HANDBOOK OF VISUAL OPTICS

VOLUME II

Handbook of Visual Optics

Handbook of Visual Optics: Fundamentals and Eye Optics, Volume One
Handbook of Visual Optics: Instrumentation and Vision Correction, Volume Two

HANDBOOK OF VISUAL OPTICS
Instrumentation and Vision Correction
VOLUME II

edited by
Pablo Artal

CRC Press
Taylor & Francis Group
Boca Raton London New York

CRC Press is an imprint of the
Taylor & Francis Group, an **informa** business

MATLAB' is a trademark of The MathWorks, Inc. and is used with permission. The MathWorks does not warrant the accuracy of the text or exercises in this book. This book's use or discussion of MATLAB' software or related products does not constitute endorsement or sponsorship by The MathWorks of a particular pedagogical approach or particular use of the MATLAB' software.

CRC Press
Taylor & Francis Group
6000 Broken Sound Parkway NW, Suite 300
Boca Raton, FL 33487-2742

First issued in paperback 2019

© 2017 by Taylor & Francis Group, LLC
CRC Press is an imprint of Taylor & Francis Group, an Informa business

No claim to original U.S. Government works

ISBN-13: 978-1-4822-3792-4 (hbk)
ISBN-13: 978-0-367-86993-9 (pbk)

Library of Congress Cataloging-in-Publication Data

Names: Artal, Pablo, editor.
Title: Handbook of visual optics / [edited by] Pablo Artal.
Description: Boca Raton : Taylor & Francis, [2017] | Includes bibliographical references.
Identifiers: LCCN 2016030030| ISBN 9781482237856 (hbk : alk. paper) | ISBN 9781315373034 (ebk) | ISBN 9781315355726 (epub) | ISBN 9781315336664 (mobi/Kindle) | ISBN 9781482237863 (web PDF)
Subjects: | MESH: Vision, Ocular--physiology | Optical Phenomena | Vision Tests--instrumentation | Eye Diseases--therapy
Classification: LCC QP475 | NLM WW 103 | DDC 612.8/4--dc23
LC record available at https://lccn.loc.gov/2016030030

Visit the Taylor & Francis Web site at
http://www.taylorandfrancis.com

and the CRC Press Web site at
http://www.crcpress.com

Contents

Preface

For many years, first as a student and later as a senior researcher in the area of physiological optics, I have wanted a comprehensive resource for frequently arising questions. Although the situation in today's Internet era is different than before, still I believe there is need for a reliable single source of encyclopedic knowledge. Finally, the dream of my youth—a handbook in visual optics—is a reality and in your hands (or on your screen). I hope this will help interested readers for a long time to come.

At the beginning of this adventure of compiling the handbook, I wanted to accomplish a number of goals (probably, too many!). Among others, I wanted to provide general useful information for beginners, or for those approaching the field from other disciplines, and the latest research presented from the most recent experiments in laboratories. As with most activities in life, success depends on the quality of individuals involved. In this regard, I was tremendously fortunate to have such an exceptional group of contributors. If we can apply the optical equivalence, this handbook is the result of a *coherent superposition* of exceptional expertise.

This handbook builds from the fundamentals to the current state of the art of the field of visual optics. The eye as an optical instrument plays a limiting role in the quality of our vision. A better understanding of the optics of the eye is required both for ophthalmic instrumentation and vision correction. The handbook covers the physics and engineering of instruments together with procedures to correct the ocular optics and its impact on visual perception. The field of physiological, or visual, optics is a classic area in science, an arena where many new practical technologies have been tested and perfected. Many of the most brilliant scientists in history were interested in the eye. Based in well-established physical and physiological principles, the area was described as nearly completed in the second part of the twentieth century. However, from the 1980s onward, a tremendous new interest in this field appeared. This was driven in part by new technology, such as lasers and electronic cameras, which allowed the introduction of new instrumentation. For example, the use of wave-front sensors and adaptive optics concepts on the eye completely changed the field. In relatively few years, these ideas expanded to the clinical areas of ophthalmology and optometry. Today, research in new aspects of vision correction and instruments is extremely active, with many groups working on it around the world. This area is a mixture of fundamentals and applications, and is at the crossroad of many disciplines: physics, medicine, biology, psychology, and engineering. I tried to find an equilibrium among the different approaches and sensibilities to serve all tastes. This book can be accessed sequentially, but also by individual parts whenever a particular topic is required.

The handbook is organized in two volumes, with five total parts. Volume One begins with an introductory part that gives an exceptional appetizer by two giants of the field: Gerald Westheimer presents an historical account of the field, and David Williams explores the near past and the future. Part II covers background and fundamental information on optical principles, ocular anatomy and physiology, and the eye and ophthalmic instruments. Each chapter is self-contained but oriented to provide the proper background for the rest of the handbook. Basic optics is covered by Schwiegerling (geometrical optics), Malacara (wave optics), and Sasián (aberrations). The concepts of photometry and colorimetry are summarized in Chapter 6 (Ohno). The basics and limits of the generation of visual stimuli are described in Chapter 7 (Farrell et al.). Furlan provides a complete revision on the main ophthalmic instruments, and Dainty an introduction on adaptive optics. While the first chapters of this part are devoted to the more technical aspects, the three next chapters have a different orientation to provide the physiological basis for the eye and the visual system. Choh and Sivak describe the anatomy and embryology of the eye in Chapter 10. Freed reviews the retina, and Winawer the architecture of the visual system. In the final chapter in this part, Pelli and Solomon describe psychophysical methods. Part II sets the foundation for the various principles that follow in the rest of the handbook.

Part III covers the current state of the art on the understanding of the optics of the eye and the retina. Collins et al. and Manns describe, respectively, what we know today about the optical properties of the cornea and the lens. Atchison reviews in Chapters 16 and 17 the different schematics eyes and the definitions and implications of the axes and angles in ocular optics. The optics of the retina is detailed in Chapter 18 (Vohnsen). Once the different components are evaluated, the next chapters concentrate on the impact of optical quality. Refractive errors (Wilson) and monochromatic (Marcos et al.) aberrations are described. Although traditionally most attention has been paid to optical characteristics of the eye in the fovea, the important role of peripheral optics is described in Chapter 21 (Lundström and Rosén). Tabernero describes personalized eye models in Chapter 22. Beyond refractive errors and aberrations, scattering in the eye affects image quality. van den Berg exhaustively reviews the state of the art of the impact and measurements of this phenomenon (Chapter 23). The eye in young subjects has the ability to focus objects placed at different distances efficiently. Bharadwaj provides a review of the accommodative mechanism (Chapter 24), and Winn and Gray describe its dynamics (Chapter 25). The eyes are continually moving to place the fovea on the area of interest. This dynamic behavior has important implications described in Chapter 26 (Anderson). Although the human eye is very robust, serving us over many years, aging obviously affects its optics. In Chapter 27, Charman reviews how the eye changes with age. Several species are able to detect the state of polarization of light. Although our visual system is not capable of something similar, polarization plays a role in optical properties as described in Chapter 28 (Bueno).

Volume Two focuses on the important topics of instrumentation and vision correction. Part I is dedicated to novel ophthalmic instrumentation for imaging, including the anterior segment and the retina, and for visual testing. An introductory chapter is dedicated to reviewing the concepts of light safety (Barat). Molebny presents a complete description of different wavefront sensors and aberrometers in Chapter 2. Hitzenberger reviews the principle of low-coherence interferometry (Chapter 3). This was the basis for one of the most successful techniques in ophthalmology: optical coherence tomography (OCT). Grulkowski concentrates on the current state of the art in OCT applied to the anterior segment (Chapter 4). Popovic (Chapter 5) and Doble (Chapter 6) present how adaptive optics implemented in ophthalmoscopes has changed the field in recent years. A different application of adaptive optics is its use for visual testing. Fernandez (Chapter 7) shows the history, present, and future potential of this technology. Imaging of the ocular media using multiphoton microscopy is a recent scientific frontier. Jester (Chapter 8) and Hunter (Chapter 9) cover, respectively, the applications of this emerging technology for the cornea and the retina.

Part II describes the different devices and techniques for surgical and nonsurgical visual correction procedures, from traditional to futuristic approaches. Ophthalmic lenses are still the most widely used approach and clearly deserve to be well recognized. Malacara (Chapter 10) presents a complete overview of this topic. Contact lenses are described in depth in Chapter 11 (Cox). The specific case of correcting highly aberrated eyes is addressed in Chapter 12 (Marsack and Applegate). A particularly relevant type of correcting devices is intraocular lenses (IOLs), implanted to substitute the crystalline lens after cataract surgery. Two emerging types of IOLs, accommodating and adjustable, are reported in Chapters 13 (Findl and Himschall) and 14 (Sandstedt). Chapter 15 (Alio and El Bahrawy) presents a review of refractive surgical approaches for the cornea. The potential for nonlinear manipulation of the ocular

tissues may open the door to new reversible future treatments. Chapter 17 (van de Pol) presents the state of the art of using cornea onlays and inlays for vision correction.

Part III reviews the relationship between the ocular optics and visual perception. Aspects related to optical visual metrics (Chapter 18, Guirao) and the prediction of visual acuity (Chapter 19, Navarro) are included. Adaptation is a key element in vision and may have significant clinical implications. Chapters 20 (Webster and Marcos) and 21 (Shaeffel) describe adaptation to blur and contrast. Visual functions change with age. A description of these characteristics is a useful resource for those interested in any practical application. Chapter 22 (Wood) reviews age-related aspects of vision. Finally, Chapter 23 (Jimenez) explores the impact of the eye's optics in stereovision.

I thank the many people who contributed to this handbook: of course, all the authors for providing accurate and up-to-date chapters; Carmen Martinez for helping me with secretarial work, and Luna Han from Taylor & Francis Group for her guidance and patience. I am also indebted to the financial help received by my lab, which allowed dedication to this endeavor: the European Research Council, the Spanish Ministry of Science, and the Fundacion Seneca, Murcia region, Spain.

Pablo Arta
Universidad de Murcia
Murcia, Spain

Editor

Pablo Artal was born in Zaragoza (Spain) in 1961. He studied Physics at the University of Zaragoza. In 1984, he moved to Madrid with a predoctoral fellowship to work at the CSIC "Instituto de Optica." He was a postdoctoral research fellow, first at Cambridge University (UK) and later at the Institut d'Optique in Orsay, France. After his return to Spain, he obtained a permanent researcher position at the CSIC in Madrid. In 1994 he became the first full professor of optics at the University of Murcia, Spain, where he founded his "Laboratorio de Optica."

Prof. Artal was secretary of the Spanish Optical Society from 1990 to 1994, associated dean of the University of Murcia Science Faculty from 1994 to 2000, and director of the Physics Department at Murcia University from 2001 to 2003. From 2004 to 2007 he was in charge of the reviewing grants panel in physics at the Spanish Ministry of Science in Madrid. Since 2006 he is the founding director of the Center for Research in Optics and Nanophysics at Murcia University. He was president of the Academy of Science of the Murcia Region from 2010 to 2015. From 2015 he is the president of the "Fundación de Estudios Medicos," an outreach organization dedicated to promote science. During his career he often spent periods doing collaborative research in laboratories in Europe, Australia, Latin America, and the United States. This included two sabbatical years in Rochester (USA) and Sydney (Australia).

Dr. Artal's research interests are centered on the optics of the eye and the retina and the development of optical and electronic imaging techniques to be applied in vision, ophthalmology, and biomedicine. He has pioneered a number of highly innovative and significant advances in the methods for studying the optics of the eye and has contributed substantially to our understanding of the factors that limit human visual resolution. In addition, several of his results and ideas in the area of ophthalmic instrumentation over the last years have been introduced in instruments and devices currently in use in clinical ophthalmology.

He has published more than 200 reviewed papers that received more than 7600 citations with an H-index of 45 (12700 and 60 in Google scholar) and presented more than 200 invited talks in international meetings and around 150 seminars in research institutions around the world. He was elected fellow member of the Optical Society of America (OSA) in 1999, fellow of the Association for Research in Vision and Ophthalmology in 2009 and 2013 (gold class), and fellow of the European Optical Society in 2014.

In 2013, he received the prestigious award "Edwin H. Land Medal" for his scientific contributions to the advancement of diagnostic and correction alternatives in visual optics. This award was established by the OSA and the Society for Imaging Science and Technology to honor Edwin H. Land. This medal recognizes pioneering work empowered by scientific research to create inventions, technologies, and products. In 2014, he was awarded with a prestigious "Advanced Grant" of the European Research Council. In 2015, he received the "King Jaime I Award on New Technologies" (applied research). This is one of the most prestigious awards for researchers in all areas in Spain. It consists of a medal, mention, and 100000€ cash prize.

He is a coinventor of 22 international patents in the field of optics and ophthalmology. Twelve of them extended to different countries and in some cases expanded to complete families of patents covering the world. Several of his proposed solutions and instruments are currently in use in the clinical practice. Dr. Artal is the cofounder of three spin-off companies developing his concepts and ideas.

He has been the mentor of many graduate and postdoctoral students. His personal science blog is followed by readers, mostly graduate students and fellow researchers, from around the world. He has been editor of the *Journal of the Optical Society of America A* and the *Journal of Vision*.

Contributors

Jorge L. Alió
Vissum Corporación
and
Division of Ophthalmology
Universidad Miguel Hernández
Alicante, Spain

Raymond A. Applegate
Visual Optics Institute
University of Houston
Houston, Texas

Ken Barat
Laser Safety Solutions
Phoenix, Arizona

Alex Black
School of Optometry and Vision Science
Queensland University of Technology
Brisbane, Queensland, Australia

Donald J. Brown
Gavin Herbert Eye Institute
and
Department of Biomedical Engineering
University of California Irvine
Irvine, California

Ian Cox
Center for Visual Science
University of Rochester
Rochester, New York

Nathan Doble
College of Optometry
The Ohio State University
Columbus, Ohio

Mohamed El Bahrawy
Vissum Corporación Alicante
Universidad Miguel Hernández
Alicante, Spain

Enrique Josua Fernández
Laboratorio de Óptica
Universidad de Murcia
Murcia, Spain

Oliver Findl
Vienna Institute for Research in Ocular Surgery
Hanusch Hospital
Vienna, Austria

and

Moorfields Eye Hospital
NHS Foundation Trust
London, United Kingdom

Ireneusz Grulkowski
Institute of Physics
Nicolaus Copernicus University
Toruń, Poland

Antonio Guirao
Department of Physics
University of Murcia
Murcia, Spain

Nino Hirnschall
Vienna Institute for Research in Ocular Surgery
Hanusch Hospital
Vienna, Austria

Christoph K. Hitzenberger
Center for Medical Physics and Biomedical Engineering
Medical University of Vienna
Vienna, Austria

Jennifer J. Hunter
Flaum Eye Institute
University of Rochester
Rochester, New York

James V. Jester
Gavin Herbert Eye Institute
and
Department of Biomedical Engineering
University of California Irvine
Irvine, California

José Ramón Jiménez
Department of Optics
University of Granada
Granada, Spain

Holger Lubatschowski
ROWIAK GmbH
Hannover, Germany

Daniel Malacara
Centro de Investigación en Optica
León, Mexico

Susana Marcos
Instituto de Optica
Consejo Superior de Investigaciones Científicas
Madrid, Spain

Jason D. Marsack
Visual Optics Institute
University of Houston
Houston, Texas

Vasyl Molebny
Department of Optics
School of Physics
National University of Kiev
Kiev, Ukraine

Rafael Navarro
Materials Science Institute of Aragón
National Council for Scientific Research &
    University of Zaragoza
Zaragoza, Spain

Zoran Popovic
Department of Ophthalmology
University of Gothenburg
Gothenburg, Sweden

Christian A. Sandstedt
Calhoun Vision, Inc.
Pasadena, California

Frank Schaeffel
Section of Neurobiology of the Eye
Ophthalmic Research Institute
Tuebingen, Germany

Robin Sharma
The Institute of Optics and Center for Visual Science
University of Rochester
Rochester, New York

Corina van de Pol
Southern California College of Optometry
Marshall B. Ketchum University
Fullerton, California

Michael A. Webster
Department of Psychology
University of Nevada Reno
Reno, Nevada

Moritz Winkler
Gavin Herbert Eye Institute
and
Department of Biomedical Engineering
University of California Irvine
Irvine, California

Joanne Wood
School of Optometry and Vision Science
Queensland University of Technology
Brisbane, Queensland, Australia

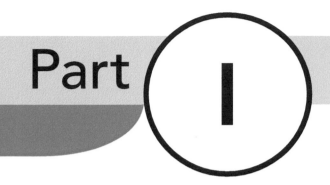

Part 1

Ophthalmic instrumentation

1 Light safety

Ken Barat

Contents

1.1 INTRODUCTION

The goal of this chapter is to discuss the hazard, in particular laser light. We will cover a number of topics that the reader may think they are familiar with, but in the case of safety, repetition is not a bad thing. In addition, there is coverage of a number of laser incidents. The author of this chapter hopes that by the end the reader will have an appreciation for the potential harm they may be exposed to. A number of resource materials will be referenced, and the author wishes to thank those individuals who allow open access to their material.

An awareness of laser safety is extremely important, not only for the benefit of users and ancillary/support staff but also to answer patient questions and concerns. With the increase in an aging population comes an increase in eye diseases. Ophthalmology has responded with technology such as mydriatic and nonmydriatic fundus cameras, phacoemulsification devices, femtosecond lasers, supercontinuum light sources, and optical coherence tomography. These advances support the need for understanding an appreciation of laser safety.

1.2 WHY ALL THE CONCERN OVER LASERS?

With all the hazards one faces during their day, why do we feel laser safety deserves special attention? While laser light or radiation has properties that distinguish it from natural light, the answer is rather simple. Visible (400–700 nm) and near-infrared wavelengths (700–1400 nm) are focused by the lens of the eye to a spot size an order of magnitude smaller than that of natural or incoherent visible light, a 300–200 μm spot size to one of 10–20 μm. Thus producing an irradiance (power per square centimeter) much higher than incoherent light. The item that gets lost sometimes is while no one expects to look directly into a laser beam, with this magnification of 100,000 to the macula even a small reflection of a laser beam between 400 and 1400 nm has the potential to cause some level of retinal injury.

See Figures 1.1 and 1.2 for a graphic display of this point. A 1 mW/cm^2 beam in is 100 W/cm^2 at ones macula.

1.3 CONTINUOUS WAVE

Laser systems that can produce uninterrupted laser energy are called continuous wave (CW) lasers. As long as they are turned on, a beam of laser light is emitted. By convention, any laser that emits light for longer than 0.25 s is called a CW laser. External shutters can be used to "chop" the beam so it has a strobe appearance, but the output power remains constant even in the chopped beam. The seemly pulse of the bar code scanner is an example of a chopped CW laser. The power output of CW lasers is measured in units of watts.

1.4 PULSED LASERS

Pulsed laser systems emit a beam that is less than 0.25 s in duration. The 0.25 s demarcation between CW and pulsed lasers is based on the approximate time for a human reflex to very intense light to work (aversion response). The pulsing in a pulsed laser system is achieved internal to the laser. Some employ an electrical "Q-switch" or are "mode locked" to achieve the shorter pulse widths. The pulse is defined by the pulse width (or emission duration) and the number of pulses per second known as pulse repetition frequency. Unlike the CW laser, all of the energy is emitted in short bursts. The total energy is compacted into a shorter time interval so the peak energy output can be very high. This makes shorter pulses more hazardous. Many pulsed lasers have pulse widths measured in nanoseconds (1 billionth of a second). Pulse rates to attosecond (10^{18} s) pulse widths have been routinely produced (but only in vacuum). A laser must output pulses at a rate faster than 1 Hz to be considered a repetitive pulsed laser. The output is measured in terms of Joules (Watts × seconds). Turnkey units are available for the generation of femtosecond laser pulses, 10^{15} s.

With the development and refinement of the laser diode, a highly reliable and convenient laser source became available for medical applications. A source that is robust and does not require an in-house service person to keep it running has few maintenance needs. Many laser systems combine the diode or a diode array with a fiber optic system. The fiber optics is used as a means to deliver the laser radiation without the use of open beam paths or beam enclosures.

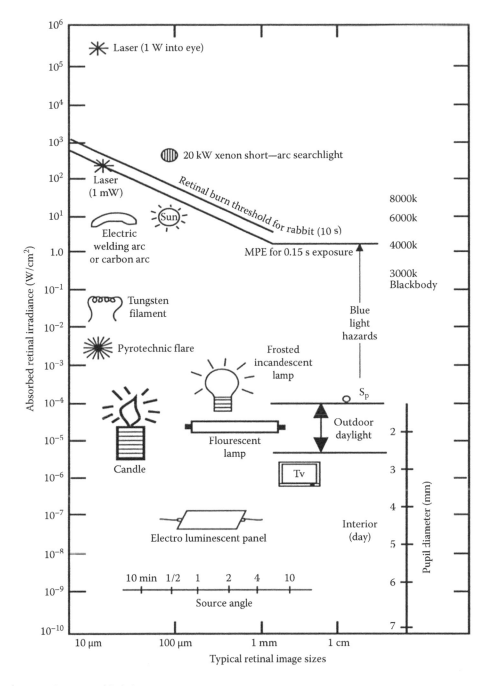

Figure 1.1 Implications from environmental lighting.

1.5 HAZARD CLASSIFICATION

An internationally agreed to hazard classification system has been developed. Its simple goal is to allow one to understand the potential risk of a laser by just knowing its classification. Classes range from Class 1, 1M, 2, 2M, 3R, 3B, and 4, and one should include Class 1 product. The specific class definitions are as follows.

1.5.1 CLASS 1

Class 1 lasers and laser systems by definition are not an eye or skin hazard, their output is usually in the microwatt range.

1.5.2 CLASS 1M

Class 1M lasers do not exceed the Class 1 AEL for unaided viewing but do exceed the Class 1 AEL for optically aided viewing. Class 1M lasers do not exceed the Class 3B AEL for optically aided viewing. This does not, however, mean that the system is incapable of doing harm. ANSI Z136.1 requires that classification is based only on unaided and 5 cm aided viewing conditions. Therefore, hazards may still exist when using viewing optics with a greater optical gain than 7.14 (5 cm optics). They also have no unique safety requirements for their use (no extra engineering controls or safety devices). They also have no unique safety requirements for their use (no extra engineering controls or safety devices).

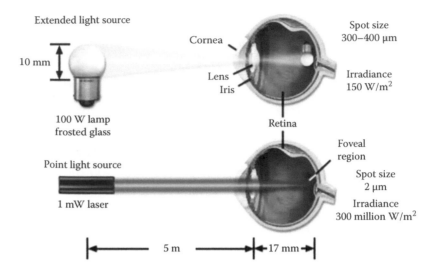

Figure 1.2 Extended and point source power density at the retina.

1.5.3 CLASS 2

Class 2 lasers are visible (400–700 nm wavelength) CW and repetitive-pulse lasers and laser systems that can emit accessible radiant energy exceeding the appropriate Class 1 AEL for the maximum duration inherent in the design or intended use of the laser or laser system, but not exceeding the Class 1 AEL for any applicable pulse (emission) duration <0.25 s and not exceeding an average radiant power of 1 mW. Laser systems with invisible beams cannot be Class 2.

1.5.4 CLASS 2M

Class 2M lasers are limited to visible wavelengths (400–700 nm) at accessible emission levels below the Class 2 AEL for the unaided eye. However, they can exceed the Class 2 AEL under optically aided viewing conditions. Class 2M lasers do not exceed the Class 3B AEL for optically aided viewing.

1.5.5 CLASS 3R (ONCE KNOWN AS 3A)

Class 3R lasers include lasers and laser systems that have an accessible output between one and five times the Class 2 AEL for visible lasers and the Class 1 AEL for all other lasers, based on the appropriate exposure duration. The output range is 1–5 mW.

1.5.6 CLASS 3B

Class 3B lasers and laser systems include the following:
  UV (180–400 nm) and IR (1400 nm to 1 mm) lasers and laser systems that emit accessible radiant power in excess of the Class 3R AEL during any emission duration within the maximum duration inherent in the design of the laser or laser system, but that (a) cannot emit an average radiant power in excess of 0.5 W (5–500 mW CW) for greater than or equal to 0.25 s or (b) cannot produce a radiant energy greater than 125 mJ in 0.25 s.
  Visible (400–700 nm) and near IR (700–1400 nm) lasers and laser systems that emit in excess of the Class 3R AEL but

that (a) cannot emit an average radiant power in excess of 0.5 W for greater than or equal to 0.25 s and (b) cannot produce a radiant energy greater than 30 mJ/pulse.
Class 3B lasers are not eye safe. It is hazardous to view the beam of a Class 3B laser under most conditions. Some may have a diffuse reflection hazard. All Class 3B lasers require some engineering controls and require a danger label/sign.

1.5.7 CLASS 4

Class 4 lasers and laser systems are those that emit radiation that exceeds the Class 3B AEL. These lasers can often be fire hazards and have diffuse reflection hazards.

They are not considered "eye safe" under any viewing conditions. Many have hazardous diffuse reflections and can be skin and combustion hazard. Class 4 lasers require many more safety devices and engineering controls.

1.5.8 CLASS 1 PRODUCT (SOMETIMES REFERRED TO AS AN EMBEDDED LASER SYSTEM)

This is a laser device or product that incorporates a laser, greater than Class 1 (usually class 3B or 4) inside the unit in such a way that during normal operation there is no potential for laser exposure to the operator. During any service or maintenance, this can all change. These also have no unique safety requirements for their use (no extra engineering controls or safety devices). It is always to ones advantage to use a class 1 product.

1.6 COMMON MEDICAL LASER WAVELENGTHS

An increasing number of wavelengths are used in medical applications. This number will only increase over time and as new medical applications are developed and wavelengths become available for use. As an example, at one time, the use of femtosecond lasers would have seen as out of reach in a local clinic. This would be from an operator knowledge perspective, maintenance requirements, and cost. But now, femtosecond laser systems have become available as turnkey systems. Much of this has to do with the absorption and transmission of various layers of skin and cellular components as well as the development of new optics (i.e., nonlinear optics). The following list should be considered as basic examples of the type of lasers used in ophthalmology and not all inclusive.

LASER TYPE	WAVELENGTH (nm)
Excimer	193,308
Argon ion	488–514
Freq. doubled Nd:YAG neodymium–yttrium aluminum garnet	532
Ruby	694
Diode	780–840
Nd:YAG	1,064
Holmium	2,100
Carbon dioxide	10,600

Conditions and procedures one can apply laser technology to
- Eyelid growths, including lid cancers
- Histoplasmosis
- Central serous retinopathy
- LASIK
- Cataract surgery
- Glaucoma surgery
- Diabetic retinopathy
- Misdirected eye lashes (*trichiasis*)
- Opening up or treating blockage of the opening to the tear ducts (*lacrimal punctum*)
- Benign (*noncancerous*) growth on the tissue that covers the surface of the white part of the eye and may grow onto the cornea (*pterygium*)
- Increased eye pressure (*glaucoma*)
- Prevention and treatment of attacks of acute glaucoma (*laser iridotomy*)
- Retinal tears and detachment treatment
- Diabetic retinopathy
- Treatment of cancer of the eye (such as retinoblastoma)
- Open-angle glaucoma

1.6.1 LASER SAFETY ITEMS

Laser users need to look toward the American National Institute Standards Z136 series for user laser safety guidance. If developing a product, one needs to look at laser safety product regulations either from the FDA or IEC (outside U.S. requirements). From a user point of view, there is Z136.1 for Safe Use of Lasers, Z136.3 for Safe Use of Lasers in Health Care Facilities, and Z136.8 for Safe Use of Lasers in Research, Development, and Testing. The Laser Safety Officer (LSO) can choose from these for the control guidance that fits their use location and setting rather than just using one of these standards. Regardless when there is a potential for laser beam exposure, a number of common administrative safety steps need to be addressed.

1.6.1.1 Appointment of a laser safety officer

No facility can say they address laser safety if no one has the title and responsibilities of the LSO. Their chief role is to put in place the appropriate control measures and ensure training is delivered to staff with possible beam exposure. In the clinic setting, this is usually the physician or senior medical staff person, not the one who is judged least valuable.

1.6.1.2 Laser protective eyewear

When required, laser eyewear must be labeled with the wavelength coverage it is designed for and its optical density (OD) (optical attenuation). As important as correct OD is visual light transmission, meaning one needs to see while wearing the eyewear. While laser eyewear is ones last line of defense, its use when required is critical. This is where having an LSO or MLSO (Medical Laser Safety Officer) is critical; one should not count on vendors for eyewear selection.

Patient eye protection should never be overlooked where practical.

1.6.1.3 Signage

Posting of the laser use area while it will not prevent an accident, it is a demonstration of laser safety awareness. Check with the appropriate standard for signage requirements. The signage needs to convey what eyewear is required and the class of the laser in use. Often with mobile medical equipment, the sign and eyewear travel with the laser cart.

1.6.1.4 Operating procedures

Any time Class 3B or Class 4 lasers or laser systems are in use, standard operating procedures are required. They need to be more than a cookbook work procedure, but indicate hazards and what controls will be applied to control them. In medical settings, most SOPs are a combination of pre- and postoperation checklists. These checklists need to include safety items, that is, post sign, perform beam alignment, check eyewear, and communicate laser hazard to staff in room. Is patient safety in place?

1.7 BIOLOGICAL EFFECTS

1.7.1 MAXIMUM PERMISSIBLE EXPOSURE

From a safety perspective, maximum permissible exposure (MPE) needs to be defined. In simple terms, it is the exposure limit for exposure levels to the skin and eye. These are generally expressed in J/cm^2 or W/cm^2 measured at the cornea. Damage threshold is generally 5–20 times higher than the MPE. Exposure up to and including the particular MPE value will not cause injury. The official definition is MPE values are the level of laser radiation to which an unprotected person may be exposed without adverse biological changes in the eye or skin.

1.7.2 NOMINAL OCULAR HAZARD ZONE

The distance along the axis of the unobstructed beam from a laser, fiber end or connector to the human eye beyond which the irradiance or radiant exposure is not expected to exceed the MPE. Think of it as the starting line of the safe zone or end of the safe zone if one is closer.

1.7.3 DAMAGE MECHANISMS

1.7.3.1 Electromechanical/photodisruption/acoustic damage

Photomechanical (or photoacoustic or photodisruption) damage occurs when the laser energy is deposited faster than mechanical relaxation can occur and typically occurs for intense pulses shorter than 1 ns. As a result, a thermoelastic pressure wave is produced, and tissue is disrupted by shear forces or by cavitation. This type of damage requires beams of a power density (10^9–10^{12} W/cm²) in extremely short pulses (ns) to delivery fluences of about 100 J/cm². Such a pulse induces dielectric breakdown in tissue, resulting in a microplasma or ionized volume with a very large number of electrons. A localized mechanical rupture of tissue occurs due to the shock wave associated with the plasma expansion. Laser pulses of less than 10 µs duration can induce a shock wave in the retinal tissue that causes tissue rupture. This damage is permanent, as with a retinal burn. Acoustic damage is actually more destructive to the retina than a thermal burn. Acoustic damage usually affects a greater area of the retina, and the threshold energy for this effect is substantially lower. The MPE values are reduced for short laser pulses to protect against this effect (Figures 1.3 and 1.4).

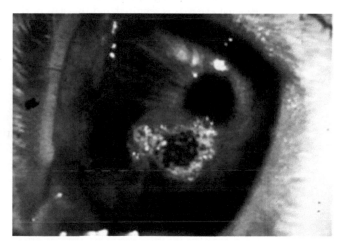

Figure 1.3 Corneal burn—rabbit.

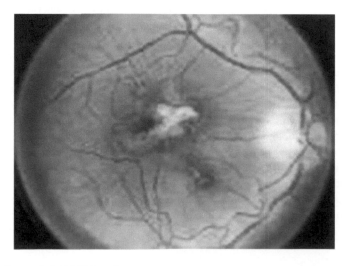

Figure 1.4 Nd:YAG pulses.

1.7.3.2 Photoablation

Photoablation is the photodissociation or direct breaking of intramolecular bonds in biopolymers, caused by absorption of incident photons and subsequent release of biological material. Molecules of collagen, for example, may dissociate by absorption of single photons in the 5–7 eV energy range. Excimer lasers at several ultraviolet (UV) wavelengths (ArF, 193 nm/6.4 eV; KrF, 248 nm/5 eV; XeCl, 308 nm/4 eV) with nanosecond pulses focused on tissue at power densities of about 108 W/cm² can produce this photoablative effect. UV radiation is extremely strongly absorbed by biomolecules, and thus absorption depths are small, of the order of a few micrometers.

1.7.3.3 Thermal damage

Thermal damage occurs because of the conversion of laser energy into heat. With the laser's ability to focus on points a few micrometers in diameter, high power densities can be spatially confined to heat target tissues. Depth of penetration into the tissue varies with wavelength of the incident radiation, determining the amount of tissue removal and bleeding control. The photothermal process occurs first with the absorption of photon energy, producing a vibrational excited state in molecules, and then in elastic scattering with neighboring molecules, increasing their kinetic energy and creating a temperature rise. Under normal conditions, the kinetic energy per molecule (kT) is about 0.025 eV. Heating effects are largely controlled by molecular target absorption such as free water, hemoproteins, melanin, and other macromolecules such as nucleic acids.

For the retina, the main absorber of visible light is the RPE. Temperature elevation of about 10°C causes protein denaturation and damage to the cells' immediate whitening (photocoagulation). Depending on how fast the laser energy is deposited, either photocoagulation or photovaporization could occur. Photocoagulation occurs when the laser energy causes the tissue temperature to produce protein denaturation, while in the case of photovaporization, the temperature rise causes the vaporization of water, generally temperatures over 100°C.

1.7.3.4 Photochemical damage

Light below 400 nm is completely absorbed by the lens and vitreous and does not therefore affect the retina. The light can be laser output, UV from the excitation pump light, or blue light from a target interaction. The effect is cumulative over a period of days. Photochemical mechanisms, in particular mechanisms that arise from illumination with blue light, are responsible for solar retinitis and for iatrogenic retinal insult from ophthalmological instruments. Further, blue light may play a role in the pathogenesis of age-related macular degeneration. Laboratory studies have suggested that photochemical damage includes oxidative events. Retinal cells die by apoptosis in response to photic injury, and the process of cell death is operated by diverse damaging mechanisms. Photochemical damage occurs when light is absorbed by a chromophore and leads to the formation of an electronically excited state of that molecule, which then undergoes either chemical transformation itself and/or interacts with other molecules leading to chemical changes of both interacting molecules or to a transfer of the excitation energy to the other molecules. Importantly, when photochemical damage occurs, there is no substantial increase in temperature of the tissue. In a particular type of photochemical damage, photosensitized damage, the photoexcited chromophore in its electronically excited singlet state undergoes intersystem crossing and forms an excited triplet state.

Note: Photochemical damage is the most common form of retinal damage caused by exposure to direct sunlight and several artificial light sources, including ophthalmic instruments.

1.7.3.5 Off-axis laser damage

The location of the exposure within the eye determines the degree of incapacitation from a retinal injury of a given degree of severity. The fovea (the central two degrees of the visual mechanism) is the region of the retina, which is most critical for vision. The rest of the retina is increasingly less sensitive to the light as one moves away from the fovea. An injury to the fovea can severely reduce visual functioning in terms of resolution. A laser lesion in the peripheral areas of the retina may not be noticed or cause a significant reduction in visual function (i.e., visual acuity) because the vision in those areas is very poor compared to vision in the fovea. Lesions in areas other than the fovea are referred to as "off-axis" damage or hits. Most are benign. In fact, a laser may be used to surgically treat diabetes-related problems in the retina. Some diabetics suffer visual dysfunction because of weakened/leaky blood vessels in the retina. A laser can be used to cauterize the smaller vessels, and visual function is improved by that procedure.

If the off-axis damage is on the nerve fibers or blood vessels in the eye, the visual functioning of the cells downstream can result in disrupted vision in much larger areas than the initial scotoma. Blood leaking into the inner chamber of the eye from an off-axis hit can also severely impact foveal vision.

Key advice: DO NOT look directly into the beam, with remaining eye.

1.7.4 ULTRAFAST LASER INJURY

Damage mechanism for ultrafast laser pulses (pico and femtosecond duration) is still being studied and does not fit the damage mechanisms from classical laser systems. Ultrafast laser systems are now turnkey and can be used with little technical support. The damage mechanism changes as pulses become less than 50 fs. A self-focusing mechanism seems to be at play as well as intense shock wave production. Two good references on this matter are

Rockwell, B., Thomas, R., and Vogel A., Ultrashort laser pulse retinal damage mechanisms and their impact on thresholds, *Med. Laser Appl.*, 25, 84–92, 2010.

Mclin, L., A case study of a bilateral femtosecond laser injury, *Proceedings of the International Laser Safety Conference*, 2013, Laser Institute of America, paper #904.

1.7.5 LASER RADIATION EFFECTS ON SKIN

Question: Why should I care about skin? A broken fiber or missed aligned laser or premature firing of the laser system could cause a skin injury. The user should know about the possible consequences.

One's skin is a larger target for potential laser exposure but is considered less serious than injury to the eye, since functional loss of the eye is more debilitating than damage to the skin. The injury thresholds for both skin and eyes are comparable (except in the retinal hazard region, 400–1400 nm). In the far-infrared and far-UV regions of the spectrum, where optical radiation is not focused on the retina, skin injury thresholds are about the same as corneal injury thresholds. The MPE for skin and extent of damage will depend on the surface area exposed. The larger the area exposed, the lower the MPE will be due to the loss of surface area to dissipate the energy.

There is great variation in depth of penetration over the range of wavelengths, with the maximum occurring around 700–1200 nm. Injury thresholds resulting from exposure of less than 10 s to the skin from far-infrared and far-UV radiation are superficial and may involve changes to the outer dead layer of the skin. A temporary skin injury may be painful if sufficiently severe, but it will eventually heal. Burns to larger areas of the skin are more serious, as they may lead to serious loss of body fluids. Hazardous exposure of large areas of the skin is unlikely to be encountered in normal laser work.

A sensation of warmth resulting from the absorption of laser energy normally provides adequate warning to prevent thermal injury to the skin from almost all lasers except for some high-power far-infrared lasers. Any irradiance of 0.1 W/cm^2 produces a sensation of warmth at diameters larger than 1 cm. On the other hand, 1/10th of this level can be readily sensed if a large portion of the body is exposed. Long-term exposure to UV lasers has been shown to cause long-term delayed effects such as accelerated skin aging and skin cancer. The layers of the skin, which are of concern in a discussion of laser hazards to the skin, are the epidermis and the dermis (Figure 1.5).

1.7.5.1 Epidermis

The epidermis is the outer layer of skin. The thickness of the epidermis varies in different types of skin. It is the thinnest on the eyelids at .05 mm and the thickest on the palms and soles at 1.5 mm.

1.7.5.2 Dermis

The dermis also varies in thickness depending on the location of the skin. It is 0.3 mm on the eyelid and 3.0 mm on the back. The dermis is composed of three types of tissue that are present throughout—not in layers. The types of tissue are collagen, elastic tissue, and reticular fibers.

1.7.5.3 Subcutaneous tissue

The subcutaneous tissue is a layer of fat and connective tissue that houses larger blood vessels and nerves. This layer is important in the regulation of temperature of the skin itself and the body. The size of this layer varies throughout the body and from person to person.

1.7.5.4 Skin effect by wavelength

To the skin, UV-A (0.315–0.400 μm) can cause hyperpigmentation and erythema. UV-B and UV-C, often collectively referred to as "actinic UV," can cause erythema and blistering, as they are absorbed in the epidermis. UV-B is a component of sunlight that is thought to have carcinogenic effects on the skin. Exposure in

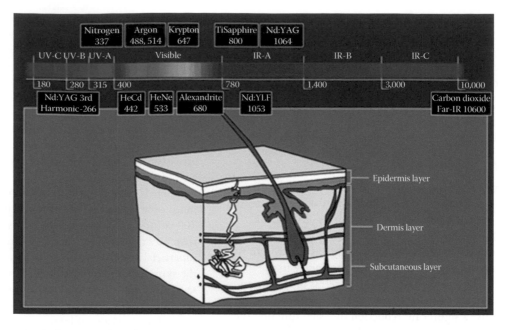

Figure 1.5 Skin penetration by laser wavelength.

the UV-B range is most injurious to skin. In addition to thermal injury caused by UV energy, there is the possibility of radiation carcinogenesis from UV-B (0.280–0.315 mm) either directly on DNA or from effects on potential carcinogenic intracellular viruses.

Exposure in the shorter UV-C (0.200–0.280 µm) and the longer UV-A ranges seems less harmful to human skin.

The shorter wavelengths are absorbed in the outer dead layers of the epidermis (stratum corium) and the longer wavelengths have an initial pigment-darkening effect followed by erythema if there is exposure to excessive levels.

IR-A wavelengths of light are absorbed by the dermis and can cause deep heating of skin tissue (Figures 1.6 through 1.8).

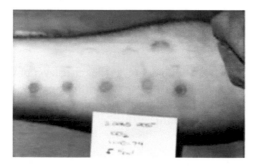
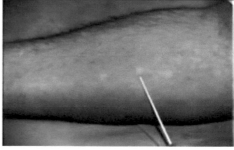

Figure 1.6 Twenty-year evaluation of CO_2 laser (5 W/cm², 1 s at 10,600 nm) exposure of human skin. *Note:* At long-term follow-up, burn region display nondescript fibrous scarring. No other symptoms were observed over the 20-year period.

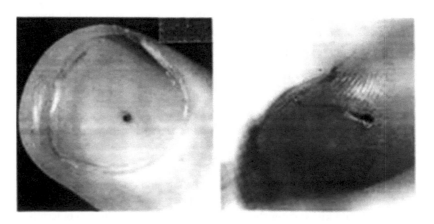

Figure 1.7 Burn through finger.

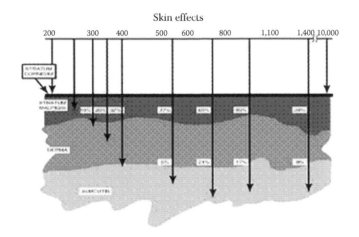

Figure 1.8 Wavelength penetration.

1.7.6 OPTICS

1.7.6.1 Increased hazards when using optics

Magnifying optics can collect more energy than the unaided eye and direct that laser energy onto the cornea and into the pupil and allow the energy to be focused on ocular tissues. However, the two types of optical aids—telescopic optics and eye-loupe magnifiers—increase ocular hazards quite differently.

Note: A hand magnifier or eye loupe cannot increase the retinal irradiance from a collimated laser beam, but a telescope generally does not.

Telescopes and binoculars collect more energy and maintain a minimal retinal image in the relaxed, that is, focused at infinity normal eye. However, the hand magnifier or eye loupe increases the retina hazard only if the eye is placed very close to a diverging source, so that the energy from a highly diverging beam can be directed into the pupil and brought to focus at the retina.

The normal viewing condition for the human eye when using an eye loupe or magnifier is the relaxed condition since the magnifying aid provides a great enhancement of the eye's ability to accommodate focus at close distance. The front focal plane of the magnifier is placed at the plane of the object under study, and rays from a point on the object surface emerge from the magnifier as effectively "collimated." The eye can be positioned over a range of distances from the magnifier optic and still be in focus provided the eye remains relaxed, that is, viewing at infinity. The refracting power of an eye loupe depends on its focal length. The magnifying power of an eye loupe or hand magnifier is standardized internationally relative to a reference focal length of 25 cm, the 4 diopter near point of accommodation of most adults, and the magnifying power, M, is the ratio of the reference focal length of 25 cm to the lens focal length, f. Therefore, a simple 53 hand magnifier has a focal length of

$$f = 25 \text{ cm}/5x = 5 \text{ cm}$$

and a 2.53 magnifier has a focal length of 10 cm, the reference distance used for laser measurements as the closest point where the eye may be able to image an object. Hence, the assessment of the increased risk of optical aids for close viewing must begin with M greater than 2.5. If the optical source is placed at the focal

plane of a 73 cyc loupe, the output power that is transmitted by a 7 mm aperture located 3.5 cm away is the power that would enter a fully dilated 7 mm pupil of the eye. The image on the retina would be seven times larger than when viewing the same source with the unaided eye at 250 mm to 25 cm. For a person who can actually focus on objects at 10 cm, the "73" eye loupe effectively increases the retinal image by a factor of 2.8 compared to unaided viewing at 10 cm. However, a sevenfold increase in retinal image diameter would only occur for those individuals capable of accommodating to 10 cm when the standard power of the loupe was 183. Higher-power viewing aides are often used in the fiber optics industry by trained technicians to provide diagnostic information about the condition of fiber tips. However, such aids are rarely provided without suitable wavelength blocking filters to protect a user if the laser source is inadvertently emitting power during the inspection. Also, higher-power optics would require firm mounting so that the head could be stabilized with respect to the object being examined such as is done with a microscope. Although such specialized eye loupes are available, their use in practical situations is limited since the position of the loupe must be held at a precise location with respect to both the eye and object viewed to obtain a clear image. Since hazards based on telescopic optics are limited to 73, although higher-power optics are available, it is also reasonable to limit the hazard evaluation of loupes or hand magnifiers to 73, except for those industries where higher-power loupes are known to be routinely used.

1.8 AIDED VIEWING FROM A FIBER

Figure 1.9 illustrates eye exposure by emission from a fiber when viewed with an eye loupe. This viewing condition assumes that the laser power was not turned off or disconnected from the cable before viewing began, that the emission is invisible so the person is unaware of being exposed, that the eye loupe provides little additional attenuation at the laser wavelength relative to visible light, and that the viewer will examine the tip under ideal conditions for maximizing the hazard for 100 s. Good laser safety practice would preclude such a condition from happening; however, the addition of the eye loupe could increase the hazard by allowing the eye to focus on the tip of the fiber optic cable at a closer distance and thus increase the energy entering the eye. However, the addition of the loupe will not increase the radiance of the source. Normally, the retinal irradiance would not be increased either; however, since the tip could be very small, the image on the retina of the core diameter, from which the emission originates, could still be on the same order as a min. Therefore, the retinal irradiance is increased since viewing the fiber tip with the unaided eye produced a minimal spot size on the retina, and viewing with

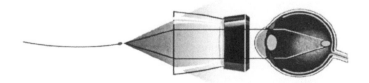

Figure 1.9 Eye exposure from viewing a fiber tip through an eye loupe.

the loupe produces an image on the retina that is not much larger, if larger at all. Therefore, the hazard would be increased since more energy is concentrated in this minimal spot on the retina.

1.9 ACCIDENTS AND RELATED EVENTS

Laser accidents and incidents including near misses are happening. More than 500 a year are reported to the Food and Drug Administration (FDA). These can be found in the MAUDE database. Other events find their way to the Rockwell Laser Industries accident database (both of these databases can be accessed on the web). In addition, a number of incidents never see the light of day but are talked about in hushed tones or kept under lock and key at the incident site institution. One of the most important services provided by the MAUDE DB is it allows tracking and trending review. It can see equipment problems that may only happen once at an institution but seem to be repeated at others, hence allowing for data to generate a product recall or modification.

1.10 MAUDE DATABASE

Each year, on average, the U.S. FDA received several hundred thousand medical device reports (MDRs) of suspected device-associated deaths, serious injuries, and malfunctions. The FDA uses MDRs to monitor device performance, detect potential device-related safety issues, and contribute to benefit-risk assessments of these products. The MAUDE database houses MDRs submitted to the FDA by mandatory reporters (manufacturers, importers, and device user facilities) and voluntary reporters such as health care professionals, patients, and consumers.

MDRs are a passive surveillance system with limitations, including the potential submission of incomplete, inaccurate, untimely, unverified, or biased data. In addition, the incidence or prevalence of an event cannot be determined from this reporting system alone due to potential underreporting of events and lack of information about frequency of device use. Because of this, MDRs cannot be accepted as the complete listing of medical device incidents and problems.

Note that certain types of report information are protected from public disclosure under the Freedom of Information Act. If a report contains trade secret or confidential business information, that text is replaced by a standard statement. MAUDE is updated monthly; the FDA seeks to include all reports received prior to the update, but the inclusion of some reports may be delayed. The following are a sampling of MDRs in the ophthalmic field report. As you review, remember the limitations of this reporting system.

1.10.1 REPORT 1: MAUDE DB IRIS MEDICAL SLIT LAMP REPORT NUMBER 2939653-1999-00002

752959 patient sequence number: On May 5, 1999, a sales rep demonstrated an iris medical occu-light glass laser system with a slit lamp adapter delivery device to three Drs at the hosp. While making practice burns on a business card, Dr 2 commented on the brightness through the oculars after test firing the laser.

Drs 2 and 3 said they saw "spots" after test firing; Dr 1 and the sales rep noticed no unusual brightness. Upon further inspection of the slit lamp adapter delivery device, *it was noticed that the safety filter frame was labeled 810 nm. The wavelength of the laser being demonstrated was 532 nm.* On May 6, 1999, it was verbally reported by the hospital risk manager to the sales rep that Dr 3 was found to have 20/50 vision and three suspected laser burns on his retina. The ophthalmologist who examined Dr 3 indicated that it may take several months for the vision to recover/stabilize.

1.10.2 REPORT 2: ALCON RESEARCH LTD/ HUNTINGTON ACRYSOF RESTOR INTRAOCULAR LENS

Model # SN60D3 event type: injury

In a Journal article, the authors presented the results of a patient who had decreased visual acuity in her left eye and disturbing concentric ring dysphotopsias in both eyes after bilateral intraocular multifocal lens implants. The pt had undergone laser treatment for a peripheral retinal tear in the left eye in 1995. Photocoagulation scars peripherally in the left eye were observed during an examination in 2007, macular epiretinal membrane reduced the beva in the left eye to 20/40. Her beva returned to 20/30 after removal of the epiretinal membrane. The pt continued to report concentric rings. There are two MDRs associated with this event.

1.10.3 REPORT 3: DEVICE PROBLEM, SELF ACTIVATION OR KEYING

A customer reported that when the unit was set on endo during a vitrectomy procedure, it switched to laser indirect ophthalmoscope (LIO) on its own.

Model LXT

1.10.4 REPORT 4: MDR REPORT KEY 4121056 DEVICE OPERATED DIFFERENTLY THAN EXPECTED

A surgeon reported that two 33° arcuate incisions were programmed before surgery, but during treatment, one of the arcuate was made at a 120° width. The surgeon used a compression suture to minimize the effect of the incorrect arcuate and the procedure was completed.

1.10.5 REPORT 5: MDR REPORT KEY 4105971 OPHTHALMIC FEMTOSECOND LASER, LENSEX LASER SYSTEM

An ophthalmic surgeon reported suction difficulties during a laser assisted cataract procedure. A bubble formed in the anterior chamber and between the patient interfaces. After several suction attempts, the procedure was completed. An anterior and posterior capsular rupture occurred, which was relayed as being contributed by air in the capsular bag.

1.10.6 REPORT 6: REPORT KEY 4076806 WAVELIGHT EX500 EXCIMER LASER

A health care professional reported a case where eye tracker was not detecting the pupil and the system stopped during LASIK treatment procedure. Reporter indicated the procedure was not completed.

1.10.7 REPORT 7: MDR REPORTING KEY 3735743 WAVELIGHT FS200 FEMTOSECOND LASER

A surgeon reported multiple cases of "spotted pattern," post LASIK surgery. This case references patient's right eye. Additional information received indicated that the patient was treated with a steroid taper and reported issue has not yet resolved. According to surgeon, the patient is not experiencing any visual disturbances. A company representative has visited the site and recommended review of user-adjusted laser energy settings, possibly reducing the treatment energy settings. Additional information has been requested. This reported event involved multiple MDRs. This report references the fourth patient's right eye.

1.10.8 REPORT 8: MDR REPORT KEY 4124364 CATALYS PRECISION LASER SYSTEM OPHTHALMIC FEMTOSECOND LASER

It was reported that a pt who underwent anterior capsulotomy and lens fragmentation with the Catalys system subsequently experienced capsular tears in the operating room (or) during the surgical procedure to remove the cataract. No additional complications and/or medical intervention were reported.

1.10.9 REPORT 9: LASER FIRE

A woman patient was burned when a fire broke out during laser beam surgery at NYU Medical Center. Patient was listed in satisfactory condition in intensive care unit, with second degree burns on her face, neck, and shoulder. The accident began when the laser beam ignited a sheet used to cover surgical instruments near the patient. The sheet was supposed to be flame retardant. The small blaze was put out by attending doctors and nurses. The operation was over when the fire started; the laser should have been off.

1.10.10 REPORT 10: MODEL NO. PUREPOINT LIO NATURE OF PROBLEM, MALFUNCTION

A nurse reported that the laser embedded in the system did not work during a vitrectomy procedure. The system was exchanged and the procedure was completed.

1.10.11 REPORT 11: EVENT, NEGLIGENCE

While a user was looking through an optical viewer to inspect the cleanliness of fiber optics, another user pulsed the laser resulting in an eye exposure.

1.10.12 REPORT 12: DEVICE PROBLEM, DEVICE INOPERABLE

A nurse reported the surgeon couldn't get the LIO to work during a photocoagulation surgery. The surgeon decided to stop operating.

1.10.13 REPORT 13: LASIK EYE SURGERY

Event type: Permanent-external eye damage

After LASIK procedure, a man suffered permanent vision damage and had to leave his job. Eye center filed to diagnose patient's keratoconus, which made him a poor candidate for LASIK surgery.

1.10.14 REPORT 14: MEDLITE C6

According to the allegedly injured party, in 2006, a laser technician practitioner was performing a procedure using a Medlite C6 laser at 6 J/cm^2, 1064 nm, 10 Hz with a 4 mm spot size while compressing the skin with a glass window to force the blood away from the treatment site to minimize the formation of purpura. She reports that during the procedure, she noticed bright spots that caused her to blink. After the procedure, she noticed that there continued to be a "black" spot or shadow in her central vision similar to what happens after you look directly into a light bulb or the sun and then look away. She further reports that 3 days later, she went to see an ophthalmologist who examined her and found bleeding and referred her to a retinal specialist. We have not yet received her formal medical records. Sixteen days later, the patient was sent an email to an independent contractor working with hoya conbio, informing her of the adverse event. The independent contractor reviewed the email for the first time and reported the alleged injury to hoya for the first time 5 days later.

1.10.14.1 Manufacturer response

Device is not evaluated as there is no defect or problem with the device. The event may be related to a lack of eyewear or the use of incorrect eyewear. The eyewear provided with the laser system is used to protect the eyes with the 1064 and 532 nm wavelengths. This eyewear has a slight amber tint. The reporter requested that a hoya representative send her a clear set of eyewear without the amber tint. Hoya sent clear eyewear to her, but sent eyewear rated only for 2940 nm. Upon receipt of the eyewear, she did not check the rating on the eyewear prior to using them. She used them for several months prior to 11/06. She claims to have been wearing this eyewear in 2006 when she allegedly received a back reflection off of the surface of the compression window she was using. The 2940 nm eyewear was removed from the treatment room immediately after the event and replacement eyewear with the correct protection was sent to institute on 12/20/06.

1.10.15 REPORT 15: ELECTRICAL SHOCK

Our service technician was working on a Medlite C6 laser at the office of the doctor. The laser was reporting error 22, which is "no end of charge," which means the hv capacitor is not getting fully charged and the flashlamp was not flashing. He evaluated the system and found the simmer supply was working, the lamp was simmering, but the lamps were not flashing. He decided the hv power supply should be replaced. He reports that he turned the system off, unplugged the system, and laid down on the floor to remove the power supply. He disconnected the hv cable from the supply that goes to the hv capacitor and the ac input cable to the supply, and the last thing he remembered was removing the control cable to the supply and then he received the shock. The office staff heard a loud bang from the room, found him bleeding from the ears, nose, and mouth, and they called the building manager who tried to resuscitate him. The emergency response team resuscitated him and took him to the hospital for treatment.

1.10.15.1 Manufactures response

They submitted a report on the site evaluation of the system and interviews with the office personnel and with service technician. The high voltage power supply, the scr board, and the charge capacitor were returned from the system for further evaluation: the scr board is intact, the high voltage discharge relay and discharge resistors are intact, and the resistance that was connected to the charge capacitor to discharge the capacitor measured 33 kΩ. This is consistent with the design that has 3100 kΩ resistors in parallel. The high voltage power supply was opened and we found that the bridge rectifier on the input ac was burned and all diodes were shorted, which would not allow the power supply to develop a high voltage output. This is consistent with report that indicated that when they turned on the system, it reported an error 22 and there was no voltage developed on the charge capacitor. In the report, service technician indicated when the system would not flash and was reporting an error 22, he also noticed a burning smell. This is consistent with the burned input bridge rectifier in the power supply. The capacitor was received in good condition. There is some evidence that the threads on one of the posts were damaged, not cross threaded, but the post had some rough spots on the threads. This is consistent with the observation in the report that the nut on the wire going from the charge cap to the scr board was slightly loose. This is the path to rapidly discharge the capacitor. Note the attached spread sheet on discharge times for the primary discharge path and the backup discharge path. It was observed that you could move the wire on the post that it was not tight, but it also was not sloppy loose. The diameter of the ring lug on the wire that goes on that post is very close to the diameter of the post, so there is little chance that the loose wire could have been really disabled from discharging the capacitor. The backup discharge resistor on the capacitor measures 2.2 MΩ that is consistent with the design. None of the evidence identifies defective components, parts, or designs that would cause the accident to happen. The only explanation that is possible is the power supply was able to deposit a charge on the capacitor as it was failing. In addition, the loose nut on the capacitor broke the connection to the fast discharge resistors on the scr board; as a consequence, the backup bleed resistor was discharging the partially charged capacitor when service technician contacted a high voltage point when he was removing the low voltage cables from the power supply. The only conclusion is if he had discharged the high voltage capacitor as he had been recently instructed to do, this accident would not have happened even if one of the system safety discharge circuits was inoperative. Because of the serious nature of the incident, they have taken additional preventative action steps to impress upon the service personnel the importance of following a specific safety regimen when working in and around high voltage. The service bulletin will be incorporated into the service manual, in addition to being sent to the service personnel. Preventative action: a service bulletin has been generated, which has been emailed to all the service engineers and distributors worldwide. This bulletin specifically addresses the safety precautions that must be observed when working in and around high voltage components. It also states the minimum wait time that should be observed before attempting to discharge the high voltage capacitor to allow the backup bleed resistor to discharge the capacitor to minimize the risk if the capacitor is not fully discharged. This service bulletin will be sent via UPS with a return receipt tracking requested to verify that all personnel and distributors have received the information. The manufacturing and engineering personnel will be trained in these safety procedures as well.

BIBLIOGRAPHY

American National Standards Institute, Z136.1. Safe use of lasers—201 version.

American National Standards Institute, Z136.8. Safe use of lasers in th research, Development and Testing Environment-2012 version.

http://www.eyecareamerica.org/eyecare/tmp/laser-surgery-of-the-eye.cfm

Wesley, J. M., Brumage, E. C., and Sliney, D. H., Intrabeam viewing of extended-source lasers with telescopes, *J. Laser Appl.*, 19(2), 89 May 2007.

Wesley, J. M. and Sliney, D. H., Methods for hazard assessment from viewing fiber optics with eye loupes, *J. Laser Appl.*, 16(3), 178, August 2004.

WORTH READING REFERENCES

Many of the references cited in the following were used to develo biological exposure levels in the existing laser standards as well a in the development of this chapter. Several of the references are review articles; their bibliographies should be used as a source of additional references. The most comprehensive and up-to-date bibliography of laser effects on the eye and skin is *Laser Hazards Bibliography* published by the U.S. Army Environmental Hygien Agency, Aberdeen Proving Ground, MD 21010-5422, and the latest version should be consulted.

Adams, D. O., Beatrice, E. S., and Bedell, R. B., Retinal ultrastructura alterations produced by extremely low levels of coherent radiatior *Science*, 177(4043), 58–60, 1972.

Barat, K., *Laser Safety: Tools and Training*, CRC Press, Boca Raton, FL, 2012.

Barat, K., Laser reference guide, Lawrence Berkeley National Laboratory, Berkeley, CA, Laser Safety web page, 2012.

Birngruber, R., Hillenkamp, F., and Gabel, V. P., Theoretical investigation of laser thermal retinal injury, *Health Phys.*, 48(6), 781–796, 1985.

Birngruber, R., Puliafito, C. A., Gawande, A., Lin, W., Schoenlein, R. T., and Fujimoto, J. G., Femtosecond laser tissue interactions: Retinal injury studies, *IEEE J. Quant. Electron.*, QE-23(10), 1836–1844, 1987.

Cain, C. P., Toth, C. A., DiCarlo, C. D., Stein, C. D., Noojin, G. D., Stolarski, D. J., and Roach, W. P., Visible retinal lesions from ultrashort laser pulses in the primate eye, *Invest. Ophthalmal. Vis. Sci.*, 36, 879–888, 1995.

Cain, C. P., Toth, C. A., Noojin, G. D., Carothers, V., Stolarski, D. J., and Rockwell, B. A., Thresholds for visible lesions in the primate eye produced by ultrashort near infrared laser pulses, *Invest. Ophthalmal. Vis. Sci.*, 40, 2343–2349, 1999.

Clark, A. M., Ocular hazards from lasers and other optical sources, *Cri Rev. Environ. Control*, 3, 307–339, November 1970.

Docchio, F. and Sacchi, C. A., Shielding properties of laser induced plasma in ocular media Irradiated by single Nd:YAG pulses of different durations, *Invest. Ophthalmal. Vis. Sci.*, 29(3), 437–443, 1988.

Farrer, D. N., Graham, E. S., Ham, W. T. Jr., Geeraets, W. J., Williams, R. C., Mueller, H. A., Cleary, S. F., and Clarke, A. M., The effect of threshold macular lesions and subthreshold macular exposures on visual acuity in the rhesus monkey, *Am. Ind. Hyg. Assn. J.*, 31(2), 198–295, 1970.

Gabel, V. P. and Birngruber, R. A., Comparative study of threshold laser lesions in the retina of human volunteers and rabbits, *Health Phys.*, 40, 238–240, 1981.

Ham, W. T. and Mueller, H. A., Phytopathology and nature of blue light and near UV retinal lesions produced by lasers and other optical sources. In: *Laser Applications in Medicine and Biology*, Wolbarsht, M. L., ed., Plenum Publishing Corporation, New York, 1989, pp. 191–246.

Lund, B. J., Laser retinal thermal damage threshold: Impact of small-scale ocular motion, *J. Biomed. Opt.*, 11(6), 064033-1, 2006.

Lund, B. J., Lund, D. J., and Edsall, P. R., Laser-induced damage threshold measurements with wavefront correction, *J. Biomed. Opt.*, 13(6), 064011-1, 2008a.

Lund, D. and Beatrice, E.S., Near infrared laser ocular bioeffects, *Health Phys.*, 56(5), 631–636, 1989.

Lund, D. J., Chapter 11: The maximum permissible exposure: A biophysical basis. In: *Laser Safety: Tools and Training*, Barat, K., ed., CRC Press, Taylor & Francis, Boca Raton, FL, 2009, pp. 139–165.

Lund, D. J., Edsall, P. R., and Stuck, B. E., Spectral dependence of retinal thermal injury, *J. Laser Appl.*, 20(2), 76–82, 2008b.

Lund, D. J., Edsall, P., Stuck, B. E., and Schulmeister, K., Variation of laser-induced retinal injury thresholds with retinal irradiated area: 0.1 s duration, 514 nm exposures, *J. Biomed. Opt.*, 12(2), 024023-1–024023-7, 2007.

Lund, D. J., Stuck, B. E., and Edsall, P., Retinal injury thresholds for blue wavelength lasers, *Health Phys.*, 90(5), 477–484, 2006.

Marshall, J., Trokel, S., Rothery, S., and Schubert, H., An ultrastructural study of corneal incisions induced by excimer laser at 193 nm, *Ophthalmology*, 92(6), 749–758, 1985.

Marshall, W. O., Bell, J., and Sliney, D., Methods for hazard assessment from viewing fiber optics with eye loupes, *J. Laser Appl.*, 16(3), 178, August 2004.

Marshall, W. O., Brumage, E. C., and Sliney, D. H., Intrabeam viewing of extended-source lasers with telescopes, *J. Laser Appl.*, 19(2), 89, 2007.

Mellerio, J., Light effects on the retina. In: *Principles and Practice of Ophthalmology*, Albert, D., Jakobiec, F. A., and Saunders, W. B., eds., WB Saunders, Philadelphia, PA, 1994, Chapter 116, pp. 1–23.

Morgan, J. I. W., Hunter, J. J., Masella, B., Wolfe, R., Gray, D. C., Merigan, W. H., Delori, F. C., and Williams, D. R., Light-induced retinal changes observed with high-resolution autofluorescence imaging of the retinal pigment epithelium, *Invest. Ophthalmol.*, 49(8), 3715–3729, 2008.

Ness, J. W., Zwick, H., Stuck, B. E., Lund, D. J., Lund, B. J., Molchany, J. W., and Sliney, D. H., Retinal image motion during deliberate fixation: Implications to laser safety for long duration viewing, *Health Phys.*, 78(2), 131–142, 2000.

Rockwell, B. A., Hammer, D. X., Hopkins, R. A., Payne, D. J., Toth, C. A., Roach, W. P., Druessel, J. J. et al., Ultrashort laser pulse bioeffects and safety, *J. Laser Appl.*, 11, 42–44, 1999.

Schulmeister, K., Husinsky, J., Seiser, B., Edthofer, F., Fekete, B., Farmer, L., and Lund, D. J., Ex vivo and computer model study on retinal thermal laser-induced damage in the visible wavelength range, *J. Biomed. Opt.*, 13(5), 054038-1–054038-13, 2008.

Schulmeister, K., Stuck, B. E., Lund D. J., and Sliney D. H., Review of thresholds and recommendations for revised exposure limits for laser and optical radiation for thermally induced retinal injury, *Health Phys.*, 100, 210–220, 2010.

Sliney, D. H., Development of laser safety criteria. In: *Laser Applications in Medicine and Biology*, Wolbarsht, M. L., ed., Plenum Press, New York, 1971, vol. 1, pp. 153–238.

Sliney, D. H., Interaction mechanisms of laser radiation with ocular tissues. In: *National Bureau of Standards, Special Publication*, 1984, pp. 355–367, NBS (ASTM STP 847).

Sliney, D. H., Aron-Rosa, D., DeLori, F., Fankhauser, F., Landry, R., Mainster, M., Marshall, J. et al., Adjustment of guidelines for exposure of the eye to optical radiation from ocular instruments: Statement from a task group of the International Commission on Non-Ionizing Radiation Protection (ICNIRP), *Appl. Opt.*, 44(11), 2162–2176, 2005.

Sliney, D. H., Dolch, B. R., Rosen, A., and DeJacma, F. W. Jr., Intraocular lens damage from ND:YAG laser pulses focused in the vitreous. Part 11: Mode-locked lasers, *J. Cataract Refract. Surg.*, 14(5), 530–532, 1988.

Sliney, D. H. and Wolbarsht, M. L., *Safety with Lasers and Other Optical Sources*, Plenum Publishing Company, New York, 1980.

Vincelette, R. L., Welch, A. J., Thomas, R. J., Rockwell, B. A., and Lund, D. J., Thermal lensing in ocular media exposed to continuous-wave near-infrared radiation: The 1150–1350 nm region, *J. Biomed. Opt.*, 13(5), 054005-1, 2008.

Vos, J. J., A theory of retinal burns, *Bull. Math. Biophys.*, 24, 115–128, 1962.

White, T. J., Mainster, M. A., Tips, I. A., and Wilson, P. W., Chorioretinal thermal behavior, *Bull. Math. Biophys.*, 32(9), 315–322, 1970.

Zuclich, J. A., Ultraviolet laser radiation injury to the ocular tissues. In: *Lasers et Nornes de Protection, First International Symposium on Laser Biological Effects and Exposure Limits*, Court, L. A., Duchene, A., and Courant, D., eds., Fontenay-aux-Roses: Commissariat al'Energic Atomique, Departement de Protection Sanitaire. Service de Documentation, pp. 256–275, 1988.

Zuclich, J. A., Edsall, P. R., Lund, D. J., Stuck, B. E., Till, S., Hollins, R. C., Kennedy, P. K., and McLin, L. N., New data on the variation of laser-induced retinal-damage threshold with retinal image size, *J. Laser Appl.*, 20(2), 83–88, 2008.

Zuclich, J. A., Lund, D. J., and Stuck, B. E., Wavelength dependence of ocular damage thresholds in the near-IR to far-IR transition region: Proposed revisions to MPEs, *Health Phys.*, 92(1), 15–23, 2007.

2 Wavefront sensors

Vasyl Molebny

Contents

2.1 INTRODUCTION

In ophthalmology, the term "wavefront sensing" is used in parallel with "aberrometry" and "spatially resolved refractometry." They have different origins, but they have the same meaning of acquiring the information on the optical system of the eye. *Wavefront sensing* originates from physics, especially from astronomy and from military applications of lasers. *Aberrometry* is a term in wide use in optics, as a means to describe the quality of optics. *Spatially resolved refractometry* was introduced to characterize nonhomogeneity of refractive properties of the eye. The acquired information on refractive imperfections of the ocular optics is presented in the form of mathematical expressions and displayed in the form of maps, other diagrams, and parameters.

Many speculations exist on Helmholtz's notice that he would never be a God's customer ordering the optics for the human eye. The reasons not to be a God's customer are in the aberrations produced by the imperfections of the ocular optics. A word-by-word translation of what he said is as follows [32]:

> Now, it is not too much to say that if an optician wanted to sell me an instrument, which had all these defects, I should think myself quite justified in blaming his carelessness in the strongest terms, and giving him back his instrument. Of course, I shall not do this with my eyes, and shall be only too glad to keep them as long as I can - defects and all.

For his time, he could not mean today's description of these defects like refraction nonhomogeneity with Zernike polynomials since Zernike invented them later.

But he certainly knew about the experiments of Scheiner [85] who looked at a candle through small separated holes demonstrating differing refractive power of the ocular optics in different zones of the eye aperture. He also knew pioneering experiments of Thomas Young who demonstrated the existence of aberrations by means of a simple experiment [113]. Young extended a thread from a point close to the eye, nearly along its visual axis, to a considerable distance away. Then, he closed the eye with a screen with four slits and observed four separate images of the thread and saw the threads intersecting differently depending on the accommodation state. Volkmann [105] modified Young's experiment looking through four small openings placed across the pupil and moving them to different distances from the center. He tested several observers and found some of them having a positive spherical aberration and some having a negative aberration. Donders moved a small perforated screen in front of the eye and noticed that the focus was not the same for different sectors. Helmholtz recognized the significance of these experiments [31] that gave him the audacity to criticize the lower-quality nature-created ocular optics as compared to the man-made optics.

2.2 EARLY STUDIES OF SPHERICAL ABERRATION

Following the approach of point-by-point (or even zone-by-zone) exploration of the eye, diversities of the techniques were proposed and tried. In the majority of studies, the goal was to find out what kind of aspherics the optical system of the eye is—myopic or hyperopic. Volkmann's studies were based on subjective patient's judgment whether the image is in or out of focus after having past along a certain path through small apertures in front of the eye. Bringing the image in focus from its out-of-focus position is less sensitive than evaluating the distances between two point images. Ames and Proctor tried this approach [5] and measured spherical aberrations along two meridians in the relaxed eye. They directed in the eye two beams of light of small cross section (Figure 2.1). One beam was passing through the center of the pupil, while the second was made to pass consecutively through a series of points. The observers had a possibility to vary the distance between the beams. Rotation resulted in changing the investigated meridian of the eye. Spherical aberration was calculated from the angle between the axial and the peripheral beams.

The investigators found that from the center of the pupil to a radius of about 1.5 mm, the eye became increasingly myopic; that is, it showed positive spherical aberration. From 1.5 mm to the pupil margin, the aberration decreased tending to become zero again at the very edge. The maximum observed values of the aberration were between 0.4 and 0.8 diopter.

Bahr [8] measured spherical aberration along the horizontal and vertical meridians of the pupil, using pairs of small pinholes to sample various zones along a meridian. He found that most of the eyes exhibited positive spherical aberration, the zones of the pupil becoming gradually more myopic from the center of the pupil outward. Negative aberration was present in few eyes. Some eyes exhibited negative aberration in one-half of a meridian only. Of all the eyes measured, there were none in which the aberration was completely symmetrical.

Otero and Duran [77] investigated the aberrations of the eye looking through each of a series of circular artificial pupils of different diameters. Koomen et al. [48] modified these measurements, subdividing the eye pupil into a central circular area of 2 mm diameter and several annular areas of increasing mean diameter. For each annular zone, a spectacle lens power was adjusted to produce the sharpest vision. The spectacle corrections for the annular zones, considered relative to the correction required for the small central area, gave the magnitude and the sign of the aberration. The measurements were made under conditions of controlled accommodation.

A simplified setup is shown in Figure 2.2. Annular apertures used for dividing the eye pupil into selected zones are placed in turn in front of the eye aperture. The annular apertures consist of transparent rings in an opaque background. The target, observed through the annular aperture, consists of a distant pair of point sources whose separation is held just above the resolution threshold for each annular zone. To fix the state of accommodation of the eye, there is a second target introduced by means of a small beam splitter.

Measurement of spherical aberration of an eye was also proposed by means of a dioptometer—a telescopic system that can be focused on objects located at different distances from its entrance pupil. One of the two special apertures limits the light entering the telescopic system under test to a ray located near

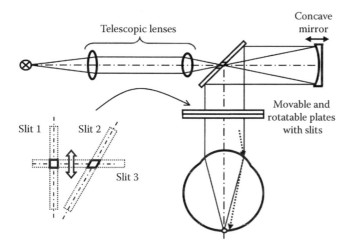

Figure 2.1 Measuring spherical aberrations with two narrow beams.

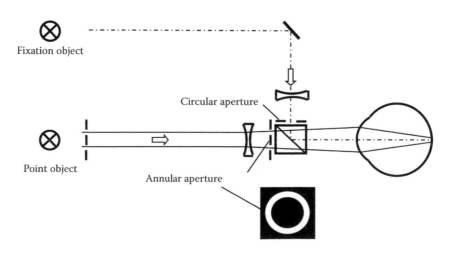

Figure 2.2 Measuring spherical aberrations with variable annular aperture.

Ophthalmic instrumentation

the paraxial region, and the other aperture limits the ray to the marginal region of the telescopic system [15]. The difference in the dioptometer focal readings obtained for these two pairs of "rays" is regarded to be a spherical aberration in diopters.

Ivanoff modified [46,47] the Ames and Proctor technique [5] using two thin vertical wires (upper and lower ones) in front of the objective illuminated from two different sources of light. When the eye has aberrations, the wires must be adjusted to be perceived as a continuation of each other. The "amount of adjustment" is a measure of the aberration. Similarly to his predecessors, Ivanoff was concentrated on the spherical aberrations along meridians. He found out that the outer zones of the average eye exhibited approximately 0.9 diopter of positive spherical aberration when in the relaxed state and 1.25 diopters of negative spherical aberration when accommodated at 3.0 diopters.

Koomen et al. [49] paid attention to the fact that the choice of the reference axis has a strong effect on the shape of the curve of spherical aberration. Since an improper choice may lead to a highly nonsymmetrical curve, Koomen et al. recommended to refer to an axis relative to which the eye exhibits the greatest degree of symmetry.

2.3 WAVEFRONT RECONSTRUCTION: PRELASER STAGE

To the middle of the past century, it became clear that it is not enough to measure spherical aberrations to describe the quality of the optical system of the eye. Even earlier, Helmholtz noticed [31] that "in most meridians of most eyes the points of intersection of the rays refracted with the central ray do not form a continuous series at all, as if the conception of spherical aberration does not apply here."

The results of pioneering studies were published by Smirnov [93] and confirmed by Brink [12]. Both of them not only measured the distribution of the optical power across certain meridians of the eye pupil but also reconstructed the maps of refraction nonhomogeneity of the human eye.

With Smirnov's instrument (Figure 2.3), the observer looks at the light-scattering shield S through a successively installed polarizer P_2 with a small central opening, through an opening in the test object (an opaque white *Screen*), crossed wires CW, and a

polarizer P_1. A rectangular grid is drawn on the test object, which is illuminated by the lamp L_1. Polarizers P_1 and P_2 are installed with orthogonal orientation of their axes. The shield S is back illuminated with the lamp L_2. With this layout, the observer can see the *Screen* through the polarizer P_2 in a wide field of view and the crossed wires CW only through the opening in the polarizer P_1 and through the opening in the *Screen*. The dimensions in this layout are as follows: distance from the eye to the polarizer L_2 is 15 cm, distance to the *Screen* is 1 m, the opening in the polarizer L_2 is 0.4 mm in diameter, and the diameter of the opening in the test object is 1 cm.

Brink's instrument was based on focusing of an image of a stripe object that is less sensitive to refraction variations than Smirnov's one.

Smirnov reconstructed the wavefront maps, having given them names of "plates of errors" (see an example in Figure 2.3), analyzed and recalculated the results of Ivanoff, explained the causes of the discrepancies in Ivanoff's conclusions, compared his own results with the results of other authors, and came to the conclusion that, generally speaking, the wavefront errors of the eye cannot be described only in terms of defocusing or simple astigmatism. They, as a rule, are essentially nonsymmetrical. He noted that the high level of aberrations correlates with the drop of visual acuity.

But for "not very high level of aberrations there does not seem to be a robust and direct relationship between the level of the aberrations and the visual acuity."

Based on his findings, Smirnov expressed a far-reaching idea: "In principle, it is possible to manufacture a lens compensating the wave aberration of the eye in the complex form of the plates of errors. The lenses must obviously be contact ones. Otherwise, even small turns of the eye will produce sharp increase in aberrations of the system." The time to implement this idea came about 40 years later [18].

As with the other techniques described earlier, Smirnov's was quite laborious: the measurement took 1–2 h, and calculations in those days took 10–12 h. For these reasons, Smirnov suggested that it is unlikely that "such detailed measurements will ever be adopted by practicing ophthalmologists." He believed that principal value of his study was the understanding of the restrictions of the visual channel of information acquisition, and for these reasons, he abandoned

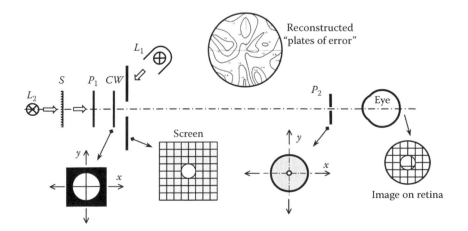

Figure 2.3 Smirnov's experimental setup.

further development of his technique (M.S. Smirnov, 2005, private communication to V. Molebny).

Instead of point-by-point investigation of the eye, Tscherning devised his "aberroscope" to see the whole picture of aberrations at a glance. He designed it as a rectangular grid of thin wires combined with a plano-convex lens [102]. With the lens in front of his eye, he observed a distant luminous point. Due to the "artificial myopia," the opaque lines produced shadows on the retina. From the direction of the curvature of the shadows, Tscherning differentiated positive and negative spherical aberrations and found that in his eyes, the aberration is positive when the eye is relaxed, tending toward the negative as his eye accommodates. Examples of distorted projections are shown in Figure 2.4.

In 1968, Howland modified Tscherning's technique of grid projection and applied it for photographic lens evaluation [37]. Instead of a grid–lens pair, a lens–grid–lens triplet was used where lenses were plus and minus plano cylinders, with the grid sandwiched between their plano surfaces (Figure 2.5). Same with Tscherning's grid–lens pair, the lens–grid–lens triplet was also called an "aberroscope." Later, this technique was applied for eye investigation [36,38], and it is known as a crossed-cylinder technique. A point source of light was generated by placing a fiber optic at the front focal plane of a microscope objective (Figure 2.6, no photocamera is used in this mode of operation). The lenses taken were +5 and –5 D. The combination forms a single test lens element that fits into the front of a trial lens frame

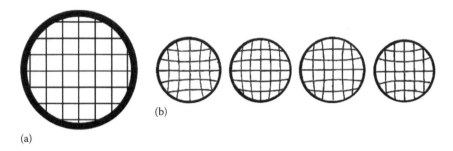

(b)

(a)

Figure 2.4 (a) Projected grid of the Tscherning aberroscope and (b) its distortions in the human eye.

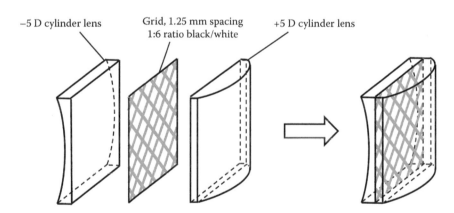

−5 D cylinder lens · Grid, 1.25 mm spacing 1:6 ratio black/white · +5 D cylinder lens

Figure 2.5 Crossed-cylinder components.

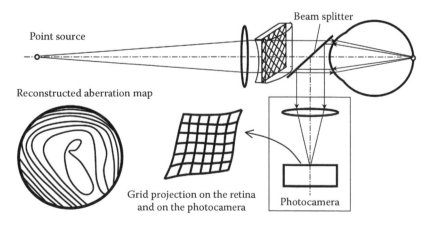

Point source · Beam splitter · Reconstructed aberration map · Grid projection on the retina and on the photocamera · Photocamera

Figure 2.6 Measuring aberrations with crossed cylinders.

The grid spacing is 1.25 mm. The diagonals of the grid coincide with the axes of the crossed-cylinder lens combination.

The subject looks at a point source of light about 1 m distant in a darkened room. The subject's near prescription is placed in the trial frame. The lines in the central squares of the grid are seen as horizontal and vertical. Any change in spherical or cylindrical power causes a tilting of the grid lines. The subject is asked to memorize the grid's pertinent features so that he or she can later sketch the grid with these features. The sketch is then analyzed quantitatively to estimate the wave aberration.

Photographing the subject's retina instead of subjective sketching was proposed by Walsh et al. [107]. Computer reconstruction of an aberroscope grid projection on the retina (and on photocamera) and topographic representation of wave aberrations (aberration map) are illustrated in Figure 2.6. The contours in the topographic maps are at 0.33 mm intervals. Large errors in measured aberrations occur from incorrect choice of grid center, from incorrect alignment, and from errors in setting up the aberroscope [95].

In the early 1990s, a next step was made in aberrometry with the introduction of a spatially resolved refractometer (SRR) [111]. Its principle is explained in Figure 2.7. Here, P_1 and P_2 are the pinholes, which are xyz movable. Lens L_0 images the pinholes at the plane of the cornea. A beam splitter BS transmits light for P_1 and reflects light for P_2. Light from a lamp shines on a steel bearing ball and is reflected in all directions. Some of this light falls on the gimballed mirror and is reflected to the lens L_1 and through the pinhole P_1 is directed into the eye. When reaching the cornea, this light is collimated. Its direction is controlled by the gimballed mirror that changes the beam's direction without changing its position at the cornea.

The P_2 is imaged at a region near the corneal apex (ideally coinciding with the "center" of the pupil), and the position of the image of P_1 is varied in the plane of the pupil (relatively the "center"). Through P_2, the subject looks at a reference image (alignment target AT) and through P_1 at a test image (a bright dot of a point source of light). The gimballed mirror is controlled by a joystick. Subject's task is to align the test image with the reference image by means of the joystick. The orientation and magnitude of the angular change of the beam defined by the tilt of the gimballed mirror are recorded; then the test aperture is

moved to a new position for another measurement. The authors took measurements in 30 separate positions on the cornea (having a dilated pupil 6–7 mm in diameter and a probe beam being 1 mm in diameter).

Authors described the displaying of the results of the refraction measurements in the form of a vector map (Figure 2.7). Each vector is anchored on a dot at a corresponding location x, y that is the position of the test aperture relative to the reference aperture at the cornea. The origin of this coordinate system is held close to the pupil center (usually within one mm) by the alignment procedure that was used, although the pupil center is not obligatory at that origin.

It was noticed by the authors that several general properties of these vector maps have simple interpretations. For example, if all vectors point toward the central point, the eye is myopic with spherical error only. If all vectors at the same radius have the same length, then there is no cylindrical error. The refraction in diopters can be estimated in regular cases by dividing the vector length in milliradians by the location radius in millimeters. The superposition of all the vectors describes the point spread function of the eye over the measured pupil. An estimate of the blur can be made by projecting the ray vectors from the whole pupil around a retinal position whose centroid is the image center and whose root-mean-square (rms) diameter is an estimate of the size of the blurred image.

In Sergienko's astigmometer [89], bringing two images together was implemented by along-axis shifting of the target. Sixteen points were chosen in the eye aperture of 3 and 5 mm in diameter.

In one of the SRR modifications [29], a laser with rotating diffuser at the exit was used as a point source of light. A chopper was inserted on the beam path to arrange a pulse mode of operation. A focusing block was also introduced to compensate for defocus. Separate channels were used for providing the test stimulus, a reference stimulus with adjustable pupil size, and a real-time view of the subject's pupil.

In another modification [60], an oscilloscope display played the role of a source of light with a pupil sampling aperture randomly selecting the positions within the aperture (37 positions in the aperture sized about 6 mm, beam diameter being about 1 mm).

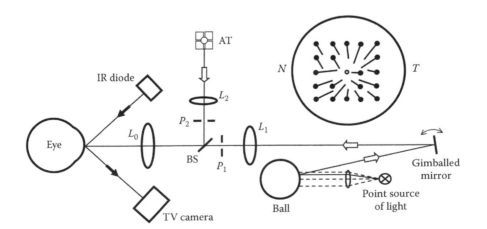

Figure 2.7 Spatially resolved aberrometer.

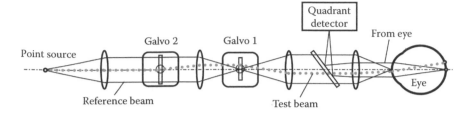

Figure 2.8 Spatially resolved aberrometer with automatic tilt adjustment of the projected laser beam.

In a later work [110], an approach was described of making the procedure of target alignment automatic. Two beams are projected into the eye alternately, the first one as a reference and another one as a test beam (Figure 2.8). A quadrant detector "looks" into the eye, measures relative positions of the test beam, and controls its positioning and tilt until the coordinates of its projection on the retina coincide with the coordinates of the reference beam.

The authors analyzed several techniques for the role of point source of light and came to the conclusion that it is impractical to use liquid crystal modulators, since they have a relaxation time of 50 ms at room temperature. If using the galvanometer-controlled mirrors having a resonance frequency above 1 kHz, an average time for large repositioning could be about 30 ms, resulting in 2 s for 65 corneal positions.

2.4 LASER-BASED WAVEFRONT SENSING: CLINICAL ABERROMETERS

The advent of laser opened new perspectives for wavefront sensing. Shack brought Platt's lenslets (made in his "wife's kitchen") from Arizona to Bille in Heidelberg, who asked his young postgraduate Liang to check them with the human eyes. It was the rebirth of Hartmann–Shack wavefront sensor for ophthalmology.

V. Molebny, a Ukrainian former star war scientist, met Pallikaris in Crete and discovered a problem for himself no less interesting than space—to study the refractive structure of the eye. His experience of spacecraft docking gave birth to three new lines of study (single- and double-beam ray tracing and holographic Hartmann–Shack wavefront sensor), the single-beam ray tracing being the most lucky one due to I. Chyzh' "wife's kitchen", to the talents of V. Sokurenko in math, S. Molebny in programming, E. Smirnov in acousto-optics, and J. Wakil in management. Fast acousto-optics of E. Smirnov's team was principal for the clinical use.

In Spain, Navarro's postgraduate Losada proposed him to try a "light pencil" for point-by-point eye investigation. They made a ray tracing instrument but missed an opportunity of fast scanning; their conclusion based on electromechanical scanning was negative for clinical use; the instrument needed about 5 s for eye investigation.

In Dresden, Seiler with the team of physicists found the way to rebuild the 100-year-old Tscherning aberroscope into Dresden aberrometer. It brought much trouble for Mierdel and Krinke to solve the problem of higher sensitivity: the first demonstrations were possible only in a dark room.

In the town of Gamagori, a young engineer Fujieda devoted to autorefractometry, inspired by Ozawa, started thinking of using

the skiascopic approach for aberrometry. The instrument entered the clinics later than three other approaches mentioned earlier, but its felicitous engineering (even with mechanical scanning) keeps it durably in the market.

2.4.1 HARTMANN–SHACK OCULAR WAVEFRONT SENSING

The program of the 1971 OSA Spring Meeting [90] left a transcript of an abstract of the Shack and Platt presentation MG2 where "a conventional perforated screen" was replaced "with an array of contiguous lenticular elements, each approximately 1 mm square and approximately 125 mm in focal length. The recording is made in the common focal plane of the lenticular elements." It was a less than common 10 min presentation and was marked with a note that the paper could be presented "only if the chairman of the session rules that time permits."

The initial idea of "a conventional perforated screen" was described by Hartmann [27] at the beginning of the twentieth century who checked the quality of large-sized optics for astronomy. On the path of a bundle of light propagating through an optical component (lens), he installed an obscure screen with a series of small apertures. These small holes in the obscure screen were replaced by a lenslet array manufactured by Platt. An attempt was made to file for a patent but the application was never filed. A complete system was designed, fabricated, and delivered to Cloudcroft, NM, in the early 1970s but was never installed, and the facility was decommissioned [79].

The results of experiments on human eyes with Platt's lenslets were published in 1994 [54]. Another group working in Ukraine manufactured the lenslets with holographic technologies and used the microscopic Fresnel lenslets to measure the wavefront distorted by the eye [73]. To exclude the manufacturing errors, an additional reference channel was incorporated into the layout. As a result, each holographic microlens produced two images: of the reference spot and of the test spot. The distance between these spots was a measure of the transverse aberration.

A simplified schematic is shown in Figure 2.9. A thin laser beam is directed into the eye and, being reflected from the retina, exits the eye. Usually, the microlenses are arranged in the form of matrix. Each microlens of the lenslet array projects its part of the beam cross section on the charge-coupled device (CCD) (complementary metal-oxide-semiconductor [CMOS] or other types) camera. Computer reconstructs the wavefront and can calculate any of the derivative parameters of the optical system of the eye. The combination of a set (a matrix) of lenses (a lenslet array) and a 2D photosensitive detector (e.g., a CCD camera) are usually called Hartmann–Shack sensor. The image in the plane of the photosensitive detector is called hartmanngram.

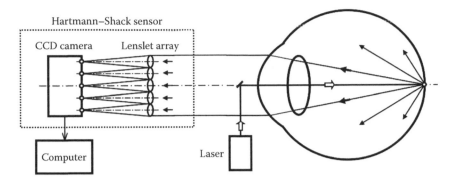

Figure 2.9 Principle of measuring the wavefront errors in the eye with Hartmann–Shack sensor.

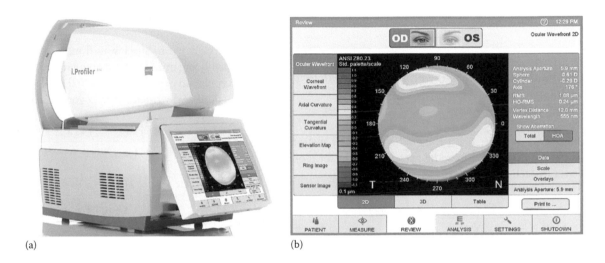

(a)   (b)

Figure 2.10 (a) iProfiler (Carl Zeiss Vision GmbH) multifunctional instrument (5-in-1) and (b) an example of the wavefront map.

Many commercial versions of clinical aberrometers were developed, based on Hartmann–Shack wavefront sensing: iDesign (Abbott Medical Optics [AMO]) [40], iProfiler (Carl Zeiss Vision GmbH) [41], KR-1W (Topcon) [42], etc. One of the marketing features of the last models of all commercial aberrometers is their multifunctionality. Zeiss' iProfiler (Figure 2.10), for example, is a combination of ocular wavefront aberrometer, autorefractometer, corneal topographer, pupillometer, and keratometer. Its specs are typical for Hartmann–Shack technique: ocular wavefront is reconstructed up to seventh-order Zernike decomposition. The instrument is provided with touch screen. The measuring range for sphere is –20 to +20 D [41]. The measuring range for cylinder is 0 to +8 D. The axis is 0°–180°. The three measurement zones are within 2.0–7.0 mm. The number of measuring points is up to 1500. Reference wavelength for the interpretation of refractive errors (referring to maximum luminosity function $V(\lambda)$ of the human eye in daylight) is 555 nm. The instrument measures both eyes automatically in approximately 30 s.

2.4.2 WAVEFRONT RECONSTRUCTION

Presentation of measured wavefront data and the procedure of the wavefront error reconstruction for ophthalmologic applications are described elsewhere [7,51,70,98]. Also, precise and detailed description is given in the U.S. and International standards [4,45]. In general, aberrometers measure the gradient of the wavefront error function or, in other words, the deflection of rays from an unaberrated direction. From each measured ray or location in the wavefront, the data contain four numbers. The first two of them are the horizontal (x) and vertical (y) coordinates of the location given. The second two numbers are the measured horizontal and vertical component values of the gradient. When reconstructing, the value of wavefront error should be calculated in each (x, y) point. Two-dimensional distribution (a map) of this error can be found as an approximation using the values of wavefront error in each point or as an interpolation.

Analytically, the approximated surface of the wavefront error contains a set of 2D polynomials that is similar to a stack of multiple layer transparencies [69]. This way of presentation is called a global one because each member of the polynomial is a component describing the whole map, not a part of it. Interpolation describing local specificities (in each zone of the map) is called a zonal presentation.

The mentioned standards recommend approximation based on Zernike polynomial presentation, which are not the only possibility. Other types of polynomials can be used: Bhatia–Zernike, Fourier, Fourier–Mellin, Taylor, Bessel, Legendre, Didon, Karhunen–Loeve, etc. [11,36,83,101]. In most cases, preference is given to Zernike polynomials, because their components, describing the lower-order aberrations, coincide with the conventional ophthalmologic description. If the wavefront error is reconstructed in Zernike polynomials, it can be transformed into the presentation as a Fourier series, and vice versa [16]. The same transformations can be done between Zernike and Taylor polynomials. The simplest way of interpolation is linear, just connecting two points by a line. More sophisticated is connecting by a curve, satisfying certain requirements. These curves are usually the splines [6]. Combination of modal (e.g., Zernike polynomials) and zonal (e.g., cubic B-splines) can be efficient in cases when polynomial description would need too long series [21].

If the reconstructed wavefront $W(x, y)$ ("plate of error"—according to Smirnov) is to be represented by Zernike polynomials, it will look like

$$W(x, y) = \sum_{m=0}^{M} C_m Z_m(x, y), \qquad (2.1)$$

where

$Z_m(x, y)$ is the mth mode of the polynomial (its mth term being a 2D function)

C_m is its coefficient ("weight"), being a real number

Since measured parameters are the gradients of the wavefront, the results of measurements can be presented in the form of the derivatives in each of N points:

$$\frac{\partial W(x_n, y_n)}{\partial x}, \frac{\partial W(x_n, y_n)}{\partial y}. \qquad (2.2)$$

With this in mind, one may rewrite Equation 2.1 in the form of two equations:

$$\frac{\partial W(x_n, y_n)}{\partial x} = \sum_{m=1}^{M} C_m \frac{\partial Z_m(x_n, y_n)}{\partial x},$$

$$\frac{\partial W(x_n, y_n)}{\partial y} = \sum_{m=1}^{M} C_m \frac{\partial Z_m(x_n, y_n)}{\partial y}. \qquad (2.3)$$

C_m are the unknowns to be calculated. Applying Equation 2.3 for all N points of the aperture will result in the following expression:

$$\begin{pmatrix} \dfrac{\partial W(x_n, y_n)}{\partial x} \\ \dfrac{\partial W(x_n, y_n)}{\partial y} \end{pmatrix} = (C_m) \times \begin{pmatrix} \dfrac{\partial Z_m(x_n, y_n)}{\partial x} \\ \dfrac{\partial Z_m(x_n, y_n)}{\partial y} \end{pmatrix}$$

$$\text{for all } n = 1, \ldots, N \text{ and } m = 1, \ldots, M. \qquad (2.4)$$

Practically, in most cases, this system is overdetermined, meaning that the number of points N, in which the data on wavefront slopes are acquired, is at least twice as large as the number of modes $M + 1$ of the set of Zernike polynomials. It means that the number of equations is at least four times larger than the number of unknowns C.

Solving Equation 2.4 means the best fit of the coefficients C, when the rms of the difference between measured and calculated

values of the gradients is minimized [33]. Zernike polynomials have some specific properties like orthogonality within the aperture, whose radius should be a unit. The solution suggests also a series of requirements to be applied to the shape of the aperture. It is not the goal of this chapter to go into the details of the wavefront reconstruction; they can be found in more specific publications [9,57,108].

Not all commercial aberrometers use Zernike decomposition. AMO, for example, prefers Fourier series. Fourier algorithms use data from pupils of any shape and reconstruct all peripheral data (data that lie beyond the circular area) unlike with the use of Zernike reconstruction. With data from about 240 points, the AMO Fourier algorithm provides precise measurement of the wavefront error that is equivalent to 20th-order Zernike measurements [39], while other instruments are restricted with seven orders.

In Hartmann–Shack wavefront sensor, the subapertures are normally organized as a rectangular grid. For this case, cubic spline interpolation can be presented [62] in the form of spline gradient functions $S(x)$ written for orthogonal directions as follows:

$$S(x) = \frac{(x_{i+1} - x)^2 [2(x - x_i) + (x_{i+1} - x_i)]}{(x_{i+1} - x_i)^3} S(x_i)$$

$$+ \frac{(x - x_i)^2 [2(x_{i+1} - x) + (x_{i+1} - x_i)]}{(x_{i+1} - x_i)^3} S(x_{i+1})$$

$$+ \frac{(x_{i+1} - x)^2 (x - x_i)}{(x_{i+1} - x_i)^2} S'(x_i) + \frac{(x - x_i)^2 (x - x_{i+1})}{(x_{i+1} - x_i)^2} S'(x_{i+1}), \qquad (2.5)$$

where

x is a generalized coordinate (both x and y)

x_i and x_{i+1} are the coordinates of ith and $(i + 1)$-th points of an eye aperture in which the wavefront tilts $S(x_i)$ and $S(x_{i+1})$ are measured

$S'(x_i)$ and $S'(x_{i+1})$ are the first partial derivatives in points with coordinates x_i and x_{i+1} correspondingly.

There is a fundamental limitation of Hartmann–Shack wavefront sensing following from the condition that each spot imaged by a lenslet array must be within a virtual region on a detector to calculate its centroid correctly. This limitation creates a trade-off between accuracy and dynamic range. The longer the focal length of the microlenses, the higher is the sensitivity. But when highly aberrated wavefronts are measured with a long-focal-length lenslet array, crossed-over spots could appear within the same virtual centroiding area due to a larger spot displacement.

The simplest way to avoid this problem is to reduce the amount of spot displacement by using a significantly shorter focal length. However, this solution also decreases measurement sensitivity due to the inability to detect small spot displacements when relatively small amounts of aberration are measured.

"Countermeasures" to subaperture violation can be software or hardware based. Software-based methods are usually more time consuming. One of the hardware-based ways is to switch the subapertures on and off, so that a definite assignment of the spots to their subapertures is possible. Yoon et al. [112] proposed the blocking of adjacent lenslets using translatable plates. Instead of mechanical switching, Lindlein et al. [56] used spatial light modulators in front of the microlenses of the sensor to switch on and off the subapertures. It was done with a coding algorithm that allows a definite assignment of the spots to their subapertures. With relaxation time 50 ms for liquid crystal modulators and 50 ms readout time for CCD device, a five-time switching and reading will result in at least 500 ms for measurement.

Switching the position of the spot itself on the retina corresponding to the neighboring lenslets was proposed by Molebny [64,72], where the switching was provided with acousto-optic deflectors. Grouping the subapertures like that with a Yoon translatable plate, each probing provides one-fourth of the information. Switching four times with different tilts of the probing beam will complete the cycle. The tilt of the beam should correspond to the pitch of the lenslet array. Sequential scanning of the retinal spot image in front of an elementary lenslet detector was proposed [99], which suggests a separate (in time) measurement in each point of the wavefront.

Not to spend time on switching, a parallel layout was proposed with astigmatic lenslets having different orientations of axes in such a way that no one neighbor of any lenslet has the same orientation [55]. The orientations change in 10° steps, the lenslets giving elongated images (Figure 2.11a). Even if there are crossover spots (e.g., see the central zone of Figure 2.11b), they can be identified by tilts of elongated images.

Groening et al. [26] described an algorithm that assigns the spots to their reference points even if they are situated far outside their subaperture. For this assignment, a spline function is extrapolated in successive steps with the iterative algorithm. Starting with a number of spots that can be assigned to their reference points unequivocally (normally, it is a central 3 × 3 matrix), the algorithm estimates the positions of the spots of neighboring microlenses by extrapolating a 2D spline function that assigns the spots to their respective reference points. If spots are found in a small area around the extrapolated spot positions, they are used to recalculate the spline function. This procedure is carried on iteratively until no spots are found at the extrapolated spot positions.

Lundström and Unsbo [58] modified this unwrapping algorithm, having used a B-spline function through a least squares estimate to the deviations of the central spots. Leroux and Dainty [53] used a similar approach but fit Zernike polynomials to the spot displacement data instead of a spline, producing a smoothing effect that suppresses noise. Lee et al. [52] and Smith and Greivenkamp [94] described algorithms in which spots are located in a predetermined order in such a way as to reduce ambiguity of the lenslet-spot correspondence.

Bedggood and Metha [10] compared these software methods for different kinds of errors and found that Leroux and Dainty ("Zernike") and Lundström and Unsbo ("B-spline") algorithms are more accurate in describing the aberrations they modeled.

2.4.3 RECONSTRUCTION OF A HARTMANNGRAM WITH RAY TRACING

In 1988, Häusler and Schneider [28] applied the principle of ray tracing known in theoretical optics to experimentally acquire a hartmanngram. Figure 2.12 demonstrates how the acquisition of the hartmanngram can be implemented with a laser, instead of multiple pinholes. A laser beam is directed into the tested optics with a translatable mirror (Figure 2.12a). Such point-by-point probing of the optics under test is done sequentially in time. Measurement of the position of the beam projections is made with the position sensing detector (PSD).

With nonmoving laser beam, translatable can be the tested optics itself. A PSD is positioned in the focal plane and measures the coordinates of the beam projection in the process of translation of the tested optics (Figure 2.12b). This approach was called an experimental ray tracing. The procedure yields directly the local

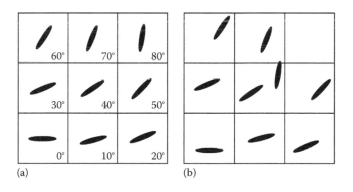

(a)   (b)

Figure 2.11 Focal spots of astigmatic lenslets: (a) no aberration; (b) with aberrations, crossover in the central zone. Projection from 80° lenslet can be identified by its tilt.

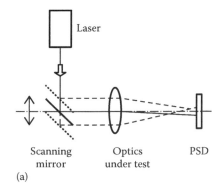

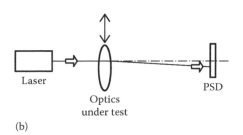

Figure 2.12 (a, b) Getting hartmanngrams with ray tracing.

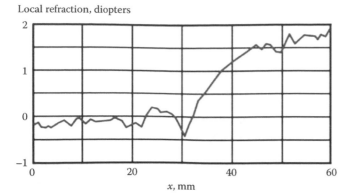

Figure 2.13 Reconstruction of a continuously varying refractive power in an aspheric spectacle lens.

refracting power of spherical and even strong aspheric optics. The tested optics was the aspheric spectacle lenses with a continuously varying refracting power from −0.5 to 2.0 D. An example of the reconstruction of the distribution of the local refractive power along a meridian in the pupil plane is given in Figure 2.13.

It was demonstrated also that the experimental ray tracing can successfully be applied for reconstruction of 3D shape of large optical objects like a car windshield whose profile was reconstructed from the data of the beam displacement along the glass plate.

2.4.4 RAY TRACING ABERROMETRY

The first ophthalmology-related publication baptized as a retina ray tracing was made in 1997 [68]. The initial solution was formulated by Molebny in his discussion with Pallikaris in 1995 and was implemented in 1996–1998. The instrument went through the first clinical tests in 1998–1999 [71,78]. After the FDA approval, the instrument is being manufactured by Tracey Technologies.

The ray tracing technique uses measurement of the position of a thin laser beam projected onto the retina. A beam of light is directed into the eye parallel to the visual axis having passed a two-directional (x, y) acousto-optic deflector and a collimating lens (Figure 2.14).

The front focal point of the collimating lens is positioned in the center of scanning C of the acousto-optic deflector. Each entrance point, one at a time, provides its own projection on the retina. The PSD measures the transverse displacement δ_x, δ_y of the laser spot on the retina. An objective lens is used to optically conjugate the retina and the detector plane.

Figure 2.15 shows a version of an iTrace aberrometer. A set of entrance points overlaid on the image of the eye is encircled by a green outer contour corresponding to the shape of the pupil. In the process of beam point-by-point repositioning over the eye aperture, data on the transverse aberration for each point of the beam entrance into the eye are collected. They represent a 2D distribution known as a retinal spot diagram (Figure 2.15, right bottom). The instrument uses a narrow (0.30 mm) laser diode beam with the wavelength 785 nm being displaced over the entrance pupil of an eye while kept parallel to the visual axis.

To measure the positions of laser spots on retina, X and Y linear arrays (each of 1024 photodetectors) are used. Examples of signals from these photodetectors in one of the 256 points of the aperture are shown in Figure 2.16. The left diagram demonstrates a signal from the laser beam focused on the retina. The signal of a defocused laser spot would be wider; its centroid would be determined with a lower accuracy. It is clearly seen that the "wings" of the signal are widespread. It is the result of light scatter in the layers of the retina.

If there is scatter in the crystalline lens, it will be added to the scatter in the retina; this phenomenon changes the shape of the "wings" (Figure 2.16, middle). The portion of scatter in the lens can be found from the difference between the scatter in the retina (red line in Figure 2.16, right) and the total scatter in the "wings" from the same retinal spot.

Examples of point spread function (PSF) and modulation transfer function (MTF) are given in Figure 2.17. The instrument allows 2D and 3D displays of both of them and includes in calculations any combination of Zernike modes. Any of these functions can be calculated for any size of the pupil.

In the process of measurement, an image of the pupil is captured; its size and position are automatically measured in each TV frame. The permission to fire the radiation is given by the software only when the positions of the pupil and visual axis are within a predetermined correspondence.

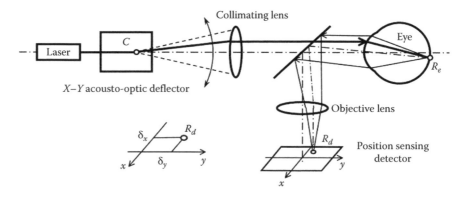

Figure 2.14 Ocular ray tracing with a fast X–Y acousto-optic deflector.

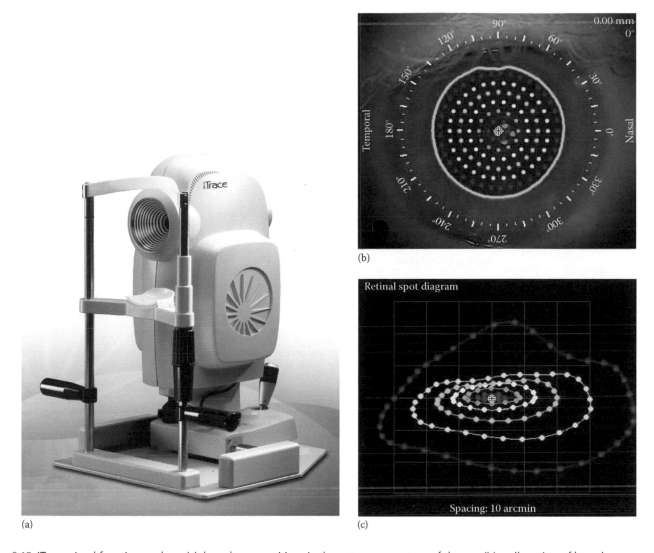

(a)

(b)

(c)

Figure 2.15 iTrace visual function analyzer (a); laser beam positions in the entrance aperture of the eye (b); collocation of laser beam projections on the retina–retinal spot diagram (c).

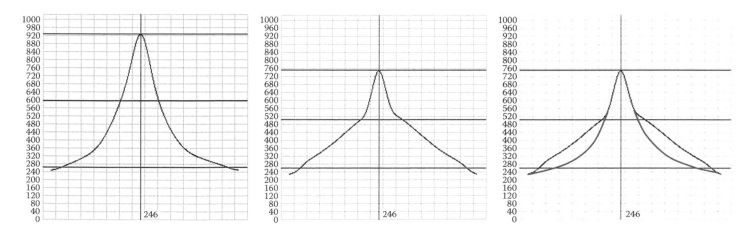

Figure 2.16 Signals from the separate retinal laser spots provide a unique opportunity to calculate the scatter in the crystalline lens and to evaluate its status.

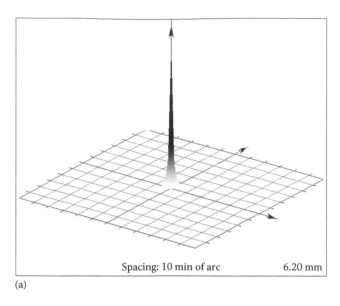

Spacing: 10 min of arc   6.20 mm

(a)

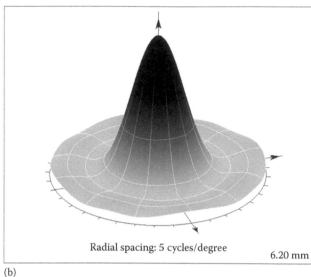

Radial spacing: 5 cycles/degree   6.20 mm

(b)

Figure 2.17 (a) 3D PSF and (b) 3D MTF.

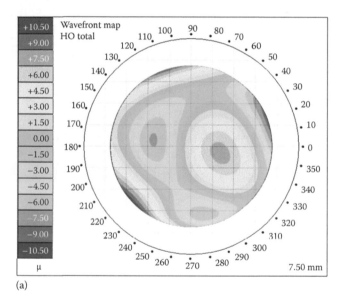

(a)

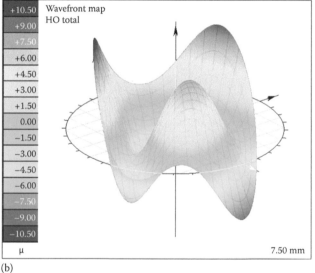

(b)

Figure 2.18 Wavefront map displayed in iTrace in (a) 2D and (b) 3D versions.

The total time of scanning for the entire aperture of the eye is within 100 ms. The duration depends on the number of test points at the eye entrance pupil (from 64 to 256). Acousto-optic deflectors are very fast devices: transition time for switching a position is about 10 μs, that is, for 256 points, less than 3 ms is spent on switching. This is an advantage of acousto-optics as compared to much slower scanners (no more than 20 points/s), used in the laboratory studies by Navarro et al. [75,76].

Examples of reconstructed wavefront maps are given in Figure 2.18; refraction maps are shown in Figure 2.19. Both are shown in 2D and 3D versions and with excluded defocus.

2.4.5 SKIASCOPIC ABERROMETER

The principle of skiascopy was used by Fujieda [22] to design the aberrometer that was combined with the corneal topographer [59]. Narrow strips of light are directed into the eye. These strips

are the projections of the light propagating from inside the rotating cylinder-shaped drum (chopper wheel) with narrow slits in the generatrix (Figure 2.20). Like in the ray tracing technique, the center of scanning (center of the drum cylinder) coincides with the front focus of the collimating lens. An objective lens conjugates the retina and a set of photodetectors. The strips running in the retina plane are imaged in the plane of this set of photodetectors. The set consists of two groups: (1) the delay measuring detector array and (2) the centering detectors. The first group is used to measure the sign and value of ametropia and the second ones to center the images of the strips.

According to the principle of skiascopy, the direction of strip's run depends on whether the eye is myopic or hyperopic. For the direction of drum rotation shown in Figure 2.20, the directions of strip movement in the myopic and hyperopic eyes are indicated by arrows for the retina plane and for the plane of photodetectors

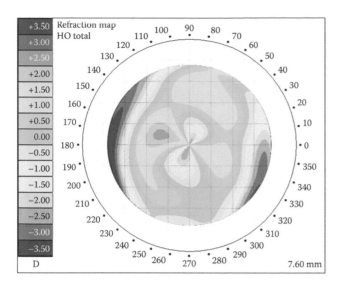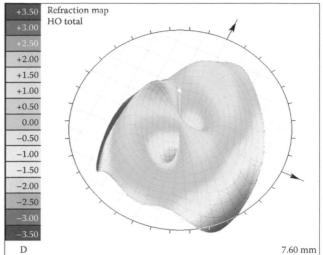

Figure 2.19 Refraction map displayed in Trace in 2D and 3D versions.

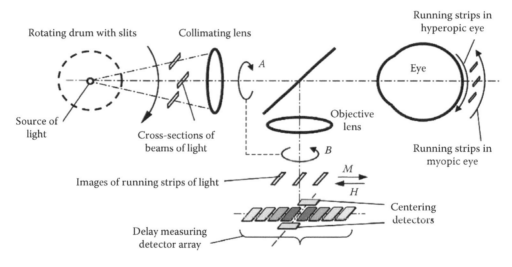

Figure 2.20 Optical layout of the aberrometer based on the skiascopy principle.

The time interval between the moments of light strip passing over individual detectors of the detector array depends on the distance between the detectors and the degree of ametropy for this given meridian of the eye. The measurements are taken between the pairs of detectors symmetrically positioned relatively to the center (in Figure 2.20, the detectors of the same pair are filled with the same color). This ranging from the center corresponds to certain diameters in the corneal plane. In one of the versions of the instrument, there are four pairs of detectors ranged as far as 2.0, 3.2, 4.4, and 5.5 mm in the plane of the cornea. The time difference in each pair of photodetectors is converted into the refractive power. In this way, four-ring data are accumulated for the reconstruction of the refraction map. After having measured the time intervals, the meridian is changed by the rotation in the direction of arrows A and B, the rotation of both axes being matched. More detector pairs would result in more dense spatial data in the radial direction.

An infrared LED is used as a source of light. The projecting system of the chopper wheel rotates 180° in 0.4 s across both semimeridians, so that 360 meridians are covered. The LED and photodetectors are conjugated with the cornea. The aperture stop (not shown in Figure 2.20) is conjugated with the retina when the eye is emmetropic. In myopia, the aperture stop is in front of the retina; in hyperopia, the aperture stop is behind the retina.

One of the last versions of Nidek/Marco OPD-Scan III is shown in Figure 2.21 [35]. Like its Hartmann–Shack competitors, it is multifunctional. Wavefront maps of the total optical system of the eye, of the cornea, and of the internal optics can be displayed with the Zernike graphs in the same window (Figure 2.21, right).

2.4.6 LASER-BASED TSCHERNING APPROACH

Seiler et al. modified the Tscherning idea of grid projection on the retina in such a way that the grid was substituted by a set of thin laser beams [74]. They formed this set of a parallel bundle of light having split it into a group of single thin parallel rays by means of a mask with a regular matrix of fine holes (Figure 2.22). These rays are focused by a low-power lens in front of the eye

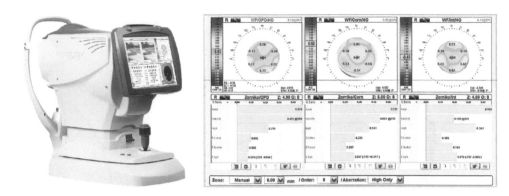

Figure 2.21 Nidek/Marco OPD-Scan III displaying higher-order wavefront map of total, corneal, and internal aberrations.

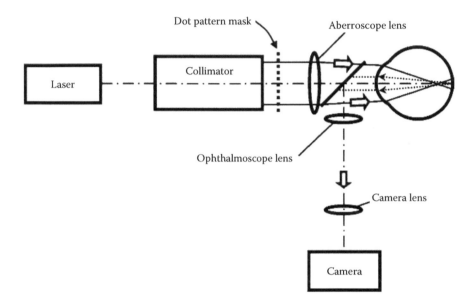

Figure 2.22 Optical layout of the aberrometer using the Tscherning principle.

so that the intraocular focus point is located about 2.5 mm in front of the retina. The purpose of the aberroscope lens is only to sufficiently enlarge the retinal spot pattern to separate and to identify the single light spots. Its optical power depends on the mean ocular spherical refraction. For example, an emmetropic eye needs this lens to be in the range from +4 to +5 D. In myopic eyes with the refraction from −6 to −9 D, this lens can be omitted. For hyperopic eyes of more than +2 D, a lens with a power of more than +5 D is needed.

The light source for the measuring rays was a green 532 nm diode-pumped solid-state laser with a beam diameter of about 2 mm and an output power of 10 mW. The measuring light is controlled by an electromechanical shutter with an opening time of about 60 ms. The laser beam is enlarged to a diameter of about 25 mm by means of a beam expander (a Keplerian telescope). Figure 2.23 (left) shows the configuration of the dot pattern mask that produces the system of test rays. The diameter of a single ray is about 0.3 mm. These rays are focused by a changeable aberroscope lens. The dot pattern masks are also changeable. They differ only in the dot spacing. Figure 2.23 (right) shows an example of a retinal light spot pattern. In an aberration-free eye, a projected spot configuration will have the same regularity

as at its origin. With real eyes, the spot positions are measured. For this purpose, the retinal light pattern is imaged onto the sensor of a low-light CCD camera using the ophthalmoscopic approach with a central near paraxial channel about 1 mm in diameter within the eye. This optical channel is assumed to be approximately free of higher-order aberrations.

According to the opinion of the authors [61], the described aberrometer does not provide exact measurements because of some specific reasons. The assumption of an approximately aberration-free paraxial ocular channel with a diameter of about 1 mm is not valid in every individual case. Furthermore, some opacities of the ocular lens or the vitreous body can considerably diminish the intensity of the rays, and some spots cannot be detected by the image processing program.

The technique allows only the measurement of eyes with astigmatism of less than 3–4 D. In eyes with greater astigmatic differences, the retinal light spot separations in one of the main axes are so reduced that clear identification of the spots is not possible. Nevertheless, practical experience shows that this aberrometer is applicable under clinical conditions. Wavelight Allegro Analyzer operates with the Alcon Allegretto wavefront-guided excimer laser [43].

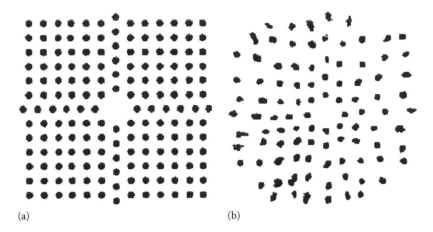

(a)   (b)

Figure 2.23 Dot pattern mask (hole diameters 0.33 mm, aberroscope lens +4 D, hole spacing 0.805 mm) and an example of images of retinal spots.

2.5 BEYOND THE COMMERCIAL ABERROMETRY

Interference, diffraction, and refraction phenomena are among those that can be used to determine the distortions of the wavefront. Light diffraction on gratings (Talbot effect), diffraction on edges (Foucault knife), shearing interferometry, and double- and triple-beam interferometry are the most known ones that use coherent light. Noncoherent techniques use beam splitting with pyramid-like (or double-roof) beam splitters, axicons, and concentric circular lenses.

2.5.1 TALBOT WAVEFRONT SENSING

Talbot effect discovered in 1836 [100] was later used for wavefront sensing [50,84] as an alternative to the Hartmann–Shack sensor [81]. It is based on the self-reconstruction of the image of a linear diffraction grating at certain distances behind the grating without any optical system, due to interference of diffracted waves, while the grating is illuminated by a monochromatic beam with a plane wavefront. The grating image is distorted when an optically inhomogeneous object is put between the grating and the self-imaging plane. This occurs because the spatial harmonics get different phase shifts while passing through different parts of the object. The degree of the image distortion describes the object quality.

Talbot wavefront sensing effect was further developed using the fractional Talbot effect [92]. Usually, the tested wavefront shape is calculated by estimating its deviation from a plane or a spherical wave.

Usage of the Talbot effect to measure ocular aberrations was experimented by Sekine et al. [88]. Several regular structures were involved that rotate relatively each other to make easier the identification of different types of aberrations [34].

The Talbot and Hartmann–Shack sensors using holographic gratings or holographic lenslets may have much in common, like optical setup for information registration, measurements of the spot shifts in the hartmanngrams or in the grating images. From the other side, grating distortions look similar to the regular structure distortions in Tscherning approach. Combination of moiré deflectometry with Talbot approach was proposed to measure ocular aberrations [20].

The wavefront sensor based on the Talbot effect correctly detects the wavefront slopes, if the principal curvature of the wavefront is substantially less than the inverse of the Talbot distance. Therefore, when analyzing the wavefront with a significant curvature, it is necessary to reduce the period of the grating accordingly that reduces the angular sensitivity of the sensor. This results in the similarity of the performance features of the Talbot and Hartmann–Shack sensors in their requirements to the period of lenslets in the Hartmann–Shack sensor and to the period of gratings in the Talbot sensor [80].

2.5.2 INTERFEROMETRIC RAY TRACING

Interferometric or frequency-shifted ray tracing technique has its roots in phase microscopy and its further development for profile measurement based on phase difference measurement is based between two beams that passed different optical paths [86,97]. In applications for ophthalmology, this technique was initially studied with frequency shifting along one of the two orthogonal axes, along which the scanning is provided [66,67]. Later, a three-beam system was proposed for microprofilometry [65].

The layout of the frequency-shifted ray tracing aberrometer is explained in Figure 2.24. Its principal difference from the simple ray tracing aberrometer consists of additional splitting of the beam into three components having the frequencies f_0, f_x, and f_y and using the phase discriminator instead of a PSD to calculate the wavefront tilt. The splitting is provided by the acousto-optic modulator (AOM) controlled by its driver. The configuration of the beams (shown in Figure 2.24) contains the central beam having the frequency f_0 and two diffracted beams having the frequencies f_x and f_y, which originate from the central frequency being shifted by ΔF_x and ΔF_y. These two beams are scanned in X and Y directions by the acousto-optic deflector (the term "scan" here means repositioning from one point to another within a certain aperture, being unmoved in each position for a prescribed time interval). The beams interfere on the retina, whose plane is conjugated with the photosensitive plane of the photodetector. Two signals of the frequencies ΔF_x and ΔF_y are filtered at the output of the photodetector. Reference signals of the same frequencies ΔF_x and ΔF_y that are used for controlling the AOMs are the reference inputs for the

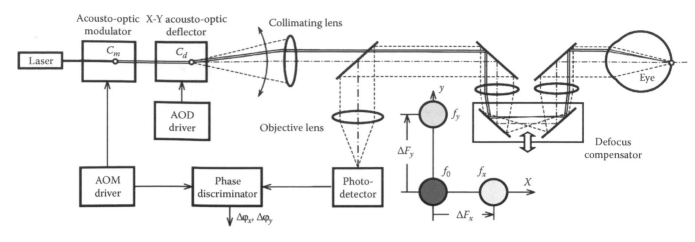

Figure 2.24 Layout of the frequency-shifted double-beam interferometric ray tracing aberrometer.

phase discriminator measuring the phase difference $\Delta\varphi_x$ and $\Delta\varphi_y$ at each of these two frequencies ΔF_x and ΔF_y.

The measured phase differences correspond to the optical path differences in X and Y directions, that is, to the X and Y components of the wavefront tilt in the given point of the eye aperture where the beam is positioned in the moment of measurement. Due to the specificity of the data processing with this technique, a defocus compensator is introduced on the beam paths into and out of the eye. This is necessary to make plane the wavefronts of the beams on the retina. Defocus compensator can be recommended in any other aberrometer, but in interferometric methods it is very important from the signal-to-noise considerations. The sensitivity of this technique is of the order of nanometers.

2.5.3 PYRAMID WAVEFRONT SENSORS

The main component of the pyramid sensor is a four-sided glass pyramidal prism (Figure 2.25). The incoming light is focused

onto the prism apex. The four facets of the pyramid split the incoming light into four beams, propagating in slightly different directions [17].

A relay lens placed behind the prism reimages the four beams, allowing adjustable sampling of the four different images I_1, I_2, I_3 and I_4 of the aperture on the CCD camera. The two measurement sets S_x and S_y are obtained from the four intensity patterns as

$$S_x = \frac{\left[I_1(x,y) + I_2(x,y)\right] - \left[I_3(x,y) + I_4(x,y)\right]}{I_0}, \quad (2.6)$$

$$S_y = \frac{\left[I_1(x,y) + I_4(x,y)\right] - \left[I_2(x,y) + I_3(x,y)\right]}{I_0}, \quad (2.7)$$

where I_0 is the average intensity per subaperture.

The four-sided pyramidal prism can be substituted with two orthogonally placed two-sided roof prisms. Each roof provides

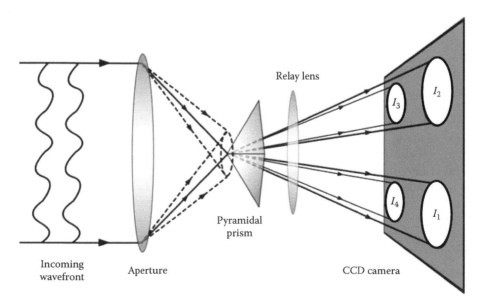

Figure 2.25 The idea of splitting the beam by a pyramidal prism.

two different images of the aperture on the detector. The two data sets S_x and S_y are obtained as the difference between the two intensity patterns. Due to the prisms' orthogonal placement with respect to each other, the two signal sets S_x and S_y are independent and contain information about the wavefront tilts only in x and y direction correspondingly.

2.5.4 ABERROMETER WITH CIRCULAR LENSLET ARRAY

A modified wavefront sensor was proposed involving a set of toroidal lenses [14]. The sensor consists of a set of concentric "half-donut" surfaces molded on an acrylic surface with a CCD located at the focal plane. When illuminated with a plane wavefront, it focuses a symmetric pattern of concentric discs on the photodetector plane, similar to Placido images from videokeratography. For a distorted wavefront, an asymmetric disc pattern is formed. From detection of shift in the radial direction, radial slopes are computed for a maximum of 2880 points and the traditional least squares procedure is used to fit these partial derivatives to a set of Zernike polynomials. A prototype contained eight discs (Figure 2.26). The central lens works as a simple spherical lenticule, which is used as a central reference for alignment.

A wavefront with aberrations hits the circular lenslet array and is projected on the photodetector (CCD camera), which is placed at the exact focal distance (F) of the "donut" lenses. One of the differences between this sensor and the Hartmann–Shack sensor is that only the radial slopes in the radial direction are available ($\Delta\rho/f$). A polar coordinate system is used to process and save disc information to computer files. All detected values of radial distance are functions of two parameters: disc number (which means the value of the radius) and the polar angle (θ).

Data containing the radial distances for both the calibration (c) and aberration (a) images are used to compute the radial slopes as

$$\frac{\partial z}{\partial \rho} = \frac{\rho_a - \rho_c}{F}, \tag{2.8}$$

where
  ρ_c and ρ_a are the radial distances for the nonaberrated and aberrated wavefront, respectively
  F is the focal distance of the toroidal lenses

Since this sensor does not form single spots on the image plane, there is no information regarding the polar slope $\partial z/\partial\theta$. Nevertheless, to the opinion of the authors, the lack of information in the tangential direction can be compensated by a greater resolution (a maximum of 2880 points). From the radial slopes and a set of Zernike polynomials, a least squares minimization function may be written and the coefficients are computed by solving a linear system.

Also, the difference in spatial resolution, from the central portion to the periphery of this sensor, is compensated by the greater amount of points when compared to the conventional Hartmann–Shack sensor. An advantage of the sensor is that less data points are lost on the edge of the pupil due to the circular shape of the optics.

Dynamic modulation of the incoming beam was proposed [103], which allows increasing of the linear range of the pyramid sensor and is also used to adjust its sensitivity [2]. The modulation can be accomplished in several ways, either by oscillating the pyramid itself [82] or with a steering mirror [13], implementing any of the two modulation scenarios—linear or circular.

Vohnsen et al. described wavefront sensing with a large-apex-angle axicon [104]. The focal length of an axicon is proportional to the off-axis distance of any ray incident parallel to its optical axis, and thus a positive axicon will focus an incident beam into an axial line image. Divergence of the incident beam shifts the image axially, suggesting a possible use in wavefront sensing. An axicon resembles the pyramid, but with an infinite number of facets forming the conical prism or axicon lens. The schematics are similar to that with pyramid, the apex serving as a scattering center, creating an interferometric reference wave.

For larger aberrations, more light intersects the axicon farther from the apex, reducing the contrast and complicating the interpretation of the interferograms due to refraction. Beyond ±3 D defocus, an image becomes noisy. Alternatively, an axicon with a larger apex area may be used, similar to the use of larger holes in the point-diffraction interferometer [1].

When aberrations are small, the focused light will be localized to a plateau near the imperfect axicon apex, and the sensor performance is similar to that of a point-diffraction interferometer. For larger aberrations, the refraction caused by the axicon becomes dominant, making a sensing application more complex

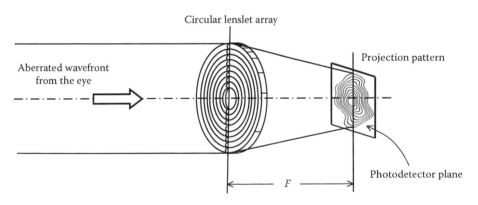

Figure 2.26 Wavefront sensing with a circular lenslet array.

but still feasible. The linear operational range of the sensor can be improved by reducing the refraction using a larger apex angle or liquid immersion or by focusing onto it more tightly.

2.6 DISCUSSION AND CONCLUSIONS

The early stages of studies of the irregular astigmatism were summarized by Helmholtz [31]. After Scheiner, among the names of those who worked on this problem in the candle era were Young, Purkinje, Péglet, Airy, Heineken, Hamilton, et al. The results of Volkmann [105] and Donders [19] created the basics for understanding the irregular astigmatism.

The nineteenth century and the prelaser era of the twentieth century brought new names in the studies of the refraction problems in the optics of the human eye: Tscherning, Ames, Proctor, Bahr, Otero, Coleman, Ivanoff, et al. And not so long ago, these became Smirnov, Brink, Sergienko, Webb, Penney, Howland, et al.

Some of the commercialized approaches in aberrometry were the direct continuation of the previous experience: the wavefront sensing with lenslet optics pioneeringly made by Platt was transfigured into Hartmann–Shack sensors, Tscherning's aberroscope was transformed into Tscherning aberrometer (initially called Dresden aberrometer), and Nidek's skiascopy has its roots in autorefractometry. The only commercialized technology developed from scratch is ray tracing.

Initially, these technologies were concentrated only on the wavefront analysis based on modal presentation with Zernike polynomials, including total, higher-order, and selected higher-order aberrations, and characteristics like MTF, PSF, and also vision simulation. Zonal presentation with splines was introduced in Tracey aberrometer. VISX (now AMO) prefers Fourier analysis.

Later on, corneal topography with Placido discs was added to commercial aberrometers. It enabled more detailed analysis of the eye components: cornea and crystalline lens can be analyzed separately. Ray tracing technique has the potential to be easily combined with the corneal topography [63,96].

The advertizing of the aberrometers as having 5-in-1 functionality became popular providing wavefront measurement, corneal topography, autorefractometry, pupillometry, and keratometry. Zeiss and Tracey added the functions necessary for IOL calculations. Tracey has an option of calculating the depth of focus and displaying the dysfunctional lens index. All companies work on simple and reliable alignment in the process of measurement, since it is critical for acquiring the correct results. Polarizing properties of the retina are a good candidate for objective measurement associated with eye position monitoring, eye fixation, and eye tracking [25,44]. New technologies of corneal inlays [109] with small apertures require very accurate positioning and, therefore, accurate determination of the visual axis [106].

Unique property of the ray tracing is that each laser spot imaged on the retina contains more information than only the position on the eye bottom. Its shape is influenced by the media of the beam propagation and by the medium of the eye bottom as well. From this shape, the scatter and opacity of the lens, as well

as their distribution over the lens can be determined that has sen for timely surgery. The technique of ray tracing can be combined with other techniques to measure the lens opacity [23].

The tendency to acquire more information about the structure of the eye in a combination of several approaches attracts the minds of researchers [3,30]. Such combination adds new measured parameters, like axial length [87] or objectively determined visual axis [114]. It allows to acquire 3D data on aberrations (aberration tomography) [24,91], etc.

REFERENCES

1. Acosta, E., S. Chamadoira, R. Blendowske, Modified point diffraction interferometer for inspection and evaluation of ophthalmic components. *J. Opt. Soc. Am. A* 23 (2006): 632.
2. Akondi, V., S. Castillo, B. Vohnsen, Digital pyramid wavefront sensor with tunable modulation. *Opt. Express* 21 (2013): 18261–18272.
3. Almeida, J. M., S. M. Franco, System to measure the topography of both corneal surfaces and corneal thickness. US Patent 6,913,358 (2005).
4. American National Standard for Ophthalmics, Methods for Reporting Optical Aberrations of Eyes. American National Standard for Ophthalmics. ANSI Z80.28-2004, American National Standards Institute, Merrifield, VA (2004).
5. Ames, A., C. A. Proctor, Dioptrics of the eye. *J. Opt. Soc. Am.* 5 (1921): 22–84.
6. Ares, M., S. Royo, Comparison of cubic B-spline and Zernike-fitting techniques in complex wavefront reconstruction. *Appl. Opt.* 45 (2006): 6954–6964.
7. Atchison, D. A., Recent advances in representation of monochromatic aberrations of human eyes. *Clin. Exp. Optom.* 87 (2004): 138–148.
8. Bahr, G. von, Investigations into the spherical and chromatic aberrations of the eye, and their influence on its refraction. *Acta Ophthalmol.* 23 (1945): 1–47.
9. Bará, S., J. Arines, J. Ares, P. Prado, Direct transformation of Zernike eye aberration coefficients between scaled, rotated and/or displaced pupils. *J. Opt. Soc. Am. A* 23 (2006): 2061–2066.
10. Bedggood, P., A. Metha, Comparison of sorting algorithms to increase the range of Hartmann-Shack aberrometry. *J. Biomed. Opt.* 15(1–7) (2010): 067004.
11. Bhatia, A. B., E. Wolf, On the circle polynomials of Zernike and related orthogonal sets. *Proc. Cambridge Philos. Soc.* 50 (1954): 40–5.
12. Brink, G. van den, Measurements of the geometrical aberrations of the eye. *Vision Res.* 2 (1962): 233–244.
13. Burvall, A., E. Daly, S. R. Chamot, C. Dainty, Linearity of the pyramid wavefront sensor. *Opt. Express* 14 (2006): 11925–11934.
14. Carvalho, L. A., J. Castro, W. Chamon, P. Schor, A new wavefront sensor with polar symmetry: Quantitative comparisons with a Shack Hartmann wavefront sensor. *J. Refract. Surg.* 22 (2006): 954–958.
15. Coleman, H. S., M. F. Coleman, D. L. Fridge, Theory and use of the dioptometer. *J. Opt. Soc. Am.* 41 (1951): 94–97.
16. Dai, G., Zernike aberration coefficients transformed to and from Fourier series coefficients for wavefront representation. *Opt. Lett.* 31 (2006): 501–503.
17. Daly, E. M., C. Dainty, Ophthalmic wavefront measurements using a versatile pyramid sensor. *Appl. Opt.* 49 (2010): G67–G77.
18. Dick, M., E. Schröder, J. Fiedler, H. Mäusezahl, V. Molebny, Method and device for completely correcting visual defects of the human eye. US Patent 6,616,275 (2003).
19. Donders, F. C., Beiträge zur Kenntniss der Refractions- und Accommodationsanomalien. *Arch. für Ophthalm.* 7 (1) (1862): 155–204.

20. Eagan, B. T., Moiré aberrometer. US Patent 7,341,348 (2008).

21. Espinosa, J., D. Mas, J. Pérez, C. Illueca, Optical surface reconstruction technique through combination of zonal and modal fitting. *J. Biomed. Opt.* 15(2) (2010): 026022.

22. Fujieda, M., Ophthalmic measurement apparatus having plural pairs of photoreceiving elements. US Patent 5,907,388 (1999).

23. Ginis, H., O. Sahin, A. Pennos, P. Artal, Compact optical integration instrument to measure intraocular straylight. *Biomed. Opt. Express* 5 (2014): 3036–3041.

24. Goncharov, A. V., M. Nowakowski, M. T. Sheehan, C. Dainty, Reconstruction of the optical system of the human eye with reverse ray-tracing. *Opt. Express* 16 (2008): 1692–1703.

25. Gramatikov, B. I., O. H. Y. Zalloum, Y. K. Wu, D. G. Hunter, D. L. Guyton, Directional eye fixation sensor using birefringence-based foveal detection. *Appl. Opt.* 46 (2007): 1809–1818.

26. Groening, S., B. Sick, K. Donner, J. Pfund, N. Lindlein, J. Schwider, Wave-front reconstruction with a Shack-Hartmann sensor with an iterative spline fitting method. *Appl. Opt.* 39 (2000): 561–567.

27. Hartmann, J., Objektivuntersuchungen. *Z. Instrum.* 24 (1904): 1–21.

28. Häusler, G., G. Schneider, Testing optics by experimental ray tracing with a lateral effect photodiode. *Appl. Opt.* 27 (1988): 5160–5164.

29. He, J. C., S. Marcos, R. H. Webb, S. A. Burns, Measurement of the wave-front aberration of the eye by a fast psychophysical procedure. *J. Opt. Soc. Am. A* 15 (1998): 2449–2456.

30. Hellmuth, T., Method and apparatus for simultaneously measuring the length and refractive error of an eye. US Patent 5,975,699 (1999).

31. Helmholtz, H. von, Optique physiologique. 1. Dioptrique de l'oeil. Edition Jacques Gabay, Paris, France (1867). Reimpression Jacques Gabay, Sceaux, France (1989).

32. Helmholtz, H. von, *Popular Scientific Lectures on Scientific Subjects*, Vol. 1, p. 194, Routledge/Thoemmes Press, London, U.K. (1996) (reprint of the 1895 edition).

33. Herrmann, J., Least-squares wave front errors of minimum norm. *J. Opt. Soc. Am.* 70 (1980): 28–35.

34. Heugten, A. Y. van, Wavefront sensor. US Patent 8,619,405 (2013).

35. Hoffman, R. S., Nidek OPD-Scan III. Pre- and post-operative diagnostic tool. http://www.finemd.com/pdfs/Nidek-OPD-Scan-III.pdf, accessed October 1, 2016.

36. Howland, B., H. C. Howland, Subjective measurement of high order aberrations of the eye. *Science* 193 (1976): 580–582.

37. Howland, B., Use of crossed cylinder lens in photographic lens evaluation. *Appl. Opt.* 7 (1968): 1587–1600.

38. Howland, H. C., B. Howland, A subjective method for the measurement of monochromatic aberrations of the eye. *J. Opt. Soc. Am.* 67 (1977): 1508–1518.

39. WaveScan. WaveFront System Home. http://www.abbottmedicaloptics.com/products/refractive/ilasik/wavescan-wavefront-system, accessed October 1, 2016.

40. iDesign. Un estudio WaveScan de avanzada. http://www.andreccorporation.com/docs/iDesign_Overview.pdf, accessed October 1, 2016.

41. iProfiler by ZEISS, http://www.hassans.com/upload/iprofiler_1161.pdf, accessed October 1, 2016.

42. KR-1W, Wavefront analyser, http://www.topcon-medical.eu/eu/products/40-kr-1w-wavefront-analyser.html#images, accessed October 1, 2016.

43. LASIK. Using the Alcon Allegretto Wavefront-Guided Excimer Laser vs AMO Visx Wavefront-Guided Excimer Laser (LASIK), https://clinicaltrials.gov/ct2/show/NCT01454843, accessed October 1, 2016.

44. Hunter, D. G., S. N. Patel, D. L. Guyton, Automated detection of foveal fixation by use of retinal birefringence scanning. *Appl. Opt.* 38 (1999): 1273–1279.

45. International Standard ISO/FDIS 24157:2008 (E), Ophthalmic optics and instruments—Reporting aberrations of the human eye. International Standard Organization, Geneva, Switzerland (2008).

46. Ivanoff, A., Les aberrations de l'oeil (Editions de la Revue de l'Optique Theorique et Instrumentale). Paris, France (1953).

47. Ivanoff, A., Sur une méthode de mesure des aberrations chromatiques et sphériques de l'oeil en lumière dirigée. *C. R. Acad. Sci.* 323 (1946): 170–172.

48. Koomen, M., R. Tousey, R. Scolnik, The spherical aberration of the eye. *J. Opt. Soc. Am.* 39 (1949): 370–376.

49. Koomen, M. J., R. Scolnik, R. Tousey, Spherical aberration of the eye and the choice of axis. *J. Opt. Soc. Am.* 46 (1956): 903–904.

50. Koryakovskiy, A., V. Marchenko, Wavefront detector based on the Talbot effect. *Sov. J. Tech. Phys.* 26 (1981): 821–825.

51. Lane, R. G., M. Tallon, Wave-front reconstruction using a Shack-Hartmann sensor. *Appl. Opt.* 31 (1992): 6902–6908.

52. Lee, J., R. V. Shack, M. R. Descour, Sorting method to extend the dynamic range of the Shack-Hartmann wave-front sensor. *Appl. Opt.* 44 (2005): 4838–4845.

53. Leroux, C., C. Dainty, A simple and robust method to extend the dynamic range of an aberrometer. *Opt. Express* 17 (2009): 19055–19061.

54. Liang, J., B. Grimm, S. Goelz, J. F. Bille, Objective measurement of wave aberrations of the human eye with the use of a Hartmann-Shack wave-front sensor. *J. Opt. Soc. Am. A* 11 (1994): 1949–1957.

55. Lindlein, N., J. Pfund, Experimental results for expanding the dynamic range of a Shack-Hartmann sensor using astigmatic microlenses. *Opt. Eng.* 41 (2002): 529–533.

56. Lindlein, N., J. Pfund, J. Schwider, Algorithm for expanding the dynamic range of a Shack-Hartmann sensor by using a spatial light modulator array. *Opt. Eng.* 40 (2001): 837–840.

57. Lundström, L., P. Unsbo, Transformation of Zernike coefficients: Scaled, translated, and rotated wavefronts with circular and elliptical pupils. *J. Opt. Soc. Am. A* 24 (2007): 569–577.

58. Lundström, L., P. Unsbo, Unwrapping Hartmann-Shack images from highly aberrated eyes using an iterative B-spline based extrapolation method. *Optom. Vis. Sci.* 81 (2004): 383–388.

59. MacRae, S., M. Fujieda, Slit skiascopic-guided ablation using the Nidek laser. *J. Refract. Surg.* 16 (2000): 576–580.

60. Marcos, S., S. A. Burns, E. Moreno-Barriuso, R. Navarro, A new approach to the study of ocular chromatic aberrations. *Vis. Res.* 39 (1999): 4309–4323.

61. Mierdel, P., M. Kämmerer, M. Mrochen, H. E. Krinke, T. Seiler, Ocular optical aberrometer for clinical use. *J. Biomed. Opt.* 6 (2001): 200–204.

62. Molebny, S., V. Molebny, L. F. Laster, Method for measuring the wave aberrations of the eye. US Patent 7,380,942 (2008).

63. Molebny, V., I. Pallikaris, Y. Wakil, S. Molebny, Method and device for synchronous mapping of the total refraction non-homogeneity of the eye and its refractive components. US Patent 6,409,345 (2002).

64. Molebny, V., Method of measurement of wave aberrations of an eye and device for performing the same. US Patent 6,715,877 (2004).

65. Molebny, V. V., G. W. Kamerman, E. M. Smirnov, L. M. Ilchenko, S. O. Kolenov, V. O. Goncharov, Three-beam scanning laser radar microprofilometer. *Proc. SPIE* 3380 (1998): 280–284.

66. Molebny, V. V., I. G. Pallikaris, L. P. Naoumidis, E. M. Smirnov, L. M. Ilchenko, V. O. Goncharov, High precision double-frequency interferometric measurement of the cornea shape. *Proc. SPIE* 2965 (1996): 121–126.

67. Molebny, V. V., I. G. Pallikaris, L. P. Naoumidis, G. W. Kamerman, E. M. Smirnov, L. M. Ilchenko, V. O. Goncharov, Dual-beam dual-frequency scanning laser radar for investigation of ablation profiles. *Proc. SPIE* 2748 (1996): 68–75.

68. Molebny, V. V., I. G. Pallikaris, L. P. Naoumidis, I. H. Chyzh, S. V. Molebny, V. M. Sokurenko, Retina ray-tracing technique for eye-refraction mapping. *Proc. SPIE* 2971 (1997): 175–183.

69. Molebny, V. V., I. H. Chyzh, V. M.Sokurenko, S. V. Molebny, I. G. Pallikaris, L. P. Naoumidis, Phase-transparency model of an eye optical system. *Proc. SPIE* 3192 (1997): 233–242.

70. Molebny, V. V., I. H. Czyzh, V. M. Sokurenko, I. G. Pallikaris, L. P. Naoumidis, Eye aberration analysis with Zernike polynomials. *Proc. SPIE* 3246 (1998): 228–237.

71. Molebny, V. V., S. I. Panagopoulou, S. V. Molebny, Y. S. Wakil, I. G. Pallikaris, Principles of ray tracing aberrometry. *J. Refract. Surg.* 16 (2000): 572–575.

72. Molebny, V. V., Scanning Shack-Hartmann wave front sensor. *Proc. SPIE* 5412 (2004): 66–71.

73. Molebny, V. V., V. N. Kurashov, I. G. Pallikaris, L. P. Naoumidis, Adaptive optics technique for measuring eye refraction distribution. *Proc. SPIE* 2930 (1996): 147–157.

74. Mrochen, M., M. Kämmerer, P. Mierdel, H. E. Krinke, T. Seiler, Principles of Tscherning aberrometry. *J. Refract. Surg.* 16 (2000): 570–571.

75. Navarro, R., E. Moreno-Barriuso, Laser ray-tracing method for optical testing. *Opt. Lett.* 24 (1999): 951–953.

76. Navarro, R., M. A. Losada, Aberrations and relative efficiency of light pencils in the living human eye. *Optom. Vis. Sci.* 74 (1997): 540–547.

77. Otero, J. M., A. Duran, Continuacion del estudio de la miopia nocturna. *An. Fis. Quim.* 38 (1942): 236.

78. Pallikaris, I. G., S. I. Panagopoulou, V. V. Molebny, Clinical experience with the Tracey Technology wavefront device. *J. Refract. Surg.* 16 (2000): 588–591.

79. Platt, B. C., R. S. Shack, History and principles of Shack-Hartmann wavefront sensing. *J. Refract. Surg.* 17 (2001): S573–S577.

80. Podanchuk, D., A. Kovalenko, V. Kurashov, M. Kotov, A. Goloborodko, V. Danko, Bottlenecks of the wavefront sensor based on the Talbot effect. *Appl. Opt.* 53 (2014): B223–B230.

81. Podanchuk, D., V. Kurashov, A. Goloborodko, V. Danko, M. Kotov, N. Goloborodko, Wavefront sensor based on the Talbot effect with the precorrected holographic grating. *Appl. Opt.* 51 (2012): C125–C132.

82. Ragazzoni, R., Pupil plane wavefront sensing with an oscillating prism. *J. Mod. Opt.* 43 (1996): 289–293.

83. Rosema, J., On the wavefront aberrations of the human eye and the search for their origins. PhD dissertation, University of Antwerp, Antwerpen, Belgium (2004).

84. Salama, N., D. Patrignani, L. De Pasquale, E. Sicre, Wavefront sensor using the Talbot effect. *Opt. Laser Technol.* 31 (1999): 269–272.

85. Scheiner, C., Oculus hoc est: Fundamentum opticum in quo ex accurata oculi. Oeniponti, Innsbruck, Austria (1619).

86. See, C. W., M. V. Iravani, H. K. Wickramasinghe, Scanning differential phase contrast optical microscope: Application to surface studies. *Appl. Opt.* 24 (1985): 2373–2379.

87. Sekine, A., I. Minegishi, H. Koizumi, Axial eye-length measurement by wavelength-shift interferometry. *J. Opt. Soc. Am. A* 10 (1993): 1651–1655.

88. Sekine, R., T. Shibuya, K. Ukai et al., Measurement of wavefront aberration of human eye using Talbot image of two-dimensional grating. *Opt. Rev.* 13 (2006): 207–211.

89. Sergienko, N. M., *Ophthalmologic Optics*. Meditsina, Moscow, Russian (1991).

90. Shack, R. V., B. C. Platt, Production and use of a lenticular Hartmann screen. *J. Opt. Soc. Am.* 61 (1971): 656.

91. Sheehan, M. T., M. Nowakowski, A. V. Goncharov, Wavefront tomography of the human eye assisted with corneal topography and optical path measurements. FIO/LS Technical Digest, Frontiers in Optics OSA Mtg., San Jose, CA (2011).

92. Siegel, C., F. Loewenthal, J. E. Balmer, A wavefront sensor based on the fractional Talbot effect. *Opt. Commun.* 194 (2001): 265–275.

93. Smirnov, M. S., Measurement of the wave aberration of the human eye. *Biophysics* 6 (1961): 687–703. Russian original: *Biofizika* 6 (1961): 776–795.

94. Smith, D. G., J. E. Greivenkamp, Generalized method for sorting Shack-Hartmann spot patterns using local similarity. *Appl. Opt.* 47 (2008): 4548–4554.

95. Smith, V., R. A. Applegate, D. A. Atchison, Assessment of the accuracy of the crossed-cylinder aberroscope technique. *J. Opt. Soc. Am. A* 15 (1998): 2477–2487.

96. Sokurenko, V., V. Molebny, Damped least-squares approach for point-source corneal topography. *Ophthalmol. Physiol. Opt.* 29 (2009): 330–337.

97. Sommargren, G. E., Optical heterodyne profilometry. *Appl. Opt.* 20 (1981): 610–618.

98. Southwell, W. H., Wave-front estimation from wave-front slope measurements. *J. Opt. Soc. Am.* 70 (1980): 998–1006.

99. Su, W., Y. Zhou, Adaptive sequential wavefront sensor with variable aperture. US Patent 8,591,027 (2013).

100. Talbot, H. F., Facts relating to optical science. *Philos. Mag.* 9 (4) (1836): 401–407.

101. Thibos, L. N., W. Wheeler, D. Horner, Power vectors: An application of Fourier analysis to the description and statistical analysis of refractive error. *Optom. Vis. Sci.* 74 (1997): 367–375.

102. Tscherning, M., Die monochromatischen Aberrationen des menschlichen Auges. *Z. Psychol. Physiol. Sinnesorg.* 6 (1894): 456–471.

103. Vérinaud, C., On the nature of the measurements provided by a pyramid wave-front sensor. *Opt. Commun.* 233 (2004): 27–38.

104. Vohnsen, B., S. Castillo, D. Rativa, Wavefront sensing with an axicon. *Opt. Lett.* 36 (2011): 846–948.

105. Volkmann, A., W. Sehen, *Wagner's Handwörterbuch der Physiologie*. Vieweg und Sohn, Brunswick, Lake Forest, IL (1846).

106. Wakil, J., S. P. Saarlos, V. Molebny, S. Molebny, Method for objectively determining the visual axis of the eye and measuring its refraction. US Patent Appl. 2014/0160438 (2014).

107. Walsh, G., W. N. Charman, H. C. Howland, Objective technique for the determination of monochromatic aberrations of the human eye. *J. Opt. Soc. Am. A* 1 (1984): 987–992.

108. Wang, J. Y., D. E. Silva, Wave-front interpretation with Zernike polynomials. *Appl. Opt.* 19 (1980): 1510–1518.

109. Waring (IV), G. O., S. D. Klyce, Corneal inlays for the treatment of presbyopia. *Int. Ophthalmol. Clin.* 51(2) (2011): 51–62.

110. Webb, R. H., C. M. Penney, J. Sobiech, P. R. Staver, S. A. Burns, SRR (spatially resolved refractometer): A null-seeking aberrometer. *Appl. Opt.* 42 (2003): 736–744.

111. Webb, R. H., C. M. Penney, K. P. Thompson, Measurement of ocular local wavefront distortion with a spatially resolved refractometer. *Appl. Opt.* 31 (1992): 3678–3686.

112. Yoon, G., S. Pantanelli, L. J. Nagy, Large-dynamic-range Shack-Hartmann wavefront sensor for highly aberrated eyes. *J. Biomed. Opt.* 11(1–3) (2006): 030502.

113. Young, T., The Bakerian lecture. On the mechanism of the eye. *Philos. Trans. R. Soc. Lond.* 91 (1801): 23–88.

114. Molebny, V., Objective designation of the visual axis of the eye. *Proc. 8 Europ. Meet. Vis. Physiol. Opt.* (2016): 208–210.

Low-coherence interferometry

Christoph K. Hitzenberger

Contents

3.1 INTRODUCTION

Modern ophthalmology has tremendously benefitted from advanced imaging methods and high-precision metrology techniques. Optical biometry is nowadays the standard technology for measuring intraocular distances as, for example, the axial eye length, which is needed for an accurate determination of the lens power of artificial intraocular lenses used to replace cataract lenses. For this application, optical biometry has largely replaced the previously used ultrasound ranging technique due to its higher precision, resolution, and contact-free application. Retinal diagnostics has been dramatically improved by the introduction of imaging techniques such as scanning laser ophthalmoscopy (SLO), fluorescein angiography, scanning laser polarimetry, and adaptive optics (AO) flood illumination cameras and SLOs. While these techniques are able to image retinal structures with high transverse resolution,

their depth resolution is limited by the numerical aperture (NA) of the eye to about 300 μm (or about 50 μm if AO technology is used). This prevents the use of these technologies for imaging and quantifying individual retinal layers.

These limitations were overcome by optical coherence tomography (OCT), a technique providing high-resolution cross-sectional and 3D images of transparent and translucent tissues (Huang et al. 1991). OCT has similarities to ultrasound B-mode imaging; however, it uses near-infrared light to probe the tissue. By exploiting coherence properties, the axial resolution is decoupled from the transversal resolution and thus is independent from NA, enabling an unprecedented depth resolution on the order of a few μm even in low-NA situations like in vivo human retinal imaging. Combined with an extraordinary sensitivity of ~100 dB and high imaging speed, OCT has truly revolutionized retinal diagnostics in the last decade (Drexler and Fujimoto 2008; Geitzenauer et al. 2011).

OCT images are synthesized from several optical depth profiles termed optical A-scans (in analogy to ultrasound A-scans). The basic ranging technique that provides the optical A-scans is low (or partial) coherence interferometry (LCI or PCI). Early demonstrations of LCI in scattering media (although with low resolution) were reported by Ivanov et al. (1977), who used a multimode argon laser as a light source. The first application of LCI to the biomedical field was demonstrated by Fercher and coworkers in the second half of the 1980s: using multimode laser diodes, they were able to measure the axial length of the human eye in vivo (Fercher and Roth 1986; Fercher et al. 1988). While these early demonstrations of LCI were rather slow and of limited sensitivity, the introduction of the heterodyne measurement principle (Fercher et al. 1991; Hitzenberger 1991) improved measurement speed and sensitivity dramatically, a prerequisite for practical applications. This heterodyne LCI technique with constantly moving reference mirror is now known as time domain (TD) optical A-scan technique (in some of our early papers, we used the term "laser Doppler interferometry" to indicate the signal readout via the heterodyne signal's Doppler frequency shift). It forms the basis of high-precision ocular biometry methods (Fercher et al. 1991; Hitzenberger 1991; Drexler et al. 1998a) and TD OCT (Huang et al. 1991; Fercher et al. 1993; Swanson et al. 1993).

TD OCT for retinal imaging has a limited speed of a few 100 A-scans/s. While this is sufficient for recording 2D cross-sectional images, it is too slow for 3D image acquisition in the living eye, where involuntary ocular motions cause image artifacts and distortions. The enormous progress and success of OCT in recent years can largely be attributed to a paradigm shift in OCT technology—the introduction of Fourier domain (FD) OCT (Fercher et al. 1995; Häusler and Lindner 1998; Wojtkowski et al. 2002b): instead of shifting the reference mirror, a stationary mirror is used. The light exiting the interferometer is spectrally dispersed by a grating onto a line scan camera, and a fast Fourier transform (FFT) of the spectral interference signal provides the depth profile, that is, the entire depth profile is recorded in a single shot. Another variant of the technology uses a rapidly swept laser source, and the spectral interferogram is recorded over time (Chinn et al. 1997; Lexer et al. 1997). Both techniques are massively parallel and achieve speeds and sensitivities that are 2–3 orders of magnitude better than that of TD OCT (Choma et al. 2003; de Boer et al. 2003; Leitgeb et al. 2003a). These advantages enabled 3D retinal imaging at speeds of ~30 kA-lines/s for commercial instruments and several 100 kA-lines/ to beyond 1 MA-lines/s in experimental systems (Potsaid et al. 2008; An et al. 2011; Klein et al. 2011).

This chapter describes the basic principles of TD and FD LCI and discusses system parameters like resolution, sensitivity, and measurement range. Furthermore, it demonstrates the use of LCI for ophthalmic applications like intraocular ranging (high-precision biometry) and OCT with applications to retinal imaging and finally introduces functional extensions of OCT like polarization-sensitive OCT and Doppler OCT.

3.2 BASIC PRINCIPLES OF LOW-COHERENCE INTERFEROMETRY

3.2.1 TIME DOMAIN LOW-COHERENCE INTERFEROMETRY

LCI is the basic technology that creates optical A-scans. LCI has two main goals: (1) determining the axial (depth) position of reflecting sites within a sample (e.g., an intraocular interface) and (2) determining the reflectivity of backscattering sites. To explain the basic concept of TD LCI, we consider the Michelson interferometer shown in Figure 3.1.

A light source emits a collimated beam with intensity I_S toward a Michelson interferometer. At first, we consider the case of a monochromatic light source, for example, a single-mode laser, that emits a wavelength λ_0. The beam is split into two components of equal intensity at a 50:50 beam splitter. One of the beam components forms the reference beam, and the other the sample beam. The reference beam is reflected at the reference mirror

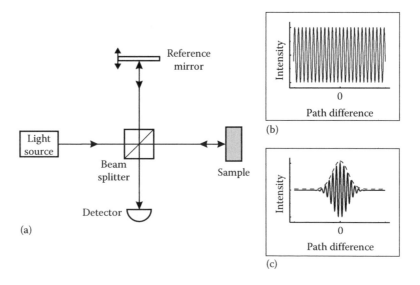

Figure 3.1 Basic principle of time domain low-coherence interferometry. (a) Sketch of Michelson interferometer. (b) Interferogram obtained with coherent light source. (c) Interferogram obtained with short-coherent light source.

(reflectivity = 1), and the sample beam is reflected at the sample surface with a reflectivity R (for simplicity, we assume a plane sample surface). Both reflected beams are recombined at the beam splitter, and 50% of each beam intensity is directed toward the photodetector. The light intensity at the detector is given by

$$I_D = \frac{I_S}{4}\left[1 + R + 2\sqrt{R}\cos\Delta\Phi\right], \qquad (3.1)$$

where $\Delta\Phi$ is the phase difference between reference and sample beam path:

$$\Delta\Phi = 2\pi\frac{2z}{\lambda_0}, \qquad (3.2)$$

with z being the interferometer arm length difference and $2z$ the round trip path length difference for light beams travelling in the sample and the reference arm (and assuming that the light is travelling in air).

Equation 3.1 is the well-known interferometer equation. It consists of a DC term (background term) $1 + R$ and an oscillating term (interferogram) $2\sqrt{R}\cos\Delta\Phi$ (if the common term $\frac{1}{4}I_S$ is neglected).

If the reference mirror is moved with constant speed, the detector signal will oscillate with constant frequency, and the movement of the mirror can be precisely measured by counting the oscillations (interference fringes). However, since z is contained only in the cosine term, an unambiguous determination of the absolute value of z (and hence the sample surface position) is not possible.

Let us now consider the case of a partially coherent light source that has a high spatial coherence (i.e., all beam components across the beam diameter are in phase) but a low temporal coherence. Examples of such light sources are multimode lasers, superluminescent diodes (SLDs), femtosecond pulse lasers, or supercontinuum sources. The source emits light with a center wavelength λ_0 and a bandwidth $\Delta\lambda$. In this case, Equation 3.1 is modified to

$$I_D = \frac{I_S}{4}\left[1 + R + |\gamma(2z)|\,2\sqrt{R}\cos\Delta\Phi\right], \qquad (3.3)$$

where $\gamma(2z)$ is the complex degree of coherence of the light field. $|\gamma(2z)|$ modulates the oscillating part of the interferometric signal. It acts as an envelope of the interferogram and depends on the shape and width of the light spectrum ($\gamma(2z)$ is proportional to the inverse Fourier transform of the light's spectral density (Born and Wolf 1987)). If the light spectrum has a Gaussian shape, $|\gamma(2z)|$ also has a Gaussian shape with an full width at half maximum (FWHM) width equal to the coherence length (Swanson et al. 1992):

$$l_c = \frac{4\ln 2}{\pi}\cdot\frac{\lambda_0^2}{\Delta\lambda}. \qquad (3.4)$$

$|\gamma(2z)|$ has a maximum of 1 for $z = 0$ (i.e., if reference and sample arm have equal length) and falls off along a Gaussian shape with width l_c.

Figures 3.1b and c show the interferometric signals observed by the photodetector in the case of a monochromatic and a broadband light source, respectively. In the latter case, the oscillation has a maximum strength for equal lengths of reference and sample arm. Since the reference arm length is known, the position of the reflector in the sample arm can be easily determined. In many cases, it is sufficient to record the envelope of the oscillating signal, the position of the coherence signal peak immediately provides the sample surface location, and the height of the signal peak is proportional to $\sqrt{R}$.

If the sample has several interfaces (or backscattering sites) separated in depth, each of them generates a signal peak. The peaks can be resolved with a resolution of $\sim l_c/2$. To record an optical A-scan, the reference mirror is moved with constant speed over a certain depth range, and the signal envelope is recorded (Figure 3.2).

3.2.2 FOURIER DOMAIN LOW-COHERENCE INTERFEROMETRY

In the previous section, it was shown how an optical A-scan can be generated by moving the reference mirror of a time domain low-coherence interferometer while recording the light intensity in the detection arm by a single detector. This scheme has two disadvantages: (1) the mechanical scan of the reference mirror limits the scanning speed to a few 100 Hz (or a few kHz in the case of special rapid scanning optical delay lines [Tearney et al. 1997]), and (2) since only a narrow "coherence gate" is analyzed by the reference mirror at a single instant (light from other depths is rejected by the LCI mechanism), light power is wasted and the overall sensitivity is limited.

Another scheme of LCI avoids these drawbacks: FD LCI analyzes the spectrum of the interfering light at the interferometer exit. FD LCI is based on the Fourier relationship between the light field amplitude, as a function of wavenumber, $E(k)$, and the backscattering potential of the sample as a function of path length difference $F(2z)$ (Fercher et al. 1995; Fercher and Hitzenberger 2002). An (inverse) Fourier transform of the spectral amplitude would therefore directly provide a depth profile of the object structure. However, amplitude data are not directly accessible. Instead, the spectral intensity distribution $I_D(k)$ is measured in the detection arm of the interferometer.

We consider a sample containing n reflectors of reflectivity R_i located at depth positions corresponding to (round trip) path length differences $2z_i$. A modification of Equation 3.1 yields the spectral intensity distribution at the interferometer exit:

$$I_D(k) = \frac{I_S(k)}{4}\Big\{1 + \Sigma_i R_i + 2\Sigma_i\sqrt{R_i}\cos 2kz_i$$
$$+ 2\Sigma_{i\neq j}\sqrt{R_i R_j}\cos\left[2k\left(z_i - z_j\right)\right]\Big\} \qquad (3.5)$$

with $I_S(k) \sim |E_S(k)|^2$ being the power spectrum of the source. Furthermore, we assume for simplicity that the reference mirror has a wavenumber-independent reflectivity = 1 and the sample reflectivities R_i are also independent of wavenumber.

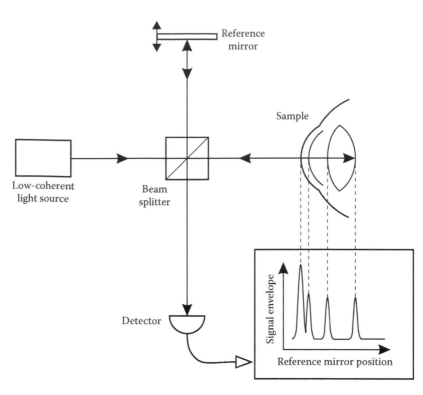

Figure 3.2 Time domain low-coherence interferometry for intraocular ranging. Principle of optical A-scan of anterior eye segment.

An inverse Fourier transform of the measured intensity $I_D(k)$ yields not a depth profile of the object structure but its autocorrelation (Fercher et al. 1995):

$$FT^{-1}\{I_D(k)\} \sim \Gamma(2z) + \Sigma_i R_i \Gamma(2z)$$
$$+ \Sigma_{i \neq j} \sqrt{R_i R_j} \Gamma \{[2z + 2(z_i - z_j)] + \Gamma[2z - 2(z_i - z_j)]\}$$
$$+ \Sigma_i \sqrt{R_i} \{\Gamma[2(z + z_i)] + \Gamma[2(z - z_i)]\}, \qquad (3.6)$$

where Γ represents the first-order electric field correlation function. It is a complex function: $\Gamma(2z) = A(2z)\exp[i2kz]$, where the modulus $A(2z)$ corresponds to the envelope of the signal, that is, plays the same role as $|\gamma(2z)|$ in TD LCI. The width of $A(2z)$ corresponds to the resolution in FD LCI.

The first two terms on the right-hand side of Equation 3.6 are DC terms, representing the intensity of light directly reflected by the reference mirror and the sample interfaces. The sum of the third term represents autocorrelation terms that are caused by mutual interference of waves scattered within the sample. This term can be considered as coherent noise that, similar to the DC term, obscures the true object structure. The sum of the last term contains the real object structure. This object structure term is symmetric about the zero position which can be seen by the fact that the correlation function Γ appears twice in the sum, at positive and negative signs of z_i. The reason for this is that the Fourier transform of the real-valued spectral intensity is Hermitian (Bracewell 2000). As a consequence, the reconstructed depth profile contains a mirror image of the original signal, with symmetry about the zero position. For unambiguous distance measurements and OCT imaging by FD LCI, the sample has to be placed at a position so that $z > z_i - z_j$, that is, the sample

surface has to be located far enough from the zero position so tha an overlap with the autocorrelation terms is avoided. Because of the mirror term, only half of the imaging distance can be used in order to avoid an overlap of the depth profile with its mirror signal. It should be mentioned that there are ways to solve these problems by full-range complex FD LCI techniques that record complex signals by means of phase shifting interferometry. However, these techniques are beyond the scope of this chapter. For more information, see, for example, Fercher et al. (1999), Wojtkowski et al. (2002a), Leitgeb et al. (2003b), and Baumann et al. (2007).

Two different variants of FD LCI have been developed and used extensively for applications like FD OCT. They are described in the following sections.

3.2.2.1 Spectrometer-based Fourier domain low-coherence interferometry

Spectrometer-based FD LCI, or spectral domain (SD) LCI, was the first FD LCI scheme to be applied to the biomedical field and demonstrated for 1D (Fercher et al. 1995) and 2D SD OCT (Häusler and Lindner 1998). A few hardware modifications are needed to convert the basic TD LCI interferometer of Figure 3.1 into an SD LCI interferometer. Figure 3.3 shows a basic sketch.

The scanning reference mirror is replaced by a static mirror (although, because of the limited depth range, a mounting on a movable stage is helpful to coarsely adjust the zero position). The single detector arm is replaced by a spectrometer, consisting essentially of a diffraction grating, a lens, and a line scan camera (CCD or CMOS). The same broadband light source can be used as in TD LCI. Instead of moving the reference mirror, the interfering light beams are spectrally dispersed by the spectrometer whose data are read out by a frame grabber board.

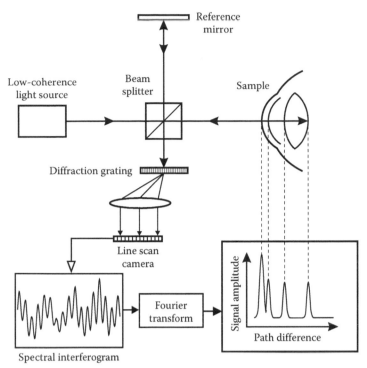

Figure 3.3 Basic principle of spectral domain low-coherence interferometry. Michelson interferometer configuration.

The spectral interferogram obtained in this way shows oscillations whose frequencies are characteristic for the path length differences to be measured (i.e., characteristic for the sample's depth profile). An FFT of the spectral data by the PC provides the depth scan. Since no moving parts are needed and the method

is inherently very sensitive (see Section 3.2.6), SD LCI can be very fast. The typical speed of first-generation commercial SD OCT systems is on the order of 20–40 kA-scans/s. Experimental systems have been demonstrated with even higher speed: 70–300 kHz A-scan rates are now available (Potsaid et al. 2008) and up to 500 kHz have been demonstrated with interleaved operation of two high-speed line scan cameras (An et al. 2011).

An important point that has to be considered for SD OCT is the spectral characteristics of the spectrum provided by the spectrometer. The grating generates a spectrum that is linear in wavelength. However, the FT relationship is between distance and wavenumber. To avoid resolution loss and signal distortion, the spectrum has to be rescaled into a version that is linear in k-space before FT. This is usually done numerically, although hardware solutions have been demonstrated (Hu and Rollins 2007).

3.2.2.2 Swept source–based Fourier domain low-coherence interferometry

Swept source (SS)–based FD LCI, also known as wavelength-tuning LCI or optical frequency domain ranging, was applied to intraocular ranging (Hitzenberger et al. 1997; Lexer et al. 1997) and 2D OCT imaging (Chinn et al. 1997) shortly after SD LCI. To eliminate excess noise caused by the tunable light source, a somewhat more complicated interferometric scheme based on a Mach–Zehnder interferometer and dual balanced detection is commonly used.

Figure 3.4 shows a sketch of a fiber-optic version of such an SS LCI system. Instead of a broadband light source, a tunable laser of narrow instantaneous linewidth $\delta\lambda$, whose wavelength can be rapidly swept over a large bandwidth $\Delta\lambda$, is used. The beam is split by a first fiber-optic beam splitter (fiber coupler) into the

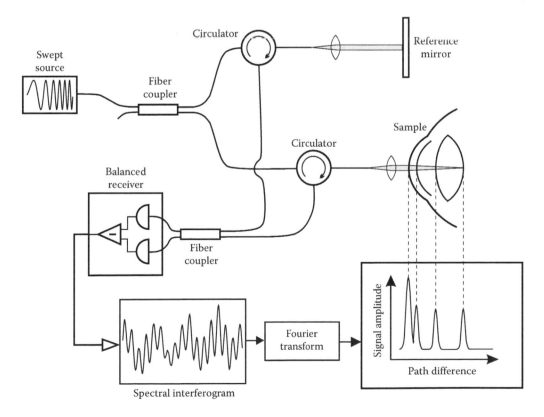

Figure 3.4 Basic principle of swept source low-coherence interferometry. Fiber-optic Mach–Zehnder interferometer configuration.

sample and the reference path. In each path, a circulator guides the beam to a free-space path containing sample and reference mirror, respectively, collects the reflected beams again and directs them to a second fiber-optic beam splitter where the two beams interfere. At both exits of the second beam splitter, interfering beams are available, which are 180° out of phase. They enter into a dual balanced detector that generates a difference signal. This doubles the amplitude of the interferometric oscillations while eliminating common intensity fluctuations and noise generated by the source.

Contrary to SD LCI, the sample is not illuminated by all wavelengths simultaneously but subsequently by rapidly tuning the wavenumber of the laser. Since SS LCI has a similar sensitivity advantage over TD LCI as the SD variant, very fast sweeping can be used while good signal quality is maintained. Commercial swept sources with tuning speeds on the order of ~100 kHz are presently the standard. However, much higher speeds beyond 1 MHz have been demonstrated for experimental in vivo retinal SS OCT systems (Klein et al. 2011, 2013). Similar to the case of SD LCI, care has to be taken that the detected signal is sampled linearly in k-space. Since the direct sweep is typically nonlinear, additional measures have to be taken, for example, a reference sweep can be used to generate a rescaling function that is applied prior to Fourier transform, or the laser is equipped with a "k-clock" that generates a trigger signal that is linear in k-space.

3.2.3 LIGHT SOURCES AND WAVELENGTHS

One of the most important components of an LCI or OCT system is the light source. Via scattering and absorption coefficients, the central wavelength λ_0 determines the light penetration into the tissue. The bandwidth (or sweep range) determines (in combination with λ_0) the axial resolution (cf. Section 3.2.5), and the power determines the sensitivity of the system (cf. Section 3.2.7).

LCI light sources have to fulfill very special criteria: good spatial coherence is required, while the temporal coherence has to be very low (short l_c) in the case of TD LCI and SD LCI. For SS LCI, on the other hand, the instantaneous spectral linewidth has to be narrow (i.e., high instantaneous temporal coherence) and rapidly tunable over a wide range. A detailed discussion of the different light sources used for LCI and OCT is beyond the scope of this chapter; only a brief summary of commonly used sources can be given here. For more details, the reader is referred to literature, for example, Shidlovski (2008), Unterhuber et al. (2008), Yun and Bouma (2008), and Drexler et al. (2014).

Among the broadband sources used in TD and SD LCI and OCT, femtosecond pulse lasers, supercontinuum sources, multimode laser diodes, and SLDs should be named here. Femtosecond pulse lasers and supercontinuum sources achieve the largest bandwidths and therefore the best axial resolution. However, they are very expensive and partly difficult to operate. Therefore, their use for OCT is presently restricted to research laboratories; they are not yet implemented in commercial OCT systems.

Multimode laser diodes were used before SLDs were readily available; they are of historic interest since the first LCI ocular biometry experiments were made with this type of source.

The most widely used commercial LCI biometry system, the Carl Zeiss IOLMaster, also uses such a multimode laser diode with a central wavelength of ~780 nm and a bandwidth of $\Delta\lambda$ ~2.5 nm. The advantage of this light source is that it is very cheap; however disadvantages are the poor axial resolution of ~100 μm and rather strong side lobes of the coherence function (generating multiple signal peaks that can obscure the detection of weakly reflecting interfaces in the vicinity of a strong reflector).

SLDs are most commonly used in OCT. They are comparatively cheap, lightweight, easy to operate, and available with a large selection of wavelengths. Available center wavelength range from ~650 to ~1600 nm and output powers up to 30 mW from single-mode fiber (providing optimum spatial coherence) are available. By combining several sources with different center wavelengths, very broad bandwidths of up to 300 nm are available (though with lower output power), providing axial resolutions down to the μm range.

Swept sources based on a large variety of operating principles and with a wide range of scanning speeds were reported. The wavelength tuning can be based on various mechanisms like polygon scanners, MEMS-tunable filters, and FD mode locking in combination with a tunable filter. Central wavelengths for swept sources useful for LCI and OCT applications are presently limited (with few exceptions) to ~1050 and ~1300 nm. While typical sweep rates of commercially available MEMS filter-based swept sources are of the order of 100 kHz, achieving output powers of ~20 mW and sweep ranges of ~100 nm, higher speeds have been reported for prototype MEMS-tunable VCSEL lasers (Grulkowski et al. 2012) and especially for experimental FD mode locked lasers that can operate at speeds beyond 1 MHz (Klein et al. 2013).

Ophthalmic applications of LCI and OCT operate in the near-infrared range. The reasons are that the retina tolerates more intensity in that wavelength range than in the visible range and that better light sources are available in the near infrared. Three wavelength regimes are used for ocular applications:

1. Central wavelengths near 800–860 nm are used most commonly for retinal OCT imaging. Rather cheap SLDs with excellent coherence properties are available in this range and reasonably priced silicon-based line scan cameras can be used in the detection unit. Most commercial retinal OCT scanners operate in this wavelength range, and SD technology is presently exclusively used for commercial systems (for LCI biometry, multimode laser diodes centered at 780 nm are still frequently used in a dual-beam TD LCI configuration).

2. The wavelength regime around λ_0 ~1050 nm is becoming increasingly popular for experimental retinal OCT systems. Recently, a first commercial instrument operating at this wavelength was introduced. Drawing advantage from the "water absorption window" (~1000–1100 nm) (Unterhuber et al. 2005) and lower scattering, this wavelength penetrates better into deeper layers beyond the retinal pigment epithelium (RPE), making this wavelength the ideal choice for imaging tissues like choroid and sclera. The dominating technology here is SS OCT since good swept sources are available for this wavelength and InGaAs line scan cameras that would be needed for SD versions are rather expensive.

3. Center wavelengths around 1300 nm can be used for anterior segment OCT. Even higher powers can be used at this wavelength; however, water absorption prevents the use of this wavelength for retinal imaging in the human eye.

Finally, it should be mentioned that light source development is a very active field of research. It can be expected that also the 800 nm regime will switch to swept source technology once stable, good quality tunable lasers with suitable parameters become available at a competitive price.

3.2.4 QUANTITATIVE RANGING WITHIN OPTICAL MEDIA

So far, we have considered the ideal case of interferometric paths within air, that is, light beams travelling within a medium of refractive index $n = 1$. (The case of a fiber-optic interferometer is similar if the fiber length traversed by the light beams is exactly equal for reference and sample arms and if similar fibers are used in both arms.) However, for imaging and ranging within a medium of refractive index $n > 1$ (e.g., ocular media), the optical properties of the media have to be taken into account. LCI measures optical distances (the product of geometric distance and refractive index). To convert optical distances into geometric distances, the optical distances have to be divided by the refractive index. In a real medium, the refractive index is a function of wavelength (i.e., $n = n(\lambda)$), an effect called dispersion. In this case, the wave groups emitted by the light source travel with the group velocity $v_g = c/n_g$ (c, vacuum light speed; n_g, group refractive index) through the medium, and n_g instead of the phase index n has to be used to convert the measured optical distances to geometric distances (Hitzenberger 1991). The group index of ocular media is in the range of ~1.34–1.41 for the 800–860 nm wavelength range most commonly used for LCI and OCT applications (Drexler et al. 1998c).

3.2.5 RESOLUTION

Contrary to other optical imaging techniques like microscopy, axial and lateral resolutions are decoupled in LCI and OCT.

3.2.5.1 Axial resolution

One of the main advantages of LCI and OCT is the fact that the axial resolution is neither influenced by the NA of the eye nor by its optical aberrations. It depends only on the coherence properties of the light field. According to Equation 3.4, the axial resolution is proportional to $\lambda_0^2/\Delta\lambda$, that is, at a given central wavelength λ_0, the resolution is inversely proportional to the source bandwidth (or sweep range in the case of SS LCI) $\Delta\lambda$. At a wavelength of $\lambda_0 = 840$ nm as it is typically used for intraocular ranging or retinal OCT, and a bandwidth $\Delta\lambda \cong 50$ nm, the axial resolution is $\cong 6$ μm (in air, assuming a Gaussian shape of the source spectrum). Using multiplexed SLDs or femtosecond Ti:Al$_2$O$_3$ lasers, broader bandwidths of 100 nm and beyond are available, improving the axial resolution into the μm range (Drexler et al. 2001; Drexler 2004; Wojtkowski et al. 2004; Ko et al. 2004).

These considerations hold for measuring interfaces in air. However, for intraocular ranging and imaging, effects of media have to be considered (cf. Section 3.2.3). For a nondispersive medium, the coherence length of the light field within the medium is shortened by $1/n$, that is, the resolution is improved by the same factor. In the case of a dispersive medium, the different spectral components of the light field travel with different speed within the medium, leading to a broadening of the coherence function and a degradation of resolution (Hitzenberger et al. 1999). The width of the coherence envelope, after double passing a medium of length L with a group index n_g and group dispersion $GD = \mathrm{d}n_g/\mathrm{d}\lambda$, is broadened by a factor (Hitzenberger et al. 1999):

$$\beta = \frac{1}{l_c}\sqrt{l_c^2 + (GD \cdot 2L \cdot \Delta\lambda)^2}, \qquad (3.7)$$

as compared to the nondispersive case, or β/n_g as compared to measurements in air. As can be seen, especially very broadband light sources can lead to large signal broadening and resolution degradation. The effect of dispersion broadening can be compensated either by carefully matching the dispersion in the sample arm by placing appropriate dispersive material of similar effect in the reference arm (Hitzenberger et al. 1999) or by numerical methods (Fercher et al. 2001; Wojtkowski et al. 2004).

3.2.5.2 Transversal resolution

The transversal resolution of OCT is limited by similar criteria as in the case of confocal microscopy. The smallest distance δx that can be resolved in transverse direction can be defined, in analogy to the axial resolution, as the FWHM diameter of the electric field amplitude of the beam scanning the sample (assuming a sufficient transversal sampling density). For a Gaussian field amplitude distribution of the sampling beam, the FWHM diameter of the focal spot is given by (Fercher et al. 2003)

$$\delta x = 4\ln(2) \cdot \frac{f\lambda_0}{n\pi d}, \qquad (3.8)$$

where
  f is the focal length of the lens
  n is the refractive index of the medium in which the lens is working
  d is the FWHM diameter of the collimated beam

If $1/e^2$ intensity distribution diameters are used instead of FWHM electric field amplitude diameters, the factor $\ln(2)$ in Equation 3.8 has to be omitted. With a beam diameter of ~1 mm, a transverse resolution of ~10–15 μm can be expected at the retina of a human eye, the same order as the transversal resolution obtained by SLOs. Increasing the beam diameter in healthy eyes with low aberration can improve the resolution (Pircher et al. 2006a); however for beam diameters beyond ~3–4 mm, the aberrations of the optical elements of the eye degrade the transversal resolution (for eyes with moderate to high aberrations, the useful limit of d is ~1–2 mm). Furthermore, a larger beam diameter reduces the depth of focus, requiring a careful focusing of the beam. Solutions to these problems are dynamic focusing (Lexer et al. 1999; Pircher et al. 2006b) and the use of AO that compensate the aberrations (Liang et al. 1997; Fernandez et al. 2001; Roorda et al. 2002; Zawadzki et al. 2005; Zhang et al. 2006; Felberer et al. 2014).

3.2.6 IMAGING DEPTH

While in TD LCI the imaging depth is only limited by the travel range of the reference mirror, FD LCI is more restricted. Since the distances are encoded in the frequencies of the oscillations of the spectral interferogram, with higher frequencies being equivalent to larger distances, the spectral resolution limits the maximum possible distances that can be measured. If we consider the Nyquist criterion, the maximum imaging depth is given by

$$z_{max} = \frac{1}{4} \frac{\lambda_0^2}{n \cdot \delta\lambda}, \tag{3.9}$$

where $\delta\lambda$ is the spectral resolution.

If we consider the case of SD LCI, $\delta\lambda = \Delta\Lambda/N$, where $\Delta\Lambda$ is the spectral range covered by the spectrometer with N pixels. Typically, a spectral range $\Delta\Lambda$ about twice the FWHM spectral width $\Delta\lambda$ of the source is chosen. For typical values of $\lambda_0 = 840$ nm, $\Delta\Lambda = 100$ nm, and a line scan camera of 2048 pixels, $z_{max} \cong 3.6$ mm (in air) is obtained. The case of SS LCI is similar; instead of the spectrometer resolution, the spectral resolution is determined by the laser sweep rate and range, the electric bandwidth of the detector, and the sampling rate of the AD converter.

An additional effect that affects imaging depth is the sensitivity roll-off with depth. Maximum sensitivity (as discussed in Section 3.2.7) is only achieved at zero position. The sensitivity degrades at larger depths (even if sample-dependent effects like scattering and absorption are ignored). In the case of SD LCI, the rectangular shape of the camera pixels is convolved with the spectral interferogram, causing a multiplication of the depth-dependent signal with a sinc function after the Fourier transform (Leitgeb et al. 2003a). This leads to a depth-dependent signal loss (cf. Figure 3.5). Moreover, other factors like uneven sampling of longer and shorter wavelengths in wavenumber domain, limited spot size, and camera pixel cross talk further enhance the signal decay with depth (Bajraszewski et al. 2008). Typical signal roll-offs for SD LCI and OCT are ~10–15 dB over the image depth range.

In the case of SS LCI, the signal decay is less pronounced, depending essentially on the instantaneous linewidth of the laser. A narrow linewidth leads to a long instantaneous coherence

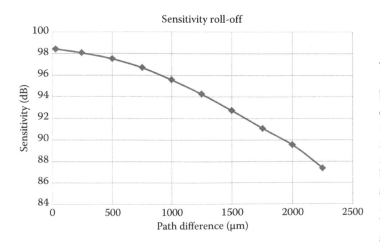

Figure 3.5 Sensitivity roll-off of an SD LCI system.

length and a low roll-off. For commercial swept source lasers used for ocular OCT applications, 6 dB roll-off imaging depths >4 mm have been reported (Potsaid et al. 2010). However, new experimental sources can achieve much larger ranging values beyond 100 mm and even beyond 1 m (with reduced scanning speed and resolution) (Grulkowski et al. 2013; Bonesi et al. 2014).

3.2.7 SENSITIVITY

An important parameter of LCI ranging and OCT imaging is detection sensitivity. Sensitivity can be defined via the smallest signal just discernible from noise. Since LCI signals are generated by backscattered light, a suitable definition of sensitivity S is the ratio between the reflectivity of a perfectly reflecting mirror ($R = 1$) and the sample reflectivity $R_{S,min}$ that yields a signal power equal to noise (or a signal-to-noise ratio $SNR = 1$):

$$S = 1/R_{S,min}, \tag{3.10}$$

or, if defined in dB,

$$S \,(dB) = 10\log\left(1/R_{S,min}\right). \tag{3.11}$$

In TD LCI, the SNR is defined as the ratio of the mean square signal photocurrent to the mean square photocurrent fluctuations

$$SNR = \frac{2\alpha^2 P_S P_R}{\left\langle \Delta i_{sh}^2 \right\rangle + \left\langle \Delta i_{eh}^2 \right\rangle + \left\langle \Delta i_{re}^2 \right\rangle}, \tag{3.12}$$

where

P_S and P_R are the powers of sample and reference beams at the detector, respectively

$\alpha = \eta q_e/(h\nu)$ is a conversion factor for converting optical power to current (η, detector quantum efficiency; q_e, electron charge; h, Planck's constant; ν, light frequency)

The three terms in the denominator of Equation 3.12 are the mean square photocurrent fluctuations caused by shot noise, excess noise, and receiver noise. For a Michelson interferometer configuration with a 50:50 beam splitter and a reference mirror reflectivity of 1, and from the definition of $S = 1/R_{S,min}$ at an $SNR = 1$ we obtain for the sensitivity of a TD LCI system

$$S_{TD} = \frac{\alpha^2}{8} \frac{P_0^2}{\left\langle \Delta i_{sh}^2 \right\rangle + \left\langle \Delta i_{ex}^2 \right\rangle + \left\langle \Delta i_{re}^2 \right\rangle}. \tag{3.13}$$

where P_0 is the source power.

A detailed discussion of the individual noise current terms is beyond the scope of this chapter; more details can be found, for example, in Rollins and Izatt (1999) and Podoleanu and Jackson (1999). In short, we can say that at low light power levels, receiver noise will dominate while at very high power levels, excess noise dominates. At intermediate power levels, shot noise will be dominating, and this is the optimum condition that can be achieved in a well-designed LCI or OCT system. Shot noise is inevitable; it is caused by the inherent quantum nature of light. With $\left\langle \Delta i_{sh}^2 \right\rangle = 2q_e B \langle i \rangle$ (B, electrical bandwidth of the detector) and $\langle i \rangle = \alpha P_0/4$ (where we assume that the mean DC detector current $\langle i \rangle$ is caused by the reference light and its sample light component can be neglected), we arrive at the sensitivity of a shot noise limited LCI system:

$$S_{TD,sh} = \frac{\alpha P_0}{4q_e B}. \tag{3.14}$$

$S_{TD,sh}$ is linearly proportional to the source power and inversely proportional to the detector bandwidth, that is, inversely proportional to the measurement speed. As a rule of thumb, a sensitivity of >90–95 dB should be achieved for acceptable to good quality retinal OCT imaging.

It has been shown, both theoretically and experimentally, that FD LCI (both SD and SS LCI versions) can achieve much higher sensitivity (Choma et al. 2003; de Boer et al. 2003; Leitgeb et al. 2003a). In the case of FD LCI, Equation 3.14 is replaced by

$$S_{FD,sh} = \frac{\alpha P_0}{4q_e LR}, \tag{3.15}$$

where LR is the line rate (A-scan rate) of the measurement (and we have assumed that the readout time of the line scan camera is negligible compared to the integration time τ, i.e., $LR = 1/\tau$).

It can be seen from Equations 3.14 and 3.15 that the sensitivity of FD LCI is improved by a factor of B/LR as compared to TD LCI. If we assume a TD LCI detection bandwidth $B \cong \left(4\Delta\lambda z_{max}/\lambda_0^2\right) LR$ (which considers the broadening of the detected heterodyne interference fringe frequency caused by the various wavelengths emitted by the broadband source), we arrive at

$$S_{FD,sh} \simeq \frac{N}{2} S_{TD,sh}. \tag{3.16}$$

FD LCI has a sensitivity advantage over TD LCI that is proportional to the number of pixels of the line scan camera. Equation 3.16 resembles an ideal case; practically, sensitivity advantages on the order of 20–30 dB have been demonstrated. This huge sensitivity advantage has led to a paradigm shift, largely replacing TD OCT by FD variants in the last decade.

Different approaches have been taken to explain this advantage (see, e.g.,[Choma et al. 2003; de Boer et al. 2003; Leitgeb et al. 2003a]). However, probably the most intuitive explanation is that TD LCI gates out all of the light from the detection process that is not within the narrow coherence gate at any given instant. That is, only a fraction of $l_c/2z_{max}$ contributes to the signal at any instant. In FD LCI, on the other hand, light backscattered from all depths is simultaneously collected and contributes to the signal generation.

3.3 APPLICATIONS OF LOW-COHERENCE INTERFEROMETRY

3.3.1 INTRAOCULAR RANGING

3.3.1.1 Axial eye length measurement

The measurement of the axial length of the human eye was historically the first application of LCI to the biomedical field (Fercher and Roth 1986; Fercher et al. 1988), and it is still an important application. The axial length of the eye is needed

for a precise determination of the required power of artificial intraocular lenses that replace the natural crystalline lens after cataract surgery (Drexler et al. 1998a; Olsen 2007; Norrby 2008). Since the introduction of heterodyne LCI (Fercher et al. 1991; Hitzenberger 1991), this technology has largely replaced the ultrasound biometry technique because of its higher precision and contact-free application. Commercial instruments are available, for example, from Carl Zeiss Meditec, Jena, Germany; Haag-Streit, Koeniz, Switzerland; and Tomey, Nagoya, Japan.

Because of the long distance to be measured—the geometric length of the eye is about 24 mm, equivalent to an optical length of ~32–33 mm—the Michelson interferometer setups shown in Figures 3.1 through 3.3 cannot be directly used: the measurement range of SD LCI is too low, and for TD LCI, the required travel range of the reference mirror is too large (axial eye motions between the recording of the corneal reflex and the retinal reflex would yield erroneous results).

To solve this problem, the interferometric setup is modified (cf. Figure 3.6): the eye is not placed in the sample arm of the interferometer; instead, an external interferometer is used that generates two coaxial beams that have a path delay whose value is close to the path delay within the eye. This dual beam illuminates the eye via a beam splitter. Both beam components are reflected at the ocular interfaces and detected via the beam splitter. We now consider four beam components: C_1, C_2, R_1, and R_2, where C indicates beam components reflected at the cornea, R indicates beam components reflected at the retina, and 1 and 2 indicate beams reflected by interferometer mirrors 1 and 2, respectively. To measure the axial length of the eye, the interferometer arm length difference d is changed by moving interferometer mirror 2. If d is equal to the optical length of the eye OL_{eye}, two of the four beam components, C_2 and R_1, will travel over the same total distance and hence cause an LCI signal (the other two components just add an incoherent background).

This scheme has two advantages: instead of the whole optical eye length, only a few millimeters have to be scanned by the interferometer mirror (spanning the range of natural eye lengths). Moreover, this scheme is completely insensitive to axial eye motions since they affect both interfering beam components in a similar way.

The first measurements with this technology, both in healthy and cataract eyes (Fercher et al. 1991; Hitzenberger 1991; Hitzenberger et al. 1993), were performed using a rather narrow-band multimode laser diode as the light source (also the first commercial LCI biometry instrument used this type of light source). While the width of the (round trip) coherence envelope $l_c/2$ was rather broad (~110 μm), the measurement precision was much better, ~20–25 μm, since the signal peak position can be measured with much better precision than the coherence length (which determines the resolution of two closely spaced surfaces). Figure 3.7 shows a plot of the LCI signal envelope recorded during an eye length measurement obtained in a healthy human eye with the first experimental heterodyne LCI system (Hitzenberger 1991). Two signal peaks can be observed, corresponding to the distances from the corneal surface to the inner limiting membrane and the posterior side of the retina (comprising the photoreceptor layer and the RPE, which cannot be further resolved in these early data because of the low axial resolution). The position of the latter peak

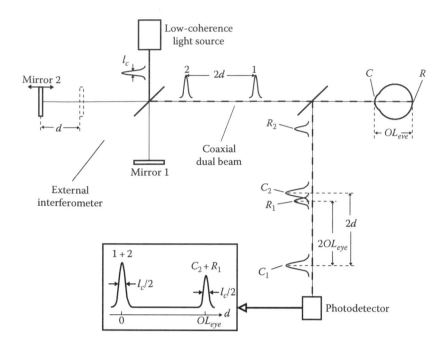

Figure 3.6 Dual-beam time domain LCI system for ocular biometry. C, cornea; R, retina; d, interferometer arm length difference; l_c, coherence length; OL_{eye}, optical length of the eye.

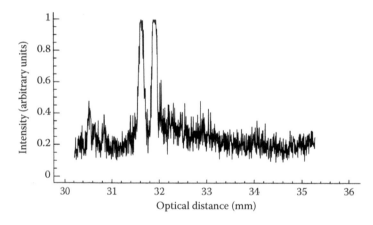

Figure 3.7 TD LCI signal envelope recorded during axial eye length measurement in a healthy eye. The two signal peaks correspond to the positions of inner limiting membrane and retinal pigment epithelium. (Reproduced from Hitzenberger, C.K., *Invest. Ophthalmol. Vis. Sci.*, 32(3), 616, 1991, by permission from the Association for Research in Vision and Ophthalmology.)

corresponds to OL_{eye} (the first peak is not always visible in these LCI biometry measurements; it depends on a strong specular reflex at the inner limiting membrane).

As mentioned before, the eye length provided by LCI is the optical length OL_{eye}, which has to be converted into the geometric length. This is achieved by division by the group index n_g. Ideally, each intraocular optical distance (corneal thickness, anterior chamber depth, lens thickness, vitreous length) should be measured separately and divided by the group index of the respective medium (cf. Table 3.1). The individual geometric distances can then be added to obtain the total geometric axial eye length. However, since measurements of anterior chamber depth and lens thickness by LCI are somewhat

more complicated (see Section 3.3.1.2), simpler methods are frequently used, like division of OL_{eye} by a mean group index or a method based on the assumption that the main difference in eye lengths is caused by differences of vitreous lengths (Hitzenberger 1991). These simplified methods can lead to a somewhat lower accuracy.

Comparisons of LCI and ultrasound biometry demonstrated a very good correlation between the two techniques (with superior precision of LCI), both in healthy (Hitzenberger 1991) and cataract eyes (Hitzenberger et al. 1993) (cf. Figure 3.8). An observation made in these and other studies was a systematic difference between the axial eye length results obtained by LCI and ultrasound. The latter yields eye lengths that are on average ~460 µm smaller than those obtained by LCI. The differences were attributed to two reasons: (1) LCI measures the distance from the cornea to the RPE, while ultrasound measures the distance to the inner limiting membrane, that is, the two technologies should differ by the retinal thickness, and (2) for the ultrasound measurements, the applanation technique is commonly used where the direct contact between ultrasound transducer and cornea generates a small indentation of the cornea leading to a slightly reduced eye length. If the more elaborate immersion ultrasound technique is used, the difference between ultrasound and LCI results is reduced to approximately the value of the retinal thickness (cf. Figure 3.8b). Another factor might be the unknown group index of the cataract lens that might differ slightly from that of the healthy lens.

The results of these studies laid the foundation of the nowadays widespread use and commercial success of LCI ocular biometry. Measurement precision could later be improved to below 10 µm by replacing the multimode laser diode by an SLD with broader bandwidth (Drexler et al. 1998a; Findl et al. 1998), which was also a prerequisite for high-precision biometry of the

Table 3.1 **Group refractive indices of ocular media**

CENTRAL WAVELENGTH (NM)	CORNEA	AQUEOUS HUMOR	VITREOUS HUMOR	LENS (MEAN VALUE)	TOTAL EYE (MEAN VALUE)
814	1.3854	1.3457	1.3443	1.4068	1.3546
855	1.3851	1.3454	1.3440	1.4065	1.3543

Source: Drexler, W. et al., *Exp. Eye Res.*, 66, 25, 1998c. Permission from Academic Press.

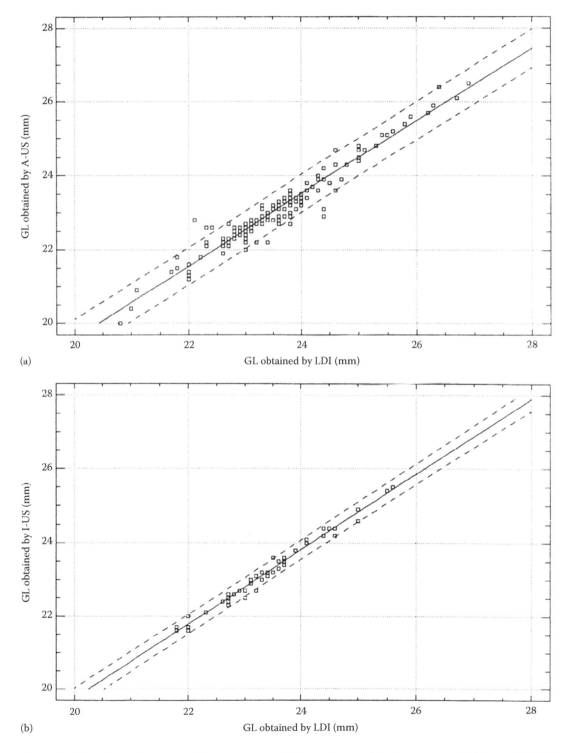

(a)

(b)

Figure 3.8 Comparison of axial eye lengths measured in cataract eyes by low-coherence interferometry and ultrasound. (a) LCI versus applanation ultrasound. (b) LCI versus immersion ultrasound. (Reproduced from Hitzenberger, C.K. et al., *Invest. Ophthalmol. Vis. Sci.*, 34(6), 1886, 1993. With permission from the Association for Research in Vision and Ophthalmology.)

anterior eye segment (cf. next section) and for high-resolution OCT imaging in the retina (cf. Section 3.3.2).

3.3.1.2 Anterior segment measurement

The approach of the dual beam generated by the external interferometer (Figure 3.6) was later extended for measuring distances in the anterior segment (Drexler et al. 1997a,b, 1998a; Hitzenberger 1992; Hitzenberger et al. 1994). Measurements of corneal thickness are comparatively simple because the two corneal surfaces are (nearly) parallel to each other and generate overlapping reflections that can easily be used to generate LCI signals. Measurements of the anterior chamber depth and the lens thickness are somewhat more difficult. These measurements involve light beams reflected at the lens surfaces. In this case, measurements have to be performed parallel to the optical axis of the eye instead of parallel to the vision axis, as is done in the case of axial eye length measurements. The angle between the two axes is ~5°, and if specular reflections are used, the angle tolerance is only ~±2° (Drexler et al. 1997b). Therefore, an additional fixation light is needed and subjects need to have sufficient fixation capabilities. (For this reason, e.g., the Carl Zeiss "IOLMaster," the most widely used commercial LCI biometry system, measures only the axial eye length by LCI; the anterior segment is measured by conventional slit lamp–based technology. It should be mentioned, however, that with the advent of new high-sensitive FD OCT systems with long measurement range, specular reflections are no longer needed; instead, weaker backscattered light can be used, alleviating these conditions (Grulkowski et al. 2009a; Jungwirth et al. 2009; Li et al. 2014).)

Figure 3.9 shows an example of an anterior segment LCI scan taken in a cataract eye. An SLD with a center wavelength of 855 nm was used as the light source. The LCI signal intensity is plotted as a function of optical distance to the anterior corneal surface (ACS). Four sharp peaks are observed, corresponding to the positions of the ACS, the posterior corneal surface (PCS), the anterior lens surface (ALS), and the posterior lens surface (PLS). Two additional broader signal peaks are observed within the lens, labeled 1 and 2; they are probably generated by lens opacities near the cortex–nucleus interfaces. The distances between ACS and PCS, PCS and ALS, and ALS and PLS correspond to the optical thickness of cornea, anterior chamber, and lens, respectively. A division by the respective group index (Table 3.1) provides the geometric distances.

LCI biometry achieves a very high precision, both for the axial eye length and the anterior segment. While sub-μm precision can be achieved for central corneal thickness (Drexler

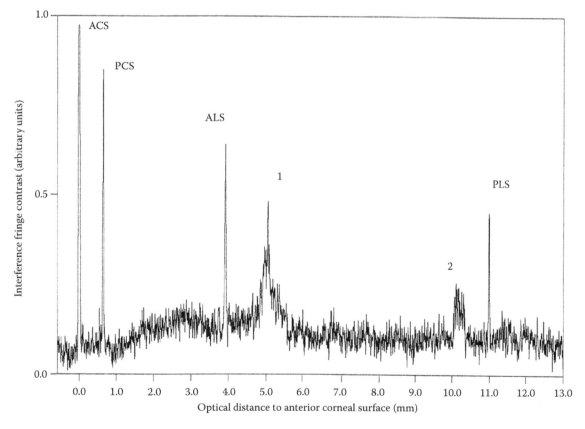

Figure 3.9 TD LCI signal envelope recorded during anterior segment biometry in a cataract eye. The signal peaks correspond to ACS, anterior corneal surface; PCS, posterior corneal surface; ALS, anterior lens surface; 1 and 2, lens opacities near cortex–nucleus interface; PLS, posterior lens surface. (Reprinted from *Am. J. Ophthalmol.*, 126(4), Drexler, W., Findl, O., Menapace, R., Rainer, G., Vass, C., Hitzenberger, C.K., and Fercher, A.F., Partial coherence interferometry: A novel approach to biometry in cataract surgery, 524–534, Copyright 1998a, with permission from Elsevier.)

Table 3.2 **Biometric precision of ultrasound and low-coherence interferometry determined in 85 cataract eyes (values in μm)**

	ULTRASOUND		LOW COHERENCE INTERFEROMETRY	
	MEAN ± SDEV	RANGE	MEAN ± SDEV	RANGE
Axial eye length	88.3 ± 49.5	10–310	8.8 ± 3.2	3.8–14.3
Ant. chamber depth	69.7 ± 33.0	35–206	4.5 ± 2.1	0.5–12.0
Lens thickness	94.1 ± 106.5	35–710	4.6 ± 2.3	2.0–10.8
Vitreous length	89.6 ± 89.2	35–674	6.0 ± 2.3	2.5–11.2
Corneal thickness	—	—	0.82 ± 0.2	0.10–1.2

Source: Reprinted from *Am. J. Ophthalmol.*, 126(4), Drexler, W., Findl, O., Menapace, R., Rainer, G., Vass, C., Hitzenberger, C.K., and Fercher, A.F., Partial coherence interferometry: A novel approach to biometry in cataract surgery, 524–534, Copyright 1998a, with permission from Elsevier.

et al. 1997b), precision figures on the order of 5–10 μm were reported for the anterior chamber depth, lens thickness, vitreous length, and axial eye length. Table 3.2 compares precision results obtained by repeated LCI and ultrasound measurements in 85 cataract eyes (Drexler et al. 1998b). As can be seen, the precision of LCI is an order of magnitude (or more) better than that of ultrasound biometry. By eliminating residual accommodation using cycloplegia, the precision of anterior chamber depth and lens thickness measurements could be further improved to values below 2 μm in healthy volunteers (Drexler et al. 1997b). In addition to standard biometry, the high precision and resolution of LCI enabled detailed studies of, for example, anterior chamber depth and lens thickness changes during accommodation (Drexler et al. 1997a), eye elongation during accommodation in emmetropes and myopes (Drexler et al. 1998b), and the lens–capsule distance in pseudophakic eyes (Findl et al. 1998).

3.3.2 OPTICAL COHERENCE TOMOGRAPHY

3.3.2.1 Two- and three-dimensional imaging

An LCI A-scan can be regarded as a 1D image: it provides the distribution and scattering potential of backscattering sites along the sampling beam. To obtain 2D cross-sectional images (OCT B-scan images), multiple A-scans are recorded at adjacent sample positions. The LCI signal intensities are converted to gray levels or color values, and the individual scans are mounted to form a 2D gray level or false color image that shows a cross section of the sample (cf. Figure 3.10). The concept can easily be extended to 3D imaging by recording multiple B-scans at adjacent positions. To interrogate the different positions on the sample, a pair of scanning mirrors is typically used that deflects the sample beam along a raster scan pattern of predefined geometry. Present commercial FD OCT systems have typical raster scan patterns of, for example, 512(*x*) × 128(*y*) or 200(*x*) × 200(*y*) A-scans, which are distributed over a square-shaped scan field and recorded within few seconds. However, other patterns with, for example, denser sampling in the fast direction or circular scan patterns (around the optic nerve head [ONH]) are frequently used. With the introduction of even faster SS OCT systems, volume acquisitions within less than 1 s become available.

3.3.2.2 Image contrast

Conventional OCT is based on the intensity of backreflected or backscattered light (according to Equations 3.3 and 3.6, the LCI signal envelope (or the object term in FD OCT) is proportional to the square root of the reflectivity). Backreflection occurs at interfaces within the sample that separate tissues of different refractive index. Such an interface can only be observed in OCT images if the light is retroreflected (i.e., if the interface is perpendicular to the incident beam). In this case, the interface is displayed as a bright line in the B-scan image. On the other hand, backscattering occurs in all directions. It is caused by particles whose refractive index is different from that of the surrounding matrix. The backscattered intensity varies with angle and depends on particle size, shape, orientation, and refractive index mismatch (Schmitt and Kumar 1998). Backscattering structures are visible in OCT independently of the orientation of the backscattering structure. Backscattering surfaces appear as bright boundaries, and volume backscatterers (particles dispersed in a surrounding matrix) appear as "speckled" structures.

Scattering attenuates the light, leading to an exponential decay of backscattered signal with depth. Absorption also attenuates the signal with depth (although this effect is considerably weaker in most tissues for the wavelengths used for OCT imaging). Strongly scattering (or absorbing) material can shadow tissue structures beneath; examples are, in the case of retinal imaging, the RPE, which frequently hinders imaging of deeper structures like the choroid (at least in the 800 nm wavelength regime), or thick blood vessels, which cast shadows on deeper retinal layers.

Because of the exponential attenuation of the probing light with depth and because of the large variations of backscattering (and backreflecting) coefficients with tissue type, the amount of backscattered light can vary strongly within an OCT image. Therefore, the images are usually displayed on logarithmic intensity (or false color) scales, instead of linear scales.

Finally, it should be mentioned that LCI and OCT have band-pass characteristics. As a consequence, only structures where the refractive index changes within a short distance (on the order of the wavelength or smaller) give rise to measurable backscattered or reflected light. Structures with slowly varying

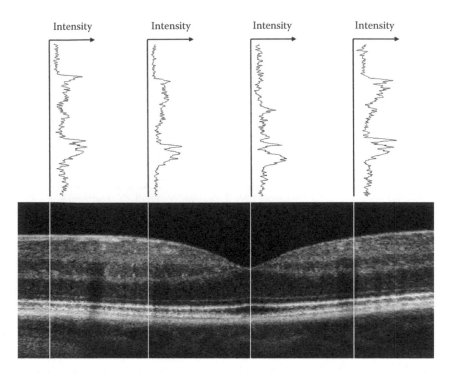

Figure 3.10 Schematic of synthesis of OCT B-scan from several A-scans.

refractive index remain invisible in intensity-based OCT. For a more detailed discussion of this feature, see, for example, Fercher and Hitzenberger (2002). Advanced contrasting techniques like polarization-sensitive OCT or phase-sensitive OCT can overcome this limit in some cases.

3.3.2.3 Applications to retinal imaging

Retinal imaging was historically the first application of OCT (Huang et al. 1991; Fercher et al. 1993) and is still the most important of applications, with presently about 10 companies selling retinal OCT scanners, and 10s of millions of OCT procedures taken each year. Given the huge body of literature in that field, a comprehensive overview is beyond the scope of this chapter. Instead, only few examples of healthy and diseased retinal OCT scans will be presented and briefly discussed, as well as safety precautions considered.

3.3.2.3.1 Laser safety

Since OCT uses spatially coherent light beams (with exception of certain full-field OCT systems (Grieve et al. 2004), which are, however, not used for in vivo retinal imaging), emitted by lasers or SLDs, laser safety regulations have to be obeyed. This is especially important since the retina is the organ most sensitive to light damage and because light in the near infrared is used that is barely visible (800 nm regime) or totally invisible (1050 nm regime) to the human eye.

Two laser safety standards regulate the use of lasers (American National Standards for Safe Use of Lasers 2000; International Electrotechnical Commission 2001; they have to be applied depending on the geographical region of use. Fortunately, the two regulations are rather similar as far as OCT imaging is concerned. For the case of OCT imaging (and LCI biometry), the maximum permissible exposure (MPE) is calculated based on the assumption that the beam is collimated (corresponding

Table 3.3 **Maximum permissible exposure (MPE) values for intrabeam viewing > 10 s for various wavelengths used in ophthalmic LCI and OCT**

Wavelength (nm)	780	840	1050
MPE (mW)	0.56	0.73	1.93

to a small source subtending a visual angle ≤1.5 mrad) and the deposited power is averaged over a 7 mm diameter entrance pupil. Table 3.3 shows MPE values calculated for typical LCI and OCT wavelengths for illumination times >10 s. These values are based on the assumption that the beam spot size on the retina has a diameter of ~20–30 µm and that natural ocular motions smear out the beam energy over larger areas in the case of extended illumination times. This implies that no additional means are taken to stabilize the beam on the retina (by using a retinal tracker) and that higher-order ocular aberrations are not corrected (e.g., by use of AO). Since, for OCT, the beam is scanned over the retina, the energy is distributed over a larger area and, in principle, higher powers as in Table 3.3 could be used, depending on a detailed analysis of the scan pattern size and timing, and a careful observation of the safety limits. In this case, a safety shutter that blocks the beam in the case of scanner malfunction is mandatory. To be on the safe side, most companies presently prefer to use powers not exceeding those of Table 3.3.

3.3.2.3.2 Examples of retinal OCT imaging

Figure 3.11 shows results from a 3D OCT data set recorded in the right eye of a healthy human volunteer. A 40° × 40° scan field was imaged with a raster scan pattern consisting of 1024(x) × 250(y) A-scans. An SD OCT system operating at 70 kA-scans/s was used (Zotter et al. 2012). Figure 3.11a shows an en face intensity projection image (or pseudo-SLO image), which is generated by adding up intensity values along the A-line (z)

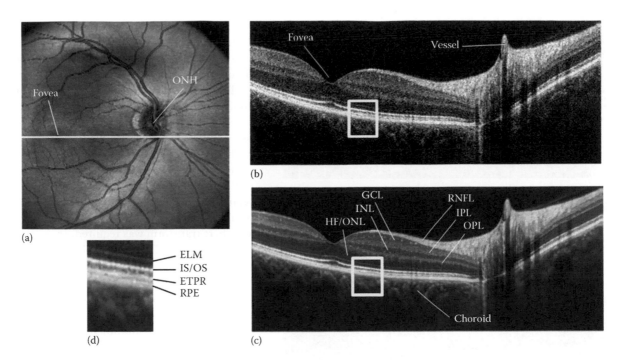

Figure 3.11 Example of retinal OCT imaging in a healthy eye. (a) Intensity projection image (pseudo-SLO). Scan field size: 40° × 40°. (b) B-scan image taken at the position of yellow line in (a). (c) Average of 50 consecutive B-scans taken at the same location as (b). (d) Magnification of area within yellow rectangle in (b) and (c). ONH, optic nerve head; RNFL, retinal nerve fiber layer; GCL, ganglion cell layer; IPL, inner plexiform layer; INL, inner nuclear layer; OPL, outer plexiform layer; HF/ONL, Henle fiber/outer nuclear layer; ELM, external limiting membrane; IS/OS, junction between inner and outer photoreceptor segments; ETPR, end tips of photoreceptors; RPE, retinal pigment epithelium. (Adapted from Zotter, S., Pircher, M., Torzicky, T., Baumann, B., Yoshida, H., Hirose, F., Roberts, P. et al., Large-field high-speed polarization sensitive spectral domain OCT and its applications in ophthalmology, *Biomed. Opt. Express*, 3(11), 2720–2732, 2012. With permission of Optical Society of America.)

direction (Jiao et al. 2005). The image covers the macula, the ONH, and the major arcuate retinal vessels. Figure 3.11b shows a B-scan at the position indicated by the yellow line in Figure 3.11a. This papillomacular scan covers the fovea centralis and the inferior rim of the ONH. In the area of the ONH, the thick hyperreflective retinal nerve fiber layer (RNFL) can be clearly observed, as well as a cross section through a major retinal vessel. The large vessel and some smaller vessels on the nasal side cast shadows on the deeper tissue. Figure 3.11c shows an average of 50 consecutive B-scans recorded at the same position as Figure 3.11b. This averaging improves the signal-to-noise ratio and reduces speckle noise. The retinal layer structure, though already well visible in the single frame image (Figure 3.11b), is especially clear and well discernible in this noise-reduced image. The individual retinal layers are labeled in Figure 3.11c and in the magnified zoom-in Figure 3.11d.

As an example for retinal imaging in diagnostic applications, Figure 3.12 shows several images recorded in patients with age-related macular degeneration (AMD) at various stages of the disease. AMD is a multifactorial disease that leads to progressive vision loss in the macula. It is the leading cause of blindness in the industrialized world.

Figure 3.12a shows an early stage of AMD, a B-scan through a retina that contains several drusen (marked by arrows) (Ahlers et al. 2010). Drusen are caused by deposits of metabolic end products, accumulated beneath the RPE. They are identified as local elevations of the RPE (detachments of the RPE from Bruch's membrane). At later stages, AMD can progress into two forms that cause severe damage to the retina, leading to severe vision loss: dry AMD and wet (neovascular) AMD.

Figure 3.12b shows a B-scan obtained in a patient with dry AMD (Sugita et al. 2014). The retina is heavily damaged; the most striking observation is an absence of the RPE in the central area of the image, extending over major parts of the macula. In the area of this so-called geographic atrophy (GA), the layers associated with the photoreceptors are also missing, leading to a complete vision loss in this region of the retina. A further observation is the increased penetration of the sampling light into deeper areas (choroid and sclera) since the heavily scattering and absorbing RPE no longer obstructs the light. Figure 3.12c shows a B-scan of a retina with neovascular AMD (Ahlers et al. 2010). The retina is heavily distorted and large cysts are visible.

These few images can only provide a first impression of the wealth of information provided by OCT for retinal diagnostics. Retinal OCT is nowadays an indispensable tool for diagnostics and monitoring of a large variety of retinal diseases like AMD, diabetic retinopathy, and glaucoma.

3.3.2.4 Functional extensions of OCT

Despite all the success of conventional, intensity-based OCT, it still has some limitations: it just provides structural information and does not allow a direct differentiation of tissues or functional information. To overcome these issues, functional extensions of OCT are under development. Among the most promising are polarization-sensitive (PS) OCT and Doppler (D) OCT. These two extensions shall be briefly discussed in the next two subsections. It should be mentioned, however, that several variants of these technologies are currently under development and evaluation. Because of the limited space of this chapter, only basic concepts and few examples can be presented here.

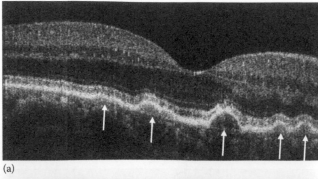

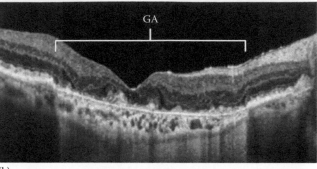

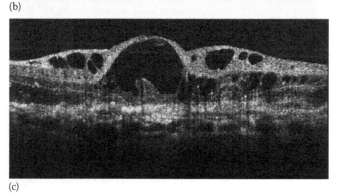

Figure 3.12 Examples of retinal imaging in eyes affected by age-related macular degeneration (AMD). (a) Drusen in early AMD (marked by arrows). (b) Geographic atrophy (GA) in late-stage dry AMD. (c) Cysts in late-stage neovascular AMD. ([a,c]: Reproduced from Ahlers, C. et al., *Invest. Ophthalmol. Vis. Sci.*, 51(4), 2149, 2010. With permission from the Association for Research in Vision and Ophthalmology; [b] From Sugita, M., Zotter, S., Pircher, M., Makihira, T., Saito, K., Tomatsu, N., Sato, M., Roberts, P., Schmidt-Erfurth, U., and Hitzenberger, C.K., Motion artifact and speckle noise reduction in polarization sensitive optical coherence tomography by retinal tracking, *Biomed. Opt. Express*, 5(1), 106–122, 2014. With permission of Optical Society of America).

3.3.2.4.1 Polarization-sensitive OCT

Intensity-based OCT—both TD and FD variants—cannot directly differentiate tissues. While retinal layers in healthy eyes can readily be identified by their known sequence, this is frequently difficult in diseases that distort the layers or where layers are missing. Pure intensity contrast can be insufficient in these cases. However, some tissues can change the light's polarization state. This effect can be used by PS OCT for tissue identification and for quantitative measurements. With PS OCT, the sample is typically illuminated either by circularly polarized light (Hee et al. 1992; de Boer et al. 1997; Hitzenberger et al. 2001; Götzinger et al. 2005) or successively with different

polarization states (Park et al. 2001; Yamanari et al. 2006, 2008; Cense et al. 2007). The backscattered light is decomposed into two orthogonal polarization states that are measured simultaneously.

The second detection channel and some polarizing components in the beam paths add complexity and costs to the system. In addition, data acquisition and processing are more complex. While for intensity-based OCT the acquisition of the signal amplitude (or envelope of the fringe pattern) is sufficient, PS OCT also requires measurement of the phase of the signal (except for a very basic form that just provides retardation data). While in TD OCT either a demodulation of the signal or an analysis by a more complex method like a Hilbert transform was required, the situation is simpler in the case of FD OCT. The Fourier transform of the real-valued spectral interference signal directly provides a complex signal from which the phase information can easily be derived. The spatially resolved amplitude and phase enables calculation of several parameters simultaneously. For example, in the case of a system that illuminates the sample directly by circularly polarized light, reflectivity, retardation, birefringent axis orientation, and Stokes vectors can be derived. More information on the procedures and equations required can be found in a recent review (Pircher et al. 2011).

Two polarization-changing light–tissue interaction mechanisms can be used for ocular imaging: birefringence, which is found in fibrous tissues (form birefringence), and depolarization, which can be caused by multiple light scattering or scattering at large, nonspherical particles (Schmitt and Xiang 1998). Applications of PS OCT to ocular imaging were reported for the anterior and the posterior eye segment. For the sake of brevity, we only consider retinal applications in this section. The structures of the ocular fundus can be classified into polarization preserving (e.g., photoreceptor layer), birefringent (e.g., RNFL, Henle's fiber layer, fibrotic tissue), and depolarizing tissue (e.g., RPE) (Pircher et al. 2011). Two promising applications shall be mentioned here: analysis of the RNFL for glaucoma diagnostics and segmentation of the RPE for AMD diagnostics.

The RNFL is a fibrous tissue that shows birefringence (Dreher et al. 1992). Birefringence introduces a phase delay between orthogonally polarized beam components that traverse the RNFL perpendicularly to the fiber orientation. This phase retardation can be quantified by PS OCT (Cense et al. 2002). Thereby, RNFL thickness and birefringence can be measured (Cense et al. 2002, 2004; Zotter et al. 2012, 2013). Glaucoma, one of the leading causes of blindness in the industrialized world, damages the RNFL, leading to a thinning of the tissue, a reduction of retardation (Quigley et al. 1982; Weinreb et al. 1995), and, according to recent studies in nonhuman primates, a reduction of birefringence at an even earlier state (Fortune et al. 2013). Therefore, analysis of the RNFL is an interesting diagnostic application of PS OCT.

Figure 3.13 shows RNFL retardation maps derived from 3D PS OCT data sets acquired with a wide-field SD PS OCT instrument (Zotter et al. 2012). Figure 3.13a was recorded in a healthy eye. The increased retardation (greenish-yellowish color) caused by the thick RNFL bundles superior and inferior to the ONH, stretching along the major arcuate vessels, can be clearly

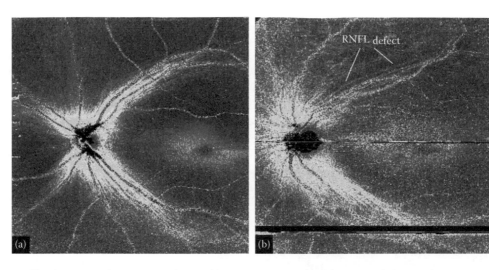

Figure 3.13 Retinal nerve fiber layer retardation maps obtained by PS OCT in (a) a healthy eye and (b) in a glaucomatous eye. (From Zotter, S., Pircher, M., Torzicky, T., Baumann, B., Yoshida, H., Hirose, F., Roberts, P. et al., Large-field high-speed polarization sensitive spectral domain OCT and its applications in ophthalmology, *Biomed. Opt. Express*, 3(11), 2720–2732, 2012. With permission of the Optical Society of America.)

observed, as well as the (somewhat lower) retardation caused by Henle's fiber layer (doughnut-shaped ring centered at the fovea centralis). Figure 3.13b was obtained in a patient with early glaucoma. An RNFL bundle defect along the superior arcuate bundle is clearly visible.

The RPE scrambles the polarization state of backscattered light, equivalent to depolarization. This effect can be quantified by measuring spatially resolved Stokes vectors (that completely describe the light's polarization state) and analyzing them within a small, sliding evaluation window (kernel). Thereby, a quantity called degree of polarization uniformity (DOPU) can be calculated (Götzinger et al. 2008). DOPU is ~1 in tissues that backscatter light with a well-defined polarization state and smaller in depolarizing tissue like the RPE. By generating a DOPU image, the RPE can be directly identified by its intrinsic, tissue-specific polarization contrast. This effect can further be used to segment the RPE. Since the integrity of the RPE is decisive for the metabolism of the photoreceptors, a segmentation of the RPE and associated lesions is very important for diagnosing and monitoring diseases that affect the RPE like AMD.

Figure 3.14 illustrates the segmentation of the RPE by PS OCT (Götzinger et al. 2008). Figure 3.14a shows an intensity image. The three bright lines at the posterior side of the retina, labeled IS/OS, ETPR, and RPE, show nearly similar intensity. Figure 3.14b is a DOPU image. The RPE can clearly be recognized by the reduced DOPU value (blue-green color). This layer was extracted by a thresholding algorithm, finally generating an overlay image showing the segmented RPE in red on top of the intensity image (Figure 3.14c).

Lesions of the RPE can be segmented by advanced postprocessing algorithms that use the RPE that was segmented by its DOPU value as a backbone. For example, algorithms that automatically segment drusen and GAs have been developed (Baumann et al. 2010) and were successfully compared to manual segmentation (Schlanitz et al. 2011; Schutze et al. 2013). Figure 3.15 shows a result of drusen segmentation by PS OCT, and Figure 3.16 an example of GA segmentation (Baumann et al. 2010).

More information on PS OCT methods and applications to ocular imaging can be found in a recent review (Pircher et al. 2011).

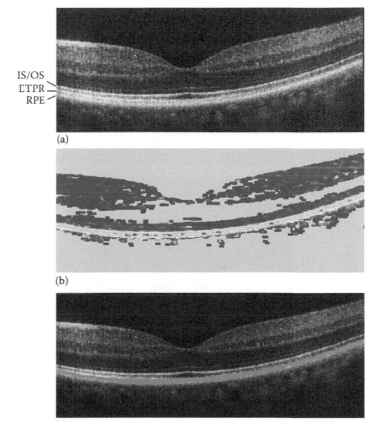

Figure 3.14 Segmentation of the retinal pigment epithelium (RPE) by polarization-sensitive OCT in a healthy eye. (a) Intensity B-scan. (b) Degree of polarization uniformity (DOPU) image. Color bar: cf. Figure 3.15 (black: DOPU = 0; red: DOPU = 1). Polarization preserving tissue shown in red, depolarizing tissue shown in blue–green. (c) Overlay image. The segmented RPE is shown in red on top of the intensity image. IS/OS, junction between inner and outer photoreceptor segments; ETPR, end tips of photoreceptors. (From Götzinger, E., Pircher, M., Geitzenauer, W., Ahlers, C., Baumann, B., Michels, S., Schmidt-Erfurth, U., and Hitzenberger, C.K., Retinal pigment epithelium segmentation by polarization sensitive optical coherence tomography, *Opt. Express*, 16(21), 16410–16422, 2008. With permission of Optical Society of America.)

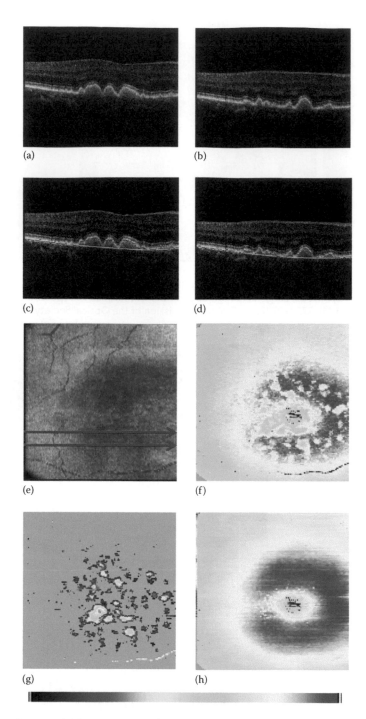

(a)   (b)

(c)   (d)

(e)   (f)

(g)   (h)

Figure 3.15 Segmentation of drusen in an eye affected by early AMD. (a, b) Intensity B-scan images. (c, d) Overlaid positions of inner limiting membrane (ILM) (blue), retinal pigment epithelium (RPE) (red), and Bruch's membrane (BM) (green). (e) Intensity projection image. (f) Retinal thickness map showing the distance between ILM and RPE position (color scale: 70–350 μm). (g) Drusen thickness map (distance between RPE and BM; color scale: 0–128 μm). (h) Retinal thickness map (distance between ILM and BM; color scale: 70–350 μm). (Reproduced from Baumann, B. et al., *J. Biomed. Opt.*, 15, 061704, 2010. With permission from SPIE.)

3.3.2.4.2 Doppler OCT

Similar to imaging techniques like ultrasound, OCT can also provide velocity information by exploiting the Doppler effect and related phase-based mechanisms. Thereby, blood flow can be measured, and vessel contrast (angiography-like images) can be

generated. These methods are presently very active research fields; a great variety of different implementations have been reported. A comprehensive overview of these techniques is beyond the scope of this chapter; however, it can be found in recent reviews (Mahmud et al. 2013; Leitgeb et al. 2014). In the following paragraphs, two methods shall be briefly presented: quantitative velocity measurements and vessel contrast generation.

The first applications of D-OCT were quantitative measurement of velocity and flow. Early TD variants of the technique measured the Doppler shift of the carrier frequency of the A-scan signal caused by moving particles (Wang et al. 1995; Izatt et al. 1997). To obtain velocity information in a depth-resolved manner, the A-scan signal was evaluated by short-time Fourier transform. As with other OCT applications, a major boost to the technology came with the switch to FD-based techniques (Leitgeb et al. 2003c). In this case, two successive A-scans are taken at the same position, and the phase difference between the two signals is calculated along the A-scan length (as with PS OCT, advantage is taken from the fact that a complex-valued signal is available after the Fourier transform of the spectral interferogram, directly providing phase values).

The flow speed of the moving scatterers (at each location along the A-scan) is proportional to the phase difference between the A-scans at the corresponding position and inversely proportional to the time delay between the two A-scans. Instead of measuring the phase difference between A-scans at exactly the same location (which would require a stop of the galvo scanner at each sampling position), measurements can also be taken at adjacent A-scans during transverse scanning of the sample beam, as long as a significant overlap of the beam spots at adjacent positions is warranted to provide a meaningful phase relation between the two signals. The minimum measurable speed is determined by the phase noise of the signal, and the maximum speed by the 2π ambiguity of phase measurements.

It should be mentioned that D-OCT, as described earlier, only provides the velocity component along the measuring beam. Measurements of the absolute velocity require either knowledge of the vessel geometry (Wang et al. 2008) or more complicated two- or three-beam approaches (Werkmeister et al. 2008; Trasischker et al. 2013) that record different velocity vector components.

Figure 3.17 shows an example of velocity profiles recorded in retinal vessels by D-OCT (Bachmann et al. 2007). Figure 3.17a shows an intensity B-scan in the vicinity of the ONH. Within the thick RNFL, cross sections through vessels can be observed. Figure 3.17b shows the corresponding velocity image. The vessels are clearly recognizable by their blue or red color (indicating opposite flow directions). Figure 3.17c shows the velocity profiles extracted from four vessels. An important clinical application of such flow measurements by D-OCT might be glaucoma diagnostics, since there is increasing evidence that ocular blood flow plays an important role in this disease.

The second application to be briefly discussed here is the use of D-OCT and related techniques to generate vessel contrast that can be used for noninvasive angiography. The basic idea is to record two or more spatially correlated data sets of the tissue. While static tissue generates equal data sets, moving scatterers can be identified by changes from one data set to the next.

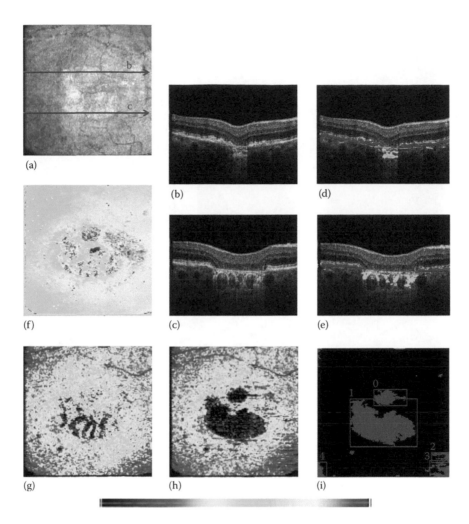

Figure 3.16 Segmentation of geographic atrophies in an eye affected by advanced AMD. (a) Intensity projection image. (b, c) Intensity B-scan images. (d, e) Overlay of depolarizing structures: RPE, red; choroidal pigments, green. (f) Retinal thickness map. (g) Map of overall number of depolarizing pixels per A-line. Zones of RPE atrophy are masked by choroidal depolarization. (h) Map displaying the thickness of the depolarizing RPE. (i) Binary map of atrophic zones. The color map scales from 70 to 325 µm for (f) and from 0 to 39 pixels for (g) and (h). (Reproduced from Baumann, B. et al., *J. Biomed. Opt.*, 15, 061704, 2010. With permission from SPIE.)

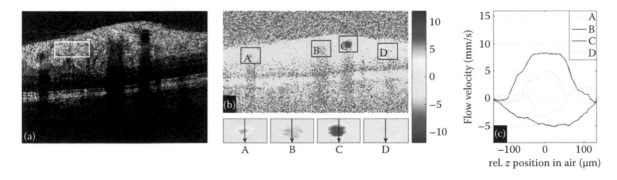

Figure 3.17 Doppler OCT imaging in a healthy eye. (a) Intensity B-scan. (b) Velocity map (top) with zoomed and average-filtered vessel regions (bottom). (c) Velocity profiles extracted along the indicated lines in the vessel regions. (From Bachmann, A.H., Villiger, M.L., Blatter, C., Lasser, T., and Leitgeb, R.A., Resonant Doppler flow imaging and optical vivisection of retinal blood vessels, *Opt. Express*, 15(2), 408–422, 2007. With permission of Optical Society of America.)

For example, the phase difference between adjacent A-scans can be exploited, as in D-OCT, to generate vessel contrast (Makita et al. 2006). Static tissue would exhibit near-zero phase difference, while vessels, containing moving blood cells as scatterers, can be differentiated by nonzero phase differences. The use of phase differences between adjacent A-scans, however, requires rather fast motion of the scatterers to generate a measurable signal well discernible from phase noise (because of the high A-scan rate of modern FD OCT systems). Therefore, only larger retinal vessels can be imaged by this method. An alternative is to measure phase differences between B-scans (Grulkowski et al. 2009). The larger time interval between two B-scans allows visualization of the

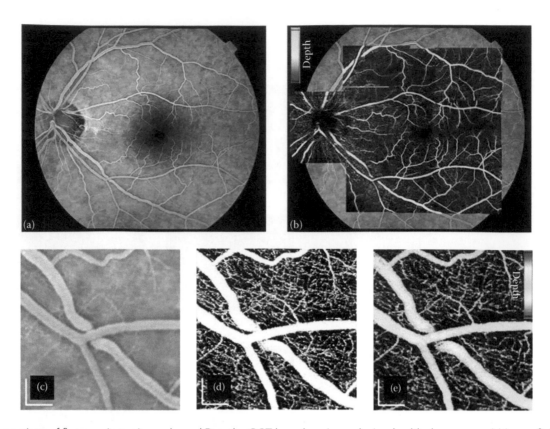

Figure 3.18 Comparison of fluorescein angiography and Doppler OCT-based angiography in a healthy human eye. (a) Large-field fluorescein angiography. (b) Fluorescein angiography overlaid with stitched phase variance OCT vessel maps (color coding shows vessel depth). (c–e) Zoom in's. (c) Fluorescein angiography; (d) phase variance OCT capillary map; (e) color-coded phase variance OCT capillary map (color corresponds to depth). (From Kim, D.Y., Fingler, J., Werner, J.S., Schwartz, D.M., Fraser, S.E., and Zawadzki, R.J., In vivo volumetric imaging of human retinal circulation with phase-variance optical coherence tomography, *Biomed. Opt. Express*, 2(6), 1504–1513, 2011. With permission of Optical Society of America.)

slower moving blood cells in smaller vessels and capillaries. Other methods calculate the phase variance between several data sets (Fingler et al. 2009), but also intensity differences or variances between scans can be exploited (Blatter et al. 2012).

Figure 3.18 shows an example of OCT-based angiography in the fundus of a healthy human eye (Kim et al. 2011). Color coding is used to show the depth location of different vascular beds. Important diagnostic applications of these technologies could be diabetic retinopathy, where alterations of the perifoveal capillary network occur at an early stage, or retinal vein occlusions.

3.4 OUTLOOK

LCI and OCT have revolutionized ocular biometry and imaging, as discussed in the previous sections. However, these are rapidly evolving fields and further improvements of technology and applications can be expected in the near future. Commercial biometry instruments are still largely based on the older, less-sensitive TD LCI technique, mainly because of the lower depth range of SD LCI. However, with the introduction of new tunable lasers with very long coherence length, biometry of the full axial eye length will become available. Experimental prototypes were already demonstrated successfully; it seems only a matter of availability of these sources at a competitive price that SS LCI will replace TD LCI. The higher sensitivity of FD-based technologies

should also enable a direct measurement of the anterior segment, including the lens, by LCI (using backscattered light instead of reflected light), so the use of conventional, less precise slit lamp–based technology for anterior segment measurements can likely be abandoned in the near future.

Also for OCT, a switch to SS technology can be expected in the near future, provided swept sources become cheaper. SS OCT has a lower sensitivity decay with depth than SD OCT, and the 1050 nm wavelength, for which sources with high sweep rate are available, has a better penetration into deeper tissue layers like the choroid and sclera. A first commercial SS OCT retinal scanner operating at 100 kA-scans/s is already available; more are expected for the near future. Further increases of imaging speed to a few 100 kHz should be possible maintaining good image quality, reducing the time for a high-resolution volume raster scan to well below one second. Other hardware technology improvements, for example, multibeam OCT, PS OCT, and Doppler OCT (which, in its simplest form, can also be realized as a pure software add-on), are presently under investigation and might become available in future generation OCT systems, provided they can be built at a competitive price.

Another area where future improvements can be expected is software: automated segmentation of retinal layers is one of the main goals of this field of research. While software algorithms exist that can automatically segment up to ~10 layers in the healthy retina, their reliability is poorer in diseased retinas where retinal layers are heavily distorted, are interrupted, or are missing. Interdisciplinary research combining the expertise of various fields (ophthalmology, optics, informatics, signal processing, etc.) is underway to provide improved segmentation algorithms for reliable diagnostics and treatment monitoring.

ACKNOWLEDGMENTS

The author acknowledges the permission of several authors and publishers to reproduce figures from their work and their journals.

Furthermore, the author wishes to thank A.F. Fercher for the stimulation to work in the field of biomedical optics and for many fruitful discussions. Cooperation and fruitful discussions with several colleagues and coworkers at the Center for Medical Physics and Biomedical Engineering, Medical University of Vienna, are acknowledged: B. Baumann, A. Baumgartner, W. Drexler, E. Götzinger, R. Leitgeb, M. Pircher, H. Sattmann, L. Schmetterer, M. Sticker, T. Torzicky, W. Trasischker, R. Werkmeister, S. Zotter, and many more. Equally important, the cooperation and discussions with our clinical partners are acknowledged: C. Dolezal, H. Gnad, M. Juchem, and F. Skorpik, Lainz Hospital, contributed in the early days of LCI biometry; O. Findl, B. Kiss, R. Menapace, and others continued this work at the Department of Ophthalmology and Optometry, Medical University of Vienna; U. Schmidt-Erfurth, Department of Ophthalmology and Optometry, Medical University of Vienna, and her team: C. Ahlers, M. Bolz, J. Lammer, S. Michels, M. Ritter, P. Roberts, F. Schlanitz, C. Schütze, C. Vass, and many more, are acknowledged for the cooperation in the field of OCT imaging.

Part of the work reported in this chapter was financially supported by the Austrian Science Fund (FWF grants P7300, P09781, P14103, P16776, P19624, P26553), by the European Commission (project FUN OCT, FP7 HEALTH, contract no. 201880), and by Canon Inc., Tokyo.

REFERENCES

Ahlers, C., E. Gotzinger, M. Pircher, I. Golbaz, F. Prager, C. Schutze, B. Baumann, C. K. Hitzenberger, and U. Schmidt-Erfurth. 2010. Imaging of the retinal pigment epithelium in age-related macular degeneration using polarization-sensitive optical coherence tomography. *Investigative Ophthalmology and Visual Science* 51 (4):2149–2157.

American National Standards for Safe Use of Lasers. 2000. In *ANSI Z 136.1*, ed. American National Standards. Institute. Orlando, FL: Laser Institute of America.

An, L., P. Li, T. T. Shen, and R. K. Wang. 2011. High speed spectral domain optical coherence tomography for retinal imaging at 500,000 A-lines per second. *Biomedical Optics Express* 2:2770–2783.

Bachmann, A. H., M. L. Villiger, C. Blatter, T. Lasser, and R. A. Leitgeb. 2007. Resonant Doppler flow imaging and optical vivisection of retinal blood vessels. *Optics Express* 15 (2):408–422.

Bajraszewski, T., M. Wojtkowski, M. Szkulmowski, A. Szkulmowska, R. Huber, and A. Kowalczyk. 2008. Improved spectral optical coherence tomography using optical frequency comb. *Optics Express* 16 (6):4163–4176.

Baumann, B., E. Götzinger, M. Pircher, H. Sattman, C. Schütze, F. Schlanitz, C. Ahlers, U. Schmidt-Erfurth, and C. K. Hitzenberger. 2010. Segmentation and quantification of retinal lesions in age-related macular degeneration using polarization-sensitive optical coherence tomography. *Journal of Biomedical Optics* 15:061704.

Baumann, B., M. Pircher, E. Götzinger, and C. K. Hitzenberger. 2007. Full range complex spectral domain optical coherence tomography without additional phase shifters. *Optics Express* 15 (20):13375–13387.

Blatter, C., T. Klein, B. Grajciar, T. Schmoll, W. Wieser, R. Andre, R. Huber, and R. A. Leitgeb. 2012. Ultrahigh-speed non-invasive widefield angiography. *Journal of Biomedical Optics* 17 (7):070505.

Bonesi, M., M. P. Minneman, J. Ensher, B. Zabihian, H. Sattmann, P. Boschert, E. Hoover, R. A. Leitgeb, M. Crawford, and W. Drexler. 2014. Akinetic all-semiconductor programmable swept-source at 1550 nm and 1310 nm with centimeters coherence length. *Optics Express* 22 (3):2632–2655.

Born, M. and E. Wolf. 1987. *Principles of Optics*, 6th edn. Oxford, U.K.: Pergamon Press.

Bracewell, R. N. 2000. *The Fourier Transform and Its Application*, 3rd edn. New York: McGraw-Hill.

Cense, B., T. C. Chen, B. H. Park, M. C. Pierce, and J. F. de Boer. 2002. In vivo depth-resolved birefringence measurements of the human retinal nerve fiber layer by polarization-sensitive optical coherence tomography. *Optics Letters* 27 (18):1610–1612.

Cense, B., T. C. Chen, B. H. Park, M. C. Pierce, and J. F. de Boer. 2004. Thickness and birefringence of healthy retinal nerve fiber layer tissue measured with polarization-sensitive optical coherence tomography. *Investigative Ophthalmology and Visual Science* 45 (8):2606–2612.

Cense, B., M. Mujat, T. C. Chen, B. H. Park, and J. F. de Boer. 2007. Polarization-sensitive spectral-domain optical coherence tomography using a single line scan camera. *Optics Express* 15 (5):2421–2431.

Chinn, S. R., E. A. Swanson, and J. G. Fujimoto. 1997. Optical coherence tomography using a frequency-tunable optical source. *Optics Letters* 22 (5):340–342.

Choma, M. A., M. V. Sarunic, C. H. Yang, and J. A. Izatt. 2003. Sensitivity advantage of swept source and Fourier domain optical coherence tomography. *Optics Express* 11 (18):2183–2189.

de Boer, J. F., B. Cense, B. H. Park, M. C. Pierce, G. J. Tearney, and B. E. Bouma. 2003. Improved signal-to-noise ratio in spectral-domain compared with time-domain optical coherence tomography. *Optics Letters* 28 (21):2067–2069.

de Boer, J. F., T. E. Milner, M. J. C. van Gemert, and J. S. Nelson. 1997. Two-dimensional birefringence imaging in biological tissue by polarization-sensitive optical coherence tomography. *Optics Letters* 22 (12):934–936.

Dreher, A. W., K. Reiter, and R. N. Weinreb. 1992. Spatially resolved birefringence of the retinal nerve-fiber layer assessed with a retinal laser ellipsometer. *Applied Optics* 31 (19):3730–3735.

Drexler, W. 2004. Ultrahigh-resolution optical coherence tomography. *Journal of Biomedical Optics* 9 (1):47–74.

Drexler, W., A. Baumgartner, O. Findl, C. K. Hitzenberger, and A. F. Fercher. 1997a. Biometric investigation of changes in the anterior eye segment during accommodation. *Vision Research* 37 (19):2789–2800.

Drexler, W., A. Baumgartner, O. Findl, C. K. Hitzenberger, H. Sattmann, and A. F. Fercher. 1997b. Submicrometer precision biometry of the anterior segment of the human eye. *Investigative Ophthalmology and Visual Science* 38 (7):1304–1313.

Drexler, W., O. Findl, R. Menapace, G. Rainer, C. Vass, C. K. Hitzenberger, and A. F. Fercher. 1998a. Partial coherence interferometry: A novel approach to biometry in cataract surgery. *American Journal of Ophthalmology* 126 (4):524–534.

Drexler, W., O. Findl, L. Schmetterer, C. K. Hitzenberger, and A. F. Fercher. 1998b. Eye elongation during accommodation in humans: Differences between emmetropes and myopes. *Investigative Ophthalmology and Visual Science* 39 (11):2140–2147.

Drexler, W. and J. G. Fujimoto. 2008. State-of-the-art retinal optical coherence tomography. *Progress in Retinal and Eye Research* 27 (1):45–88.

Drexler, W., C. K. Hitzenberger, A. Baumgartner, O. Findl, H. Sattmann, and A. F. Fercher. 1998c. Investigation of dispersion effects in ocular media by multiple wavelength partial coherence interferometry. *Experimental Eye Research* 66 (1):25–33.

Drexler, W., M. Y. Liu, A. Kumar, T. Kamali, A. Unterhuber, and R. A. Leitgeb. 2014. Optical coherence tomography today: Speed, contrast, and multimodality. *Journal of Biomedical Optics* 19 (7):071412.

Drexler, W., U. Morgner, R. K. Ghanta, F. X. Kartner, J. S. Schuman, and J. G. Fujimoto. 2001. Ultrahigh-resolution ophthalmic optical coherence tomography. *Nature Medicine* 7 (4):502–507.

Felberer, F., J. S. Kroisamer, B. Baumann, S. Zotter, U. Schmidt-Erfurth, C. K. Hitzenberger, and M. Pircher. 2014. Adaptive optics SLO/OCT for 3D imaging of human photoreceptors in vivo. *Biomedical Optics Express* 5 (2):439–456.

Fercher, A. F., W. Drexler, C. K. Hitzenberger, and T. Lasser. 2003. Optical coherence tomography–Principles and applications. *Reports on Progress in Physics* 66 (2):239–303.

Fercher, A. F., C. Hitzenberger, and M. Juchem. 1991. Measurement of intraocular optical distances using partially coherent laser-light. *Journal of Modern Optics* 38 (7):1327–1333.

Fercher, A. F. and C. K. Hitzenberger. 2002. Optical coherence tomography. *Progress in Optics* 44:215–302.

Fercher, A. F., C. K. Hitzenberger, W. Drexler, G. Kamp, and H. Sattmann. 1993. In-vivo optical coherence tomography. *American Journal of Ophthalmology* 116 (1):113–115.

Fercher, A. F., C. K. Hitzenberger, G. Kamp, and S. Y. El-Zaiat. 1995. Measurement of intraocular distances by backscattering spectral interferometry. *Optics Communications* 117 (1–2):43–48.

Fercher, A. F., C. K. Hitzenberger, M. Sticker, R. Zawadzki, B. Karamata, and T. Lasser. 2001. Numerical dispersion compensation for partial coherence interferometry and optical coherence tomography. *Optics Express* 9 (12):610–615.

Fercher, A. F., R. Leitgeb, C. K. Hitzenberger, H. Sattmann, and M. Wojtkowski. 1999. Complex spectral interferometry OCT. *Proceedings of the SPIE* 3564:173–178.

Fercher, A. F., K. Mengedoht, and W. Werner. 1988. Eye-length measurement by interferometry with partially coherent-light. *Optics Letters* 13 (3):186–188.

Fercher, A. F. and E. Roth. 1986. Ophthalmic laser interferometry. *Proceeding of the SPIE* 658:48–51.

Fernandez, E. J., I. Iglesias, and P. Artal. 2001. Closed-loop adaptive optics in the human eye. *Optics Letters* 26 (10):746–748.

Findl, O., W. Drexler, R. Menapace, C. K. Hitzenberger, and A. F. Fercher. 1998. High precision biometry of pseudophakic eyes using partial coherence interferometry. *Journal of Cataract and Refractive Surgery* 24 (8):1087–1093.

Fingler, J., R. J. Zawadzki, J. S. Werner, D. Schwartz, and S. E. Fraser. 2009. Volumetric microvascular imaging of human retina using optical coherence tomography with a novel motion contrast technique. *Optics Express* 17 (24):22190–22200.

Fortune, B., C. F. Burgoyne, G. Cull, J. Reynaud, and L. Wang. 2013. Onset and progression of peripapillary retinal nerve fiber layer (RNFL) retardance changes occur earlier than RNFL thickness changes in experimental glaucoma. *Investigative Ophthalmology and Visual Science* 54 (8):5653–5660.

Geitzenauer, W., C. K. Hitzenberger, and U. M. Schmidt-Erfurth. 201 Retinal optical coherence tomography: Past, present and future perspectives. *British Journal of Ophthalmology* 95 (2):171–177.

Götzinger, E., M. Pircher, W. Geitzenauer, C. Ahlers, B. Baumann, S. Michels, U. Schmidt-Erfurth, and C. K. Hitzenberger. 2008. Retinal pigment epithelium segmentation by polarization sensitive optical coherence tomography. *Optics Express* 16 (21):16410–16422.

Götzinger, E., M. Pircher, and C. K. Hitzenberger. 2005. High speed spectral domain polarization sensitive optical coherence tomography of the human retina. *Optics Express* 13 (25):10217–10229.

Grieve, K., M. Paques, A. Dubois, J. Sahel, C. Boccara, and J. F. Le Gargasson. 2004. Ocular tissue imaging using ultrahigh-resolution, full-field optical coherence tomography. *Investigative Ophthalmology and Visual Science* 45 (11):4126–4131.

Grulkowski, I., M. Gora, M. Szkulmowski, I. Gorczynska, D. Szlag, S. Marcos, A. Kowalczyk, and M. Wojtkowski. 2009a. Anterior segment imaging with Spectral OCT system using a high-speed CMOS camera. *Optics Express* 17 (6):4842–4858.

Grulkowski, I., I. Gorczynska, M. Szkulmowski, D. Szlag, A. Szkulmowska, R. A. Leitgeb, A. Kowalczyk, and M. Wojtkowski 2009b. Scanning protocols dedicated to smart velocity ranging i Spectral OCT. *Optics Express* 17 (26):23736–23754.

Grulkowski, I., J. J. Liu, B. Potsaid, V. Jayaraman, J. Jiang, J. G. Fujimoto, and A. E. Cable. 2013. High-precision, high-accuracy ultralong-range swept-source optical coherence tomography usin vertical cavity surface emitting laser light source. *Optics Letters* 38 (5):673–675.

Grulkowski, I., J. J. Liu, B. Potsaid, V. Jayaraman, C. D. Lu, J. Jiang, A. E. Cable, J. S. Duker, and J. G. Fujimoto. 2012. Retinal, anterior segment and full eye imaging using ultrahigh speed swept source OCT with vertical-cavity surface emitting lasers. *Biomedical Optics Express* 3 (11):2733–2751.

Häusler, G., and M. W. Lindner. 1998. "Coherence radar" and "spectra radar"—New tools for dermatological diagnosis. *Journal of Biomedical Optics* 3 (1):21–31.

Hee, M. R., D. Huang, E. A. Swanson, and J. G. Fujimoto. 1992. Polarization-sensitive low-coherence reflectometer for birefringence characterization and ranging. *Journal of the Optical Society of America B-Optical Physics* 9 (6):903–908.

Hitzenberger, C. K. 1991. Optical measurement of the axial eye length by laser Doppler interferometry. *Investigative Ophthalmology and Visual Science* 32 (3):616–624.

Hitzenberger, C. K. 1992. Measurement of corneal thickness by low-coherence interferometry. *Applied Optics* 31 (31):6637–6642.

Hitzenberger, C. K., A. Baumgartner, W. Drexler, and A. F. Fercher. 1994 Interferometric measurement of corneal thickness with micrometer precision. *American Journal of Ophthalmology* 118 (4):468–476.

Hitzenberger, C. K., A. Baumgartner, W. Drexler, and A. F. Fercher. 1999. Dispersion effects in partial coherence interferometry: Implications for intraocular ranging. *Journal of Biomedical Optics* 4 (1):144–151.

Hitzenberger, C. K., W. Drexler, A. Baumgartner, F. Lexer, H. Sattma M. Eßlinger, M. Kulhavy, and A. F. Fercher. 1997. Optical measurement of intraocular distances: A comparison of methods. *Lasers and Light in Ophthalmology* 8:85–95.

Hitzenberger, C. K., W. Drexler, C. Dolezal, F. Skorpik, M. Juchem, A. F. Fercher, and H. D. Gnad. 1993. Measurement of the axial length of cataract eyes by laser-doppler interferometry. *Investigative Ophthalmology and Visual Science* 34 (6):1886–1893.

Hitzenberger, C. K., E. Götzinger, M. Sticker, M. Pircher, and A. F. Fercher. 2001. Measurement and imaging of birefringence and optic axis orientation by phase resolved polarization sensitive optical coherence tomography. *Optics Express* 9 (13):780–790.

Hu, Z. and A. M. Rollins. 2007. Fourier domain optical coherence tomography with a linear-in-wavenumber spectrometer. *Optics Letters* 32 (24):3525–3527.

Huang, D., E. A. Swanson, C. P. Lin, J. S. Schuman, W. G. Stinson, W. Chang, M. R. Hee et al. 1991. Optical coherence tomography. *Science* 254 (5035):1178–1181.

International Electrotechnical Commission. 2001. Safety of laser products—Part 1: Equipment classification and requirements. In *IEC 60825 -1 Ed. 2*, International Electrotechnical Commission, Geneva.

Ivanov, A. P., A. P. Chaikovskii, and A. A. Kumeisha. 1977. New method for high-range resolution measurements of light scattering in optically dense inhomogeneous media. *Optics Letters* 1 (6):226.

Izatt, J. A., M. D. Kulkami, S. Yazdanfar, J. K. Barton, and A. J. Welch. 1997. In vivo bidirectional color Doppler flow imaging of picoliter blood volumes using optical coherence tomography. *Optics Letters* 22 (18):1439–1441.

Jiao, S. L., R. Knighton, X. R. Huang, G. Gregori, and C. A. Puliafito. 2005. Simultaneous acquisition of sectional and fundus ophthalmic images with spectral-domain optical coherence tomography. *Optics Express* 13 (2):444–452.

Jungwirth, J., B. Baumann, M. Pircher, E. Gotzinger, and C. K. Hitzenberger. 2009. Extended in vivo anterior eye-segment imaging with full-range complex spectral domain optical coherence tomography. *Journal of Biomedical Optics* 14 (5):050501.

Kim, D. Y., J. Fingler, J. S. Werner, D. M. Schwartz, S. E. Fraser, and R. J. Zawadzki. 2011. In vivo volumetric imaging of human retinal circulation with phase-variance optical coherence tomography. *Biomedical Optics Express* 2 (6):1504–1513.

Klein, T., W. Wieser, C. M. Eigenwillig, B. R. Biedermann, and R. Huber. 2011. Megahertz OCT for ultrawide-field retinal imaging with a 1050 nm Fourier domain mode-locked laser. *Optics Express* 19 (4):3044–3062.

Klein, T., W. Wieser, L. Reznicek, A. Neubauer, A. Kampik, and R. Huber. 2013. Multi-MHz retinal OCT. *Biomedical Optics Express* 4 (10):1890–1908.

Ko, T. H., D. C. Adler, J. G. Fujimoto, D. Mamedov, V. Prokhorov, V. Shidlovski, and S. Yakubovich. 2004. Ultrahigh resolution optical coherence tomography imaging with a broadband superluminescent diode light source. *Optics Express* 12 (10):2112–2119.

Leitgeb, R., C. K. Hitzenberger, and A. F. Fercher. 2003a. Performance of Fourier domain vs. time domain optical coherence tomography. *Optics Express* 11 (8):889–894.

Leitgeb, R. A., C. K. Hitzenberger, A. F. Fercher, and T. Bajraszewski. 2003b. Phase-shifting algorithm to achieve high-speed long-depth-range probing by frequency-domain optical coherence tomography. *Optics Letters* 28 (22):2201–2203.

Leitgeb, R. A., L. Schmetterer, W. Drexler, A. F. Fercher, R. J. Zawadzki, and T. Bajraszewski. 2003c. Real-time assessment of retinal blood flow with ultrafast acquisition by color Doppler Fourier domain optical coherence tomography. *Optics Express* 11 (23):3116–3121.

Leitgeb, R. A., R. M. Werkmeister, C. Blatter, and L. Schmetterer. 2014. Doppler optical coherence tomography. *Progress in Retinal and Eye Research* 41:26–43.

Lexer, F., C. K. Hitzenberger, W. Drexler, S. Molebny, H. Sattmann, M. Sticker, and A. F. Fercher. 1999. Dynamic coherent focus OCT with depth-independent transversal resolution. *Journal of Modern Optics* 46 (3):541–553.

Lexer, F., C. K. Hitzenberger, A. F. Fercher, and M. Kulhavy. 1997. Wavelength-tuning interferometry of intraocular distances. *Applied Optics* 36 (25):6548–6553.

Li, P., M. Johnstone, and R. K. K. Wang. 2014. Full anterior segment biometry with extended imaging range spectral domain optical coherence tomography at 1340 nm. *Journal of Biomedical Optics* 19 (4):046013.

Liang, J. Z., D. R. Williams, and D. T. Miller. 1997. High resolution imaging of the living human retina with adaptive optics. *Investigative Ophthalmology and Visual Science* 38 (4):55–55.

Mahmud, M. S., D. W. Cadotte, B. Vuong, C. Sun, T. W. H. Luk, A. Mariampillai, and V. X. D. Yang. 2013. Review of speckle and phase variance optical coherence tomography to visualize microvascular networks. *Journal of Biomedical Optics* 18 (5):050901.

Makita, S., Y. Hong, M. Yamanari, T. Yatagai, and Y. Yasuno. 2006. Optical coherence angiography. *Optics Express* 14 (17):7821–7840.

Norrby, S. 2008. Sources of error in intraocular lens power calculation. *Journal of Cataract and Refractive Surgery* 34 (3):368–376.

Olsen, T. 2007. Calculation of intraocular lens power: A review. *Acta Ophthalmologica Scandinavica* 85 (5):472–485.

Park, B. H., C. Saxer, S. M. Srinivas, J. S. Nelson, and J. F. de Boer. 2001. In vivo burn depth determination by high-speed fiber-based polarization sensitive optical coherence tomography. *Journal of Biomedical Optics* 6 (4):474–479.

Pircher, M., B. Baumann, E. Götzinger, and C. K. Hitzenberger. 2006a. Retinal cone mosaic imaged with transverse scanning optical coherence tomography. *Optics Letters* 31 (12):1821–1823.

Pircher, M., E. Götzinger, and C. K. Hitzenberger. 2006b. Dynamic focus in optical coherence tomography for retinal imaging. *Journal of Biomedical Optics* 11 (5):054013.

Pircher, M., C. K. Hitzenberger, and U. Schmidt-Erfurt. 2011. Polarization sensitive optical coherence tomography in the human eye. *Progress in Retinal and Eye Research* 30:431–451.

Podoleanu, A. G. and D. A. Jackson. 1999. Noise analysis of a combined optical coherence tomograph and a confocal scanning ophthalmoscope. *Applied Optics* 38 (10):2116–2127.

Potsaid, B., B. Baumann, D. Huang, S. Barry, A. E. Cable, J. S. Schuman, J. S. Duker, and J. G. Fujimoto. 2010. Ultrahigh speed 1050 nm swept source/Fourier domain OCT retinal and anterior segment imaging at 100,000 to 400,000 axial scans per second. *Optics Express* 18 (19):20029–20048.

Potsaid, B., I. Gorczynska, V. J. Srinivasan, Y. L. Chen, J. Jiang, A. Cable, and J. G. Fujimoto. 2008. Ultrahigh speed Spectral/Fourier domain OCT ophthalmic imaging at 70,000 to 312,500 axial scans per second. *Optics Express* 16 (19):15149–15169.

Quigley, H. A., E. M. Addicks, and W. R. Green. 1982. Optic-nerve damage in human glaucoma. 3. Quantitative correlation of nerve-fiber loss and visual-field defect in glaucoma, ischemic neuropathy, papilledema, and toxic neuropathy. *Archives of Ophthalmology* 100 (1):135–146.

Rollins, A. M. and J. A. Izatt. 1999. Optimal interferometer designs for optical coherence tomography. *Optics Letters* 24 (21):1484–1486.

Roorda, A., F. Romero-Borja, W. J. Donnelly, H. Queener, T. J. Hebert, and M. C. W. Campbell. 2002. Adaptive optics scanning laser ophthalmoscopy. *Optics Express* 10 (9):405–412.

Schlanitz, F. G., B. Baumann, T. Spalek, C. Schutze, C. Ahlers, M. Pircher, E. Gotzinger, C. K. Hitzenberger, and U. Schmidt-Erfurth. 2011. Performance of automated drusen detection by polarization-sensitive optical coherence tomography. *Investigative Ophthalmology and Visual Science* 52 (7):4571–4579.

Schmitt, J. M. and G. Kumar. 1998. Optical scattering properties of soft tissue: a discrete particle model. *Applied Optics* 37 (13):2788–2797.

Schmitt, J. M. and S. H. Xiang. 1998. Cross-polarized backscatter in optical coherence tomography of biological tissue. *Optics Letters* 23 (13):1060–1062.

Schutze, C., M. Bolz, R. Sayegh, B. Baumann, M. Pircher, E. Gotzinger, C. K. Hitzenberger, and U. Schmidt-Erfurth. 2013. Lesion size detection in geographic atrophy by polarization-sensitive optical coherence tomography and correlation to conventional imaging techniques. *Investigative Ophthalmology and Visual Science* 54 (1):739–745.

Shidlovski, V. R. 2008. Superluminescent diode light sources for OCT. In *Optical Coherence Tomography: Technology and Applications*, eds. W. Drexler and J. G. Fujimoto. Berlin, Germany: Springer.

Sugita, M., S. Zotter, M. Pircher, T. Makihira, K. Saito, N. Tomatsu, M. Sato, P. Roberts, U. Schmidt-Erfurth, and C. K. Hitzenberger. 2014. Motion artifact and speckle noise reduction in polarization sensitive optical coherence tomography by retinal tracking. *Biomedical Optics Express* 5 (1):106–122.

Swanson, E. A., D. Huang, M. R. Hee, J. G. Fujimoto, C. P. Lin, and C. A. Puliafito. 1992. High-speed optical coherence domain reflectometry. *Optics Letters* 17 (2):151–153.

Swanson, E. A., J. A. Izatt, M. R. Hee, D. Huang, C. P. Lin, J. S. Schuman, C. A. Puliafito, and J. G. Fujimoto. 1993. In-vivo retinal imaging by optical coherence tomography. *Optics Letters* 18 (21):1864–1866.

Tearney, G. J., B. E. Bouma, and J. G. Fujimoto. 1997. High-speed phase- and group-delay scanning with a grating-based phase control delay line. *Optics Letters* 22 (23):1811–1813.

Trasischker, W., R. M. Werkmeister, S. Zotter, B. Baumann, T. Torzicky, M. Pircher, and C. K. Hitzenberger. 2013. In vitro and in vivo three-dimensional velocity vector measurement by three-beam spectral-domain Doppler optical coherence tomography. *Journal of Biomedical Optics* 18 (11):116010.

Unterhuber, A., B. Povazay, A. Aguirre, Y. Chen, F. X. Kaertner, J. G. Fujimoto, and W. Drexler. 2008. Broad bandwidth laser and nonlinear optical light sources for OCT. In *Optical Coherence Tomography: Technology and Applications*, eds. W. Drexler and J. G. Fujimoto. Berlin, Germany: Springer.

Unterhuber, A., B. Povazay, B. Hermann, H. Sattmann, A. Chavez-Pirson, and W. Drexler. 2005. In vivo retinal optical coherence tomography at 1040 nm-enhanced penetration into the choroid. *Optics Express* 13 (9):3252–3258.

Wang, X. J., T. E. Milner, and J. S. Nelson. 1995. Characterization of fluid-flow velocity by optical Doppler tomography. *Optics Letters* 20 (11):1337–1339.

Wang, Y. M., B. A. Bower, J. A. Izatt, O. Tan, and D. Huang. 2008. Retinal blood flow measurement by circumpapillary Fourier domain Doppler optical coherence tomography. *Journal of Biomedical Optics* 13 (6):064003.

Weinreb, R. N., S. Shakiba, and L. Zangwill. 1995. Scanning laser polarimetry to measure the nerve-fiber layer of normal and glaucomatous eyes. *American Journal of Ophthalmology* 119 (5):627–636.

Werkmeister, R. M., N. Dragostinoff, M. Pircher, E. Gotzinger, C. K. Hitzenberger, R. A. Leitgeb, and L. Schmetterer. 2008. Bidirectional Doppler Fourier-domain optical coherence tomography for measurement of absolute flow velocities in human retinal vessels. *Optics Letters* 33 (24):2967–2969.

Wojtkowski, M., A. Kowalczyk, R. Leitgeb, and A. F. Fercher. 2002a. Full range complex spectral optical coherence tomography technique in eye imaging. *Optics Letters* 27 (16):1415–1417.

Wojtkowski, M., R. Leitgeb, A. Kowalczyk, T. Bajraszewski, and A. F. Fercher. 2002b. In vivo human retinal imaging by Fourier domain optical coherence tomography. *Journal of Biomedical Optics* 7 (3):457–463.

Wojtkowski, M., V. J. Srinivasan, T. H. Ko, J. G. Fujimoto, A. Kowalczyk, and J. S. Duker. 2004. Ultrahigh-resolution, high-speed, Fourier domain optical coherence tomography and methods for dispersion compensation. *Optics Express* 12 (11):2404–2422.

Yamanari, M., S. Makita, V. D. Madjarova, T. Yatagai, and Y. Yasuno. 2006. Fiber-based polarization-sensitive Fourier domain optical coherence tomography using B-scan-oriented polarization modulation method. *Optics Express* 14 (14):6502–6515.

Yamanari, M., S. Makita, and Y. Yasuno. 2008. Polarization-sensitive swept-source optical coherence tomography with continuous source polarization modulation. *Optics Express* 16 (8):5892–5906.

Yun, S. H. and B. E. Bouma. 2008. Wavelength swept lasers. In *Optical Coherence Tomography: Technology and Applications*, eds. W. Drexler and J. G. Fujimoto. Berlin, Germany: Springer.

Zawadzki, R. J., S. M. Jones, S. S. Olivier, M. T. Zhao, B. A. Bower, J. A. Izatt, S. Choi, S. Laut, and J. S. Werner. 2005. Adaptive-optics optical coherence tomography for high-resolution and high-speed 3D retinal in vivo imaging. *Optics Express* 13 (21):8532–8546.

Zhang, Y., B. Cense, J. Rha, R. S. Jonnal, W. Gao, R. J. Zawadzki, J. S. Werner, S. Jones, S. Olivier, and D. T. Miller. 2006. High-speed volumetric imaging of cone photoreceptors with adaptive optics spectral-domain optical coherence tomography. *Optics Express* 14 (10):4380–4394.

Zotter, S., M. Pircher, E. Götzinger, T. Torzicky, H. Yoshida, F. Hirose, S. Holzer et al. 2013. Measuring retinal nerve fiber layer birefringence, retardation, and thickness using wide-field, high-speed polarization sensitive spectral domain OCT. *Investigative Ophthalmology and Visual Science* 54 (1):72–84.

Zotter, S., M. Pircher, T. Torzicky, B. Baumann, H. Yoshida, F. Hirose, P. Roberts et al. 2012. Large-field high-speed polarization sensitive spectral domain OCT and its applications in ophthalmology. *Biomedical Optics Express* 3 (11):2720–2732.

Ophthalmic instrumentation

4 Anterior segment OCT

Ireneusz Grulkowski

Contents

4.1 INTRODUCTION

The human eye is a complex and dynamic optical structure enabling visual perception, that is, conversion of light into electrical signals that are later processed by the brain. Anatomically, the front of the eye constitutes its anterior segment (AS) and includes the elements spanning from the cornea to the posterior surface of the crystalline lens. The main function of the AS of the eye is to perform and control light focusing on the retina since it contains all refracting surfaces leading to the sharp image of the object on the retina. Therefore, from the physical (optical) point of view, the AS of the eye is composed of the optical elements of different refractive

properties, thus forming a compound optical system (Atchison and Smith, 2000, Grosvenor, 2007).

Inspection of AS structures is an integral part of ophthalmic examination. In particular, imaging of the AS of the eye is a common clinical procedure in the diagnosis and surgical management of ocular disorders such as glaucoma and cataract (Low and Nolan, 2009, See, 2009, Ursea and Silverman, 2010, Leung and Weinreb, 2011). It also aims at evaluation of clinically relevant features of the AS of the eye in a variety of eye conditions (Konstantopoulos et al., 2007). Moreover, visualization and quantitative description of the AS of the eye reveals refractive status of the eye, thus playing an important role in efficient vision correction (e.g., lens fitting) or refractive surgery (Wang et al., 2011).

In clinical practice, imaging of the AS has been traditionally carried out with slit-lamp biomicroscopy. However, AS imaging is currently a rapidly advancing field of ophthalmology and provides tools that supplement well-established modalities. Those modern techniques include ultrasound biomicroscopy (UBM), confocal microscopy, Scheimpflug imaging, and corneal topography (Wolffsohn and Peterson, 2006, Konstantopoulos et al., 2007, Wolffsohn and Davies, 2007, Guthoff and Stachs, 2011). Apart from instrumentation development, a tremendous effort has been made to implement image analysis techniques that allow obtaining quantitative information on the state of the cornea, anterior chamber, corneoscleral angle, and crystalline lens (Wolffsohn and Peterson, 2006, Heur and Dupps, 2009, Pinero, 2015). Consequently, application of new technologies for in vivo imaging significantly improves clinical procedures in ophthalmology.

One of the main breakthrough developments in eye diagnostics was the introduction of optical coherence tomography (OCT) that is a noninvasive imaging modality generating micrometer resolution, 2-D cross-sectional images and 3-D volumetric data on the internal structure of optically scattering and reflective tissues (Fujimoto, 2001, 2003, Drexler and Fujimoto, 2008a,b). The idea of OCT dates back to early studies on interferometric measurement of intraocular distances in the eye that were performed in the 1980s and the early 1990s (Fujimoto et al., 1986, Fercher et al., 1988, 1991, Hitzenberger, 1991, 1992, Swanson et al., 1992). OCT was first demonstrated in 1991 in J.G. Fujimoto laboratory at Massachusetts Institute of Technology for imaging of the human retina (Huang et al., 1991). OCT imaging of the AS of the eye, referred to as anterior segment OCT (AS-OCT), was first demonstrated 3 years later (Izatt et al., 1994).

Although OCT is now a standard clinical imaging modality in ophthalmology for the detection and treatment monitoring of macular degeneration, retinopathy, and glaucoma, scanning the anterior eye is still considered investigational.

In this chapter, we will present state-of-the-art AS-OCT technology and its applications in visualization of both structure and functions of the AS of the eye. We will review clinical OCT systems for AS imaging. We will demonstrate also the main research directions in the field of AS-OCT including the results of clinical studies. Most of the results shown in this chapter were obtained by the author and his colleagues at the Nicolaus Copernicus University in Toruń (Poland) and at the Massachusetts Institute of Technology (Cambridge, MA).

4.2 OPTICAL COHERENCE TOMOGRAPHY

The idea of OCT is analogous to that of ultrasound imaging. OCT detects photons that are back reflected or backscattered at the interfaces of the imaged object (Figure 4.1a) (Huang et al., 1991, Drexler and Fujimoto, 2008a). The effect of light backscattering is caused by the nonhomogeneity of the optical properties (i.e., refractive index) of the object. Depth profiles (axial scans [A-scans] are generated by measuring the amplitude and echo time delay of backscattered light. Furthermore, scanning the incident optical beam over the object in one and/or two lateral directions enables obtaining cross-sectional images (B-scans) and/or 3-D volumetric data (Figure 4.1b). The advantage of having access to 3-D data is that one can extract any virtual cross-sectional image of the object along any plane. Since the speed of light is much higher than sound, the measurement of echo time delay is not performed directly but can be extracted from the interferometric fringes, whe

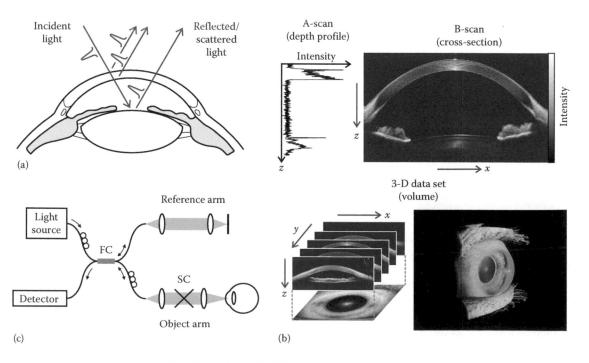

Figure 4.1 Optical coherence tomography (OCT). (a) Principle of OCT operation. (b) OCT image (B-scan) is composed of adjacent axial scans. Volumetric image enables 3-D reconstruction of the internal structure of the object. The nomenclature of the images is similar to that used in ultrasonography. (c) Scheme of simple OCT setup (FC, fiber coupler; SC, scanners).

the effect of superposition of both reference and object beams is analyzed. Typical optical fiber-based interferometric configuration in OCT system is shown in Figure 4.1c.

The most important parameters that drive the OCT performance include imaging speed, depth range, resolution, and sensitivity (Fercher et al., 2003, Drexler et al., 2014).

The imaging (acquisition) speed in OCT is specified as the A-scan rate or number of A-scans acquired per unit time. Typical imaging speeds in current clinical OCT instruments achieve the range of tens to hundreds of kHz (kA-scans/s) (Wojtkowski, 2010, Drexler et al., 2014). Imaging speed is also linked and inversely proportional to the system sensitivity, that is, the ability to detect weak signals trades off with increased imaging speed (Choma et al., 2003, de Boer et al., 2003, Leitgeb et al., 2003).

Maximum optical path difference z_{max} that can be measured by the interferometric system defines the axial (imaging depth) range (Figure 4.2). Detailed discussion of this parameter is described in Section 4.4. The imaging depth range is usually determined by the design of the instrument. However, it should not be confused with the penetration depth of the light into tissue. The penetration depth depends on the light scattering properties of the tissue and it may impact the effective imaging depth (Fercher et al., 2003). In the case of highly scattering tissues like sclera, the light attenuation prevents efficient light delivery and detection at deeper depths, and the effective imaging depth can be actually limited by the penetration depth.

Different physical phenomena determine the resolution in OCT images. The axial resolution is governed by the coherence length of the light source. Assuming the Gaussian shape of the spectrum of light source used in OCT, it is possible to estimate the axial resolution:

$$\Delta z = l_c = \frac{2 \ln 2}{\pi} \cdot \frac{\lambda_0^2}{\Delta \lambda}, \qquad (4.1)$$

where
  λ_0 is the central wavelength of the light source
  $\Delta \lambda$ stands for its bandwidth defined as full width at half maximum of wavelength spectrum

Therefore, OCT systems utilize broadband light sources (i.e., with low temporal coherence but high spatial coherence) to achieve micrometer-scale axial resolution. In practice, the light sources of ~100 nm bandwidths in near infrared region are utilized, which enables obtaining axial resolutions 3–10 μm in tissue.

On the other hand, transverse (lateral) resolution is determined by the beam spot size incident on the object. Hence, it depends on the optical design of the object beam:

$$\Delta x = \frac{4\lambda_0}{\pi} \cdot \frac{f}{d} = \frac{2\lambda_0}{\pi} \cdot \frac{1}{NA}, \qquad (4.2)$$

where
  f is the focal length of the objective lens
  d is the beam diameter incident on the objective
  NA is the numerical aperture of the optical system (Figure 4.2)

The advantage of OCT comes from the fact that it decouples axial and transverse resolutions. Although relatively low numerical aperture can be used in AS-OCT systems, coherence gating provides high axial resolution along the axial direction (Drexler and Fujimoto, 2008a). We emphasize here that the axial and transverse resolution should not be confused with the digital resolution of the OCT image, that is, the pixel size. Specific processing of the acquired interferometric signal and high lateral sampling density allow for obtaining very high digital image resolution. However, it does not indicate that it is possible to resolve object located by a distance defined by the image pixel size.

The available axial imaging range should be matched to the depth of focus of the object beam to ensure optimum photon collection efficiency in the whole sampled volume (Figure 4.2). The depth of focus is defined by the Rayleigh range of the beam focused on the object:

$$b = 2z_R = \frac{\pi (\Delta x)^2}{2\lambda_0}, \qquad (4.3)$$

where z_R is the Rayleigh range. Equation 4.3 shows that there is a trade-off between transverse resolution Δx and depth of focus b of illuminating beam.

4.3 CHALLENGES OF ANTERIOR SEGMENT OCT IMAGING

The AS of the eye has specific properties that make the OCT imaging challenging. In this section, we describe different aspects of OCT imaging, specific for in vivo visualization of the AS of the eye, which pose technical limitations and determine the development of laboratory and clinical AS-OCT systems.

Clinical applications can be enabled only for in vivo imaging that aims at visualization of the eye in its natural state. The movements of the eye facilitated by the extraocular muscles make the eyeball possible to be both shifted in all three directions and rotated. Additionally, some ocular structures such as the crystalline lens change their positions and shapes during accommodation. Since the eye is in a constant motion, imaging of the ocular structures requires the acquisition time to be short enough to

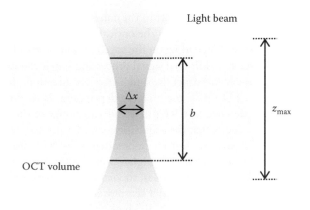

Figure 4.2 Light beam incident on the object in optical coherence tomography (OCT) (Δx, beam spot size; b, depth of focus; z_{max}, imaging depth range in OCT).

minimize the impact of ocular motility. This can be realized by high imaging speed defined by high A-scan rate so that the data acquisition can be shortened. Consequently, the advances in imaging speed are critical for a precise and accurate extraction of quantitative parameters of the AS (Drexler et al., 2014). This issue can be also addressed by the implementation of specific algorithms for motion artifact removal in the post-processing (Maintz and Viergever, 1998, Potsaid et al., 2008, Ricco et al., 2009, Kraus et al., 2012). Clinical practice shows that the acquisition of OCT data should not exceed 2–3 s in order to assure both acceptable image quality and comfort of the patient during scanning.

Another important aspect is associated with the AS spatial dimensions that define the optically sampled volume for proper 3-D AS visualization. AS is relatively large morphological structure compared to the retina, especially in terms of axial direction. If we consider intraocular distances as well as the average refractive indices of human ocular media (cornea and aqueous), the axial imaging range of the AS-OCT setup should be at least 6 mm. What is more, average anterior chamber width of the eye determines transverse scan range to be at least 13 mm. The estimations mentioned earlier indicate that scanning ca. ~13 mm × 13 mm × 6 mm volume is necessary to visualize the entire anterior chamber in all three dimensions spanning from the limbus to limbus and from the apex to at least the front of the crystalline lens. This in turn leads to very large data sets containing information on the AS structures (Gora et al., 2009). Keeping the transverse scanning density constant, imaging of the AS of the eye with standard scan protocol means at least 4× larger scan area than in retinal imaging. It also indicates 4× bigger data set. This challenge can be partially solved by development of specific scanning protocols that account for the symmetry of the front part of the eye. In some clinical AS-OCT instruments, radial scans in meridional directions are implemented (Li et al., 2006).

Optical imaging modalities require light to penetrate the tissue in order to generate the image. The AS of the eye is composed of tissues that have different scattering properties, and optimum light delivery to deeper layers depends on the wavelength of the light source. A proper design of the OCT system for AS imaging should be equipped with the light source that provides sufficient light penetration. The choice of operating wavelength is a primary importance if one wants to image through highly scattering tissues like sclera. Light scattering dominates in the near infrared spectral region, which limits penetration depth. Since light absorption and scattering of most tissue components decrease with the illumination wavelength, higher penetration depth can be achieved with light at 1310 nm compared with light at 830 nm. On the other hand, the absorption of the vitreous humor, mainly composed of water, increases very rapidly for near infrared light (Boettner and Wolter, 1962, van den Berg and Spekreijse, 1997, Sardar et al., 2007, Jacques, 2013). Consequently, it does not allow light at longer wavelengths to reach the retina. Therefore, safety standards enable higher illumination powers to be sent to the eye when longer wavelengths are used (Radhakrishnan et al., 2001). Although AS imaging prefers wavelength region around 1310 nm, the challenge of light source choice comes from the fact that other factors coming from the specific applications of AS-OCT imaging should be also taken into account (e.g., resolution).

Optical inhomogeneity of the AS components causes the OCT images to be distorted. Those distortions come from the fact that OCT measures optical rather than geometric distances. Moreover, light refraction occurs due to different values of the refraction indices of the cornea, the aqueous, and the crystalline lens (Atchison and Smith, 2000). Therefore, physical properties of the eye imply that the OCT images do not reveal the AS morphology correctly. The true morphology of the AS of the eye can be obtained from the OCT data when additional complex image post-processing steps are applied (Steinert and Huang, 2008). Biometric information on the corneal topography requires light refraction correction of the OCT data (Westphal et al., 2002, Podoleanu et al., 2004, Ortiz et al., 2010, Zhao et al., 2010, Siedlecki et al., 2012).

To conclude, the AS-OCT instrument should feature the following parameters: high acquisition speed, sufficient penetration depth, relatively long imaging depth range, homogeneous resolution (beam spot size) over the entire sampled volume, and ability to generate the maps of internal surfaces of ocular structures (quantitative 3-D imaging). Imaging the AS of the eye requires a combination of those capabilities to obtain clinically useful information. We will also discuss those issues in the following sections.

4.4 DEVELOPMENT OF OCT TECHNOLOGY

There are different ways of detecting the interferometric signals, which is a key step in OCT image generation. Those approaches define different OCT technology generations that have been developed over the course of historic development.

4.4.1 TIME-DOMAIN OCT

First generation OCT systems detect the time delay of optical echoes by mechanically scanning the path length of an interferometer reference arm (Figure 4.3a). Since the method encodes the location of each reflection in the time information while the reference path length is scanned, this approach is referred to as time-domain OCT (TD-OCT) (Huang et al., 1991). The interference signal appears only when the optical path difference between light reflected from the object and from the reference mirror is within the coherence length of the light source. The signal recorded by the optical detector is modulated with the frequency depending on the speed of optical path length scanning. Therefore, additional signal processing step (demodulation) is necessary to extract the signal envelope (depth profile).

In TD-OCT, the imaging depth range is set directly by the reference scan length. The disadvantage of time-domain detection is that the acquisition speed is limited by the speed of the reference mirror. In practice, TD-OCT instruments are unable to reach speeds higher than ~4000 A-scans/s due to mechanical limitations in the reference arm as well as limitations in detection sensitivity (Tearney et al., 1997). It is sufficient only to generate few individual cross-sectional images in a single acquisition procedure. First demonstrations of AS imaging using TD-OCT included evaluation of laboratory instruments in clinical practice and served as a basis for the next developments leading to the introduction of the first commercial instruments

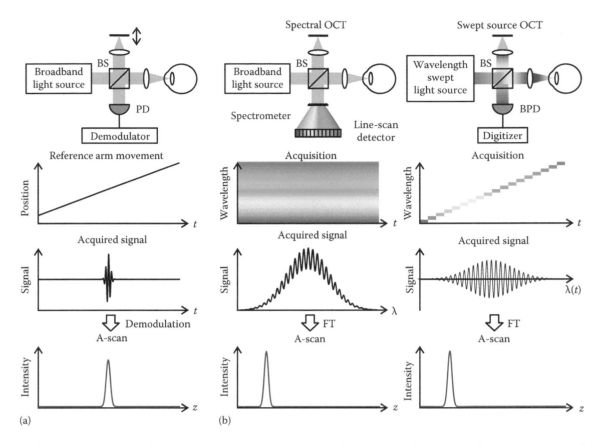

Figure 4.3 Time-domain (a) and Fourier-domain optical coherence tomography (b). PD, photodetector; BS, beam splitter; BPD, balanced photodetector; FT, Fourier transformation.

(Izatt et al., 1994, Koop et al., 1997, Hoerauf et al., 2000, Wirbelauer et al., 2000, Hirano et al., 2001, Radhakrishnan et al., 2001, Huang et al., 2004).

4.4.2 FOURIER-DOMAIN/SPECTRAL OCT

The development of Fourier-domain detection enabled a breakthrough in OCT imaging sensitivity and speed. Fourier-domain detection is based on capturing the spectral interference fringes in wavelength (wavenumber) domain by acquisition of multiple (usually of the order of thousands) spectral channels with no need for scanning the reference mirror (Figure 4.3b, left panel). The information on the depth positions of particular reflection/scattering events (echo time delays) is coded in the oscillation frequencies of the obtained interference spectrum. The echo magnitude and time delay can be extracted from the received interferometric signal by its Fourier transformation (Fercher et al., 1995). The highest frequency that can be measured by the detection system defines the imaging depth range of OCT. The performance advantages of Fourier-domain OCT were recognized independently by several research groups in 2003 (Choma et al., 2003, de Boer et al., 2003, Leitgeb et al., 2003). In general, Fourier-domain OCT achieves ~20 dB higher sensitivity than previous generation OCT systems with time-domain detection enabling ~50–100 times faster imaging. This feature makes OCT more feasible in clinical applications (Wojtkowski et al., 2002b).

The first implementation of Fourier-domain detection is based on a spectrometer with line-scan detector (Figure 4.3b,

middle panel). In this approach, called spectral/Fourier-domain OCT (SD-OCT), 1-D array detector enables simultaneous (parallel in time) measurement of all spectral components (Wojtkowski et al., 2002b).

The spectrometer design is responsible for the performance of this second generation of OCT technology. The imaging depth range in SD-OCT depends on the spectral resolution of the spectrometer (defined by a spectral range of a single spectral component), and the acquisition speed is determined by the repetition rate of the line-scan camera (i.e., reciprocal of a single camera exposition time). SD-OCT imaging of the AS of the eye was demonstrated in 2002 with open air interferometer operating at the speed 7.8 kA-scans/s (Kaluzny et al., 2002, Kaluzny et al., 2006, p. 463). Since then, significant progress in line-scan camera technology allowed increases in speed exceeding 100 kHz rates (Potsaid et al., 2008, Sarunic et al., 2008, Stehouwer et al., 2010). Nowadays, most current ophthalmic OCT instruments utilize SD-OCT technology (Wojtkowski et al., 2012).

4.4.3 FOURIER-DOMAIN/SWEPT SOURCE OCT

The newest (third) generation of Fourier-domain OCT instruments is based on wavelength swept light sources (swept source/Fourier-domain OCT [SS-OCT], known also as optical frequency domain imaging) (Chinn et al., 1997, Golubovic et al., 1997, Choma et al., 2003, Yun et al., 2003b). The swept laser is a form of tunable light source, in which periodic tuning of spectrally narrow wavelength is provided. Such a laser generates instantaneous temporally coherent electromagnetic radiation.

However, the emitted light is temporally incoherent when the entire sweep is considered, as demanded by OCT. Wavelength-swept lasers utilize different tuning mechanisms and are considered to be a key technology for SS-OCT (Chinn et al., 1997, Golubovic et al., 1997, Yun et al., 2003a, 2004a, Choma et al., 2005, Huber et al., 2005, 2006, Yasuno et al., 2005, Oh et al., 2010, Okabe et al., 2012, Wieser et al., 2012). Recent advances include miniaturization of laser cavity, which improves both the tuning speed and coherence properties (Fujiwara et al., 2008, Potsaid et al., 2010, Totsuka et al., 2010, Jayaraman et al., 2011, 2012, 2013, Minneman et al., 2011, Potsaid et al., 2012).

In SS-OCT, the interferometric signal (fringe) is acquired in time when the laser is sweeping the wavelength. The light is detected by a high-speed point photodetector and digitized by a data acquisition board (analog-to-digital card), which reconstruct the spectral fringes acquired over one frequency sweep of the light source (Figure 4.3b, right panel). In other words, consecutive (serial in time) measurement of spectral channels is performed. Similar to SD-OCT, the echo delays or A-scans can be generated by Fourier transforming the detector signal.

The maximum range that can be measured by SS-OCT system depends on laser and hardware specifications:

$$z_{max} \cong \frac{1}{4} \cdot \frac{f_{ADC} \cdot d_c}{f_{laser}} \cdot \frac{\lambda_0^2}{\Delta\lambda}, \tag{4.4}$$

where

f_{ADC} is the sampling rate (bandwidth) of the acquisition (A/D card)
f_{laser} is the sweep rate of the swept light source
d_c is the duty cycle of the laser source
λ_0 is the central wavelength of the light source
$\Delta\lambda$ is the wavelength sweep range

On the other hand, the imaging speed in SS-OCT is determined by the sweep repetition rate of the swept light source. First demonstration of AS-OCT imaging with swept source technology was shown in 2005 using the instrument operating at 20 kA-scans/s, but modern systems can achieve even MHz A-scan rates (Sarunic et al., 2005, Yasuno et al., 2005, Kerbage et al., 2007, Potsaid et al., 2012, Wieser et al., 2012).

4.4.4 INTERFACE DESIGNS

Independently on the detection advances, different illumination schemes have been developed. The most standard configuration is based on the flying spot, in which a single beam is incident on the eye. The beam scanning is performed by a scanning system (e.g., a pair of galvanometric mirrors) in the sample arm (Figure 4.4a) (Huang et al., 1991, Hee et al., 1995). Raster lateral scanning enables acquisition of cross-sectional images or 3-D data sets. The extension of this standard approach involves scanning with more beams to multiply the effective acquisition speed or to extract additional contrast parameters from OCT data (Figure 4.4b). Dual-beam scanning was demonstrated (Iftimia et al., 2008, Potsaid et al., 2010, Makita et al., 2011, Zotter et al., 2011, Jeong et al., 2012, Klein et al., 2013, Nankivil et al., 2014, Fan et al., 2015). The limitations are the safety standards of light exposure on the eye.

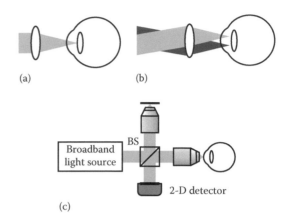

Figure 4.4 Illumination schemes in optical coherence tomography (OCT): (a) flying spot, (b) multiple beams, and (c) full-field OCT.

The depth information can be also extracted by the alternative technique, known as full-field OCT (Figure 4.4c). In this approach, Michelson interferometer with microscope objectives in both arms (Linnik interferometer) is used for full-field illumination. Full-field OCT was introduced in 2002 (Dubois et al., 2002, Vabre et al., 2002), and the first images of the cornea were demonstrated very soon (Grieve et al., 2005, Akiba et al., 2007). Full-field OCT enables obtaining images in the en-face orientation so that there is no need to scan the beam. Full-field OCT directly registers 2-D en-face images with megapixel camera to reconstruct 3-D image with the isotropic resolutions of ~1 μm in all three directions (Grieve et al., 2005). Another example of a parallelization of signal acquisition is line-field (linear) OCT (Grajciar et al., 2005). However, the sensitivity of systems in those two approaches is limited (Fercher et al., 2003, Fercher, 2010).

4.5 COMPARISON OF OCT WITH OTHER ANTERIOR SEGMENT IMAGING MODALITIES

AS imaging modalities give access to both qualitative and quantitative assessment of the eye. Although those modalities provide usually equivalent information, it is important to note that they are not always interchangeable. Each imaging method possesses its strengths and weaknesses. Generally, the advantage of optical techniques used in AS imaging is that the examination is performed relatively fast and in a noncontact way. However, the light penetration can be limited by the clarity of the cornea or by the pupil diameter.

Confocal microscopy can be used for high-resolution imaging of ocular tissues and provides en-face images of corneal layers (Chiou et al., 2006). More advanced version of confocal microscopy with scanning slit enables video rate imaging. Confocal microscopy requires an objective with short focal distance, and the imaging session is limited to a small area of the eye (Bohnke and Masters, 1999). In contrast to confocal microscopy, axial and transverse resolutions are decoupled in OCT, which means that the axial resolution does not depend on the numerical aperture of the objective lens and even high-resolution images can be obtained in low-NA systems (cf. Equations 4.1 and 4.2). Contrary to OCT, confocal microscopy images at cellular level from a particular depth

Scheimpflug imaging provides cross-sectional images of the whole AS, from the anterior cornea to the posterior lens, with large depth of focus. By rotating the Scheimpflug camera, it is possible to extract radial AS sections at different angles (meridionals) as well as to perform corneal topography and phakometry (Rosales and Marcos, 2009). However, the resolution of the images is much lower than that of OCT, and the field of view of posterior lens is limited by the pupil aperture (Ishibazawa et al., 2011). What is more, the obtained images suffer from geometric and optical distortion (like in OCT). Consequently, a correction algorithm is necessary to obtain quantitative information on topography of cornea and crystalline lens (Rosales et al., 2006).

The properties similar to Scheimpflug imaging are observed in all slit illumination imaging techniques. Slit scanning tomography provides a series of images by changing the angle of rotating slit. The images are later used to reconstruct the anterior and posterior surface of the cornea as well as corneal thickness and anterior surface of the lens (Oliveira et al., 2011). Due to the limited depth of focus, assessment of the posterior lens is not possible.

UBM enables obtaining cross-sectional images with the depth range up to 5 mm. Standard UBM instruments operating at ultrasound frequencies ranging between 50 and 100 MHz can image with the resolution of 25–50 μm (Nolan, 2008b). Therefore, compared with UBM, OCT technology is characterized by a higher axial resolution. Higher image resolutions in ultrasound imaging can be achieved by using sound of higher frequency. However, it also considerably increases attenuation of ultrasonic waves, which limits penetration into tissue and imaging depth. Moreover, although ultrasound probe does not contact the eye directly, the measurement is relatively invasive since it involves coupling medium to launch ultrasound (i.e., water immersion). Accordingly, special eyecup is placed at the cornea under topical anesthesia. The contact nature of this time-consuming examination along with a supine position of the patient can generate some sense of discomfort for the patient as well as may impact the extracted quantitative data (e.g., ocular compression). Additionally, UBM requires an experienced operator to avoid possible corneal abrasion. Unlike UBM, OCT scanning in the clinic is in most cases performed in a seated, upright position. However, ultrasound penetration through optically nontransparent iris makes UBM superior in the imaging through opacities and in the imaging the structures posterior to the iris such as ciliary body, zonules, and the periphery of the crystalline lens (Radhakrishnan et al., 2005, Dada et al., 2007).

The limitation of penetration depth does not exist in magnetic resonance imaging (MRI). MRI has a large field of view compared to other ocular imaging modalities. Thus, it is well suited to the imaging of the whole eye globe and diagnosis of the orbit diseases. However, the resolution of MR images is rather poor, which makes visualization of AS structures harder. MRI scanner for ophthalmic imaging can be equipped with special magnetic coils to increase resolution (up to ~100–200 μm) (Langner et al., 2010). Another disadvantage of the MRI procedure is that it takes several minutes and is even more expensive than OCT. Consequently, patients have to maintain eye and head positions for the duration of imaging to obtain MR images of good quality with no motion artifacts. Therefore, MRI is usually used in the assessment of the eye in specific cases (Townsend et al., 2008, Fanea and Fagan, 2012).

4.6 CLINICAL OCT SYSTEMS FOR ANTERIOR SEGMENT IMAGING

The widespread acceptance of OCT technique resulted in development of the market that is fed by ca. 40 companies offering OCT instruments for different research and clinical applications (Swanson, 2015). The instrumentation for AS-OCT imaging includes several devices that implement both time- and Fourier-domain detection schemes. A list of selected instrumentation for AS-OCT with their basic parameters is presented in Table 4.1.

Currently, there are three OCT systems strictly dedicated to AS imaging (Li et al., 2014). The first commercially used AS-OCT system is the Visante system (Carl Zeiss Meditec, United States), which operates at the central wavelength of 1300 nm and at the acquisition speed 2 kHz (Tang et al., 2006). Longer wavelengths used in that instrument enable better light penetration through the sclera and iris. Visante can generate images of the entire anterior chamber and anterior chamber angle (Radhakrishnan et al., 2007). The same TD-OCT technology and operating wavelengths are used in SL-OCT developed by Heidelberg Engineering (Germany) (Wirbelauer et al., 2002). This device is actually an OCT incorporated with slit-lamp biomicroscope. The newest clinical system is CASIA introduced by Tomey (Japan) in 2008 (Leung and Weinreb, 2011). The device can be regarded as the first clinical SS-OCT instrument. It employs wavelength-tunable laser at 30 kHz developed by Santec (Japan). Imaging at longer wavelengths (1300 nm) and longer imaging depth range make 3-D biometry of the anterior eye possible.

The OCT market includes also instruments dedicated to retinal imaging with AS imaging mode as a complementary (additional) mode. Primary function of the device determines imaging depth range that results in the ability to visualize only the cornea or anterior chamber angle at high resolution rather than the entire AS of the eye. Since clinical practice can accept only small changes between operational modes, the AS imaging mode employs light sources centered at ~830 nm. Moreover, optical system of the eye requires modification of the sample arm (interface) if one wants to switch between both imaging modes. Therefore, an adaptor lens is usually added to the retinal configuration. This solution is applied in the clinical SD-OCT systems such as Stratus and Cirrus (Carl Zeiss, United States), RTVue (Optovue, United States), Spectralis (Heidelberg Engineering, Germany), SOCT Copernicus (Optopol Technology, Poland), HS-100 (Canon, Japan), Envisu (Bioptigen, United States), 3D OCT-2000 (Topcon, Japan), and RS-3000 Advance (Nidek, United States) (Kalev-Landoy et al., 2007, Tang et al., 2010a, Alonso-Caneiro et al., 2012, Rodrigues et al., 2012, Shapiro et al., 2013). The SD-OCT devices mentioned earlier can acquire images at speeds up to 70 kA-scans/s with the imaging range of 2–3 mm. The adaptor lens must be used in SS-OCT instrument Triton offered by Topcon (Japan). It provides anterior and posterior imaging at 1050 nm wavelength with additional options such as fluorescein angiography and fundus autofluorescence (Park et al., 2014). However, full FDA clearance procedure has not been completed yet.

AS-OCT has been also implemented in the instruments to support surgical procedures. The integration of SD-OCT imaging

Table 4.1 **Examples of clinical optical coherence tomography instruments for anterior segment imaging**

INSTRUMENT/ COMPANY	TECHNOLOGY	LIGHT SOURCE/ WAVELENGTH (nm)	ACQUISITION SPEED	RESOLUTION (μm)	DEPTH RANGE (mm)	COMMENTS
Stratus OCT 3000/ Carl Zeiss Meditec (2002)	TD-OCT	SLD/820	400 Hz	10 (z) 20 (x)	2	Anterior segment adaptor; cornea and angle visualization; CCT measurement
SL-OCT/Heidelberg Engineering (2003)	TD-OCT	SLD/1310	200 Hz	25 (z) 20–100 (x)	7	Combined with slit lamp
Visante/Carl Zeiss Meditec (2005)	TD-OCT	SLD/1300	2 kHz	18 (z) 60 (x)	3 or 6	For anterior segment imaging only
SOCT Copernicus/ Optopol Technology (2006)	SD-OCT	SLD/830	27 kHz	5 (z) 12–18 (x)	2	Anterior module
RTVue Premier/ Optovue (2006)	SD-OCT	SLD/840	26 kHz	5 (z) 8 (x)	2–2.3	Corneal and retinal imaging; corneal power mapping; epithelium thickness mapping
Cirrus HD-OCT/Carl Zeiss Meditec (2007)	SD-OCT	SLD/840	27 kHz	5 (z) 15 (x)	2–5.8	Two interchangeable lenses for corneal, anterior chamber and wide angle-to-angle imaging
Spectralis/Heidelberg Engineering (2007)	SD-OCT	SLD/870	40 kHz (85 kHz)	7 (z) 14 (x)	1.8	Combined with cSLO platform; anterior segment module (add-on lens and software)
Casia SS-1000/Tomey (2008)	SS-OCT	HSL-200–30/1310	30 kHz	10 (z) 30 (x)	6	For anterior segment imaging only
3D OCT-2000/ Topcon (2009)	SD-OCT	SLD/840	50 kHz	6 (z) 20 (x)	2.3	For retinal and anterior segment imaging
RS-3000 Advance/ Nidek (2009)	SD-OCT	SLD/880	53 kHz	7 (z) 20 (x)	2.1	Optional anterior segment module
iOCT/OPMedT; Haag-Streit Surgical (2010)	SD-OCT	SLD/840	10 kHz	10 (z)	4.2	An OCT camera for the operating microscope
Envisu/Bioptigen (2011)	SD-OCT	SLD/830	32 kHz	2.4 (z) 12–18 (x)	2.5	Handheld system; for retinal and corneal/ anterior lens imaging; intraoperative imaging platform
HS-100/Canon (2012)	SD-OCT	SLD/855	70 kHz	3 (z) 20 (x)	2	Anterior segment adapter (ASA-1)
DRI OCT Triton/ Topcon (2013)	SS-OCT	Swept light source/1050	100 kHz	8 (z) 20 (x)	2.6	Anterior segment attachment kit (wide-field lens adapter, AA1)
Rescan 700 OCT/Carl Zeiss Meditec (2014)	SD-OCT	SLD/840	27 kHz	5.5 (z)	2	Integrated intraoperative OCT engine

into ophthalmic microscope was realized recently in commercial Rescan 700 OCT (Carl Zeiss Meditec, United States) and in iOCT (Haag-Streit Surgical, Wedel, Germany) to enhance and control surgical manipulations (Steven et al., 2013, 2014, Ehlers et al., 2014a). Intraoperative OCT is currently an extensively explored area of application (Ehlers et al., 2013, 2014b, Tao et al., 2014). Moreover, OCT-based guidance is applied in the platforms for femtosecond laser–assisted cataract and refractive surgery (Lubatschowski, 2008, He et al., 2011, Tomita et al., 2012). The systems like Catalys (Abbott Medical Optics, United States), LenSx (Alcon-Novartis, Germany), Femto LDV Z models (Ziemer Ophthalmic Systems, Germany), or Amaris (Schwind, Germany) utilize high-resolution OCT imaging of the AS (including crystalline lens) for precise design laser light delivery during incisions (Arbelaez and Samuel Arba Mosquera, 2009, Nagy et al., 2009, 2013, Friedman et al., 2011, Bali et al., 2012, Tomita et al., 2012).

4.7 LABORATORY OCT SYSTEMS FOR ANTERIOR SEGMENT IMAGING

We developed several OCT laboratory systems for AS imaging with Fourier-domain detection (Gora et al., 2009, Grulkowski et al., 2009, 2012, Alonso-Caneiro et al., 2011, Karnowski et al., 2011). The schemes of fiber-optic-based interferometers are shown in Figure 4.5. The fiber configurations as well as detection

were adjusted to available optimized components at particular wavelength range and optimized to obtain the highest OCT signal. The acquisition units in SS-OCT instruments (Figure 4.5b and c) are not shown for the sake of scheme clarity.

SS-OCT system operating at the wavelength 1050 nm (Figure 4.5b) was equipped with either a short external cavity tunable laser (Axsun Technologies, United States) or a prototype of MEMS-tunable vertical cavity surface emitting laser (VCSEL) (Praevium/Thorlabs, United States). The light sources emitting light centered around 1310 nm used in laboratory SS-OCT instruments were a short external cavity tunable laser (Axsun Technologies, United States) and a custom-made Fourier-domain mode locking (FDML) laser. The systems are characterized by different performance parameters, as shown in Table 4.2. The laboratory systems are usually more flexible in adjusting the performance. The results presented in subsequent sections were obtained with those instruments.

4.8 IN VIVO SD-OCT IMAGING

Standard SD-OCT operating at the 830 nm wavelength region is well suited to the imaging of the posterior segment of the eye. Retinal SD-OCT instrument adopted to AS imaging can image only relatively shallow structures and therefore might be oriented on specific components of the AS of the eye such as cornea

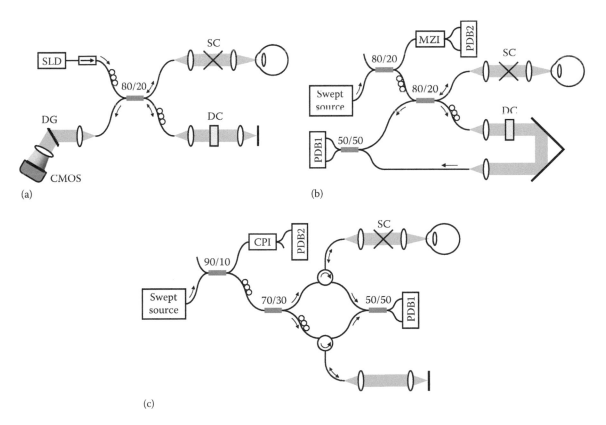

Figure 4.5 Schematics of the prototype Fourier-domain anterior segment optical coherence tomography (OCT) systems: (a) spectral/Fourier-domain OCT instrument at 840 nm (SLD, superluminescent diode; SC, galvanometric scanners; DC, dispersion compensation; DG, diffraction grating; CMOS, line-scan camera); (b) swept source/Fourier-domain OCT (SS-OCT) instrument at 1050 nm (SC, galvanometric scanners; DC, dispersion compensation; MZI, Mach–Zehnder interferometer; PBD1, balanced photodiode in the imaging interferometer; PBD2, balanced photodiode in the calibration interferometer); (c) SS-OCT instrument at 1310 nm (SC, galvanometric scanners; CPI, common path interferometer; PBD1, balanced photodiode in the imaging interferometer; PBD2, balanced photodiode in the calibration interferometer).

Table 4.2 **Parameters of laboratory systems for anterior segment imaging**

PARAMETER	SD-OCT	SS-OCT	SS-OCT	SS-OCT	SS-OCT
Light source technology	SLD (broadlighter)	MEMS-based short external cavity tunable laser	MEMS-tunable VCSEL	MEMS-based short external cavity tunable laser	FDML
Wavelength (nm)	$\lambda_0 = 840$	$\lambda_0 = 1040$	$\lambda_0 = 1060$	$\lambda_0 = 1310$	$\lambda_0 = 1300$
	$\Delta\lambda = 50$	$\Delta\lambda = 100$	$\Delta\lambda = 45$–85	$\Delta\lambda = 100$	$\Delta\lambda = 35$–135
Acquisition speed (kA-scans/s)	14–135	100	50–100	50	108–200
Axial resolution (in tissue) (μm)	6.9	6.0	9.0–12.4	6.7	9.0–25.0
Transverse resolution (μm)	28	48	73	20	32
Depth range (in air) (mm)	7.0	5.8	18–50	8.2	2–8
Sensitivity	102 dB at 25 kHz	102 dB at 100 kHz	100 dB at 50 kHz	109 dB at 50 kHz	103 dB at 200 kHz
Sensitivity drop (–6 dB depth)	19.5 dB (2.5 mm)	15 dB (2.5 mm)	12 dB (40 mm)	9 dB (7.3 mm)	15 dB (5.5 mm)

(Wang et al., 2011). Early studies with high-resolution imaging systems demonstrated applicability of SD-OCT in the imaging of different corneal pathologies (Christopoulos et al., 2007, Kaluzny et al., 2008, 2009, 2010, Wylegala et al., 2009). Figure 4.6a and b presents exemplary cross sections of the human corneas obtained with ultrahigh-resolution laboratory system equipped with a femtosecond laser. Ultrashort pulses represent very broadband spectrum in Fourier domain, which gave axial resolution of 2 μm in tissue. High axial resolution of the OCT images permits distinguishing layer-like microarchitecture of the cornea with epithelium, Bowman membrane, stroma, and endothelium (Figure 4.6a). This feature is especially useful in the pre- and postsurgical assessment of the cornea (Kaluzny et al., 2014). Figure 4.6b shows remodeling of a new epithelial cell layer after epi-Bowman keratectomy, which appears as a hyper-reflective layer covering corneal stroma. High-resolution imaging enables also visualization of pathologic states of the cornea.

Figure 4.6c and d demonstrates deposits on the cornea and the dystrophy of corneal endothelium, respectively.

Anterior chamber imaging with SD-OCT requires optimization of the spectrometer design to achieve sufficient depth range. The laboratory prototype SD-OCT system was developed with custom design spectrometer equipped with complementary metal-oxide-semiconductor (CMOS) sensor array as a detector (Grulkowski et al., 2009). Using CMOS technology provides with more flexible solution by enabling adjustment of the number of readout pixels, which consequently enables changing the acquisition speed and axial resolution of OCT images.

The most important feature of SD-OCT is that it enables much faster acquisition times than TD-OCT. Moreover, SD-OCT offers more efficient detection of photons reflected from the sample, which significantly enhances sensitivity of the method, that is, its ability to detect very low light intensities. High speeds along with better sensitivity allow for decreasing the

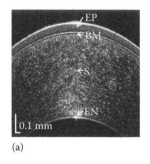
(a)

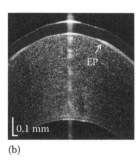
(b)

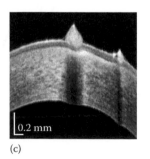
(c)

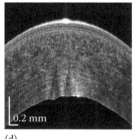
(d)

Figure 4.6 High-resolution spectral/Fourier-domain optical coherence tomography imaging of the cornea (EP, epithelium; BM, Bowman's membrane; S, stroma; EN, endothelium). (a) B-scan of the healthy cornea imaged with a prototype instrument at 810 nm. (b) B-scan of the cornea a week after epi-Bowman keratectomy imaged with a prototype instrument at 810 nm. Remodeling of the epithelium after the procedure is visible as more scattering interface. Therapeutic contact lens is placed on the cornea. (c) Corneal deposits imaged with RTVue system. (d) Cornea with Fuchs' endothelial dystrophy imaged with RTVue system. ([a]: Reproduced from Grulkowski, I. et al., *Opt. Express*, 17, 4842, 2009; [b]: Courtesy of M. Wojtkowski, Nicolaus Copernicus University, Torun, Poland; [c,d]: Courtesy of W. Fojt.)

acquisition time for in vivo imaging to reduce motion artifacts and improve reproducibility of quantitative measurements. Faster acquisition means also that more data can be obtained during a given scan duration, which results in a denser sampling and/or a wider field of view. In particular, SD-OCT enables 3-D imaging with improved scan coverage and better image quality.

Figure 4.7a and b shows cross-sectional high-definition images as well as 3-D reconstruction of the anterior chamber and the crystalline lens (Grulkowski et al., 2009). High-definition imaging was performed at 14 kHz. The renderings of volumetric data sets cover a 15 × 15 mm² area of the anterior chamber. Since the imaging was performed at 73 kHz, the volume consisting of 1000 × 50 A-scans was acquired in 0.68 s. On the other hand, volumetric image of the crystalline lens was acquired from the 9 × 9 mm² area. The reconstruction is based on 80 B-scans, each consisting of 1000 A-scans. Central cross sections demonstrate general architecture of the AS of the eye and include the components such as cornea, limbus, and sclera. The myopic subject had a soft contact lens at its cornea (Figure 4.7a). The crystalline lens exhibits nonhomogeneous

optical properties due to the differences in the refractive index. As a result, it is possible to distinguish the nucleus and the cortex in Figure 4.7b.

AS-OCT enables also imaging of the corneoscleral angle, visualization of which is important in the glaucoma assessment (Li et al., 2012). Figure 4.8 shows the images of the anterior chamber angle of the healthy subject.

Limited imaging depth range in SD-OCT can be extended using the fundamental property of the Fourier-domain detection. Fourier transformation of the interferometric signals actually generates two complex conjugate images that carry the same structural information. Several approaches have been developed to provide complex conjugate removal (Wojtkowski et al., 2002a, Sarunic et al., 2005, 2006, 2008, Baumann et al., 2007b, Leitgeb et al., 2007, Tao et al., 2007, Hofer et al., 2009, Kim et al., 2010, Dhalla and Izatt, 2011, Dhalla et al., 2012b). In particular, complex Fourier-domain OCT method is based on the hardware modification to assure constant optical path difference change between both arms of the interferometer when the sample is scanned (Wojtkowski et al., 2002a). Specific post-processing

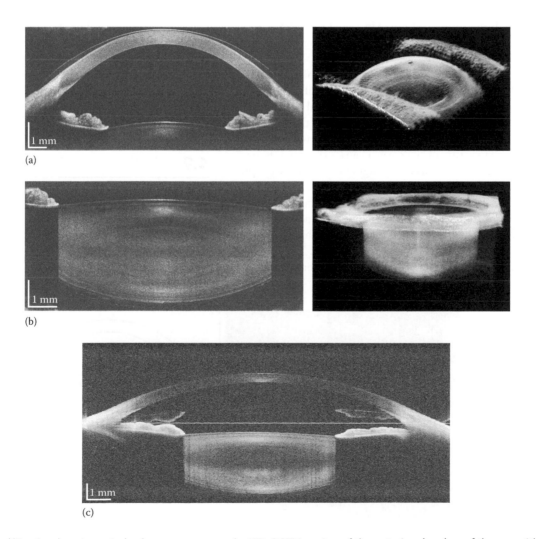

(a)

(b)

(c)

Figure 4.7 Spectral/Fourier-domain optical coherence tomography (SD-OCT) imaging of the anterior chamber of the eye with a prototype SD-OCT system at 840 nm. (a) High-definition cross-sectional image of the anterior chamber of the myopic eye with a contact lens. 3-D rendering of volumetric data set. (b) Cross-sectional image of the crystalline lens. 3-D rendering of volumetric data set. (c) Complex conjugate removal method allows for extension of the depth range of the SD-OCT systems. Cross-sectional image of the whole anterior segment. (Reproduced from Grulkowski, I. et al., *Opt. Express*, 17, 4842, 2009.)

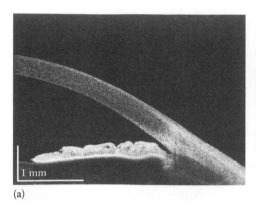
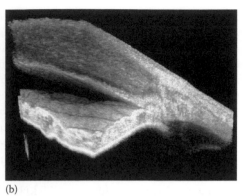

(a)   (b)

Figure 4.8 Anterior chamber angle imaging with spectral/Fourier-domain optical coherence tomography operating at 840 nm and at the speed of 20 kA-scans/s. (a) High-definition cross section of the angle. (b) Volumetric image of the anterior chamber angle of healthy subject. (Reproduced from Grulkowski, I. et al., *Opt. Express*, 17, 4842, 2009.)

procedure enables separation of complex conjugate images that in turn leads to the extension of the depth range by a factor of two. This technique was applied to AS-OCT to generate cross section of the entire AS spanning from the corneal apex to the posterior surface of the crystalline lens (Grulkowski et al., 2009). Figure 4.7c shows high-definition full-range image of the entire AS obtained by our prototype SD-OCT instrument operating at the speed of 14 kA-scans/s.

The main drawback of SD-OCT is relatively high signal drop with the depth. Although it is inherent for all systems with Fourier-domain detection scheme, the signal at the end of imaging range can be even ~20 dB lower than the signal at zero path difference in spectrometer-based systems. Moreover, relatively large curvature of the cornea causes the signal to be washed out. As a consequence, both signal drop-off with depth and signal washout become important limitations in AS imaging with SD-OCT. The impact of those two factors mentioned earlier can be minimized in AS imaging by a proper placement of the eye with respect to the focal plane of the objective lens and zero optical path difference of the interferometer. The imaging procedure here takes advantage of the fact that components

of the eye are characterized by different optical properties in terms of light scattering. The aqueous is practically transparent, whereas cornea and crystalline lens exhibit moderate scattering. However, sclera and iris demonstrate very high scattering, which manifests in high signal in OCT images. Signal drop-off and low corneal scattering can be compensated by adjustment the cornea near zero path difference and putting the focal plane deeper into the anterior chamber. This methodology is similar to enhanced depth imaging in retinal OCT imaging, when one wants to increase the signal coming from the choroid (Spaide et al., 2008). As shown in Figure 4.9, it allows for obtaining the OCT images of better quality.

4.9 IN VIVO SS-OCT IMAGING

SS-OCT demonstrates distinct features compared to SD-OCT. SS-OCT uses high-speed, balanced photodetectors instead of a spectrometer and line-scan camera that is used in SD-OCT. SS-OCT detection losses are reduced compared with SD-OCT because of the higher detection efficiency of photodetectors. In addition, the signal roll-off with imaging depth, which is

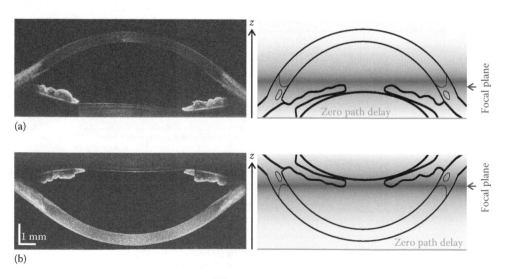

(a)

(b)

Figure 4.9 Optimization of spectral/Fourier-domain optical coherence tomography image quality of the anterior segment of the eye. (a) Zero delay behind the pupil plane. (b) Zero delay in front of the eye. (Reproduced from Grulkowski, I. et al., *Opt. Express*, 17, 4842, 2009.)

present in SD-OCT, can be significantly reduced in SS-OCT due to the high instantaneous coherence properties of tunable lasers and the ability to detect high-frequency signals with broad bandwidth detection and data acquisition systems (Grulkowski et al., 2013b). In general, three main factors contribute to the effective signal roll-off in SS-OCT: instantaneous linewidth (coherence width), bandwidth of the photodetector and digitizer, as described by the following relation:

$$S_{SS-OCT}(z) = C(z) \otimes S_{PDB}(z) \otimes S_{ADC}(z), \qquad (4.5)$$

where

$C(z)$ is the coherence function of the swept laser
$S_{PDB}(z)$ and $S_{ADC}(z)$ are the frequency characteristics (bandwidths) of the photodetector and acquisition board, respectively

Wavelength-tunable light source technology includes several laser configurations with different inherent light coherence properties (Grulkowski et al., 2012, Drexler et al., 2014). It is not easy to manufacture wideband swept laser operating in ~830 nm wavelength region due to the technological limitations. Therefore, most swept lasers can emit the light at longer wavelengths (with central wavelength 1050 or 1310 nm). The measured coherence functions $C(z)$ of few popular swept light sources used in our AS studies are presented in Figure 4.10a. The impact of the previously mentioned factors on signal roll-off in SS-OCT is demonstrated in Figure 4.10b. The signal drop for AS-OCT with spectral-domain detection is also included for comparison. Different signal roll-offs demonstrate the ability to image deeper structures, thus defining particular applications of AS imaging.

High-speed and reduced parasitic sensitivity roll-off is important for 3-D imaging of the anterior eye. The SS-OCT systems with short external cavity tunable lasers were used here

(cf. Table 4.2). Figure 4.11a shows a rendering of a volumetric data set covering a 13 × 13 mm² area of the anterior eye. The data were acquired with SS-OCT system operating at 1050 nm. The volume consists of 500 × 500 A-scans acquired in 2.6 s. Figure 4.11b presents a volumetric data set consisting of 500 × 100 A-scans obtained with SS-OCT with 1300 nm central wavelength at 50 kA-scans/s. The lower resolution for longer wavelength (as given by Equation 4.1) results in less visible details of the AS morphology. Compared to SD-OCT, signal washout at the steepest parts of the cornea is not as severe in SS-OCT.

SS-OCT imaging can be also applied in clinical practice as a high-resolution imaging modality for the diagnosis and surgical treatment. Figure 4.11 shows pathologic features associated with the AS of the eye imaged with SS-OCT. The noninvasive visualization of AS architecture can be also helpful in the pre- and postsurgical evaluation of the eye. The examples in Figure 4.11c and d show the eyes of the patients after cataract surgery (obtained with VCSEL-based SS-OCT instrument) and after penetrating keratoplasty (obtained with FDML-based SS-OCT instrument). The surfaces of intraocular lens (IOL) implant are visible in those two cases.

SS-OCT systems imaging at 1 and 1.3 µm wavelengths (longer than typical wavelengths of 840 nm that are used in SD-OCT) enable better penetration of light and improved imaging depths in scattering tissues. The wavelength-dependent scattering tissue properties can be visualized in the imaging of the anterior chamber angle and the ciliary body. Figure 4.12 demonstrates the images of anterior chamber angle of the same healthy subject measured with three generations of OCT systems at different wavelengths: 840, 1050, and 1310 nm. Deeper penetration for longer wavelengths enables more effective visualization of the ciliary body, which is located deeper than highly reflective sclera.

Recent advances in swept laser technology resulted in the development of novel light sources whose specifications including long coherence length, high sweeping rate, and wide tuning range

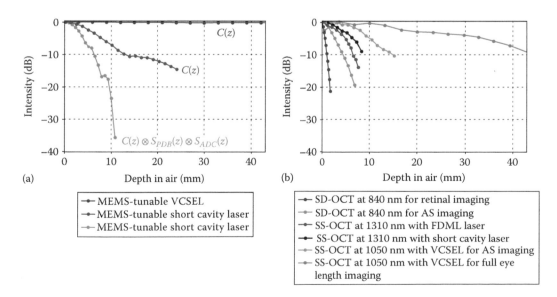

Figure 4.10 Coherence properties of swept light sources used in swept source/Fourier-domain optical coherence tomography (OCT) imaging. (a) Coherence functions for two light sources operating at central wavelength of 1050 nm. The impact of the frequency characteristics of both photodetector and acquisition card is provided. (b) Signal roll-off for laboratory anterior segment OCT systems with Fourier-domain detection. Typical signal drop for retinal spectral/Fourier-domain OCT instrument is given for comparison.

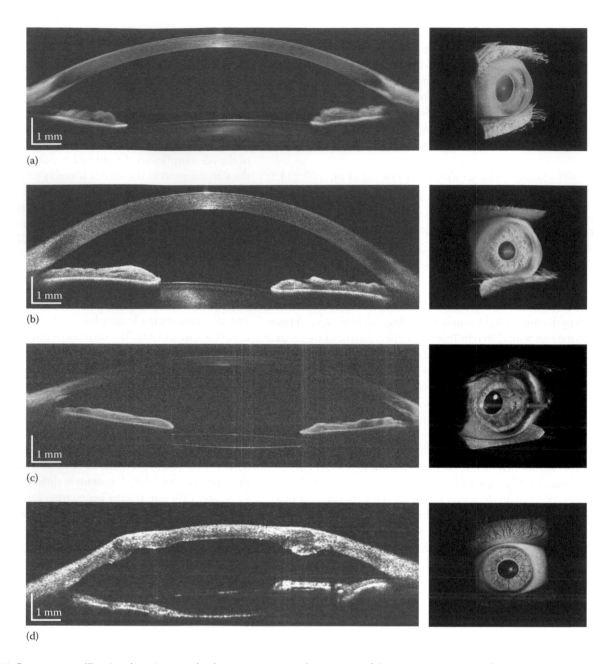

Figure 4.11 Swept source/Fourier-domain optical coherence tomography imaging of the anterior segment of the eye—cross-sectional images and 3-D reconstructions. (a) Eye of a healthy subject measured with a system operating at 1050 nm and the speed 100 kA-scans/s. (b) Eye of a healthy subject measured with a system operating at 1310 nm and the speed 50 kA-scans/s. (c) Eye of a patient after cataract surgery and intraocular lens implantation measured with a system operating at 1050 nm and the speed 100 kA-scans/s. (d) Eye of a patient after penetrating keratoplasty measured with a system operating at 1310 nm and the speed 108 kA-scans/s. Sutures are visible after data rendering ([a,c]: Courtesy of J.G. Fujimoto, MIT, Cambridge, MA; [b,d]: Courtesy of K. Karnowski and M. Wojtkowski, Nicolaus Copernicus University, Toruń, Poland.)

are optimized for SS-OCT. One example of such development is the MEMS-tunable VCSEL technology that offers many advantages for OCT imaging (Jayaraman et al., 2011, 2012, 2013, Grulkowski et al., 2012, Potsaid et al., 2012). The micron-scale cavity length of the VCSEL and the rapid MEMS response enable high imaging speeds. Moreover, the VCSEL operates with a single longitudinal mode instead of multiple modes and therefore has an extremely narrow instantaneous linewidth. This yields a very long coherence length and enables long imaging ranges.

Many attempts have been performed to perform imaging of the whole AS, from the front of the cornea to the back of the crystalline lens. The approaches in SD-OCT technology are usually complex and limit the performance of long-range SD-OCT systems (Grulkowski et al., 2009, Jungwirth et al., 2009, Zhou et al., 2009, Shen et al., 2010, Du et al., 2012, Ruggeri et al., 2012, Li et al., 2013a, 2014, Tao et al., 2013, Dai et al., 2014, Zhong et al., 2014, Fan et al., 2015). Nowadays, SS-OCT offers the opportunity to achieve centimeter depth ranges without any complicated depth extension procedure (Potsaid et al., 2012). Figure 4.13a demonstrates OCT image of the whole AS including the cornea, iris, and entire crystalline lens and spans the entire transverse width of the anterior

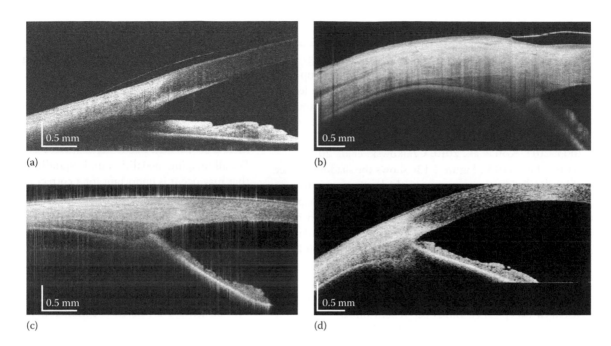

(a) (b)

(c) (d)

Figure 4.12 Imaging of ciliary body of the same subject using different optical coherence tomography technologies and different wavelengths: (a) laboratory SS-OCT system at 840 nm, (b) laboratory SS-OCT system at 1050 nm, (c) laboratory SS-OCT system at 1310 nm, (d) commercial TD-OCT system at 1310 nm. ([a]: Reproduced from Grulkowski, I. et al., *Opt. Express*, 17, 4842, 2009; [b]: Reproduced from Grulkowski, I. et al., *Photon. Lett. Poland*, 3, 132, 2011.)

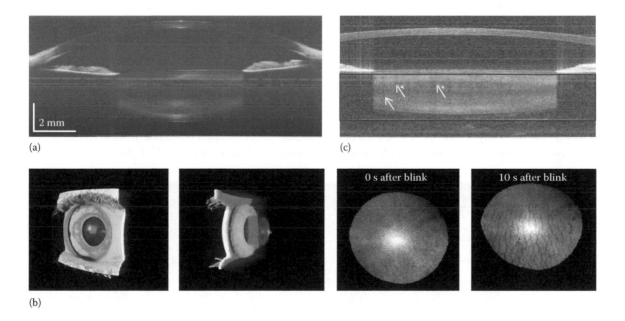

(a) (c)

(b)

Figure 4.13 Anterior segment imaging with swept source/Fourier-domain optical coherence tomography based on vertical cavity surface emitting laser light source. (a) Cross-sectional image of the anterior segment of the eye. (b) 3-D reconstruction of the eye: front and back view. (c) High sensitivity of the system enables ocular surface assessment. The shadows in the crystalline lens signal are due to the tear film breakup. The en-face views of the tear film integrity immediately after blink and 10 s after blink. ([a,b]: Reproduced from Grulkowski, I. et al., *Biomed. Opt. Express*, 3, 2733, 2012; [c]: Reproduced from Liu, J.J. et al., *Proc. SPIE*, 8567, 85670X, 2013.)

chamber, from limbus to limbus. The cross-sectional image in Figure 4.13a shows also the crystalline lens that is not an optically homogeneous structure since the nucleus and cortex can be distinguished. The imaging was performed at 1050 nm. The acquisition of two orthogonal volumes consisting of 400 × 400 A-scans at the speed of 100 kHz took ~3.2 s. After registration, the 3-D reconstruction of the volumetric data provides insight into morphology of the AS of the eye from different viewpoints (Figure 4.13b). Such a direct visualization of the entire AS in a simple scanning session does not require full-range technique due to the long coherence length of the VCSEL light source and virtually no parasitic sensitivity roll-off.

High sensitivity achieved in SS-OCT can enable indirect but noninvasive imaging of tear film dynamics. The tear film impacts the optical quality of the eye and contributes to the etiology of dry eye associated with contact lens wear. Tear film breakup time is usually measured with fluorescein dye. Several studies demonstrated that high-resolution OCT method has the ability to noninvasively quantify the dynamics of tear film characteristics, such as thickness, integrity, and meniscus dimensions (Wang et al., 2003, 2011, Koh et al., 2010, Czajkowski et al., 2012, Werkmeister et al., 2013). Figure 4.13c shows the shadows caused by tear film disruption that appear over the lens region. This feature can be applied to generate the tear film integrity map by summing the detected OCT signal over the lens region where optical shadowing from tear film breakup appears. The en-face images in Figure 4.13c present how drying cornea after blink affects the tear film.

SS-OCT using new tunable MEMS-VCSELs has the advantage that the MEMS mirror dynamics can be easily controlled so that the wavelength sweep range and sweep repetition rate can be adjusted in order to tailor the image resolution, imaging range, and imaging speed to match specific applications. This unique operational feature is specific for SS-OCT and provides means to meet the requirements of different imaging modes with versatile functionalities. Using high bandwidth detection, it is possible to achieve imaging ranges that are not accessible with SD-OCT (Grulkowski et al., 2012, 2013b,c). According to Equation 4.5, by trading off wavelength tuning range and sweep rate, one can increase the available axial measurement range. The example of this approach has been used in development of a novel imaging mode called full eye length imaging. The imaging from the anterior to posterior pole of the eyeball is demonstrated in Figure 4.14. VCSEL operated at 50 kA-scans/s and wavelength tuning range of 30 nm gave the axial resolution of 20 μm. The eye was scanned telecentrically, which means that the light refracted on the AS components converges into a single point of the retina.

As shown in Figure 4.14c, refraction correction of the image reveals that issue.

The operational mode of long-range imaging can be very useful in performing OCT-based ocular biometry. Ocular biometry aims at the measurement of intraocular distances of the eye to determine proper IOL before cataract surgery (see Section 4.11.2).

4.10 ARTIFACTS IN AS-OCT IMAGES

For all imaging modalities, understanding possible artifacts that may occur is important for a correct image interpretation. Recognizing artifacts and understanding how they were created are critical in clinical practice to differentiate artifacts from pathology (Chhablani et al., 2014). Some examples of artifacts present in AS-OCT imaging are shown in Figure 4.15 and are discussed shortly in this section.

4.10.1 MOTION ARTIFACTS

Motion artifacts are the most common in clinical practice. In vivo imaging is the subject to motion artifacts due to the eye, head, and respiratory movements as well as involuntary blinking (Yun et al., 2004b). Although current OCT systems can achieve high imaging speeds and reduce scan acquisition times considerably, it is not possible to fully eliminate motion-related artifacts. The cross section of the AS extracted from 3-D data set in slow scan direction can easily demonstrate the distortion of the corneal shape due to axial and lateral eye motion (Figure 4.15a). Different hardware and software approaches have been proposed in laboratory systems and, or implemented in commercial instruments to generate motion-free data (Maintz and Viergever, 1998, Jorgensen et al., 2007, Potsaid et al., 2008, Ricco et al., 2009, Kraus et al., 2012).

4.10.2 MIRROR ARTIFACT

A complex conjugate (mirror) image is a unique feature of all Fourier-domain OCT devices. Fourier transformation of the

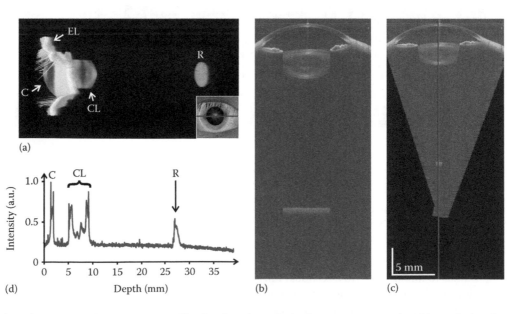

(a)

(d)

(b)

(c)

Figure 4.14 Full eye length imaging using swept source/Fourier-domain optical coherence tomography with vertical cavity surface emitting laser. (a) 3-D rendering of the volumetric data set. C, Cornea; EL, eyelid; CL, crystalline lens; R, retinal signal are visible. (b) Central cross section before refraction correction. (c) Central cross section after refraction correction. (d) Central depth profile enables performing ocular biometry. (Reproduced from Grulkowski, I. et al., *Biomed. Opt. Express*, 3, 2733, 2012.)

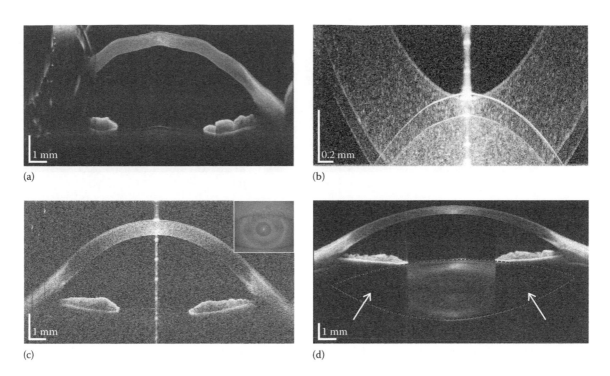

(a)

(b)

(c)

(d)

Figure 4.15 Examples of image artifacts in anterior segment optical coherence tomography: (a) motion artifact (rippled cornea) visible in 3-D imaging visible in cross section extracted along slow scan axis; (b) mirror artifact in Fourier-domain optical coherence tomography; (c) specular reflection and corneal signal washout; and (d) shadowing of the crystalline lens by the iris. ([a,d]: Courtesy of J.G. Fujimoto, MIT, Cambridge, MA.)

measured fringes actually generates two OCT images that are symmetric around the zero-delay line and carry the same structural information (Wojtkowski et al., 2002a, Ho et al., 2010). Therefore, visualization of one OCT image is sufficient, and the other image is truncated. If the AS image physically crosses the zero-delay line, the corresponding symmetric mirror image on the truncated side superimposes on the scanning range. A "ghost" mirror artifact will appear covering the details of actual image or producing false interpretation of AS morphology (Figure 4.15b). The complex conjugate effect can suppressed by applying phase changing in the reference arm, which results in doubling the axial imaging range.

4.10.3 CORNEAL SPECULAR REFLECTION

When the cross section of an AS-OCT image is on a corneal meridian, a vertical white beam (central vertical flare) appears in the anterior chamber and a small hyper-reflective area appears on the corneal surface in OCT images (Figure 4.15b and c). This type of artifacts is caused by strong specular back reflection from the cornea. Specular reflection saturates the detector during acquisition that results in a very high intensity in the corresponding A-scans. Although specular reflection hints useful signal from the eye, it may be practically used to center the scan area.

4.10.4 SHADOWING AND SIGNAL WASHOUT

Another type of the artifacts is linked to the limited penetration of light into the ocular tissue. AS of the eye is composed of light attenuating tissues. Sclera and limbus exhibit high scattering even for near-infrared radiation. Therefore, they prevent from penetration of light into deeper layers and make imaging the ciliary body difficult due to the multiple scattering.

As a result, ciliary body can be more effectively imaged with longer wavelengths and when side illumination of the eye (tilted eyeball) is used (Figure 4.12). Other example of artifact generating structures are the eye lashes and sutures that strongly reflect the light causing shadows to appear in the OCT image (Figure 4.11d). In addition, the iris limits the numerical aperture of the eye and it is also highly pigmented element. This in turn blocks the light beam to penetrate deeper and generates shadows in OCT images in the regions that are posterior to the iris. Consequently, it is impossible to visualize the peripheral parts of the crystalline lens in OCT (Figure 4.15d).

The cornea is relatively steep structure. When the beam is incident on a steep interface, different parts of the beam are reflected from slightly different depths. The interference fringe averaging (washout) that occurs in such a situation decreases the OCT signal displayed in OCT images. The signal washout can be severe in SD-OCT leading to the degraded image quality of the steepest parts of the cornea. However, SS-OCT is less sensitive to that kind of signal degradation. Another source of decreased interference signal from the cornea is its birefringent properties. The amplitude of interference does not only depend on the light intensity of both sample and reference components but also on the polarization states of both interfering waves. The light passing through the cornea changes the polarization, which impacts the OCT signal.

4.10.5 IMAGE DISTORTION CAUSED BY LIGHT REFRACTION

In the eye, light refraction at intraocular interfaces (i.e., air–tear film, tear film–cornea, cornea, aqueous, aqueous–crystalline lens) and light propagation through media of different refractive indices determine both the direction of beam propagation and

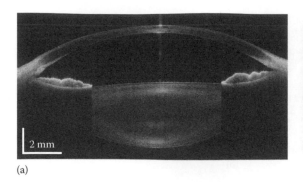
(a)

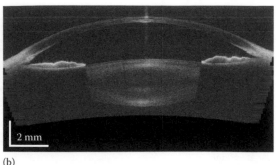
(b)

Figure 4.16 Light refraction correction. Optical coherence tomography cross-sectional image before (a) and after (b) refraction correction. (Reproduced from Grulkowski, I. et al., *Biomed. Opt. Express*, 3, 2733, 2012.)

the distances measured from OCT images that display "optical paths." Consequently, AS-OCT images do not represent true morphology of the eye. The refraction distortion in OCT images must be corrected to obtain anatomically correct structural images. In general, the correct physical shape of AS morphology can only be obtained after fan distortion and light refraction correction algorithms. Simple 2-D and advanced 3-D refraction correction (dewarping) algorithms for OCT data are usually based on ray-tracing methods that utilize basic Snell's law of light refraction and assumed refraction indices of ocular structures (Westphal et al., 2002, Podoleanu et al., 2004, Ortiz et al., 2010, Zhao et al., 2010, Siedlecki et al., 2012). Ray tracing is also primarily based on segmentation of ocular interfaces from OCT images. The refraction correction of OCT images is compulsory prior to performing biometric measurements of the eye based on OCT data. We developed a 3-D dewarping software that enabled correcting AS-OCT images. The impact of this kind of distortion is shown in Figure 4.16.

4.11 ANTERIOR SEGMENT BIOMETRY BASED ON OCT

The possibility to perform quantitative description by dedicated image analysis tools is regarded as an essential feature of all modern medical imaging modalities. In particular, AS-OCT revealing morphology of the eye is often clinically applied in qualitative assessment of the AS by looking at the scans, determining the anatomical landmarks in pathologic cases and checking the change of the images after treatment. However, the diagnostic potential of AS-OCT can be fully explored by the extraction of relevant quantitative information on AS from OCT data sets (Li et al., 2014). The main advantage of quantitative imaging in ophthalmology is that it makes the eye diagnosis more objective. Moreover, noncontact and noninvasive nature of OCT examination together with high-speed acquisition ensures that biometry of the eye based on OCT data is not affected by the scanning procedure itself or altered by motion artifacts. Consequently, OCT can demonstrate sufficient reliability and reproducibility of biometric assessment. In addition to that, access to 3-D eye morphology provides comprehensive information on the shape of ocular structures at micrometer resolution, which significantly increases diagnostic sensitivity of detecting small morphologic changes. Commercial instruments for AS-OCT

have semiautomatic or automatic analytical tools (software) enabling extraction of the most common biometric parameters.

Quantitative evaluation of the AS of the eye includes measurements of several clinically relevant parameters. The image in Figure 4.17 demonstrates light refraction-corrected cross section of the eye along with the basic biometric parameters that are usually extracted from OCT data.

4.11.1 CORNEAL AND CRYSTALLINE LENS TOPOGRAPHY

Corneal topography can be revealed by mapping the corneal thickness (pachymetry), surface elevation, or curvature/power mapping (keratometry) and plays an important role in the assessment of keratorefractive surgical procedures, corneal transplantation, or screening the corneal degenerations and dystrophies (e.g., keratoconus). There are several techniques enabling topographic description of the eye (Mejia-Barbosa and Malacara-Hernandez, 2001, Pinero, 2015). Access to volumetric data sets enables mapping the corneal thickness (corneal pachymetry). Several studies have been performed to show reproducibility of OCT pachymetry and agreement with standard instruments (Wirbelauer et al., 2002, Li et al., 2006, 2008, 2010, Pinero et al., 2008). The comparison of corneal pachymetry for normal subject and patient with keratoconus is presented in Figure 4.18 (Karnowski et al., 2011). The maps were calculated from corresponding 3-D data sets after segmentation of corneal interfaces and correction for light refraction.

Keratometry with standard instruments has the disadvantage that the modalities cannot measure the curvature of posterior corneal surface. This issue was solved with the introduction of OCT. The measurement of central corneal power has been implemented in commercial SD-OCT instrument (Tang et al., 2006, 2010a,b, Sorbara et al., 2010).

Since the cornea and the crystalline lens contribute to refractive properties of the eye, topographic measurements of the crystalline lens are also crucial in vision science for understanding the optical quality of the eye as well as aberrations in certain conditions such as accommodation or presbyopia. The difficulty in correct description of the crystalline lens comes from the fact that there is no access to the individual in vivo measurement of the gradient index distribution (de Castro et al., 2010, Siedlecki et al., 2012, de Freitas et al., 2013, Gambra et al., 2013, Sun et al., 2014). Phakometric measurements provide valuable information

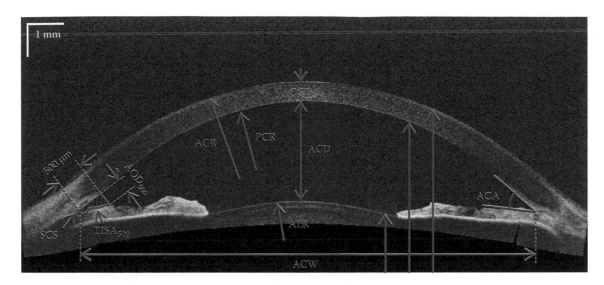

Figure 4.17 Biometric parameters of the eye. CCT, central corneal thickness; ACD, anterior chamber depth; ALR, anterior corneal radius of curvature; PCR, posterior corneal radius of curvature; ALR, anterior lens radius of curvature; ACA, anterior chamber angle; SCS, scleral spur; AOD_{500}, angle opening distance 500 μm anterior to sclera spur; $TISA_{500}$, trabecular iris space area; ACW, anterior chamber width.

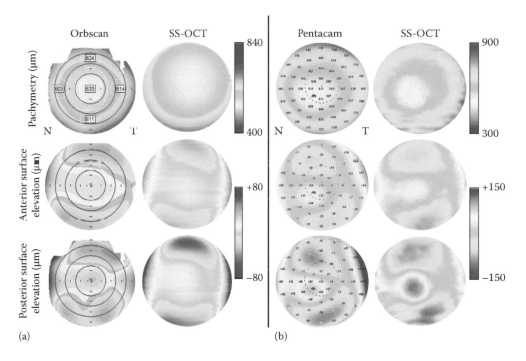

Figure 4.18 Pachymetric maps and elevation maps of both corneal surfaces. (a) Left eye of a healthy subject was measured with swept source/ Fourier-domain optical coherence tomography (SS-OCT) instrument at 1050 nm and with Orbscan topographer (scanning slit lamp). (b) Left eye of a patient with keratoconus was measured with SS-OCT instrument at 1310 nm and with Pentacam topographer (rotating Scheimpflug camera). ([b]: Reproduced from Karnowski, K. et al., *Biomed. Opt. Express*, 2, 2709, 2011.)

used during planning and evaluation of IOL implantation in cataract surgery.

The measurements of anterior chamber depth and anterior chamber angle are used in diagnosing glaucoma and planning of surgical interventions. OCT goniometry is based on measurements of several parameters characterizing the anterior chamber angle (Karandish et al., 2004, 2006, Radhakrishnan et al., 2005, Meinhardt et al., 2006, Nolan, 2008a, Fukuda et al., 2011, Leung and Weinreb, 2011, Tian et al., 2011, Radhakrishnan and Yarovoy, 2014).

4.11.2 OCT-BASED OCULAR BIOMETRY

Novel long-range applications of SS-OCT enable determination of intraocular distances, known as ocular biometry. OCT-based ocular biometry is essential for accurate outcomes in cataract and keratorefractive surgeries. Precisely measured axial intraocular distances are important for proper IOL power calculation (Fledelius, 1997, Olsen, 2007). Currently, ocular biometry with ultrasound is a gold standard in ophthalmology. Ultrasound devices can perform axial length measurements with a resolution of ~100 μm. However, ultrasound techniques require contact of the eye by a transducer in order to measure axial distance. The optical biometric methods based on partial coherence interferometry were also demonstrated, and commercial optical devices such as IOL Master (Zeiss) and LensStar (Haag-Streit) were developed (Perkins et al., 1976, Fujimoto et al., 1986, Fercher et al., 1988, Drexler et al., 1996, Haigis et al., 2000, Vogel, 2001, p. 130, Packer et al., 2002, Rajan et al., 2002, Buckhurst et al., 2009, Rohrer et al., 2009, Mylonas et al., 2011, Shammas and Hoffer, 2012). Simultaneous visualization of anterior and posterior eye structures as well as extraction of intraocular distances from OCT data is a developing field of application (Grajciar et al., 2008, Chong et al., 2009, Dhalla et al., 2012a, Ruggeri et al., 2012). Optical biometry is noncontact and provides higher resolution (10–20 μm) than ultrasound.

Advances in swept laser technology and high-speed data acquisition systems enable long-range OCT imaging that is extremely useful in ocular biometry. Three-dimensional full eye length images contain comprehensive biometric information and permit extraction of an averaged depth profile of the eye enabling calculation of the intraocular distances (Figure 4.19a). SS-OCT-based biometry demonstrates excellent reproducibility and repeatability with the axial length measurement precision of 16 μm (Grulkowski et al., 2013c). We measured the intraocular distances of both eyes of 10 healthy subjects with no ocular surface pathology. Comparison with clinical ocular optical biometers (IOL Master, Carl Zeiss Meditec) and immersion A-scan ultrasound biometer (Axis II PR, Quantel Medical) shows also very good correlation of axial eye

length measurements between SS-OCT, the IOL Master, and immersion ultrasound (Figure 4.19b).

4.12 IMAGING OF THE DYNAMICS OF THE AS OF THE EYE

Different functions of the eye require that organ and its components to change dynamically, and the processes are characterized by different time scales. The studies of in vivo dynamic processes in the eye are extremely challenging and involve combination of the technological advances in imaging speed, ultrafast acquisition, and real-time data processing to process and visualize the vast amounts of data. The acquisition of rapidly repeated volumetric data sets in time defines emerging imaging mode called 4-D (3-D + time) imaging. Four-dimensional OCT imaging has been recently demonstrated in a number of biomedical applications (Rey et al., 2009, Jenkins et al., 2010, Kobler et al., 2010, Klein et al., 2011, Sylwestrzak et al., 2012, Wieser et al., 2014). Some applications of imaging of dynamic processes are described in the following.

4.12.1 BLINKING

High speed offers the potential to visualize motion of the eye. The most obvious example of the movement of the ocular structures is blinking. Figure 4.20a demonstrates volumetric image of the eye in few selected phases of the eyelid closing along with corresponding en-face images. The imaging was performed by the SD-OCT laboratory instrument operating at

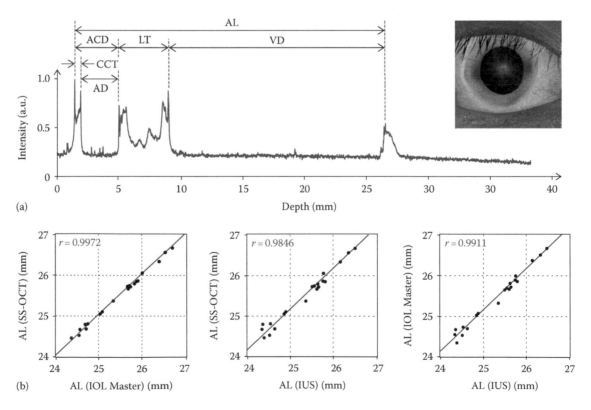

Figure 4.19 Optical coherence tomography (OCT) ocular biometry. (a) Extraction of intraocular distances from averaged central axial scan (A-scan). (b) Comparison of swept source/Fourier-domain optical coherence tomography ocular biometry with clinical biometric devices: IOL Master and immersion A-scan ultrasound biometer (IUS). Correlation coefficients are given as a measure of comparison. (Reproduced from Grulkowski, I. et al., *Ophthalmology*, 120, 2184, 2013c.)

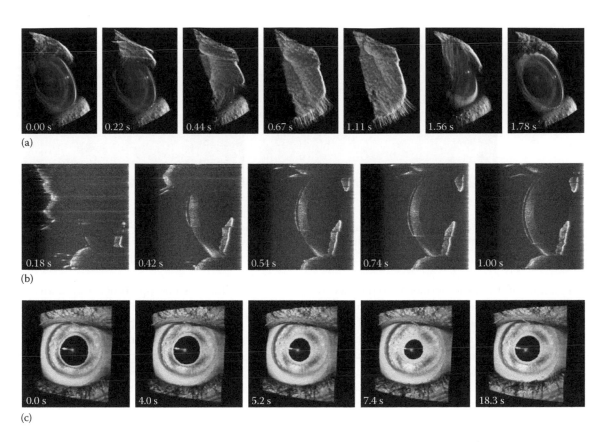

Figure 4.20 Imaging the dynamic processes in the eye using high-speed optical coherence tomography (OCT). (a) 4-D spectral/Fourier-domain OCT of the eye blink measured in vivo with the system operating at 840 nm. (b) Vertical cross sections of the eye wearing rigid contact lens after blink (imaging with swept source/Fourier-domain OCT [SS-OCT] at 1310 nm). (c) 4-D OCT imaging of eye reaction on light stimulus (SS-OCT at 1050 nm). ([a]: Reproduced from Grulkowski, I. et al., *Opt. Express*, 17, 4842, 2009; [b]: Reproduced from Gora, M. et al., *Opt. Express*, 17, 14880, 2009; [c]: Reproduced from Liu, J.J. et al., *Proc. SPIE*, 8567, 85670X, 2013.)

an A-scan rate of 135 kA-scans/s, and a 4-D scan protocol was used with 300×100 A-scan volumes over 15×15 mm² area at a rate of 4.5 volumes/s, which gave total imaging time of 2.7 s (Grulkowski et al., 2009). Evaluation of blink-induced contact lens movement was also visualized using SS-OCT with FDML laser operating at 200 kA-scans/s with the central wavelength of 1310 nm (Gora et al., 2009). To better elucidate the dynamics of that process, repeated vertical scans consisting of 4000 A-scans were acquired for 2 s. The acquisition speed enabled obtaining scan rate of 50 B-scans/s. Sequential images presented in Figure 4.20b show that rigid contact lens slides down after the blink. Both the position of the bottom edge of the lens and the alignment quality on corneal surface can be easily examined. Such experiments are useful in studying contact lens fitting.

4.12.2 PUPILLARY RESPONSE

The iris is a dynamic structure whose configuration regularly changes in response to light and during accommodation. Dynamic changes in intraocular structures caused by illumination or dilation are suggested to be risk factors for glaucoma development. Studying the dynamic response of the pupil to dark–light stimulus with OCT provided a more comprehensive assessment of risks to primary angle closure development and may help understand the pathophysiology of angle closure glaucoma (Leung et al., 2007, Cheung et al., 2010, Li et al., 2014). We demonstrated 4-D imaging of the pupil

response to light stimulus from a light emitting diode, where sequential 150×150 A-scan volumes over 17×17 mm² were acquired with SS-OCT system at ~8 volumes/s for 5 s with an A-scan rate of 200 kHz enabling the visualization of changes in the iris in time and in three dimensions (Figure 4.20c) (Liu et al., 2013). Additionally, this process can be quantified by measuring the pupil area at each time instance. As shown in Figure 4.20c, the pupil area decreased drastically when light stimulus was applied.

4.12.3 ACCOMMODATION

Another important dynamic process occurring in everyday life is the eye accommodation, which enables sharp vision of the objects located at different distances from the eye. Anatomically, the accommodation manifests in changing the shape and relative position of the crystalline lens due to the action of ciliary muscles. Studies of the lenticular accommodation processes are crucial for understanding the accommodation mechanism itself as well as for identifying the backgrounds of presbyopia development. Several theories have been proposed to explain the physiology of human eye accommodation (Von Helmholtz, 1855, Tscherning, 1899, Schachar, 1994). However, the verification possibilities of those theories were limited due to the lack of noninvasive and high-resolution imaging modalities. The advent of OCT provided a tool for fast imaging of the eye in different accommodation states and may shed a new light on the debate

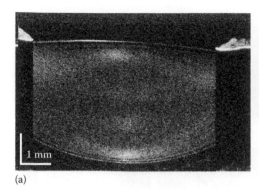
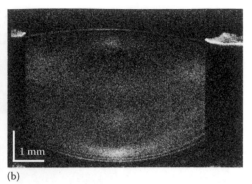

(a)   (b)

Figure 4.21 Spectral/Fourier-domain optical coherence tomography imaging of accommodating eye at 840 nm system. (a) Eye in the relaxed state. (b) Eye in the accommodated state. (Reproduced from Grulkowski, I. et al., *Opt. Express*, 17, 14880, 2009.)

over the validity of the eye accommodation theories. The response of the eye to accommodation can be studied either with dynamically varying demand or with a series of static demands measured consecutively. In both cases, comparison of biometric parameters with respect to the relaxed state can be performed. Multiple studies have been recently performed to visualize the accommodating eye with high-speed OCT (Drexler et al., 1997, Baikoff et al., 2004, Richdale et al., 2008, Zhou et al., 2009, Ruggeri et al., 2012, Zhong et al., 2014). We used high-speed SD-OCT system to visualize the morphologic changes in the crystalline lens during accommodation. The sections shown in Figure 4.21 demonstrate that the crystalline lens becomes thicker and the radii of curvature decrease.

The availability OCT systems with extended depth ranges allows to obtain more comprehensive information (Shen et al., 2010, Du et al., 2012, Shi et al., 2012, Gambra et al., 2013, Shao et al., 2013, Leng et al., 2014). In another study, the interface of the long-range SS-OCT system to perform study with variable accommodation was equipped with the additional channel, that is, Badal optometer, to provide desired accommodation demand (Atchison et al., 1995). We imaged the accommodating eyes in vivo with SS-OCT instrument for selected accommodation demands ranged from 0 (relaxed state) to 6 D (diopters) (Grulkowski et al., 2013a). We performed simultaneous topographic measurements of the cornea and the crystalline lens as well as ocular biometry. The quantitative data indicate that the crystalline lens parameters vary when a subject focuses at a near object, while the cornea and the eye globe are relatively rigid structure and do not change with accommodation. The crystalline lens moves toward the cornea making the anterior chamber shallower.

4.12.4 OCT ELASTOGRAPHY OF THE EYE

In most applications, OCT images provide structural contrast due to variations of light scattering properties based on refractive index inhomogeneity in the tissue. However, access to interferometric fringes enable also visualization additional parameters such as the changes of polarization (polarization-sensitive OCT) or motion (Doppler OCT) (Götzinger et al., 2004, Baumann et al., 2007a, Miyazawa et al., 2009, Kagemann et al., 2011, Li et al., 2011, 2013b, Lim et al., 2011, Pircher et al., 2011, Sun et al., 2015).

The dynamic aspects of AS structures depend on the biomechanics properties of the ocular tissues and the intraocular pressure. Therefore, the assessment of elastic properties might be of great importance in the detection and treatment monitoring of corneal abnormalities affecting its biomechanical properties, for example, keratoconus. In most cases, elastographic studies are based on examination of corneal response to mechanical stimulu like sound or air puff. A clinically approved device called Ocular Response Analyzer (Reichert Ophthalmic Instruments, United States) enables determination on intraocular pressure as well as parameters like corneal hysteresis and resistance factor that are associated with biomechanical properties of the cornea (Luce, 2005). Corneal deformation can be also visualized with other commercial instrument Corvis ST (Oculus, Germany) based on Scheimpflug imaging (Hon and Lam, 2013). The measurement of elastic properties of the cornea has been recently demonstrated with shear wave imaging or Brillouin microscopy (Scarcelli and Yun, 2008, Tanter et al., 2009).

OCT-based elastography was first presented in 1998 (Schmitt, 1998, Kennedy et al., 2014). OCT was also recently applied in the extraction of biomechanical properties of the corneal tissue using different excitation and detection schemes (Alonso-Caneiro et al., 2011, Ford et al., 2011, Dorronsoro et al., 2012, Nahas et al., 2013, Wang and Larin, 2014, Kling et al., 2014a,b). The advantage of application of OCT in elasticity imaging of the eye comes from high spatiotemporal resolution and noninvasiveness of this modality. We developed the SS-OCT system operating at 1310 nm integrated with clinical air puff that can image at the speed of 50 kA-scans/s (Figure 4.22a) (Alonso-Caneiro et al., 2011). The M-scan consisting of consecutively acquired A-scans in time in the corneal apex reveal deformation of the cornea and the crystalline lens caused by the applied air puff (Figure 4.22b). The quantification of that highly dynamic process is based on segmentation of corneal interfaces and analysis of their dynamic behavior. However, full interpretation of the parameters is still the goal of undergoing studies. Figure 4.22c demonstrates reaction and restoration of the anterior chamber.

4.13 SUMMARY AND PERSPECTIVES

Modern OCT systems serve as efficient instrumentation for noninvasive high-speed and high-resolution AS imaging. Commercial and research OCT systems enable qualitative and

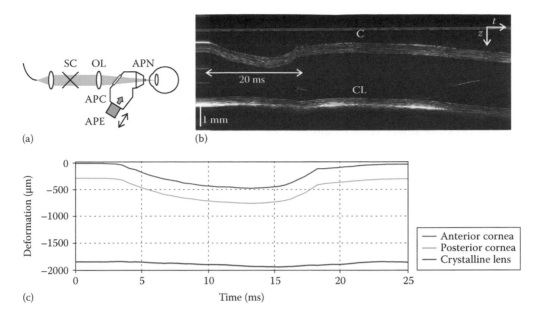

Figure 4.22 Air-puff optical coherence tomography (OCT) for determination of elastic properties of the cornea. (a) Interface of an air-puff OCT instrument. (b) M-scan showing anterior chamber dynamics during air pulse (C, cornea; CL, crystalline lens). (c) Deformation curves of the corneal and lenticular surfaces. (Reproduced from Alonso-Caneiro, D. et al., *Opt. Express*, 19, 14188, 2011.)

quantitative study of AS morphology in various conditions of the eye. Recent advances in OCT technology address the challenges of AS imaging and allow for development of the flexible platforms enabling comprehensive in vivo visualization and biometry of the eye for both research and clinical applications. High imaging speeds enable rapid acquisition times as well as 4-D imaging of dynamical processes. Novel light sources designed for SS-OCT provide imaging ranges not available before due to its long coherence length and have applications for biometry of the anterior eye and full eye length imaging.

The research will still focus on the evaluation of OCT for AS imaging in various clinical cases. New emerging applications such as real-time surgical guidance (intraoperative imaging) will be a cutting edge of clinical AS-OCT in the nearest future. Additionally, development of new sophisticated software tools for extraction of quantitative 3-D information from OCT data may advance customized innovative treatments of the front part of the human eye. Finally, the extension of OCT including corneal elastography will enhance diagnostic power of OCT technique to provide information complementary to structural imaging.

ACKNOWLEDGMENTS

The author would like to acknowledge colleagues with whom he had an opportunity to work at the Nicolaus Copernicus University in Toruń as well as at the Massachusetts Institute of Technology: Prof. M. Wojtkowski, Prof. J.G. Fujimoto, Prof. B. Kałużny, Prof. J.S. Duker, Prof. D. Huang, Prof. S. Marcos, Dr. hab. M. Szkulmowski, Dr. K. Karnowski, Dr. M. Góra, Dr. I. Gorczyńska, Dr. D. Bukowska, Dr. D. Szlag, Dr. J.J. Liu, Dr. B. Baumann, Dr. B. Potsaid, Dr. Y. Tao, Dr. M. Kraus, Dr. W.J. Choi, Dr. Mehreen Ahdi, Dr. J.Y. Zhang, Dr. V. Jayaraman, and C.D. Lu.

REFERENCES

Akiba, M., Maeda, N., Yumikake, K., Soma, T., Nishida, K., Tano, Y., and Chan, K. P. 2007. Ultrahigh-resolution imaging of human donor cornea using full-field optical coherence tomography. *Journal of Biomedical Optics*, 12, 041202.

Alonso-Caneiro, D., Karnowski, K., Kaluzny, B. J., Kowalczyk, A., and Wojtkowski, M. 2011. Assessment of corneal dynamics with high-speed swept source Optical Coherence Tomography combined with an air puff system. *Optics Express*, 19, 14188–14199.

Alonso-Caneiro, D., Shaw, A. J., and Collins, M. J. 2012. Using optical coherence tomography to assess corneoscleral morphology after soft contact lens wear. *Optometry and Vision Science*, 89, 1619–1626.

Arbelaez, M. C. and Samuel Arba Mosquera, M. 2009. The Schwind Amaris total-tech laser as an all-rounder in refractive surgery. *Middle East African Journal of Ophthalmology*, 16, 46–53.

Atchison, D. A., Bradley, A., Thibos, L. N., and Smith, G. 1995. Useful variations of the Badal optometer. *Optometry and Vision Science*, 72, 279–284.

Atchison, D. A. and Smith, G. (eds.) 2000. *Optics of the Human Eye*, Edinburgh, Scotland: Butterworth-Heinemann.

Baikoff, G., Lutun, E., and Ferraz, C. 2004. Static and dynamic analysis of the anterior segment with optical coherence tomography. *Journal of Cataract and Refractive Surgery*, 30, 1843–1850.

Bali, S. J., Hodge, C., Lawless, M., Roberts, T. V., and Sutton, G. 2012. Early experience with the femtosecond laser for cataract surgery. *Ophthalmology*, 119, 891–899.

Baumann, B., Götzinger, E., Pircher, M., and Hitzenberger, C. K. 2007a. Single camera based spectral domain polarization sensitive optical coherence tomography. *Optics Express*, 15, 1054–1063.

Baumann, B., Pircher, M., Gotzinger, E., and Hitzenberger, C. K. 2007b. Full range complex spectral domain optical coherence tomography without additional phase shifters. *Optics Express*, 15, 13375–13387.

Boettner, E. A. and Wolter, J. R. 1962. Transmission of the Ocular Media. *Investigative Ophthalmology and Visual Science*, 1, 776–783.

Bohnke, M. and Masters, B. R. 1999. Confocal microscopy of the cornea. *Progress in Retinal and Eye Research*, 18, 553–628.

Buckhurst, P. J., Wolffsohn, J. S., Shah, S., Naroo, S. A., Davies, L. N., and Berrow, E. J. 2009. A new optical low coherence reflectometry device for ocular biometry in cataract patients. *British Journal of Ophthalmology*, 93, 949–953.

Cheung, C. Y.-L., Liu, S., Weinreb, R. N., Liu, J., Li, H., Leung, D. Y.-L., Dorairaj, S., Liebmann, J., Ritch, R., and Lam, D. S. C. 2010. Dynamic analysis of iris configuration with anterior segment optical coherence tomography. *Investigative Ophthalmology and Visual Science*, 51, 4040–4046.

Chhablani, J., Krishnan, T., Sethi, V., and Kozak, I. 2014. Artifacts in optical coherence tomography. *Saudi Journal of Ophthalmology*, 28, 81–87.

Chinn, S. R., Swanson, E. A., and Fujimoto, J. G. 1997. Optical coherence tomography using a frequency-tunable optical source. *Optics Letters*, 22, 340–342.

Chiou, A. G. Y., Kaufman, S. C., Kaufman, H. E., and Beuerman, R. W. 2006. Clinical corneal confocal microscopy. *Survey of Ophthalmology*, 51, 482–500.

Choma, M. A., Hsu, K., and Izatt, J. A. 2005. Swept source optical coherence tomography using an all-fiber 1300-nm ring laser source. *Journal of Biomedical Optics*, 10, 044009.

Choma, M. A., Sarunic, M. V., Yang, C. H., and Izatt, J. A. 2003. Sensitivity advantage of swept source and Fourier domain optical coherence tomography. *Optics Express*, 11, 2183–2189.

Chong, C. H., Suzuki, T., Totsuka, K., Morosawa, A., and Sakai, T. 2009. Large coherence length swept source for axial length measurement of the eye. *Applied Optics*, 48, D144–D150.

Christopoulos, V., Kagemann, L., Wollstein, G., Ishikawa, H. R., Gabriele, M. L., Wojtkowski, M., Srinivasan, V. et al. 2007. In vivo corneal high-speed, ultra-high-resolution optical coherence tomography. *Archives of Ophthalmology*, 125, 1027–1035.

Czajkowski, G., Kaluzny, B. J., Laudencka, A., Malukiewicz, G., and Kaluzny, J. J. 2012. Tear meniscus measurement by spectral optical coherence tomography. *Optometry and Vision Science*, 89, 336–342.

Dada, T., Sihota, R., Gadia, R., Aggarwal, A., Mandal, S., and Gupta, V. 2007. Comparison of anterior segment optical coherence tomography and ultrasound biomicroscopy for assessment of the anterior segment. *Journal of Cataract and Refractive Surgery*, 33, 837–840.

Dai, C., Fan, S., Chai, X., Li, Y., Ren, Q., Xi, P., and Zhou, C. 2014. Dual-channel spectral-domain optical-coherence tomography system based on 3 × 3 fiber coupler for extended imaging range. *Applied Optics*, 53, 5375–5379.

de Boer, J. F., Cense, B., Park, B. H., Pierce, M. C., Tearney, G. J., and Bouma, B. E. 2003. Improved signal-to-noise ratio in spectral-domain compared with time-domain optical coherence tomography. *Optics Letters*, 28, 2067–2069.

de Castro, A., Ortiz, S., Gambra, E., Siedlecki, D., and Marcos, S. 2010. Three-dimensional reconstruction of the crystalline lens gradient index distribution from OCT imaging. *Optics Express*, 18, 21905–21917.

de Freitas, C., Ruggeri, M., Manns, F., Ho, A., and Parel, J.-M. 2013. In vivo measurement of the average refractive index of the human crystalline lens using optical coherence tomography. *Optics Letters*, 38, 85–87.

Dhalla, A. and Izatt, J. A. 2011. Complete complex conjugate resolved heterodyne swept-source optical coherence tomography using a dispersive optical delay line. *Biomedical Optics Express*, 2, 1218–1232.

Dhalla, A.-H., Nankivil, D., Bustamante, T., Kuo, A., and Izatt, J. A. 2012a. Simultaneous swept source optical coherence tomography of the anterior segment and retina using coherence revival. *Optics Letters*, 37, 1883–1885.

Dhalla, A.-H., Nankivil, D., and Izatt, J. A. 2012b. Complex conjugate resolved heterodyne swept source optical coherence tomography using coherence revival. *Biomedical Optics Express*, 3, 633–649.

Dorronsoro, C., Pascual, D., Pérez-Merino, P., Kling, S., and Marcos, S. 2012. Dynamic OCT measurement of corneal deformation by an air puff in normal and cross-linked corneas. *Biomedical Optics Express*, 3, 473–487.

Drexler, W., Baumgartner, A., Findl, O., Hitzenberger, C. K., and Fercher, A. F. 1997. Biometric investigation of changes in the anterior eye segment during accommodation. *Vision Research*, 37, 2789–2800.

Drexler, W. and Fujimoto, J. G. (ed.) 2008a. *Optical Coherence Tomography: Technology and Applications*, Berlin, Germany: Springer-Verlag.

Drexler, W. and Fujimoto, J. G. 2008b. State-of-the-art retinal optical coherence tomography. *Progress in Retinal and Eye Research*, 27, 45–88.

Drexler, W., Hitzenberger, C. K., Baumgartner, A., Findl, O., Strenn, K., Rainer, G., Menapace, R., and Fercher, A. F. 1996. (Sub)micrometer precision biometry of the human eye by optical coherence tomography and topography. *Investigative Ophthalmology and Visual Science*, 37, 4374.

Drexler, W., Liu, M. Y., Kumar, A., Kamali, T., Unterhuber, A., and Leitgeb, R. A. 2014. Optical coherence tomography today: Speed, contrast, and multimodality. *Journal of Biomedical Optics*, 19, 071412.

Du, C., Shen, M., Li, M., Zhu, D., Wang, M. R., and Wang, J. 2012. Anterior segment biometry during accommodation imaged with ultralong scan depth optical coherence tomography. *Ophthalmology*, 119, 2479–2485.

Dubois, A., Vabre, L., Boccara, A.-C., and Beaurepaire, E. 2002. High-resolution full-field optical coherence tomography with a Linnik microscope. *Applied Optics*, 41, 805–812.

Ehlers, J. P., Kaiser, P. K., and Srivastava, S. K. 2014a. Intraoperative optical coherence tomography using the Rescan 700: Preliminary results from the Discover study. *British Journal of Ophthalmology*, 98, 1329–1332.

Ehlers, J. P., Srivastava, S. K., Feiler, D., Noonan, A. I., Rollins, A. M., and Tao, Y. K. 2014b. Integrative advances for OCT-guided ophthalmic surgery and intraoperative OCT: Microscope Integration, surgical instrumentation, and heads-up display surgeon feedback. *PLoS ONE*, 9, e105224.

Ehlers, J. P., Tao, Y. K., Farsiu, S., Maldonado, R., Izatt, J. A., and Toth, C. A. 2013. Visualization of real-time intraoperative maneuvers with a microscope-mounted spectral domain optical coherence tomography system. *Retina*, 33, 232–236.

Fan, S. H., Li, L., Li, Q., Dai, C. X., Ren, Q. S., Jiao, S. L., and Zhou, C. Q. 2015. Dual band dual focus optical coherence tomography for imaging the whole eye segment. *Biomedical Optics Express*, 6, 2481–2493.

Fanea, L. and Fagan, A. J. 2012. Review: Magnetic resonance imaging techniques in ophthalmology. *Molecular Vision*, 18, 2538–2560.

Fercher, A. F. 2010. Optical coherence tomography—Development, principles, applications. *Zeitschrift für Medizinische Physik*, 20, 251–276.

Fercher, A. F., Drexler, W., Hitzenberger, C. K., and Lasser, T. 2003. Optical coherence tomography—Principles and applications. *Reports on Progress in Physics*, 66, 239–303.

Fercher, A. F., Hitzenberger, C., and Juchem, M. 1991. Measurement of intraocular optical distances using partially coherent laser light. *Journal of Modern Optics*, 38, 1327–1333.

Fercher, A. F., Hitzenberger, C. K., Kamp, G., and Elzaiat, S. Y. 1995. Measurement of intraocular distances by backscattering spectral interferometry. *Optics Communications*, 117, 43–48.

Fercher, A. F., Mengedoht, K., and Werner, W. 1988. Eye-length measurement by interferometry with partially coherent light. *Optics Letters*, 13, 186–188.

Fledelius, H. C. 1997. Ultrasound in ophthalmology. *Ultrasound in Medicine and Biology*, 23, 365–375.

Ford, M. R., Dupps, W. J., Rollins, A. M., Roy, A. S., and Hu, Z. 2011. Method for optical coherence elastography of the cornea. *Journal of Biomedical Optics*, 16, 016005.

Friedman, N. J., Palanker, D. V., Schuele, G., Andersen, D., Marcellino, G., Seibel, B. S., Batlle, J. et al. 2011. Femtosecond laser capsulotomy. *Journal of Cataract and Refractive Surgery*, 37, 1189–1198.

Fujimoto, J. G. 2001. Optical coherence tomography. *Comptes Rendus de l'Academie des Sciences—Series IV—Physique-Astrophysique*, 2, 1099–1111.

Fujimoto, J. G. 2003. Optical coherence tomography for ultrahigh resolution in vivo imaging. *Nature Biotechnology*, 21, 1361–1367.

Fujimoto, J. G., Desilvestri, S., Ippen, E. P., Puliafito, C. A., Margolis, R., and Oseroff, A. 1986. Femtosecond optical ranging in biological systems. *Optics Letters*, 11, 150–152.

Fujiwara, N., Yoshimura, R., Kato, K., Ishii, H., Kano, F., Kawaguchi, Y., Kondo, Y., Ohbayashi, K., and Oohashi, H. 2008. 140-nm quasi-continuous fast sweep using SSG-DBR lasers. *IEEE Photonics Technology Letters*, 20, 1015–1017.

Fukuda, S., Kawana, K., Yasuno, Y., and Oshika, T. 2011. Repeatability and reproducibility of anterior chamber volume measurements using 3-dimensional corneal and anterior segment optical coherence tomography. *Journal of Cataract and Refractive Surgery*, 37, 461–468.

Gambra, E., Ortiz, S., Perez-Merino, P., Gora, M., Wojtkowski, M., and Marcos, S. 2013. Static and dynamic crystalline lens accommodation evaluated using quantitative 3-D OCT. *Biomedical Optics Express*, 4, 1595–1609.

Golubovic, B., Bouma, B. E., Tearney, G. J., and Fujimoto, J. G. 1997. Optical frequency-domain reflectometry using rapid wavelength tuning of a Cr4+: Forsterite laser. *Optics Letters*, 22, 1704–1706.

Gora, M., Karnowski, K., Szkulmowski, M., Kaluzny, B. J., Huber, R., Kowalczyk, A., and Wojtkowski, M. 2009. Ultra high-speed swept source OCT imaging of the anterior segment of human eye at 200 kHz with adjustable imaging range. *Optics Express*, 17, 14880–14894.

Götzinger, E., Pircher, M., Sticker, M., Fercher, A. F., and Hitzenberger, C. K. 2004. Measurement and imaging of birefringent properties of the human cornea with phase-resolved, polarization-sensitive optical coherence tomography. *Journal of Biomedical Optics*, 9, 94–102.

Grajciar, B., Pircher, M., Fercher, A., and Leitgeb, R. 2005. Parallel Fourier domain optical coherence tomography for in vivo measurement of the human eye. *Optics Express*, 13, 1131–1137.

Grajciar, B., Pircher, M., Hitzenberger, C. K., Findl, O., and Fercher, A. F. 2008. High sensitive measurement of the human axial eye length in vivo with Fourier domain low coherence interferometry. *Optics Express*, 16, 2405–2414.

Grieve, K., Dubois, A., Simonutti, M., Paques, M., Sahel, J., Le Gargasson, J. F., and Boccara, C. 2005. In vivo anterior segment imaging in the rat eye with high speed white light full-field optical coherence tomography. *Optics Express*, 13, 6286–6295.

Grosvenor, T. (ed.) 2007. *Primary Care Optometry*, St Louis, MI: Butterworth-Heinemann Elsevier.

Grulkowski, I., Gora, M., Szkulmowski, M., Gorczynska, I., Szlag, D., Marcos, S., Kowalczyk, A., and Wojtkowski, M. 2009. Anterior segment imaging with Spectral OCT system using a high-speed CMOS camera. *Optics Express*, 17, 4842–4858.

Grulkowski, I., Liu, J. J., Baumann, B., Potsaid, B., Lu, C., and Fujimoto, J. G. 2011. Imaging limbal and scleral vasculature using Swept Source Optical Coherence Tomography. *Photonics Letters of Poland*, 3, 132–134.

Grulkowski, I., Liu, J., Potsaid, B., Jayaraman, V., Cable, A., Kraus, M., Hornegger, J., Duker, J., Huang, D., and Fujimoto, J. 2013a. Three-dimensional biometric measurements of accommodation using full-eye-length swept-source OCT. *Investigative Ophthalmology & Visual Science*, 54, 381.

Grulkowski, I., Liu, J. J., Potsaid, B., Jayaraman, V., Jiang, J., Fujimoto, J. G., and Cable, A. E. 2013b. High-precision, high-accuracy ultralong-range swept-source optical coherence tomography using vertical cavity surface emitting laser light source. *Optics Letters*, 38, 673–675.

Grulkowski, I., Liu, J. J., Potsaid, B., Jayaraman, V., Lu, C. D., Jiang, J., Cable, A. E., Duker, J. S., and Fujimoto, J. G. 2012. Retinal, anterior segment and full eye imaging using ultrahigh speed swept source OCT with vertical-cavity surface emitting lasers. *Biomedical Optics Express*, 3, 2733–2751.

Grulkowski, I., Liu, J. J., Zhang, J. Y., Potsaid, B., Jayaraman, V., Cable, A. E., Duker, J. S., and Fujimoto, J. G. 2013c. Reproducibility of a long-range swept-source optical coherence tomography ocular biometry system and comparison with clinical biometers. *Ophthalmology*, 120, 2184–2190.

Guthoff, R. F. and Stachs, O. 2011. Anterior segment imaging—Present and future. *Klinische Monatsblatter für Augenheilkunde*, 228, 1051.

Haigis, W., Lege, B., Miller, N., and Schneider, B. 2000. Comparison of immersion ultrasound biometry and partial coherence interferometry for intraocular lens calculation according to Haigis. *Graefes Archive for Clinical and Experimental Ophthalmology*, 238, 765–773.

He, L., Sheehy, K., and Culbertson, W. 2011. Femtosecond laser-assisted cataract surgery. *Current Opinion In Ophthalmology*, 22, 43–52.

Hee, M. R., Izatt, J. A., Swanson, E. A., Huang, D., Schuman, J. S., Lin, C. P., Puliafito, C. A., and Fujimoto, J. G. 1995. Optical coherence tomography of the human retina. *Archives of Ophthalmology*, 113, 325–332.

Heur, M. and Dupps, W., Jr. 2009. Anterior Segment Imaging. *In:* Kohnen, T. and Koch, D. (eds.) *Cataract and Refractive Surgery*. Berlin, Germany: Springer.

Hirano, K., Ito, Y., Suzuki, T., Kojima, T., Kachi, S., and Miyake, Y. 2001. Optical coherence tomography for the noninvasive evaluation of the cornea. *Cornea*, 20, 281–289.

Hitzenberger, C. K. 1991. Optical measurement of the axial eye length by laser Doppler interferometry. *Investigative Ophthalmology and Visual Science*, 32, 616–624.

Hitzenberger, C. K. 1992. Measurement of corneal thickness by low-coherence interferometry. *Applied Optics*, 31, 6637–6642.

Ho, J., Castro, D. P. E., Castro, L. C., Chen, Y., Liu, J., Mattox, C., Krishnan, C., Fujimoto, J. G., Schuman, J. S., and Duker, J. S. 2010. Clinical assessment of mirror artifacts in spectral-domain optical coherence tomography. *Investigative Ophthalmology and Visual Science*, 51, 3714–3720.

Hoerauf, H., Wirbelauer, C., Scholz, C., Engelhardt, R., Koch, P., Laqua, H., and Birngruber, R. 2000. Slit-lamp-adapted optical coherence tomography of the anterior segment. *Graefes Archive for Clinical and Experimental Ophthalmology*, 238, 8–18.

Hofer, B., Povazay, B., Hermann, B., Unterhuber, A., Matz, G., and Drexler, W. 2009. Dispersion encoded full range frequency domain optical coherence tomography. *Optics Express*, 17, 7–24.

Hon, Y. and Lam, A. K. 2013. Corneal deformation measurement using Scheimpflug noncontact tonometry. *Optometry and Vision Science*, 90, e1–e8.

Huang, D., Li, Y., and Radhakrishnan, S. 2004. Optical coherence tomography of the anterior segment of the eye. *Ophthalmology Clinics of North America*, 17, 1–6.

Huang, D., Swanson, E. A., Lin, C. P., Schuman, J. S., Stinson, W. G., Chang, W., Hee, M. R. et al. 1991. Optical coherence tomography. *Science*, 254, 1178–1181.

Huber, R., Wojtkowski, M., and Fujimoto, J. G. 2006. Fourier Domain Mode locking (FDML): A new laser operating regime and applications for optical coherence tomography. *Optics Express*, 14, 3225–3237.

Huber, R., Wojtkowski, M., Taira, K., Fujimoto, J. G., and Hsu, K. 2005. Amplified, frequency swept lasers for frequency domain reflectometry and OCT imaging: Design and scaling principles. *Optics Express*, 13, 3513–3528.

Iftimia, N. V., Hammer, D. X., Ferguson, R. D., Mujat, M., Vu, D., and Ferrante, A. A. 2008. Dual-beam Fourier domain optical Doppler tomography of zebrafish. *Optics Express*, 16, 13624–13636.

Ishibazawa, A., Igarashi, S., Hanada, K., Nagaoka, T., Ishiko, S., Ito, H., and Yoshida, A. 2011. Central corneal thickness measurements with Fourier-domain optical coherence tomography versus ultrasonic pachymetry and rotating Scheimpflug camera. *Cornea*, 30, 615–619.

Izatt, J. A., Hee, M. R., Swanson, E. A., Lin, C. P., Huang, D., Schuman, J. S., Puliafito, C. A., and Fujimoto, J. G. 1994. Micrometer-scale resolution imaging of the anterior eye in-vivo with optical coherence tomography. *Archives of Ophthalmology*, 112, 1584–1589.

Jacques, S. L. 2013. Optical properties of biological tissues: A review. *Physics in Medicine and Biology*, 58, R37–R61.

Jayaraman, V., Cole, G. D., Robertson, M., Burgner, C., John, D., Uddin, A., and Cable, A. 2012. Rapidly swept, ultra-widely-tunable 1060 nm MEMS-VCSELs. *Electronics Letters*, 48, 1331–1333.

Jayaraman, V., Jiang, J., Li, H., Heim, P. J. S., Cole, G. D., Potsaid, B., Fujimoto, J. G., and Cable, A. 2011. OCT imaging up to 760 kHz axial scan rate using single-mode 1310 nm MEMS-tunable VCSELs with >100 nm tuning range. *2011 Conference on Lasers and Electro-Optics*, Baltimore, MD. Institute of Electrical and Electronics Engineers, pp. 1–2.

Jayaraman, V., Potsaid, B., Jiang, J., Cole, G. D., Robertson, M. E., Burgner, C. B., John, D. D. et al. 2013. High-speed ultra-broad tuning MEMS-VCSELs for imaging and spectroscopy. *Proceedings of SPIE*, 8763, 87630H.

Jenkins, M. W., Peterson, L., Gu, S., Gargesha, M., Wilson, D. L., Watanabe, M., and Rollins, A. M. 2010. Measuring hemodynamics in the developing heart tube with four-dimensional gated Doppler optical coherence tomography. *Journal of Biomedical Optics*, 15, 066022.

Jeong, H.-W., Lee, S.-W., and Kim, B.-M. 2012. Spectral-domain OCT with dual illumination and interlaced detection for simultaneous anterior segment and retina imaging. *Optics Express*, 20, 19148–19159.

Jorgensen, T. M., Thomadsen, J., Christensen, U., Soliman, W., and Sander, B. 2007. Enhancing the signal-to-noise ratio in ophthalmic optical coherence tomography by image registration—Method and clinical examples. *Journal of Biomedical Optics*, 12, 041208.

Jungwirth, J., Baumann, B., Pircher, M., Goetzinger, E., and Hitzenberger, C. K. 2009. Extended in vivo anterior eye-segment imaging with full-range complex spectral domain optical coherence tomography. *Journal of Biomedical Optics*, 14, 050501.

Kagemann, L., Wollstein, G., Ishikawa, H., Sigal, I. A., Folio, L. S., Xu, J., Gong, H. Y., and Schuman, J. S. 2011. 3D visualization of aqueous humor outflow structures in-situ in humans. *Experimental Eye Research*, 93, 308–315.

Kalev-Landoy, M., Day, A. C., Cordeiro, M. F., and Migdal, C. 2007. Optical coherence tomography in anterior segment imaging. *Acta Ophthalmologica Scandinavica*, 85, 427–430.

Kaluzny, B. J., Gora, M., Karnowski, K., Grulkowski, I., Kowalczyk, A., and Wojtkowski, M. 2010. Imaging of the lens capsule with an ultrahigh-resolution spectral optical coherence tomography prototype based on a femtosecond laser. *British Journal of Ophthalmology*, 94, 275–277.

Kaluzny, B. J., Kaluzny, J. J., Szkulmowska, A., Gorczynska, I., Szkulmowski, M., Bajraszewski, T., Wojtkowski, M., and Targowski, P. 2006. Spectral optical coherence tomography—A novel technique for cornea imaging. *Cornea*, 25, 960–965.

Kaluzny, B. J., Szkulmowska, A., Szkulmowski, M., Bajraszewski, T., Kowalczyk, A., and Wojtkowski, M. 2009. Fuchs' endothelial dystrophy in 830-nm spectral domain optical coherence tomography. *Ophthalmic Surgery Lasers and Imaging*, 40, 198–200.

Kaluzny, B. J., Szkulmowska, A., Szkulmowski, M., Bajraszewski, T., Wawrocka, A., Krawczynski, M. R., Kowalczyk, A., and Wojtkowski, M. 2008. Granular corneal dystrophy in 830-nm spectral optical coherence tomography. *Cornea*, 27, 830–832.

Kaluzny, B. J., Szkulmowski, M., Bukowska, D. M., and Wojtkowski, M. 2014. Spectral OCT with speckle contrast reduction for evaluation of the healing process after PRK and transepithelial PRK. *Biomedical Optics Express*, 5, 1089–1098.

Kaluzny, J. J., Wojtkowski, M., and Kowalczyk, A. 2002. Imaging of the anterior segment of the eye by spectral optical coherence tomography. *Optica Applicata*, 32, 581–589.

Karandish, A., Wirbelauer, C., Haberle, H., and Pham, D. T. 2004. Reproducibility of goniometry with slitlamp-adapted optical coherence tomography. *Ophthalmologe*, 101, 608–613.

Karandish, A., Wirbelauer, C., Haberle, H., and Pham, D. T. 2006. Goniometry with optical coherence tomography in angle-closure glaucoma. *Ophthalmologe*, 103, 35–39.

Karnowski, K., Kaluzny, B. J., Szkulmowski, M., Gora, M., and Wojtkowski, M. 2011. Corneal topography with high-speed swept source OCT in clinical examination. *Biomedical Optics Express*, 2, 2709–2720.

Kennedy, B. F., Kennedy, K. M., and Sampson, D. D. 2014. A review of optical coherence elastography: Fundamentals, techniques and prospects. *IEEE Journal of Selected Topics in quantum Electronics*, 20, 272–288.

Kerbage, C., Lim, H., Sun, W., Mujat, M., and De Boer, J. F. 2007. Large depth-high resolution full 3D imaging of the anterior segments of the eye using high speed optical frequency domain imaging. *Optics Express*, 15, 7117–7125.

Kim, D. Y., Werner, J. S., and Zawadzki, R. J. 2010. Comparison of phase-shifting techniques for in vivo full-range, high-speed Fourier-domain optical coherence tomography. *Journal of Biomedical Optics*, 15, 056011.

Klein, T., Wieser, W., Eigenwillig, C. M., Biedermann, B. R., and Huber, R. 2011. Megahertz OCT for ultrawide-field retinal imaging with a 1050 nm Fourier domain mode-locked laser. *Optics Express*, 19, 3044–3062.

Klein, T., Wieser, W., Reznicek, L., Neubauer, A., Kampik, A., and Huber, R. 2013. Multi-MHz retinal OCT. *Biomedical Optics Express*, 4, 1890–1908.

Kling, S., Akca, I. B., Chang, E. W., Scarcelli, G., Bekesi, N., Yun, S.-H., and Marcos, S. 2014a. Numerical model of optical coherence tomographic vibrography imaging to estimate corneal biomechanical properties. *Journal of The Royal Society Interface*, 1 20140920.

Kling, S., Bekesi, N., Dorronsoro, C., Pascual, D., and Marcos, S. 2014b. Corneal viscoelastic properties from finite-element analysis of in vivo air-puff deformation. *PLoS ONE*, 9, e104904.

Kobler, J. B., Chang, E. W., Zeitels, S. M., and Yun, S.-H. 2010. Dynamic imaging of vocal fold oscillation with four-dimensional optical coherence tomography. *Laryngoscope*, 120, 1354–1362.

Koh, S., Tung, C., Aquavella, J., Yadav, R., Zavislan, J., and Yoon, G. 2010. Simultaneous measurement of tear film dynamics using wavefront sensor and optical coherence tomography. *Investigative Ophthalmology and Visual Science*, 51, 3441–3448.

Konstantopoulos, A., Hossain, P., and Anderson, D. F. 2007. Recent advances in ophthalmic anterior segment imaging: A new era for ophthalmic diagnosis? *British Journal of Ophthalmology*, 91, 551–557.

Koop, N., Brinkmann, R., Lankenau, E., Flache, S., Engelhardt, R., and Birngruber, R. 1997. Optical coherence tomography of cornea and anterior segment of the eye. *Ophthalmologe*, 94, 481–486.

Kraus, M. F., Potsaid, B., Mayer, M. A., Bock, R., Baumann, B., Liu, J. J., Hornegger, J., and Fujimoto, J. G. 2012. Motion correction in optical coherence tomography volumes on a per A-scan basis using orthogonal scan patterns. *Biomedical Optics Express*, 3, 1182–1199.

Langner, S., Martin, H., Terwee, T., Koopmans, S. A., Kruger, P. C., Hosten, N., Schmitz, K. P., Guthoff, R. F., and Stachs, O. 2010. 7.1 T MRI to assess the anterior segment of the eye. *Investigative Ophthalmology and Visual Science*, 51, 6575–6581.

Leitgeb, R., Hitzenberger, C. K., and Fercher, A. F. 2003. Performance of fourier domain vs. time domain optical coherence tomography. *Optics Express*, 11, 889–894.

Leitgeb, R. A., Michaely, R., Lasser, T., and Sekhar, S. C. 2007. Complex ambiguity-free Fourier domain optical coherence tomography through transverse scanning. *Optics Letters*, 32, 3453–3455.

Leng, L., Yuan, Y. M., Chen, Q., Shen, M. X., Ma, Q. K., Lin, B. B., Zhu, D. X., Qu, J., and Lu, F. 2014. Biometry of anterior segment of human eye on both horizontal and vertical meridians during accommodation imaged with extended scan depth optical coherence tomography. *PLoS ONE*, 9, 0104775.

Leung, C. K.-S., Cheung, C. Y. L., Li, H., Dorairaj, S., Yiu, C. K. F., Wong, A. L., Liebmann, J., Ritch, R., Weinreb, R., and Lam, D. S. C. 2007. Dynamic analysis of dark–light changes of the anterior chamber angle with anterior segment OCT. *Investigative Ophthalmology and Visual Science*, 48, 4116–4122.

Leung, C. K. S. and Weinreb, R. N. 2011. Anterior chamber angle imaging with optical coherence tomography. *Eye*, 25, 261–267.

Li, H., Jhanji, V., Dorairaj, S., Liu, A., Lam, D. S., and Leung, C. K. 2012. Anterior segment optical coherence tomography and its clinical applications in glaucoma. *Journal of Current Glaucoma Practice*, 6, 68–74.

Li, P., An, L., Lan, G., Johnstone, M., Malchow, D., and Wang, R. K. 2013a. Extended imaging depth to 12 mm for 1050-nm spectral domain optical coherence tomography for imaging the whole anterior segment of the human eye at 120-kHz A-scan rate. *Journal of Biomedical Optics*, 18, 016012.

Li, P., An, L., Reif, R., Shen, T. T., Johnstone, M., and Wang, R. K. 2011. In vivo microstructural and microvascular imaging of the human corneo-scleral limbus using optical coherence tomography. *Biomedical Optics Express*, 2, 3109–3118.

Li, P., Johnstone, M., and Wang, R. K. K. 2014. Full anterior segment biometry with extended imaging range spectral domain optical coherence tomography at 1340 nm. *Journal of Biomedical Optics*, 19, 046013.

Li, P., Shen, T. T., Johnstone, M., and Wang, R. K. 2013b. Pulsatile motion of the trabecular meshwork in healthy human subjects quantified by phase-sensitive optical coherence tomography. *Biomedical Optics Express*, 4, 2051–2065.

Li, Y., Meisler, D. M., Tang, M., Lu, A. T. H., Thakrar, V., Reiser, B. J., and Huang, D. 2008. Keratoconus diagnosis with optical coherence tomography pachymetry mapping. *Ophthalmology*, 115, 2159–2166.

Li, Y., Shekhar, R., and Huang, D. 2006. Corneal pachymetry mapping with high-speed optical coherence tomography. *Ophthalmology*, 113, 792–799.

Li, Y., Tang, M. L., Zhang, X. B., Salaroli, C. H., Ramos, J. L., and Huang, D. 2010. Pachymetric mapping with Fourier-domain optical coherence tomography. *Journal of Cataract and Refractive Surgery*, 36, 826–831.

Lim, Y., Yamanari, M., Fukuda, S., Kaji, Y., Kiuchi, T., Miura, M., Oshika, T., and Yasuno, Y. 2011. Birefringence measurement of cornea and anterior segment by office-based polarization-sensitive optical coherence tomography. *Biomedical Optics Express*, 2, 2392–2402.

Liu, J. J., Grulkowski, I., Potsaid, B., Jayaraman, V., Cable, A. E., Kraus, M. F., Hornegger, J., Duker, J. S., and Fujimoto, J. G. 2013. 4D dynamic imaging of the eye using ultrahigh speed SS-OCT. *Proceedings of SPIE*, 8567, 85670X.

Low, S. and Nolan, W. 2009. Anterior segment imaging for glaucoma: Where are we and what next? *Clinical and Experimental Ophthalmology*, 37, 431–433.

Lubatschowski, H. 2008. Overview of commercially available femtosecond lasers in refractive surgery. *Journal of Refractive Surgery*, 24, S102–S107.

Luce, D. A. 2005. Determining in vivo biomechanical properties of the cornea with an ocular response analyzer. *Journal of Cataract and Refractive Surgery*, 31, 156–162.

Maintz, J. B. and Viergever, M. A. 1998. A survey of medical image registration. *Medical Image Analysis*, 2, 1–36.

Makita, S., Jaillon, F., Yamanari, M., Miura, M., and Yasuno, Y. 2011. Comprehensive in vivo micro-vascular imaging of the human eye by dual-beam-scan Doppler optical coherence angiography. *Optics Express*, 19, 1271–1283.

Meinhardt, B., Stachs, O., Stave, J., Beck, R., and Guthoff, R. 2006. Evaluation of biometric methods for measuring the anterior chamber depth in the non-contact mode. *Graefes Archive for Clinical and Experimental Ophthalmology*, 244, 559–564.

Mejia-Barbosa, Y. and Malacara-Hernandez, D. 2001. A review of methods for measuring corneal topography. *Optometry and Vision Science*, 78, 240–253.

Minneman, M. P., Ensher, J., Crawford, M., and Derickson, D. 2011. All-semiconductor high-speed akinetic swept-source for OCT. *Proceedings of SPIE*, 8311, 831116.

Miyazawa, A., Yamanari, M., Makita, S., Miura, M., Kawana, K., Iwaya, K., Goto, H., and Yasuno, Y. 2009. Tissue discrimination in anterior eye using three optical parameters obtained by polarization sensitive optical coherence tomography. *Optics Express*, 17, 17426–17440.

Mylonas, G., Sacu, S., Buehl, W., Ritter, M., Georgopoulos, M., and Schmidt-Erfurth, U. 2011. Performance of three biometry devices in patients with different grades of age-related cataract. *Acta Ophthalmologica*, 89, e237–e241.

Nagy, Z., Takacs, A., Filkorn, T., and Sarayba, M. 2009. Initial clinical evaluation of an intraocular femtosecond laser in cataract surgery. *Journal of Refractive Surgery*, 25, 1053–1060.

Nagy, Z. Z., Filkorn, T., Takács, Á. I., Kránitz, K., Juhasz, T., Donnenfeld, E., Knorz, M. C., and Alio, J. L. 2013. Anterior segment OCT imaging after femtosecond laser cataract surgery. *Journal of Refractive Surgery*, 29, 110–112.

Nahas, A., Bauer, M., Roux, S., and Boccara, A. C. 2013. 3D static elastography at the micrometer scale using Full Field OCT. *Biomedical Optics Express*, 4, 2138–2149.

Nankivil, D., Dhalla, A., Gahm, N., Shia, K., Farsiu, S., and Izatt, J. A. 2014. Coherence revival multiplexed, buffered swept source optical coherence tomography: 400 kHz imaging with a 100 kHz source. *Optics Letters*, 39, 3740–3743.

Nolan, W. 2008a. Anterior segment imaging: Identifying the landmarks. *British Journal of Ophthalmology*, 92, 1575–1576.

Nolan, W. 2008b. Anterior segment imaging: Ultrasound biomicroscopy and anterior segment optical coherence tomography. *Current Opinion in Ophthalmology*, 19, 115–121.

Oh, W. Y., Vakoc, B. J., Shishkov, M., Tearney, G. J., and Bouma, B. E. 2010. >400 kHz repetition rate wavelength-swept laser and application to high-speed optical frequency domain imaging. *Optics Letters*, 35, 2919–2921.

Okabe, Y., Sasaki, Y., Ueno, M., Sakamoto, T., Toyoda, S., Yagi, S., Naganuma, K. et al. 2012. 200 kHz swept light source equipped with KTN deflector for optical coherence tomography. *Electronics Letters*, 48, 201–202.

Oliveira, C. M., Ribeiro, C., and Franco, S. 2011. Corneal imaging with slit-scanning and Scheimpflug imaging techniques. *Clinical and Experimental Optometry*, 94, 33–42.

Olsen, T. 2007. Calculation of intraocular lens power: A review. *Acta Ophthalmologica Scandinavica*, 85, 472–485.

Ortiz, S., Siedlecki, D., Grulkowski, I., Remon, L., Pascual, D., Wojtkowski, M., and Marcos, S. 2010. Optical distortion correction in Optical Coherence Tomography for quantitative ocular anterior segment by three-dimensional imaging. *Optics Express*, 18, 2782–2796.

Packer, M., Fine, I. H., Hoffman, R. S., Coffman, P. G., and Brown, L. K. 2002. Immersion A-scan compared with partial coherence interferometry—Outcomes analysis. *Journal of Cataract and Refractive Surgery*, 28, 239–242.

Park, H.-Y. L., Lee, N. Y., Choi, J. A., and Park, C. K. 2014. Measurement of scleral thickness using swept-source optical coherence tomography in patients with open-angle glaucoma and myopia. *American Journal of Ophthalmology*, 157, 876–884.

Perkins, E. S., Hammond, B., and Milliken, A. B. 1976. Simple method of determining axial length of eye. *British Journal of Ophthalmology*, 60, 266–270.

Pinero, D. P. 2015. Technologies for anatomical and geometric characterization of the corneal structure and anterior segment: A review. *Seminars in Ophthalmology*, 30, 161–170.

Pinero, D. P., Plaza, A. B., and Alio, J. L. 2008. Anterior segment biometry with 2 imaging technologies: Very-high-frequency ultrasound scanning versus optical coherence tomography. *Journal of Cataract & Refractive Surgery*, 34, 95–102.

Pircher, M., Hitzenberger, C. K., and Schmidt-Erfurth, U. 2011. Polarization sensitive optical coherence tomography in the human eye. *Progress in Retinal and Eye Research*, 30, 431–451.

Podoleanu, A., Charalambous, I., Plesea, L., Dogariu, A., and Rosen, R. 2004. Correction of distortions in optical coherence tomography imaging of the eye. *Physics in Medicine and Biology*, 49, 1277–1294.

Potsaid, B., Baumann, B., Huang, D., Barry, S., Cable, A. E., Schuman, J. S., Duker, J. S., and Fujimoto, J. G. 2010. Ultrahigh speed 1050 nm swept source/Fourier domain OCT retinal and anterior segment imaging at 100,000 to 400,000 axial scans per second. *Optics Express*, 18, 20029–20048.

Potsaid, B., Gorczynska, I., Srinivasan, V. J., Chen, Y., Jiang, J., Cable, A., and Fujimoto, J. G. 2008. Ultrahigh speed Spectral/Fourier domain OCT ophthalmic imaging at 70,000 to 312,500 axial scans per second. *Optics Express*, 16, 15149–15169.

Potsaid, B., Jayaraman, V., Fujimoto, J. G., Jiang, J., Heim, P. J. S., and Cable, A. E. 2012. MEMS tunable VCSEL light source for ultrahigh speed 60 kHz-1 MHz axial scan rate and long range centimeter class OCT imaging. *Proceedings of SPIE*, 8213, 82130M.

Radhakrishnan, S., Goldsmith, J., Huang, D., Westphal, V., Dueker, D. K., Rollins, A. M., Izatt, J. A., and Smith, S. D. 2005. Comparison of optical coherence tomography and ultrasound biomicroscopy for detection of narrow anterior chamber angles. *Archives of Ophthalmology*, 123, 1053–1059.

Radhakrishnan, S., Rollins, A. M., Roth, J. E., Yazdanfar, S., Westphal, V., Bardenstein, D. S., and Izatt, J. A. 2001. Real-time optical coherence tomography of the anterior segment at 1310 nm. *Archives of Ophthalmology*, 119, 1179–1185.

Radhakrishnan, S., See, J., Smith, S. D., Nolan, W. P., Ce, Z., Friedman, D. S., Huang, D., Li, Y., Aung, T., and Chew, P. T. 2007. Reproducibility of anterior chamber angle measurements obtained with anterior segment optical coherence tomography. *Investigative Ophthalmology and Visual Science*, 48, 3683–3688.

Radhakrishnan, S. and Yarovoy, D. 2014. Development in anterior segment imaging for glaucoma. *Current Opinion in Ophthalmology*, 25, 98–103.

Rajan, M. S., Keilhorn, I., and Bell, J. A. 2002. Partial coherence laser interferometry vs conventional ultrasound biometry in intraocular lens power calculations. *Eye*, 16, 552–556.

Rey, S. M., Povazay, B., Hofer, B., Unterhuber, A., Hermann, B., Harwood, A., and Drexler, W. 2009. Three- and four-dimensional visualization of cell migration using optical coherence tomography. *Journal of Biophotonics*, 2, 370–379.

Ricco, S., Chen, M., Ishikawa, H., Wollstein, G., and Schuman, J. 2009. Correcting motion artifacts in retinal spectral domain optical coherence tomography via image registration. *In:* Yang, G. Z., Hawkes, D., Rueckert, D., Nobel, A., and Taylor, C. (eds.), *Proceedings of Medical Image Computing and Computer-Assisted Intervention—MICCAI 2009: Part I*, Berlin, Germany. Berlin, Germany: Springer-Verlag, pp. 100–107.

Richdale, K., Bullimore, M. A., and Zadnik, K. 2008. Lens thickness with age and accommodation by optical coherence tomography. *Ophthalmic and Physiological Optics*, 28, 441–447.

Rodrigues, E. B., Johanson, M., and Penha, F. M. 2012. Anterior segment tomography with the cirrus optical coherence tomography. *Journal of Ophthalmology*, 2012, 806989.

Rohrer, K., Frueh, B. E., Walti, R., Clemetson, I. A., Tappeiner, C., and Goldblum, D. 2009. Comparison and evaluation of ocular biometry using a new noncontact optical low-coherence reflectometer. *Ophthalmology*, 116, 2087–2092.

Rosales, P., Dubbelman, M., Marcos, S., and Van der Heijde, R. 2006. Crystalline lens radii of curvature from Purkinje and Scheimpflug imaging. *Journal of Vision*, 6, 1057–1067.

Rosales, P. and Marcos, S. 2009. Pentacam Scheimpflug quantitative imaging of the crystalline lens and intraocular lens. *Journal of Refractive Surgery*, 25, 421–428.

Ruggeri, M., Uhlhorn, S. R., De Freitas, C., Ho, A., Manns, F., and Parel, J. M. 2012. Imaging and full-length biometry of the eye during accommodation using spectral domain OCT with an optical switch. *Biomedical Optics Express*, 3, 1506–1520.

Sardar, D., Swanland, G.-Y., Yow, R., Thomas, R., and Tsin, A. C. 2007. Optical properties of ocular tissues in the near infrared region. *Lasers in Medical Science*, 22, 46–52.

Sarunic, M. V., Applegate, B. E., and Izatt, J. A. 2006. Real-time quadrature projection complex conjugate resolved Fourier domain optical coherence tomography. *Optics Letters*, 31, 2426–2428.

Sarunic, M. V., Asrani, S., and Izatt, J. A. 2008. Imaging the ocular anterior segment with real-time, full-range Fourier-Domain optical coherence tomography. *Archives of Ophthalmology*, 126, 537–542.

Sarunic, M. V., Choma, M. A., Yang, C. H., and Izatt, J. A. 2005. Instantaneous complex conjugate resolved spectral domain and swept-source OCT using 3 × 3 fiber couplers. *Optics Express*, 13, 957–967.

Scarcelli, G. and Yun, S. H. 2008. Confocal Brillouin microscopy for three-dimensional mechanical imaging. *Nature Photonics*, 2, 39–43.

Schachar, R. A. 1994. Zonular function: A new hypothesis with clinical implications. *Annals of Ophthalmology*, 26, 36–38.

Schmitt, J. 1998. OCT elastography: Imaging microscopic deformation and strain of tissue. *Optics Express*, 3, 199–211.

See, J. L. S. 2009. Imaging of the anterior segment in glaucoma. *Clinical and Experimental Ophthalmology*, 37, 506–513.

Shammas, H. J. and Hoffer, K. J. 2012. Repeatability and reproducibility of biometry and keratometry measurements using a noncontact optical low-coherence reflectometer and keratometer. *American Journal of Ophthalmology*, 153, 55–61.

Shao, Y. L., Tao, A. Z., Jiang, H., Shen, M. X., Zhong, J. G., Lu, F., and Wang, J. H. 2013. Simultaneous real-time imaging of the ocular anterior segment including the ciliary muscle during accommodation. *Biomedical Optics Express*, 4, 466–480.

Shapiro, B. L., Cortes, D. E., Chin, E. K., Li, J. Y., Werner, J. S., Redenbo, E., and Mannis, M. J. 2013. High-resolution spectral domain anterior segment optical coherence tomography in type 1 boston keratoprosthesis. *Cornea*, 32, 951–955.

Shen, M., Wang, M. R., Yuan, Y., Chen, F., Karp, C. L., Yoo, S. H., and Wang, J. 2010. SD-OCT with prolonged scan depth for imaging the anterior segment of the eye. *Ophthalmic Surgery Lasers and Imaging*, 41, S65–S69.

Shi, G., Wang, Y., Yuan, Y., Wei, L., Lv, F., and Zhang, Y. 2012. Measurement of ocular anterior segment dimension and wavefront aberration simultaneously during accommodation. *Journal of Biomedical Optics*, 17, 120501.

Siedlecki, D., De Castro, A., Gambra, E., Ortiz, S., Borja, D., Uhlhorn, S., Manns, F., Marcos, S., and Parel, J.-M. 2012. Distortion correction of OCT images of the crystalline lens: Gradient index approach. *Optometry and Vision Science*, 89, E709–E718.

Sorbara, L., Maram, J., Fonn, D., Woods, C., and Simpson, T. 2010. Metrics of the normal cornea: Anterior segment imaging with the Visante OCT. *Clinical and Experimental Optometry*, 93, 150–156.

Spaide, R. F., Koizumi, H., and Pozonni, M. C. 2008. Enhanced depth imaging spectral-domain optical coherence tomography. *American Journal of Ophthalmology*, 146, 496–500.

Stehouwer, M., Verbraak, F. D., De Vries, H., Kok, P. H. B., and van Leeuwen, T. G. 2010. Fourier Domain Optical Coherence Tomography integrated into a slit lamp; a novel technique combining anterior and posterior segment OCT. *Eye*, 24, 980–984.

Steinert, R. F. and Huang, D. (eds.) 2008. *Anterior Segment Optical Coherence Tomography*, Thorofare, NJ: SLACK Incorporated.

Steven, P., Le Blanc, C., Lankenau, E., Krug, M., Oelckers, S., Heindl, L. M., Gehlsen, U., Huettmann, G., and Cursiefen, C. 2014. Optimising deep anterior lamellar keratoplasty (DALK) using intraoperative online optical coherence tomography (iOCT). *British Journal of Ophthalmology*, 98, 900–904.

Steven, P., Le Blanc, C., Velten, K., Lankenau, E., Krug, M., Oelckers, S., Heindl, L. M., Gehlsen, U., Hüttmann, G., and Cursiefen, C. 2013. Optimizing descemet membrane endothelial keratoplasty using intraoperative optical coherence tomography. *JAMA Ophthalmology*, 131, 1135–1142.

Sun, M., Birkenfeld, J., De Castro, A., and Ortiz, S. 2014. OCT 3-D surface topography of isolated human crystalline lenses. *Biomedical Optics Express*, 5, 3547–3561.

Sun, Y.-C., Li, P., Johnstone, M., Wang, R. K., and Shen, T. T. 2015. Pulsatile motion of trabecular meshwork in a patient with iris cyst by phase-sensitive optical coherence tomography: A case report. *Quantitative Imaging in Medicine and Surgery*, 5, 171–173.

Swanson, E. A. 2015. OCT technology transfer and the OCT market. *In:* Drexler, W. and Fujimoto, J. G. (eds.) *Optical Coherence Tomography*, Cham, Switzerland: Springer International Publishing.

Swanson, E. A., Huang, D., Hee, M. R., Fujimoto, J. G., Lin, C. P., and Puliafito, C. A. 1992. High-speed optical coherence domain reflectometry. *Optics Letters*, 17, 151–153.

Sylwestrzak, M., Szlag, D., Szkulmowski, M., Gorczynska, I., Bukowska, D., Wojtkowski, M., and Targowski, P. 2012. Four-dimensional structural and Doppler optical coherence tomography imaging on graphics processing units. *Journal of Biomedical Optics*, 17, 100502.

Tang, M. L., Chen, A., Li, Y., and Huang, D. 2010a. Corneal power measurement with Fourier-domain optical coherence tomography. *Journal of Cataract and Refractive Surgery*, 36, 2115–2122.

Tang, M. L., Li, Y., Avila, M., and Huang, D. 2006. Measuring total corneal power before and after laser in situ keratomileusis with high-speed optical coherence tomography. *Journal of Cataract and Refractive Surgery*, 32, 1843–1850.

Tang, M. L., Li, Y., and Huang, D. 2010b. An intraocular lens power calculation formula based on optical coherence tomography: A pilot study. *Journal of Refractive Surgery*, 26, 430–437.

Tanter, M., Touboul, D., Gennisson, J. L., Bercoff, J., and Fink, M. 2009. High-resolution quantitative imaging of cornea elasticity using supersonic shear imaging. *IEEE Transactions on Medical Imaging*, 28, 1881–1893.

Tao, A., Shao, Y., Zhong, J., Jiang, H., Shen, M., and Wang, J. 2013. Versatile optical coherence tomography for imaging the human eye. *Biomedical Optics Express*, 4, 1031–1044.

Tao, Y. K., Srivastava, S. K., and Ehlers, J. P. 2014. Microscope-integrated intraoperative OCT with electrically tunable focus and heads-up display for imaging of ophthalmic surgical maneuvers. *Biomedical Optics Express*, 5, 1877–1885.

Tao, Y. K., Zhao, M., and Izatt, J. A. 2007. High-speed complex conjugate resolved retinal spectral domain optical coherence tomography using sinusoidal phase modulation. *Optics Letters*, 32, 2918–2920.

Tearney, G. J., Bouma, B. E., and Fujimoto, J. G. 1997. High-speed phase- and group-delay scanning with a grating-based phase control delay line. *Optics Letters*, 22, 1811–1813.

Tian, J., Marziliano, P., Baskaran, M., Wong, H. T., and Aung, T. 2011. Automatic anterior chamber angle assessment for HD-OCT images. *IEEE Transactions on Biomedical Engineering*, 58, 3242–3249.

Tomita, M., Chiba, A., Matsuda, J., and Nawa, Y. 2012. Evaluation of LASIK treatment with the Femto LDV in patients with corneal opacity. *Journal of Refractive Surgery*, 28, 25–30.

Totsuka, K., Isamoto, K., Sakai, T., Morosawa, A., and Chong, C. H. 2010. MEMS scanner based swept source laser for optical coherence tomography. *Proceedings of SPIE*, 7554, 75542Q.

Townsend, K. A., Wollstein, G., and Schuman, J. S. 2008. Clinical application of MRI in ophthalmology. *NMR in Biomedicine*, 21, 997–1002.

Tscherning, M. 1899. The theory of accommodation. *Ophthalmic Review*, 18, 91–99.

Ursea, R. and Silverman, R. H. 2010. Anterior-segment imaging for assessment of glaucoma. *Expert Review of Ophthalmology*, 5, 59–74.

Vabre, L., Dubois, A., and Boccara, A. C. 2002. Thermal-light full-field optical coherence tomography. *Optics Letters*, 27, 530–532.

van den Berg, T. J. T. P. and Spekreijse, H. 1997. Near infrared light absorption in the human eye media. *Vision Research*, 37, 249–253.

Vogel, A., Dick, H. B., and Krummenauer, F. 2001. Reproducibility of optical biometry using partial coherence interferometry—Intraobserver and interobserver reliability. *Journal of Cataract and Refractive Surgery*, 27, 1961–1968.

von Helmholtz, H. 1855. Über die Akkommodation des Auges. *Albrecht von Graefes Archiv für Klinische und Experimentelle Ophthalmologie*, 1, 1–74.

Wang, J., Shousha, M. A., Perez, V. L., Karp, C. L., Yoo, S. H., Shen, M., Cui, L. et al. 2011. Ultra-high resolution optical coherence tomography for imaging the anterior segment of the eye. *Ophthalmic Surgery Lasers and Imaging*, 42, S15–S27.

Wang, J. H., Fonn, D., Simpson, T. L., and Jones, L. 2003. Precorneal and pre- and postlens tear film thickness measured indirectly with optical coherence tomography. *Investigative Ophthalmology and Visual Science*, 44, 2524–2528.

Wang, S. and Larin, K. V. 2014. Noncontact depth-resolved micro-scale optical coherence elastography of the cornea. *Biomedical Optics Express*, 5, 3807–3821.

Werkmeister, R. M., Alex, A., Kaya, S., Unterhuber, A., Hofer, B., Riedl, J., Bronhagl, M. et al. 2013. Measurement of tear film thickness using ultrahigh-resolution optical coherence tomography. *Investigative Ophthalmology and Visual Science*, 54, 5578–5583.

Westphal, V., Rollins, A. M., Radhakrishnan, S., and Izatt, J. A. 2002. Correction of geometric and refractive image distortions in optical coherence tomography applying Fermat's principle. *Optics Express*, 10, 397–404.

Wieser, W., Draxinger, W., Klein, T., Karpf, S., Pfeiffer, T., and Huber, R. 2014. High definition live 3D-OCT in vivo: Design and evaluation of a 4D OCT engine with 1 GVoxel/s. *Biomedical Optics Express*, 5, 2963–2977.

Wieser, W., Klein, T., Adler, D. C., Trépanier, F., Eigenwillig, C. M., Karpf, S., Schmitt, J. M., and Huber, R. 2012. Extended coherence length megahertz FDML and its application for anterior segment imaging. *Biomedical Optics Express*, 3, 2647–2657.

Wirbelauer, C., Scholz, C., Hoerauf, H., Engelhardt, R., Birngruber, R., and Laqua, H. 2000. Corneal optical coherence tomography before and immediately after excimer laser photorefractive keratectomy. *American Journal of Ophthalmology*, 130, 693–699.

Wirbelauer, C., Scholz, C., Hoerauf, H., Thoai Pham, D., Laqua, H., and Birngruber, R. 2002. Noncontact corneal pachymetry with slit lamp-adapted optical coherence tomography. *American Journal of Ophthalmology*, 133, 444–450.

Wojtkowski, M. 2010. High-speed optical coherence tomography: Basics and applications. *Applied Optics*, 49, D30–D61.

Wojtkowski, M., Kaluzny, B., and Zawadzki, R. J. 2012. New directions in ophthalmic optical coherence tomography. *Optometry and Vision Science*, 89, 524–542.

Wojtkowski, M., Kowalczyk, A., Leitgeb, R., and Fercher, A. F. 2002a. Full range complex spectral optical coherence tomography technique in eye imaging. *Optics Letters*, 27, 1415–1417.

Wojtkowski, M., Leitgeb, R., Kowalczyk, A., Fercher, A. F., and Bajraszewski, T. 2002b. In vivo human retinal imaging by Fourier domain optical coherence tomography. *Journal of Biomedical Optics*, 7, 457–463.

Wolffsohn, J. S. and Davies, L. N. 2007. Advances in anterior segment imaging. *Current Opinion in Ophthalmology*, 18, 32–38.

Wolffsohn, J. S. and Peterson, R. C. 2006. Anterior ophthalmic imaging. *Clinical and Experimental Optometry*, 89, 205–214.

Wylegala, E., Teper, S., Nowinska, A. K., Milka, M., and Dobrowolski, D. 2009. Anterior segment imaging: Fourier-domain optical coherence tomography versus time-domain optical coherence tomography. *Journal of Cataract and Refractive Surgery*, 35, 1410–1414.

Yasuno, Y., Madjarova, V. D., Makita, S., Akiba, M., Morosawa, A., Chong, C., Sakai, T., Chan, K. P., Itoh, M., and Yatagai, T. 2005. Three-dimensional and high-speed swept-source optical coherence tomography for in vivo investigation of human anterior eye segments. *Optics Express*, 13, 10652–10664.

Yun, S. H., Boudoux, C., Pierce, M. C., de Boer, J. F., Tearney, G. J., and Bouma, B. E. 2004a. Extended-cavity semiconductor wavelength-swept laser for biomedical imaging. *IEEE Photonics Technology Letters*, 16, 293–295.

Yun, S. H., Boudoux, C., Tearney, G. J., and Bouma, B. E. 2003a. High-speed wavelength-swept semiconductor laser with a polygon-scanner-based wavelength filter. *Optics Letters*, 28, 1981–1983.

Yun, S. H., Tearney, G. J., de Boer, J. F., and Bouma, B. E. 2004b. Motion artifacts in optical coherence tomography with frequency domain ranging. *Optics Express*, 12, 2977–2998.

Yun, S. H., Tearney, G. J., de Boer, J. F., Iftimia, N., and Bouma, B. E. 2003b. High-speed optical frequency-domain imaging. *Optics Express*, 11, 2953–2963.

Zhao, M. T., Kuo, A. N., and Izatt, J. A. 2010. 3D refraction correction and extraction of clinical parameters from spectral domain optical coherence tomography of the cornea. *Optics Express*, 18, 8923–8936.

Zhong, J., Tao, A., Xu, Z., Jiang, H., Shao, Y., Zhang, H., Liu, C., and Wang, J. 2014. Whole eye axial biometry during accommodation using ultra-long scan depth optical coherence tomography. *American Journal of Ophthalmology*, 157, 1064–1069.

Zhou, C. Q., Wang, J. H., and Jiao, S. L. 2009. Dual channel dual focus optical coherence tomography for imaging accommodation of the eye. *Optics Express*, 17, 8947–8955.

Zotter, S., Pircher, M., Torzicky, T., Bonesi, M., Gotzinger, E., Leitgeb, R. A., and Hitzenberger, C. K. 2011. Visualization of microvasculature by dual-beam phase-resolved Doppler optical coherence tomography. *Optics Express*, 19, 1217–1227.

5 Adaptive optics ophthalmoscopes

Zoran Popovic

Contents

5.1 INTRODUCTION

The optics of the human eye have been studied by learned men for many centuries, one of which was Ibn al-Haytham (known in the West as Alhazen), the most influential figure in the study of optics and vision in the Middle Ages (Alhazen and Sabra 1989). Although eyeglasses were introduced in the thirteenth century, it was not until the seventeenth century that Johannes Kepler formulated an accurate theory and the correction of myopia (positive defocus or nearsightedness) and hyperopia (negative defocus or farsightedness). Two centuries later, George Airy devised a cylindrical lens for the correction of astigmatism (Airy 1827; Levene 1966).

All optical systems are affected by optical aberrations that degrade imaging performance. Correcting the lower-order aberrations of the eye's optics such as defocus and astigmatism provides most people with good vision. However, in order to better understand the function of the normal retina and the

pathophysiology of retinal disease, one needs to look into the eye. This can, of course, be accomplished by studying histological preparations *ex vivo*, but the retina has to be studied *in vivo* in order to diagnose and monitor retinal disease and evaluate efficacy of medical treatment. This was facilitated by the independent inventions of the ophthalmoscope by Charles Babbage in 1847 and the indirect ophthalmoscope by Hermann von Helmholtz in 1851. The functionality and usability of the ophthalmoscope were greatly advanced by the invention of the handheld direct-illuminating ophthalmoscope by Francis Welch and William Allyn in 1915. Modern-day versions of the Welch–Allyn ophthalmoscope are routinely used in eye examinations today.

The fundus camera, introduced in the early twentieth century by the Zeiss company in Germany (Van Cader 1978), essentially combined an ophthalmoscope with a camera and allowed ophthalmologists to image the living human retina. Today, several other modalities, for example, scanning laser ophthalmoscopy (SLO) and optical coherence tomography (OCT), are also used to routinely image the living retina in clinical practice. Because the optics of the eye also induce higher-order aberrations and the majority of these modalities only correct for defocus, they are limited in their ability to image the finer details of the living retina. Retinal structures such as cones, rods, and the smallest retinal blood vessels (capillaries) cannot be resolved, and significant structural damage has often occurred by the time pathology becomes visible. It is therefore necessary to measure and compensate for both lower- and higher-order aberrations in order to overcome this resolution limit and image the smallest retinal structures that are of the order of a few microns in size. The introduction of adaptive optics (AO) into ophthalmic photography has made this possible.

A vision science AO system, similar to AO systems used in astronomy and communications, comprises four main components: (1) a retinal beacon light source, (2) a wavefront sensor (WFS), (3) a control computer, and (4) a wavefront corrector (WFC). A system with these four components can dynamically measure and correct ocular aberrations. Enhanced functionality can be obtained by adding, for example, a digital camera for retinal imaging or a display for visual function testing.

AO has been successfully integrated with three major ophthalmic imaging modalities: (1) flood-illumination fundus cameras that take short-exposure images of the retina, (2) confocal SLO systems that acquire an image by rapidly scanning a point source across the retina, and (3) OCT systems that also scan a point source across the retina but generate a 3D image using low-coherence interferometry.

The focus of this chapter is the design and functionality of flood-illumination AO ophthalmoscopes. Only a brief overview will be given of AO-SLO and AO-OCT systems and the reader will be referred to other chapters in this book for detailed discussions of these modalities.

5.2 AO OPHTHALMOSCOPE DESIGN

The first closed-loop flood-illuminated AO fundus camera capable of correcting higher-order aberrations in the eye was constructed in David Williams lab at the University of Rochester (Liang et al. 1997). This system was soon followed by automated real-time AO systems capable of aberration measurement and correction (Fernandez et al. 2001), flood-illumination imaging (Hofer et al. 2001), AO-SLO imaging (Roorda et al. 2002), and AO-OCT imaging (Hermann et al. 2004). This successful integration of AO with the major ophthalmic imaging modalities has since led to a plethora of custom-built laboratory-based research systems. There are several publications available that explain the basic concepts of AO systems and also give guidance on engineering tools, improved optical designs, and tricks of the trade that are needed to develop your own AO system (Glanc et al. 2004; Gomez-Vieyra et al. 2009; Hampson 2008; Miller et al. 2005; Porter 2006).

5.2.1 BRIEF OVERVIEW OF DIFFERENT AO INSTRUMENT DESIGNS

This section will briefly discuss different designs of AO imaging instruments, each offering different benefits, such as flood-illuminated AO, AO-SLO, and AO-OCT systems.

5.2.1.1 Flood-illumination AO

A flood-illumination AO ophthalmoscope (Liang et al. 1997) has two main advantages over scanning systems such as AO-SLO and AO-OCT; it is conceptually the simplest one with generally lower optical and mechanical system complexity, and image artifacts caused by natural eye movements can be mitigated by using short-exposure times. A potential drawback of flood-illumination AO ophthalmoscopes using a flash as the imaging light source is that the charge time of the flash can limit imaging frame rates. This can be avoided by using a pulsed or shutter-controlled light source, for example, a superluminescent diode (SLD), a light-emitting diode (LED), or a laser diode (LD) in combination with a high-speed camera (Glanc et al. 2007; Jonnal et al. 2007; Rha et al. 2006, 2009).

5.2.1.2 AO-SLO

The retinal image in an AO-SLO (Roorda et al. 2002) is created by scanning a small point source of light over the retina in a raster fashion, typically at video rates (25–30 Hz), and recording the reflected scattered light with a photomultiplier tube. The benefit of AO in such a system is that the focused spot of light can be made very small by correcting optical aberrations over a large entrance pupil. Image contrast is improved by placing a confocal spatial filter (pinhole) in a plane confocal to the retinal plane just in front of the detector. Light returning from the focal plane will pass through the pinhole and light returning from out of focus planes will be blocked, leading to improved axial resolution and contrast. The pinhole also blocks scattered light, yielding further contrast improvement. The combination of AO and confocal filtering in an AO-SLO thus enables axial sectioning of the retina and allows for direct visualization of different retinal structural layers, for example, the photoreceptor, blood vessel, and nerve fiber layers. Lateral and axial resolution in an AO-SLO can be tuned by modifying the size of the confocal pinhole (Romero-Borja et al. 2005).

Another benefit of the AO-SLO is that the raster scan feature can be used to deliver stimuli to the retina by turning off the scanning beam at predetermined scan positions (Poonja et al. 2005). Single cone stimulus delivery has been achieved by using the AO-SLO as a high-frequency eye tracker (Arathorn et al. 2007; Yang et al. 2010).

5.2.1.3 AO-OCT

The retinal image in an AO-OCT (Hermann et al. 2004) is also created by scanning a small point source of light over the retina. Correcting aberrations using AO will decrease the size of the point source and yield improved lateral resolution and sensitivity. The AO-OCT, instead of creating a point-by-point intensity image like an AO-SLO, uses time-domain, spectral-domain (SD), or swept-source low-coherence interferometry to generate a high-axial-resolution 3D image of the retina. Current state-of-the-art ultrahigh-resolution AO SD-OCT demonstrate an isotropic 3D resolution of $3 \times 3 \times 3\ \mu m^3$ in retinal tissue (Miller et al. 2011). The combined benefit of both high axial and high lateral resolution in AO-OCT thus enables 3D cellular-resolution imaging of retinal structures (Fernandez et al. 2005; Hermann et al. 2004; Miller et al. 2011; Pircher and Zawadzki 2007; Zawadzki et al. 2005; Zhang et al. 2005).

Axial and lateral resolutions in an AO-OCT are decoupled. Axial resolution is determined by the spectral bandwidth of the light source. A broader bandwidth yields better axial resolution, but a significantly broader bandwidth is needed to achieve the same axial resolution with increasing central wavelength (Drexler 2004). Dispersion compensation is therefore important to fully benefit from the increased bandwidth due to the significant longitudinal chromatic aberration of the human eye (Drexler 2004; Fernandez et al. 2006; Zawadzki et al. 2008). Lateral resolution and sensitivity is limited by the lateral diameter of the point source, which is significantly affected by ocular aberrations (Drexler 2004). The reader is referred to Chapter 6 of *Handbook of Visual Optics: Fundamentals and Eye Optics, Volume Two* for a detailed discussion of AO-OCT systems.

5.2.2 DESIGN ASPECTS OF FLOOD-ILLUMINATED AO SYSTEMS

5.2.2.1 AO ophthalmoscope optical components

It is essential that the design of an AO ophthalmoscope is modeled using an optical design software such as Zemax, CODE V, OSLO, or FRED, to properly evaluate and optimize system components and overall design in order to minimize the static optical aberrations of the ophthalmoscope and maximize system performance.

Various optical components are used in AO ophthalmoscopes to relay light from the eye to the WFS and imaging camera. These components can be lenses, plane or curved mirrors, and a variety of beam splitters (e.g., pellicle, plate, cube, or dichroic).

Lenses in AO ophthalmoscopes are usually off-the-shelf precision achromats that have been corrected for various aberrations, for example, spherical aberration and coma, as well as longitudinal chromatic aberration. The lenses are normally arranged in afocal pairs (i.e., telescopes) with effective focal lengths equal to infinity that change the magnification of incoming parallel beams. The use of afocal telescopes generates pupil and retinal conjugate planes in which various system components such as apertures, deformable mirrors (DMs), and WFSs can be placed. However, an achromatic lens is designed to bring only two wavelengths into focus in the same plane, typically blue and red for visible wavelengths, independent of it being a doublet or triplet lens. Use of wavelengths much different from the two design wavelengths will induce chromatic aberration in the system. An alternative, albeit expensive, option is to use custom-designed lenses with application-specific design wavelengths.

A major drawback of using lenses in an AO ophthalmoscope is that a properly aligned lens will generate back reflections along the optical axis of the system that can interfere with the true retinal signal at the WFS or science camera. The negative effects of reflected light can be somewhat mitigated by choosing appropriate antireflection coatings, but it can still be an issue since the intensity of the reflected light is normally greater than the intensity of reflected light from the retina. Crossed polarizers can also be used to reduce back reflections, but this depends on the application. An effective solution to reduce back reflections is to minimize the number of lenses in the affected optical path(s) of an AO ophthalmoscope.

The problems mentioned earlier associated with lenses can be avoided entirely by using curved mirrors. Off-axis parabolic mirrors offer several benefits over lenses since they do not induce back reflections or chromatic aberrations. Off-axis parabolic mirrors also avoid the issue of inducing astigmatism associated with spherical mirrors, which for nonnormal incidence decrease the focal length by a factor of the cosine of the angle of incidence in the tangential direction and increase the focal length by a factor of the inverse cosine of that angle in the sagittal direction. Another useful feature of a mirror-based design is that the size of the instrument can be reduced by folding the beam.

Reflective components such as pellicle, plate, or cube beam splitters should be of sufficient optical quality to have minimal impact on system performance. Other design parameters are reflection/transmission ratios, thickness, and the option of a wedge in plate beam splitters to avoid ghost reflections.

5.2.2.2 Wavefront sensors

The dominating WFS used in ophthalmic AO is by far the Shack–Hartmann WFS (Shack and Platt 1971). A Shack–Hartmann WFS combines a lenslet array, an array of small lenses with the same focal length, with an imaging sensor, typically a charge-coupled device (CCD) camera. The lenslet array is placed in a plane conjugate to the pupil so that each illuminated lenslet provides a local sample of an incoming wavefront by focusing the incoming light into a focal spot on the sensor. The local slope of the wavefront for each lenslet can then be calculated from the shift of the corresponding focal spot on the sensor. This allows the wavefront in the pupil plane to be approximated by a combination of slopes from all illuminated lenslets.

The reader is referred to Chapter 2 of *Handbook of Visual Optics: Fundamentals and Eye Optics, Volume Two* for a detailed discussion of WFSs.

5.2.2.3 Wavefront correctors

A WFC in an AO ophthalmoscope is positioned in a plane that is conjugate to the pupil of the eye. The rationale for this is that since (1) the cornea and crystalline lens are the primary aberration sources in the eye, (2) aberrations caused by the cornea and lens can essentially be described by pure phase variations (i.e., a distorted wavefront) at the pupil of the eye, and (3) WFCs are phase-only correctors, the maximum effectiveness of a WFC will be at (or near) the pupil of the eye (Roorda et al. 2005).

A WFC in an AO system compensates for aberrations by changing the optical path length (OPL), and thus the shape, of an incoming wavefront. The most common WFC used in AO ophthalmoscopes is the continuous facesheet DM. Other WFC devices are segmented DMs, bimorph DMs, and liquid crystal spatial light modulators (LC-SLMs). An extensive comparison of different DMs was published by Devaney et al. (2008). It provides an in-depth comparison of various DMs and designs available at the time of publication, although new DMs have appeared on the market since then.

Many parameters affect DM performance: total number of actuators, single actuator stroke, interactuator stroke, actuator pitch, actuator coupling, actuator influence function, response and settling times, hysteresis, and creep. The total number of actuators, in combination with the number of lenslets in the WFS, determines the correctable number of modes of an AO system. The maximum single actuator stroke imposes a limit on the maximum wavefront amplitude that can be corrected, while the correction of higher-order aberrations is limited by the maximum interactuator stroke. Actuator pitch, that is, the distance between neighboring actuators, determines actuator coupling, which is a measure of how much the movement of one actuator will displace neighboring actuators. Influence functions represent DM shapes that either describe continuous mirror displacements (modal influence functions) or localized responses on the mirror surface (zonal influence functions). The response time determines how long it takes for a DM to respond to a new measurement and the settling time is the time it takes for the DM to assume a new shape after receiving a new command. The nonlinear effects of hysteresis, which results in different actuator displacements depending on the direction of movement, and creep, which is the slow and time-delayed deformation of the DM surface as a response to an applied command, mainly affect piezoelectric and bimorph DMs. Creep is not an issue for AO systems running in closed loop but can affect performance in open-loop control.

Wavefront correction with continuous facesheet DMs is accomplished by deforming the surface of the mirror using an array of discrete, electrostatic, magnetic, or piezoelectric actuators that produce local deformations of the DM surface. Depending on mirror type, the actuators exert a physical, electrical, or magnetic force to push or pull the mirror surface into a shape that corresponds to half the amplitude of the wavefront that is to be corrected. The discrete actuators of segmented DMs, capable of piston (up/down) or piston/tip/tilt motion, are independently moved to approximate the desired wavefront shape.

Bimorph DMs are made of layers of electrically active and inactive materials where one or more active layers contain piezoelectric or electrostrictive actuator patterns. This allows for the implementation of different actuator functionalities; for example, one set of actuators can be used for curvature deformations of the mirror surface while another set can be used for local slope deformations. The actuators respond to an applied voltage signal and thus deform the mirror surface into the desired wavefront shape.

The LC-SLM is a segmented WFC device made of an array of liquid crystals (pixels) that function in either reflective or transparent mode. Wavefront correction is accomplished by changing the refractive index, and thus the OPL, through electrical or optical addressing of individual pixels. The total number of pixels determines the spatial resolution of an LC-SLM and the stroke is typically around one wavelength, but this can be extended using modulo-2π phase wrapping. LC-SLMs can only be used with a linearly polarized light source as the liquid crystal molecules only modulate light along their polarization axis. The consequence of this is that a large portion of reflected light from the retina is lost due to depolarization by the optics of the eye and the retina.

The reader is referred to Chapter 2 of *Handbook of Visual Optics: Fundamentals and Eye Optics, Volume One* for a detailed discussion of WFCs.

5.2.2.4 Control computer

The requirements on a control computer for an AO ophthalmoscope are modest compared to those in, for example, astronomy, in that the control software that is used for

closed-loop control of the AO ophthalmoscope can be run on a workstation PC. The control software runs the real-time control algorithm that converts wave aberration measurements from the WFS into actuator commands that are applied to the WFC and handles timing synchronization between system components.

5.2.2.5 Multi-WFC AO

Although the vast majority of AO ophthalmoscopes implement one WFC and one WFS, other concepts using two WFCs and one or more WFSs have been proposed to extend either the dynamic range of aberration correction or the size of the corrected field of view (FOV).

In a woofer–tweeter arrangement, two DMs, both controlled by the same WFS, are placed in separate planes conjugate to the pupil plane to extend the dynamic range of aberration correction. A high-stroke woofer DM is used for correction of lower-order aberrations and a low-stroke tweeter DM is used for correction of higher-order aberrations. This approach adds a layer of complexity to the AO control loop, since the correction algorithm is required to sort the wavefront aberrations into one group for the woofer correction and one for the tweeter correction. Several control algorithms have been proposed on this matter (Cense et al. 2009; Chen et al. 2007; Hu et al. 2006; Lavigne and Véran 2008; Li et al. 2010; Zawadzki et al. 2005, 2007; Zou et al. 2008). The woofer–tweeter approach has to date only been implemented in AO-SLO and AO-OCT systems (Cense et al. 2009; Chen et al. 2007; Ferguson et al. 2010; Li et al. 2010; Zawadzki et al. 2007).

Another approach, based on the concept of multiconjugate AO (MCAO) in astronomy, has been investigated theoretically (Bedggood et al. 2006, 2008; Bedggood and Metha 2010; Thaung et al. 2009) and implemented in an AO ophthalmoscope by Thaung et al. (2009). The benefit of the MCAO approach is that it overcomes the limitation of the small isoplanatic patch size of approximately 2° in the human eye (Bedggood et al. 2006; Dubini et al. 2008; Goncharov et al. 2008; Thaung et al. 2009) and allows for AO-corrected imaging of larger retinal patches. A corrected FOV of approximately 7° has been demonstrated in the human eye (Thaung et al. 2009). Aberrations are measured in different angular directions using five displaced retinal beacons and five WFS patterns, one for each beacon, on a single WFS CCD camera. Correction is performed with two DMs, one in a plane conjugate to the pupil and one conjugate to a plane within the eye, and the concept is consequently called dual-conjugate AO (DCAO).

5.2.2.6 Imaging light sources

Several factors need to be taken into consideration when it comes to selecting a light source for flood-illuminated AO retinal imaging, and it can prove difficult to find a commercially available product that fulfills all the criteria. The light source should have a narrow spectral bandwidth; otherwise, imaging quality will be compromised due to the large inherent chromatic aberration of the eye of approximately 2.5 diopters across the visible spectrum (Ames and Proctor 1921; Westheimer 2006). A light source with a broad spectral bandwidth can be used in combination with a narrow band-pass optical filter, and other solutions using chromatic aberration correctors have been proposed (Benny et al. 2007; Fernandez et al. 2006).

The light source should be able to deliver enough light to uniformly illuminate the retina, allow for short-exposure images on the order of milliseconds to eliminate image dither, and also exhibit low spatial coherence to reduce speckle noise in the image. Another important factor that determines the region on the retina that can be effectively illuminated is the étendue of the illumination channel. The étendue is constant in an optical system and is related to the Lagrange and optical invariants. It can be calculated as the product of the area of the source and the solid angle of the entrance pupil of the system or as the product of the area of the exit pupil and the solid angle of the image. The optimal size of the illuminating source can thus be determined if one knows the size of the illuminated region on the retina and the size of the pupil of the eye.

Common imaging light sources in flood-illuminated AO ophthalmoscopes are flash lamps, SLDs, or LEDs. A flash lamp is filled with a noble gas (e.g., xenon or krypton) that generates a plasma to produce very short pulses of incoherent full-spectrum white light. Since the flash distributes light in all directions, only a small fraction is utilized in the illumination channel. Flash lamps are used in combination with narrow band-pass filters to constrain the effect of chromatic aberration, thus further reducing the amount of light that is actually used for imaging. Long charge times on the order of several seconds limit the repetition rate during imaging.

SLDs are more efficient light sources that emit highly directional and quasi-monochromatic light. Unwanted speckle noise can be attenuated by passing the light through a multimode optical fiber (Mims 1982) to effectively reduce the spatial coherence of the SLD. Directional high-power LEDs offer many benefits when compared to SLDs; they are smaller, cheaper, less coherent, and less directional, and they require less power, have longer lifetimes, and offer fast switching times.

5.2.2.7 Other factors affecting high-resolution retinal imaging

Several other factors have to be considered in order to facilitate high-resolution retinal imaging. Most of the incoming light is absorbed by the retina, and only approximately 0.1%–10% is reflected depending on the wavelength (Delori and Pflibsen 1989; Van Norren and Tiemeijer 1986). An easy solution would be to increase measurement and imaging light intensities to improve WFS and retinal imaging performance. However, these cannot exceed the maximum amount of light that can be safely coupled into the eye as determined by safety standards (Landry et al. 2011). This coupling between retinal attenuation and light intensity has to be taken into account in order to maximize both WFS and retinal imaging performance.

Small eye movements and structures within the eye also affect the quality of retinal imaging. Microscopic eye movements that occur even during fixation (Havermann et al. 2014) induce lateral shifts of the retina that can blur the image. Microfluctuations of the crystalline lens induce changes in ocular power (Charman and Heron 1988), and changes in retinal illuminance affect the size of the eye's pupil (Barbur et al. 1992). Short exposures on the order of a few milliseconds are normally used to avoid detrimental eye movement effects.

Two types of eye drops prevent changes in pupil size, where one type stimulates the contraction of the eye muscles that widen the pupil and the other type relaxes the muscles thatconstrict the pupil. The latter has the additional benefit of paralyzing accommodation, thus preventing fluctuations of ocular power. Pupil dilation can be maximized by simultaneous administration of both types of eye drops.

5.3 AO OPHTHALMOSCOPE EXAMPLES

The vast majority of current flood-illuminated AO ophthalmoscopes are custom-built laboratory-based research systems. It is outside the scope of this chapter to give a comprehensive review of this field. The following section will give a brief description of a few systems and highlight some of the results that have been obtained with these and other flood-illuminated AO ophthalmoscopes.

5.3.1 LABORATORY-BASED FLOOD-ILLUMINATED AO SYSTEMS

Many different flood-illumination AO ophthalmoscope designs have been presented over the years. They all conform to the basic principle described in Figure 5.1 but have introduced different solutions for wavefront sensing light and imaging light delivery as well as various implementations of other components such as off-axis parabolic mirrors, DMs, and retinal cameras. Many of the drawbacks of the first AO ophthalmoscopes have been addressed and other improvements have been developed. Only a brief overview of these instruments will be given in the text and the reader is referred to the referenced articles for detailed descriptions.

5.3.1.1 Rochester second-generation AO ophthalmoscope

The Rochester second-generation flood illuminated AO ophthalmoscope (Putnam et al. 2005) was based on the modified version (Hofer et al. 2001) of the original AO system described by Liang et al. (1997). The main differences from the previous two systems were that the second-generation system implemented a DM (Xinetics, Devens, MA) with 97 instead of 37 actuators and two off-axis parabolic mirrors instead of lenses to image the pupil onto the DM. The large 75 mm diameter active area of the DM required long focal length off-axis parabolic mirrors to fully illuminate the surface (Figure 5.2).

Wavefront sensing with a Shack–Hartmann WFS was performed at rates up to 30 Hz using 820 nm SLD light and a 17×17 square lenslet array (f = 24 mm, pitch = 0.4 mm) over a 6.8 mm pupil. The fixation target was a Maltese cross presented at 550 nm. Retinal illumination at a wavelength of 550 nm was delivered using a 4 ms flash from a spectrally filtered broadband krypton flash lamp. The flash was triggered automatically when a desired root-mean-square wavefront aberration was achieved (~0.1 μm) or after 10 cycles of measurement and correction, whichever occurred first. Retinal images of approximately 1° in diameter were acquired with a cooled CCD camera (Roper Scientific, Trenton, NJ) over the central 6 mm of the pupil to reduce edge artifacts in aberration correction.

5.3.1.2 Medical College of Wisconsin AO ophthalmoscope

The flood-illuminated AO ophthalmoscope at the Medical College of Wisconsin (Rha et al. 2009) measured monochromatic aberrations at 13.3 Hz with a Shack–Hartmann WFS using light

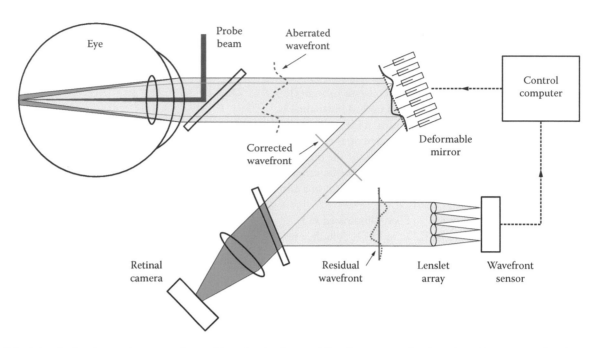

Figure 5.1 Principle of adaptive optics correction and imaging in the eye. Wavefront correction is an iterative process that is accomplished by focusing a probe beam of light onto the retina of the eye. The reflected beam of light from the focused spot (yellow beam) is aberrated by the optical media of the eye and exits through the pupil as an aberrated wavefront (red dashed line). It is reflected by a wavefront corrector (black dotted line), here a deformable mirror (DM), placed in a pupil conjugate plane. The aberrated beam is finally imaged by a lenslet array, which samples the wavefront in a pupil conjugate plane, as an array of spots on a wavefront sensor (WFS) camera. The WFS image is analyzed by a control computer that calculates a correction signal that is applied to the DM. After the DM has assumed its new shape, a new measurement is made and the residual wavefront (red dotted line) is again sampled by the WFS. The process is repeated until the DM has assumed a shape (solid black line) that generates a flat residual wavefront (straight red line) and a flat corrected wavefront (solid green line). Imaging light that is reflected from the retina (green beam and lines) will then be corrected by the DM and allow for diffraction-limited imaging with a retinal camera.

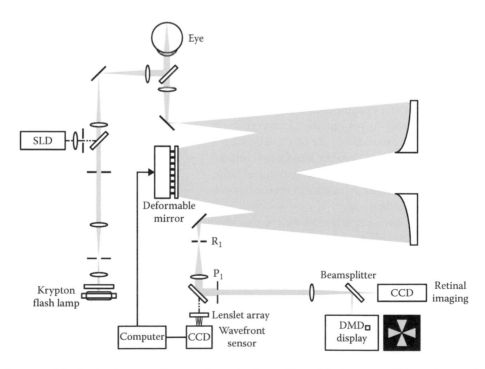

Figure 5.2 Schematic diagram of the Rochester second-generation adaptive optics ophthalmoscope. Please observe that the vertical orientation of both off-axis parabolic mirrors should be reversed. The correct orientation is shown in Figure 5.1 in the paper by Doble et al. (2002). *Abbreviations:* CCD, charge-coupled device camera; DMD, dot-matrix display; P_1, pupil conjugate plane; R_1, retinal conjugate plane; SLD, superluminescent diode. (Originally published in Putnam, N.M. et al., *J. Vis.*, 5(7), 632, 2005.)

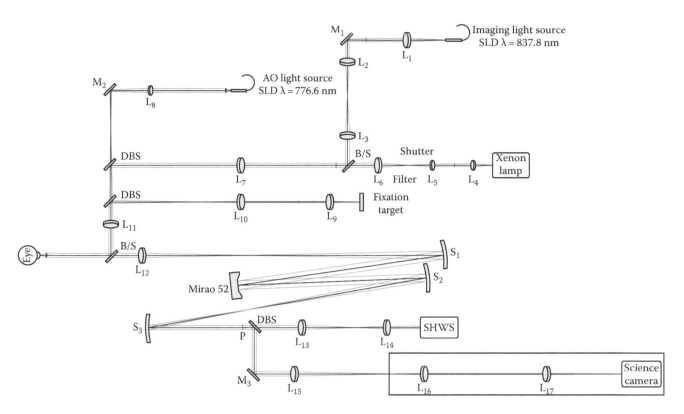

Figure 5.3 Schematic diagram of the Medical College of Wisconsin adaptive optics ophthalmoscope. *Abbreviations*: BS, beam splitter; DBS, dichroic beam splitter; SHWS, Shack–Hartmann wavefront sensor; P, pupil plane; L_1–L_{17}, achromatic lenses; M_1–M_3, flat mirrors; S_1–S_3, spherical mirrors. (Originally published in Rha, J. et al., *Opt. Lett.*, 34(24), 3782, 2009.)

from a 776.6 nm ($\Delta\lambda$ = 19.2 nm) SLD (Superlum, Cork, Ireland). Aberrations were corrected with a 52 actuator Mirao DM (Imagine Eyes, Orsay, France; Figure 5.3).

A 1.8° diameter retinal patch was illuminated using a 500 ms flash from a fiber coupled 837.8 nm ($\Delta\lambda$ = 14.1 nm) SLD (Superlum, Cork, Ireland) that had been passed through 110 m of multimode step index fiber (Fiberguide Industries, Stirling, NJ) to reduce the spatial coherence of the light and thus speckle noise in the image. Continuous 6 ms exposures were captured at a frame rate of 167 fps during each 500 ms flash with a back-illuminated scientific-grade 12-bit Cam1M100-SFT CCD camera (Sarnoff Corporation, Princeton, NJ).

5.3.1.3 Indiana high-speed AO ophthalmoscope

The Indiana high-speed flood-illumination AO ophthalmoscope (Rha et al. 2006) comprises a Shack–Hartmann WFS and 37 actuator DM (Xinetics, Devens, MA). The WFS employs a 0.75 mW pigtailed single-mode SLD (exposure level at the cornea was 5 μW) operating at 788 nm ($\Delta\lambda$ = 20 nm) and a 17 × 17 lenslet array (f = 24 mm, pitch = 0.4 mm) that samples the wavefront across a 6.8 mm pupil (Figure 5.4).

Two light sources were used for retinal illumination, a 10 mW pigtailed single-mode SLD (λ = 679 nm, $\Delta\lambda$ = 10.8 nm) coupled to 25 m of multimode step index optical fiber (NA = 0.22, d_{core} = 105 μm, n_{core} = 1.457 at 633 nm [Lucent Technologies, Murray Hill, NJ]), and a 200 mW multimode LD (λ = 670 nm) coupled to a 300 m multimode step index optical fiber (NA = 0.39, d_{core} = 200 μm, n_{core} = 1.457 at 633 nm [Lucent Technologies, Murray Hill, NJ]). The lengths of the fibers were sufficient to reduce source spatial coherence and thus speckle noise

in the image. The fiber tips were placed in a plane conjugate to the retina and illuminated retinal patches of 1° (SLD) and 1.8° (LD).

Retinal images were captured in a synchronized manner to the strobing SLD and LD using a high-speed back-illuminated scientific-grade 12-bit Quantix 57 CCD camera (Roper Scientific, Tucson, AZ). Higher frame rates were achieved by designating a smaller region of interest on the CCD camera, that is, a smaller retinal FOV: 10 Hz (full frame), 30 Hz, and 60 Hz provided a 1.8°, 0.8°, and 0.4° FOV, respectively. Short-burst imaging sequences of up to 500 Hz were achieved by temporarily storing sequences of four 256 × 530 pixel or eight 128 × 530 pixel 1 ms images on the CCD for total exposure times of 7 and 15 ms, respectively.

5.3.1.4 Wide-field DCAO prototype

The current wide-field flood-illuminated DCAO ophthalmoscope (Popovic et al. 2012) is based on the first implementation of the MCAO concept in an ophthalmic instrument by Thaung et al. (2009). Wavefront correction is accomplished using five retinal beacons from an 835 ± 10 nm SLD (Superlum, Cork, Ireland), a multireference WFS with five separate 13 × 13 spot patterns on a single CCD camera, and two DMs (ALPAO SAS, Montbonnot, France). A 52 actuator DM (DM1, Figure 5.5) is positioned in a plane conjugate to the pupil and corrects for an average of the wavefronts measured with the five WFSs. A 97 actuator DM (DM2, Figure 5.5), used to correct for field-dependent aberrations, is positioned in a plane conjugate to a plane within the eye where the footprints of the five beacons are separated but still exhibit a certain amount of overlap.

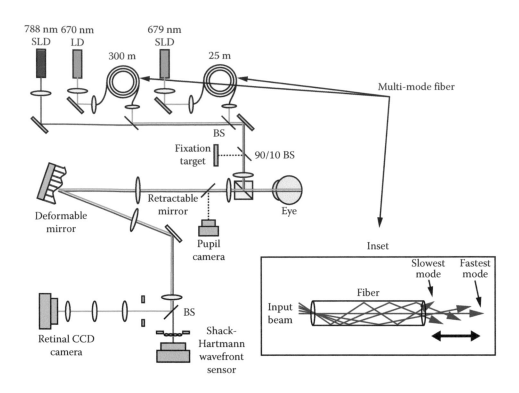

Figure 5.4 Schematic diagram of the Indiana high-speed adaptive optics (AO) ophthalmoscope. The camera comprises three subsystems: a pupil retroillumination and fixation channel to align the eye of the subject to the camera, an AO channel to measure and compensate the wave aberrations of the eye, and a retinal imaging channel with a fiber-based light source and scientific-grade CCD. (Inset) The light from an LD or SLD that is coupled into a multimode fiber is distributed among the fiber modes that propagate along the fiber length at different velocities, thus reducing the spatial coherence of the light at the output of the fiber. *Abbreviations*: BS, beam splitter; CCD, charge-coupled device camera; LD, laser diode; SLD, superluminescent diode. (Originally published in Rha, J. et al., *Opt. Express*, 14(10), 4552, 2006.)

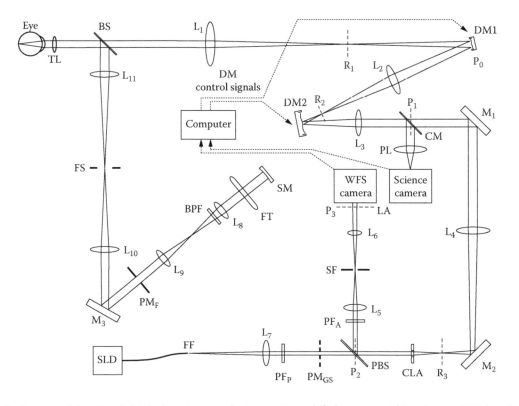

Figure 5.5 Schematic diagram of the wide-field dual-conjugate adaptive optics ophthalmoscope. *Abbreviations*: BPF, band-pass filter; BS, beam splitter; CLA, collimating lens array; CM, cold mirror; DM1, pupil DM; DM2, field DM; FS, field stop; FT, flash tube; LA, lenslet array; M, mirror; P, pupil conjugate plane; PBS, pellicle beam splitter; PFP/PFA, polarization filters; PL, photographic lens; PMF, flash pupil mask; PMGS, GS pupil mask; R, retinal conjugate plane; SF, spatial filter; SM, spherical mirror; SLD, superluminescent diode; TL, trial lens. (Originally published in Popovic, Z. et al., In *Adaptive Optics Progress*, R. Tyson (Ed.), pp. 1–21. InTech, Rijeka, Croatia, 2012; under CC BY 3.0 license.)

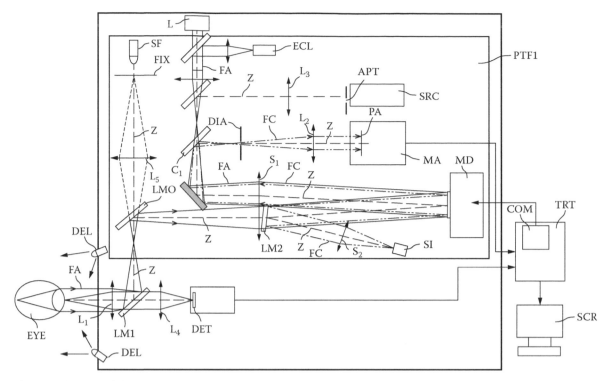

Figure 5.6 Schematic diagram of the Imagine Eyes rtx1 AO ophthalmoscope. *Abbreviations:* APT, illumination aperture; C_1, beam splitters; COM, control unit; DEL, diode; DET, pupil camera; DIA, measurement diaphragm; ECL, retinal illumination unit; FA, main beam; FC, wavefront sensing measurement light beam; FIX, fixation image; L, retinal camera; L_1–L_5, lenses; LM1, LM2, LMO, beam splitters; MA, wavefront sensor; MD, deformable mirror; PA, wavefront sensor measurement plane; PTF1, mobile platform; S_1, S_2, optical systems; SCR, monitor; SF, fixation light source; SI, calibration light source; SRC, wavefront sensing light source; TRT, processing unit; Z, optical axis. (Originally published in Levecq, X., Phase modulation device for an ophthalmic instrument, ophthalmic instruments equipped with such device, and related calibration method, Patent Application No. US20110001930A1, United States Patent and Trademark Office, 2011.)

Light from a spectrally filtered (575 ± 10 nm) broadband xenon flash lamp is used to illuminate a 10° × 10° retinal field, of which the central 7° × 7° are used for imaging to avoid nonuniform illumination artifacts at the border of the field. Short-exposure (5–7 ms) images are captured with a scientific-grade 14-bit monochromatic Stingray F-504B CCD camera (Allied Vision Technologies GmbH, Stadtroda, Germany).

5.3.2 IMAGINE EYES RTX1 FLOOD-ILLUMINATED AO SYSTEM

The rtx1 (Imagine Eyes, Orsay, France) is the only commercially available AO ophthalmoscope designed for clinical research. The instrument features a pupil tracking system, an SLD with a center wavelength of 750 nm and a Haso 32-eye WFS (Imagine Eyes, Orsay, France) for wavefront sensing, and a 52 actuator Mirao 52-e DM (Imagine Eyes, Orsay, France) for wavefront correction. An LED with a center wavelength of 850 nm (Δλ = 30 nm) delivers 9 ms pulses at a frequency of 9.5 Hz over a 4° × 4° FOV that is imaged by a low-noise high-resolution 12-bit CCD camera (Roper Scientific, Tucson, AZ; Figure 5.6).

5.4 APPLICATIONS OF FLOOD-ILLUMINATED AO RETINAL IMAGING

Many flood-illumination AO ophthalmoscope studies have focused on imaging the cone photoreceptor mosaic to analyze the variation of density (Bidaut Garnier et al. 2014; Carroll

et al. 2008; Choi et al. 2006, 2008, 2011; Dabir et al. 2014; Hofer et al. 2005; Lombardo et al. 2012, 2013a,c, 2014a,b; Michaelides et al. 2011; Mrejen et al. 2014; Murhiah et al. 2014; Obata and Yanagi 2014; Putnam et al. 2005; Stepien et al. 2009; Tojo et al. 2013a,b; Wagner-Schuman et al. 2010; Werner et al. 2011; Wolfing et al. 2006; Xue et al. 2007), spacing (Bidaut Garnier et al. 2014; Choi et al. 2011; Dabir et al. 2014; Hofer et al. 2001; Kitaguchi et al. 2007; Liang et al. 1997; Lombardo et al. 2013c; Tojo et al. 2013b), waveguide properties (Choi et al. 2005; Roorda and Williams 2002), and reflectance properties (Bedggood and Metha 2012b; Jacob et al. 2014; Jonnal et al. 2007; Mrejen et al. 2014; Pallikaris et al. 2003) of cone photoreceptors in healthy and diseased retina. Investigators have established normal ranges that can be used to differentiate healthy from pathological photoreceptors even in early stages of retinal disease. Recent studies have also studied properties of the retinal microvasculature (Bedggood and Metha 2012a, 2014; Koch et al. 2014; Lombardo et al. 2013b; Popovic et al. 2011; Rha et al. 2006) and nerve fiber bundles (Ramaswamy et al. 2014).

5.5 SUMMARY

Due to the typically slow progression of most retinal diseases, it is estimated that a clinical loss of visual function is only seen years after the onset of the disease, although currently subclinical changes may have been detectable during that time. The introduction of AO in ophthalmic instrumentation has created exciting new opportunities to image the living human retina at a microscopic level. The improvement in contrast and resolution in AO retinal images facilitates direct observation of retinal microstructures such as photoreceptors, retinal microvasculature, and nerve fiber bundles. This allows for *in vivo* analysis of the structural integrity of normal retinal tissue and pathological abnormalities caused by retinal disease. Although AO retinal imaging has shown its potential in applications that cannot be addressed with other imaging technologies, it has not yet gained widespread clinical acceptance.

Many of the drawbacks of early flood-illuminated AO ophthalmoscopes have been addressed through the implementation of, for example, optimized optical designs, real-time AO correction, high-speed cameras, eye tracking, and fast switching retinal illumination sources. But a remaining issue in most instruments is the relatively small FOV on the order of 1°. This is often compensated by performing repeated imaging of neighboring retinal areas, a time-consuming operation that generates a significant amount of imaging data in need of extensive, often manual, postprocessing. The development of automated image processing routines for alignment, stitching, and analysis, possibly in combination with a larger FOV, is therefore a very important step in order to gain clinical acceptance for AO retinal imaging.

REFERENCES

Airy, G. B. 1827. On a peculiar defect in the eye, and a mode of correcting it. *Edinburgh J Sci* 7:322–326.

Alhazen, and A. I. Sabra. 1989. *The Optics of Ibn al-Haytham*. Books I–III, On Direct Vision. 2 vols., Studies of the Warburg Institute. London, U.K.: Warburg Institute, University of London.

Ames, A. and C. A. Proctor. 1921. Dioptrics of the eye. *J Opt Soc Am* 5(1):22–84. doi: 10.1364/JOSA.5.000022.

Arathorn, D. W., Q. Yang, C. R. Vogel, Y. Zhang, P. Tiruveedhula, and A. Roorda. 2007. Retinally stabilized cone-targeted stimulus delivery. *Opt Express* 15(21):13731–13744.

Barbur, J. L., A. J. Harlow, and A. Sahraie. 1992. Pupillary responses to stimulus structure, colour and movement. *Ophthalmic Physiol Opt* 12(2):137–141.

Bedggood, P. A., R. Ashman, G. Smith, and A. B. Metha. 2006. Multiconjugate adaptive optics applied to an anatomically accurate human eye model. *Opt Express* 14(18):8019–8030.

Bedggood, P., M. Daaboul, R. Ashman, G. Smith, and A. Metha. 2008. Characteristics of the human isoplanatic patch and implications for adaptive optics retinal imaging. *J Biomed Opt* 13(2):024008. doi: 10.1117/1.2907211.

Bedggood, P. and A. Metha. 2010. System design considerations to improve isoplanatism for adaptive optics retinal imaging. *J Opt Soc Am A* 27(11):A37–A47. doi: 10.1364/JOSAA.27.000A37.

Bedggood, P. and A. Metha. 2012a. Direct visualization and characterization of erythrocyte flow in human retinal capillaries. *Biomed Opt Express* 3(12):3264–3277. doi: 10.1364/boe.3.003264.

Bedggood, P. and A. Metha. 2012b. Variability in bleach kinetics and amount of photopigment between individual foveal cones. *Invest Ophthalmol Vis Sci* 53(7):3673–3681. doi: 10.1167/iovs.11-8796.

Bedggood, P. and A. Metha. 2014. Analysis of contrast and motion signals generated by human blood constituents in capillary flow. *Opt Lett* 39(3):610–613. doi: 10.1364/ol.39.000610.

Benny, Y., S. Manzanera, P. M. Prieto, E. N. Ribak, and P. Artal. 2007. Wide-angle chromatic aberration corrector for the human eye. *J Opt Soc Am A Opt Image Sci Vis* 24(6):1538–1544.

Bidaut Garnier, M., M. Flores, G. Debellemaniere, M. Puyraveau, P. Tumahai, M. Meillat, C. Schwartz, M. Montard, B. Delbosc, and M. Saleh. 2014. Reliability of cone counts using an adaptive optics retinal camera. *Clin Exp Ophthalmol* 42(9):833–840. doi: 10.1111/ceo.12356.

Carroll, J., S. S. Choi, and D. R. Williams. 2008. In vivo imaging of the photoreceptor mosaic of a rod monochromat. *Vision Res* 48(26):2564–2568. doi: 10.1016/j.visres.2008.04.006.

Cense, B., E. Koperda, J. M. Brown, O. P. Kocaoglu, W. Gao, R. S. Jonnal, and D. T. Miller. 2009. Volumetric retinal imaging with ultrahigh-resolution spectral-domain optical coherence tomography and adaptive optics using two broadband light sources. *Opt Express* 17(5):4095–4111.

Charman, W. N. and G. Heron. 1988. Fluctuations in accommodation: A review. *Ophthalmic Physiol Opt* 8(2):153–164.

Chen, D. C., S. M. Jones, D. A. Silva, and S. S. Olivier. 2007. High-resolution adaptive optics scanning laser ophthalmoscope with dual deformable mirrors. *J Opt Soc Am A Opt Image Sci Vis* 24(5):1305–1312.

Choi, S. S., J. Christou, D. R. Williams, N. Doble, and J. Lin. 2005. Effect of wavelength on in vivo images of the human cone mosaic. *J Opt Soc Am A* 22(12):2598–2605. doi: 10.1364/JOSAA.22.002598.

Choi, S. S., N. Doble, J. L. Hardy, S. M. Jones, J. L. Keltner, S. S. Olivier, and J. S. Werner. 2006. In vivo imaging of the photoreceptor mosaic in retinal dystrophies and correlations with visual function. *Invest Ophthalmol Vis Sci* 47(5):2080–2092. doi: 10.1167/iovs.05-0997.

Choi, S. S., R. J. Zawadzki, M. A. Greiner, J. S. Werner, and J. L. Keltner. 2008. Fourier-domain optical coherence tomography and adaptive optics reveal nerve fiber layer loss and photoreceptor changes in a patient with optic nerve drusen. *J Neuroophthalmol* 28(2):120–125. doi: 10.1097/WNO.0b013e318175c6f5.

Choi, S. S., R. J. Zawadzki, M. C. Lim, J. D. Brandt, J. L. Keltner, N. Doble, and J. S. Werner. 2011. Evidence of outer retinal changes in glaucoma patients as revealed by ultrahigh-resolution in vivo retinal imaging. *Br J Ophthalmol* 95(1):131–141. doi: 10.1136/bjo.2010.183756.

Dabir, S., S. Mangalesh, K. A. Kumar, M. K. Kummelil, S. A. Roy, and R. Shetty. 2014. Variations in the cone packing density with eccentricity in emmetropes. *Eye (Lond)* 28(12):1488–1493. doi: 10.1038/eye.2014.229.

Delori, F. C. and K. P. Pflibsen. 1989. Spectral reflectance of the human ocular fundus. *Appl Opt* 28(6):1061–1077. doi: 10.1364/AO.28.001061.

Devaney, N., E. Dalimier, T. Farrell, D. Coburn, R. Mackey, D. Mackey, F. Laurent, E. Daly, and C. Dainty. 2008. Correction of ocular and atmospheric wavefronts: A comparison of the performance of various deformable mirrors. *Appl Opt* 47(35):6550–6562.

Doble, N., G. Yoon, L. Chen, P. Bierden, B. Singer, S. Olivier, and D. R. Williams. 2002. Use of a microelectromechanical mirror for adaptive optics in the human eye. *Opt Lett* 27(17):1537–1539.

Drexler, W. 2004. Ultrahigh-resolution optical coherence tomography. *J Biomed Opt* 9(1):47–74. doi: 10.1117/1.1629679.

Dubinin, A., T. Cherezova, A. Belyakov, and A. Kudryashov. 2008. Human retina imaging: Widening of high resolution area. *J Modern Opt* 55(4–5):671–681. doi: 10.1080/09500340701467710.

Ferguson, R. D., Z. Zhong, D. X. Hammer, M. Mujat, A. H. Patel, C. Deng, W. Zou, and S. A. Burns. 2010. Adaptive optics scanning laser ophthalmoscope with integrated wide-field retinal imaging and tracking. *J Opt Soc Am A Opt Image Sci Vis* 27(11):A265–A277. doi: 10.1364/JOSAA.27.00A265.

Fernandez, E. J., I. Iglesias, and P. Artal. 2001. Closed-loop adaptive optics in the human eye. *Opt Lett* 26(10):746–748.

Fernandez, E. J., B. Povazay, B. Hermann, A. Unterhuber, H. Sattmann, P. M. Prieto, R. Leitgeb, P. Ahnelt, P. Artal, and W. Drexler. 2005. Three-dimensional adaptive optics ultrahigh-resolution optical coherence tomography using a liquid crystal spatial light modulator. *Vision Res* 45(28):3432–3444. doi: 10.1016/j.visres.2005.08.028.

Fernandez, E. J., A. Unterhuber, B. Povazay, B. Hermann, P. Artal, and W. Drexler. 2006. Chromatic aberration correction of the human eye for retinal imaging in the near infrared. *Opt Express* 14(13):6213–6225. doi: 10.1364/Oe.14.006213.

Glanc, M., L. Blanco, L. Vabre, F. Lacombe, P. Puget, G. Rousset, G. Chenegros et al. 2007. First adaptive optics images with the upgraded Quinze-Vingts hospital retinal imager. In *Adaptive Optics: Methods, Analysis and Applications*, Vancouver, British Columbia, Canada, June 18, 2007.

Glanc, M., E. Gendron, F. Lacombe, D. Lafaille, J.-F. Le Gargasson, and P. Lena. 2004. Towards wide-field retinal imaging with adaptive optics. *Opt Commun* 230(4–6):225–238. doi: 10.1016/J.Optcom.2003.11.020.

Gomez-Vieyra, A., A. Dubra, D. Malacara-Hernandez, and D. R. Williams. 2009. First-order design of off-axis reflective ophthalmic adaptive optics systems using afocal telescopes. *Opt Express* 17(21):18906–18919. doi: 10.1364/OE.17.018906.

Goncharov, A. V., M. Nowakowski, M. T. Sheehan, and C. Dainty. 2008. Reconstruction of the optical system of the human eye with reverse ray-tracing. *Opt Express* 16(3):1692–1703. doi: 10.1364/OE.16.001692.

Hampson, K. M. 2008. Adaptive optics and vision. *J Modern Opt* 55(21):3425–3467. doi: 10.1080/09500340802541777.

Havermann, K., C. Cherici, M. Rucci, and M. Lappe. 2014. Fine-scale plasticity of microscopic saccades. *J Neurosci* 34(35):11665–11672. doi: 10.1523/JNEUROSCI.5277-13.2014.

Hermann, B., E. J. Fernandez, A. Unterhuber, H. Sattmann, A. F. Fercher, W. Drexler, P. M. Prieto, and P. Artal. 2004. Adaptive-optics ultrahigh-resolution optical coherence tomography. *Opt Lett* 29(18):2142–2144.

Hofer, H., J. Carroll, J. Neitz, M. Neitz, and D. R. Williams. 2005. Organization of the human trichromatic cone mosaic. *J Neurosci* 25(42):9669–9679. doi: 10.1523/JNEUROSCI.2414-05.2005.

Hofer, H., L. Chen, G. Y. Yoon, B. Singer, Y. Yamauchi, and D. R. Williams. 2001. Improvement in retinal image quality with dynamic correction of the eye's aberrations. *Opt Express* 8(11):631–643.

Hu, S., B. Xu, X. Zhang, J. Hou, J. Wu, and W. Jiang. 2006. Double-deformable-mirror adaptive optics system for phase compensation. *Appl Opt* 45(12):2638–2642. doi: 10.1364/AO.45.002638.

Jacob, J., M. Paques, V. Krivosic, B. Dupas, A. Couturier, C. Kulcsar, R. Tadayoni, P. Massin, and A. Gaudric. 2014. Meaning of visualizing retinal cone mosaic on adaptive optics images. *Am J Ophthalmol* 159(1):118–123. doi: 10.1016/j.ajo.2014.09.043.

Jonnal, R. S., J. Rha, Y. Zhang, B. Cense, W. Gao, and D. T. Miller. 2007. In vivo functional imaging of human cone photoreceptors. *Opt Express* 15(24):16141–16160.

Kitaguchi, Y., K. Bessho, T. Yamaguchi, N. Nakazawa, T. Mihashi, and T. Fujikado. 2007. In vivo measurements of cone photoreceptor spacing in myopic eyes from images obtained by an adaptive optics fundus camera. *Jpn J Ophthalmol* 51(6):456–461. doi: 10.1007/s10384-007-0477-7.

Koch, E., D. Rosenbaum, A. Brolly, J. A. Sahel, P. Chaumet-Riffaud, X. Girerd, F. Rossant, and M. Paques. 2014. Morphometric analysis of small arteries in the human retina using adaptive optics imaging: Relationship with blood pressure and focal vascular changes. *J Hypertens* 32(4):890–898. doi: 10.1097/hjh.0000000000000095.

Landry, R. J., R. G. Bostrom, S. A. Miller, D. Shi, and D. H. Sliney. 2011. Retinal phototoxicity: A review of standard methodology for evaluating retinal optical radiation hazards. *Health Phys* 100(4):417–434. doi: 10.1097/HP.0b013e3181f4993d.

Lavigne, J.-F. and J.-P. Véran. 2008. Woofer-tweeter control in an adaptive optics system using a Fourier reconstructor. *J Opt Soc Am A* 25(9):2271–2279. doi: 10.1364/JOSAA.25.002271.

Levecq, X. 2011. Phase modulation device for an ophthalmic instrument, ophthalmic instruments equipped with such device, and related calibration method. Patent Application No. US20110001930A1: United States Patent and Trademark Office.

Levene, J. R. 1966. Sir George Biddell Airy, F.R.S. (1801–1892) and the discovery and correction of astigmatism. *Notes Rec R Soc Lond* 21(2):180–199.

Li, C., N. Sredar, K. M. Ivers, H. Queener, and J. Porter. 2010. A correction algorithm to simultaneously control dual deformable mirrors in a woofer-tweeter adaptive optics system. *Opt Express* 18(16):16671–16684. doi: 10.1364/OE.18.016671.

Liang, J., D. R. Williams, and D. T. Miller. 1997. Supernormal vision and high-resolution retinal imaging through adaptive optics. *J Opt Soc Am A Opt Image Sci Vis* 14(11):2884–2892.

Lombardo, M., G. Lombardo, D. Schiano Lomoriello, P. Ducoli, M. Stirpe, and S. Serrao. 2013a. Interocular symmetry of parafoveal photoreceptor cone density distribution. *Retina* 33(8):1640–1649. doi: 10.1097/IAE.0b013e3182807642.

Lombardo, M., M. Parravano, G. Lombardo, M. Varano, B. Boccassini, M. Stirpe, and S. Serrao. 2014a. Adaptive optics imaging of parafoveal cones in type 1 diabetes. *Retina* 34(3):546–557. doi: 10.1097/IAE.0b013e3182a10850.

Lombardo, M., M. Parravano, S. Serrao, P. Ducoli, M. Stirpe, and G. Lombardo. 2013b. Analysis of retinal capillaries in patients with type 1 diabetes and nonproliferative diabetic retinopathy using adaptive optics imaging. *Retina* 33(8):1630–1639. doi: 10.1097/IAE.0b013e3182899326.

Lombardo, M., S. Serrao, P. Ducoli, and G. Lombardo. 2012. Variations in image optical quality of the eye and the sampling limit of resolution of the cone mosaic with axial length in young adults. *J Cataract Refract Surg* 38(7):1147–1155. doi: 10.1016/j.jcrs.2012.02.033.

Lombardo, M., S. Serrao, P. Ducoli, and G. Lombardo. 2013c. Eccentricity dependent changes of density, spacing and packing arrangement of parafoveal cones. *Ophthalmic Physiol Opt* 33(4):516–526. doi: 10.1111/opo.12053.

Lombardo, M., S. Serrao, and G. Lombardo. 2014b. Technical factors influencing cone packing density estimates in adaptive optics flood illuminated retinal images. *PLoS ONE* 9(9):e107402. doi: 10.1371/journal.pone.0107402.

Michaelides, M., J. Rha, E. W. Dees, R. C. Baraas, M. L. Wagner-Schuman, J. D. Mollon, A. M. Dubis et al. 2011. Integrity of the cone photoreceptor mosaic in oligocone trichromacy. *Invest Ophthalmol Vis Sci* 52(7):4757–4764. doi: 10.1167/iovs.10-6659.

Miller, D. T., O. P. Kocaoglu, Q. Wang, and S. Lee. 2011. Adaptive optics and the eye (super resolution OCT). *Eye* 25(3):321–330.

Miller, D. T., L. N. Thibos, and X. Hong. 2005. Requirements for segmented correctors for diffraction-limited performance in the human eye. *Opt Express* 13(1):275–289. doi: 10.1364/Opex.13.000275.

Mims, F. M. 1982. *A Practical Introduction to Lightwave Communications*, 1st edn. Indianapolis, IN: H.W. Sams.

Mrejen, S., T. Sato, C. A. Curcio, and R. F. Spaide. 2014. Assessing the cone photoreceptor mosaic in eyes with pseudodrusen and soft Drusen in vivo using adaptive optics imaging. *Ophthalmology* 121(2):545–551. doi: 10.1016/j.ophtha.2013.09.026.

Muthiah, M. N., C. Gias, F. K. Chen, J. Zhong, Z. McClelland, F. B. Sallo, T. Peto, P. J. Coffey, and L. da Cruz. 2014. Cone photoreceptor definition on adaptive optics retinal imaging. *Br J Ophthalmol* 98(8):1073–1079. doi: 10.1136/bjophthalmol-2013-304615.

Obata, R. and Y. Yanagi. 2014. Quantitative analysis of cone photoreceptor distribution and its relationship with axial length, age, and early age-related macular degeneration. *PLoS ONE* 9(3):e91873. doi: 10.1371/journal.pone.0091873.

Pallikaris, A., D. R. Williams, and H. Hofer. 2003. The reflectance of single cones in the living human eye. *Invest Ophthalmol Vis Sci* 44(10):4580–4592.

Pircher, M. and R. J. Zawadzki. 2007. Combining adaptive optics with optical coherence tomography: Unveiling the cellular structure of the human retina in vivo. *Expert Rev Ophthalmol* 2(6):1019–1035. doi: doi:10.1586/17469899.2.6.1019.

Poonja, S., S. Patel, L. Henry, and A. Roorda. 2005. Dynamic visual stimulus presentation in an adaptive optics scanning laser ophthalmoscope. *J Refract Surg* 21(5):S575–S580.

Popovic, Z., P. Knutsson, J. Thaung, M. Owner-Petersen, and J. Sjostrand. 2011. Noninvasive imaging of human foveal capillary network using dual-conjugate adaptive optics. *Invest Ophthalmol Vis Sci* 52(5):2649–2655. doi: 10.1167/iovs.10-6054.

Popovic, Z., J. Thaung, P. Knutsson, and M. Owner-Petersen. 2012. Dual conjugate adaptive optics prototype for wide field high resolution retinal imaging. In *Adaptive Optics Progress*, R. Tyson (Ed.), pp. 1–21. InTech, Rijeka, Croatia.

Porter, J. 2006. *Adaptive Optics for Vision Science: Principles, Practices, Design, and Applications*, Wiley Series in Microwave and Optical Engineering. Hoboken, NJ: Wiley-Interscience.

Putnam, N. M., H. J. Hofer, N. Doble, L. Chen, J. Carroll, and D. R. Williams. 2005. The locus of fixation and the foveal cone mosaic. *J Vis* 5(7):632–639. doi: 10:1167/5.7.3.

Ramaswamy, G., M. Lombardo, and N. Devaney. 2014. Registration of adaptive optics corrected retinal nerve fiber layer (RNFL) images. *Biomed Opt Express* 5(6):1941–1951. doi: 10.1364/boe.5.001941.

Rha, J., R. S. Jonnal, K. E. Thorn, J. Qu, Y. Zhang, and D. T. Miller. 2006. Adaptive optics flood-illumination camera for high speed retinal imaging. *Opt Express* 14(10):4552–4569.

Rha, J., B. Schroeder, P. Godara, and J. Carroll. 2009. Variable optical activation of human cone photoreceptors visualized using a short coherence light source. *Opt Lett* 34(24):3782–3784. doi: 10.1364/OL.34.003782.

Romero-Borja, F., K. Venkateswaran, A. Roorda, and T. Hebert. 2005. Optical slicing of human retinal tissue in vivo with the adaptive optics scanning laser ophthalmoscope. *Appl Opt* 44(19):4032–4040.

Roorda, A., D. T. Miller, and J. Christou. 2005. Strategies for high-resolution retinal imaging. In *Adaptive Optics for Vision Science*, J. Porter (Ed.), pp. 235–287. John Wiley & Sons, Inc., Hoboken, NJ.

Roorda, A., F. Romero-Borja, W. Donnelly Iii, H. Queener, T. Hebert, and M. Campbell. 2002. Adaptive optics scanning laser ophthalmoscopy. *Opt Express* 10(9):405–412.

Roorda, A. and D. R. Williams. 2002. Optical fiber properties of individual human cones. *J Vis* 2(5):404–412. doi: 10:1167/2.5.4.

Shack, R. V. and B. C. Platt. 1971. Production and use of a lenticular Hartmann screen. *J Opt Soc Am* 61(5):656.

Stepien, K. E., D. P. Han, J. Schell, P. Godara, J. Rha, and J. Carroll. 2009. Spectral-domain optical coherence tomography and adaptive optics may detect hydroxychloroquine retinal toxicity before symptomatic vision loss. *Trans Am Ophthalmol Soc* 107:28–33.

Thaung, J., P. Knutsson, Z. Popovic, and M. Owner-Petersen. 2009. Dual-conjugate adaptive optics for wide-field high-resolution retinal imaging. *Opt Express* 17(6):4454–4467.

Tojo, N., T. Nakamura, C. Fuchizawa, T. Oiwake, and A. Hayashi. 2013a. Adaptive optics fundus images of cone photoreceptors in the macula of patients with retinitis pigmentosa. *Clin Ophthalmol* 7:203–210. doi: 10.2147/opth.s39879.

Tojo, N., T. Nakamura, H. Ozaki, M. Oka, T. Oiwake, and A. Hayashi. 2013b. Analysis of macular cone photoreceptors in a case of occult macular dystrophy. *Clin Ophthalmol* 7:859–864. doi: 10.2147/opth.s44446.

Van Cader, T. C. 1978. History of ophthalmic photography. *J Ophthal Photogr* 1(1):7–9.

Van Norren, D. and L. F. Tiemeijer. 1986. Spectral reflectance of the human eye. *Vision Res* 26(2):313–320.

Wagner-Schuman, M., J. Neitz, J. Rha, D. R. Williams, M. Neitz, and J. Carroll. 2010. Color-deficient cone mosaics associated with Xq28 opsin mutations: A stop codon versus gene deletions. *Vision Res* 50(23):2396–2402. doi: 10.1016/j.visres.2010.09.015.

Werner, J. S., J. L. Keltner, R. J. Zawadzki, and S. S. Choi. 2011. Outer retinal abnormalities associated with inner retinal pathology in nonglaucomatous and glaucomatous optic neuropathies. *Eye (Lond)* 25(3):279–289. doi: 10.1038/eye.2010.218.

Westheimer, G. 2006. Specifying and controlling the optical image on the human retina. *Prog Retin Eye Res* 25(1):19–42. doi: 10.1016/j.preteyeres.2005.05.002.

Wolfing, J. I., M. Chung, J. Carroll, A. Roorda, and D. R. Williams. 2006. High-resolution retinal imaging of cone-rod dystrophy. *Ophthalmology* 113(6):1019.e1. doi: 10.1016/j.ophtha.2006.01.056.

Xue, B., S. S. Choi, N. Doble, and J. S. Werner. 2007. Photoreceptor counting and montaging of en-face retinal images from an adaptive optics fundus camera. *J Opt Soc Am A* 24(5):1364–1372. doi: 10.1364/JOSAA.24.001364.

Yang, Q., D. W. Arathorn, P. Tiruveedhula, C. R. Vogel, and A. Roorda. 2010. Design of an integrated hardware interface for AOSLO image capture and cone-targeted stimulus delivery. *Opt Express* 18(17):17841–17858. doi: 10.1364/OE.18.017841.

Zawadzki, R. J., B. Cense, Y. Zhang, S. S. Choi, D. T. Miller, and J. S. Werner. 2008. Ultrahigh-resolution optical coherence tomography with monochromatic and chromatic aberration correction. *Opt Express* 16(11):8126–8143.

Zawadzki, R. J., S. S. Choi, S. M. Jones, S. S. Oliver, and J. S. Werner. 2007. Adaptive optics-optical coherence tomography: Optimizing visualization of microscopic retinal structures in three dimensions. *J Opt Soc Am A Opt Image Sci Vis* 24(5):1373–1383.

Zawadzki, R. J., S. M. Jones, S. S. Olivier, M. Zhao, B. A. Bower, J. A. Izatt, S. Choi, S. Laut, and J. S. Werner. 2005. Adaptive-optics optical coherence tomography for high-resolution and high speed 3D retinal in vivo imaging. *Opt Express* 13(21):8532–8546.

Zhang, Y., J. Rha, R. Jonnal, and D. Miller. 2005. Adaptive optics parallel spectral domain optical coherence tomography for imaging the living retina. *Opt Express* 13(12):4792–4811.

Zou, W., X. Qi, and S. A. Burns. 2008. Wavefront-aberration sorting and correction for a dual-deformable-mirror adaptive-optics system. *Opt Lett* 33(22):2602–2604. doi: 10.1364/OL.33.002602.

Ophthalmic instrumentation

6 Adaptive optics optical coherence tomography (AO-OCT)

Nathan Doble

Contents

6.1 INTRODUCTION

This chapter details the developments in combining the high axial resolution afforded by optical coherence tomography (OCT) with the benefits of improved lateral resolution provided by adaptive optics (AO) for imaging of the living retina. The discussion assumes that the reader is familiar with the concepts of OCT and AO detailed in previous chapters.

6.1.1 BRIEF HISTORY

OCT and AO developed along separate paths and it is only relatively recently that retinal imaging systems combining both technologies have been developed. Figure 6.1 describes the chronological timeline of the major milestones in both the OCT and AO fields—scanning laser ophthalmoscope (SLO) development has been added for completeness. The shaded areas show where two or more modalities were combined for the first time.

6.1.2 OPTICAL COHERENCE TOMOGRAPHY (OCT)

OCT is a powerful optical imaging technique that has provided remarkable images of the living human retina. It is a noninvasive imaging modality that uses partially coherent interferometry to achieve very high axial resolution. The decoupling of the lateral and axial resolutions has contributed to its widespread adoption

YEAR	OCT	AO	SLO	OTHER
1953		Concept of AO proposed (Babcock 1953)		
1977		First astronomical systems (Buffington et al. 1977a,b, Hardy et al. 1977)		
1980			First SLO (Webb et al. 1980)	
1986	Optical ranging of rabbit cornea in vivo (Fujimoto et al. 1986)			
1987			First confocal SLO (Webb et al. 1987)	
1988	Eye length measurement using partially coherent light (Fercher et al. 1988)			
1989		Use of a DM in an SLO, no wavefront sensing (Dreher et al. 1989)		In vivo measure of cone spacing (Artal and Navarro 1989)
1990				
1991	In vitro (TD)-OCT (Huang et al. 1991)			
1993	In vivo (TD)-OCT (Fercher et al. 1993, Swanson et al. 1993)			
1994		First SH-WFS measure in the human eye (Liang et al. 1994)		
1995	In vivo ocular length measurements with (SD)-OCT (Fercher et al. 1995)			
1996				First image of cones in the living human eye (Miller et al. 1996)
1997	First implementations of (SS)-OCT on the eye (Lexer et al. 1997); non-eye (Golubovic et al. 1997)	First AO system for the eye (Liang et al. 1997)		
1998	Combined (TS)-OCT-SLO (Podoleanu and Jackson 1998)			
2001		Real-time AO systems for the eye (Hofer et al. 2001b, Fernandez et al. 2001)		
2002	In vivo retinal imaging with (SD)-OCT (Wojtkowski et al. 2002a,b)	First AO-SLO (Roorda et al. 2002)		
2003	First AO en-face coherence-gated (TD) OCT (Miller et al. 2003)			
2004	AO-(TD)-OCT (Hermann et al. 2004)			
2005	AO-(SD)-OCT (Zawadzki et al. 2005, Zhang et al. 2005, Fernandez et al. 2005a)			
2006	AO-(TS)-OCT-SLO (Merino et al. 2006)			(TS)-OCT in vivo imaging of cones (non-AO) (Pircher et al. 2006a)
2009	AO-(SD)-OCT-SLO (Zawadzki et al. 2009)			
2010	AO-(SS)-OCT-SLO (Mujat et al. 2010)			

Figure 6.1 Charting the chronological technological milestones in the implementation of adaptive optics and optical coherence tomography to the human eye. Key scanning laser ophthalmoscope (SLO) developments are also described. TD, time domain; TS, transversal scanning; SD, spectral domain; SS, swept source. SD and SS are both considered to be Fourier domain (FD) techniques. The shaded areas show where two or more modalities were combined for the first time.

Ophthalmic instrumentation

as images of good clarity can be acquired through undilated pupils (1–2 mm) albeit with transverse resolutions limited to ~15 μm. The origin of OCT (Masters 1999) can be traced back to the work of Fujimoto et al. (1986) and their in vivo optical ranging measurements of rabbit corneas and to that of Fercher et al. (1988), who used interferometry with a partially coherent light source to measure the axial length of the human eye. Huang et al. (1991) first used the term "optical coherence tomography" and demonstrated the imaging capabilities with in vitro tissue (human retina and coronary artery) using time domain (TD)-OCT. A few years later saw the first in vivo images of the human eye taken with (TD)-OCT systems (Fercher et al. 1993, Swanson et al. 1993) and remarkably the first commercial OCT system appeared in 1996, the OCT-1 by Carl Zeiss Meditec.

OCT is unique in that it can be implemented in many forms, with the main variants aside from (TD)-OCT being Fourier techniques such as spectral domain (SD)-OCT (Fercher et al. 1995) or swept source (SS)-OCT (Lexer et al. 1997); further subdivision of these broad classes is then possible. For the advantages/disadvantages of the various OCT techniques, the reader is directed to the literature (Fercher 1996, Bouma and Tearney 2002, Fercher et al. 2003, Leitgeb et al. 2003, Wojtkowski 2010).

6.1.3 ADAPTIVE OPTICS (AO)

The concept of AO was proposed in 1953 (Babcock 1953) to address atmospheric turbulence. One of the first closed loop realizations was the correction of horizontal path atmospheric turbulence (Hardy et al. 1977). The COME-ON system was the first use on a terrestrial telescope (Rousset et al. 1990). This technology has since been transitioned into other fields such as the ophthalmic/microscopy space over the past 20 years.

Around the same time as the first OCT papers, the application of AO to the human eye began with the use of a membrane deformable mirror (DM) for the low-order correction of sphere and cylinder in an SLO (Dreher et al. 1989). This was followed by the use of a Shack–Hartmann wavefront sensor (SH-WFS) in the human eye (Liang et al. 1994, Liang and Williams 1997). The first full AO system utilizing a WFS and DM soon followed in 1997 (Liang et al. 1997). The intervening years have seen the in vivo imaging of many retinal cells along with advances in the development of the associated AO hardware. A detailed discussion is beyond the scope of this chapter, and further information is provided both within this book and other published works (Porter et al. 2006, Hampson 2008, Miller and Roorda 2009, Godara et al. 2010, Rossi et al. 2011, Williams 2011, Carroll et al. 2013, Lombardo et al. 2013). For further information on the theory and implementation of AO systems, the reader is directed to reference texts (Hardy 1998, Tyson 2010).

Note that Miller et al. (1996) showed that it was possible to see cone photoreceptors in the living human eye in a non-AO fundus camera with a careful correction of sphere and cylinder. Other authors have since reported images of cones from OCT systems without the use of AO (Pircher et al. 2006a, Potsaid et al. 2008).

6.1.4 COMBINING AO AND OCT

The first AO-OCT demonstration was the en-face coherence-gated (TD)-OCT system described by Miller et al. (2003). This approach aimed to exploit the parallel, en-face image acquisition

capability and the integration with an existing flood-illuminated AO system. Unfortunately, this first implementation suffered from low sensitivity and problematic speckle. The following year saw a (TD)-OCT system employing AO (Hermann et al. 2004), but this system was unable to resolve single retinal cells in vivo. In 2005, several (SD)-OCT systems were demonstrated, had much faster image acquisition, and provided the first in vivo single cell images. Zhang et al. (2005) used a novel line scanning approach enabling the capture of full B-scans. Their system demonstrated the improved lateral resolution by clearly showing the breakup of the inner segment (IS)/outer segment (OS) junction at several retinal eccentricities. Zawadzki et al. (2005) demonstrated three-dimensional volumetric imaging of the cone photoreceptors and retinal capillaries using the more conventional A-scan to B-scan image reconstruction. Fernandez et al. (2005a) detail the construction of an AO-(SD)-OCT system employing a liquid crystal spatial light modulator (LC-SLM) as the wavefront corrector. The following year, Zhang et al. (2006) built a faster system that showed cone photoreceptors within the reconstructed C-scans due to the reduced eye motion effects.

Merino et al. (2006) used (TD)-OCT approach with AO, but instead of employing the conventional A-scan to B-scan reconstruction, the beam was scanned in an en-face plane similar to the image acquired in an SLO. Once the first C-scan is acquired, the reference arm position is moved axially changing the position of the coherence gate and another C-scan is acquired; this is termed transverse scanning or (TS)-OCT. The technique results in single C-scan images with reduced transverse eye motion effects but is sensitive to axial eye motion (Pircher et al. 2010).

Several groups have now demonstrated AO-OCT systems using a variety of OCT and AO implementations, and these are discussed later in the chapter. Recent years have seen the extension to also including the SLO imaging modality, that is, AO-OCT-SLO systems (Merino et al. 2006, Zawadzki et al. 2009, Mujat et al. 2010, Hammer et al. 2012). Additionally, wavefront sensorless AO-OCT (Jian et al. 2014, Wong et al. 2015) and computational-based AO-OCT (Adie et al. 2012) have been reported.

While several companies have commercialized stand-alone OCT technology for ophthalmology (Bioptigen Inc., Carl Zeiss Meditec Inc., Heidelberg Engineering GmbH, Nidek Co. Ltd., Optopol Technology, Optos Plc., Optovue Inc., and Topcon Medical Systems), there are no commercial AO-OCT systems currently on the market.

The next sections detail how AO is combined with OCT, its intrinsic advantages and describes the current status of the field. Prior reviews also describe AO-OCT development (Pircher and Zawadzki 2007, Miller et al. 2011, Wojtkowski et al. 2012).

6.2 BENEFITS OF COMBINING AO AND OCT

Combining AO with OCT provides three distinct advantages: (1) increased lateral resolution, (2) smaller speckle size, and (3) increased sensitivity due to the ability to collect light over a larger pupil diameter. It should be noted, however, that the disadvantage is much higher system complexity. For example, multiple extra telescopes are required to relay the beam to the WFS and DM, and due to light losses in the longer sample arms, the sensitivity of AO-OCT is not that much higher than for simple OCT systems.

6.2.1 RESOLUTION IMPROVEMENT

Like a conventional fundus camera, the lateral (transverse) resolution, Δx, of an OCT system is described by Equation 6.1:

$$\Delta x = \frac{1.22 f \lambda}{D} \tag{6.1}$$

where

    f is the focal length of the eye
    λ is the imaging wavelength
    D is the pupil diameter

Dilating the pupil, hereby increasing D would appear to increase the lateral resolution but unfortunately the increasing ocular aberration mitigates any potential benefit. For a strictly confocal system where either the fiber input (OCT) or the pinhole (cSLO) is close to the diameter of the Airy disk, Equation 6.1 becomes (Webb 1996)

$$\Delta x = \frac{0.88 f \lambda}{D} \tag{6.2}$$

Figure 6.2 shows the point spread function (PSF) for a normal human subject as a function of the pupil diameter (2, 4, and 6 mm) as measured by a Shack–Hartmann WFS in the system described by Headington et al. (2011). The equivalent unaberrated PSFs are also shown. In the aberrated case, it is clear that the form of the PSF degrades with increasing pupil size becoming blurred with less energy in the Airy disk.

Fortunately, AO corrects for the increased aberration allowing for the use of dilated pupils. This maximizes the available resolution and allows for corrected PSFs comparable to the 6 mm unaberrated case shown in Figure 6.2.

The axial (depth, longitudinal) resolution assuming a light source with a Gaussian spectral profile, Δz, of an OCT system is given by (Fercher and Hitzenberger 2002)

$$\Delta z = \frac{2 \ln 2 \lambda_c^2}{\pi n \Delta \lambda} \tag{6.}$$

where

    λ_c is the center wavelength
    $\Delta \lambda$ is the bandwidth of the imaging light source
    n is the refractive index of the retinal tissue

Note that Δz is not dependent on the pupil diameter D, resulting in the ability to acquire images with high axial resolution through undilated pupils. Equation 6.3 however does assume that the dispersion is matched in both the reference and sample arms. The use of high bandwidth sources (>100 nm) makes this dispersion condition more challenging to achieve as elements such as beam splitters, mirror coatings, and the eye itself must be accurately matched (Wojtkowski et al. 2004).

Assume a human eye with D = 7 mm, f = 16.7 mm, and n = 1.38 being imaged with an AO-OCT system with a light source λ_c = 840 nm and $\Delta \lambda$ = 100 nm. Equations 6.1 and 6.3 give lateral and axial resolutions of Δx = 2.4 µm and Δz = 2.3 µm, respectively—comparable to the smallest cells

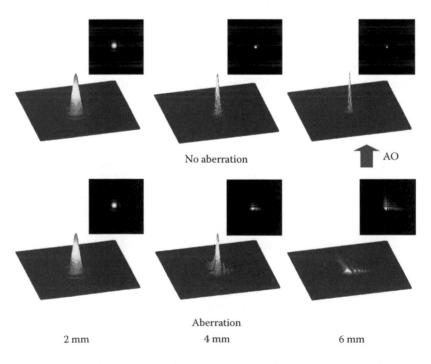

Figure 6.2 The effect of ocular aberrations on the point spread function (PSF) as a function of pupil size for a normal human subject. The top row shows the unaberrated PSF for 2, 4, and 6 mm pupil diameters. The bottom row shows the actual plots for the human subject reconstructed from SH-WFS measurements in a flood-illuminated adaptive optics system (Headington et al. 2011). Note the increasing blur of the PSF with increasing pupil size. All PSF amplitudes have been plotted to the same scale to allow for comparison. The inset figure shows the same PSF but in 2D.

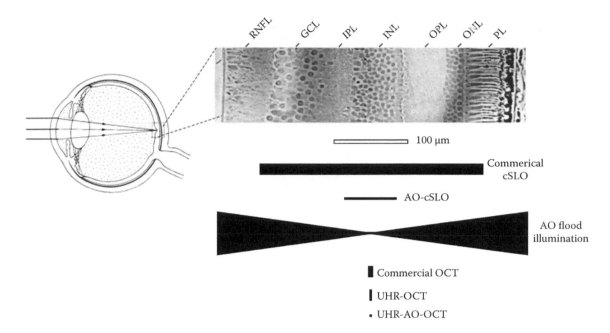

Figure 6.3 Comparison of the lateral and axial resolution for the three primary imaging modalities, (1) flood-illuminated fundus cameras, (2) cSLO, and (3) OCT, as compared to a scaled histological section. OCT provides the smallest three-dimensional resolution volume especially when used with large bandwidth light sources and lateral resolution improvement through adaptive optics (AO) (Miller et al. 2011). The commercial systems have limited lateral resolution due to the use of small pupils to limit the effect of the ocular aberrations. The AO flood-illuminated fundus cameras have good lateral resolution but accept all of the light reflected from the retina. (Reprinted by permission from Macmillan Publishers Ltd. *Eye* (*Lond.*), Miller, D.T., Kocaoglu, O.P., Wang, Q., and Lee, S., Adaptive optics and the eye (super resolution OCT), 25(3), 321–330, copyright 2011.)

found in the human eye. To further illustrate the resolution capability of AO-OCT, the lateral and axial resolutions for both commercial and research imaging instruments are shown in Figure 6.3 (Miller et al. 2011). When compared to a scaled histological section, it is clear that OCT has the best theoretical resolution when compared to other imaging modalities such as SLO and flood-illuminated fundus cameras.

6.2.2 INFLUENCE OF SPECKLE

Due to the interferometric nature of OCT, speckle is a major challenge and is observed as a random intensity distribution whenever coherent light is reflected from a rough optical surface, in this case the retina (Dainty 1984). Speckle size is proportional to $\lambda f/D$, being smallest at the diffraction limit of the system. AO decreases speckle size in the lateral dimension, while the use of a broad bandwidth sources reduces the axial size. Miller et al. (2011) estimate a 25-fold decrease in the speckle volume when combining AO with OCT. It is important that speckle is not interpreted as actual retinal structure and strategies to overcome speckle complications are described later in the chapter.

To illustrate the speckle effect, Figure 6.4 compares an image from a commercial OCT system to that from a research grade instrument. The top figure is the B-scan image at an eccentricity of 6° superior to the fovea (3° field of view [FOV] at the retina) from a 62-year-old subject taken with the Heidelberg Spectralis system. The lower figure is the same retinal area taken with the AO-OCT system at Indiana University focused at the retinal nerve fiber layer (RNFL) (Miller et al. 2011). The difference in the image quality is due in part to the reduction in speckle size.

6.2.3 SENSITIVITY

A further benefit of using AO is the ability to effectively collect light from a larger pupil. Cense et al. (2009a) report a 4–8 dB sensitivity increase when going from a 1.2 mm pupil diameter without AO to a 6 mm pupil with AO. The exact increase is dependent on the actual retinal layer under study. Figure 6.5a shows an AO-OCT scan from a human subject that shows the breakup of the photoreceptor layer due to the improved lateral resolution afforded by AO. By introducing a defocus shift, the plane of focus can be shifted from the photoreceptors to the RNFL as shown in Figure 6.5b (Kocaoglu et al. 2011a).

One of the biggest benefits of AO-OCT to ophthalmology is its ability to help interpret clinical OCT findings including interpretation of cellular origins responsible for OCT signals. As an example, Jonnal et al. (2014) discuss the origins of OCT photoreceptor layers bands, a highly relevant topic in recent years.

6.3 IMPLEMENTATION

Figure 6.6a and b shows schematic diagrams of the AO control loop and OCT subsystems, respectively, together with their integration into the combined AO-OCT system—Figure 6.6c. All of the AO-OCT systems reported in the literature have custom optical designs and implementations. These variations can be in the placement of the DM (e.g., before or after the scanners), the use of separate light sources for the AO and OCT subsystems through to the particular implementation of the OCT or AO. These differences are highlighted throughout the following discussion.

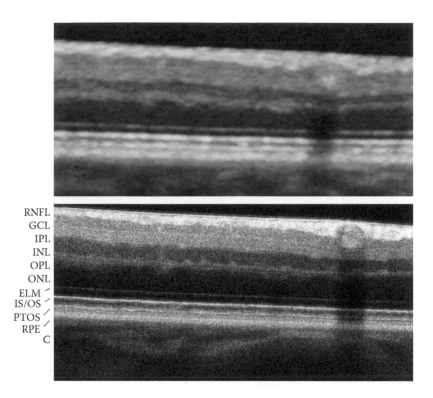

Figure 6.4 Example of the image improvement observed when adaptive optics (AO) is combined with optical coherence tomography (OCT). The top figure is the B-scan image at an eccentricity of 6° superior to the fovea (3° field of view at the retina) from a 62-year-old subject taken with the Heidelberg Spectralis system and the lower figure is the same retinal area taken with the AO-OCT system at Indiana University focused at the RNFL. The AO improves the speckle size in the lateral dimension, while the use of a broadband light source reduces the axial speckle size. C, choroid; ELM, external limiting membrane; GCL, ganglion cell layer; INL, inner nuclear layer; IPL, inner plexiform layer; IS/OS, inner segment/outer segment junction; ONL, outer nuclear layer; OPL, outer plexiform layer; PTOS, posterior tip of the outer segment; RNFL, retinal nerve fiber layer; RPE, retinal pigment epithelium. (Reprinted by permission from Macmillan Publishers Ltd. *Eye (Lond.)*, Miller, D.T., Kocaoglu, O.P., Wang, Q., and Lee, S., Adaptive optics and the eye (super resolution OCT), 25(3), 321–330, copyright 2011.)

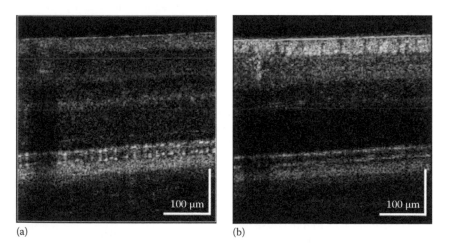

(a) (b)

Figure 6.5 (a) AO-OCT B-scans with the focal plane at the photoreceptors. The breakup of the photoreceptor layer is clear due to the improved lateral resolution. (b) Plane of focus shifted to the RNFL with the shift being introduced by the DM. (Reprinted from *Vision Res.*, 51(16), Kocaoglu, O.P., Cense, B., Jonnal, R.S. et al., Imaging retinal nerve fiber bundles using optical coherence tomography with adaptive optics, 1835–1844, Copyright 2011a, with permission from Elsevier.)

6.3.1 THE AO CONTROL LOOP

The AO control loop consists of three basic components: (1) a wavefront corrector, (2) a wavefront sensor (WFS), and (3) the wavefront reconstructor and control—Figure 6.6a. Imagine a point source located on the retinal surface resulting in diverging wavefronts, which are then distorted by aberrations in the lens and cornea of the eye. This light is then reflected off the wavefront corrector, typically a DM. The majority of the light is transmitted to the imaging camera/confocal pinhole, but a fraction (5%–20%) is reflected to the WFS where the aberrations are measured. The wavefront reconstructor and control computer generate the appropriate control signals for the wavefront corrector. The continual measurement and correction

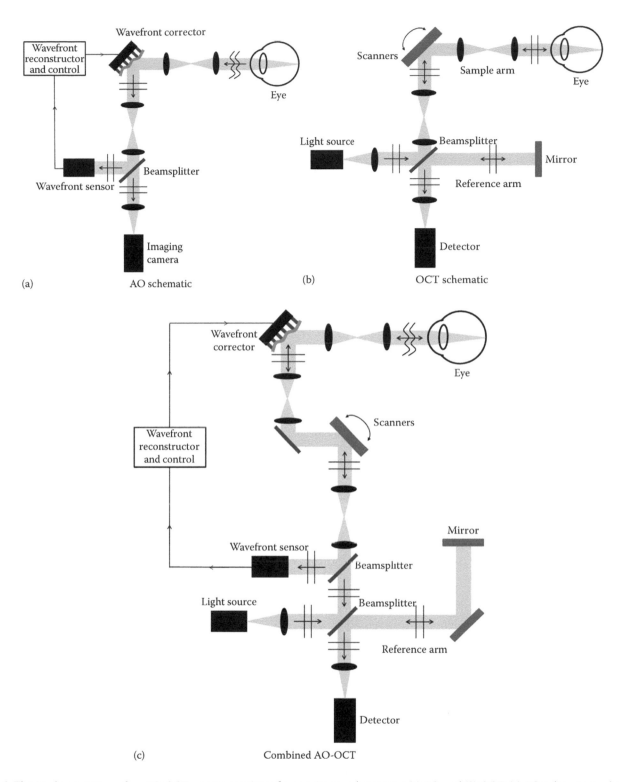

Figure 6.6 The implementation of an AO-OCT system consists of two primary subsystems: (1) AO and (2) OCT. (a) AO subsystem schematic. A point source is reflected of the retinal surface resulting in diverging wavefronts that are then distorted by aberrations in the lens and cornea of the eye. This light is reflected off the wavefront corrector, typically a DM. The majority of the light is transmitted to the imaging camera, but a small fraction (5%–20%) is reflected to the WFS where the aberrations are measured and the appropriate control commands are generated and written to the DM. (b) OCT schematic. A broadband light source (incident from the left) is split into two sample and reference arms via a beam splitter. In the sample arm, scanners (located in a pupil conjugate plane) raster the beam across the retinal surface. The reflected light is then recombined via the same beam splitter with the light that has propagated through the reference arm. These two beams interfere and the interference is measured by the detector. (c) The combined AO-OCT system. The light source is again split by the beam splitter and is reflected off the scanning assembly, the DM, and is finally incident on the retinal surface. The reflected light then follows the reverse path through the DM and scanners. Part of the light is reflected into the WFS arm, while the majority is recombined with the reference arm light to provide the OCT signal. The WFS is positioned to "see" as many of the optical components as possible to reduce the effect of noncommon path errors. Current AO-OCT systems have individual custom designs varying in the placement of the DM (e.g., before or after the scanners), the use of separate light sources for the AO and OCT subsystems through to the particular implementation of the OCT.

of the aberrations must be fast enough to correct for the temporal changes in the ocular aberration (Hofer et al. 2001a, Diaz-Santana et al. 2003), and if this condition is met, then the system is termed closed loop.

6.3.1.1 The wavefront corrector

This device provides the conjugate optical correction to the measured aberration profile. Any wavefront corrector will try to equalize the optical path length (OPL) seen by all points on the incident wavefront, resulting in a plane wavefront. The OPL is the product of the refractive index, n, and the physical path length, d, and two broad classes of wavefront corrector can be defined based on how the OPL correction is achieved.

The most popular wavefront correctors for AO-OCT are DMs whose top mirrored surface is deformed by an underlying two-dimensional pattern of electrodes, that is, it varies d. The benefit of using mirrored surfaces in AO-OCT systems is the lack of back reflections (advantageous for the WFS) and zero dispersion (important when using broad bandwidth light sources). The DM is also used to control the focal plane in such systems by introducing a defocus offset.

A variety of DMs with varying actuation mechanisms have been successfully used in AO-OCT systems; these include (1) membrane bulk MEMS DMs from OKO Technologies (Hermann et al. 2004, Merino et al. 2006), (2) AOA Xinetics DMs (Zhang et al. 2005), (3) AOptix bimorph DMs (Zawadzki et al. 2005, Zhang et al. 2006), (4) surface micromachined MEMS DMs from Boston Micromachines Corporation (BMC) (Bigelow et al. 2007) and Iris AO Inc. (Jian et al. 2013), (5) Imagine Eyes DMs (Torti et al. 2009, Kurokawa et al. 2010, Sasaki et al. 2012, Felberer et al. 2014), and (6) ALPAO SAS DMs (Zawadzki et al. 2011, Meadway et al. 2013, Kocaoglu et al. 2014a, Wells-Gray et al. 2015). Manufacturer's websites are included in the references.

Some systems have employed two DMs in a woofer-tweeter arrangement with one mirror providing high stroke compensation of low-order aberrations such as defocus and astigmatism and the second giving correction of the higher-order aberrations. Results from an AOptix/BMC combination have been published by several authors (Zawadzki et al. 2007, 2008, 2011, Cense et al. 2009a,b, Kocaoglu et al. 2011b); an Imagine Eyes/BMC pairing has also been reported (Mujat et al. 2010, Hammer et al. 2012).

A second but less common wavefront corrector for AO-OCT is LC-SLMs. These devices produce localized changes in their refractive index, n (d is fixed). The refractive index change is achieved through the application of a voltage that rotates the liquid crystal molecules. Fernandez et al. (2005a) report the use of an optically addressed LC-SLM from Hamamatsu Photonics.

Irrespective of form, any wavefront corrector must be capable of correcting the magnitude and spatial frequency profiles of the aberrations found in the human eye (Porter et al. 2001, Thibos et al. 2002). The theoretical correction performance of various wavefront correctors has been described by Miller et al. (2005) and Doble et al. (2007). The temporal requirements for vision science AO applications are modest and within the range of current wavefront corrector technologies, but the stroke requirements are demanding.

6.3.1.2 The wavefront sensor (WFS)

Almost all AO-OCT systems employ Shack–Hartmann WFS to measure the ocular aberration. The principle behind their operation is described in this book as well as in the reference texts (Hardy 1998, Tyson 2010). As with any AO system the location of the WFS should be chosen to "see" as many of the delivery optics as possible and is generally placed first in this arm, hereby reducing the effect of noncommon path errors. The larger question is the use of a separate light source for the WFS as opposed to using the same source for both the AO and OCT channels. The use of the same light source will reduce the acquisition speed of the OCT system as some of the light is redirected to the WFS rather than the fiber input. The use of sensitive WFS cameras can somewhat offset this effect, but with increasing numbers of DM actuators and hence required number of WFS subapertures, this will continue to be a concern. For systems employing a further SLO channel, there is more flexibility as these generally require two separate wavelength sources. Regardless of the system specifics, the requirement to stay within the maximum permissible exposure limits remains (Delori et al. 2007) and needs to be calculated on system-by-system basis.

6.3.2 OCT

Figure 6.6b shows the OCT schematic. In its most basic form light from a broadband light source (incident from the left) is split into the sample and reference arms via a beam splitter. In the sample arm, scanners (located in pupil conjugate planes) raster the beam across the retinal surface. The reflected light is then recombined via the same beam splitter with the light that has propagated through the reference arm. These two beams interfere and this interference is measured by the detector.

6.3.3 OVERALL INTEGRATION

Figure 6.6c shows the schematic of the AO-OCT implementation using a single light source. The light source is again split by the beam splitter and is reflected off the scanning assembly, the DM, and is finally incident onto the retinal surface. The reflected light then follows the reverse path through the DM and scanners. Part of the light is reflected into the WFS arm, while the majority is recombined with the reference arm light to provide the OCT signal. Current AO-OCT systems have individual custom designs varying in the placement of the DM (e.g., before or after the scanners), the use of separate light sources for the AO and OCT subsystems through to the particular implementation of the OCT. The AO components are always placed in the sample arm of the OCT layout, that is, the optical path to the eye. Aberrations present in the reference arm are generally small and will have no effect on the lateral resolution. The usual OCT requirements such as balancing the dispersion in both the reference and sample arms must still be followed, for example.

However, the use of AO does not preclude the need for careful optical design. The sample arm delivery optics follow the same design rules as that of AO-SLOs usually employing $4f$ optics to reply the pupil conjugate plane of the eye to the scanners, DM, and WFS. AO is not fast enough to correct for the field aberrations as the beam as it scanned across the retinal surface; however through careful system design, these

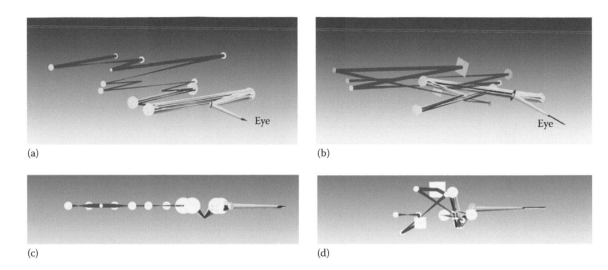

Figure 6.7 The out-of-plane aberration cancelling AO-OCT design of Lee et al. (2013). (a) and (c) show complementary views of their older design, whereas (b) and (d) show the extension to three dimensions. The out-of-plane design reduces the system aberrations and pupil wander. (Reprinted from Lee, S.H. et al., Improved visualization of outer retinal morphology with aberration cancelling reflective optical design for adaptive optics—Optical coherence tomography, *Biomed. Opt. Express*, 4(11), 2508–2517, 2013. With permission of Optical Society of America.)

aberrations can be significantly reduced. Recent AO-SLO designs (Burns et al. 2007, Gomez-Vieyra et al. 2009, Dubra and Sulai 2011) whereby the relay telescopes are folded out of the plane to minimize field-dependent aberrations have been recently applied to AO-OCT design (Lee et al. 2013). Figure 6.7 shows their out-of-plane aberration cancelling AO-OCT design, and Figure 6.7a and c shows complementary views of their older design, whereas Figure 6.7b and d shows the extension to three dimensions. Successive pairs of telescopes are arranged out of plane so that their aberrations effectively cancel; this reduces the system aberrations and pupil wander. An in-plane AO-OCT system employing toroidal mirrors (TM) to deliver a similar outcome has also been described (Liu et al. 2013).

The position of the DM can be located either side of the scanning assembly. Locating the DM between the light source and the scanners eliminates any pupil wander as the beam is descanned upon reflection from the eye. Other designs place the DM in the first pupil plane after the eye as shown in Figure 6.6c. Correcting the aberrations in the plane closest to the eye minimizes the area of the optical surfaces used in subsequent telescopes. Furthermore, by keeping the magnification as close to unity as possible maximizes the ability to cancel the astigmatism in both the retinal and pupil conjugate planes (Gomez-Vieyra et al. 2009). One disadvantage of this DM placement can be slight pupil plane wander on the DM as the beam is scanned.

An alternative but uncommon placement of the AO system is in the detection, that is, between the detector and the first beam splitter. The advantages and disadvantages of this approach are discussed in Chapter 10 of Porter et al. (2006).

6.4 AO-OCT SYSTEMS

Table 6.1 describes the operating parameters for the various AO-OCT systems in use at the current time. Many of these have gone through several design/upgrade iterations and the relevant references are provided. Performance figures in their respective high-resolution imaging modes are given, and only

the details for the AO-OCT system are described when part of a multimodality system.

The majority are (SD)-OCT systems employing an superluminscent diode (SLD) or femtosecond laser (fs laser) as their light source, operating with a center wavelength in the 800–850 nm range although longer wavelength sources up to 1020 nm have been used (Sasaki et al. 2012). The A-scan acquisition rates can be up to a few hundred kHz resulting in 1° × 1° volumes acquired in a second or less with both lateral and axial resolutions of a few micrometers. Another FD technique, (SS)-OCT, has been successfully combined with AO, and this is described in more detail later (Mujat et al. 2010).

Other groups have built (TD)-OCT-based AO systems whereby the reference mirror is rapidly scanned to give an A-scan with subsequent registration into B-scans. More recently, this has given way to transversal scanning (TS)-OCT, whereby an en-face OCT image of the retina is acquired analogous to an SLO image, the plane of focus is then adjusted, and another image is acquired (Merino et al. 2006, Felberer et al. 2014).

The majority are designed for human imaging and several also employ an SLO channel and retinal tracking in addition to AO-OCT. Common to almost all the AO subsystems, however, is the use of a SH-WFS and a DM.

Hammer et al. (2012) and Jian et al. (2013, 2014) describe systems for animal imaging. The 2014 paper by Jian is unusual in that it employs a wavefront sensorless approach. The next sections describe examples of the various AO-OCT implementations in detail.

6.4.1 SYSTEM EXAMPLE: AO-(SD)-OCT

Figure 6.8 shows a schematic layout of the AO-(SD)-OCT system at Indiana University (Kocaoglu et al. 2014a). The system also includes a wide-field line scanning ophthalmoscope (LSO) and a retinal tracker. The blue line shows the AO-OCT path, the red line is the path used for the LSO, and the green path is used for the retinal tracker.

Table 6.1 Basic operational specifications of AO-OCT systems detailed in the literature

PUPIL DIA. (mm)	LIGHT SOURCE TYPE	λ_c (nm)	$\Delta\lambda$ (nm)	POWER (µW)	A-SCAN RATE (kHz)	B-SCAN RATE (Hz)	VOL.	VOL. ACQ. TIME (S)	FOV (DEG)	LAT. (µm)	AXIAL (µm)	SEPARATE WFS SOURCE	DM (# ACT.)	SLO	RT	NOTE	REFERENCES
Spectral domain																	
6.7	fs laser	809	81	400	167	606 270 A/B	216 B-scans	4.6 (13 vols)	0.90 × 0.72		2.6	No	Alpao (97)	No	Yes	a	Kocaoglu et al. (2011b, 2014a), Liu et al. (2013), Cense et al. (2009b)
6	SLD	840	50	500		140 512 A/B			1 × 1		4.7	Yes	Alpao (97)	Yes	No	b	Meadway et al. (2013, 2014)
6.8	SLD	840	112	650	102				0.75 × 0.75	2		No	Alpao (97)	No	No	c	Lee et al. (2013)
7.2	SLD	840	140	350	36	60 601 A/B	250	4.3	0.5 × 1			No	Alpao (97)	Yes	no		Wells-Gray et al. (2015)
7.5	SLD	850			29	28 1024 A/B			1 × 1, 2 × 2			Yes	BMC (140), Alpao (97)/ Mirao (52)	Yes	Yes	d	Hammer et al. (2012), Bigelow et al. (2007)
7.4	SLD	1020	106	1280	91.9	128 A-scans	128 B-scans	5.6 vols/s	0.8 × 0.8		3.4	Yes	Mirao (52)	No	No		Sasaki et al. (2012), Kurokawa et al. (2010)
6	Ti:sapp laser	800	140	800	120	256 A-scans	256 B-scans	0.5	~1.3 × 1.3	2.7	2	No	Mirao (52)	No	No		Torti et al. (2009), Fernandez et al. (2008)
6	Ti:sapp laser	800	130		25	1024 A-scans	45 B-scans	2	~750 × 110 µm			No	Hamamatsu LC-SLM (230,400)	No	No		Fernandez et al. (2005a)
Spectral domain: Designed specifically for animal imaging																	
1	SLD	860	135	~750	90	600 360 A/B	360 B-scan	4.6 vol/s	333 × 333 µm			No WFS	Iris AO Inc (37)	No	No	e	Jian et al. (2013, 2014)
Swept source																	
7–8	SS laser	1070	79	<2600	20	20			1 × 3 mm		4.6	Yes	BMC (140), Mirao (52)	Yes	Yes	b	Mujat et al. (2010)

(Continued)

Table 6.1 (Continued) Basic operational specifications of AO-OCT systems detailed in the literature

PUPIL DIA. (mm)	LIGHT SOURCE				ACQUISITION RATE				FOV (DEG)	RESO-LUTION		SEPARATE WFS SOURCE	DM (# ACT.)	SLO	RT	NOTE	REFERENCES
	TYPE	λc (nm)	Δλ (nm)	POWER (μW)	A-SCAN RATE (kHz)	B-SCAN RATE (Hz)	VOL.	VOL. ACQ. TIME (s)		LAT. (μm)	AXIAL (μm)						
Time domain																	
3.68	Ti:sapp laser	800	130	800	125–250 Hz A-scans	600 A-scans			.25–2.8 mm	5–10	3	No	OKO (37)	No	No	f	Hermann et al. (2004)
6	SLD	840	49	325		500 A-scans			1.5 mm		6.7	Yes	PZT DM (37)	No	No	g	Shi et al. (2008)
					Frame size (pix)	Frame rate (Hz)	Scan depth (μm)	Vol. acq. time (s)									
Transversal scanning																	
7	SLD	840	50	<700	1152 × 790	20	200	6	1 × 1	~2.2	~4.5	No	Mirao (52)	Yes	No	h	Felberer et al. (2012, 2014), Pircher et al. (2008)
6	SLD	831	17	300		2			330 × 550 μm	5	13	No	OKO (37)	Yes	No		Merino et al. (2006)

Notes: Many systems have seen several iterations and only the details for the latest reported version are provided here. Figures for their respective high-resolution imaging mode are given throughout, and only the details for the AO-OCT system are described when part of a multimodality system. All systems employ Shack–Hartmann wavefront sensing except for the wavefront sensorless system described by Jian et al. (2014).

a Source was bandpass filtered, originally λc = 800 nm, Δλ = 160 nm.
b Setup does not allow for simultaneous OCT/SLO acquisition.
c Latest design incorporates out-of-plane telescopes to minimize astigmatism and pupil wander. Prior systems incorporate OCT and SLO together (Zawadzki et al. 2005, 2007, 2008, 2011).
d System can also be configured to image small animal eyes.
e Uses a wavefront sensorless approach designed for the optics of the mouse eye. Also report the use of a SH-WFS in prior system version.
f OCT system was based on the commercially available OCT-1 from Carl Zeiss Meditec AG, Dublin, CA. Did not have simultaneous AO and OCT capability. Acquired B-scans only.
g Only demonstrated on porcine eyes.
h Employs axial eye tracking.

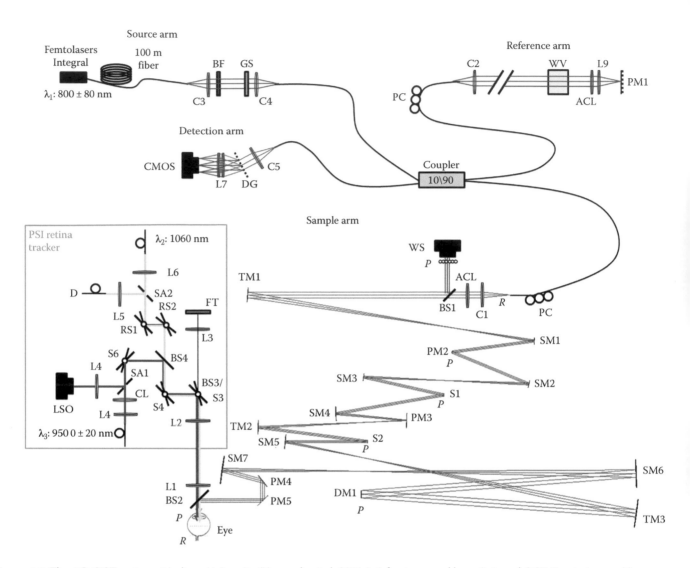

Figure 6.8 The AO-OCT system at Indiana University (Kocaoglu et al. 2014a). A femtosecond laser (Integral OCT, Femto Lasers, Vienna, Austria) with a center wavelength of 800 nm ($\Delta\lambda$ = 160 nm) is coupled into the system via a 90:10 fiber coupler. One output goes to the reference arm containing a water vial to compensate for the ocular dispersion. The other output leads to the sample arm and after initial collimatio by lens C1 the beam passes through a custom achromatizing lens to correct for the ocular chromatic aberration. The beam then passes throug a series of relay telescopes to adjust the beam diameter onto the scanning mirrors denoted by S1 and S2. A further telescope (SM5/TM3) rela the beam to a 97 actuator DM from ALPAO. The last telescope provides the correct pupil magnification at the eye (6.7 mm pupil diameter). Upon reflection from the eye, the light follows the reverse path through the optics where it is descanned by S1 and S2; a fraction of the light is coupled to the Shack–Hartmann WFS (WS). Note that the WFS is placed close to the initial collimating lens (C1) so that it "sees" all of the optics, hence minimizing any noncommon path errors. The majority of the light however is coupled back into the fiber coupler; the reflected light from the source arm and that from the reference arm interferes, is dispersed by the diffraction grating (DG), and is finally incident on a linear CCD camera (Basler Sprint). The bandwidth of the light source was limited to 81 nm using a bandpass filter giving an axial resolution of 2.6 μm (n = 1.38). This allowed an A-line rate of 167 kHz using the central 1408 pixels of the Basler camera. The field of view in high-resolution mode was 0.9° × 0.72°. The use of the toroidal mirrors allows for correction of the astigmatism in both the retinal and pupil planes due to the use of spherical mirrors (Liu et al. 2013). The system also includes an improved retinal tracker module built by Physical Sciences Inc. (Ferguson et al. 2004, Hammer et al. 2005), and this addition is discussed later in the chapter. The blue line shows the AO-OCT path, the red line is the path used for a wide-field line scanning ophthalmoscope, and the green path is used for the retinal tracking. BF, bandpass filter; BS, beam splitter; C, collimator; CL, cylindrical lens; DG, diffraction grating; DM, deformable mirror; D, detector; FT, fixation target; GS, glass slide; L, lens; PC, polarization controller; PM, planar mirror, RS, resonant scanner; SA, aperture splitting optic. (Reprinted with permission from Kocaoglu, O.P. et al., *Biomed. Opt. Express*, 5(7), 2262, 2014a.)

A femtosecond laser (Integral OCT, Femto Lasers, Vienna, Austria) with a center wavelength of 800 nm ($\Delta\lambda$ = 160 nm) is coupled into the system via a 90:10 fiber coupler. One output goes to the reference arm containing a water vial to compensate for the ocular dispersion. The other output leads to the sample arm, and after initial collimation by lens C1 the beam passes through a custom achromatizing lens (ACL) to correct for the eye's chromatic aberration. The beam then passes through a series

of relay telescopes to adjust the beam diameter onto the scannin mirrors denoted by S1 and S2. A further telescope (SM5/TM3) relays the beam to the DM (DM97, ALPAO SAS, France). The last telescope provides the correct pupil magnification at the eye (6.7 mm pupil diameter). Upon reflection from the eye, the ligh follows the reverse path through the optics where it is descanne by S1 and S2; a fraction of the light is coupled to the SH-WFS (WS) with beam splitter, BS1. Note that the WFS is placed clos to the initial collimating lens (C1) to minimize noncommon path errors. The majority of the light however is transmitted through BS1 and coupled back into the fiber coupler; the reflected light from the source arm and that from the reference arm then interferes, is dispersed by the diffraction grating (DG), and is finally incident on a linear CCD camera (Basler Sprint). The bandwidth of the light source was limited to 81 nm using a filter giving an axial resolution of 2.6 μm (n = 1.38). This allowed an A-line rate of 167 kHz using the central 1408 pixels of the Basler camera. The FOV in high-resolution mode was 0.9° × 0.72°. The use of the TM allowed for correction of the astigmatism in both the retinal and pupil planes due to the use of spherical mirrors (SM) (Liu et al. 2013). The system also includes a retinal tracker module built by Physical Sciences Inc. (Ferguson et al. 2004, Hammer et al. 2005), and this addition is discussed later in the chapter.

Figure 6.9 shows AO-OCT images of retinal nerve fiber bundles (RNFB) using an earlier version of the Indiana AO-OCT system (Kocaoglu et al. 2011a). Images from the same normal female control subject at 6° retinal eccentricity taken seven months apart are shown. Figure 6.9a shows a wide-field SLO image, while Figure 6.9b and c shows the en-face C-scans superimposed on the SLO image. The striations of the RNFL are clearly visible along with retinal vasculature. Figure 6.9d shows the B-scan image from the first imaging session (red line), and Figure 6.9e shows the same image from the second session (green line). The shape and size of the RNFB had a high degree of correlation between the two sessions.

The Indiana group has also used the same system to image the retinal capillary network close the foveal avascular zone (Wang et al. 2011) showing excellent correlation with entopic viewing. The review by Miller et al. (2011) discusses AO-OCT imaging of the nerve fiber layer and retinal capillaries in detail.

Jonnal et al. (2012) describe a clever experiment whereby phase-sensitive AO-OCT was used to measure changes in the OS length of individual cones, showing a length change sensitivity of 45 nm. More recent work examines the origin of outer retinal structures seen with clinical OCT system (Jonnal et al. 2014).

6.4.1.1 Longitudinal chromatic aberration

From Equation 6.3, the use of broad bandwidth light sources increases the axial resolution; however, this benefit can only be exploited if the longitudinal chromatic aberration (LCA) of the eye is corrected (Fernandez and Drexler 2005). The average LCA of the human eye over the range of wavelengths typically used in OCT (700–900 nm) is approximately 0.4 diopters (Fernandez et al. 2005b). A solution is the use of an ACL that in most cases is a zero optical power triplet lens that has the equal and opposite LCA profile to that of the eye. Fernandez et al. (2006) first suggested its use for OCT imaging with the first systems employing an ACL being described a few years later (Fernandez et al. 2008, Zawadzki et al. 2008). Figure 6.10 shows AO-(SD)-OCT images from a 33-year-old subject at 4.5° in the nasal retina taken without (a) and with (b) an ACL (Zawadzki et al. 2008). The plane of focus was set to the photoreceptor layer, and the figures are plotted on a linear intensity scale. Figure 6.10c and d shows an enlargement of the areas indicated by the white dashed rectangles in (a) and (b), respectively. The separation of the cones is improved with the ACL; however, the authors did not observe a visible difference in speckle size or axial resolution. Figure 6.10d clearly shows various interfaces within a single cone, namely, the IS/OS junction and the intersection with the retinal pigment epithelium (RPE) layer.

It should be noted that an ACL does not correct for the transverse chromatic aberration (TCA) and misalignment (centration and tilt) of such lenses can cause substantial increase in the TCA amplitude. Zawadzki et al. (2008) mention several

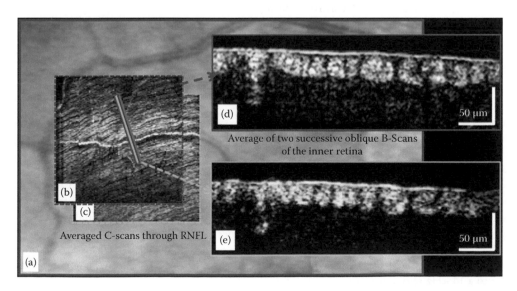

Figure 6.9 AO-OCT images from the same normal female control subject at 6° retinal eccentricity taken 7 months apart. (a) A wide-field SLO image. (b) and (c) The en-face C-scans reconstructed from the B-scans superimposed on the SLO image. The striations of the RNFL are clearly visible along with retinal vasculature. (d) The B-scan image from the first imaging session (red line) and (e) the same image from the second session (green line). The shape and size of the retinal nerve fiber bundles had a high degree of correlation between the two sessions. (Reprinted from *Vision Res.*, 51(16), Kocaoglu, O.P., Cense, B., Jonnal, R.S. et al., Imaging retinal nerve fiber bundles using optical coherence tomography with adaptive optics, 1835–1844, Copyright 2011a, with permission from Elsevier.)

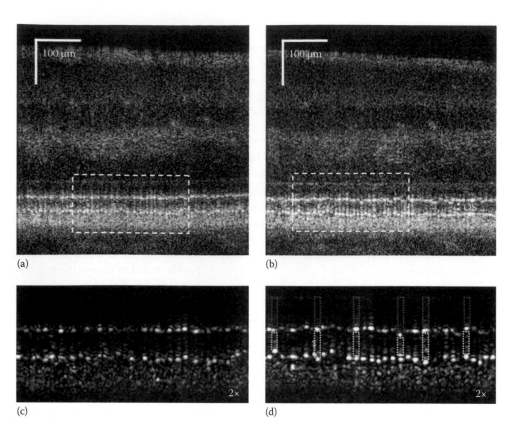

(a)

(b)

(c)

(d)

Figure 6.10 AO-(SD)-OCT images from a 33-year-old subject at 4.5° in the nasal retina with the plane of focus set to the photoreceptor layer (plotted on a linear intensity scale). (a) and (b) The same retinal location with and without the achromatizing lens, respectively. (c) and (d) An enlargement of the areas indicated by the white dashed rectangles in (a) and (b). IS, cone inner segment; OS, cone outer segment. Dashed red rectangles show the cone inner segments and the dashed yellow rectangles show the cone outer segments. (Reprinted from Zawadzki, R.J. et al., Ultrahigh-resolution optical coherence tomography with monochromatic and chromatic aberration correction, *Opt. Express*, 16(11), 8126–8143, 2008. With permission of Optical Society of America.)

aspects of AO-OCT systems that relax the designs constraints of the ACL, namely, (1) the small FOV, (2) the use of near IR wavelengths, (3) the limiting pupil diameter of the eye, (4) the use of raster scanning, and finally (5) the control of head motion due to the use of bitebars and/or forehead rests.

6.4.1.2 Inclusion of an SLO channel

It is also possible to include SLO imaging and AO-OCT capability into one system. Many of the optical components can be shared, and the simultaneous SLO image can be used for registration of the OCT frames. Zawadzki et al. (2011) detail the integration of an SLO with AO-(SD)-OCT; a challenge for this combination is the different image acquisition directions, and the OCT acquires A-scans (and hence B-scans), while the SLO scans in the en-face plane. Note that SLO integration is easier in (TS)-OCT systems—see the later discussion. In their paper separate SLD light sources are used for the OCT and SLO channels (830 and 680 nm, respectively). Both share a common galvanometric scanner operating at 27 Hz that generates the B-scan for the OCT and is also the frame scanner for the SLO. Dichroic mirrors redirect the SLO beam to a 14 kHz resonant scanner (RS) and the OCT light to a second galvo that provides the slow volume scan. Thus, one OCT B-scan and one SLO frame are acquired simultaneously allowing the use of the fast SLO frame for registration of the slower OCT B-scan/volume data. Figure 6.11 shows the registration of the OCT and SLO datasets, and Figure 6.11a is a single AO-OCT

B-scan, while Figure 6.11b shows the simultaneously acquired AO-SLO frame. The coregistered AO-OCT and AO-SLO frames are shown in Figure 6.11c.

6.4.2 SYSTEM EXAMPLE: AO-(SS)-OCT

To date the only example of an AO-OCT system using the swept source (SS) method is described by Mujat et al. (2010). In addition to being SS the system also incorporated an AO-corrected SLO, a wide-field (33°) line SLO, and a retinal tracker. The AO system was Shack–Hartmann based employing two DMs in a woofer–tweeter arrangement. The SS was manufactured by Santec Corporation, Japan, and had an output power of 11 mW with a 79 nm bandwidth centered at 1070 nm giving a theoretical axial resolution in tissue of 4.6 µm. The OCT power at the corneal plane was <2.6 mW. The OCT and SLO acquisition rates were not synchronized, and hence individual SLO frames and B-scans were acquired sequentially. OCT A-scans were acquired at 20 kHz with B-scan rates of 20 Hz. Figure 6.12 shows a comparison of AO-(SD)-OCT images at 850 nm and AO-(SS)-OCT images at 1070 nm; the deeper choroidal penetration of the 1070 nm light is clearly visible. Currently, there is a lack of suitable SS light sources in the 850 nm range.

6.4.3 SYSTEM EXAMPLE: AO-(TS)-OCT

The time domain (TD) approach was the first technique OCT method to be successfully applied to the eye and subsequently

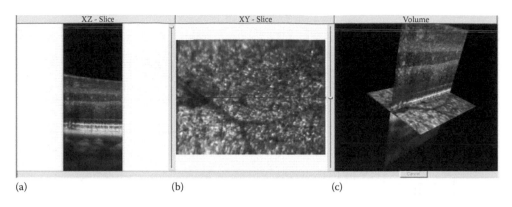

Figure 6.11 Registration of the AO-OCT and AO-SLO datasets. (a) An AO-OCT B-scan, (b) the simultaneously acquired AO-SLO frame, and (c) the coregistered AO-OCT and AO-SLO frames. (Reprinted from Zawadzki, R.J. et al., Integrated adaptive optics optical coherence tomography and adaptive optics scanning laser ophthalmoscope system for simultaneous cellular resolution in vivo retinal imaging, *Biomed. Opt. Express*, 2(6), 1674–1686, 2011. With permission of Optical Society of America.)

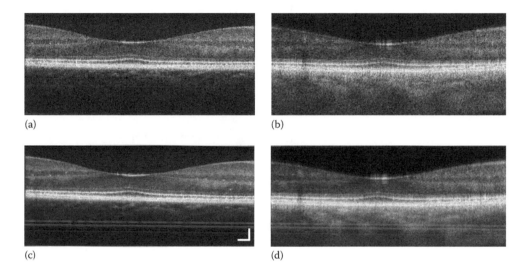

Figure 6.12 Comparison of AO-(SD)-OCT images at 850 nm (a) single frame and (c) four frame composite and AO-(SS)-OCT images at 1 μm (b) single frame and (d) four frame composite. The deeper choroidal penetration of the 1 μm light is clearly visible. The scale bar is 100 μm. (Reprinted from Mujat, M. et al., High resolution multimodal clinical ophthalmic imaging system, *Opt. Express*, 18(11), 11607–11621, 2010. With permission of Optical Society of America.)

commercialized. The general implementation rapidly scans the reference mirror in an axial direction to generate a single A-scan; this is then repeated along a series of retinal locations to build up the two-dimensional B-scan. When combined with AO this image acquisition geometry has been superseded by transversal scanning (TS); here the beam is rapidly scanned in an en-face geometry to build a single retinal image akin to an SLO frame. The reference mirror position is then adjusted shifting the coherence gate position within the sample and the process repeated hence building a three-dimensional retinal volume. One benefit of this approach is that the scanning geometry is identical to that of an SLO; thus, both an OCT and SLO image can be acquired simultaneously with pixel-to-pixel registration with added benefit of shifting focal plane together with coherence gate resulting in dynamic focus imaging. A limitation of the (TS)-OCT approach is the sensitivity to axial eye motion, which can be overcome by using an active eye tracker (Pircher et al. 2007, 2010, Cucu et al. 2010). Merino et al. demonstrated the first AO-(TS)-OCT-SLO system in 2006 (Merino et al. 2006), but image improvement was somewhat limited due to the AO technology employed.

More recently, AO-(TS)-OCT has provided retina images with resolutions high enough to show both cone and rod structures (Felberer et al. 2014). The sample arm of their AO-(TS)-OCT-SLO system is shown in Figure 6.13. The delivery arm is lens based (Felberer et al. 2012) rather than the more common SM design. The system also incorporates an SLO channel and axial tracking.

An 840 nm SLD source with a bandwidth (FWHM) of 50 nm is incident into the system through the 90:10 fiber coupler, giving a theoretical axial resolution in tissue of 4.5 μm. Due to the use of lenses instead of SM, polarization elements are used to mitigate the back reflections, which would be problematic for the WFS. A series of relay telescopes reimages the pupil onto the RS operating at 8 kHz (16 kHz line rate as both scanning directions are used) and then onto the galvo scanner (GS). Further telescopes relay the light to the DM (Mirao 52-e, Imagine Eyes) and finally to the eye. Upon reflection from the eye, the light is descanned before reflection by the polarization beam splitter; 20% goes to the WFS, 20% to the SLO, and the remaining 60% to the OCT arm. Note that the reflected sample arm light is not recoupled

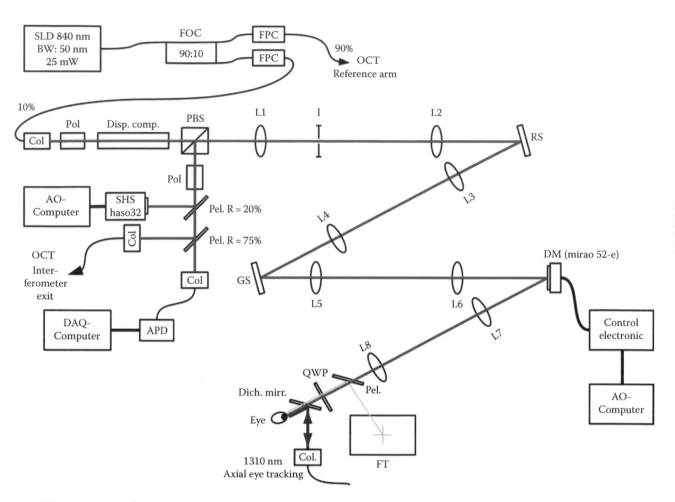

Figure 6.13 The sample arm of the AO-(TS)-OCT system described by Felberer et al. (2014). The delivery arm is lens based rather than the more common spherical mirror design. The system also incorporates an SLO channel and axial tracking. SLD, superluminescent diode; FOC, fiber-optic coupler; FPC, fiber polarization controller; Col, collimator; Pol, polarizer; Disp. comp., dispersion compensation; PBS, polarizing beam splitter; L1–L8 lenses; RS, resonant scanner; GS, galvanometer scanner; DM, deformable mirror; Pel., pellicle; QWP, quarter wave plate; SHS, Shack–Hartmann wavefront sensor; APD, avalanche photodiode; FT, fixation target; I, variable aperture stop. (Reprinted from Felberer, F et al., Adaptive optics SLO/OCT for 3D imaging of human photoreceptors in vivo, *Biomed. Opt. Express*, 5(2), 439–456, 2014. With permission of Optical Society of America.)

into the initial delivery fiber; rather it is coupled into a further single mode optical fiber, which forms one arm of a balanced detector—the other arm being the light from the reference arm.

The reference arm is more complex that typically required in a (FD)-OCT system. Ninety percent of the initial 840 nm SLD source light is directed into the reference arm via the first fiber-optic coupler. The reference light traverses polarizers to match the polarization state in the sample arm and then passes through two acousto-optic modulators (AOMs) that introduce an overall carrier frequency shift of 3 MHz. After correcting for the AOM dispersion, the beam is then input to a galvanometer-based optical delay line that can be rapidly modulated to introduce reference arm length changes to compensate for shifts in the axial position of the eye. The beam is then finally reflected off a voice coil activated mirror to provide the axial scanning before finally interfering with the sample arm light hence providing the OCT signal.

The axial tracking system uses a second light source at 1300 nm and employs low coherence interferometry to measure the position of the cornea (Pircher et al. 2006b).

The interferometer output is then used to control the position of the GS in the optical delay line; this arrangement keeps the OPL in both the sample and reference arm identical.

Imaging is performed over a 7 mm pupil with 1° × 1° images being acquired at 20 Hz, each frame comprising 1152 × 790 pixels. The scanning depth was set to 200 μm with a volume taking 6 s to acquire. One hundred twenty en-face images were included in each volume. The simultaneous acquisition of both the SLO and OCT frames allowed for post-processing to remove fixational eye motion using a strip-wise registration approach (Stevenson and Roorda 2005). Figure 6.14 shows foveal cone images from a normal human subject in their AO-(TS)-OCT-SLO system. Figure 6.14a shows the average of 120 SLO frames and Figure 6.14b shows the same area from 70 averaged OCT images. The central 50 μm indicated by the white squares in the SLO and OCT images is shown enlarged in Figure 6.14c and d, respectively. Figure 6.14e and f is the Fourier transforms of the images in Figure 6.14c and d showing Yellot's ring. By performing a radial average the nearest neighbor spacing was calculated to be 3.11 μm, in agreement with histology (Curcio et al. 1990).

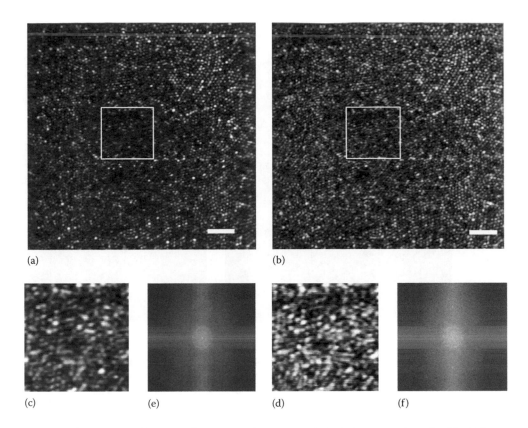

(a) (b)

(c) (e) (d) (f)

Figure 6.14 Foveal cone images from a normal human subject using the AO-(TS)-OCT-SLO system described by Felberer et al. (2014). (a) The average of 120 SLO frames and (b) The same area from 70 averaged OCT images. The central 50 μm indicated by the white squares in the SLO and OCT images is shown enlarged in (c) and (d), respectively. Figures (e) and (f) are the Fourier transforms of (c) and (d) showing Yellot's ring. The scale bar indicates 30 μm. (Reprinted from Felberer, F. et al., Adaptive optics SLO/OCT for 3D imaging of human photoreceptors in vivo, *Biomed. Opt. Express*, 5(2), 439–456, 2014. With permission of Optical Society of America.)

6.4.4 WAVEFRONT SENSORLESS AO IMPLEMENTATIONS

Wavefront sensorless AO systems have recently been demonstrated for imaging human retinas (Hofer et al. 2011) and mouse (Biss et al. 2007, Jian et al. 2014). Instead of using a WFS, the DM is controlled based on measurements of a merit function (MF). This MF can be as simple as monitoring the light through the confocal pinhole as the amplitude of various Zernike modes written to the DM is varied. Maximizing the light transmitted through the confocal pinhole is an indicator of the aberration correction state.

The principle is not new and from a historical perspective the wavefront sensorless approach was one of the first successful demonstrations of AO (Muller and Buffington 1974, Buffington et al. 1977a,b). The technique (and variants of it) has been applied successfully in AO microscopy applications (Albert et al. 2000, Booth et al. 2002) and has the advantage of removing noncommon path errors and furthermore does not require that a percentage of the light be reflected to a WFS arm.

The approach has been successfully applied to (SD)-OCT (Bonora and Zawadzki 2013) and most recently to the human eye (Wong et al. 2015). Indeed, when imaging the mouse retina, acquiring SH-WFS spots is challenging and the wavefront sensorless technique may provide a better alternative (Jian et al. 2014).

6.5 APPLICATION TO THE STUDY OF DISEASE

Various retinal conditions have been studied with AO-(SD)-OCT exploiting the increased lateral and axial resolution due to AO and the use of broader bandwidth light sources. An example of the potential clinical utility is the study of subretinal drusenoid deposits (SDD) that are associated with age-related macular degeneration (Meadway et al. 2014).

SDD lesions are found between the cone photoreceptors and the RPE layer. Figure 6.15a and b shows an SDD imaged with AO-OCT and the same area taken with the Spectralis SD-OCT system from Heidelberg Engineering but in the orthogonal direction. The improvement in lateral resolution due to the use of AO is clearly visible. Both systems show that the SDD has disturbed the ellipsoid portion of the photoreceptor IS (EZ), with the AO-OCT system showing little effect beyond the ELM.

Choi et al. (2008) examined the retinas of several patients with various forms of optic neuropathy showing that when permanent thinning of the NFL-GC-IP complex was measured, there were associated outer retinal changes—in particular, a loss of cone photoreceptor density and a less distinct OS/RPE layer in the OCT images. Further work looking at 10 glaucoma patients imaged with a variety of AO-equipped imaging modalities

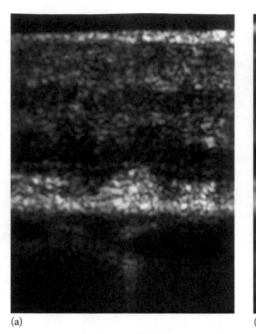
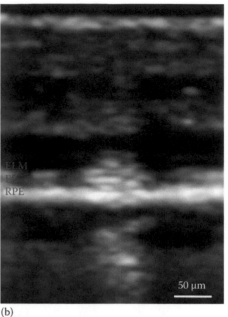

(a)   (b)

Figure 6.15 Subretinal drusenoid deposits (SDD) imaged with AO-(SD)-OCT. SSD are extracellular lesions associated with age-related macular degeneration. These lesions are found between the cone photoreceptors and the RPE layer. (a) An SDD imaged with AO-OCT (b) The same area taken with the Spectralis SD-OCT system from Heidelberg Engineering but in the orthogonal direction. The improvement in lateral resolution due to the use of AO is clearly visible. Both systems show that the SDD has disturbed the ellipsoid portion of the photoreceptor inner segment (EZ), with the AO-OCT system showing little effect beyond the ELM. The Spectralis SD OCT image is shown with logarithmic gray scale, whereas the AO-OCT image has a linear gray scale. The scale bar applies to both figures. (Reprinted from Meadway, A. et al., Microstructure of subretinal drusenoid deposits revealed by adaptive optics imaging, *Biomed. Opt. Express*, 5(3), 713–727, 2014. With permission of Optical Society of America.)

showed similar outer retinal changes (Choi et al. 2011). The retinal locations with cone loss measured with AO-(SD)-OCT correlated with visual sensitivity measurements. In particular, a shortening of the OS length was observed in the AO-OCT images; loss of photoreceptor integrity was further confirmed by examining the waveguide properties of the affected cones.

Images of the lamina cribrosa (LC), the structure of which is closely associated with the progression of glaucomatous damage, have been imaged in both healthy (Kim et al. 2012, Nadler et al. 2014a) and glaucomatous eyes (Nadler et al. 2014b) using AO-(SD)-OCT systems centered at 836 and 1050 nm.

Hammer et al. (2008) describe the use of an AO-(SD)-OCT system in the study of retinopathy of prematurity (ROP). The foveal pit depth, the retinal layer thickness, and the parafoveal retinal vasculature were examined in five ROP patients ranging in age from 14 to 26 years and five age-matched controls. They found that the foveal pit was wider and shallower with increasing retinal thickness at the fovea for the ROP patients. Furthermore, an avascular zone was not observed in the ROP subjects but was clearly present in the controls. The photoreceptor layer thickness was found to be the same across both populations.

Five geographic atrophy patients were studied using AO-(SD)-OCT, and the results were complemented with multifocal ERG and microperimetry (Panorgias et al. 2013). The AO-OCT images clearly show retinal tabulation, calcified drusen deposits, and drusenoid pigment epithelium detachment; moreover, there was good correlation with the functional testing.

High-resolution imaging has significant potential in the study of new drug/stem cell therapies for a range of ocular

conditions. Park et al. (2015) used AO-(SD)-OCT to examine the retina structure of six subjects with irreversible vision loss who were injected with human bone marrow CD34+ stem cells. For the enrolled Stargardt's patient, AO-OCT revealed new hyperreflective deposits within the retinal thickness; these deposits were too small to be seen with standard clinical grade OCT.

6.5.1 IMAGING ANIMAL RETINA

A few animal species have been imaged with AO-OCT. Hammer et al. (2012) show images of the photoreceptor layer and NFL in the albino rat. Pocock et al. (2014) used an AO-(SD)-OCT system to study laser injury to the retina of eight cynomolgus monkeys. The improved resolution allowed for the longitudinal study of the same animal over time.

Figure 6.16 shows AO-(SD)-OCT images of the living mouse retina. B-scan and en-face images are shown at several retina layers (IPL, INL, OPL, and PRL) indicated by the red brackets (Jian et al. 2013). The plane of focus was controlled by adding a defocus offset to the DM. Figure 6.16a through d shows the B-scan and en-face image with the AO loop on and Figure 6.16e through h shows the corresponding images with AO off. The line plots on the right show the AO on and AO off profiles of the areas highlighted by the yellow boxes for each of the four focal depths. The bottom panel shows the normalized line graphs of the image intensity taken across the capillaries at the locations labeled 1, 2, and 3 in Figure 6.16c and g—the green dotted boxes. The width of the capillaries decreased by a factor of 2.5 with the AO turned on. As the depth is scanned, a clear increase in contrast is apparent from the line plots.

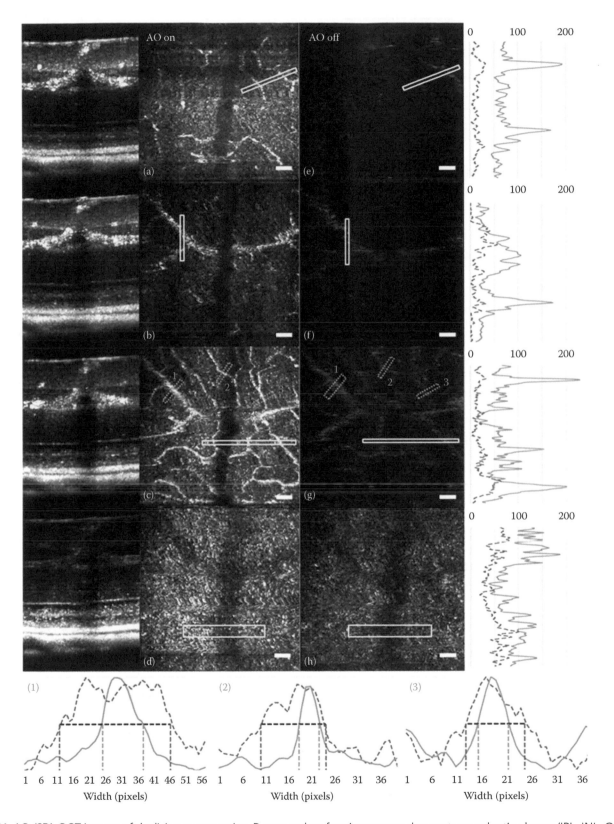

Figure 6.16 AO-(SD)-OCT images of the living mouse retina. B-scan and en-face images are shown at several retina layers (IPL, INL, OPL, and PRL) indicated by the red brackets. The plane of focus was controlled by adding a defocus offset to the DM. (a–d) The B-scan and en-face image with the AO loop on and (e–h) the corresponding images with AO off. The line plots on the right show the AO on and AO off profiles of the areas highlighted by the yellow boxes for each of the four focal depths. The bottom panel shows the normalized line graphs of the image intensity taken across the capillaries at the locations labeled 1, 2, and 3 in (c) and (g)—the green dotted boxes. The width of the capillaries decreased by a factor of 2.5 with the AO turned on. The scale bar is 30 μm. (Reprinted from Jian, Y. et al., *J. Biomed. Opt.*, 18(5), 56007, 2013. With permission of SPIE.)

6.6 FUTURE DEVELOPMENTS AND DIRECTIONS

AO-OCT imaging systems are producing some of the highest-resolution images of the living retina to date. The AO aspect is well developed and understood, the WFS can measure the aberrations at sufficient speed, and the DMs are capable of providing diffraction-limited imaging with stroke capabilities that approach those required for correction in a general human population. A limitation to commercial application is the cost; the DM in particular is still too expensive to be incorporated into clinical instruments at the current time. Other challenges include obtaining accurate WFS measurement in patients with clinical opacities, for example, in the cornea, crystalline lens, or vitreous humor. For OCT however, there still remain significant challenges in particular speckle and retinal eye motion.

6.6.1 SPECKLE REDUCTION

Speckle is a fundamental reality of OCT due to its coherent imaging nature. It has high contrast and individual speckles have sizes on the order of the theoretical lateral and axial resolutions. Therefore, care must be taken so that such features are not interpreted as actual retinal structure; hence, methods (software and hardware based) to mitigate speckle are active areas of research.

A method to reduce the effect of speckle is to simply average multiple B-scans. Assuming retinal eye motion, the speckle in each B-scan will be different and averaging minimizes their appearance. Irrespective of the particular approach, the aim is to obtain images with uncorrelated speckle patterns, and this can be achieved by introducing a diversity to the imaging procedure. This diversity might be a change in wavelength, a change in incident light angle, or the use of polarization. The review by Wojkowski et al. (2012) describes various approaches that have been described in the literature. Speckle reduction is not without cost; however, methods are time consuming and generally result in a loss in either lateral or axial resolution.

6.6.2 EYE MOTION

In any retinal imaging system, retinal eye motion can cause unacceptable blur. Flood-illuminated fundus cameras circumvent this issue by using very short exposure times (<4 ms) to freeze any eye motion. However, for imaging system employing scanning optics, the image acquisition rates are often slow enough that eye motion is problematic.

For AO-OCT, eye motion is of particular concern when acquiring volume data, which can typically take several seconds. Variation in the retinal position of individual B-scans can make the registration very challenging. There are several options to overcome this challenge: (1) increase the image acquisition speed, (2) actively track the retina and then compensate for the motion, or (3) use input from a secondary imaging source.

Kocaoglu et al. (2014b) recently demonstrated an AO-(SD)-OCT system acquiring A-scans at 1 MHz. Acquiring data at such speeds reduces eye motion artifacts allowing for improved image registration and larger retinal volumes. The system is more complex than similar systems requiring four separate spectrometers each coupled to a CMOS line scan camera. The authors state that this is the fastest ophthalmic OCT system working in the 700–900 nm range. A 500 kHz non-AO system operating at 850 nm has also been reported (An et al. 2011).

Another Fourier domain OCT technique that has great potential when combined with AO is (SS)-OCT. (SS)-OCT is capable of much higher speeds albeit limited to wavelengths above 1000 nm due to the currently available light sources. Moo and Kim (2006) report 5 MHz A-scan rates and more recently 6.7 MHz has been demonstrated (Klein et al. 2013). The only AO-(SS)-OCT system was described by Mujat et al. (2010) running at an A-scan rate of 20 kHz so there is an opportunity for huge improvement.

Another solution to eye motion is to employ retinal tracking This approach generally uses a secondary system whose sole purpose is to actively measure and correct for the retinal eye motion. Figure 6.8 shows the hardware-based retinal tracker designed by Physical Sciences Inc., currently used in the Indiana University AO-OCT system (Kocaoglu et al. 2014a). The tracker employs a 1060 nm light source that is scanned in a circular pattern on the LC using a pair of 16 kHz RS. The reflected signal can be processed to provide a signal that is proportional to the eye motion. Voltage biases can then be applied to the OCT galvanometric scanners to stabilize the image. The tracker also includes a 950 nm wide-field SLO with a 35° × 35° FOV that can be used to monitor both the AO-OC and the tracking beams. The tracker demonstrated the ability to correct eye motion up to 100 Hz and reduced residual eye motion to 10 μm RMS.

The use of hardware tracking provides an absolute measurement of retinal motion and allows for higher sampling speeds, currently 208 kHz for the Indiana system, but does add to the system complexity. An alternative approach is to use image-based stabilization, and this has been successfully used in AO-SLOs (Vogel et al. 2006, Sheehy et al. 2012) and also in OCT (Vienola et al. 2012, Braaf et al. 2013). Here, an SLO frame is broken up into a series of horizontal strips and each strip is registered to a reference frame. The horizontal direction is scanned rapidly using a fast RS and hence would see minimal eye motion. The correlation of each strip with the reference gives a measure of the eye motion during the time that strip was acquired; sampling speeds up to 960 Hz have been demonstrated (Vogel et al. 2006, Sheehy et al. 2012). The choice of reference frame for this method is however critical. In a dual AO-OCT-SLO system, the strip offsets generated by the SLO channel can then be used to realign the OCT B-scans—Figure 6.11. A review of the technique is provided by Zawadzki et al. (2014).

There are other challenges associated with AO-OCT; one of these is the limited depth of field (30 μm) compared to the 500 μm thickness of the retina. For (FD)-OCT this means that only a narrow retina thickness will be in focus; therefore, it is necessary to apply a defocus to the DM to scan through the entire retina depth. With (TS)-OCT systems this is of less concern as it is possible to arrange that focal plane is shifted in conjunction with the coherence gate keeping the retinal layer in continual focus.

A further limitation of any AO system for the eye is the limited FOV. Typically AO retinal images have an FOV of 1° × 1° to 2° × 2° in size due to the aberration profile of the eye (the isoplanatic angle). In scanning retinal imaging systems the WFS essentially integrates the field-dependent aberrations over the FOV as the AO control loop is not fast enough to correct at every point in the scan. The solution to the small FOV has been to montage multiple smaller images together to give a larger picture. This requires good fixation for the subject or increased system complexity if implemented through the optical hardware. A similar problem exists in astronomy and the solution is to use multiple DMs and guide stars termed multiconjugate AO (Esposito 2005); however, this introduces considerable system complexity and expense.

6.7 CONCLUSIONS

OCT is particularly suited to imaging the human retina due to its layered structure and overall ocular transmission properties. Being the only direct view of the central nervous system, the ability to image three-dimensional cellular level retinal structure noninvasively has important implications for a range of retinal and also systemic diseases. While published AO-OCT work in diseased eyes is currently limited, one can expect a significant increase in this area over the coming years.

There is also considerable room for improving the technology, while DMs and WFS meet the majority of the human eye requirements; the DM, in particular, remains expensive. The use of wavefront sensorless approaches reduces the cost somewhat, but a DM is still required.

Computational AO (CAO) may provide an alternative to the traditional hardware-based implementations of AO allowing for post-data-acquisition correction of images without the need for a DM or WFS. Such methods may be particularly suited to OCT as here both intensity and phase are recorded in the tomogram. Kam et al. (2001) used a space-variant deconvolution approach to correct for sample aberrations. Using Nomarski differential interference contrast microscopy to measure the refractive index profile, they generated the PSF used for deconvolution, successfully imaging a fluorescent bead immersed under an oil droplet. An alternative CAO implementation modified the pupil phase profile allowing post-data-acquisition aberration correction in an SD-OCT tomogram. Successful CAO results were demonstrated for a tissue phantom and also for ex vivo rat lung tissue (Adie et al. 2012).

Perhaps the most exciting development in the field will be advances in AO-(SS)-OCT. The much higher OCT acquisition rates will minimize the effects of eye motion while allowing for larger retinal volumes, increasing the clinical and commercial utility. However, current systems are limited to wavelengths above 1000 nm due to the lack of suitable light sources in the 800–900 nm range.

ACKNOWLEDGMENTS

The author thanks R. J. Zawadzki for helpful comments in the preparation of this work. This work was supported by grants from (1) the National Institutes of Health grant EY020901 and (2) the Department of Defense Telemedicine and Advanced Technology Research Center W81XWH-10-1-0738.

REFERENCES

Adie, S. G., B. W. Graf, A. Ahmad, P. S. Carney, and S. A. Boppart. 2012. Computational adaptive optics for broadband optical interferometric tomography of biological tissue. *Proc Natl Acad Sci USA* 109 (19):7175–7180.

Albert, O., L. Sherman, G. Mourou, T. B. Norris, and G. Vdovin. 2000. Smart microscope: An adaptive optics learning system for aberration correction in multiphoton confocal microscopy. *Opt Lett* 25 (1):52–54.

ALPAO SAS. Available from www.alpao.com/.

An, L., P. Li, T. T. Shen, and R. Wang. 2011. High speed spectral domain optical coherence tomography for retinal imaging at 500,000 Alines per second. *Biomed Opt Express* 2 (10):2770–2783.

AOA Xinetics—Northrop Grumman Corporation. Available from http://www.northropgrumman.com/BusinessVentures/AOAXinetics/Pages/default.aspx.

AOptix. Available from www.aoptix.com/.

Artal, P. and R. Navarro. 1989. High-resolution imaging of the living human fovea: Measurement of the intercenter cone distance by speckle interferometry. *Opt Lett* 14 (20):1098–1100.

Babcock, H. W. 1953. The possibility of compensating astronomical seeing. *Publ Astron Soc Pac* 65:229.

Bigelow, C. E., N. V. Iftimia, R. D. Ferguson et al. 2007. Compact multimodal adaptive-optics spectral-domain optical coherence tomography instrument for retinal imaging. *J Opt Soc Am A Opt Image Sci Vis* 24 (5):1327–1336.

Bioptigen Inc. Available from http://www.leica-microsystems.com/products/optical-coherence-tomography-oct/.

Biss, D. P., R. H. Webb, Y. Zhou et al. 2007. An adaptive optics biomicroscope for mouse retinal imaging. *Proc Lasers Ophthalmol II* 6467:46703.

Bonora, S. and R. J. Zawadzki. 2013. Wavefront sensorless modal deformable mirror correction in adaptive optics: Optical coherence tomography. *Opt Lett* 38 (22):4801–4804.

Booth, M. J., M. A. A. Neil, R. Juskaitis, and T. Wilson. 2002. Adaptive aberration correction in a confocal microscope. *Proc Natl Acad Sci USA* 99 (9):5788–5792.

Boston Micromachines Corporation. Available from www.bostonmicromachines.com/.

Bouma, B. E. and G. J. Tearney. 2002. *Handbook of Optical Coherence Tomography*. New York: Marcel Dekker.

Braaf, B., K. V. Vienola, C. K. Sheehy et al. 2013. Real-time eye motion correction in phase-resolved OCT angiography with tracking SLO. *Biomed Opt Express* 4 (1):51–65.

Buffington, A., F. S. Crawford, R. A. Muller, and C. D. Orth. 1977a. 1st observatory results with an image-sharpening telescope. *J Opt Soc Am* 67 (3):304–305.

Buffington, A., F. S. Crawford, R. A. Muller, A. J. Schwemin, and R. G. Smits. 1977b. Correction of atmospheric distortion with an image-sharpening telescope. *J Opt Soc Am* 67 (3):298–303.

Burns, S. A., R. Tumbar, A. E. Elsner, D. Ferguson, and D. X. Hammer. 2007. Large-field-of-view, modular, stabilized, adaptive-optics-based scanning laser ophthalmoscope. *J Opt Soc Am A* 24 (5):1313–1326.

Carl Zeiss Meditec Inc. Available from http://www.zeiss.com/meditec/en_us/home.html.

Carroll, J., D. B. Kay, D. Scoles, A. Dubra, and M. Lombardo. 2013. Adaptive optics retinal imaging—Clinical opportunities and challenges. *Curr Eye Res* 38 (7):709–721.

Cense, B., W. Gao, J. M. Brown et al. 2009a. Retinal imaging with polarization-sensitive optical coherence tomography and adaptive optics. *Opt Express* 17 (24):21634–21651.

Cense, B., E. Koperda, J. M. Brown et al. 2009b. Volumetric retinal imaging with ultrahigh-resolution spectral-domain optical coherence tomography and adaptive optics using two broadband light sources. *Opt Express* 17 (5):4095–4111.

Choi, S. S., R. J. Zawadzki, J. L. Keltner, and J. S. Werner. 2008. Changes in cellular structures revealed by ultra-high resolution retinal imaging in optic neuropathies. *Invest Ophthalmol Vis Sci* 49 (5):2103–2119.

Choi, S. S., R. J. Zawadzki, M. C. Lim et al. 2011. Evidence of outer retinal changes in glaucoma patients as revealed by ultrahigh-resolution in vivo retinal imaging. *Br J Ophthalmol* 95 (1):131–141.

Cucu, R. G., M. W. Hathaway, A. G. Podoleanu, and R. B. Rosen. 2010. Variable lateral size imaging of the human retina in vivo by combined confocal/en face optical coherence tomography with closed loop OPD-locked low coherence interferometry based active axial eye motion. *Proceeding of the Optical Coherence Tomography and Coherence Domain Optical Methods in Biomedicine XIV*, Vol. 7554, Bellingham, WA.

Curcio, C. A., K. R. Sloan, R. E. Kalina, and A. E. Hendrickson. 1990. Human photoreceptor topography. *J Comp Neurol* 292 (4):497–523.

Dainty, J. C. 1984. *Laser Speckle and Related Phenomena* (Topics in Applied Physics). Ed. J. C. Dainty. 2nd edn. New York: Springer-Verlag.

Delori, F. C., R. H. Webb, and D. H. Sliney. 2007. Maximum permissible exposures for ocular safety (ANSI 2000), with emphasis on ophthalmic devices. *J Opt Soc Am A Opt Image Sci Vis* 24 (5):1250–1265.

Diaz-Santana, L., C. Torti, I. Munro, P. Gasson, and C. Dainty. 2003. Benefit of higher closed-loop bandwidths in ocular adaptive optics. *Opt Express* 11 (20):2597–2605.

Doble, N., D. T. Miller, G. Yoon, and D. R. Williams. 2007. Requirements for discrete actuator and segmented wavefront correctors for aberration compensation in two large populations of human eyes. *Appl Opt* 46 (20):4501–4514.

Dreher, A. W., J. F. Bille, and R. N. Weinreb. 1989. Active optical depth resolution improvement of the laser tomographic scanner. *Appl Opt* 28 (4):804–808.

Dubra, A. and Y. Sulai. 2011. Reflective afocal broadband adaptive optics scanning ophthalmoscope. *Biomed Opt Express* 2 (6):1757–1768.

Esposito, S. 2005. Introduction to multi-conjugate adaptive optics systems. *Comptes Rendus Physique* 6 (10):1039–1048.

Felberer, F., J. S. Kroisamer, B. Baumann et al. 2014. Adaptive optics SLO/OCT for 3D imaging of human photoreceptors in vivo. *Biomed Opt Express* 5 (2):439–456.

Felberer, F., J. S. Kroisamer, C. K. Hitzenberger, and M. Pircher. 2012. Lens based adaptive optics scanning laser ophthalmoscope. *Opt Express* 20 (16):17297–17310.

Fercher, A. F. 1996. Optical coherence tomography. *J Biomed Opt* 1 (2):157–173.

Fercher, A. F., W. Drexler, C. K. Hitzenberger, and T. Lasser. 2003. Optical coherence tomography—Principles and applications. *Rep Progress Phys* 66 (2):239–303.

Fercher, A. F. and C. K. Hitzenberger. 2002. Optical coherence tomography. In *Progress in Optics*, ed. E. Wolf, pp. 215–302. Amsterdam, the Netherlands: Elsevier Science & Technology.

Fercher, A. F., C. K. Hitzenberger, W. Drexler, G. Kamp, and H. Sattmann. 1993. In vivo optical coherence tomography. *Am J Ophthalmol* 116 (1):113–114.

Fercher, A. F., C. K. Hitzenberger, G. Kamp, and S. Y. Elzaiat. 1995. Measurement of intraocular distances by backscattering spectral interferometry. *Opt Commun* 117 (1–2):43–48.

Fercher, A. F., K. Mengedoht, and W. Werner. 1988. Eye-length measurement by interferometry with partially coherent light. *Opt Lett* 13 (3):186–188.

Ferguson, R. D., D. X. Hammer, L. A. Paunescu, S. Beaton, and J. S. Schuman. 2004. Tracking optical coherence tomography. *Opt Lett* 29 (18):2139–2141.

Fernandez, E. and W. Drexler. 2005. Influence of ocular chromatic aberration and pupil size on transverse resolution in ophthalmic adaptive optics optical coherence tomography. *Opt Express* 13 (20):8184–8197.

Fernandez, E. J., B. Hermann, B. Povazay et al. 2008. Ultrahigh resolution optical coherence tomography and pancorrection for cellular imaging of the living human retina. *Opt Express* 16 (15):11083–11094.

Fernandez, E. J., I. Iglesias, and P. Artal. 2001. Closed-loop adaptive optics in the human eye. *Opt Lett* 26 (10):746–748.

Fernandez, E. J., B. Povazay, B. Hermann et al. 2005a. Three-dimensional adaptive optics ultrahigh-resolution optical coherence tomography using a liquid crystal spatial light modulator. *Vision Res* 45 (28):3432–3444.

Fernandez, E. J., A. Unterhuber, B. Povazay et al. 2006. Chromatic aberration correction of the human eye for retinal imaging in the near infrared. *Opt Express* 14 (13):6213–6225.

Fernandez, E., A. Unterhuber, P. Prieto et al. 2005b. Ocular aberrations as a function of wavelength in the near infrared measured with a femtosecond laser. *Opt Express* 13 (2):400–409.

Flexible Optical B.V. (OKO Technologies). Available from www.okotech.com/.

Fujimoto, J. G., S. De Silvestri, E. P. Ippen et al. 1986. Femtosecond optical ranging in biological systems. *Opt Lett* 11 (3):150.

Godara, P., A. M. Dubis, A. Roorda, J. L. Duncan, and J. Carroll. 2010. Adaptive optics retinal imaging: Emerging clinical applications. *Optom Vis Sci* 87 (12):930–941.

Golubovic, B., B. E. Bouma, G. J. Tearney, and J. G. Fujimoto. 1997. Optical frequency-domain reflectometry using rapid wavelength tuning of a Cr^{4+}:forsterite laser. *Opt Lett* 22 (22):1704–1706.

Gomez-Vieyra, A., A. Dubra, D. Malacara-Hernandez, and D. R. Williams. 2009. First-order design of off-axis reflective ophthalmic adaptive optics systems using afocal telescopes. *Opt Express* 17 (21):18906–18919.

Hamamatsu Photonics. Available from http://www.hamamatsu.com.

Hammer, D., R. D. Ferguson, N. Iftimia et al. 2005. Advanced scanning methods with tracking optical coherence tomography. *Opt Express* 13 (20):7937–7947.

Hammer, D. X., R. D. Ferguson, M. Mujat et al. 2012. Multimodal adaptive optics retinal imager: Design and performance. *J Opt Soc Am A Opt Image Sci Vis* 29 (12):2598–2607.

Hammer, D. X., N. V. Iftimia, R. D. Ferguson et al. 2008. Foveal fine structure in retinopathy of prematurity: An adaptive optics Fourier domain optical coherence tomography study. *Invest Ophthalmol Vis Sci* 49 (5):2061–2070.

Hampson, K. M. 2008. Adaptive optics and vision. *J Mod Opt* 55 (21):3425–3467.

Hardy, J. W. 1998. *Adaptive Optics for Astronomical Telescopes*. Oxford, U.K.: Oxford University Press.

Hardy, J. W., J. E. Lefebvre, and C. L. Koliopoulos. 1977. Real-time atmospheric compensation. *J Opt Soc Am* 67 (3):360–369.

Headington, K., S. S. Choi, D. Nickla, and N. Doble. 2011. Single cell, in vivo imaging of the chick retina with adaptive optics. *Curr Eye Res* 36 (10):947–957.

Heidelberg Engineering GmbH. Available from www.heidelbergengineering.com.

Hermann, B., E. J. Fernandez, A. Unterhuber et al. 2004. Adaptive-optics ultrahigh-resolution optical coherence tomography. *Opt Lett.* 29 (18):2142–2144.

Hofer, H., P. Artal, B. Singer, J. L. Aragon, and D. R. Williams. 2001a. Dynamics of the eye's wave aberration. *J Opt Soc Am A Opt Image Sci Vis* 18 (3):497–506.

Hofer, H., L. Chen, G. Y. Yoon et al. 2001b. Improvement in retinal image quality with dynamic correction of the eye's aberrations. *Opt Express* 8 (11):631–643.

Hofer, H., N. Sredar, H. Queener, C. Li, and J. Porter. 2011. Wavefront sensorless adaptive optics ophthalmoscopy in the human eye. *Opt Express* 19 (15):14160–14171.

Huang, D., E. A. Swanson, C. P. Lin et al. 1991. Optical coherence tomography. *Science* 254 (5035):1178–1181.

Imagine Eyes. Available from http://www.imagine-eyes.com.

Iris AO Inc. Available from www.irisao.com.

Jian, Y., J. Xu, M. A. Gradowski et al. 2014. Wavefront sensorless adaptive optics optical coherence tomography for in vivo retinal imaging in mice. *Biomed Opt Express* 5 (2):547–559.

Jian, Y., R. J. Zawadzki, and M. V. Sarunic. 2013. Adaptive optics optical coherence tomography for in vivo mouse retinal imaging. *J Biomed Opt* 18 (5):56007.

Jonnal, R. S., O. P. Kocaoglu, Q. Wang, S. Lee, and D. T. Miller. 2012. Phase-sensitive imaging of the outer retina using optical coherence tomography and adaptive optics. *Biomed Opt Express* 3 (1):104–124.

Jonnal, R. S., O. P. Kocaoglu, R. J. Zawadzki et al. 2014. The cellular origins of the outer retinal bands in optical coherence tomography images. *Invest Ophthalmol Vis Sci* 55 (12):7904–7918.

Kam, Z., B. Hanser, M. G. Gustafsson, D. A. Agard, and J. W. Sedat. 2001. Computational adaptive optics for live three-dimensional biological imaging. *Proc Natl Acad Sci USA* 98 (7):3790–3795.

Kim, D. Y., J. S. Werner, and R. J. Zawadzki. 2012. Complex conjugate artifact-free adaptive optics optical coherence tomography of in vivo human optic nerve head. *J Biomed Opt* 17 (12):126005.

Klein, T., W. Wieser, L. Reznicek et al. 2013. Multi-MHz retinal OCT. *Biomed Opt Express* 4 (10):1890–1908.

Kocaoglu, O. P., B. Cense, R. S. Jonnal et al. 2011a. Imaging retinal nerve fiber bundles using optical coherence tomography with adaptive optics. *Vis Res* 51 (16):1835–1844.

Kocaoglu, O. P., R. D. Ferguson, R. S. Jonnal et al. 2014a. Adaptive optics optical coherence tomography with dynamic retinal tracking. *Biomed Opt Express* 5 (7):2262–2284.

Kocaoglu, O. P., S. Lee, R. S. Jonnal et al. 2011b. Imaging cone photoreceptors in three dimensions and in time using ultrahigh resolution optical coherence tomography with adaptive optics. *Biomed Opt Express* 2 (4):748–763.

Kocaoglu, O. P., T. L. Turner, Z. Liu, and D. T. Miller. 2014b. Adaptive optics optical coherence tomography at 1 MHz. *Biomed Opt Express* 5 (12):4186–4200.

Kurokawa, K., K. Sasaki, S. Makita et al. 2010. Simultaneous high-resolution retinal imaging and high-penetration choroidal imaging by one-micrometer adaptive optics optical coherence tomography. *Opt Express* 18 (8):8515–8527.

Lee, S. H., J. S. Werner, and R. J. Zawadzki. 2013. Improved visualization of outer retinal morphology with aberration cancelling reflective optical design for adaptive optics—Optical coherence tomography. *Biomed Opt Express* 4 (11):2508–2517.

Leitgeb, R., C. Hitzenberger, and A. Fercher. 2003. Performance of Fourier domain vs. time domain optical coherence tomography. *Opt Express* 11 (8):889–894.

Lexer, F., C. K. Hitzenberger, A. F. Fercher, and M. Kulhavy. 1997. Wavelength-tuning interferometry of intraocular distances. *Appl Opt* 36 (25):6548–6553.

Liang, C. and D. R. Williams. 1997. Aberrations and retinal image quality of the normal human eye. *J Opt Soc Am A Opt Image Sci Vis* 14:2873–2883.

Liang, J., B. Grimm, S. Goelz, and J. F. Bille. 1994. Objective measurement of wave aberrations of the human eye with the use of a Hartmann-Shack wave-front sensor. *J Opt Soc Am A Opt Image Sci Vis* 11 (7):1949–1957.

Liang, J., D. R. Williams, and D. T. Miller. 1997. Supernormal vision and high-resolution retinal imaging through adaptive optics. *J Opt Soc Am A Opt Image Sci Vis* 14 (11):2884–2892.

Liu, Z., O. P. Kocaoglu, and D. T. Miller. 2013. In-the-plane design of an off-axis ophthalmic adaptive optics system using toroidal mirrors. *Biomed Opt Express* 4 (12):3007–3029.

Lombardo, M., S. Serrao, N. Devaney, M. Parravano, and G. Lombardo. 2013. Adaptive optics technology for high-resolution retinal imaging. *Sensors* 13 (1):334–366.

Masters, B. R. 1999. Early development of optical low-coherence reflectometry and some recent biomedical applications. *J Biomed Opt* 4 (2):236–247.

Meadway, A., C. A. Girkin, and Y. Zhang. 2013. A dual-modal retinal imaging system with adaptive optics. *Opt Express* 21 (24):29792–29807.

Meadway, A., X. Wang, C. A. Curcio, and Y. Zhang. 2014. Microstructure of subretinal drusenoid deposits revealed by adaptive optics imaging. *Biomed Opt Express* 5 (3):713–727.

Merino, D., C. Dainty, A. Bradu, and A. G. Podoleanu. 2006. Adaptive optics enhanced simultaneous en-face optical coherence tomography and scanning laser ophthalmoscopy. *Opt Express* 14 (8):3345–3353.

Miller, D. T., O. P. Kocaoglu, Q. Wang, and S. Lee. 2011. Adaptive optics and the eye (super resolution OCT). *Eye (Lond)* 25 (3):321–330.

Miller, D. T., J. L. Qu, R. S. Jonnal, and K. Thorn. 2003. Coherence gating and adaptive optics in the eye. *Proceedings of the Coherence Domain Optical Methods and Optical Coherence Tomography in Biomedicine VII*, Vol. 4956, Bellingham, WA, pp. 65–72.

Miller, D. T. and A. Roorda. 2009. Adaptive optics in retinal microscopy and vision. In *Handbook of Optics*, eds. M. Bass, C. Decusatis, J. M. Enoch, V. Lakshminarayanan, G. Li, V. C. Macdonald, N. Mahajan, and E. Van Stryland. New York: McGraw-Hill.

Miller, D. T., D. R. Williams, G. M. Morris, and J. Liang. 1996. Images of cone photoreceptors in the living human eye. *Vis Res* 36 (8):1067–1079.

Miller, D., L. Thibos, and X. Hong. 2005. Requirements for segmented correctors for diffraction-limited performance in the human eye. *Opt Express* 13 (1):275–289.

Moon, S. and D. Y. Kim. 2006. Ultra-high-speed optical coherence tomography with a stretched pulse supercontinuum source. *Opt Express* 14 (24):11575–11584.

Mujat, M., R. D. Ferguson, A. H. Patel et al. 2010. High resolution multimodal clinical ophthalmic imaging system. *Opt Express* 18 (11):11607–11621.

Muller, R. A. and A Buffington. 1974. Real-time correction of atmospherically degraded telescope images through image sharpening. *J Opt Soc Am* 64 (9):1200–1210.

Nadler, Z., B. Wang, J. S. Schuman et al. 2014a. In vivo three-dimensional characterization of the healthy human lamina cribrosa with adaptive optics spectral-domain optical coherence tomography. *Invest Ophthalmol Vis Sci* 55 (10):6459–6466.

Nadler, Z., B. Wang, G. Wollstein et al. 2014b. Repeatability of in vivo 3D lamina cribrosa microarchitecture using adaptive optics spectral domain optical coherence tomography. *Biomed Opt Express* 5 (4):1114–1123.

Nidek Co. Ltd. Available from http://www.nidek-intl.com/.

Optopol Technology. Available from http://www.optopol.com/.

Optos Plc. Available from http://www.optos.com/en-us/Products/.

Optovue Inc. Available from http://optovue.com/.

Panorgias, A., R. J. Zawadzki, A. G. Capps et al. 2013. Multimodal assessment of microscopic morphology and retinal function in patients with geographic atrophy. *Invest Ophthalmol Vis Sci* 54 (6):4372–4384.

Park, S. S., G. Bauer, M. Abedi et al. 2015. Intravitreal autologous bone marrow CD34+ cell therapy for ischemic and degenerative retinal disorders: Preliminary phase 1 clinical trial findings. *Invest Ophthalmol Vis Sci* 56 (1):81–89.

Physical Sciences Inc. Available from www.psicorp.com/.

Pircher, M., B. Baumann, E. Gotzinger, and C. K. Hitzenberger. 2006a. Retinal cone mosaic imaged with transverse scanning optical coherence tomography. *Opt Lett* 31 (12):1821–1823.

Pircher, M., B. Baumann, E. Gotzinger, H. Sattmann, and C. K. Hitzenberger. 2007. Simultaneous SLO/OCT imaging of the human retina with axial eye motion correction. *Opt Express* 15 (25):16922–16932.

Pircher, M., E. Gotzinger, and C. K. Hitzenberger. 2006b. Dynamic focus in optical coherence tomography for retinal imaging. *J Biomed Opt* 11 (5):054013.

Pircher, M., E. Gotzinger, H. Sattmann, R. A. Leitgeb, and C. K. Hitzenberger. 2010. In vivo investigation of human cone photoreceptors with SLO/OCT in combination with 3D motion correction on a cellular level. *Opt Express* 18 (13):13935–13944.

Pircher, M. and R. J. Zawadzki. 2007. Combining adaptive optics with optical coherence tomography: Unveiling the cellular structure of the human retina in vivo. *Exp Rev Ophthalmol* 2 (6):1019–1035.

Pircher, M., R. J. Zawadzki, J. W. Evans, J. S. Werner, and C. K. Hitzenberger. 2008. Simultaneous imaging of human cone mosaic with adaptive optics enhanced scanning laser ophthalmoscopy and high-speed transversal scanning optical coherence tomography. *Opt Lett* 33 (1):22–24.

Pocock, G. M., J. W. Oliver, C. S. Specht et al. 2014. High-resolution in vivo imaging of regimes of laser damage to the primate retina. *J Ophthalmol* 2014:516854.

Podoleanu, A. G. and D. A. Jackson. 1998. Combined optical coherence tomograph and scanning laser ophthalmoscope. *Electron Lett* 34 (11):1088–1090.

Porter, J., A. Guirao, I. G. Cox, and D. R. Williams. 2001. Monochromatic aberrations of the human eye in a large population. *J Opt Soc Am A Opt Image Sci Vis* 18 (8):1793–1803.

Porter, J., H. Queener, J. Lin, K. E. Thorn, and A. Awwal. 2006. *Adaptive Optics for Vision Science*. Hoboken, NJ: Wiley-Interscience.

Potsaid, B., I. Gorczynska, V. J. Srinivasan et al. 2008. Ultrahigh speed spectral/Fourier domain OCT ophthalmic imaging at 70,000 to 312,500 axial scans per second. *Opt Express* 16 (19):15149–15169.

Roorda, A., F. Romero-Borja, W. Donnelly III et al. 2002. Adaptive optics scanning laser ophthalmoscopy. *Opt Express* 10 (9):405–412.

Rossi, E. A., M. Chung, A. Dubra et al. 2011. Imaging retinal mosaics in the living eye. *Eye* (*Lond*) 25 (3):301–308.

Rousset, G., J. C. Fontanella, P. Kern, P. Gigan, and F. Rigaut. 1990. First diffraction-limited astronomical images with adaptive optics. *Astron Astrophys* 230:L29–L32.

Sasaki, K., K. Kurokawa, S. Makita, and Y. Yasuno. 2012. Extended depth of focus adaptive optics spectral domain optical coherence tomography. *Biomed Opt Express* 3 (10):2353–2370.

Sheehy, C. K., Q. Yang, D. W. Arathorn et al. 2012. High-speed, image-based eye tracking with a scanning laser ophthalmoscope. *Biomed Opt Express* 3 (10):2611–2622.

Shi, G. H., Y. Dai, L. Wang et al. 2008. Adaptive optics optical coherence tomography for retina imaging. *Chin Opt Lett* 6 (6):424–425.

Stevenson, S. B. and A. Roorda. 2005. Correcting for miniature eye movements in high resolution scanning laser ophthalmoscopy. *Ophthalmic Technol XV* 5688:145–151.

Swanson, E. A., J. A. Izatt, M. R. Hee et al. 1993. In vivo retinal imaging by optical coherence tomography. *Opt Lett* 18 (21):1864–1866.

Thibos, L. N., X. Hong, A. Bradley, and X. Cheng. 2002. Statistical variation of aberration structure and image quality in a normal population of healthy eyes. *J Opt Soc Am A Opt Image Sci Vis* 19 (12):2329–2348.

Topcon Medical Systems Inc. Available from http://www.topconmedical.com.

Torti, C., B. Povazay, B. Hofer et al. 2009. Adaptive optics optical coherence tomography at 120,000 depth scans/s for non-invasive cellular phenotyping of the living human retina. *Opt Express* 17 (22):19382–19400.

Tyson, R. K. 2010. *Principles of Adaptive Optics*, 3rd edn. Boca Raton, FL: CRC Press.

Vienola, K. V., B. Braaf, C. K. Sheehy et al. 2012. Real-time eye motion compensation for OCT imaging with tracking SLO. *Biomed Opt Express* 3 (11):2950–2963.

Vogel, C. R., D. W. Arathorn, A. Roorda, and A. Parker. 2006. Retinal motion estimation in adaptive optics scanning laser ophthalmoscopy. *Opt Express* 14 (2):487–497.

Wang, Q., O. P. Kocaoglu, B. Cense et al. 2011. Imaging retinal capillaries using ultrahigh-resolution optical coherence tomography and adaptive optics. *Invest Ophthalmol Vis Sci* 52 (9):6292–6299.

Webb, R. H. 1996. Confocal optical microscopy. *Rep Progress Phys* 59 (3):427–471.

Webb, R. H., G. W. Hughes, and F. C. Delori. 1987. Confocal scanning laser ophthalmoscope. *Appl Opt* 26 (8):1492–1499.

Webb, R. H., G. W. Hughes, and O. Pomerantzeff. 1980. Flying spot TV ophthalmoscope. *Appl Opt* 19 (17):2991–2997.

Wells-Gray, E. M., R. J. Zawadzki, S. C. Finn et al. 2015. Performance of a combined optical coherence tomography and scanning laser ophthalmoscope with adaptive optics for human retinal imaging applications. *Proceedings of the Adaptive Optics and Wavefront Control for Biological Systems*, SPIE, Vol. 9335, Bellingham, WA.

Williams, D. R. 2011. Imaging single cells in the living retina. *Vis Res* 51 (13):1379–1396.

Wojtkowski, M. 2010. High-speed optical coherence tomography: Basi and applications. *Appl Opt* 49 (16):D30–D61.

Wojtkowski, M., B. Kaluzny, and R. J. Zawadzki. 2012. New direction in ophthalmic optical coherence tomography. *Optom Vis Sci* 89 (5):524–542.

Wojtkowski, M., A. Kowalczyk, R. Leitgeb, and A. F. Fercher. 2002a. Full range complex spectral optical coherence tomography technique in eye imaging. *Opt Lett* 27 (16):1415–1417.

Wojtkowski, M., R. Leitgeb, A. Kowalczyk, T. Bajraszewski, and A. F. Fercher. 2002b. In vivo human retinal imaging by Fourier domain optical coherence tomography. *J Biomed Opt* 7 (3):457–463.

Wojtkowski, M., V. Srinivasan, T. Ko et al. 2004. Ultrahigh-resolution high-speed, Fourier domain optical coherence tomography and methods for dispersion compensation. *Opt Express* 12 (11):2404–2422.

Wong, K. S. K., Y. Jian, M. Cua et al. 2015. In vivo imaging of human photoreceptor mosaic with wavefront sensorless adaptive optics optical coherence tomography. *Biomed Opt Express* 6 (2):580–590.

Zawadzki, R. J., A. G. Capps, D. Y. Kim et al. 2014. Progress on developing adaptive optics-optical coherence tomography for retinal imaging: Monitoring and correction of eye motion artifacts. *IEEE J Sel Top Quantum Electron* 20 (2):pii: 7100912.

Zawadzki, R. J., B. Cense, Y. Zhang et al. 2008. Ultrahigh-resolution optical coherence tomography with monochromatic and chromatic aberration correction. *Opt Express* 16 (11):8126–8143.

Zawadzki, R. J., S. S. Choi, S. M. Jones, S. S. Oliver, and J. S. Werner. 2007. Adaptive optics-optical coherence tomography: Optimizing visualization of microscopic retinal structures in three dimensions. *J Opt Soc Am A Opt Image Sci Vis* 24 (5):1373–1383.

Zawadzki, R. J., S. M. Jones, D. Chen et al. 2009. Combined adaptive optics—Optical coherence tomography and adaptive optics—Scanning laser ophthalmoscopy system for retinal imaging. *Ophthalmic Technologies XIX* 7163:71630F.

Zawadzki, R. J., S. M. Jones, S. S. Olivier et al. 2005. Adaptive-optics optical coherence tomography for high-resolution and high-speed 3D retinal in vivo imaging. *Opt Express* 13 (21):8532–8546.

Zawadzki, R. J., S. M. Jones, S. Pilli et al. 2011. Integrated adaptive optics optical coherence tomography and adaptive optics scanning laser ophthalmoscope system for simultaneous cellular resolution in vivo retinal imaging. *Biomed Opt Express* 2 (6):1674–1686.

Zhang, Y., B. Cense, J. Rha et al. 2006. High-speed volumetric imaging of cone photoreceptors with adaptive optics spectral-domain optical coherence tomography. *Opt Express* 14 (10):4380–4394.

Zhang, Y., J. Rha, R. Jonnal, and D. Miller. 2005. Adaptive optics parallel spectral domain optical coherence tomography for imaging the living retina. *Opt Express* 13 (12):4792–4811.

7 Adaptive optics for visual testing

Enrique Josua Fernández

Contents

7.1 INTRODUCTION

Since time immemorial, vision has attracted the curiosity of human beings. Almost from the early beginnings of the development of science, the study of the eye and the sense of vision have occupied an eminent place. The tremendous advance of the technology in the last decades has changed how classic questions about vision are addressed. The conventional division of vision into three stages, that is, optical, neural, and psychological stages, has now been fortunately surpassed. Interdisciplinary research employing methods from distinct fields is mandatory in many cases to understand the complex puzzle of vision as a whole. In this context, one of the most successful examples of merging different disciplines to understand vision as a whole is adaptive optics (AO) visual testing. In this modality, vision and optics are beautifully combined. AO visual testing enables to fully characterize the impact of the optical quality of the retinal images in vision. AO visual testing provides the most powerful tool for the design of new optical solutions to correct our vision from refractive problems. In this chapter, a comprehensive review of some of the most important results obtained in AO visual testing, together with some important aspects of the required technology, is presented. Binocular AO visual testing, a new and promising modality emerging in this fascinating field, is presented in deeper detail.

7.1.1 WHY WE USE ADAPTIVE OPTICS

AO can be simply defined as the optical technology allowing the measurement and subsequent correction of optical aberrations. In spite of the apparent simplicity of this definition, it is composed of the two fundamental concepts sustaining AO. The first key point is the measurement of aberrations. The operation must be performed in a robust manner, and faster than the typical temporal variation that is intended to be later corrected. The other

pillar of AO is the correction of aberrations. The AO system must allow controlling or manipulating the wavefront once it is estimated. The AO aberration correction concept inherently includes the measurement of the effect of the correction over the wavefront in closed loop. There is an alternative concept generally known as active optics, which performs the correction after the measurement of aberrations, with no possibility of measuring the effect of the correction over the wavefront in closed loop.

In recent years, our knowledge of the optics of the eye has notably improved. Simplistic models to study the formation of images on the retina (El Hage and Le Grand, 1980) provide explanations of the generation of blurred retinal images in the presence of refractive errors, rough estimation of the image size, etc. They are useful to understand the effect of classic optical corrections on the blurred retinal image affected by standard refractive errors as myopia, hyperopia, and astigmatism. All these robust models have provided the necessary baseline to step further toward more sophisticated and realistic models. In the real eye, none of the refractive surfaces can be generally described by a simple radius of curvature, not even by a single conic surface. The optics of the eye is complex and shows a certain level of internal aberrations compensation (Artal et al., 2006; Artal and Tabernero, 2008; Guirao and Artal, 1998; Guirao et al., 2001; Kelly et al., 2004; Tabernero et al., 2007). The different surfaces separating media of distinct refractive indexes are not aligned, nor their hypothetical centers of curvature lay on a common axis. That makes the different surfaces to appear misaligned to each other in the eye (Nishi et al., 2010; Rosales et al., 2010; Tabernero et al., 2006). In addition, the optics of the eye exhibits continuous fluctuations in time (Chin et al., 2008; Diaz-Santana et al., 2003; Fernández and Artal, 2007; Hofer et al., 2001a), which are produced mainly by changes in the crystalline lens. The tear film, the intraocular pressure, and the changes and movement associated with the ocular humors are other factors preventing a steady situation in the eye. It can be summarized that retinal images are degraded dynamically by the individual combination of ocular aberrations in both eyes in addition to classic refractive errors such as myopia, hypermetropia, and astigmatism (Cagigal et al., 2002; Castejón-Mochón et al., 2002; Guirao and Artal, 1998; Porter et al., 2001; Thibos et al., 2002). Consequently, modern techniques for the correction of vision must take into account, and perhaps benefit from, all these aberrations affecting the retinal image. This is precisely what AO visual testing can accomplish and the reason we use it.

7.2 VISUAL SIMULATION AND THE IMPACT OF ABERRATIONS IN VISION

The use of AO to provide a virtually perfect vision from an optical perspective was first demonstrated in 1997 (Liang et al., 1997). Correcting ocular aberrations should permit reaching the perceptual limits of vision once the constraints imposed by the optics of the eye are overcome (Fernández et al., 2001; Hofer et al., 2001b; Li et al., 2009; Yoon and Williams, 2002). The experiments conducted by Liang showed an increase in visual performance when aberrations were corrected as expected. However, it must be noted that the correction of higher-order aberrations, those beyond defocus and astigmatism, exhibited a modest impact in vision in the normal young eye. Some eyes affected by pathological corneas would possibly be the best candidates to benefit from the AO compensation of higher-order aberrations. A typical example of highly aberrated optics can be found in those patients with keratoconus. This is a relatively well-studied condition producing a severe deformation of the anterior and posterior surfaces of the cornea. Several works have studied the impact in vision of the optical correction of keratoconus with AO (Rocha et al., 2010; Sabesan et al., 2007a; Sabesan and Yoon, 2009, 2010). Unfortunately, the current state of the art prevents the use of AO for improving vision. The complexity of the optical systems and their cost make it unreasonable, at present, to integrate AO technology into everyday vision. Nevertheless, AO can be successfully employed to manipulate aberrations and perform experiments to better understand their impact on vision. This use and concept was first demonstrated by Fernandez et al. (2002). The AO system was coupled with an optical relay to permit the display of visual stimuli simultaneously to the manipulation of aberrations. This apparatus was presented as a visual simulator, a term that is being used in the literature since then. The system incorporated a membrane deformable mirror as correcting device with 37 independent actuators (Fernández et al., 2001; Paterson et al., 2000). This type of mirror presented a limitation in the amplitude of deformation, so certain aberrations as defocus or astigmatism had to be corrected by other means, for instance, use of an additional optometer. A comprehensive study about the capability of electrostatic deformable mirrors in the context of vision can be found elsewhere (Fernandez and Artal, 2003). In this seminal experiment with the visual simulator, a subject was asked to complete visual acuity tests while some aberrations were manipulated. The subject presented a natural coma aberration,

which was the selected aberration to be manipulated. The natur coma aberration produced a significant directionality in the subject's point spread function due to the absence of other aberrations. Some amounts of controlled coma aberrations were added to the natural aberrations of the subject producing the effect of rotating the point spread function. Visual acuity was obtained under three cases in the study: vision with natural aberrations; addition of coma aberration in the same orientatior of the one existing in the subject's eye; and coma aberration rotated 90° regarding the natural orientation of the subject's on The analysis of the modulation transfer function (MTF) reveale that from an optical perspective, the qualities of the natural anc the rotated cases were similar, so an acuity test obtained from tł average of different orientations should not exhibit substantial differences. In the case of addition of coma aberration, the area of the MTF presented a significant decrease at every frequency, so the poorest acuity was expected in this case. The results revealed that adding coma aberration in the direction of the naturally existing one in the eye degraded visual performance less than with other orientations. This result suggested that vision could be adapted to the natural aberrations of the eye. Tł larger tolerance of vision when retinal images are degraded in a particular direction might be due to a kind of adaptation of the visual system to its own aberrations.

The adaptation of vision to the aberrations has been investigated in different experiments within the context of AO. The first work where the vision of a number of subjects was systematically tested with manipulated wavefronts appeared in 2004 (Artal et al., 2004). In that study, the perception of blur was evaluated using a random pattern showing irregular black spots on a bright background. Monochromatic illumination at 543 nm was used in order to avoid the possible effects of chromatic aberrations. The ocular aberrations of each subject were recorded. Using the retrieved wavefront, an algorithm generated the appropriate shape for the deformable mirror so that the combination of ocular and mirror wavefronts produced a rotated version of the original ocular aberration. The procedure was performed in closed loop in real time. This operation guaranteed that the subject was continuously viewin the test through the modified aberration pattern. It must be stressed that the experiment forced that the optical quality of the retinal images projected on the observer was exactly the same, but in the orientation of the optical distortion. A two-alternative forced-choice test was programmed. Two images were displayed. One through the natural aberrations of the eye and the other through the rotated version of those aberrations. Subjects were asked to change the level of perceived blur in the manipulated version of the images until equalization of blur with natural aberrations was achieved. For accomplishing the task, the subjects multiplied the rotated version of the wavefront by a factor that globally reduced the level of aberrations, simultaneously in all terms. The results were a systematic perception of a larger blur when viewing through a rotated version of the subjects' own aberrations. Since the optical quality of the retinal images was similar in all cases, the different perceptions of blur had to be produced by neural factors. The proposed explanation to this novel effect was a nev kind of adaptation to the own individual aberration pattern.

Essentially, the brain could be used or adapted to extract information from images degraded in a constant manner, imposed by the subject's ocular aberrations. Any change away from this natural shape degrades perception, even if the optical quality levels are similar. The experiment was interesting since it opened the door to a novel perspective for understanding vision. Many other experiments using AO since then have explored a number of open questions related to the adaptation of vision to ocular aberrations in the eye (Chen et al., 2007; Murray et al., 2010; Sawides et al., 2011a,b, 2012, 2013; Vinas et al., 2012).

7.3 AO VISUAL TESTING FOR ADVANCED OPHTHALMIC DESIGN

Some of the first uses of AO for visual testing were connected with the study of the impact of spherical aberration in vision. Spherical aberration (Ivanoff, 1956) is one of the few optical distortions with a predictable value in the human eye. Spherical aberration is known to show low values in the young eye. It progressively increases with age toward positive values. Spherical aberration compensation mechanism in the human eye is naturally disrupted in the aging eye mostly because of the changes occurring in the crystalline lens (Artal et al., 2001, 2002, 2006; Berrio et al., 2010; Guirao et al., 2000; Tabernero et al., 2007). In addition to this, spherical aberration can be easily compensated by using aspheric surfaces implemented in visual corrections. Among the different alternatives, some very convenient and advantageous candidates to incorporate spherical aberration compensation are intraocular lenses. It can be mentioned that they are implanted close to the exit pupil of the eye, their position and centering are usually steady over time, and the implementation of asphericity on their surfaces is technically feasible with moderate cost. Piers and collaborators (2004) employed AO technology in order to study to what extent the correction of spherical aberration in the eye, specifically in the context of IOL design, could affect vision. The correction of natural spherical aberration in the phakic eye, produced by the cornea, should increase the optical quality of retinal images. However, this slight increase in optical quality might not be translated into better vision. In the case of spherical aberration, naturally appearing in the aged eye, there is an additional issue

with the hypothetical benefits of its correction connected with the depth of field. Spherical aberration, although degrading the images at the paraxial focal plane, produces an extension in the depth of field. That increase in the aged eye can be a useful passive mechanism permitting to ameliorate the loss of accommodation. This fact is of critical importance in the pseudophakic eye, whose capability to accommodate at different distances by changing the power of the lens no longer exists. The incorporation of spherical aberration correction in the IOL could render a better image in the best plane for distant vision, impairing vision at any other closer distance more severely than the natural phakic eye. AO visual testing allowed studying the problem without the necessity of implantation of any IOL. It was accomplished by optically simulating the effect of the spherical aberration correction. In the work of Piers and collaborators (2004), the AO system incorporated an electrostatic deformable mirror with 37 actuators driven in closed loop from the measures obtained in a Hartmann–Shack wavefront sensor (Fernández et al., 2001). Due to the modest amplitude of deformation of the electrostatic mirror, additional elements were required to cover the typical magnitude of the aberrations found in the human eye. Defocus was precompensated by using a motorized optometer, and spherical aberration was implemented in the system by aspheric plates located in a conjugated plane of the subject's pupil. The deformable mirror compensated the remaining optical aberrations in real time, including their temporal variations. The system incorporated a relay for displaying optical stimuli to the subject through manipulated wavefronts. Two different conditions of spherical aberrations were programmed in the experiment. The visual acuity was tested with full compensation of spherical aberration and with the value typically found in the pseudophakic eye: 0.15 μm 4.8 mm pupil diameter (Guirao et al., 2000; Holladay et al., 2002; Mester et al., 2003). The measurements were accomplished on four young subjects. Contrast sensitivity at certain frequencies and visual acuity were obtained under the two conditions. Some of the results are summarized in Figure 7.1. In the plot, the decimal average visual acuity from the four subjects is depicted as a function of defocus in order to study the effect of the correction of spherical aberration on depth of focus. The correction of spherical aberration (blue curve in Figure 7.1) produced a benefit in the visual acuity in the best focus.

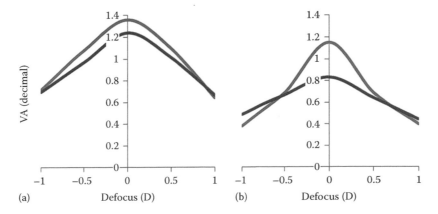

Figure 7.1 Average visual acuity through focus obtained in white light (a) and in green light (b). Blue curve corresponds to the case with full spherical aberration correction. The red curve shows the results obtained with 0.15 μm of spherical aberration and 4.8 mm of pupil diameter.

The enhancement was progressively lost when some additional defocus was included as compared with the situation emulating the typical values of spherical aberration in the aged eye. In the case of monochromatic green light centered at 540 nm, the tails of the function are even below those corresponding to the case with additional spherical aberration (panel b in Figure 7.1). The effect of chromatic aberration seemed to reduce not only the benefit of correcting spherical aberration but also the due degradation out of focus. Still, in view of the results customized and total correction of spherical aberration to improve vision is relevant. The application of AO permitted to answer a practical question without the necessity of implanting IOL with different spherical aberrations. This work showed how AO visual testing is a powerful tool in the design of novel concepts, especially in the early stages of the procedure, for testing the hypothetical visual benefits of any new optical solution.

Since spherical aberration is present in the normal eye, it could be possible that it really plays a role in enhancing vision when combined with other ocular aberrations. Therefore, the paradigm of incorporating solely the correction of spherical aberration in new ophthalmic designs, as in the intraocular lenses, must be first experimentally demonstrated if the rest of the higher-order aberrations are kept. In the absence of other higher-order aberrations, the correction of spherical aberration is known to increase vision (Piers et al., 2004). The question about the interaction of spherical aberration with typical values of higher-order aberrations in natural eyes was studied with an AO visual simulator endowed with a deformable mirror (Piers et al., 2007). In that work, a contrast sensitivity function was obtained with distinct values of spherical aberration both with natural and corrected higher-order aberrations. The individual results exhibited significant intersubject variability. That indicated complex and, in general, not-easy-to-predict interactions across the different aberrations naturally occurring in the eye. The correction of spherical aberration did not produce a peak in the contrast sensitivity function in all cases. However, when analyzing the results on average, it was found that correcting spherical aberration produced the best vision. The results of this work support the use of average values of spherical aberration intending its total compensation in the eye, although it also shows the benefits of using customized corrections. Similarly, using AO visual testing, the impact of spherical aberration in the presence of a high amount of scattering was simulated in another good example of the application of visual simulators (Pérez et al., 2009). Surprisingly, the results showed that under certain conditions, spherical aberration could act as a protection in terms of visual quality when scattering is also present. That is precisely the situation occurring in the normal-aged eye, where both spherical aberration and scattering are naturally high.

The different systems and results presented so far in this chapter were obtained with different types of deformable mirrors used as correcting elements in AO simulators. Deformable mirrors, though an established technology in managing ocular aberrations, present some inherent limitations. A constraint for practical applications in the field of visual testing is the failure of such correctors to produce diffractive profiles. Diffractive profiles are useful to be incorporated in intraocular lenses to produce multifocality (Hayashi et al., 2015; Ortiz et al., 2008; Ravikumar et al., 2014; Weeber and Piers, 2012). Another possible use under development is the extension of ocular depth of focus through more sophisticated diffractive profiles (Cagigal et al., 2004; Flores et al., 2004; Liu et al., 2010; Mas et al., 2007; Rao et al., 2011; Zhang et al., 2008), as a method to alleviate presbyopia. Diffractive profiles, and in general any phase showing discontinuities, can be programmed with liquid crystal–based phase modulators. The use of liquid crystals for ophthalmic applications has been reported in the past, demonstrating their capability to correct ocular aberrations (Vargas-Martín et al., 1998). Some other works have explored and characterized these elements for visually oriented applications (Fernández et al., 2009b; Prieto et al., 2004). These phase modulators have also been used as aberration correctors for AO retinal imaging in the field of optical coherence tomography (OCT) (Fernández et al., 2005). The work of Manzanera et al. (2007) specifically employed a liquid crystal phase modulator for visual testing in the context of ophthalmic optics design. The experimental apparatus incorporates a relay for the recording of point spread functions through the system. Some phase profiles intended for multifocality were tested, isolated and in combination with real eyes. In the latter case, subjects performed visual tests through focus to better characterize the effects of the proposed phase profiles. These could be eventually incorporated into intraocular lenses, for instance. The visual simulator provided an excellent platform for the design of new solutions, and, equally importantly, for testing the proposed solutions on real eyes. This is of particular importance when sophisticated diffractive profiles are tested. The natural higher-order aberrations occurring in the normal eye, in spite of their relatively low amplitude, can nevertheless interact with the programmed profiles altering or even neutralizing the theoretical outcomes of the diffractive mask. Some examples were provided in the work of Manzanera et al. (2007) illustrating this concept.

Liquid crystal phase modulators present some limitations in the temporal response. Phase variations are typically in the order of 50 Hz. When operating in closed loop aberration correction, they are at the peak of their performance to compensate ocular dynamics. Therefore, if temporal stabilization of the ocular aberrations is required during AO operation, liquid crystal technology might be insufficient. To overcome this limitation, a hybrid experimental setup incorporating both a deformable mirror and a liquid crystal phase modulator was demonstrated for advanced visual testing (Cánovas et al., 2010). The principle of operation of the systems consisted in the use of a deformable mirror to compensate dynamically ocular aberrations, while the liquid crystal phase modulator generated diffractive or discontinuous phase profiles. The subjects experienced the effect of both the correction and the phase mask simultaneously with the performance of visual testing. The system successfully merged two technologies for ocular aberration correction, but the complexity and the cost of the setup probably limit the application of such an approach in regular clinical practice. However, the system perfectly meets the requirement of the ophthalmic industry as an advanced visual simulation station to test and design new and customized solutions to correct vision.

7.4 BINOCULAR AO VISUAL TESTING

During its early development, and so far in this chapter, AO visual testing was systematically applied under monocular vision. It was not employed in binocular studies until the technique proved mature enough to face the increase in complexity of managing both eyes simultaneously. Characterizing vision from the data obtained in a single eye provides useful and valuable information, but normal vision is not monocular vision. Therefore, any technique aiming to fully understand vision must take into account binocularity. Binocular vision provides a number of advantages in perception, for instance, in contrast sensitivity (Campbell and Green, 1965). Superior perception of the reality is accomplished by binocular vision. This fact has driven the need for new instruments and techniques able to objectively measure binocular refraction, for instance, binocular optometers. Some of them were developed in the recent past, and they can be considered the necessary antecedents to first binocular wavefront sensors (Clark and Crane, 1978; Heron et al., 1989; Heron and Winn, 1989; Okuyama et al., 1993). Kobayashi et al. (2008) reported a system endowed with two Hartmann–Shack wavefront sensors specifically designed for binocular estimation of higher-order aberrations from both eyes. The system operated in open view. With that apparatus, the subjects could undergo visual testing while their ocular aberrations were retrieved. An evolution of the system was later employed in the estimation of clinical refraction (Mihashi et al., 2010). The approach of using a separated wavefront sensor for each eye brings about the duplication of the cost of the system, together with an increase in the complexity of the control of the setup. During the same time, another alternative to this first solution for binocular objective estimation of the ocular aberrations was presented in the work of Hampson et al. (2007; Chin et al., 2008; Hampson et al., 2007). In this other approach, a design allowing the estimation of the aberrations from both eyes employing a single sensor was reported. Using a Hartmann–Shack-based apparatus, the light emerging from each pupil was directed into the system, keeping the two beams spatially resolved along the optical relays. The exit pupils of both eyes were projected on the surface of the wavefront sensor, so that the camera could obtain in a single frame the spots from both pupils. Appropriate postprocessing allowed the retrieval of the ocular aberrations from both eyes. The advantages of this apparatus were evident in terms of cost and complexity. A schematic representation of this concept is shown in Figure 7.2. The diagram does not include the lenses and mirrors required to conjugate the subject's pupils and the plane of the Hartmann–Shack wavefront sensor. The system incorporated a pair of beam splitters for open view operation, so that the subjects could perform visual testing or just fixation during the measurements of their aberrations. The images from the two pupils were conserved along the setup spatially resolved right up to the plane of the wavefront sensor, enabling the formation of two separate sets of spots corresponding to each pupil. Appropriate software calculated the aberrations from each pupil appearing in a single camera frame.

An interesting alternative setup for open view binocular wavefront sensing has been reported using IR light in the 1050 nm,

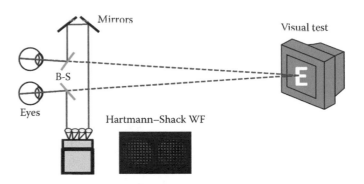

Figure 7.2 Schematic diagram of the setup allowing the retrieval of aberrations from both eyes simultaneously in a Hartmann–Shack wavefront (WF) sensor. The key feature in the system is the projection of the two pupils of the subject on the surface of a single WF sensor spatially resolved. A pair of beam splitters (B-S) enabled open view operation.

at intensities low enough to make it undetectable (Chirre et al., 2014). A remarkable feature of that instrument is the capability to study accommodation while simultaneously retrieving pupil sizes, vergence, and ocular aberrations during changes in focus.

The use of a single wavefront sensor for simultaneously measuring both pupils of a subject was developed in parallel within the context of AO aberration correction as well. In 2009, the first binocular AO system was reported and successfully applied on real eyes (Fernández et al., 2009a). The system enabled the manipulation of measured aberrations from the eyes simultaneously. One of the most remarkable features of this system was the employment of a single correcting device in combination with a single wavefront sensor. The latter employed am approach similar to that reported by Chin et al. (2008) in the wavefront, separating the two pupils on the surface of the detector (Figure 7.3).

The correcting device employed an analogous principle for managing the two pupils. The pupils were simultaneously projected on its surface, still spatially resolved, and controlled by the computer. A fundamental issue arising from this configuration was procuring enough resolution on the corrector for manipulating aberrations, including higher-order aberrations, from both pupils with sufficient accuracy. In order to overcome this obstacle, a liquid crystal–based correcting device was incorporated in the system. Such correctors have been key to understand the current state of the art of binocular AO visual testing. One of their most important and attractive characteristics is possibly their enormous resolution, particularly as compared to deformable mirrors. In the first binocular visual simulator, a commercial spatial light modulator (LCOS-SLM X10468-04, Hamamatsu Photonics, Japan) with an SVGA resolution (800 × 600) was incorporated. The total number of pixels was 480,000, therefore dedicating a maximum of 240,000 pixels for each pupil. The device allowed keeping a large number of pixels dedicated for the control of aberrations from each eye, in any case far beyond the range of deformable mirrors. It is interesting to note that the possibility of using AO under binocular conditions had been previously suggested in different works and patents by duplicating the apparatus designed for a single eye. However, in most of the cases, it proved to be an impractical solution with low efficiency as compare to the possibility of operation with a

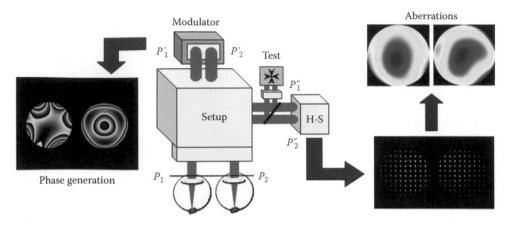

Figure 7.3 Schematic diagram of the binocular AO visual simulator. The pupils of the two eyes are conjugated simultaneously on a single phase modulator, on a single Hartmann–Shack wavefront sensor, and on the stimuli presentation relay.

single corrector and sensor. The capability of this novel approach was demonstrated by testing the impact of different combinations of spherical aberration on vision. For this purpose, the contrast sensitivity of a subject at 7.8 c/deg in green light was measured under different conditions. The contrast sensitivity was obtained by using a two-alternative forced-choice test: a panel with target grating and different contrast and another with a homogenous background. The subject identified which displayed the grating. The time for displaying the stimuli was 500 ms. The grating subtended 1°, assuring isoplanatic conditions at the retina. The contrast of the grating was set randomly. A psychometric curve was obtained for inferring the value of the contrast sensitivity, which was estimated as the detection threshold at 75% of confidence. The experiment was repeated under monocular and binocular conditions to discern the influence of binocular summation. The value of spherical aberration was ±0.2 μm for a pupil of diameter 4 mm. This experiment demonstrated the potential of the technique. Although a single subject underwent contrast sensitivity measurements, some interesting results were obtained, which are shown in Figure 7.4.

The plot shows the contrast sensitivity obtained for different combinations of spherical aberration. The + and – signs indicate the value of the spherical aberration added on the eyes.

The position of the value in the brackets represents which eye was affected for each value, the first value for the left eye and the second for the right eye. The subject reported his dominant eye to be the left one. Systematically, keeping the dominant eye with the natural aberrations caused a better visual acuity. That was particularly evident when the results were compared with those corresponding to the reversal cases, where spherical aberration was added to the nondominant eye. Another interesting result was that the degradation of visual acuity was more dramatic when positive spherical aberration was added to the dominant eye, in absolute values. This particular eye exhibited a larger tolerance to negative spherical aberration.

An advantage of binocular vision is stereopsis. Stereopsis is connected to the visual perception of depth (Fielder and Moseley, 1996; Howard and Brian, 1996; O'Connor et al., 2010; Reading, 1983). It is defined as the capability for perceiving the depth in a scene originated exclusively for the distinct position of the images on each retina or more commonly referred to as retinal disparity. This kind of retinal parallax produced by the relatively shifted perspective of the object in each eye has been widely studied in the context of visual perception. Stereopsis involves a delicate neural processing in addition to the purely optical or geometrical stage. Actually, stereopsis was traditionally studied in the context of psychology of perception. Some attempts to introduce the possible impact of the ocular optical quality, or ocular aberrations, on stereopsis have been reported in the past (Castro et al., 2009; Jiménez et al., 2008). Binocular AO provide the ideal tool for characterizing the actual impact of aberrations on stereopsis, not only lower-order aberrations, as defocus and astigmatism can be tested, but any aberration in general. In this direction, the first work where stereopsis was evaluated degrading the wavefront with different aberrations was reported in 2010 (Fernández et al., 2010). The experimental system incorporated two different internal displays for projecting distinct retinal images over each retina. This permitted the generation of retinal disparity, indistinguishable from an optical perspective to that created with a real scene in front of the subject. The stereopsis was tested through the measurement of the stereoacuity. Stereoacuity provides a numerical estimate of the capability of the subject for detecting changes in depth associated exclusively with retinal disparity. Actually, stereoacuity is the minimum retinal disparity

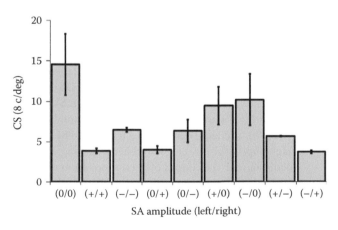

Figure 7.4 Contrast sensitivity for 7.68 c/deg obtained under different values of spherical aberration for the left and right eye. The amplitude of the aberration was 0 or ±0.2 μm.

given as the subtended angle, causing the perception of depth. Using separate displays for each eye guaranteed that no other monocular cue was involved in the experiment. Random dot stereograms were used for obtaining stereoacuity. Random dot stereograms are a useful tool for understanding pure stereopsis, particularly global stereopsis. They were introduced by Julesz in the 1970s (Julesz, 1971), and since then have been widely employed for binocular vision research. In Julesz's work, pure defocus was first added under different conditions. Selecting a pupil diameter of 4 mm, stereoacuity was retrieved for natural vision with lower-order refractive correction (including defocus and astigmatism), addition of 1 D of defocus in both eyes, and 1 D of defocus in one of the eyes. In addition to defocus, for the first time the impact on stereopsis of other higher-order aberrations generated by AO was evaluated. In particular, the trefoil aberration was selected and applied to both binocular and monocular visions, while simultaneous measurement of the stereoacuity was performed. The resulting trend was similar to that typically occurring for defocus. The addition of trefoil in a single eye produced a larger degradation on stereopsis than the bilateral case. The obtained values of stereoacuity were 4, 13, and 18 s for natural vision (with no additional aberrations) and unilateral and bilateral addition of pure trefoil, respectively. Figure 7.5 shows graphically the results obtained under different conditions.

An excellent work related to the impact of ocular aberrations on binocular vision appeared in 2011, where stereopsis was also obtained in some subjects with natural conditions and through static aberration correction (Vlaskamp et al., 2011). The results of that work showed an unexpected lack of benefit on stereopsis when ocular aberrations were compensated. Although the number of subjects participating in the study was relatively modest, the results were consistent enough to demonstrate this relevant effect. The authors hypothesized eye movements as the probable cause for this lack of increase in stereopsis when the retinal images are

corrected from aberrations. Further experiments might confirm that explanation and help better understand these findings. Those results dramatically exhibit our ignorance on several aspects of binocular vision and its connection with ocular optics.

An interesting issue of practical importance with binocular vision concerns the summation in the presence of asymmetric pupils (Tabernero et al., 2011). A straightforward approach to expand depth of field consists in imaging through small pupils. In the case of the eye, constraining the pupils brings about a significant loss of irradiance in the retinal images, whose more direct consequence is the drop in contrast sensitivity. A possible alternative to ameliorate this effect is the use of asymmetric pupils. This concept was explored with the help of AO (Tabernero et al., 2011) as an alternative to compensate presbyopia. The idea resembles monovision, where one eye is set for far vision, while the other is endowed with appropriate near refraction. In the case of asymmetric pupils, one eye is looking through its natural pupil and the other through a small aperture. The latter eye is expected to show larger depth of field as compared with the other. The open question then is how binocular vision is affected by this approach. The experimental setup was based on previous works (Fernández et al., 2010), but it also incorporated two asymmetric entrance pupils of 4 and 1.5 mm of diameter optically projected on the subjects' pupils, limiting the amount of light from the stimulus relay. In the experiment, binocular visual acuity was systematically enhanced as compared with any of the monocular situations for corrected or distant vision. However, when closer distances were considered, summation did not occur, and consequently, the obtained visual acuity resembled that measured under monocular conditions through the small pupil. A possible explanation provided by the authors was the significant differences in the images' blur across the two eyes, producing suppression of the worst information channel. Still, the effect is not well understood and further experiments are needed. A particular valuable aspect of this study is the possibility to measure vision through the selected conditions without the need for surgically implanting asymmetric pupils or manufacturing any optical element. An evolution of this system was reported by Schwarz et al. (2011). The apparatus incorporated an additional liquid crystal operating in transmission for the manipulation of the pupils. This other liquid crystal acted as an amplitude modulation device, enabling dynamic manipulation of both pupils' size and position during the measurements, retaining the capability for phase modulation with the other correcting device. The modulation in the amplitude of the entrance pupils opened the door to a new family of experiments.

A completely different experimental approach to study binocular vision with manipulation of ocular aberrations was reported by Sabesan et al. (2012). They presented a binocular AO system as a duplication of a previous monocular AO setup (Sabesan and Yoon, 2009; Sabesan et al., 2007). This system had separate phase modulator and wavefront sensor for each eye. Another important difference as compared with others binocular AO systems previously reported (Fernández et al., 2009a) was the technology selected to control and manipulate aberrations. In this case, a magnetic deformable mirror was selected to perform the corrections. These kinds of mirrors have been successfully demonstrated for ocular applications

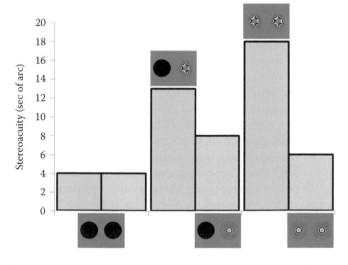

Figure 7.5 Stereoacuity in seconds of arc obtained from experimental psychometric functions under different conditions. Bars in light blue show the results measured with trefoil aberration, monocular, or binocular. The pale orange bars are the results obtained with pure defocus. The first two bars exhibit the stereoacuity under natural vision.

(Fernández et al., 2006a). They exhibit high potential in the ability to compensate the aberrations occurring in the eye. In this work, the authors surveyed the behavior of visual acuity and contrast sensitivity in the absence of aberrations. Binocular summation was obtained in a variety of cases to better understand how vision responds under aberration correction. Some curious and apparently paradoxical results were reported. For instance, in view of the results, it was found that correcting native optical aberrations decreased binocular advantage. In spite of a clear increase in both contrast sensitivity and visual acuity with aberration correction, summation was reduced when the subjects performed the tests through near-perfect optics. The increase of summation might be a kind of protection against the slight blur produced by higher-order aberrations. However, in the absence of aberrations, binocular vision becomes more similar to monocular vision in terms of visual quality. The effect was particularly evident at higher spatial frequencies. These results need to be studied further to understand how binocularity enriches our vision in the presence of natural aberrations. The same experimental setup was employed in another work to study the addition of spherical aberration in combination with monovision to ameliorate the impact of presbyopia (Zheleznyak et al., 2013). Spherical aberration has the potential to increase depth of focus, as it has been explained before in the context of AO visual testing under monocular conditions. On the other hand, monovision has been demonstrated as one of the most simplistic and effective passive solutions to provide reasonable visual quality at different distances (Erickson, 1988; Evans, 2007; Jain et al., 1996). In the work of Zheleznyak et al. (2013), binocular performance through focus was investigated in monovision (1.5 D difference across eyes) with some spherical aberration; in particular, they selected the values of ±0.2 and ±0.4 μm in a 4 mm pupil diameter. Vision was tested for visual acuity through focus and contrast sensitivity at 10 c/deg. Few subjects participated in the study, probably due to the complexity in the experimental measurements. Besides, the results were valuable and can provide some information about the most efficient method to combine spherical aberration with monovision. It was found that monovision with positive spherical aberration presented a larger benefit for intermediate distances than that with negative spherical aberration. However, negative values of spherical aberrations exhibited larger enhancements in visual acuity at near distances. Binocular summation was absent at all object distances except 0.5 D, intermediate, where it slightly improved by about 20% over monocular vision. AO visual testing can help to optimize new solutions and aids to correct vision with no need of real patients either true manufacturing or implantations of optical corrections.

Directly connected with the quest for new passive corrections to presbyopia, an evolution of the AO system with full control of pupils described before (Schwarz et al., 2011) in this chapter was employed to study an alternative method to increase depth of field in presbyopic eyes (Fernández et al., 2013). In this work, the impact of using asymmetric pupils in combination with some defocus on stereoacuity is confronted to traditional monovision. Stereoacuity is defined as the minimum distance, or equivalently change in vergence or retinal disparity, that the

Figure 7.6 Picture of the AO system showing the light paths from each of the microdisplays to the subject's eyes.

subject can perceive in depth. In the experiment, a three-needle test was programmed to obtain stereoacuity, where vertical lines are projected on the patient's retinae. The experimental system incorporated two different projectors to show the stimuli. The central wire can be displaced, generating disparity on the retinae, to produce the illusion of depth. The apparatus operated in open loop. Once the aberrations were estimated, the programmable phase modulator modified accordingly the aberrations of the two pupils. This operation was possible with liquid crystal–based correctors (due to their high fidelity) after proper calibration. A picture of the actual system is presented in Figure 7.6.

The five different conditions tested in the experiment were as follows:

1. *Natural*: Subjects performed through their natural optics and 4 mm pupil size.
2. *Monovision*: Pupil diameter of 4 mm and addition of 1.5 D in one of the eyes.
3. *Micromonovision*: Maintaining symmetric pupils of 4 mm diameter, and 0.75 D for the near refraction.
4. *Small aperture*: One of the eyes performed through a pupil 1.6 mm diameter, the other used 4 mm diameter, and natural aberrations in both eyes.
5. *Small aperture and micromonovision*: The eye with the near refraction was also enforced to perform through the 1.6 mm diameter pupil.

In all cases, lower-order aberrations were fully corrected and accommodation was paralyzed by using drugs. The summary of the results is presented in Figure 7.7.

Figure 7.7 does not include the case of pure monovision with the addition of 1.5 D. Subjects were not able to perform the experiment for that particular condition owing to the large difference between retinal blurs. The difference in the retinal images is very likely to cause suppression in the eye set for near vision, so monocular vision can be assumed for far and near vision under monovision. On the other hand, alleviating presbyopia by using monovision is tempting due to the simplicity in its practical implementation. For those patients who require stereoacuity, this simple and effective approach should not be recommended. As an alternative, the results of the work show that the combination of micromonovision and small aperture does not degrade this particular visual function, at least under photopic conditions. In practice, the small aperture

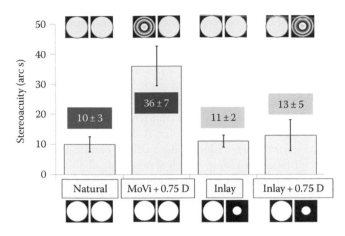

Figure 7.7 Average stereoacuity obtained with different combinations of phases and pupils. The latter are depicted at the bottom, while that corresponding phases are shown at the top of the panel.

can be implanted in the presbyopic eye by using intrastromal corneal inlays (Dexl et al., 2011, 2012; Seyeddain et al., 2012; Yilmaz et al., 2008, 2011). The reduction of light irradiance due to the use of a small pupil did not show any apparent effect under photopic conditions, but it is very likely that under mesopic and scotopic conditions, visual performance could be impaired by this solution.

In this direction, the question of how and to what extent aberration correction could affect binocular performance under conditions of different illumination was also studied with binocular AO technology (Schwarz et al., 2014b). In the experiment, subjects performed visual acuity and contrast sensitivity tests with different luminances with static aberration correction and natural conditions. Luminance was controlled by using appropriate neutral filters, selecting values from 0.2 to 0.002 cd/m² in green light (543 nm). The results showed that the benefit in visual performance of correcting aberrations increases as the luminance decreases. The fact that for photopic conditions the correction of higher-order aberrations produced solely a modest benefit as compared with that under mesopic and scotopic conditions is of practical importance for the design of advanced optical solutions. Using customized corrections implemented in contact lenses, intraocular lenses, or even programmed in corneal refractive surgery is specially indicated for those patients requiring good visual performance under low luminance. Another interesting result somehow supporting previous findings (Sabesan et al., 2012) is that the binocular benefit is more pronounced in those eyes exhibiting poorer optical quality. Binocular vision appears as a protection against higher-order aberrations in normal eyes.

In all the preceding studies, ocular aberrations are implicitly assumed to be monochromatic. However, the eye presents chromatic aberrations as well. In addition to the visible, the near infrared range has been also characterized in recent decades (Atchison and Smith, 2005; Fernández and Artal, 2008; Fernández et al., 2006b; Thibos et al., 1992). Longitudinal chromatic aberration has been proved to be an important factor affecting optical quality of the retinal images (Fernández and Drexler, 2005; Fernández et al., 2006a; Ravikumar et al., 2008). How binocular vision is affected by chromatic aberration was studied by an AO binocular simulator (Schwarz et al., 2014a). In particular, the interest of chromatic aberration correction emerges because of the state-of-the-art technology for manufacturing optical solutions, as intraocular lenses, capable to incorporate such correction. The combined correction of longitudinal chromatic aberration and spherical aberration in the eye was demonstrated theoretically to improve vision (Weeber and Piers, 2012b). The concept was previously tested (Artal et al., 2010). In both these works, monocular vision was considered. In the work of Schwarz et al. (2014a), binocular visual acuity was obtained under different conditions for spherical aberration in polychromatic and monochromatic light (550 nm) to assess the impact of chromatic aberration in vision. Spherical aberration was corrected, to emulate those intraocular lenses with such compensation in some cases, while a value of 0.15 m was induced in other cases. That amount of spherical aberration corresponds to the typical value obtained in pseudophakic eyes implanted with spherical intraocular lenses through a realistic pupil of 4.8 mm diameter (Guirao et al., 2000; Mester et al., 2003). Figure 7.8 shows the average visual acuity obtained from three subjects under the different conditions tested in the experiment. The highest visual acuity was obtained with monochromatic illumination and full correction of spherical aberration for both the monocular and binocular cases.

Conversely, monocular visual acuity obtained with polychromatic illumination and spherical aberration scored the lowest in the experiment. The visual acuity obtained under monochromatic, monocular, and spherical aberration exhibited a larger value than the binocular and polychromatic illumination conditions, indicating that chromatic aberration degrades vision more severely than the selected value of spherical aberration. From the plot, the benefit of performing the binocular test as compared to monocular test shows intriguing differences as a function of the distinct conditions set in the experiment, demanding a deeper attention. A possible way to characterize

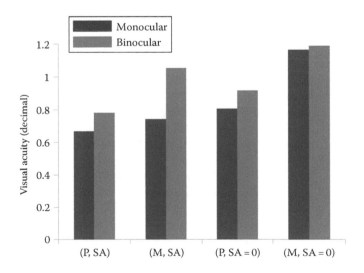

Figure 7.8 Average visual acuity obtained with different combinations of illumination and spherical aberration under monocular and binocular conditions. P indicates polychromatic illumination, while M corresponds to monochromatic light. The value programmed for spherical aberration (SA) was 0.15 µm for a 4.8 mm pupil. SA = 0 indicated full correction of spherical aberration.

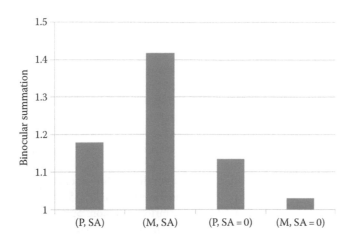

Figure 7.9 Average binocular summation coefficient obtained under the different conditions.

the binocular advantage is to obtain the summation coefficient. Figure 7.9 presents the binocular summation under the distinct conditions tested in the experiment. It seems that starting from high optical quality in the monocular case significantly decreases the binocular advantage in vision, as it is shown in the figure for monochromatic illumination and no spherical aberration. Conversely, the largest summation occurred in the presence of spherical aberration and monochromatic illumination. From the results, it could be inferred that our visual system is able to handle spherical aberration quite successfully under binocular conditions, partially alleviating impairment over the optical quality of the retinal images. When polychromatic illumination is included in the experiment, the summation is reduced. Therefore, apparently our capability for enhancing vision affected by spherical aberrations is slightly reduced.

7.5 CONCLUSIONS

Visual testing with AO technology is a relatively young field. In spite of this, a vast number of applications and studies have proliferated in recent years. In this chapter, I have presented a possible classification of applications to show the results with some organization. The division between basic and applied research is especially diffuse in this field. In principle, trying to understand how aberrations affect our vision, something AO visual testing is particularly suited for, could be taken as a basic science question, but the applications of every result in the design of better corrections are evident. Perhaps this is one of the beauties of AO visual testing: every advance can be almost immediately used to improve vision. The reader is possibly aware how presbyopia, its alleviation, is a recurrent topic in this chapter. Actually, it has been one of the engines moving the field forward, and I expect it will continue starring the history from some time more. Its correction with passive solutions, such as multifocal intraocular lenses, diffractive phase masks to extend depth of field, traditional monovision, asymmetric employment of pupils, or spherical aberration, just to cite a few, is easy to implement in practice. Some of them are quite old, but we still do not know all the details, and what is more important is how

real vision performs under such corrections. Probably, the smart combination of some already known techniques will produce desirable results in the near future to alleviate presbyopia. In this context, the amount of studies and efforts dedicated to understand the role and possible uses of spherical aberration in the eye is also significant.

I have devoted special attention to binocular AO visual simulators in this chapter. At last, we have a tool to fully characterize vision under realistic conditions. Understanding binocular vision, which is otherwise the natural case in most situations, will possibly change the way we correct vision in the future. I am convinced that progressively the field will adopt such binocular simulators as a standard for studying vision, in particular, for the optimized design of new corrections. With this technology, we can investigate and compare different solutions with no need to manufacture them, not requiring true patients, but just volunteers willing participate in the experiments with the simulator. I think that this concept will speed up the development of new advanced corrections. I can imagine a near future where AO visual testing stations would be available for patients to choose an optical solution that suits their requirements, before acquiring the correction, undergoing surgery, or even customizing and adapting their corrections to their own aberrations. I am convinced that the most important applications of AO visual testing are still to come.

ACKNOWLEDGMENTS

This work was supported by the Ministerio de Ciencia e Innovación, Spain (grants FIS2010-14926, CSD2007-00013, FIS2013-41237-R), and Fundación Séneca-Agencia de Ciencia y Tecnología de la Región de Murcia (Spain), Programa Jóvenes Líderes en Investigación (grant 18964/JLI/2013), and Programa Grupos de Excelencia (grant 4524/GERM/06).

REFERENCES

Artal, P., Benito, A., and Tabernero, J. (2006). The human eye is an example of robust optical design. *Journal of Vision*, 6(1), 1–7. doi:10.1167/6.1.1.

Artal, P., Berrio, E., Guirao, A., and Piers, P. (2002). Contribution of t cornea and internal surfaces to the change of ocular aberrations with age. *Journal of the Optical Society of America. A, Optics, Imag Science, and Vision*, 19(1), 137–143. Retrieved from http://www.ncbi.nlm.nih.gov/pubmed/11778716.

Artal, P., Chen, L., Fernández, E. J., Singer, B., Optica, L. De, F., De, D., and Williams, D. R. (2004). Neural compensation for the eye's optical aberrations. *Journal of Vision*, 16, 281–287. doi:10.1167/4.4.4.

Artal, P., Guirao, A., Berrio, E., and Williams, D. R. (2001). Compensation of corneal aberrations by the internal optics in the human eye. *Journal of Vision*, 1(1), 1–8. doi:10:1167/1.1.1.

Artal, P., Manzanera, S., Piers, P., and Weeber, H. (2010). Visual effect of the combined correction of spherical and longitudinal chromatic aberrations. *Optics Express*, 18(2), 1637–1648. Retrieve from http://www.ncbi.nlm.nih.gov/pubmed/20173991.

Artal, P. and Tabernero, J. (2008). The eye's aplanatic answer. *Nature Photonics*, 2(October), 586–589.

Ophthalmic instrumentation

Atchison, D. and Smith, G. (2005). Chromatic dispersions of the ocular media of human eyes. *Journal of the Optical Society of America A*, *22*(1), 29–37. Retrieved from http://www.opticsinfobase.org/abstract.cfm?uri=JOSAA-22-1-29.

Berrio, E., Tabernero, J., and Artal, P. (2010). Optical aberrations and alignment of the eye with age. *Journal of Vision*, *10*(14), 34. doi:10.1167/10.14.34.

Cagigal, M. P., Canales, V. F., Castejón-Mochón, J. F., Prieto, P. M., López-Gil, N., and Artal, P. (2002). Statistical description of wave-front aberration in the human eye. *Optics Letters*, *27*(1), 37–39. Retrieved from http://www.ncbi.nlm.nih.gov/pubmed/18007708.

Cagigal, M. P., Oti, J. E., Canales, V. F., and Valle, P. J. (2004). Analytical design of superresolving phase filters. *Optics Communications*, *241*(4–6), 249–253. doi:10.1016/j.optcom.2004.07.024.

Campbell, F. W. and Green, D. G. (1965). Monocular versus binocular visual acuity. *Nature*, *208*(5006), 191–192. doi:10.1038/208191a0.

Cánovas, C., Prieto, P. M., Manzanera, S., Mira, A., and Artal, P. (2010). Hybrid adaptive-optics visual simulator. *Optics Letters*, *35*(2), 196–198. Retrieved from http://www.ncbi.nlm.nih.gov/pubmed/20081966.

Castejón-Mochón, J. F., López-Gil, N., Benito, A., and Artal, P. (2002). Ocular wave-front aberration statistics in a normal young population. *Vision Research*, *42*(13), 1611–1617. Retrieved from http://www.ncbi.nlm.nih.gov/pubmed/12079789.

Castro, J. J., Jiménez, J. R., Hita, E., and Ortiz, C. (2009). Influence of interocular differences in the Strehl ratio on binocular summation. *Ophthalmic and Physiological Optics: The Journal of the British College of Ophthalmic Opticians (Optometrists)*, *29*(3), 370–374. doi:10.1111/j.1475-1313.2009.00643.x.

Chen, L., Artal, P., Gutierrez, D., and Williams, D. R. (2007). Neural compensation for the best aberration correction. *Journal of Vision*, *7*(10), 9.1–9.9. doi:10.1167/7.10.9.

Chin, S. S., Hampson, K. M., and Mallen, E. A. (2008). Binocular correlation of ocular aberration dynamics. *Optics Express*, *16*(19), 14731–14745. Retrieved from http://www.ncbi.nlm.nih.gov/pubmed/18795011.

Chirre, E., Prieto, P. M., and Artal, P. (2014). Binocular open-view instrument to measure aberrations and pupillary dynamics. *Optics Letters*, *39*(16), 4773–4775.

Clark, M. R. and Crane, H. D. (1978). Dynamic interaction in binocular vision. In R. A. Monty, J. W. Senders, and D. F. Fisher (Eds.), *Eye Movement and the Higher Psychological Functions* (pp. 77–88). New York: Erlbaum.

Dexl, A. K., Seyeddain, O., Riha, W., Hohensinn, M., Hitzl, W., and Grabner, G. (2011). Reading performance after implantation of a small-aperture corneal inlay for the surgical correction of presbyopia: Two-year follow-up. *Journal of Cataract and Refractive Surgery*, *37*(3), 525–531. doi:10.1016/j.jcrs.2010.10.044.

Dexl, A. K., Seyeddain, O., Riha, W., Hohensinn, M., Rückl, T., Reischl, V., and Grabner, G. (2012). One-year visual outcomes and patient satisfaction after surgical correction of presbyopia with an intracorneal inlay of a new design. *Journal of Cataract and Refractive Surgery*, *38*(2), 262–269. doi:10.1016/j.jcrs.2011.08.031.

Diaz-Santana, L., Torti, C., Munro, I., Gasson, P., and Dainty, C. (2003). Benefit of higher closed-loop bandwidths in ocular adaptive optics. *Optics Express*, *11*(20), 2597–2605. Retrieved from http://www.ncbi.nlm.nih.gov/pubmed/19471373.

El Hage, S. G. and Le Grand, Y. (1980). *Physiological Optics* (Vol. 13). Berlin, Germany: Springer. doi:10.1007/978-3-540-39053-4.

Erickson, P. (1988). Potential range of clear vision in monovision. *Journal of the American Optometric Association*, *59*(3), 203–205. Retrieved from http://www.ncbi.nlm.nih.gov/pubmed/3351187.

Evans, B. J. W. (2007). Monovision: A review. *Ophthalmic and Physiological Optics: The Journal of the British College of Ophthalmic Opticians (Optometrists)*, *27*(5), 417–439. doi:10.1111/j.1475-1313.2007.00488.x.

Fernandez, E. and Artal, P. (2003). Membrane deformable mirror for adaptive optics: Performance limits in visual optics. *Optics Express*, *11*(9), 1056–1069. Retrieved from http://www.ncbi.nlm.nih.gov/pubmed/19465970.

Fernández, E. J. and Artal, P. (2007). Dynamic eye model for adaptive optics testing. *Applied Optics*, *46*(28), 6971. doi:10.1364/AO.46.006971.

Fernández, E. J. and Artal, P. (2008). Ocular aberrations up to the infrared range: From 632.8 to 1070 nm. *Optics Express*, *16*(26), 21199–21208. Retrieved from http://www.ncbi.nlm.nih.gov/pubmed/19104549.

Fernández, E. J. and Drexler, W. (2005). Influence of ocular chromatic aberration and pupil size on transverse resolution in ophthalmic adaptive optics optical coherence tomography. *Optics Express*, *13*(20), 8184. doi:10.1364/OPEX.13.008184.

Fernández, E. J., Iglesias, I., and Artal, P. (2001). Closed-loop adaptive optics in the human eye. *Optics Letters*, *26*(10), 746. doi:10.1364/OL.26.000746.

Fernández, E. J., Laurent, V., Hermann, B., Unterhuber, A., Považay, B., and Drexler, W. (2006a). Adaptive optics with a magnetic deformable mirror: Applications in the human eye. *Optics Express*, *14*(20), 631–643. doi:10.1167/4.4.4.E.

Fernández, E. J., Manzanera, S., Piers, P., and Artal, P. (2002). Adaptive optics visual simulator. *Journal of Refractive Surgery (Thorofare, N.J.: 1995)*, *18*(5), 634–638. Retrieved from http://www.ncbi.nlm.nih.gov/pubmed/12361172.

Fernández, E. J., Povazay, B., Hermann, B., Unterhuber, A., Sattmann, H., Prieto, P. M. et al. (2005). Three-dimensional adaptive optics ultrahigh-resolution optical coherence tomography using a liquid crystal spatial light modulator. *Vision Research*, *45*(28), 3432–3444. doi:10.1016/j.visres.2005.08.028.

Fernández, E. J., Prieto, P. M., and Artal, P. (2009a). Binocular adaptive optics visual simulator. *Optics Letters*, *34*(17), 2628–2630. Retrieved from http://www.ncbi.nlm.nih.gov/pubmed/21045890.

Fernández, E. J., Prieto, P. M., and Artal, P. (2009b). Wave-aberration control with a liquid crystal on silicon (LCOS) spatial phase modulator. *Optics Express*, *17*(13), 11013–11025. Retrieved from http://www.ncbi.nlm.nih.gov/pubmed/19550501.

Fernández, E. J., Prieto, P. M., and Artal, P. (2010). Adaptive optics binocular visual simulator to study stereopsis in the presence of aberrations. *Journal of the Optical Society of America. A, Optics, Image Science, and Vision*, *27*(11), A48–A55. Retrieved from http://www.ncbi.nlm.nih.gov/pubmed/21045890.

Fernández, E. J., Schwarz, C., Prieto, P. M., Manzanera, S., and Artal, P. (2013). Impact on stereo-acuity of two presbyopia correction approaches: Monovision and small aperture inlay. *Biomedical Optics Express*, *4*(6), 822–830. doi:10.1364/BOE.4.000822.

Fernández, E. J., Unterhuber, A., Povazay, B., Hermann, B., Artal, P., and Drexler, W. (2006b). Chromatic aberration correction of the human eye for retinal imaging in the near infrared. *Optics Express*, *14*(13), 6213–6225. Retrieved from http://www.ncbi.nlm.nih.gov/pubmed/19516794.

Fielder, A. R. and Moseley, M. J. (1996). Does stereopsis matter in humans? *Eye (London, England)*, *10*(Pt 2), 233–238. doi:10.1038/eye.1996.51.

Flores, A., Wang, M. R., and Yang, J. J. (2004). Achromatic hybrid refractive-diffractive lens with extended depth of focus. *Applied Optics*, *43*(30), 5618–5630. Retrieved from http://www.ncbi.nlm.nih.gov/pubmed/15534993.

Guirao, A. and Artal, P. (1998). Off-axis monochromatic aberrations estimated from double pass measurements in the human eye. *Vision Research*, *26*(2), 321–325. Retrieved from http://www.ncbi.nlm.nih.gov/pubmed/10326131.

Guirao, A., Redondo, M., and Artal, P. (2000). Optical aberrations of the human cornea as a function of age. *Journal of the Optical Society of America. A, Optics, Image Science, and Vision*, *17*(10), 1697–1702. Retrieved from http://www.ncbi.nlm.nih.gov/pubmed/11028517.

Hampson, K. M., Chin, S. S., and Mallen, E. A. H. (2007). Binocular Shack-Hartmann sensor for the human eye. *Journal of Modern Optics*, *55*(4–5), 703–716. Retrieved from http://cat.inist.fr/?aModele=afficheNandcpsidt=20317258.

Hayashi, K., Masumoto, M., and Takimoto, M. (2015). Comparison of visual and refractive outcomes after bilateral implantation of toric intraocular lenses with or without a multifocal component. *Journal of Cataract and Refractive Surgery*, *41*(1), 73–83. doi:10.1016/j.jcrs.2014.04.032.

Heron, G. and Winn, B. (1989). Binocular accommodation reaction and response times for normal observers. *Ophthalmic and Physiological Optics: The Journal of the British College of Ophthalmic Opticians (Optometrists)*, *9*(2), 176–183. Retrieved from http://www.ncbi.nlm.nih.gov/pubmed/2622653.

Heron, G., Winn, B., Pugh, J. R., and Eadie, A. S. (1989). Twin channel infrared optometer for recording binocular accommodation. *Optometry and Vision Science: Official Publication of the American Academy of Optometry*, *66*(2), 123–129. Retrieved from http://www.ncbi.nlm.nih.gov/pubmed/2710510.

Hofer, H., Artal, P., Singer, B., Aragón, J. L., and Williams, D. R. (2001a). Dynamics of the eye's wave aberration. *Journal of the Optical Society of America A*, *18*(3), 497. doi:10.1364/JOSAA.18.000497.

Hofer, H., Chen, L., Yoon, G. Y., Singer, B., Yamauchi, Y., and Williams, D. R. (2001b). Improvement in retinal image quality with dynamic correction of the eye's aberrations. *Optics Express*, *8*(11), 631–643. Retrieved from http://www.ncbi.nlm.nih.gov/pubmed/19421252.

Holladay, J. T., Piers, P. A., Koranyi, G., van der Mooren, M., and Norrby, N. E. S. 2002. A new intraocular lens design to reduce spherical aberration of pseudophakic eyes. *Journal of Refractive Surgery (Thorofare, N.J.: 1995)*, *18*(6), 683–691. Retrieved from http://www.ncbi.nlm.nih.gov/pubmed/12458861.

Howard, I. P. and Brian, J. R. (1996). *Binocular Vision and Stereopsis* (p. 736). New York: Oxford University Press. doi:10.1093/acprof:oso/9780195084764.001.0001.

Ivanoff, A. (1956). About the spherical aberration of the eye. *Journal of the Optical Society of America*, *46*(10), 901–903. Retrieved from http://www.ncbi.nlm.nih.gov/pubmed/13367938.

Jain, S., Arora, I., and Azar, D. T. 1996. Success of monovision in presbyopes: Review of the literature and potential applications to refractive surgery. *Survey of Ophthalmology*, *40*(6), 491–499. Retrieved from http://www.ncbi.nlm.nih.gov/pubmed/8724641.

Jiménez, J. R., Castro, J. J., Jiménez, R., and Hita, E. (2008). Interocular differences in higher-order aberrations on binocular visual performance. *Optometry and Vision Science: Official Publication of the American Academy of Optometry*, *85*(3), 174–179. doi:10.1097/OPX.0b013e31816445a7.

Julesz, B. (1971). *Foundations of Cyclopean Perception* (March 2006, p. 428). Cambridge, MA: The MIT Press.

Kelly, J. E., Mihashi, T., and Howland, H. C. (2004). Compensation of corneal horizontal/vertical astigmatism, lateral coma, and spherical aberration by internal optics of the eye. *Journal of Vision*, *4*(4), 262–271. doi:10.1167/4.4.2.

Kobayashi, M., Nakazawa, N., Yamaguchi, T., Otaki, T., Hirohara, Y., and Mihashi, T. (2008). Binocular open-view Shack-Hartmann wavefront sensor with consecutive measurements of near triad and spherical aberration. *Applied Optics*, *47*(25), 4619–4626.

Li, S., Xiong, Y., Li, J., Wang, N., Dai, Y., Xue, L. et al. (2009). Effects of monochromatic aberration on visual acuity using adaptive optics. *Optometry and Vision Science: Official Publication of the American Academy of Optometry*, *86*(7), 868–874. doi:10.1097/OPX.0b013e3181adfdff.

Liang, J., Williams, D. R., and Miller, D. T. (1997). Supernormal vision and high-resolution retinal imaging through adaptive optics. *Journal of the Optical Society of America. A, Optics, Image Science, and Vision*, *14*(11), 2884–2892. Retrieved from http://www.ncbi.nlm.nih.gov/pubmed/9379246.

Liu, H., Lu, Z., Sun, Q., and Zhang, H. (2010). Design of multiplexed phase diffractive optical elements for focal depth extension. *Optics Express*, *18*(12), 12798–12806. Retrieved from http://www.ncbi.nlm.nih.gov/pubmed/20588408.

Manzanera, S., Prieto, P. M., Ayala, D. B., Lindacher, J. M., and Artal, P. (2007). Liquid crystal Adaptive Optics Visual Simulator: Application to testing and design of ophthalmic optical elements. *Optics Express*, *15*(24), 16177. doi:10.1364/OE.15.016177.

Mas, D., Espinosa, J., Perez, J., and Illueca, C. (2007). Three dimensional analysis of chromatic aberration in diffractive elements with extended depth of focus. *Optics Express*, *15*(26), 17842–17854. Retrieved from http://www.ncbi.nlm.nih.gov/pubmed/19551079.

Mester, U., Dillinger, P., and Anterist, N. (2003). Impact of a modified optic design on visual function: Clinical comparative study. *Journal of Cataract and Refractive Surgery*, *29*(4), 652–660. Retrieved from http://www.ncbi.nlm.nih.gov/pubmed/12686231.

Mihashi, T., Kobayashi, M., Nakazawa, N., Yamaguchi, T., Hirohara, Y., and Otaki, T. (2010). Refraction measurements with an open-view binocular Shack-Hartmann wavefront sensor. *Journal of Vision*, *6*(13), 57. doi:10.1167/6.13.57.

Murray, I. J., Elliott, S. L., Pallikaris, A., Werner, J. S., Choi, S., and Tahir, H. J. (2010). The oblique effect has an optical component: Orientation-specific contrast thresholds after correction of high-order aberrations. *Journal of Vision*, *10*(11), 10. doi:10.1167/10.11.10.

Nishi, Y., Hirnschall, N., Crnej, A., Gangwani, V., Tabernero, J., Artal, P., and Findl, O. (2010). Reproducibility of intraocular lens decentration and tilt measurement using a clinical Purkinje meter. *Journal of Cataract and Refractive Surgery*, *36*(9), 1529–1535. doi:10.1016/j.jcrs.2010.03.043.

O'Connor, A. R., Birch, E. E., Anderson, S., and Draper, H. (2010). The functional significance of stereopsis. *Investigative Ophthalmology and Visual Science*, *51*(4), 2019–2023. doi:10.1167/iovs.09-4434.

Okuyama, F., Tokoro, T., and Fujieda, M. (1993). Binocular infrared optometer for measuring accommodation in both eyes simultaneously in natural-viewing conditions. *Applied Optics*, *32*(22), 4147–4154.

Ortiz, D., Alió, J. L., Bernabéu, G., and Pongo, V. (2008). Optical performance of monofocal and multifocal intraocular lenses in the human eye. *Journal of Cataract and Refractive Surgery*, *34*(5), 755–762. doi:10.1016/j.jcrs.2007.12.038.

Paterson, C., Munro, I., and Dainty, J. (2000). A low cost adaptive optics system using a membrane mirror. *Optics Express*, *6*(9), 175. doi:10.1364/OE.6.000175.

Pérez, G. M., Manzanera, S., and Artal, P. (2009). Impact of scattering and spherical aberration in contrast sensitivity. *Journal of Vision*, 9(3), 19.1–19.10. doi:10.1167/9.3.19.

Piers, P. A, Fernandez, E. J., Manzanera, S., Norrby, S., and Artal, P. (2004). Adaptive optics simulation of intraocular lenses with modified spherical aberration. *Investigative Ophthalmology and Visual Science*, 45(12), 4601–4610. doi:10.1167/iovs.04-0234.

Piers, P. A., Manzanera, S., Prieto, P. M., Gorceix, N., and Artal, P. (2007). Use of adaptive optics to determine the optimal ocular spherical aberration. *Journal of Cataract and Refractive Surgery*, 33(10), 1721–1726. doi:10.1016/j.jcrs.2007.08.001.

Porter, J., Guirao, A., Cox, I. G., and Williams, D. R. (2001). Monochromatic aberrations of the human eye in a large population. *Journal of the Optical Society of America*, 18(8), 1793–1803.

Prieto, P., Fernández, E., Manzanera, S., and Artal, P. (2004). Adaptive optics with a programmable phase modulator: Applications in the human eye. *Optics Express*, 12(17), 4059–4071. Retrieved from http://www.ncbi.nlm.nih.gov/pubmed/19483947.

Rao, F., Wang, Z.-Q., Liu, Y.-J., and Wang, Y. (2011). A novel approach to design intraocular lenses with extended depth of focus in a pseudophakic eye model. *Optik—International Journal for Light and Electron Optics*, 122(11), 991–995. doi:10.1016/j.ijleo.2010.06.035.

Ravikumar, S., Bradley, A., and Thibos, L. N. (2014). Chromatic aberration and polychromatic image quality with diffractive multifocal intraocular lenses. *Journal of Cataract and Refractive Surgery*, 40(7), 1192–1204. doi:10.1016/j.jcrs.2013.11.035.

Ravikumar, S., Thibos, L. N., and Bradley, A. (2008). Calculation of retinal image quality for polychromatic light. *Journal of the Optical Society of America. A, Optics, Image Science, and Vision*, 25(10), 2395–2407.

Reading, R. W. (1983). *Binocular Vision: Foundations and Applications* (p. 384). Boston, MA: Butterworth-Heinemann.

Rocha, K. M., Vabre, L., Chateau, N., and Krueger, R. R. (2010). Enhanced visual acuity and image perception following correction of highly aberrated eyes using an adaptive optics visual simulator. *Journal of Refractive Surgery (Thorofare, N J · 1995)*, 26(1), 52–56. doi:10.3928/1081597X-20101215-08.

Rosales, P., De Castro, A., Jiménez-Alfaro, I., and Marcos, S. (2010). Intraocular lens alignment from Purkinje and Scheimpflug imaging. *Clinical and Experimental Optometry: Journal of the Australian Optometrical Association*, 93(6), 400–408. doi:10.1111/j.1444-0938.2010.00514.x.

Sabesan, R., Ahmad, K., and Yoon, G. (2007a). Correcting highly aberrated eyes using large-stroke adaptive optics. *Journal of Refractive Surgery (Thorofare, N.J.: 1995)*, 23(9), 947–952. Retrieved from http://www.ncbi.nlm.nih.gov/pubmed/18041252.

Sabesan, R., Jeong, T. M., Carvalho, L., Cox, I. G., Williams, D. R., and Yoon, G. (2007b). Vision improvement by correcting higher-order aberrations with customized soft contact lenses in keratoconic eyes. *Optics Letters*, 32(8), 1000. doi:10.1364/OL.32.001000.

Sabesan, R. and Yoon, G. (2009). Visual performance after correcting higher order aberrations in keratoconic eyes. *Journal of Vision*, 9(5), 6.1–6.10. doi:10.1167/9.5.6.

Sabesan, R. and Yoon, G. (2010). Neural compensation for long-term asymmetric optical blur to improve visual performance in keratoconic eyes. *Investigative Ophthalmology and Visual Science*, 51(7), 3835–3839. doi:10.1167/iovs.09-4558.

Sabesan, R., Zheleznyak, L., and Yoon, G. (2012). Binocular visual performance and summation after correcting higher order aberrations. *Biomedical Optics Express*, 3(12), 3176–3189. doi:10.1364/BOE.3.003176.

Sawides, L., de Gracia, P., Dorronsoro, C., Webster, M. A., and Marcos, S. (2011a). Vision is adapted to the natural level of blur present in the retinal image. *PLoS ONE*, 6(11), e27031. doi:10.1371/journal.pone.0027031.

Sawides, L., de Gracia, P., Dorronsoro, C., Webster, M., and Marcos, S. (2011b). Adapting to blur produced by ocular high-order aberrations. *Journal of Vision*, 11(7), 21. doi:10.1167/11.7.21.

Sawides, L., Dorronsoro, C., de Gracia, P., Vinas, M., Webster, M., and Marcos, S. (2012). Dependence of subjective image focus on the magnitude and pattern of high order aberrations. *Journal of Vision*, 12(8), 4. doi:10.1167/12.8.4.

Sawides, L., Dorronsoro, C., Haun, A. M., Peli, E., and Marcos, S. (2013). Using pattern classification to measure adaptation to the orientation of high order aberrations. *PLoS ONE*, 8(8), e70856. doi:10.1371/journal.pone.0070856.

Schwarz, C., Cánovas, C., Manzanera, S., Weeber, H., Prieto, P. M., Piers, P., and Artal, P. (2014a). Binocular visual acuity for the correction of spherical aberration in polychromatic and monochromatic light. *Journal of Vision*, 14(2), 8. doi:10.1167/14.2.8.

Schwarz, C., Manzanera, S., and Artal, P. (2014b). Binocular visual performance with aberration correction as a function of light level. *Journal of Vision*, 14(14), 6. doi:10.1167/14.14.6.

Schwarz, C., Prieto, P. M., Fernández, E. J., and Artal, P. (2011). Binocular adaptive optics vision analyzer with full control over the complex pupil functions. *Optics Letters*, 36(24), 4779–4781. Retrieved from http://www.ncbi.nlm.nih.gov/pubmed/22179881.

Seyeddain, O., Hohensinn, M., Riha, W., Nix, G., Rückl, T., Grabner, G., and Dexl, A. K. (2012). Small-aperture corneal inlay for the correction of presbyopia: 3-year follow-up. *Journal of Cataract and Refractive Surgery*, 38(1), 35–45. doi:10.1016/j.jcrs.2011.07.027.

Tabernero, J., Benito, A., Alcón, E., and Artal, P. (2007). Mechanism of compensation of aberrations in the human eye. *Journal of the Optical Society of America. A, Optics, Image Science, and Vision*, 24(10), 3274–3283. Retrieved from http://www.ncbi.nlm.nih.gov/pubmed/17912320.

Tabernero, J., Benito, A., Nourrit, V., and Artal, P. (2006). Instrument for measuring the misalignments of ocular surfaces. *Optics Express*, 14(22), 10945. doi:10.1364/OE.14.010945.

Tabernero, J., Schwarz, C., Fernández, E. J., and Artal, P. (2011). Binocular visual simulation of a corneal inlay to increase depth of focus. *Investigative Ophthalmology and Visual Science*, 52(8), 5273–5277. doi:10.1167/iovs.10-6436.

Thibos, L. N., Hong, X., Bradley, A., and Cheng, X. (2002). Statistical variation of aberration structure and image quality in a normal population of healthy eyes. *Journal of the Optical Society of America. A, Optics, Image Science, and Vision*, 19, 2329–2348. doi:10.1364/JOSAA.19.002329.

Thibos, L. N., Ye, M., Zhang, X., and Bradley, A. (1992). The chromatic eye: A new reduced-eye model of ocular chromatic aberration in humans. *Applied Optics*, 31(19), 3594–3600.

Vargas-Martín, F., Prieto, P. M., and Artal, P. (1998). Correction of the aberrations in the human eye with a liquid-crystal spatial light modulator: Limits to performance. *Journal of the Optical Society of America. A, Optics, Image Science, and Vision*, 15(9), 2552–2562. Retrieved from http://www.ncbi.nlm.nih.gov/pubmed/9729868.

Vinas, M., Sawides, L., de Gracia, P., and Marcos, S. (2012). Perceptual adaptation to the correction of natural astigmatism. *PLoS ONE*, 7(9), e46361. doi:10.1371/journal.pone.0046361.

Vlaskamp, B. N. S., Yoon, G., and Banks, M. S. (2011). Human stereopsis is not limited by the optics of the well-focused eye. *The Journal of Neuroscience: The Official Journal of the Society for Neuroscience*, 31(27), 9814–9818. doi:10.1523/JNEUROSCI.0980-11.2011.

Weeber, H. A. and Piers, P. A. (2012). Theoretical performance of intraocular lenses correcting both spherical and chromatic aberration. *Journal of Refractive Surgery (Thorofare, N.J.: 1995)*, *28*(1), 48–52. doi:10.3928/1081597X-20111103-01.

Yılmaz, O. F., Alagöz, N., Pekel, G., Azman, E., Aksoy, E. F., Cakır, H. et al. (2011). Intracorneal inlay to correct presbyopia: Long-term results. *Journal of Cataract and Refractive Surgery*, *37*(7), 1275–1281. doi:10.1016/j.jcrs.2011.01.027.

Yilmaz, O. F., Bayraktar, S., Agca, A., Yilmaz, B., McDonald, M. B., and van de Pol, C. (2008). Intracorneal inlay for the surgical correction of presbyopia. *Journal of Cataract and Refractive Surgery*, *34*(11), 1921–1927. doi:10.1016/j.jcrs.2008.07.015.

Yoon, G.-Y. and Williams, D. R. (2002). Visual performance after correcting the monochromatic and chromatic aberrations of the eye. *Journal of the Optical Society of America. A, Optics, Image Science, and Vision*, *19*(2), 266–275. Retrieved from http://www.ncbi.nlm.nih.gov/pubmed/11822589.

Zhang, H., Liu, H., Lu, Z., and Zhang, H. (2008). Modified phase function model for kinoform lenses. *Applied Optics*, *47*(22), 4055–4060. Retrieved from http://www.ncbi.nlm.nih.gov/pubmed/18670562.

Zheleznyak, L., Sabesan, R., Oh, J.-S., MacRae, S., and Yoon, G. (2013). Modified monovision with spherical aberration to improve presbyopic through-focus visual performance. *Investigative Ophthalmology and Visual Science*, *54*(5), 3157–3165. doi:10.1167/iovs.12-11050.

8 Multiphoton imaging of the cornea

Moritz Winkler, Donald J. Brown, and James V. Jester

Contents

8.1 INTRODUCTION

The human cornea fulfills a dual role, acting as both a protective cover to maintain ocular integrity and as the primary refractive element responsible for transmitting and focusing light onto the retina. As part of the eye's outer tunic, it contains the outward force exerted by intraocular pressure (IOP) and shields the inner eye from mechanical, biological, and chemical insults. Due to its curvature and the difference in refractive indices between the surrounding air and itself, the cornea is also a powerful focusing lens, providing nearly twice the refractive power of the crystalline lens.

This dual role poses unique challenges, requiring the cornea to be highly transparent, mechanically stable, and capable of maintaining a precise curvature. Indeed, the transparency of the cornea provides a "window to eye," allowing evaluation of intraocular structures, including the iris, lens, and posterior retina. Corneal transparency also provides unique opportunities to evaluate corneal function and pathology at the cellular level using various imaging approaches ranging from biomicroscopy using a slit lamp to optical coherence tomography to high magnification in vivo confocal microscopy (for reviews, please see Ramos et al. 2009 and Masters 2009).

Importantly, corneal shape is a functional requirement of the eye, as the refractive properties of the cornea are a function of its curvature. Visual acuity is thus directly linked to corneal shape, and deviations from an optically ideal curvature result

in aberrations that prevent the focusing of a sharp image on the retina. This functionally crucial shape is controlled by the structure and biomechanics of the cornea.

Measuring the mechanical properties of the cornea has been attempted numerous times using a host of different techniques (Hjortdal 1995, 1996, Liu and Roberts 2005, Zeng et al. 2001). However, determining even the most basic biomechanical properties, particularly the elastic modulus, has proven to be challenging. Depending on the type of measurement, reported elastic moduli for normal human corneas differ by as much as six orders of magnitude, ranging from as little as tens of kilopascals well into the gigapascal range (Hjortdal 1995, 1996, Last et al. 2009). This is in part due to the anisotropy of the elements that make up the cornea. In particular, fibrous collagen, which gives rise to the cornea's structural and biomechanical properties, possesses very high longitudinal tensile strength but is comparatively weak along the other axes. Collagen fiber orientation greatly influences the mechanical properties of tissues (Martin and Boardman 1993, Martin and Ishida 1989) and leads to a wide range of different effects depending on the orientation and type of mechanical strains encountered.

Corneal structure has been studied extensively using a variety of imaging modalities (Abahussin et al. 2009, Aghamohammadzadeh et al. 2004, Daxer et al. 1998, Han et al. 2005, Komai and Ushiki 1991, Meek et al. 1987, Morishige et al. 2006). Despite a large volume of research being devoted to exploring the precise nature of the structure of the cornea, the mechanisms that control corneal shape remain poorly understood. This is in part due to the lack of a comprehensive structural/biomechanical model of the entire eye. As noted by Kokott in 1938, understanding the mechanisms controlling corneal shape will require a structural "blueprint" of the cornea that has yet to be elucidated (Kokott 1938).

8.1.1 CURRENT UNDERSTANDING OF CORNEAL STRUCTURE

The cornea consists of five distinct layers; however, the mechanical strength and shape of the cornea are largely dependent on the corneal stroma, which makes up over 90% of the corneal thickness. Collagen accounts for approximately 70% of the corneal total dry weight and is the major structural element of the corneal stroma (Abahussin et al. 2009).

Collagen assembles to form long fibrils that in the human cornea show a uniform diameter of approximately 31–34 nm (Daxer et al. 1998). Fibrillar size, spacing, and stability are regulated by nonfibrillar collagen and proteoglycans, found in the interfibrillar matrix (Zimmermann et al. 1986). Collagen fibrils are generally organized into independent bundles or fibers, which in the cornea have been referred to as "lamellae." Collagen lamellae are approximately 1–2 µm thick and 10–200 µm wide and are conventionally thought to traverse the entire cornea from limbus to limbus (Abahussin et al. 2009). The biomechanical properties of the cornea and hence corneal shape are thought to be controlled principally by collagen fiber organization although other factors may contribute (Ruberti and Zieske 2008).

The corneal nanostructure has been studied using electron microscopy (Komai and Ushiki 1991), which has the capability to resolve individual collagen fibrils. X-ray diffraction has also been used to analyze bulk collagen alignment across the entire cornea and measure collagen fibril diameter and spacing (Aghamohammadzadeh et al. 2004). The combined findings of these studies have led to a proposed comprehensive model of collagen distribution, which has been the major basis for our understanding of corneal biomechanics. In this proposed model, collagen fibrils are bundled into broad ribbons or lamellae that run in-plane across the width of the cornea and are stacked vertically in approximately 200 separate planes. Lamellae within each plane are preferentially aligned along the horizontal or vertical meridians of the cornea, and each layer has a preferential alignment vector that is rotated about 90° relative to its neighbors giving an overall orthogonal arrangement to the collagen structure (Aghamohammadzadeh et al. 2004). Peripherally, lamellae are thought to have a more circumferential orientation that extends around the circumference of the cornea and forms a boundary between corneal and scleral curvature (Aghamohammadzadeh et al. 2004). In conjunction with collagen inserting from the limbus, these patterns are thought to determine the shape of the cornea with stromal lamellae existing as separate entities; their interactions limited to interweaving with adjacent lamellae (Han et al. 2005).

While our current biomechanical modeling of the human cornea is based on our understanding of these details, how this microstructure is assembled into the larger corneal organization that provides mechanical strength and controls corneal shape is not clear. What has been lacking is a larger understanding of the macrostructural details and the hierarchical collagen organizational patterns throughout the corneal stroma that is difficult to construct from these high-resolution electron microscopic and x-ray diffraction data. While extensive repetitive sampling of the cornea has been used by some laboratories to build up a larger "blueprint" of stromal collagen structure using x-ray diffraction (Aghamohammadzadeh et al. 2004, Meek and Boote 2009), a wider field of view is needed to bridge the gap between the collagen structural organization and the topographical maps of corneal shape and refractive power.

8.1.2 NONLINEAR OPTICAL IMAGING OF THE CORNEA

The ability of a microscope to resolve small structures is fundamentally limited by diffraction. As early as 1873, Karl Ernst Abbe found that the smallest possible spot that light can be focused into is directly proportional to its wavelength (Abbe 1873). Approaching this "diffraction limit" has resulted in technically more and more advanced microscopes; however, it was not until the advent of laser scanning microscopes that the resolution of these systems came close to and more recently exceeded their theoretical limits.

The first such system was the confocal laser scanning microscope (CLSM). Although the original patent dates back to 1957 (Minsky 1961), it was not until the development and widespread availability of lasers that the first 3-D CLSM was developed by Cremer in 1978 (Cremer and Cremer 1974). CLSM uses a system of pinholes to suppress out-of-plane light scatter, yielding very high axial and lateral resolution. This high resolution comes at the price of a decrease in signal-to-noise ratio. The smaller the pinhole, the higher the axial resolution; however, at the same time, a smaller pinhole permits less light to get through, making faint signals difficult to detect, increasing scan times, and lowering overall image quality.

A breakthrough in optical imaging occurred with the first demonstration of nonlinear optical (NLO) imaging by Denk et al. (1990) using a subpicosecond pulsed red laser to molecularly excite ultraviolet photoactivated fluorophores by two photon excitation (Denk et al. 1990). A unique feature of NLO imaging is that multiphoton interactions are limited to the focal volume of the objective leading to increased axial resolution. In addition to two photon-excited fluorescence, NLO microscopy can take advantage of other laser-tissue interactions including second and third harmonic generation (Juhasz et al. 1996, Olivier et al. 2010). Of particular interest in the cornea is second harmonic generation (SHG) imaging, which has emerged as a powerful new tool for investigating collagen organization (Han et al. 2005).

8.1.2.1 SHG imaging

Unlike fluorescent microscopy, SHG imaging is an absorption-free process. The underlying principle is commonly referred to as frequency doubling and was first demonstrated by Franken et al. (1961). The mathematical description of this process is as elegant as it is complex and is described in more detail by Kleinman (1962).

A more application-focused review of SHG microscopy has been compiled by Campagnola and Dong (2011).

Briefly, in SHG, the electric field of the incoming light wave serves to displace the subatomic charges that make up molecules relative to one another. Negatively charged electrons are displaced slightly, relative to the positively charged nucleus. As a result, a polarization is induced. As the electric field oscillates with the frequency of the incoming light, the polarization follows suit. The result is an induced oscillating polarization, rather than a static one.

This polarization can be expressed mathematically as the sum of linear and nonlinear terms. At low field strengths, such as those associated with low light intensities, the nonlinear terms are near zero, and the linear term dominates:

$$P = \chi E \tag{8.1}$$

where

P is the polarization vector
E is the incident electric field
χ is the electrical susceptibility tensor

However, this approximation no longer holds true in the high intensity regime, where nonlinear effects emerge. A component-wise Taylor expansion results in

$$P_k = \chi_{ik}^{(1)} E_i + \chi_{ijk}^{(2)} E_i E_j + \cdots \tag{8.2}$$

where $\chi^{(2)}$ is the second-order NLO susceptibility tensor. Its elements sum to zero in materials with inversion symmetry, canceling out this second-order effect. It can be shown here that components of the polarization field oscillate 2ω, where ω is the frequency of the incident electric field. Maxwell's laws stipulate that such an oscillating polarization itself is the source of radiation at a frequency of 2ω—the second harmonic.

As SHG is based on this induced polarization, there is no absorption of the incoming photons as is the case with fluorescent processes. Instead, a virtual two-state system can be used to illustrate the SHG process. Here, two photons of frequency ω are simultaneously destroyed while a single photon of frequency 2ω is created. Unlike fluorescent radiation, the frequency-doubled photons are not emitted isotropically, but rather in characteristic lobes.

To generate SHG signals, ultrashort laser pulses, usually no longer than several hundred femtoseconds and with wavelengths in the near-infrared regime, are focused into a specimen through microscope optics. The optics generate a small focal spot, which in conjunction with the short pulse length effectively focuses photons both spatially and temporally into a small volume. The energy of each pulse is small, typically on the order of a few hundred or dozen nanojoules. Because a phase-matching condition has to be met, only materials that lack central symmetry are capable of emitting SHG light. In the case of the cornea, fibrillar collagen, when viewed from the side or longitudinal axis, is the only structure without such symmetry and thus the only corneal element that emits SHG light. When viewed head-on in cross section, the triple-helical structure of collagen possesses central symmetry and does not generate SHG signals.

Since collagen fibrils have a cross-sectional diameter of only 31–34 nm, SHG microscopy does not resolve individual fibrils; SHG microscopy thus detects bundles of collagen fibrils or fibers and lamellae, as long as they run roughly perpendicular to the direction of the scanning beam. When imaging corneal cross sections, this means that collagen running in-plane with the section is visible, while fibers running toward the observer do not generate a signal and therefore show as black space in SHG images (Figure 8.1).

Using this imaging paradigm, SHG studies have shown that collagen is arranged in a more complex fashion than previously thought. Rather than forming highly parallel, distinct fibers that run largely uninterrupted, collagen appears to be arranged in lamellae that can change direction and interact with adjacent lamellae in ways beyond mere interweaving, such as branching and fusing. SHG studies have also revealed the presence of "sutural" lamellae that run upward and insert into Bowman's layer (Morishige et al. 2006). These lamellae were only found in the anterior portion of the cornea, indicating that the collagen organization in the anterior cornea is more complex and thus pointing toward an axial heterogeneity in collagen lamellar interconnectivity. Interestingly, keratoconus corneas show reduced or absent sutural lamellae suggesting that collagen bundle interconnectivity influences mechanical stability and possibly shape (Morishige et al. 2007, 2011).

In this chapter, we present an SHG-based imaging paradigm called nonlinear optical high-resolution macroscopy (NLO-HRMac), which allows for the generation of large-scale,

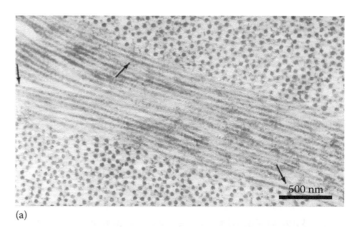

(a)

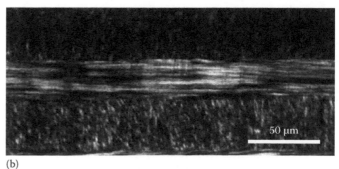

(b)

Figure 8.1 Collagen lamellae imaged using transmission electron microscopy (a) and SHG (b). Note that longitudinally section collagen lamellae show prominent SHG signal, while orthogonally arranged lamellae above and below show a weak SHG signal.

high-resolution, 3-D reconstructions of corneal cross sections. Using HRMac, we have characterized the lamellar architecture of the corneal stroma and quantified lamellar interconnectivity in human corneas as a function of stromal depth by measuring the density of fiber branching points. These studies have shown that there exists a link between the axial heterogeneity of lamellar interconnectivity and corneal mechanical stiffness and that regional variation in collagen architecture and stiffness may define corneal shape.

8.2 SHG HRMac FOR 3-D COLLAGEN MAPPING

The structure of the cornea can be described on several different levels of scale. At the molecular level, the corneal stroma is primarily made up of water, proteoglycans, and collagen molecules. The fibrils assembled from collagen molecules are several dozen nanometers in size. These scales are beyond the resolution limit for optical microscopy but can be visualized using electron or atomic force microscopy. Collagen fibers, fiber bundles, or lamellae are on a microscopic rather than a nanoscopic scale and thus can be imaged using optical microscopy. While these bundles are only 2 μm thick, we have observed them to run for a length of several millimeters, and they may even extend for several centimeters by forming a continuous loop that would connect corneal and scleral collagen.

To better understand the relationship between structure, biomechanics, and corneal shape, an imaging approach is required that can collect data on not just individual lamellae or small regions of interest but that can track lamellae across large portions of a corneal cross section. Further, such an approach needs to include the ability to compare different regions that may not necessarily be in close proximity to one another, such as the central and the peripheral cornea. A final requirement is the ability to trace lamellae in three dimensions. These prerequisites present a challenge when attempting to map the corneal ultrastructure.

Using SHG imaging, individual lamellae can be imaged at submicron resolution. A 40×/1.1 NA objective is capable of resolving the microstructure of the cornea with ease, creating submicron resolution images where each pixel measures 0.44 μm. However, this high resolution comes at the price of a vastly reduced field of view, which is merely 225 × 225 μm. Conversely, when using lower magnification, larger portions of the cornea become visible; however, the resolution is reduced to the point where fiber bundles can no longer be resolved. We have addressed this problem, dubbed the "resolution gap," by using NLO-HRMac.

8.2.1 PRINCIPLE OF HRMac

The HRMac processing pipeline is shown in Figure 8.2. NLO-HRMac combines the high-resolution and 3-D imaging capability of SHG with a field of view large enough to image entire corneal cross sections spanning some 14 mm. This is achieved by sequentially acquiring individual, overlapping image stacks to cover the desired imaging area and then digitally combining them into a single, large mosaic.

In preparation for HRMac imaging, samples were mounted and placed on a motorized, computer-controlled stage.

The imaging area was defined by marking the coordinates of the four corners of a rectangle with sufficient size to image the entire sample. Stage coordinates were then loaded into MultiTime, a commercially available plug-in for the Zeiss LSM imaging suite (Carl Zeiss Inc., Thornwood, NY). The plug-in then automatically divided the imaging area into a grid of overlapping image stacks and created a list of 3-D coordinates for each stack (1A). Each stack or "block" overlapped its neighbors by about 10% to compensate for stage positioning inaccuracies. The software would then acquire blocks and move the stage autonomously. Blocks were saved as individual files in Zeiss' native .LSM format. Following image acquisition, the individual stacks were loaded into ImageJ (National Institutes of Health, http://rsb.info.nih.gov/ij/) and saved as a numbered series of .TIFF files, one for each plane. Depending on the objective used and the imaging volume, a single HRMac scan could consist of up to 100,000 individual images.

Due to small inaccuracies in stage positioning, simply placing tiles onto a grid according to their stage coordinates resulted in visible discontinuities between adjacent tiles. To compensate, files for a single plane were stitched together using ImageStitch (Preibisch et al. 2009). This off-the-shelf plug-in used a combination of image processing algorithms to determine the best fit between neighbors and generated a list of adjusted coordinates for each image within a single plane (Figure 8.2b). A custom-written script then applied these coordinates on a plane-by-plane basis for the remainder of the tiles. This generated a single, large-scale, multiplane mosaic of the tissue (Figure 8.2c) which could then be further processed, analyzed, or rendered in 3-D (Figure 8.2d).

A typical HRMac image of a human corneal cross section is shown in Figure 8.3. This particular sample had a corneal arc length of approximately 12 mm. The original HRMac scan was 25,000 × 9,000 pixels for each of the 60 planes at a resolution of 0.44 μm/pixel and was made up of over 80,000 individual images. Figure 8.3a shows a view of the central portion of the cornea at full resolution (0.44 μm/pixel), revealing complex interactions between collagen lamellae.

8.3 3-D RECONSTRUCTION OF COLLAGEN LAMELLAE IN HUMAN CORNEAS

NLO-HRMac has been used to reconstruct three-dimensionally the organizational pattern of collagen lamellae within the human cornea at different depths for the anterior to posterior central cornea (Winkler et al. 2011). Corneas from autopsy eyes are generally much thicker (greater than 800 μm) than normal corneas (~500 μm), due to corneal swelling following the loss of corneal endothelial pump function. This swelling, in conjunction with mechanical unloading due to loss of IOP leading to increased "waviness" of lamellae. To counteract these effects, corneas used for SHG NLO HRMac were thinned prior to fixation by infusing phosphate-buffered saline (PBS) solution (pH 7.4) into the intact eye at a pressure of 50 mmHg for 30 min. Eyes were then fixed under similar pressure by perfusion with

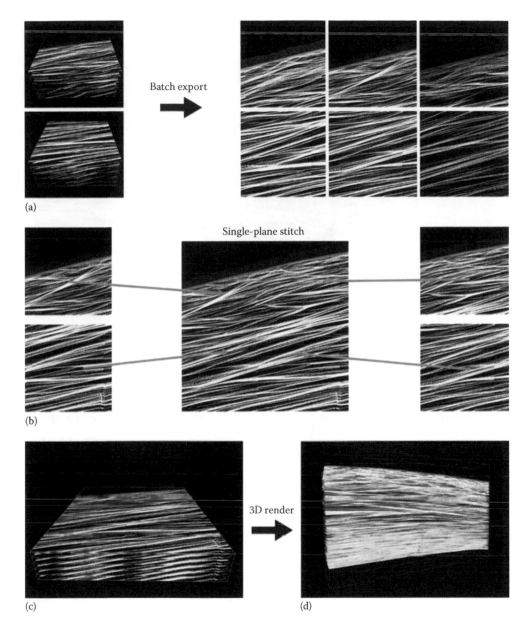

Batch export

Single-plane stitch

3D render

(a)

(b)

(c)

(d)

Figure 8.2 HRMac processing pipeline. (a) Consecutive overlapping 3-D SHG image stacks are acquired using a Zeiss LSM 510 microscope and saved as .LSM files. Planes from each stack are then batch exported and saved as individual .TIFF files. (b) All images from one plane are then stitched together to form a single-plane mosaic using ImageStitch. This process is repeated for all planes, resulting in a large-scale, 3-D stack (c) that can then be rendered in 3-D (d).

4% paraformaldehyde in PBS. Corneal orientation was then determined and the superior region identified and marked based on the insertion of the inferior oblique muscle prior to removing from the globes. Entire corneas with 2 mm of sclera were then embedded in low melting point agarose and thick 250 μm slices extending from limbus to limbus along the nasal–temporal meridian obtained using a vibratome.

For NLO-HRMac imaging, 3-D image stacks, 100 μm thick, were acquired using a 40×/1.1 NA Zeiss Apochromat objective with a step interval of 2 μm and each stack had a voxel resolution of 0.44 × 0.44 × 2 μm at 512 × 512 × 50 voxels ($x \times y \times z$). To image across the full corneal diameter, consecutive stacks were acquired, the total number of stacks varying with corneal size and curvature. Scanning a complete segment took approximately 12 h.

To reconstruct stromal lamellae, digital image processing and mosaic generation were carried out as described above. Mosaics were then loaded into Amira 5.2 (Visage Imaging, Carlsbad, CA) for 3-D reconstruction. Collagen lamellae were then individually segmented using the label field module by manually highlighting them across adjacent slices.

8.3.1 STRUCTURAL HETEROGENEITY OF CORNEAL COLLAGEN LAMELLAE

Three-dimensional reconstructions of individual collagen lamellae using the Amira surface rendering module showed branching occurred three-dimensionally with some lamellae continuing laterally after branching, whereas others showed branching in the anterior–posterior direction. Figure 8.4 shows a 3-D view of branching collagen lamellae in the anterior, mid, and

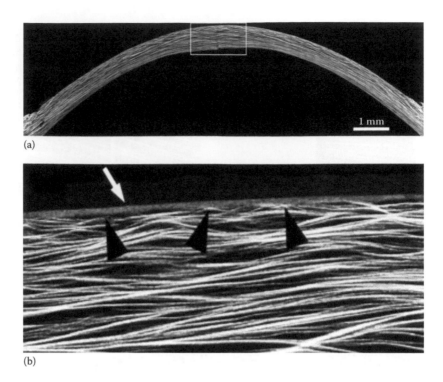

(a)

(b)

Figure 8.3 Sample HRMac image. (a) Single-plane, zoomed-out HRMac image of a full-diameter corneal cross section along the vertical meridian. The full scan comprises 60 such planes and is made up of over 80,000 individual images with a total size of 25,000 × 6,500 pixels per plane shown here at a resolution of 3 μm/pixel. (b) Shows the anterior central cornea at full resolution (0.44 μm/pixel). Note the presence of the ALL (white arrow) and the insertion of collagen fibers (black arrowheads).

posterior corneal stroma that have been segmented from the other lamellae in those regions. Extraction of the anterior segmented collagen lamellae (Figure 8.4b) shows a markedly higher degree of branching than the segmented lamellae from the deeper stroma, whereas those in the posterior part of the stroma exhibited almost no branching at all.

Three-dimensional reconstructions also identified other distinct collagen fiber structures, the first of which were long, prominent lamellae originating at or near the limbus that extended for several millimeters across large portions of the cornea (Figure 8.5). These fibers did not follow the corneal curvature from limbus to limbus; instead they traversed upward for 100s of μm before terminating at or near Bowman's layer. Interestingly, evaluation of the apparent complexity of these lamellae suggests that as these lamellae traversed toward the anterior stroma, the number and complexity of lamellar branching increased, such that in the anterior stroma the individual lamellae could no longer be identified as they joined with the highly branched anterior stromal lamellar meshwork.

A second type of structure identified was the "bow spring"–like lamellae that originated from the highly intertwined lamellae directly beneath Bowman's layer and arced upward, fusing with the Bowman's layer before arcing back down (Figure 8.6; blue, bow spring lamellae gold Bowman's layer). "Bow spring" lamellae were therefore characterized by a near-parabolic shape, the apex of which appeared fused with Bowman's layer. In part, these "bow spring" lamellae represent a dual "suture" lamellae identified earlier by Morishige et al. that are lost in keratoconus and are thought to provide mechanical stability to the corneal stroma (Morishige et al. 2007).

8.3.2 QUANTIFICATION OF LAMELLAR BRANCHING

To determine whether there were significant differences in the lamellar branching pattern as a function of depth and location within the cornea, measurements of the branching density and angle distribution of the lamellae have been performed both in the central cornea and in the entire cornea along the major meridians. In order to accomplish this task, initial studies measured the lamellar branching point density (BPD) as a function of depth in the central cornea. To accurately assess BPD in different corneas that vary in both thickness and shape, a mathematical approach was used for collecting data relative to the shape of each individual cornea as described by Winkler et al. (2011). In general, within an NLO-HRMac 3-D data set, a group of five lines perpendicular to Bowman's layer, centered around the apex of the cornea and spaced 500 μm apart, were drawn across the stroma, connecting the anterior and posterior surfaces. Fifteen segments of equal length were then created along each line (Figure 8.7a). Lamellae closest to each segment were then followed in three dimensions (x, y, and z), and the locations of the two closest branching points on either side of the line were determined and logged to a spreadsheet. By measuring the distance between branching points, an average BPD per millimeter length of lamellae as a function of stromal depth was obtained. For each cornea, the BPD for the 5 locations at the same depth were averaged and the mean and standard deviation recorded for each of the 15 depth locations.

Since measurement of BPD relied on manual tracing of lamella to find branching and fusing points, which was time consuming, labor intensive, and subject to input error, an improved, automated

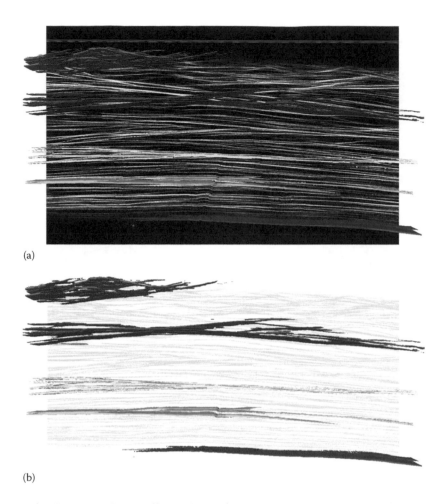

(a)

(b)

Figure 8.4 3-D reconstructions of collagen lamellae at different depths from Bowman's layer. (a) Section from a single HRMac plane of the central cornea overlaid with representative segmented lamellae at different depths. (b) This panel shows the segmented lamellae overlaying the single HRMac plane at reduced opacity.

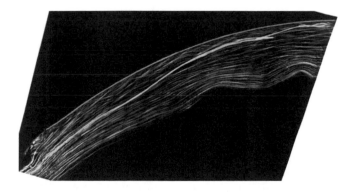

Figure 8.5 3-D reconstruction of an anchoring lamellae (green) that extends from the limbus before fusing with the highly intertwined anterior lamellar meshwork (gold).

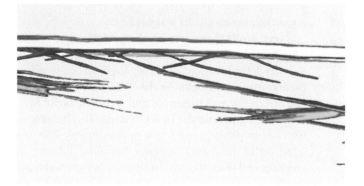

Figure 8.6 3-D reconstruction of "bow spring" lamellae (blue) that fuse with Bowman's layer (gold).

approach to measure the collagen fiber angle distribution relative to the corneal surface was developed as reported by Winkler et al. (2013). This measurement takes advantage of the fact that the lamellar branching creates two lamellae that diverge at different angles from each and hence from the corneal surface and provides distinct vector forces that may contribute to the mechanical strength of tissue as recently discussed by Petsche and Pinsky (2013). This new approach allowed for the rapid generation of large

numerical data sets that could be later incorporated into finite element models of the cornea.

In this approach, the collagen lamellar angle relative to the corneal surface was measured as a function of radial position (distance from the limbus) and stromal depth (distance from Bowman's layer). Using Metamorph software (Molecular Devices, Sunnyvale, CA), approximately 200 landmarks each were placed along the anterior and posterior surface of the corneal reconstruction from the SHG NLO HRMac 3-D data set.

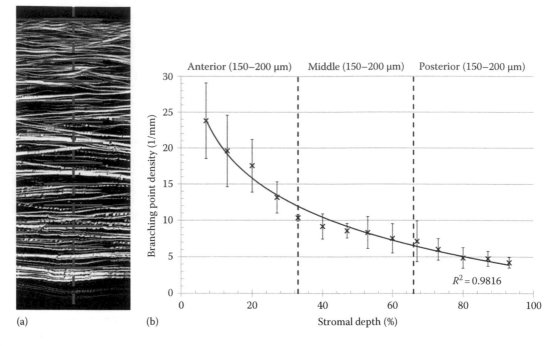

(a) (b)

Figure 8.7 Branching point densities as a function of stromal depth. (a) Fifteen segments of equal length were created along perpendicular lines to identify lamellae that were then followed and the distance to the next branching point measured three-dimensionally. (b) BPD averaged over five eyes as a function of stromal depth. The solid line shows an exponential curve fit with $R^2 > 0.98$.

Based on these landmarks, second-order polynomial curves were fitted to the anterior and posterior surface of HRMac image with high degrees of accuracy ($R^2 > 0.99$). Using custom-written software, virtual perpendicular guidelines along the cornea's anterior and posterior surfaces at 25 μm intervals along the entire length of the reconstruction were then generated. To measure lamellar angle as a function of depth, an OrientationJ algorithm developed by Sage et al. (Rezakhaniha et al. 2012) that uses structure tensors to determine the orientation and isotropy of a region of interest (ROI) was used.

Next, an ROI centered on the guideline and measuring 25 × 4 μm (width by height, corresponding to radial position by stromal depth) was created along each perpendicular, and the predominant orientation of the SHG signal as well as its energy and coherency was measured and recorded as the ROI moved down the perpendicular in 3 μm steps. This process was repeated for each guideline, resulting in approximately 200 subimages per quadrant of the cornea (superior, temporal, inferior, and nasal), yielding well over 30,000 data points per corneal quadrant.

8.3.3 DEPTH-DEPENDENT DIFFERENCES IN CORNEAL LAMELLAR ORGANIZATION

While there are considerable individual differences in the BPD as a function of depth, all of the five corneas evaluated using this approach exhibited a nonlinear decrease in BPD with increasing distance from Bowman's layer as shown in Figure 8.7b. Based on the nature of the decrease, an exponential curve was fitted to the data with a high degree of accuracy ($R^2 > 0.98$) using the following equation for the fitted curve:

$$BPD(z) = -7.766 \cdot \ln(z) + 39.084 \qquad (8.3)$$

where z is the stromal depth measured in percent. Using this equation, the data could be then interpolated to determine BPD at points of interest throughout the stroma. Importantly, the initial BPD in the anterior 5% of the stromal depth averaged 26.6 branches/mm or one branch for every 38 μm length along the lamellae. By comparison, BPD dropped significantly to approximately 8.7/mm halfway through the cornea, representing a 66% decrease in collagen lamellar branching. While lamellae in the posterior stroma also showed a significant decrease in branching compared to the anterior stroma, branching density was not significantly different from the middle stroma.

By comparison, measuring the average range in collagen lamellar angles in three human donor eyes showed a similar high lamellar fiber angle range in the anterior stroma that was significantly greater ($p < 0.02$) than the lamellar angles within the deeper stroma. There was also a corresponding reduction in the percentage of angled lamellae compared to the middle stroma. While lamellae generally ran at shallow angles, with some steeper angles found closer to the anterior surface, when analyzed as a function of depth, 95% of angles were found to be within ±11° (anterior), ±9.1° (anterior–mid), and ±8.1° (mid) relative to the corneal surface. To compare angle range and percentage of angled lamellae within different regions and quadrants of the cornea, results from the anterior stroma were plotted by region for each quadrant of the cornea (superior, temporal, inferior, nasal). While fiber angle range varied between eyes, quadrants and regions, no statistically significant differences were identified over the entire corneal reconstruction. The only significant difference other than that observed for the depth dependence regardless of region or quadrant was a difference between the superior quadrants, which showed a greater percentage of angled lamellae compared to the inferior quadrant. However, the difference was small (<8%) and not likely to be structurally relevant

8.3.4 COMPLEXITY OF CORNEAL LAMELLAR ORGANIZATION

Isolation and characterization of stromal lamellar structure using NLO-HRMac dramatically demonstrate the complexity of collagen organization in cornea that is not clearly depicted by conventional biomechanical models. Additionally, 3-D reconstructions highlight significant axial heterogeneity in the interconnectivity and branching patterns of different lamellae. Furthermore, some lamellae have unique structural features forming what appears to be "bow spring–like" lamellae that insert and exit Bowman's layer and "anchor-like" fibers that enter the cornea from the sclera in the mid to posterior stroma, traverse anteriorly, and then branch and interconnect with the anterior stromal lamellae.

It should be noted that different forms of corneal axial heterogeneity have been previously reported by multiple investigators using a variety of imaging modalities. Meek et al. describe a decrease in preferred collagen orientation in the anterior cornea using x-ray diffraction imaging (Abahussin et al. 2009). Swelling studies conducted by Müller et al. showed that corneal swelling is limited to the posterior and middle stroma (Müller et al. 2001) suggesting a structural and biomechanical heterogeneity. Transmission electron microscopy images show a larger degree of interconnectivity between lamellae in the anterior stroma (Bergmanson et al. 2005) and that the collagen lamellae occasionally bifurcate (Komai and Ushiki 1991). Finally, confocal microscopic studies show an axial heterogeneity in keratocyte density, which has been found to decrease with increasing stromal depth both in rabbit (Petroll et al. 1995) and human corneas (Patel et al. 2001).

It should also be noted that the insertions of "bow spring" lamellae into Bowman's layer observed in this study have been reported as early as 1849 by Bowman himself. Drawings of light microscopic sections in his 1849 monograph show the presence of collagen fibers extending from the anterior limiting lamina (Bowman's layer) into the stroma, where they appear to then fuse with orthogonally arranged lamellae (Bowman 1849). More recently, transmission electron microscopy studies have shown insertion of collagen lamellae into Bowman's layer at electron dense plaques with short extensions into the underlying stroma (Bergmanson et al. 2005). In this study, we observed fibers within the densely intertwined anterior lamellar meshwork just beneath Bowman's layer, arcing upward, fusing, and then returning into the anterior stroma, forming a near-parabolic arc. These "bow spring" lamellae are highly reminiscent of the load-bearing elements of girders and bridges. This population of fibers are also similar to the sutural fibers described by Morishige et al. (2006, 2007, 2011) that insert into Bowman's layer and terminate there, rather than arcing back to the anterior lamellar meshwork. Since in the former study individual lamellae were not segmented, it is likely that both sutural fibers and "bow spring" lamellae are the same and play an important role for maintaining mechanical stability. It should also be mentioned that Bowman, in considering the functional role of lamellae that inserted into Bowman's layer, suggested that they would likely play an important role in controlling corneal shape based on mechanical principles.

Another novel structural element that was identified was "anchor-like" fibers that extend from the limbus about midway between Bowman's and Descemet's layer and traverse anteriorly for several millimeters before branching and fusing with the anterior lamellar meshwork just underneath the "bow spring" fibers. These "anchor-like" fibers directly connect the most highly intertwined areas of the cornea with the limbus and bear a resemblance to the anchoring fibers described by Meek et al. (Aghamohammadzadeh et al. 2004) that also extend from the limbus and extended out toward the central cornea. Their connection to the anterior lamellar meshwork underlying the "bow spring" fibers suggests that they may also serve an anchoring function important in controlling corneal shape.

8.4 COLLAGEN STRUCTURE TO CORNEAL MECHANICS

To determine whether differences in the structural organization of the cornea influenced the mechanical properties of the cornea, LASIK tissue flaps of the anterior, middle, and posterior stroma were created using a modified Intralase© as described by Winkler et al. (2011). Corneal flaps were then mechanically tested using indentation testing to measure the effective elastic modulus. This method is used to measure compliance in biological soft tissues and is a well-established way to characterize the nanoelastic properties of various ocular tissues (Last et al. 2009, 2011). A more detailed description of indentation testing is provided by McKee et al. (2011).

In a study of seven eyes, indentation testing revealed that anterior flaps were markedly less compliant or more stiff in all eyes that were evaluated, while there was no difference in flap thickness between the samples. When the effective modulus was averaged over all seven eyes (Figure 8.8a), the anterior flap showed a significantly higher modulus and was therefore stiffer than either the middle flap ($E_{anterior}/E_{mid}$ = 1.9, $p < 0.008$) or posterior flap ($E_{anterior}/E_{posterior}$ = 2.2, $p < 0.003$). Additionally, there was no significant difference between the middle and the posterior flaps ($E_{anterior}/E_{posterior}$ = 1.18, $p = 0.6$).

To compare these results with the lamellar structural measurements, the BPD data previously discussed were grouped into anterior, mid, and posterior layers, each comprising 1/3 of total stromal thickness similar to that of the mechanical data from the corneal flaps. As shown in Figure 8.8b, the anterior third of the cornea was significantly more intertwined than the middle ($BPD_{anterior}/BPD_{mid}$ = 2.05, $p < 0.01$) and the posterior thirds ($BPD_{anterior}/BPD_{posterior}$ = 3.5, $p < 0.01$). There was no statistically significant difference between the middle and the posterior stroma ($BPD_{anterior}/BPD_{posterior}$ = 1.6, $p = 0.05$). These are identical results that were obtained for the differences in the effective modulus in these same regions, suggesting a direct association between mechanical stiffness and lamellar structure in the cornea.

Overall, the findings indicate that the highly intertwined anterior third of the cornea is significantly stiffer than the much less intertwined posterior two-thirds. This observation matches the data published by Kohlhaas et al. (2006) obtained through extensometry measurements of corneal strips, which also shows

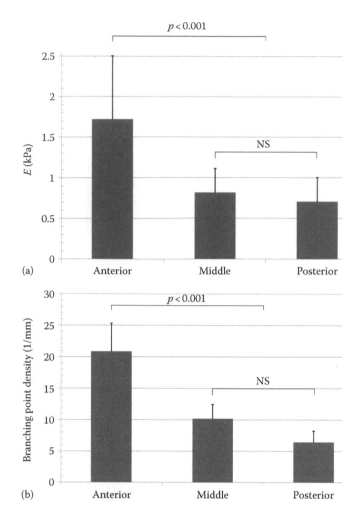

Figure 8.8 (a) Elastic modulus averaged over all seven corneas grouped by flap position. NS indicates no statistically significant difference between values ($p = 0.6$). (b) Branching point density (BPD) averaged over five HRMac-imaged corneas. BPD values from across the stroma were grouped into three layers to facilitate comparison with biomechanical. NS indicates no statistically significant difference between values ($p = 0.05$).

a stiffer anterior stroma. Interestingly, despite using a vastly different approach and measuring tensile rather than compressive modulus, Kohlhaas et al. also report a ratio of approximately 2:1 between anterior and posterior flap rigidity.

Increase stiffness of the anterior stroma is likely explained by the mechanical linking of neighboring stromal layers by lamellar branching, making it more difficult for interlamellar spacing to increase. Since the tensile strength of a collagen fiber is greatest along its long axis (Buehler 2006), lamellae that branch off and fuse with other lamellae form strong links between layers, vastly increasing axial stiffness and the force required to separate adjacent layers. Conversely, in the posterior stroma, the amount of intertwining is minimal, thereby allowing lamellae to swell in reaction to water moving into the interlamellar space, their expansion being only minimally inhibited by the presence of neighboring lamellae. These mechanical features of lamellar branching are likely to explain the interesting finding by Müller et al. regarding the resistance of the anterior stroma to corneal swelling (Müller et al. 2001).

8.5 STROMA AS A SELF-CONTAINED, CONTINUOUS STRUCTURE

Classical models often depict the corneal stroma as a nematic (liquid crystal) stack of broad lamellae, oriented along two preferential directions. In these models, lamellae form about 200 individual layers, which may interact with their nearest neighbor through interweaving but which otherwise remain separate entities.

The data presented here and elsewhere (Abahussin et al. 2009, Petsche et al. 2012, Petsche and Pinsky 2013) indicate that this model belies the more complex nature of the corneal stroma. En face SHG images, as well as x-ray diffraction studies, show that fiber bundles generally run in many different directions, rather than running in parallel and forming sheets or layers as found in nonmammalian corneas. Cross-sectional reconstructions have revealed a high density of out-of-plane transverse fibers, as well as large amounts of fiber branching and anastomosing, especially in the anterior stroma. While the amount of branching and transverse fibers is reduced with increasing stromal depth, they remain nonzero throughout the whole stroma. Branching and inclined fibers are found even in the posteriormost regions of the stroma, though less frequently than in the anterior stroma.

With these findings, we can expand upon the current model of corneal collagen organization as described by Meek et al. (Abahussin et al. 2009). First, orthogonally arranged lamellae with preferential superior–inferior and nasal–temporal orientation appear to be limited to the posterior stroma. Moving anteriorly in the stroma toward Bowman's layer, the degree of lamellar branching increases exponentially, resulting in a densely intertwined meshwork of fibers in the anterior stroma underlying Bowman's layer that exhibits a random organizational pattern. This structural heterogeneity matches the axial heterogeneity in elastic modulus; highly intertwined areas are more rigid than the orthogonally arranged lamellae that make up the posterior portion of the stroma. Second, "anchor-like" fibers extending from the sclera, rather than following the curvature of the cornea at a fixed depth, traverse several layers and terminate in an anterior lamellar meshwork underlying Bowman's layer. These fibers create mechanical links between the limbus and the central cornea and may help distribute loads. Additionally, they connect spatially separate fibers across multiple layers, which prevents slipping of layers and may serve to counteract swelling forces in the middle stroma. Third, interactions between the anterior stroma and Bowman's layer in the form of sutural or "bow spring fibers," which themselves are connected to the lower stromal regions and to the limbus through anchoring fibers, stabilize corneal shape. The lack of "bow spring" fibers in keratoconus corneas, which do not maintain normal curvature, strongly suggests that these fibers play an important role in maintaining mechanical stability and shape (Morishige et al. 2007).

This leads to an interesting conclusion: the human corneal stroma is not a layered structure in the conventional sense. Rather, it is a continuous arrangement, where every lamella is connected to every other lamellae (Figure 8.9, gold), albeit not always directly. Therefore, it is possible to trace a path of interconnected collagen lamellae from Bowman's layer through

Figure 8.9 The stroma as a continuous structure. Fiber branching and transverse fibers are present throughout the full thickness of the stroma. Therefore, through branching and anastomosing, Bowman's layer and Descemet's membrane are linked by collagen fibers. In this 3-D reconstruction of a human corneal cross section, collagen fibers (orange) forming a path between anterior and posterior surfaces have been highlighted (green).

the full thickness of the stroma back to Descemet's membrane simply by following fibers and branches (Figure 8.9, green). Lamellae arranged in such a way, rather than as a stack of separate layers, should exhibit increased shear stiffness and increased resistance to slippage, suggesting that the corneal stroma is essentially a self-stabilizing structure. While some have postulated a peripheral collagenous annulus or peripheral corneal ligament necessary to serve as a boundary condition for establishing a corneal curvature different from that of the sclera (Kokott 1938), the current 3-D reconstructions of the stroma have thus far not identified lamellae that could form such a structure. As an alternative to a peripheral ligament, the continuous and highly interconnected organization of lamellae suggests that human corneas are self-stabilizing and capable of maintaining the difference in curvature between itself and the sclera without additional peripheral stabilization. How this self-stabilized structure is built on a cellular and molecular level is not known.

ACKNOWLEDGMENT

This work was supported by a grant from the NIH NEI EY018665 and EY019719 and a challenge grant from Research to Prevent Blindness, Inc.

REFERENCES

Abahussin, M., S. Hayes, N. E. Knox Cartwright, C. S. Kamma-Lorger, Y. Khan, J. Marshall, and K. M. Meek. 2009. 3D collagen orientation study of the human cornea using x-ray diffraction and femtosecond laser technology. *Investigative Ophthalmology and Visual Science* 50 (11):5159–5164.

Abbe, E. 1873. Beiträge zur Theorie des Mikroskops und der mikroskopischen Wahrnehmung. *Archiv für mikroskopische Anatomie* 9 (1):413–418.

Aghamohammadzadeh, H., R. H. Newton, and K. M. Meek. 2004. X-ray scattering used to map the preferred collagen orientation in the human cornea and limbus. *Structure* 12 (2):249–256.

Bergmanson, J. P. G., J. Horne, M. J. Doughty, M. Garcia, and M. Gondo. 2005. Assessment of the number of lamellae in the central region of the normal human corneal stroma at the resolution of the transmission electron microscope. *Eye and Contact Lens* 31 (6):281–287.

Bowman, S. W. 1849. Lectures on the parts concerned in the operations on the eye, and on the structure of the retina, delivered at the Royal London Ophthalmic Hospital, Moorfields, June 1847: To which are added, a paper on the vitreous humor; and also a few cases of ophthalmic disease. Longman, Brown, Green, and Longmans, London, UK.

Buehler, M. J. 2006. Nature designs tough collagen: Explaining the nanostructure of collagen fibrils. *Proceedings of the National Academy of Sciences of the United States of America* 103 (33):12285–12290.

Campagnola, P. J. and C.-Y. Dong. 2011. Second harmonic generation microscopy: Principles and applications to disease diagnosis. *Laser and Photonics Reviews* 5 (1):13–26.

Cremer, C. and T. Cremer. 1974. Considerations on a laser-scanning-microscope with high resolution and depth of field. *Microscopica Acta* 81:31–44.

Daxer, A., K. Misof, B. Grabner, A. Ettl, and P. Fratzl. 1998. Collagen fibrils in the human corneal stroma: Structure and aging. *Investigative Ophthalmology and Visual Science* 39 (3):644.

Denk, W., J. H. Strickler, and W. W. Webb. 1990. Two-photon laser scanning fluorescence microscopy. *Science* 248 (4951):73–76.

Franken, P. A., A. E. Hill, C. W. Peters, and G. Weinreich. 1961. Generation of optical harmonics. *Physical Review Letters* 7 (4):118–119.

Han, M., G. Giese, and J. F. Bille. 2005. Second harmonic generation imaging of collagen fibrils in cornea and sclera. *Optics Express* 13 (15):5791–5797.

Hjortdal, J. Ø. 1995. Extensibility of the normo hydrated human cornea. *Acta Ophthalmologica Scandinavica* 73 (1):12–17.

Hjortdal, J. Ø. 1996. Regional elastic performance of the human cornea. *Journal of Biomechanics* 29 (7):931–942.

Juhasz, T., G. A. Kastis, C. Suarez, Z. Bor, and W. E. Bron. 1996. Time-resolved observations of shock waves and cavitation bubbles generated by femtosecond laser pulses in corneal tissue and water. *Lasers in Surgery Medicine* 19 (1):23–31.

Kleinman, D. A. 1962. Theory of second harmonic generation of light. *Physical Review* 128 (4):1761.

Kohlhaas, M., E. Spoerl, T. Schilde, G. Unger, C. Wittig, and L. E. Pillunat. 2006. Biomechanical evidence of the distribution of cross-links in corneastreated with riboflavin and ultraviolet A light. *Journal of Cataract and Refractive Surgery* 32 (2):279–283.

Kokott, W. 1938. Über mechanisch-funktionelle Strukturen des Auges. *Graefe's Archive for Clinical and Experimental Ophthalmology* 138 (4):424–485.

Komai, Y. and T. Ushiki. 1991. The three-dimensional organization of collagen fibrils in the human cornea and sclera. *Investigative Ophthalmology and Visual Science* 32 (8):2244.

Last, J. A., S. J. Liliensiek, P. F. Nealey, and C. J. Murphy. 2009. Determining the mechanical properties of human corneal basement membranes with atomic force microscopy. *Journal of Structural Biology* 167 (1):19–24.

Last, J. A., T. Pan, Y. Ding, C. M. Reilly, K. Keller, T. S. Acott, M. P. Fautsch, C. J. Murphy, and P. Russell. 2011. Elastic modulus determination of normal and glaucomatous human trabecular meshwork. *Investigative Ophthalmology and Visual Science* 52 (5):2147–2152.

Liu, J. and C. J. Roberts. 2005. Influence of corneal biomechanical properties on intraocular pressure measurement: Quantitative analysis. *Journal of Cataract and Refractive Surgery* 31 (1):146–155.

Martin, R. B. and D. L. Boardman. 1993. The effects of collagen fiber orientation, porosity, density, and mineralization on bovine cortical bone bending properties. *Journal of Biomechanics* 26 (9):1047–1054.

Martin, R. B. and J. Ishida. 1989. The relative effects of collagen fiber orientation, porosity, density, and mineralization on bone strength. *Journal of Biomechanics* 22 (5):419–426.

Masters, B. R. 2009. Correlation of histology and linear and nonlinear microscopy of the living human cornea. *Journal of Biophotonics* 2 (3):127–139.

McKee, C. T., J. A. Last, P. Russell, and C. J. Murphy. 2011. Indentation versus tensile measurements of Young's modulus for soft biological tissues. *Tissue Engineering Part B: Reviews* 17 (3):155–164.

Meek, K. M., T. Blamires, G. F. Elliott, T. J. Gyi, and C. Nave. 1987. The organisation of collagen fibrils in the human corneal stroma: A synchrotron x-ray diffraction study. *Current Eye Research* 6 (7):841–846.

Meek, K. M. and C. Boote. 2009. The use of x-ray scattering techniques to quantify the orientation and distribution of collagen in the corneal stroma. *Progress in Retinal and Eye Research* 28 (5):369–392.

Minsky, M. 1961. Microscopy apparatus. US Patent 3013467A.

Morishige, N., W. M. Petroll, T. Nishida, M. C. Kenney, and J. V. Jester. 2006. Noninvasive corneal stromal collagen imaging using two-photon-generated second-harmonic signals. *Journal of Cataract and Refractive Surgery* 32 (11):1784–1791.

Morishige, N., Y. Takagi, T. Chikama, A. Takahara, and T. Nishida. 2011. Three-dimensional analysis of collagen lamellae in the anterior stroma of the human cornea visualized by second harmonic generation imaging microscopy. *Investigative Ophthalmology and Visual Science* 52 (2):911.

Morishige, N., A. J. Wahlert, M. C. Kenney, D. J. Brown, K. Kawamoto, T. Chikama, T. Nishida, and J. V. Jester. 2007. Second-harmonic imaging microscopy of normal human and keratoconus cornea. *Investigative Ophthalmology and Visual Science* 48 (3):1087.

Müller, L. J., E. Pels, and G. F. J. M. Vrensen. 2001. The specific architecture of the anterior stroma accounts for maintenance of corneal curvature. *British Journal of Ophthalmology* 85 (4):437–443.

Olivier, N., F. Aptel, K. Plamann, M. C. Schanne-Klein, and E. Beaurepaire. 2010. Harmonic microscopy of isotropic and anisotropic microstructure of the human cornea. *Optics Express* 18 (5):5028–5040.

Patel, S. V., J. W. McLaren, D. O. Hodge, and W. M. Bourne. 2001. Normal human keratocyte density and corneal thickness measurement by using confocal microscopy in vivo. *Investigative Ophthalmology and Visual Science* 42 (2):333–339.

Petroll, W. M., K. Boettcher, P. Barry, H. D. Cavanagh, and J. V. Jester. 1995. Quantitative assessment of anteroposterior keratocyte density in the normal rabbit cornea. *Cornea* 14 (1):3–9.

Petsche, S. J., D. Chernyak, J. Martiz, M. E. Levenston, and P. M. Pinsky. 2012. Depth-dependent transverse shear properties of the human corneal stroma. *Investigative Ophthalmology and Visual Science* 53 (2):873–880.

Petsche, S. J. and P. M. Pinsky. 2013. The role of 3-D collagen organization in stromal elasticity: A model based on x-ray diffraction data and second harmonic-generated images. *Biomechanics and Modeling in Mechanobiology* 12 (6):1101–1113.

Preibisch, S., S. Saalfeld, and P. Tomancak. 2009. Globally optimal stitching of tiled 3D microscopic image acquisitions. *Bioinformatics* 25 (11):1463–1465.

Ramos, J. L., Y. Li, and D. Huang. 2009. Clinical and research applications of anterior segment optical coherence tomography—A review. *Clinical and Experimental Ophthalmology* 37 (1):81–89.

Rezakhaniha, R., A. Agianniotis, J. T. Christiaan Schrauwen, A. Griffa, D. Sage, C. V. C. Bouten, F. N. Van de Vosse, M. Unser, and N. Stergiopulos. 2012. Experimental investigation of collagen waviness and orientation in the arterial adventitia using confocal laser scanning microscopy. *Biomechanics and Modeling in Mechanobiology* 11 (3–4):461–473.

Ruberti, J. W. and J. D. Zieske. 2008. Prelude to corneal tissue engineering-gaining control of collagen organization. *Progress in Retinal and Eye Research* 27 (5):549–577.

Winkler, M., D. Chai, S. Kriling, C. J. Nien, D. J. Brown, B. Jester, T. Juhasz, and J. V. Jester. 2011. Nonlinear optical macroscopic assessment of 3-D corneal collagen organization and axial biomechanics. *Investigative Ophthalmology and Visual Science* 52 (12):8818–8827.

Winkler, M., G. Shoa, Y. Xie, S. J. Petsche, P. M. Pinsky, T. Juhasz, D. J. Brown, and J. V. Jester. 2013. Three-dimensional distribution of transverse collagen fibers in the anterior human corneal stroma. *Investigative Ophthalmology and Visual Science* 54 (12):7293–7301.

Zeng, Y., J. Yang, K. Huang, Z. Lee, and X. Lee. 2001. A comparison of biomechanical properties between human and porcine cornea. *Journal of Biomechanics* 34 (4):533–537.

Zimmermann, D. R., B. Trueb, K. H. Winterhalter, R. Witmer, and R. W. Fischer. 1986. Type VI collagen is a major component of the human cornea. *FEBS Letters* 197 (1–2):55–58.

9 Multiphoton imaging of the retina

Robin Sharma and Jennifer J. Hunter

Contents

9.1 MOTIVATION

Helmholtz invented the first ever ophthalmoscope for imaging the retina (Helmholtz, 1851). Ophthalmoscopy has since been the beneficiary of technical advances in microscopy along with other fields such as interferometry (optical coherence tomography) and astronomy (adaptive optics [AO]) (Huang et al., 1991; Liang et al., 1997). In the living eye, the combination of AO with scanning laser ophthalmoscopy systems has resulted in improved visualization of individual cone and rod photoreceptors (Dubra et al., 2011; Roorda et al., 2002), nerve fiber bundles (Roorda et al., 2002), vascular flow (Tam et al., 2011), lamina cribrosa in the optic nerve head (Vilupuru et al., 2007), horizontal cells (Guevara-Torres et al., 2015), and intrinsic fluorescence from retinal pigment epithelial cells (Gray et al., 2006). Extrinsic fluorescence labeling can provide the contrast necessary to image otherwise transparent individual cells in the inner retina (Geng et al., 2012, 2009; Gray et al., 2006) and monitor function, such as calcium fluctuations with visual stimulation (Yin et al., 2014, 2013). Such cellular-scale imaging is important for understanding the structure and function of the retina in health and disease. However, even with these advances, there are some scientific pursuits that current retinal imaging systems are not equipped to tackle.

First, new methodologies are needed to image intrinsic functional activity at a cellular scale. Vision begins at the level of light absorption by photopigment in retinal photoreceptors. The photopigments necessary for continued vision are regenerated after isomerization and phototransduction through the visual cycle. The kinetics of photopigment regeneration has been studied for several decades using a technique called rhodopsin densitometry that measures changes in the intensity of backscattered light over time (Hood and Rushton, 1971; van Norren and van der Kraats, 1981). Unfortunately, these measurements suffer from ambiguities related to the source of light reflected from the retina and this complicates the implementation of this technique for assessing photoreceptor health (DeLint et al., 2000; Grieve and Roorda, 2008; Masella et al., 2014). The ability to monitor the rate of formation and decay of intermediate components of the visual cycle may provide a powerful tool for monitoring photoreceptor health, since many retinal diseases can manifest when the cycle is disrupted (Travis et al., 2007).

Second, light has to propagate from the anterior optics to the photoreceptors through the inner retina. Except for recent results in horizontal cell imaging (Guevara-Torres et al., 2015), typically inner retinal layers remain hidden during ophthalmic imaging because they are naturally translucent. Not only do inner retinal cells play a crucial role in retinal function and neural processing, they can get severely affected during disease progression. For instance, ganglion cells die during prevalent eye diseases such as glaucoma (Weinreb et al., 2014) that can lead to an irreversible vision loss and is one of the most common causes of blindness in the world. While individual photoreceptors and retinal pigment epithelium (RPE) cells are routinely imaged using conventional AO ophthalmoscopy, these techniques have not been able to resolve individual ganglion cells or other cell classes in the inner retina without the use of exogenous fluorophores.

An imaging technique that has the potential to bridge this gap is nonlinear microscopy. Two or more long-wavelength photons can be combined to produce conditions similar to traditional short wavelength fluorescence excitation (Göppert-Mayer, 1931). Nonlinear fluorescence imaging has transformed microscopy in biology because it can take advantage of autofluorescence from naturally occurring, physiologically relevant molecules, has increased penetration depth, reduced photobleaching and intrinsic axial sectioning, and can provide information about the function and health of individual cells (Denk et al., 1990).

All cells in the retina contain well-known autofluorescent molecules such as NADH, NADPH (Chen et al., 2005), and flavoproteins such as FAD (Zhao et al., 2007) that are involved in cellular metabolism. Photoreceptors and RPE contain retinoids (Chen et al., 2005; Imanishi et al., 2004a; Sears and Kaplan, 1989) and lipofuscin (Bindewald-Wittich et al., 2006) that are known to be autofluorescent and are a part of the visual cycle. Additionally, collagen is present in basement membranes and extracellular matrix and is an integral component of vessel walls, along with elastin (Zoumi et al., 2002). Together, these molecules are directly associated with retinal health and physiology. In particular, recording autofluorescence from retinoids and NADH has the potential to track functional activity at a cellular scale (Chen et al., 2005; Imanishi et al., 2004a; Sears and Kaplan, 1989). Moreover, the presence and distribution of these fluorophores offers the possibility of extracting intrinsic optical contrast from otherwise invisible cells in the inner retina (Gualda et al., 2010).These molecules can be excited using ultraviolet (UV) or short-wavelength visible light (Figure 9.1). However, the advantages of nonlinear excitation make it practical, especially for *in vivo* applications where the anterior optics of the eye limit the transmission of UV and blue light (<400 nm) to the retina (Dillon et al., 2007).

This chapter discusses multiphoton imaging of the retina, including the theory of two-photon fluorescence, properties of some of the relevant endogenous fluorophores, and a summary of results

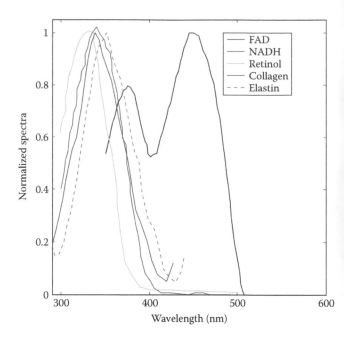

Figure 9.1 Normalized single-photon excitation spectra for some of the most common fluorophores found in the retina that can be excited with UV light. (Data adapted from Chen, C. et al., *Biophys. J.*, 88(3), 2278, 2005; Huang, S. et al., *Biophys. J.*, 82(5), 2811, 2002; Pu, et al., *J. Biomed. Opt.*, 15(4), 047008, 2010.)

from *ex vivo* and *in vivo* imaging of retinal structure and function. Two-photon fluorescence imaging of extrinsic fluorophores (Xu and Webb, 1996) is outside the scope of this chapter.

9.2 THEORY OF MULTIPHOTON IMAGING

9.2.1 TWO-PHOTON FLUORESCENCE MICROSCOPY

The branch of imaging that studies light emitted by a molecule undergoing electronic transitions is known as luminescence. Energy from the incident light is first absorbed by the molecule, causing the transition from ground (lower energy) to the excited (higher energy) state. When the molecule is in the excited state, there are three possible outcomes as shown in Figure 9.2:

1. The molecule can nonradiatively decay back to ground electronic state and energy is released as heat (not shown).
2. The molecule can spontaneously emit a photon as it decays back to ground electronic state. This is fluorescence.

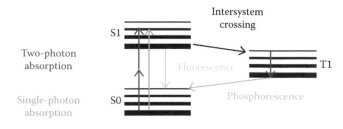

Figure 9.2 Schematic of some of the kinds of electronic transitions that are of interest in biological imaging, namely, single-photon absorption (blue), two-photon absorption (red), fluorescence emissio (green), intersystem crossing (black), and phosphorescence (orange). S0 and S1 are singlet states while T1 is a triplet system.

3. Under certain special conditions, some molecules undergo another spontaneous electronic transition to a lower excited state, through a process known as intersystem crossing, before eventually decaying to ground state and releasing a photon. This is known as phosphorescence and is not very common at room temperatures.

Typically, the absorption of a single high-energy photon provides the excitation from the ground state. A two-photon absorption event relies on the probability of two longer wavelength photons arriving at the molecule within a time duration that is shorter than 10–100 as (Diaspro et al., 2005). As the excited electron relaxes back to its electronic ground state, a photon is emitted. Fluorescence images are generated by capturing these emitted photons.

9.2.1.1 Probability of two-photon excitation

The probability of two-photon absorption is quadratic with the intensity of incident light (Göppert-Mayer, 1931). The quadratic dependence on light intensity is what makes this imaging modality nonlinear. After the invention of the laser, it was experimentally demonstrated for the first time in crystals (Kaiser and Garrett, 1961) and later in biological tissue (Denk et al., 1990). The use of ultrashort pulses ranging from a few picoseconds (Bradley et al., 1972) to a few femtoseconds (Denk et al., 1990) boosts the instantaneous intensity at the focal plane and increases the likelihood that two photons within the same pulse will arrive at the target molecules simultaneously (Diaspro et al., 2005). This also implies that the shorter the temporal width of the pulses used, the greater the probability of causing a two-photon absorption event. Even then, the efficiency of two-photon excitation is several orders of magnitude lower than for single-photon excitation (Brakenhoff et al., 1996) and is confined to a small region where the light is sharply focused. The intensity of two-photon fluorescence is related to the number of photons absorbed per fluorophore (n_{abs}), which depends on multiple factors, such as the square of the power of incident light (P_{avg}) and the two-photon absorption cross-section for the fluorophore (σ), the pulse width (τ_p), and the pulse repetition rate of the laser (f_p). Light is usually focused onto the sample through an objective lens and parameters related to spatial confinement such as the numerical aperture (NA) and wavelength of incident light (λ_{inc}) also determine the efficiency of two-photon absorption. Thus, the number of photons absorbed per unit pulse per fluorophore is given by (Denk et al., 1990)

$$n_{abs} \approx \sigma \frac{P_{avg}^2}{\tau_p f_p^2} \left(\frac{NA^2}{hc\lambda_{inc}} \right)^2$$

where

h is Planck's constant
c is the speed of light

The time-averaged two-photon fluorescence for every available molecule within the sample is given by

$$\langle I_{2P} \rangle \approx \eta \sigma \frac{P_{avg}^2}{\tau_p f_p} \left(\frac{NA^2}{hc\lambda_{inc}} \right)^2$$

where η is related to the quantum yield of collected fluorescence (Diaspro et al., 2005). Thus, emitted two-photon fluorescence is nonlinear, varying with the square of the incident power, and the numerical aperture to the fourth power. It is also inversely related to pulse width and repetition rate.

9.2.1.2 Experimental considerations

For optimal two-photon fluorescence imaging, the central wavelength of the imaging beam should to be close to the peak of the excitation spectrum for the sample being targeted. Many commercially available ultrashort pulsed lasers such as the titanium–sapphire crystal–based lasers (Lamb et al., 1994; Spence et al., 1991) provide access to a range of wavelengths from 680 to 1100 nm. This works well for nonlinear imaging of molecules whose excitation spectra lie in the UV and visible excitation regime (as shown in Figure 9.1). For biomedical imaging, the use of infrared wavelengths reduces the risk of light damage and provides deeper penetration due to reduced scatter, but the high intensities required for two-photon excitation have potential for thermal, chemical, and mechanical damage. By virtue of their temporal profile, ultrashort pulses are naturally broadband. The phase and amplitude profile of the spectrum of the pulse determine the width of the pulse in time ($\Delta\tau$). The full width at half maximum (FWHM) of the shortest achievable pulse width is inversely related to the spectral bandwidth ($\Delta\omega$) of the pulse through the time-bandwidth product:

$$\Delta\tau\Delta\omega \cong \text{time bandwidth product}$$

This parameter is related to the shape of the spectral amplitude distribution. For example, for pulses with Gaussian spectral profiles, the time-bandwidth product is 0.44. While shorter pulses improve the probability of two-photon absorption, there is a trade-off associated with the broader spectral bandwidth. The refractive indices of optical media are dependent on the wavelength of light propagating through them; this is known as dispersion. Dispersion causes different wavelengths to propagate at different velocities that spread the pulse in time and in turn reduces the probability of two-photon absorption. Broadband pulses could also suffer from chromatic aberration, which can cause different spectral components to focus at different focal planes. Additionally, the combination of chromatic aberration and dispersion can affect the spatiotemporal profile of ultrashort pulses near the focal plane, a phenomenon known as propagation time delay (Bor, 1988; Kempe et al., 1992). For singlet lenses, radially dependent dispersion delays radial components of the beam differently, consequently preventing light at the exit pupil from arriving at the focal region at the same time. This can affect the realistically achievable lateral and axial resolution from such imaging systems.

9.2.1.3 Optical resolution in two-photon microscopy

Single-photon fluorescence confocal microscopy relies on a pinhole for axial sectioning that results in losses in quantum efficiency due to chromatic aberration and scattering from turbid samples (Oheim et al., 2001). The maximal probability of two-photon excitation is restricted to the focal volume and results in intrinsic axial and lateral sectioning and removes the requirement of a pinhole. Consequently, two-photon fluorescence emitted by the sample can be coupled directly into a point detector.

Analytical expressions for the theoretical resolution for two-photon imaging systems have been estimated by using wave propagation analysis to calculate the three-dimensional intensity point spread function (IPSF) and optical transfer function (Gu, 1996; Min, 1995; Nakamura, 1993; Sheppard and Gu, 1990). Expressions for lateral and axial resolution are related to the square of IPSF. Based on fits to the analytical expressions for a Gaussian intensity profile, the lateral resolution (FWHM) is given by (Zipfel et al., 2003b)

$$\omega_{xy} = \begin{cases} 2\sqrt{\ln 2}\,\dfrac{0.320}{\sqrt{2}\,NA} & NA \leq 0.7 \\[2mm] 2\sqrt{\ln 2}\,\dfrac{0.325}{\sqrt{2}\,NA^{0.91}} & NA > 0.7 \end{cases}$$

The axial resolution (FWHM) is given by

$$\omega_z = 2\sqrt{\ln 2}\,\frac{0.532\lambda}{\sqrt{2}\,NA}\left[\frac{1}{n - \sqrt{n^2 - NA^2}}\right]$$

where

λ is the wavelength
NA is the numerical aperture
n is the image space refractive index

Theoretically, for the same emitted fluorescence wavelength, the best achievable axial and lateral resolution in single-photon confocal microscopy is almost a factor of 2 better than conventional (no pinhole) two-photon fluorescence microscopy, in large part due to the fact that the wavelength of light used for two-photon imaging is twice as long. With the use of a confocal pinhole, the resolution of two-photon imaging systems can be improved. A thorough comparison between the theoretically estimated axial and lateral resolution in single- and two-photon fluorescence imaging is available elsewhere (Gu, 1996; Min, 1995; Nakamura, 1993; Sheppard and Gu, 1990).

9.2.2 SECOND HARMONIC GENERATION IMAGING

In the previous section, we discussed the case of multiphoton absorption when two (or three or more) photons at frequencies ω_i focused at the sample can get absorbed simultaneously resulting in emission of fluorescence (and/or phosphorescence) at frequencies $<2\omega_i$ (or $<3\omega_i$ and so on). Alternatively, light of double the frequency, $2\omega_i$ can also be generated within the sample through another multiphoton process, wherein the energy of the incident photons is not absorbed, but instead is scattered by the sample via a process called harmonic up-conversion (Bloembergen, 1965). This phenomenon is known as second harmonic generation (SHG). These concepts can be extended to the cases where light at higher-order harmonics can also be created to define third harmonic generation and so on.

SHG was first demonstrated experimentally in inorganic crystals soon after the invention of the laser (Franken et al., 1961). Within the domain of electromagnetic theory, the interaction of light with matter is defined in terms of the induced dipole moment per unit volume, also known as the induced polarization. The extent of this induced polarization is dependent on (1) the electric field of incident light and (2) innate property of the material known as susceptibility (χ) (Boyd, 2008).

For linear imaging, the induced polarization depends linearly on the applied electric field via a dimensional constant known as susceptibility. For nonlinear imaging, second order–induced polarization depends on the square of the electric field and second-order material susceptibility $\chi^{(2)}$, a tensor. This tensor contains information about the ordering and arrangement of matter within the material. Fundamentally second-order polarizability is nonexistent for symmetric media, that is, materials that contain a center of inversion symmetry (also known as centrosymmetric media). Examples of such material include glass and liquids. For materials that are noncentrosymmetric, within which matter is arranged in a somewhat orderly manner, such as in crystals (e.g., $LiNbO_3$) and fibrous proteins such as collagen, second-order polarizability is nonzero. Consequently, the probability of inducing second-order nonlinear processes such as SHG is greater.

Under the right conditions, for light incident on a noncentrosymmetric material at a particular wavelength, SHG results in the production of photons at half the wavelength, or double the frequency of incident light by the destruction of two-photons of lower energy and the instantaneous creation of one photon of double the energy. Consequently, phase of higher harmonic radiation is tightly coupled with the incident light.

Apart from symmetry considerations, there are other parameters that determine the efficiency of this process, such as orientation and distribution of molecules with respect to the polarization of the incident light, refractive index for incident wavelength, and the beam propagation properties of incident light. For practical SHG, microscopy incident light has to be focused on the sample, and the theoretical treatment for this scenario has been investigated in detail (Mertz and Moreaux, 2001; Moreaux et al., 2001). The role of the size and ordered distribution of supramolecular components of tissue (e.g., collagen fibrils) has also been investigated theoretically (LaComb et al., 2008). SHG imaging in a scanning setup was first proposed by Sheppard and colleagues (Gannaway and Sheppard, 1978) and was later demonstrated in biological samples (Freund et al., 1986). Since then many advances in instrumentation have made SHG microscopy a feasible biomedical imaging tool (Campagnola, 2011; Campagnola and Loew, 2003; Cicchi et al., 2013). By virtue of the coherent nature of signal generation, SHG is primarily measured in transmission mode, also known as forward scatter. Light can also be scattered backward (e.g., under Mie scattering or multiple scattering regimes), and the ratio of forward to backscattered SHG signal (F/B ratio) is considered to be a biomarker for tissue health, although the efficiency of backscattered SHG signal is appreciably lower than forward scattered light (Rao et al., 2009; Williams et al., 2005).

9.3 MULTIPHOTON MICROSCOPY AND OPHTHALMOSCOPY

9.3.1 MICROSCOPE IMPLEMENTATION

In this section, we will discuss the simplest possible and most commonly employed implementation scenario for multiphoton imaging, although many variations and deviations exist. As mentioned previously, the efficiency of multiphoton processes depends on the square of the intensity of light incident upon the sample. To increase the density of photons at the same point in space and time, ultrashort pulsed lasers are commonly employed. Consequently, most of the commonly available two-photon microscopes are point-scanning imaging systems. Dispersion and chromatic aberration have to be minimized for efficient imaging. Excitation spectra of many common fluorophores of interest usually lie in the UV or visible, and therefore the wavelengths of light used for two-photon excitation are in the near infrared region. Emission spectra for such fluorophores are also in the UV or visible and can be easily separated from the excitation light by using dichroic mirrors. Signals are usually detected using very sensitive instruments such as avalanche photodiodes or photomultiplier tubes (PMTs). In order to limit the bleed through of incident excitation light, additional band-pass filters can be employed in the detection channel to selectively permit the transmission of fluorescence signal. Detailed description of such schematics are described in many textbooks and review articles (Denk et al., 1995, 1990; Denk and Svoboda, 1997; Diaspro et al., 2005; Diaspro and Chirico, 2003; Masters and So, 2008; So et al., 2000; Svoboda and Yasuda, 2006; Zipfel et al., 2003b).

In addition to the conventional approach to multiphoton microscopy, many other sophisticated approaches have been developed for niche applications. For example, in order to suppress background signals, investigators have deployed a scheme called simultaneous spatial and temporal focusing, where the excitation beam's spectral components are split up spatially into a rainbow beam and then focused through an objective lens onto the sample. This way, the shortest pulse is restricted to the focal volume (Oron et al., 2005; Zhu et al., 2005). Similarly, through the use of rotating microlens disks, cascaded beam splitters and microlens arrays, investigators have demonstrated higher-than-conventional-speed multiphoton microscopy across multiple focal planes for volumetric scanning in thick samples (Andresen et al., 2001; Bewersdorf et al., 1998; Buist et al., 1998; Egner and Hell, 2000; Fittinghoff et al., 2000). More examples are provided in recent review articles on the subject (Carriles et al., 2009; Hoover and Squier, 2013; Oheim et al., 2006).

Using *ex vivo* preparations, two-photon fluorescence microscopy of intrinsic fluorophores has been demonstrated throughout the retina of many species including humans, monkeys, pigs, frogs, mice, and chickens (Chen et al., 2005; Gualda et al., 2010; Han et al., 2007; Imanishi et al., 2004a; Lu et al., 2011; Palczewska et al., 2014b; Zhao, 2009). Images across different species appear very similar. The appearance of retinal structures is also similar for fixed, flat-mounted preparations and live tissue within a few hours of enucleation (Palczewska et al., 2014b). The source of fluorescence, discussed in sections 4–6, is dependent on the types of cells in the layer being imaged and on the choice of excitation wavelength and emission filter bandwidth.

9.3.2 *IN VIVO* INSTRUMENTATION

There are several challenges for two-photon imaging through the pupil of the living eye. The poor optical quality and low numerical aperture of the eye (~0.22 in primate, ~0.49 in mouse) blur the focus of the imaging beam and reduce the probability of fluorescence excitation. That, coupled with the low efficiency of two-photon imaging and limitations on the intensity of light that can be safely used in the eye, suggests that very little fluorescence emission can be expected. To generate an image, long integration times might be required, but are made difficult by the continual motion of the eye.

Measurement and correction of the eye's optical imperfections by AO (Liang et al., 1997) has allowed high resolution imaging through the pupil of the eye with a scanning light ophthalmoscope (Roorda et al., 2002). Light is raster scanned across the retina using horizontal and vertical scanners, aberrations are measured using a Hartmann–Shack wavefront sensor and corrected through AO such as deformable mirrors or spatial light modulators. A high density of photons in space that arrive at the same time are essential for efficient two-photon excitation, and thus point-scanning systems such as the AO scanning light ophthalmoscope (AOSLO) are used for the deployment of two-photon imaging in the eye (Figure 9.3).

AOSLO systems have previously been used for fluorescence imaging and images are generated by averaging the emission signal across many individual frames (Gray et al., 2006). Dual registration methods developed for single-photon AOSLO fluorescence imaging compensate for low signal levels and eye movement by tracking ocular motion using a simultaneously acquired high signal-to-noise ratio reflectance image. Dual focusing, by varying the vergence of the reflectance and fluorescence imaging beams, may be used to compensate for ocular chromatic aberration and select specific layers of interest for each imaging modality (Gray et al., 2006). These techniques are also necessary for effective two-photon imaging in the eye. The principles for two-photon microscopy described in the previous sections have been combined with the techniques that have been developed for multichannel AOSLO for imaging the retinas of living primate and mouse eyes through the ocular pupil (Hunter et al., 2011; Sharma et al., 2013, 2016a,b). Alternatively, the combination of a two-photon microscope with wavefront-sensorless AO and a flat-surfaced contact lens was employed in mouse, but is not practical for imaging in large animals and humans (Palczewska et al., 2014a). The ability to image the living eye provides the possibility to noninvasively observe both structure and function across the macula.

To avoid the optics of the eye, Imanishi et al. imaged through the sclera of the mouse eye using a commercial two-photon microscope (Imanishi et al., 2004a). Although feasible *in vivo*, imaging through the sclera requires orbital surgery and does not provide access to the central retina. It is not clear whether it would be possible in primates, which have a thicker sclera.

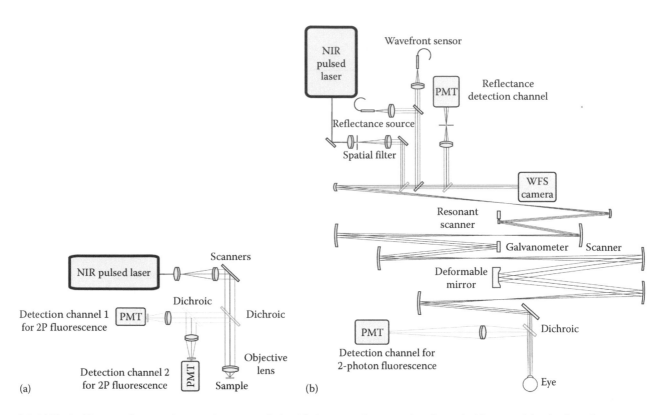

Figure 9.3 (a) Typical layout of a two-photon microscope. Pulsed light source is scanned and coupled into an objective lens. In many cases, the emitted fluorescence is collected through a dichroic and then again through another dichroic for detection using one or more detectors, typically PMTs. (b) Layout of a two-photon adaptive optics scanning laser ophthalmoscope (TPAOSLO). The schematic of the TPAOSLO is similar to that described elsewhere (Dubra and Sulai, 2011; Gray et al., 2006; Hunter et al., 2011; Roorda et al., 2002). Light from a pulsed light source is coupled into the system using appropriate dichroics. Emitted fluorescence need not be descanned and may be collected directly through a dichroic and coupled into the PMT, as shown.

9.4 IMAGING CELLULAR METABOLISM

9.4.1 METABOLITES AS FLUOROPHORES

Coenzymes such as NAD(P)H and flavoproteins like FAD are crucial for mitochondria-based cellular respiration pathways such as oxidative phosphorylation. Many retinal layers are densely packed with mitochondria because the retina is more metabolically active than the brain per unit weight (Wong-Riley, 2010). These molecules are of critical importance for cell survival and play the following roles during cellular respiration:

1. NADH and NADPH are the reduced forms of the coenzymes nicotinamide adenine dinucleotide (NAD+) and nicotinamide adenine dinucleotide phosphate (NADP+), respectively, and serve as reducing agents. NADH, a key component of cellular respiration, is generated in the cytoplasm during glycolysis. It is also produced in the mitochondria by the Krebs cycle (also known as the citric acid cycle) during aerobic respiration. NADH donates electrons in complex I, the first stage of the electron transport chain (oxidative phosphorylation), which generates adenosine triphosphate (ATP), a unit of intracellular energy. Thus, glycolysis and the Krebs cycle produce NADH, while the electron transport chain consumes NADH.

2. Flavin adenine dinucleotide (FAD) is also strongly fluorescent. FAD moves electrons between the Krebs cycle and complex II, the second stage of the electron transport chain. FAD is used during the Krebs cycle and produced during the electron transport chain.

These critical molecules are naturally fluorescent and are commonly imaged in tissue using two-photon fluorescence microscopy (Gualda et al., 2010; Huang et al., 2002; Schweitzer et al., 2007; Zipfel et al., 2003a). NADH and NADPH have identical excitation spectra (shown in Figure 9.1) and are collectively referred to as NAD(P)H. At 730 nm, the two-photon cross section of FAD is almost double that of NAD(P)H and follows a similar action spectrum until it peaks again near 900 nm (Huang et al., 2002). The emission peak of FAD is ~60 nm longer than NAD(P)H (Huang et al., 2002; Zipfel et al., 2003a). Thus, the fluorescence from these two fluorophores can be measured separately.

The ratio of FAD to NADH fluorescence intensity, known as the reduction-oxidation (redox) ratio, indicates the oxidation-reduction state in the cell and is a measure of cellular health (Skala and Ramanujam, 2010). It is related to the metabolic rate and oxygen supply within a cell (Harrison and Chance, 1970). A higher redox ratio usually indicates more active cellular metabolism and a more aerobic state. The redox ratio is increased in unhealthy tissue with abnormal metabolism, such as cancerous tumors (Li, 2012).

9.4.2 *EX VIVO* IMAGING OF METABOLITES

All retinal layers, each with a unique appearance, have been imaged using two-photon fluorescence microscopy. The primary fluorophores imaged in the inner retina were spectrally identified as mitochondrial FAD and NAD(P)H (Gualda et al., 2010; Peters et al., 2011; Schweitzer et al., 2007), although others may be present.

The nerve fiber layer appears as bright bands of nerve fiber bundles (Figure 9.4a). The fluorescence probably originates from mitochondria that reside on the nerve fibers themselves. Located between the nerve fiber bundles are processes of Müller cells that span axially from the inner limiting membrane to the photoreceptor inner segments. In a retinal cross section, the Müller cells were fluorescent throughout the entire thickness of the retina (Lu et al., 2011).

In the ganglion cell layer, fluorophores within or surrounding the cell somas provide contrast against the nonfluorescent nuclei, making it possible to isolate individual cells (Figure 9.4b). Focusing through the ganglion cell layer reveals fluorescence within the cell above and below the nucleus. Multiple layers of cell soma can be identified. Bueno and colleagues (Bueno et al., 2014) used this imaging method to demonstrate that ganglion cell bodies were larger in form-deprivation-induced ~10 D myopic chicken eyes compared to normally developing fellow eyes. The depth and pattern of individual dendrites have not been tracked using two-photon imaging of intrinsic fluorophores.

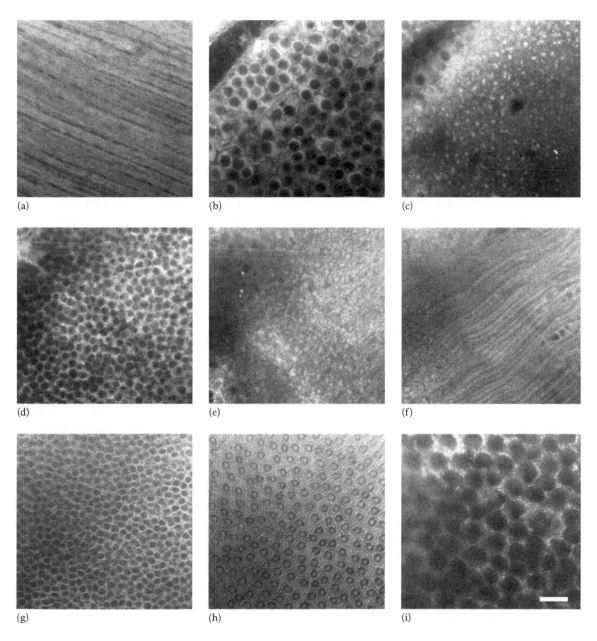

(a) (b) (c)

(d) (e) (f)

(g) (h) (i)

Figure 9.4 Two-photon fluorescence images of fixed *ex vivo* macaque retina imaged using 730 nm excitation and emission collected over the range 390 nm to 560 nm. Each image is the average of three focus depths acquired in 2 μm intervals. Multiple retinal layers were imaged: (a) nerve fiber layer, (b) ganglion cell layer, (c) inner plexiform layer, (d) inner nuclear layer, (e) outer plexiform layer, (f) Henle fiber layer (g) outer nuclear layer, (h) photoreceptor layer, and (i) retinal pigment epithelial layer. Scale bar: 20 μm.

Within the inner plexiform layer, there is a pattern of weak fluorescence that is thought to correspond to the synaptic connections between ganglion cell dendrites, bipolar cells, and amacrine cells (Figure 9.4c). In the outer plexiform layer, there is a pattern of large and small fluorescent disks that likely correspond to cone pedicles and rod spherules (Figure 9.4e). It is in this layer that they connect with bipolar and horizontal cells. In primate retina, streaks of fluorescence radiating from the fovea center are visible between the outer plexiform layer and the outer nuclear layer (Figure 9.4f). These are possibly the Henle fibers.

In the inner and outer nuclear layers, fluorescence originates from between many layers of dark spheres, corresponding to cell nuclei (Figure 9.4d and g). The surrounding fluorescence may originate from narrow regions within the cell, the cell membranes, surrounding Müller cells or from molecules in the extracellular matrix.

The mosaic of rods and cones is clearly delineated using two-photon fluorescence microscopy (Figure 9.4h). A principle photoreceptor inner segment fluorophore was spectrally identified as NAD(P)H in *ex vivo* frog rods (Chen et al., 2005). This was also observed in *ex vivo* primate inner segments as a shoulder in the emission spectrum that was not present for the outer segments (Palczewska et al., 2014b). Using single-photon excitation, a peak consistent with FAD (Zhao et al, 2007) was found for fluorescence from porcine rod and cone outer segments, but FAD would also be expected to reside in the mitochondria of the inner segments.

Functional information, such as changes in the concentration of molecules involved in cellular metabolism, could be used as a marker for retina health. In healthy mouse rods, the metabolic activity (total energy consumption) of the cell in bright light is only a quarter that in darkness (Okawa et al., 2008). In cones, the metabolic activity in bright light is at least equivalent to its demand in darkness and may even be higher. Using a combination of single- and two-photon fluorescence techniques in isolated frog rods and mouse retinal sections, Chen and colleagues (2005) measured an increase in fluorescence from the inner segments in response to visible light. The exact source of this fluorescence is unclear and may be linked to mitochondrial metabolic activity, retinoids (Section 9.5), or an unknown fluorophore.

9.5 IMAGING THE VISUAL CYCLE

9.5.1 PRODUCTS OF THE VISUAL CYCLE AS FLUOROPHORES

The first stages in vision depend on the absorption of a photon of light by a photopigment molecule that is located on the membrane in the photoreceptor outer segment. This begins the phototransduction cascade for generating a photocurrent and the visual (or retinoid) cycle for regenerating photopigment. The visual cycle begins with the photoisomerization of the chromophore 11-*cis*-retinal (vitamin A) and the release of all-*trans*-retinal from the photopigment. This is quickly converted into all-*trans*-retinol through a reaction that uses NADPH (Palczewski et al., 1994). This fluorescent molecule is then moved to the RPE where it is esterified to all-*trans*-retinyl ester. The retinyl ester may be stored in liposomes known as retinyl-ester storage particles or retinosomes (Golczak et al., 2005; Imanishi et al., 2004a,b).

These retinosomes are spread throughout the RPE cell and are highly fluorescent. When needed to regenerate photopigment, 11-*cis*-retinol is isomerized from the retinyl ester. It is then oxidized into 11-*cis*-retinal before being transported into the photoreceptor outer segments during what is thought to be the rate limiting step of the visual cycle (Lamb and Pugh, 2006). The chromophore, 11-*cis*-retinal, binds with rhodopsin or the cone opsins to form the photopigment. The aldehyde form of vitamin A exhibits very little fluorescence compared to the alcohol and ester forms. For detailed reviews of the visual cycle, see (Garwin and Saari, 2000; Lamb and Pugh, 2004; McBee et al., 2001; Palczewski, 2014). In addition to this traditional visual cycle, cone photopigment may be regenerated through an alternate faster pathway involving Müller cells (Mata et al., 2002; Muniz et al., 2007; Wang and Kefalov, 2011, 2009). In this case, 11-*cis*-retinol is transported from the Müller cells into cone inner segments, which are able to convert it into 11-*cis*-retinal. Evidence suggests that this conversion is not possible in rods (Ala-Laurila et al., 2009; Jones et al., 1989). As the concentration of retinoids changes with visual stimulation, the intensity of emitted fluorescence will vary accordingly. Disruption of the visual cycle may be measured as a variation in the fluorescence intensity of the photoreceptors or RPE in response to light stimulation. In healthy humans, the recovery of sensitivity when light levels are abruptly reduced (e.g., temporarily reduced vision when walking into a darkened theater) is partly governed by the rate of the visual cycle and is known to become slower with age (Jackson et al., 1999). Many blindness-causing diseases are linked to defects in the visual cycle, such as Stargardt macular degeneration, Leber congenital amaurosis, and retinitis pigmentosa (Palczewski and Baehr, 2005).

Not every all-*trans*-retinal molecule completes the visual cycle and is regenerated back into photopigment. In the outer segment, a molecule of all-*trans*-retinal can react with phosphatidylethanolamine (PE) and then a second all-*trans*-retinal to form A2-PE (Sparrow et al., 2010). This in turn leads to the formation of bisretinoids, such as A2E. Through the visual cycle and phagocytosis, the bisretinoids accumulate with age in the RPE. They combine with modified lipids to form lipofuscin granules located in the lysosomes of the RPE (Bazan et al., 1990; Ng et al., 2008; Sparrow et al., 2010). Apart from all-*trans*-retinal, it has been shown that 11-*cis*-retinal molecules can also combine to eventually form lipofuscin (Boyer et al., 2012). These granules are fluorescent mostly due to the presence of bisretinoids, including A2E, isomers of A2E, A2-DHP-PE (Wu et al., 2009), and the all-*trans*-retinal dimer series (Fishkin et al., 2005; Parish et al., 1998).

Retinoids and bisretinoids have differing fluorescence spectra. Two-photon fluorescence from lipofuscin is often excited using 900–950 nm, with emission collection above 600 nm. With single-photon fluorescence, the *in vivo* human excitation peaks near 510 nm (Delori et al., 1995). The peak emission for lipofuscin is near 630 nm with a broad 167 nm FWHM (Delori et al., 1995). The excitation and emission spectra for retinol and retinyl ester peak at shorter wavelengths and are narrower. Retinol and retinyl ester have identical excitation and emission spectra. The two-photon excitation cross section decreases from 700 nm with a very weak secondary maximum near 980 nm (Zipfel et al., 2003a). The emission spectrum has a peak at ~490 nm. NAD(P)H and retinol have similar two-photon excitation cross sections between 700 and 850 nm (Zipfel et al., 2003a) but differ for longer excitation wavelengths. Their emission spectra are also similar; the peak for NAD(P)H is shifted toward blue by about 20 nm relative to the peak of retinol fluorescence, making it difficult but not impossible to spectrally separate their signals (Chen et al., 2005).

9.5.2 *EX VIVO* IMAGING OF RETINOIDS AND LIPOFUSCIN IN HEALTHY RETINA

The strongest source of fluorescence in the frog retina comes from the photoreceptor layer (Lu et al., 2011) and similar findings have been observed in the primate retina (unpublished data). *En face*, each rod and cone in the mosaic can be isolated by a dark outline (Figure 9.4h). At deeper focal planes, cone outer segments were visible and there was no fluorescence from the surrounding space. In frog cones, the connecting cilium between the inner and outer segments was imaged as distinct bright spots by Lu and colleagues (2011). The autofluorescence from cones is brighter than that from rods although this could be a feature of transient photopigment density (Gualda et al., 2010; Lu et al., 2011; Palczewska et al., 2014b).

As discussed in Sections 9.4 and 9.5.1, there are many possible sources of photoreceptor autofluorescence including NAD(P)H (Chen et al., 2005; Han et al., 2007), FAD (Zhao et al., 2007), A2-PE (Zhao et al., 2007), and all-*trans*-retinol (Chen et al., 2005; Sears and Kaplan, 1989). Emission spectra showed peaks consistent with all-*trans*-retinol in outer segments of rods (Chen et al., 2005; Sears and Kaplan, 1989) and cones (Palczewska et al., 2014b). Macaque inner and outer segments excited with 730 nm light showed a peak in the emission spectra at 480 nm (Palczewska et al., 2014b). This is likely because the alternate cone visual cycle involves an influx into the cones of 11-*cis*-retinol, which is also fluorescent (Thomson, 1969). In addition, because the inner segments are packed with mitochondria, a shoulder in the emission spectral profile at 456 nm indicated a contribution from NAD(P)H (Palczewska et al., 2014b).

In the RPE, the distribution of fluorescence has the characteristic honeycomb structure (Figure 9.4i) that has been seen with *ex vivo* (Zinn and Marmor, 1979) and *in vivo* imaging (Gray et al., 2006; Morgan et al., 2009). The nuclei appear dark surrounded by strong fluorescence from granular-like structures of varying dimensions and distribution (Han et al., 2006; Imanishi et al., 2004a). The source of this fluorescence is dependent on excitation wavelength.

By exciting at 730 nm, retinyl ester located in retinosomes (Imanishi et al., 2004a) was imaged in mice. Retinosomes are cylindrical in shape (~7 μm long, <1 μm diameter), oriented axially, and cluster near the cell membrane. Palczewska et al. (2010) measured an emission spectrum that indicated retinoids as the primary source of fluorescence in mouse RPE (Palczewska et al., 2010). A similar spectrum was measured in *ex vivo* unfixed bovine RPE (Orban et al., 2011).

With 800 nm or longer excitation, lipofuscin granules are likely to be a source of fluorescence in RPE cells. Lipofuscin granules in RPE are typically ~1 μm in diameter and reside throughout the cytoplasm, often clustering near the cell membrane in young eyes. Several groups have confirmed in

different species (mouse, human) that the emission spectrum is consistent with that of lipofuscin (445–760 nm, peak near 575 nm) (Bindewald-Wittich et al., 2006; Han et al., 2007, 2006; La Schiazza and Bille, 2008; Palczewska et al., 2010; Zhao et al., 2007). In aged human retinas (>55 years), abnormally large (>2 μm diameter) lipofuscin granules with blue-shifted emission spectra (peak at ~520 nm) were observed (Bindewald-Wittich et al., 2006; Han et al., 2006, 2007). These may correspond to granules comprised of lipofuscin and melanin (Han et al, 2007).

9.5.3 *IN VIVO* TWO-PHOTON IMAGING OF RETINOIDS AND LIPOFUSCIN

Using a two-photon fluorescence microscope to image through the sclera of the living mouse with 730 nm excitation, Imanishi and colleagues (2004b) imaged retinosomes in the RPE of albino mice. Consistent with *ex vivo* images, the retinosomes appeared as distinct columns of fluorescence near the RPE cell membrane and the nuclei appeared dark.

In mice, imaging through the pupil has been performed with the use of a Leica DM6000 microscope and contact lens to null the ocular aberrations introduced by the corneal surface (Palczewska et al., 2014a). Wavefront sensorless AO was used to further correct the eyes aberrations and improve image quality. Lipofuscin and retinosomes using 850 and 730 nm excitation, respectively, were imaged in different mutant mouse models of disease.

In 2011, Hunter and co-authors demonstrated the first two-photon fluorescence retinal imaging through the pupil of the living eye. They imaged autofluorescence from the cone mosaic in macaques. Based on *ex vivo* imaging, they hypothesized that the origin was in the cone inner segments. Although not identified experimentally, the fluorophores were likely retinoids and/or NAD(P)H. At that time, 12 min of integration time, corresponding to over 16,500 frames, were required to generate a single averaged image. The fluorescence emission collection corresponded to approximately 12 photons per frame. With recent improvements in imaging instrumentation, the same group has reduced the acquisition time to approximately 2 min (Sharma et al., 2016a,b). Autofluorescence images resolving both rods and cones can now be routinely acquired with 730 nm excitation (Figure 9.5).

9.5.4 IMAGING OF VISUAL CYCLE FUNCTION IN HEALTHY AND DISEASED RETINA

Beyond imaging structure, two-photon fluorescence imaging has been used to track functional changes in photoreceptors and RPE. In response to visual stimulation, retinoid fluorescence in photoreceptor outer segments has been reported to initially increase (Chen and colleagues, 2005). Upon exposure to 730 nm light, fluorescence intensity from photoreceptors has been shown to increase in the living macaque eye (Hunter et al., 2011; Sharma et al., 2016a). This result has been confirmed in *ex vivo* monkey retina (Palczewska et al., 2014b) where a 25% increase in rod emission and doubling of cone autofluorescence was reported. Imanishi and colleagues (2004) observed changes in the size and fluorescence intensity of RPE retinosomes over the course of 30 min following photoreceptor bleaching. This corresponded to the time course of all-*trans*-retinol and all-*trans*-retinyl ester production measured using HPLC (Imanishi et al., 2004a).

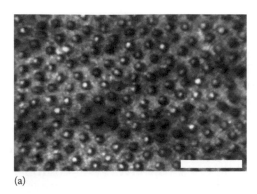 

(a)   (b)

Figure 9.5 (a) Small patch of photoreceptor mosaic imaged in the two-photon AOSLO. (a) Image of rods and cones in the peripheral retina, collected in reflectance using 790 nm light. (b) Two-photon autofluorescence image of the same mosaic of cells. Scale bar: 50 μm.

Retinal diseases involving the visual cycle can impact the concentration and distribution of retinoids and lipofuscin in the photoreceptors and RPE. Manipulating the visual cycle in mice either with drugs or transgenic models has been shown to alter the amount and type of fluorophore. In a model of Leber congenital amaurosis (RPE65$^{-/-}$), mice are unable to synthesize 11-*cis*-retinol. These mice showed enlarged retinosomes compared to wild-type mice, no change in fluorescence in response to flashes of visible light, and no lipofuscin fluorescence (Imanishi et al, 2004; Palczewska et al, 2014). A mouse model of Stargardt disease and AMD (Abca4$^{-/-}$Rdh8$^{-/-}$) presents insufficient clearance of all-*trans*-retinal and an excess accumulation of lipofuscin (Maeda et al., 2005). Days following exposure to bright light, two-photon fluorescence imaging showed enlargement of rod outer segments and overaccumulation of fluorescent granules in RPE (Maeda et al., 2005; Palczewska et al., 2014). Pretreatment with retinylamine, a retinoid cycle inhibitor, prevented this lipofuscin accumulation in RPE of these mice (Palczewska et al., 2014). Such investigations may accelerate the drug development cycle as retinoids have been suggested as a potential therapeutic target (Travis et al., 2007).

9.6 IMAGING COLLAGEN AND ELASTIN

9.6.1 COLLAGEN AND ELASTIN

The eye, just like many other organs, is comprised of a large number of highly organized scaffolds of collagens and collagen-related matrix that allow the eye to maintain its specific shape. The size and spacing between the collagen fibrils in the anatomy is related to their specific function within the tissue (Myllyharju and Kivirikko, 2001; Prockop and Kivirikko, 1995). Formation of collagen in the body is a sequential, multistep process. The smallest molecular subunits of collagen fibers are amino acids, which make up polypeptide chains, which are synthesized within cells. Three such chains combine to form a triple-stranded helix known as tropocollagen. Outside the cell, many such tropocollagen molecules assemble in a staggered configuration to create the various types of collagens that are known to exist. This has been reviewed in more detail elsewhere (Bornstein, 1974).

More than 20 different types of collagens are known to exist, and in the eye, they are present primarily in the cornea, sclera,

lens, trabecular meshwork, choroid, and the vitreous (Bailey, 1987; Ihanamäki et al., 2004; Marshall et al., 1993). In the retina collagen is present in the basement membranes near the RPE and in the walls of small and large vessels. Collagen is also present in Bruch's membrane, in the choroid and in the optic nerve head in the lamina cribrosa, and the central retinal artery.

The inner lining of all blood vessels is composed of endothelia cells, which make up the intima (or innermost) layer. This layer is ensheathed by the basement membrane and the extracellular matrix. The basement membrane contains collagen (type IV, among others) as well as glycoproteins and proteoglycans such as laminin and fibronectin. The extracellular matrix of the media is made up of elastin fibers and smooth muscle cells. The media is ensheathed by the adventitia that is composed of collagen fibers that provide tensile strength and elasticity. Immunohistochemistry studies have shown that microvessels contain collagens of type I and IV, as well as laminin, whereas larger vessels are composed of I, III, IV and V collagen, laminin, and fibronectin (Das et al., 1990; Essner and Lin, 1988; Jerdan and Glaser, 1986).

Collagens play an extremely important role in maintaining morphological integrity and defects in their production can lead to pathological changes (Bailey, 1987; Myllyharju and Kivirikko, 2001; Prockop and Kivirikko, 1995). For the eye, changes in collagen production in basement membrane near the RPE and in Bruch's membrane in the macula during aging have been reported and are potentially linked to age-related changes in the macula (Marshall et al., 1994).

For vessels in particular, collagen plays a crucial role in providing mechanical support. In diseases such as diabetic retinopathy, a thickening of the basement membrane is observed around the vessel walls (Bloodworth et al., 1970; Steinle et al., 2009; Yamashita and Becker, 1961). This might be linked to the breakdown of the blood–brain barrier and increased vessel permeability.

Apart from collagen, there are several other molecules that comprise the extracellular matrix in soft tissue such as elastin, fibrillin, fibronectin, and laminin (Halper and Kjaer, 2014; Mecham, 2001). Elastin lamellas are present in the medial layer in vessel walls and are more malleable. Elastin is a major component of the elastic fiber present in the medial layer in arteries and is present in greater quantities than other such molecules especially in large elastic arteries, where elastin and collagen are the dominant components of the extracellular matrix. Elastin provides the flexibility for the vessel to deform in response to blood flow, while collagen provides the necessary stiffness to keep this deformation under check (Wagenseil and Mecham, 2012). They also play a role in cell adhesion, migration, and survival (Halper and Kjaer, 2014).

Due to their highly organized and regular structure, SHG imaging of collagen fibers has been shown in many types of connective tissues (Freund et al., 1986) although collagen is also known to be a source of autofluorescence (Fujimoto, 1977). Elastin is also autofluorescent (Blomfield and Farrar, 1969) and the emission spectral band is the same as that for collagen. Multiphoton imaging of arteries enables segregation of these two different types of proteins because collagen offers strong SHG signals while elastin does not (Zoumi et al., 2002).

Spectroscopic absorption studies conducted in elastin and collagen powder have shown that their single-photon fluorescence excitation spectra have multiple peaks, due to their complex molecular structure. The excitation maximas for collagen are 265, 280, 330, and 450 nm, while elastin peaks at 350, 410, and 450 nm (Richards-Kortum and Sevick-Muraca, 1996). In tissue, two-photon fluorescence from collagen/elastin has been excited at different wavelengths ranging from 730 to 880 nm (Tseng et al., 2010; Zoumi et al., 2002). In blood vessels, the media has been shown to exhibit stronger fluorescence than the adventitia, due to the presence of elastin (Zoumi et al., 2002).

Vessel walls contain other possible sources of autofluorescence as well. Advanced glycation end-products (AGEs) form as a result of glycation of proteins that are present within the vessels. This is a natural process that happens with age and can be accelerated in disease. These AGEs are also known to be fluorescent and the increase in fluorescence with glycation is more severe for collagen than for elastin (Tseng et al., 2010). Collagen is more responsive to the formation of AGEs and thus a brighter signal from the collagen layer in vessel walls is indicative of potentially harmful glycation.

9.6.2 EX VIVO IMAGING OF COLLAGEN AND ELASTIN IN RETINA

Throughout the inner retina, there is strong autofluorescence from the walls of blood vessels (Figure 9.4b). Their 3D profile can be reconstructed by imaging at multiple retinal depths (Gualda et al., 2010). Focusing on the apical and distal regions of the vessel, the en face view shows fluorescence across the entire vessel. By focusing within the vessel, the walls are well-defined with reduced fluorescence inside the vessel.

Multiphoton imaging of collagen has been conducted in other parts of the retina as well. The choroid and sclera exhibit long wavelength (560–700 nm) emission two-photon fluorescence (Imanishi et al., 2004a). In the choroid, fluorescence, possibly from elastin and extracellular deposits, revealed the choriocapillaris and intercapillary pillars (Han et al., 2007). As seen in ex vivo human retina, two-photon fluorescence from Bruch's membrane appears as an indistinct, modeled pattern. The source of fluorescence is likely elastin (Han et al., 2007).

In the retina, SHG imaging of the network of collagenous fibers that comprise the lamina cribrosa (Agopov et al., 2009; Brown et al., 2007; Winkler et al., 2010) as well as axon fiber bundles in the nerve fiber layer (Lim and Danias, 2012) have been shown. SHG microscopic imaging of the collagen matrix constituting the lamina cribrosa in the optic nerve head clearly delineates the pores through which the ganglion cell axons leave the eye (Agopov et al., 2009; Brown et al., 2007; Winkler et al., 2010). SHG imaging (880 nm light) of collagen in mouse sclera has also been reported (Imanishi et al., 2007).

9.7 FUTURE DIRECTIONS

Both in vivo and ex vivo nonlinear imaging of retinal tissue have the potential to provide insight into normal visual processes and disease mechanisms. By further enhancing the capabilities for two-photon fluorescence imaging, greater insight into the source and functional response of intrinsic fluorophores can be gained. Such knowledge will allow development of nonlinear methods for monitoring disease progression that can aid in the development of new treatment strategies.

The utility of nonlinear imaging will be greatly enhanced when it is possible to deploy it in the living human eye. The greatest obstacle to achieving in vivo two-photon fluorescence imaging in human subjects is ensuring the use of safe light levels. With femtosecond lasers, there is concern for photochemical, mechanical, and thermal damage (reviewed in Boulton et al., 2001; Hunter et al., 2012). Photochemical damage could result as a consequence of two-photon absorption. Given the low efficiency of these effects, except in long duration exposures, it is unlikely to be a major source of retinal damage. Mechanical damage from femtosecond lasers results from laser-induced breakdown as the retinal tissue is ionized. This can lead to formation of plasma, shock waves that burst cells and capillaries, and cause cavitation bubbles. All retinal layers can be affected over an area that is much larger than the area of illumination (Rockwell et al., 2010). The thresholds for mechanical damage are affected by whether or not dispersion within the eye is compensated (Cain et al., 2004). In primates, there has been no visible evidence of mechanical damage as a consequence of typical in vivo two-photon imaging (unpublished data). Thermal damage is a consequence of heating of the retina beyond 10°C, usually via absorption of light by melanin. The use of low light levels will be necessary to avoid such temperature increases. Photoreceptors and RPE are more susceptible to thermal damage than inner retina; therefore, the plane of focus for the two-photon excitation plays an important role in the possibility of retinal damage. Advances in technology are currently being developed to allow two-photon imaging with increased efficiency using reduced light levels. This will be necessary before images can be acquired safely in humans.

In conjunction with the development of high-resolution AO-based ophthalmoscopy, two-photon fluorescence imaging allows access to a variety of key endogenous fluorophores in the living eye. The ability to excite fluorescence is just the first step in the implementation of many downstream, higher-level imaging techniques such as spectroscopic imaging, fluorescence lifetime imaging, and Förster resonance energy transfer imaging. Although not discussed in this review, the use of extrinsic fluorophores can also provide important measures of retinal function. For example, changes in two-photon fluorescence intensity of genetically encoded calcium indicators can be used to measure the activation of retinal neurons (Borghuis et al., 2013, 2011). In the future, the development of such two-photon fluorescence imaging-based systems could provide a unique and detailed way of studying retinal physiology in normal and diseased eyes.

ACKNOWLEDGMENTS

The authors thank David Williams, Grazyna Palczewska, Krzysztof Palczewski, James Feeks, Christina Schwarz, Joynita Sur, Sarah Walters, Tracy Bubel, and Maria Jepson and Linda Callahan of the University of Rochester Multiphoton Core Facility. The funding was provided by the National Institutes of Health under award numbers, R01EY022371, R44AG043645, and P30EY001319, and by Research to Prevent Blindness.

REFERENCES

Agopov, M., L. Lomb, O. La Schiazza, and J. F. Bille. 2009. Second harmonic generation imaging of the pig lamina cribrosa using a scanning laser ophthalmoscope-based microscope. *Lasers in Medical Science* 24 (5): 787–792. doi:10.1007/s10103-008-0641-4.

Ala-Laurila, P., M. Carter Cornwall, R. K. Crouch, and M. Kono. 2009. The action of 11-cis-retinol on cone opsins and intact cone photoreceptors. *The Journal of Biological Chemistry* 284 (24): 16492–164500. doi:10.1074/jbc.M109.004697.

Alberto, D. and G. Chirico. 2003. Two-photon excitation microscopy. *Advances in Imaging and Electron Physics* 126: 202–212. doi:10.1016/S1076-5670(03)80016-2.

Andresen, V., A. Egner, and S. W. Hell. 2001. Time-multiplexed multifocal multiphoton microscope. *Optics Letters* 26 (2): 75–77. doi:10.1364/OL.26.000075.

Bailey, A. J. 1987. Structure, function and ageing of the collagens of the. *Eye* 1 (2): 175–183. doi:10.1038/eye.1987.34.

Bazan, H. E., N. G. Bazan, L. Feeney-Burns, and E. R. Berman. 1990. Lipids in human lipofuscin-enriched subcellular fractions of two age populations. Comparison with rod outer segments and neural retina. *Investigative Ophthalmology and Visual Science* 31 (8): 1433–1443.

Bewersdorf, J., R. Pick, and S. W. Hell. 1998. Multifocal multiphoton microscopy. *Optics Letters* 23 (9): 655–657. doi:10.1364/OL.23.000655.

Bindewald-Wittich, A., M. Han, S. Schmitz-Valckenberg, S. R. Snyder, G. Giese, J. F. Bille, and F. G. Holz. 2006. Two-photon-excited fluorescence imaging of human rpe cells with a femtosecond Ti:sapphire laser. *Investigative Ophthalmology and Visual Science* 47 (10): 4553–4557. doi:10.1167/iovs.05-1562.

Bloembergen, N. 1965. *Nonlinear Optics.* Addison-Wesley Pub. Co., Redwood City, CA.

Blomfield, J. and J. F. Farrar. 1969. The fluorescent properties of maturing arterial elastin. *Cardiovascular Research* 3 (2): 161–170.

Bloodworth, J. M., R. L. Engerman, R. A. Camerini-Dávalos, and K. L. Powers. 1970. Variations in capillary basement membrane width produced by aging and diabetes mellitus. *Advances in Metabolic Disorders* 1 (Suppl 1):279.

Borghuis, B. G., J. S. Marvin, L. L. Looger, and J. B. Demb. 2013. Two-photon imaging of nonlinear glutamate release dynamics at bipolar cell synapses in the mouse retina. *The Journal of Neuroscience* 33 (27): 10972–10985. doi:10.1523/JNEUROSCI.1241-13.2013.

Borghuis, B. G., L. Tian, Y. Xu et al. 2011. Imaging light responses of targeted neuron populations in the rodent retina. *The Journal of Neuroscience* 31 (8): 2855–2867. doi:10.1523/JNEUROSCI.6064-10.2011.

Bornstein, P. 1974. The biosynthesis of collagen. *Annual Review of Biochemistry* 43 (1): 567–603. doi:10.1146/annurev. bi.43.070174.003031.

Bor, Z. 1988. Distortion of femtosecond laser pulses in lenses and lens systems. *Journal of Modern Optics* 35 (12): 1907–1918. doi:10.1080/713822325.

Boulton, M., M. Różanowska, and B. Różanowski. 2001. Retinal photodamage. *Journal of Photochemistry and Photobiology B: Biology, ESP Conference on Photoprotection* 64 (2–3): 144–161. doi:10.1016/S1011-1344(01)00227-5.

Boyd, R. W. 2008. *Nonlinear Optics*, 3rd edn. Academic Press, San Diego, CA.

Boyer, N. P., D. Higbee, M. B. Currin et al. 2012. Lipofuscin and N-retinylidene-N-retinylethanolamine (A2E) accumulate in retinal pigment epithelium in absence of light exposure: Their origin Is 11-cis-retinal. *The Journal of Biological Chemistry* 287 (26): 22276–22286. doi:10.1074/jbc.M111.329235.

Bradley, D. J., M. H. R. Hutchinson, H. Koetser, T. Morrow, G. H. C. New, and M. S. Petty. 1972. Interactions of picosecond laser puls with organic molecules. I. Two-photon fluorescence quenching ar singlet states excitation in rhodamine dyes. *Proceedings of the Roya Society of London A: Mathematical, Physical and Engineering Scien* 328 (1572): 97–121. doi:10.1098/rspa.1972.0071.

Brakenhoff, G. J., M. Müller, and R. I. Ghauharali. 1996. Analysis of efficiency of two-photon versus single-photon absorption for fluorescence generation in biological objects. *Journal of Microscop* 183 (2): 140–144. doi:10.1046/j.1365-2818.1996.870647.x.

Brown, D. J., N. Morishige, A. Neekhra, D. S. Minckler, and J. V. Jester. 2007. Application of second harmonic imaging microscopy to assess structural changes in optic nerve head structure ex vivo. *Journal of Biomedical Optics* 12 (2): 024029-1–024029-5. doi:10.1117/1.2717540.

Bueno, J. M., R. Palacios, A. Giakoumaki, E. J. Gualda, F. Schaeffel, and P. Artal. 2014. Retinal cell imaging in myopic chickens usin adaptive optics multiphoton microscopy. *Biomedical Optics Expre* 5 (3): 664–674. doi:10.1364/BOE.5.000664.

Buist, A. H., M. Müller, J. A. Squier, and G. J. Brakenhoff. 1998. Real time two-photon absorption microscopy using multi point excitation. *Journal of Microscopy* 192 (2): 217–226. doi:10.1046/j.1365-2818.1998.00431.x.

Cain, C. P., R. J. Thomas, G. D. Noojin et al. 2004. Sub-50-fs laser retinal damage thresholds in primate eyes with group velocity dispersion, self-focusing and low-density plasmas. *Graefe's Archi for Clinical and Experimental Ophthalmology* 243 (2): 101–112. doi:10.1007/s00417-004-0924-9.

Campagnola, P. 2011. Second harmonic generation imaging microscop Applications to diseases diagnostics. *Analytical Chemistry* 83 (9): 3224–3231. doi:10.1021/ac1032325.

Campagnola, P. J. and L. M. Loew. 2003. Second-harmonic imaging microscopy for visualizing biomolecular arrays in cells, tissues ar organisms. *Nature Biotechnology* 21 (11): 1356–1360. doi:10.103! nbt894.

Carriles, R., D. N. Schafer, K. E. Sheetz et al. 2009. Invited review article: imaging techniques for harmonic and multiphoton absorption fluorescence microscopy. *The Review of Scientific Instruments* 80 (8): 081101. doi:10.1063/1.3184828.

Chen, C., E. Tsina, M. Carter Cornwall, R. K Crouch, S. Vijayaraghavan, and Y. Koutalos. 2005. Reduction of all-trans retinal to all-trans retinol in the outer segments of frog and mouse rod photoreceptors. *Biophysical Journal* 88 (3): 2278–2287 doi:10.1529/biophysj.104.054254.

Cicchi, R., N. Vogler, D. Kapsokalyvas, B. Dietzek, J. Popp, and F. S. Pavone. 2013. From molecular structure to tissue architectur Collagen organization probed by SHG microscopy. *Journal of Biophotonics* 6 (2): 129–142. doi:10.1002/jbio.201200092.

Das, A., R. N. Frank, N. L. Zhang, and T. J. Turczyn. 1990. Ultrastructural localization of extracellular matrix components in human retinal vessels and Bruch's membrane. *Archives of Ophthalmology* 108 (3): 421–429. doi:10.1001/archopht.1990.01070050119045.

DeLint, P. J., T. T. Berendschot, J. van de Kraats, and D. van Norren. 2000. Slow optical changes in human photoreceptors induced by light. *Investigative Ophthalmology and Visual Science* 41 (1): 282–289.

Delori, F. C., C. K. Dorey, G. Staurenghi, O. Arend, D. G. Goger, and J. J. Weiter. 1995. In vivo fluorescence of the ocular fundus exhibits retinal pigment epithelium lipofuscin characteristics. *Investigative Ophthalmology and Visual Science* 36 (3): 718–729.

Denk, W., D. W. Piston, and W. W. Webb. 1995. Two-photon molecular excitation in laser-scanning microscopy. In *Handbook of Biological Confocal Microscopy*, ed. J. B. Pawley, pp. 445–458. Springer, New York. http://link.springer.com/chapter/10.1007/978-1-4757-5348-6_28.

Denk, W. and K. Svoboda. 1997. Photon upmanship: Why multiphoton imaging is more than a gimmick. *Neuron* 18 (3): 351–357. doi:10.1016/S0896-6273(00)81237-4.

Denk, W., J. H. Strickler, and W. W. Webb. 1990. Two-photon laser scanning fluorescence microscopy. *Science* 248 (4951): 73–76.

Diaspro, A., G. Chirico, and M. Collini. 2005. Two-photon fluorescence excitation and related techniques in biological microscopy. *Quarterly Reviews of Biophysics* 38 (2): 97–166. doi:10.1017/S0033583505004129.

Dillon, J., L. Zheng, J. C. Merriam, and E. R. Gaillard. 2007. Transmission spectra of light to the mammalian retina. *Photochemistry and Photobiology* 71 (2): 225–229. doi:10.1562/0031-8655(2000)0710225TSOLTT2.0.CO2.

Dubra, A. and Y. Sulai. 2011. reflective afocal broadband adaptive optics scanning ophthalmoscope. *Biomedical Optics Express* 2 (6): 1757–1768. doi:10.1364/BOE.2.001757.

Dubra, A., Y. Sulai, J. L Norris et al. 2011. Noninvasive imaging of the human rod photoreceptor mosaic using a confocal adaptive optics scanning ophthalmoscope. *Biomedical Optics Express* 2 (7): 1864–1876. doi:10.1364/BOE.2.001864.

Egner, A. and S. W. Hell. 2000. Time multiplexing and parallelization in multifocal multiphoton microscopy. *Journal of the Optical Society of America A* 17 (7): 1192–1201. doi:10.1364/JOSAA.17.001192.

Essner, E. and W.-L. Lin. 1988. Immunocytochemical localization of laminin, type IV collagen and fibronectin in rat retinal vessels. *Experimental Eye Research* 47 (2): 317–327. doi:10.1016/0014-4835(88)90014-0.

Fishkin, N. E., J. R. Sparrow, R. Allikmets, and K. Nakanishi. 2005. Isolation and characterization of a retinal pigment epithelial cell fluorophore: An all-trans-retinal dimer conjugate. *Proceedings of the National Academy of Sciences of the United States of America* 102 (20): 7091–7096. doi:10.1073/pnas.0501266102.

Fittinghoff, D., P. Wiseman, and J. Squier. 2000. Widefield multiphoton and temporally decorrelated multifocal multiphoton microscopy. *Optics Express* 7 (8): 273–279. doi:10.1364/OE.7.000273.

Franken, P. A., A. E. Hill, C. W. Peters, and G. Weinreich. 1961. Generation of optical harmonics. *Physical Review Letters* 7 (4): 118–119. doi:10.1103/PhysRevLett.7.118.

Freund, I., M. Deutsch, and A. Sprecher. 1986. connective tissue polarity. optical second-harmonic microscopy, crossed-beam summation, and small-angle scattering in rat-tail tendon. *Biophysical Journal* 50 (4): 693–712. doi:10.1016/S0006-3495(86)83510-X.

Fujimoto, D. 1977. Isolation and characterization of a fluorescent material in bovine Achilles tendon collagen. *Biochemical and Biophysical Research Communications* 76 (4): 1124–1129.

Gannaway, J. N. and C. J. R. Sheppard. 1978. Second-harmonic imaging in the scanning optical microscope. *Optical and Quantum Electronics* 10 (5): 435–439. doi:10.1007/BF00620308.

Garwin, G. G. and J. C. Saari. 2000. High-performance liquid chromatography analysis of visual cycle retinoids. *Methods in Enzymology* 316: 313–324.

Geng, Y., A. Dubra, L. Yin et al. 2012. Adaptive optics retinal imaging in the living mouse eye. *Biomedical Optics Express* 3 (4): 715–734. doi:10.1364/BOE.3.000715.

Geng, Y., K. P. Greenberg, R. Wolfe et al. 2009. In vivo imaging of microscopic structures in the rat retina. *Investigative Ophthalmology and Visual Science* 50 (12): 5872–5879. doi:10.1167/iovs.09-3675.

Golczak, M., Y. Imanishi, V. Kuksa, T. Maeda, R. Kubota, and K. Palczewski. 2005. Lecithin:retinol acyltransferase is responsible for amidation of retinylamine, a potent inhibitor of the retinoid cycle. *The Journal of Biological Chemistry* 280 (51): 42263–42273. doi:10.1074/jbc.M509351200.

Göppert-Mayer, M. 1931. Über elementarakte mit zwei quantensprüngen. *Annalen Der Physik* 401 (3): 273–294. doi:10.1002/andp.19314010303.

Gray, D. C., W. Merigan, J. I. Wolfing et al. 2006. In vivo fluorescence imaging of primate retinal ganglion cells and retinal pigment epithelial cells. *Optics Express* 14 (16): 7144–7158. doi:10.1364/OE.14.007144.

Grieve, K. and A. Roorda. 2008. Intrinsic signals from human cone photoreceptors. *Investigative Ophthalmology and Visual Science* 49 (2): 713–719. doi:10.1167/iovs.07-0837.

Gualda, E. J., J. M. Bueno, and P. Artal. 2010. Wavefront optimized nonlinear microscopy of ex vivo human retinas. *Journal of Biomedical Optics* 15 (2): 026007. doi:10.1117/1.3369001.

Guevara-Torres, A., D. R. Williams, and J. B. Schallek. 2015. Imaging translucent cell bodies in the living mouse retina without contrast agents. *Biomedical Optics Express* 6 (6): 2106. doi:10.1364/BOE.6.002106.

Gu, M. 1996. *Principles of Three Dimensional Imaging in Confocal Microscopes*. World Scientific, Singapore.

Halper, J. and M. Kjaer. 2014. Basic components of connective tissues and extracellular matrix: Elastin, fibrillin, fibulins, fibrinogen, fibronectin, laminin, tenascins and thrombospondins. *Advances in Experimental Medicine and Biology* 802: 31–47. doi:10.1007/978-94-007-7893-1_3.

Han, M., A. Bindewald-Wittich, F. G. Holz et al. 2006. Two-photon excited autofluorescence imaging of human retinal pigment epithelial cells. *Journal of Biomedical Optics* 11 (1): 010501. doi:10.1117/1.2171649.

Han, M., G. Giese, S. Schmitz-Valckenberg et al. 2007. Age-related structural abnormalities in the human retina-choroid complex revealed by two-photon excited autofluorescence imaging. *Journal of Biomedical Optics* 12 (2): 024012. doi:10.1117/1.2717522.

Harrison, D. E. and B. Chance. 1970. Fluorimetric technique for monitoring changes in the level of reduced nicotinamide nucleotides in continuous cultures of microorganisms. *Applied Microbiology* 19 (3): 446–450.

Helmholtz, H. 1851. *Beschreibung Eines Augen-Spiegels*. Springer, Berlin, Germany. http://link.springer.com/10.1007/978-3-662-41295-4.

Hood, C. and W. A. H. Rushton. 1971. The florida retinal densitometer. *The Journal of Physiology* 217 (1): 213–229.

Hoover, E. E. and J. A. Squier. 2013. Advances in multiphoton microscopy technology. *Nature Photonics* 7 (2): 93–101. doi:10.1038/nphoton.2012.361.

Huang, D., E. A. Swanson, C. P. Lin et al. 1991. Optical coherence tomography. *Science* 254 (5035): 1178–1181. doi:10.1126/science.1957169.

Huang, S., A. A. Heikal, and W. W. Webb. 2002. Two-photon fluorescence spectroscopy and microscopy of NAD(P)H and flavoprotein. *Biophysical Journal* 82 (5): 2811–2825. doi:10.1016/S0006-3495(02)75621-X.

Hunter, J. J., B. Masella, A. Dubra et al. 2011. Images of photoreceptors in living primate eyes using adaptive optics two-photon ophthalmoscopy. *Biomedical Optics Express* 2 (1): 139–148.

Hunter, J. J., J. I. W. Morgan, W. H. Merigan, D. H. Sliney, J. R. Sparrow, and D. R. Williams. 2012. The susceptibility of the retina to photochemical damage from visible light. *Progress in Retinal and Eye Research* 31 (1): 28–42. doi:10.1016/j.preteyeres.2011.11.001.

Ihanamäki, T., L. J. Pelliniemi, and E. Vuorio. 2004. Collagens and collagen-related matrix components in the human and mouse eye. *Progress in Retinal and Eye Research* 23 (4): 403–434. doi:10.1016/j.preteyeres.2004.04.002.

Imanishi, Y., M. L Batten, D. W Piston, W. Baehr, and K. Palczewski. 2004a. Noninvasive two-photon imaging reveals retinyl ester storage structures in the eye. *The Journal of Cell Biology* 164 (3): 373–383. doi:10.1083/jcb.200311079.

Imanishi, Y., V. Gerke, and K. Palczewski. 2004b. Retinosomes: New insights into intracellular managing of hydrophobic substances in lipid bodies. *The Journal of Cell Biology* 166 (4): 447–453. doi:10.1083/jcb.200405110.

Imanishi, Y., K. H. Lodowski, and Y. Koutalos. 2007. Two-photon microscopy: shedding light on the chemistry of vision. *Biochemistry* 46 (34): 9674–9684. doi:10.1021/bi701055g.

Jackson, G. R., C. Owsley, and G. McGwin. 1999. Aging and dark adaptation. *Vision Research* 39 (23): 3975–3982.

Jerdan, J. A. and B. M. Glaser. 1986. Retinal microvessel extracellular matrix: An immunofluorescent study. *Investigative Ophthalmology and Visual Science* 27 (2): 194–203.

Jones, G. J., R. K. Crouch, B. Wiggert, M. C. Cornwall, and G. J. Chader. 1989. Retinoid requirements for recovery of sensitivity after visual-pigment bleaching in isolated photoreceptors. *Proceedings of the National Academy of Sciences of the United States of America* 86 (23): 9606–9610.

Kaiser, W. and C. Garrett. 1961. Two-photon excitation in CaF_2:Eu$_2$+. *Physical Review Letters* 7 (6): 229–231. doi:10.1103/PhysRevLett.7.229.

Kempe, M., U. Stamm, B. Wilhelmi, and W. Rudolph. 1992. Spatial and temporal transformation of femtosecond laser pulses by lenses and lens systems. *Journal of the Optical Society of America B* 9 (7): 1158–1165. doi:10.1364/JOSAB.9.001158.

LaComb, R., O. Nadiarnykh, S. S. Townsend, and P. J. Campagnola. 2008. Phase matching considerations in second harmonic generation from tissues: Effects on emission directionality, conversion efficiency and observed morphology. *Optics Communications* 281 (7): 1823–1832. doi:10.1016/j.optcom.2007.10.040.

Lamb, K., D. E. Spence, J. Hong, C. Yelland, and W. Sibbett. 1994. All-solid-state self-mode-locked Ti:sapphire laser. *Optics Letters* 19 (22): 1864.

Lamb, T. D. and E. N. Pugh. 2004. Dark adaptation and the retinoid cycle of vision. *Progress in Retinal and Eye Research* 23 (3): 307–380. doi:10.1016/j.preteyeres.2004.03.001.

Lamb, T. D. and E. N. Pugh. 2006. Phototransduction, dark adaptation, and rhodopsin regeneration the proctor lecture. *Investigative Ophthalmology and Visual Science* 47 (12): 5138–5152. doi:10.1167/iovs.06-0849.

La Schiazza, O. and J. F. Bille. 2008. High-speed two-photon excited autofluorescence imaging of ex vivo human retinal pigment epithelial cells toward age-related macular degeneration diagnostic. *Journal of Biomedical Optics* 13 (6): 064008. doi:10.1117/1.2999607.

Liang, J., D. R. Williams, and D. T. Miller. 1997. Supernormal vision and high-resolution retinal imaging through adaptive optics. *Journal of the Optical Society of America A* 14 (11): 2884–2892. doi:10.1364/JOSAA.14.002884.

Li, L. Z. 2012. Imaging mitochondrial redox potential and its possible link to tumor metastatic potential. *Journal of Bioenergetics and Biomembranes* 44 (6): 645–653. doi:10.1007/s10863-012-9469-5.

Lim, H. and J. Danias. 2012. Effect of axonal micro-tubules on the morphology of retinal nerve fibers studied by second-harmonic generation. *Journal of Biomedical Optics* 17 (11). doi:10.1117/1.JBO.17.11.110502.

Lu, R.-W., Y.-C. Li, T. Ye, C. Strang, K. Keyser, C. A. Curcio, and X.-C. Yao. 2011. Two-photon excited autofluorescence imaging of freshly isolated frog retinas. *Biomedical Optics Express* 2 (6): 1494–1503. doi:10.1364/BOE.2.001494.

Marshall, G. E., A. G. Konstas, and W. R. Lee. 1993. Collagens in ocular tissues. *The British Journal of Ophthalmology* 77 (8): 515–524.

Marshall, G. E., A. G. Konstas, G. G. Reid, J. G. Edwards, and W. R. Lee. 1994. Collagens in the aged human macula. *Graefe's Archive for Clinical and Experimental Ophthalmology = Albrecht Von Graefes Archiv Für Klinische Und Experimentelle Ophthalmologie* 232 (3): 133–140.

Masella, B. D., J. J. Hunter, and D. R. Williams. 2014. New wrinkles in retinal densitometry. *Investigative Ophthalmology and Visual Science* 55 (11): 7525–7534. doi:10.1167/iovs.13-13795.

Masters, B. R. and P. So. 2008. *Handbook of Biomedical Nonlinear Optical Microscopy.* 1st edn. Oxford University Press, New York.

Mata, N. L., R. A. Radu, R. C. Clemmons, and G. H. Travis. 2002. Isomerization and oxidation of vitamin a in cone-dominant retinas: A novel pathway for visual-pigment regeneration in daylight. *Neuron* 36 (1): 69–80.

McBee, J. K., K. Palczewski, W. Baehr, and D. R. Pepperberg. 2001. Confronting complexity: The interlink of phototransduction and retinoid metabolism in the vertebrate retina. *Progress in Retinal and Eye Research* 20 (4): 469–529.

Mecham, R. P. 2001. Overview of extracellular matrix. *Current Protocol in Cell Biology.* Chapter 10, Unit 1. More information here: https://www.ncbi.nlm.nih.gov/pubmed/18228295.

Mertz, J. and L. Moreaux. 2001. Second-harmonic generation by focused excitation of inhomogeneously distributed scatterers. *Optics Communications* 196 (1–6): 325–330. doi:10.1016/S0030-4018(01)01403-1.

Min, G. and C. J. R. Sheppard. 1995. Comparison of three-dimensional imaging properties between two-photon and single-photon fluorescence microscopy. *Journal of Microscopy* 177 (2): 128–137. doi:10.1111/j.1365-2818.1995.tb03543.x.

Moreaux, L., O. Sandre, S. Charpak, M. Blanchard-Desce, and J. Mertz. 2001. Coherent scattering in multi-harmonic light microscopy. *Biophysical Journal* 80 (3): 1568–1574. doi:10.1016/S0006-3495(01)76129-2.

Morgan, J., A. Dubra, R. Wolfe, W. H. Merigan, and D. R. Williams. 2009. In vivo autofluorescence imaging of the human and macaque retinal pigment epithelial cell mosaic. *Investigative Ophthalmology and Visual Science* 50 (3): 1350–1359. doi:10.1167/iovs.08-2618.

Ophthalmic instrumentation

Muniz, A., E. T. Villazana-Espinoza, A. L. Hatch, S. G. Trevino, D. M. Allen, and A. T. C. Tsin. 2007. A novel cone visual cycle in the cone-dominated retina. *Experimental Eye Research* 85 (2): 175–184. doi:10.1016/j.exer.2007.05.003.

Myllyharju, J. and K. I. Kivirikko. 2001. Collagens and collagen-related diseases. *Annals of Medicine* 33 (1): 7–21.

Nakamura, O. 1993. Three-dimensional imaging characteristics of laser scan fluorescence microscopy: Two-photon excitation vs single-photon excitation. *Optik* 93 (1): 39–42.

Ng, K.-P., B. Gugiu, K. Renganathan et al. 2008. Retinal pigment epithelium lipofuscin proteomics. *Molecular and Cellular Proteomics* 7 (7): 1397–1405. doi:10.1074/mcp. M700525-MCP200.

Oheim, M., E. Beaurepaire, E. Chaigneau, J. Mertz, and S. Charpak. 2001. Two-photon microscopy in brain tissue: Parameters influencing the imaging depth. *Journal of Neuroscience Methods* 111 (1): 29–37. doi:10.1016/S0165-0270(01)00438-1.

Oheim, M., D. J. Michael, M. Geisbauer, D. Madsen, and R. H. Chow. 2006. Principles of two-photon excitation fluorescence microscopy and other nonlinear imaging approaches. *Advanced Drug Delivery Reviews* 58 (7): 788–808. doi:10.1016/j. addr.2006.07.005.

Okawa, H., A. P. Sampath, S. B. Laughlin, and G. L. Fain. 2008. ATP consumption by mammalian rod photoreceptors in darkness and in light. *Current Biology* 18 (24): 1917–1921. doi:10.1016/j. cub.2008.10.029.

Orban, T., G. Palczewska, and K. Palczewski. 2011. Retinyl ester storage particles (retinosomes) from the retinal pigmented epithelium resemble lipid droplets in other tissues. *The Journal of Biological Chemistry* 286 (19): 17248–17258. doi:10.1074/jbc. M110.195198.

Oron, D., E. Tal, and Y. Silberberg. 2005. Scanningless depth-resolved microscopy. *Optics Express* 13 (5): 1468–1476. doi:10.1364/ OPEX.13.001468.

Palczewska, G., Z. Dong, M. Golczak et al. 2014a. Noninvasive two-photon microscopy imaging of mouse retina and retinal pigment epithelium through the pupil of the eye. *Nature Medicine* 20 (7): 785–789. doi:10.1038/nm.3590.

Palczewska, G., M. Golczak, D. R. Williams, J. J. Hunter, and K. Palczewski. 2014b. Endogenous fluorophores enable two-photon imaging of the primate eye. *Investigative Ophthalmology and Visual Science* 55 (7): 4438–4447. doi:10.1167/iovs.14-14395.

Palczewska, G., T. Maeda, Y. Imanishi et al. 2010. noninvasive multiphoton fluorescence microscopy resolves retinol and retinal condensation products in mouse eyes. *Nature Medicine* 16 (12): 1444–1449. doi:10.1038/nm.2260.

Palczewski, K. 2014. Chemistry and biology of the initial steps in vision: The friedenwald lecture. *Investigative Ophthalmology and Visual Science* 55 (10): 6651–6672. doi:10.1167/iovs.14-15502.

Palczewski, K. and W. Baehr. 2005. The retinoid cycle and retinal diseases. In *Encyclopedia of Life Sciences*. John Wiley & Sons, Ltd., Chichester, U.K.

Palczewski, K., S. Jager, J. Buczylko et al. 1994. Rod outer segment retinol dehydrogenase: Substrate specificity and role in phototransduction. *Biochemistry* 33 (46): 13741–13750. doi:10.1021/bi00250a027.

Parish, C. A., M. Hashimoto, K. Nakanishi, J. Dillon, and J. Sparrow. 1998. Isolation and one-step preparation of A2E and Iso-A2E, fluorophores from human retinal pigment epithelium. *Proceedings of the National Academy of Sciences of the United States of America* 95 (25): 14609–14613.

Peters, S., M. Hammer, and D. Schweitzer. 2011. Two-photon excited fluorescence microscopy application for ex vivo investigation of ocular fundus samples. *Proceedings of SPIE* 8086:808605-1– 808605-10. doi:10.1117/12.889807.

Prockop, D. J. and K. I. Kivirikko. 1995. Collagens: Molecular biology, diseases, and potentials for therapy. *Annual Review of Biochemistry* 64: 403–434. doi:10.1146/annurev.bi.64.070195.002155.

Pu, Y., W. Wang, G. Tang, and R. R. Alfano. 2010. Changes of collagen and nicotinamide adenine dinucleotide in human cancerous and normal prostate tissues studied using native fluorescence spectroscopy with selective excitation wavelength. *Journal of Biomedical Optics* 15 (4): 047008-1–047008-5. doi:10.1117/1.3463479.

Rao, R. A., M. R. Mehta, S. Leithem, and K. C. Toussaint. 2009. Quantitative analysis of forward and backward second-harmonic images of collagen fibers using Fourier transform second-harmonic-generation microscopy. *Optics Letters* 34 (24): 3779. doi:10.1364/OL.34.003779.

Richards-Kortum, R. and E. Sevick-Muraca. 1996. Quantitative optical spectroscopy for tissue diagnosis. *Annual Review of Physical Chemistry* 47: 555–606. doi:10.1146/annurev. physchem.47.1.555.

Rockwell, B. A., R. J. Thomas, and A. Vogel. 2010. Ultrashort laser pulse retinal damage mechanisms and their impact on thresholds. *Medical Laser Application* 25: 84–92.

Roorda, A., F. Romero-Borja, III Donnelly, H. Queener, T. Hebert, and M. Campbell. 2002. Adaptive optics scanning laser ophthalmoscopy. *Optics Express* 10 (9): 405–412.

Schweitzer, D., S. Schenke, M. Hammer, F. Schweitzer, S. Jentsch, E. Birckner, W. Becker, and A. Bergmann. 2007. Towards metabolic mapping of the human retina. *Microscopy Research and Technique* 70 (5): 410–419. doi:10.1002/jemt.20427.

Sears, R. C. and M. W. Kaplan. 1989. Axial diffusion of retinol in isolated frog rod outer segments following substantial bleaches of visual pigment. *Vision Research* 29 (11): 1485–1492.

Sharma, R., C. Schwarz, G. Palczewska, K. Palczewski, D. R. Williams, and J. J. Hunter. 2015. In vivo two-photon fluorescence kinetics of primate rods and cones during light and dark adaptation. *Investigative Ophthalmology and Visual Science* 56 (7): 5968.

Sharma, R., C. Schwarz, D. R. Williams, G. Palczewska, K. Palczewski, and J. J. Hunter. 2016a. In vivo two-photon fluorescence kinetics of primate rods and cones. *Investigative Ophthalmology and Visual Science* 57(2): 647–657. doi:10.1167/iovs.15-17946.

Sharma, R., D. R. Williams, G. Palczewska, K. Palczewski, and J. J. Hunter. 2016b. Two-photon autofluorescence imaging reveals cellular structures throughout the retina of the living primate eye. *Investigative Ophthalmology and Visual Science* 57 (2): 632–646. doi:10.1167/iovs.15-17961.

Sheppard, C. J. R. and M. Gu. 1990. Image formation in two-photon fluorescence microscopy. *Optik—International Journal for Light and Electron Optics* 86 (3): 104–106.

Skala, M. and N. Ramanujam. 2010. Multiphoton redox ratio imaging for metabolic monitoring in vivo. *Methods in Molecular Biology* 594: 155–162. doi:10.1007/978-1-60761-411-1_11.

So, P. T. C., C. Y. Dong, B. R. Masters, and K. M. Berland. 2000. Two-photon excitation fluorescence microscopy. *Annual Review of Biomedical Engineering* 2 (1): 399–429. doi:10.1146/annurev. bioeng.2.1.399.

Sparrow, J. R., Y. Wu, C. Y. Kim, and J. Zhou. 2010. phospholipid meets all-trans-retinal: The making of RPE bisretinoids. *Journal of Lipid Research* 51 (2): 247–261. doi:10.1194/jlr.R000687.

Spence, D. E., P. N. Kean, and W. Sibbett. 1991. 60-Fsec pulse generation from a self-mode-locked Ti:sapphire laser. *Optics Letters* 16 (1): 42–44. doi:10.1364/OL.16.000042.

Steinle, J. J., T. S. Kern, S. A. Thomas, L. S. McFadyen-Ketchum, and C. P. Smith. 2009. Increased basement membrane thickness, pericyte ghosts, and loss of retinal thickness and cells in dopamine beta hydroxylase knockout mice. *Experimental Eye Research* 88 (6): 1014–1019. doi:10.1016/j.exer.2008.12.015.

Svoboda, K. and R. Yasuda. 2006. Principles of two-photon excitation microscopy and its applications to neuroscience. *Neuron* 50 (6): 823–839. doi:10.1016/j.neuron.2006.05.019.

Tam, J., P. Tiruveedhula, and A. Roorda. 2011. Characterization of single-file flow through human retinal parafoveal capillaries using an adaptive optics scanning laser ophthalmoscope. *Biomedical Optics Express* 2 (4): 781–793. doi:10.1364/BOE.2.000781.

Thomson, A. J. 1969. Fluorescence spectra of some retinyl polyenes. *The Journal of Chemical Physics* 51 (9): 4106–4116.

Travis, G. H., M. Golczak, A. R. Moise, and K. Palczewski. 2007. Diseases caused by defects in the visual cycle: Retinoids as potential therapeutic agents. *Annual Review of Pharmacology and Toxicology* 47: 469–512. doi:10.1146/annurev.pharmtox.47.120505.105225.

Tseng, J.-Y., A. A. Ghazaryan, W. Lo et al. 2010. Multiphoton spectral microscopy for imaging and quantification of tissue glycation. *Biomedical Optics Express* 2 (2): 218–230. doi:10.1364/BOE.2.000218.

van Norren, D. and J. van der Kraats. 1981. A continuously recording retinal densitometer. *Vision Research* 21 (6): 897–905.

Vilupuru, A. S., N. V. Rangaswamy, L. J. Frishman, E. L. Smith, R. S. Harwerth, and A. Roorda. 2007. Adaptive optics scanning laser ophthalmoscopy for in vivo imaging of lamina cribrosa. *Journal of the Optical Society of America. A, Optics, Image Science, and Vision* 24 (5): 1417–1425.

Wagenseil, J. E. and R. P. Mecham. 2012. Elastin in large artery stiffness and hypertension. *Journal of Cardiovascular Translational Research* 5 (3): 264–273. doi:10.1007/s12265-012-9349-8.

Wang, J.-S. and V. J. Kefalov. 2009. An alternative pathway mediates the mouse and human cone visual cycle. *Current Biology* 19 (19): 1665–1669. doi:10.1016/j.cub.2009.07.054.

Wang, J.-S. and V. J. Kefalov. 2011. The cone-specific visual cycle. *Progress in Retinal and Eye Research* 30 (2): 115–128. doi:10.1016/j.preteyeres.2010.11.001.

Weinreb, R. N., T. Aung, and F. A. Medeiros. 2014. The pathophysiology and treatment of glaucoma: A review. *JAMA* 311 (18): 1901–1911. doi:10.1001/jama.2014.3192.

Williams, R. M., W. R. Zipfel, and W. W. Webb. 2005. Interpreting second-harmonic generation images of collagen I fibrils. *Biophysical Journal* 88 (2): 1377–1386. doi:10.1529/biophysj.104.047308.

Winkler, M., B. Jester, C. Nien-Shy et al. 2010. High resolution three-dimensional reconstruction of the collagenous matrix of the human optic nerve head. *Brain Research Bulletin, Advances in Corneal and Retinal Research* 81 (2–3): 339–348. doi:10.1016/j.brainresbull.2009.06.001.

Wong-Riley, M. T. T. 2010. Energy metabolism of the visual system. *E and Brain* 2: 99–116. doi:10.2147/EB.S9078.

Wu, Y., N. E. Fishkin, A. Pande, J. Pande, and J. R. Sparrow. 2009. Novel lipofuscin bisretinoids prominent in human retina and in a model of recessive stargardt disease. *The Journal of Biological Chemistry* 284 (30): 20155–20166. doi:10.1074/jbc.M109.021345.

Xu, C. and W. W. Webb. 1996. Measurement of two-photon excitation cross sections of molecular fluorophores with data from 690 to 1050 nm. *Journal of the Optical Society of America B* 13 (3): 481. doi:10.1364/JOSAB.13.000481.

Yamashita, T. and B. Becker. 1961. The basement membrane in the human diabetic eye. *Diabetes* 10 (3): 167–174. doi:10.2337/diab.10.3.167.

Yin, L., Y. Geng, F. Osakada et al. 2013. Imaging light responses of retinal ganglion cells in the living mouse eye. *Journal of Neurophysiology* 10 (9):2415–2421. February. doi:10.1152/jn.01043.2012.

Yin, L., B. Masella, D. Dalkara et al. 2014. Imaging light responses of foveal ganglion cells in the living macaque eye. *The Journal of Neuroscience* 34 (19): 6596–6605. doi:10.1523/JNEUROSCI.4438-13.2014.

Zhao, L.-L. 2009. Layered-resolved autofluorescence imaging of photoreceptors using two-photon excitation. *Journal of Biomedical Science and Engineering* 02 (05): 363–365. doi:10.4236/jbise.2009.25052.

Zhao, L., J. Qu, and H. Niu. 2007. Identification of endogenous fluorophores in the photoreceptors using autofluorescence spectroscopy. *Proceedings of SPIE* 6826:682614-1–682614-7. doi:10.1117/12.760139.

Zhu, G., J. van Howe, M. Durst, W. Zipfel, and C. Xu. 2005. Simultaneous spatial and temporal focusing of femtosecond pulses. *Optics Express* 13 (6): 2153–2159. doi:10.1364/OPEX.13.002153.

Zinn, K. M. and M. F. Marmor (eds.). 1979. *The Retinal Pigment Epithelium*. Cambridge, MA: Harvard University Press.

Zipfel, W. R., R. M. Williams, R. Christie, A. Yu Nikitin, B. T. Hyman, and W. W Webb. 2003a. Live tissue intrinsic emission microscopy using multiphoton-excited native fluorescence and second harmonic generation. *Proceedings of the National Academy of Sciences of the United States of America* 100 (12): 7075–7080. doi:10.1073/pnas.0832308100.

Zipfel, W. R., R. M. Williams, and W. W. Webb. 2003b. Nonlinear magic: Multiphoton microscopy in the biosciences. *Nature Biotechnology* 21 (11): 1369–1377. doi:10.1038/nbt899.

Zoumi, A., A. Yeh, and B. J. Tromberg. 2002. Imaging cells and extracellular matrix in vivo by using second-harmonic generation and two-photon excited fluorescence. *Proceedings of the National Academy of Sciences of the United States of America* 99 (17): 11014–11019. doi:10.1073/pnas.172368799.

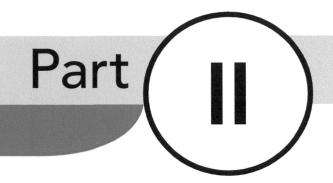

Part **II**

Vision correction

10
Ophthalmic lenses

Daniel Malacara

Contents

10.1 OPHTHALMIC LENSES

Ophthalmic lenses had been used to correct refraction defects in the eye since many centuries ago. The oldest proof of the use of eyeglasses appears in a painting by Tommaso da Modena, an Italian painter of the mid-fourteenth century (ca 1325–1379). The painting represents Cardinal Hugo of Provence (Figure 10.1) who died in 1260. Probably he never used spectacles, but the painting shows that they were used in times of the painter.

The main purpose of an ophthalmic lens is to correct ammetropies, like myopia and hypermetropia, frequently combined with astigmatism, as illustrated in Figure 10.2. They are also used to compensate for deficiencies in the accommodation to observe objects close to the eye. These lenses have the shape of a thin meniscus lens placed in front of the eye, as shown in Figure 10.3. The lens is mounted in a frame that fixes the lens position at a short distance from the eye, such that the vertex of the concave surface of the lens is 14 mm in front of the cornea. The lens forms a virtual image of the object being observed, at the proper distance from the eye to observe it clearly focused. The most important parameter in an ophthalmic lens is the back focal length, which is the distance from the vertex of the concave surface to the focus of the lens. The power of the lens in diopters is the inverse of the back focal length, measured in meters. According to the American Optometric Association Standard, the tolerance in the power of an ophthalmic lens is ± 0.06 diopters. The International Organization for Standardization (ISO) has a complete list of norms for spectacle lenses (ISO 8980-1). They have specified a more relaxed power tolerance that is different for every lens power, where the highest the diopter power, the larger the tolerance can be.

As illustrated in Figure 10.4, an eye without a refraction defect focuses a distant object on the retina of the eye, without eye accommodation, that is, with the eye lens focused for a distant object. If the eye has a refractive defect, the object is focused on the retina only if the distance from the cornea to the image is L, which is a function of the magnitude of the refractive error. If the distance L is positive (behind the eye), the eye is said to be hypermetropic, and if this distance is negative (in front of the eye), the eye is said to be myopic. To form the image at the proper place, a lens with a back focal length F_V is used, such that $F_V = L + d_V$, where d_V is the distance from the vertex of the back surface of the lens to the cornea.

Figure 10.1 Painting by Tommaso da Modena, an Italian painter of the mid-14th century. The painting represents Cardinal Hugo of Provence, who died in 1260.

If the radii of curvature of the surfaces of the lens are measured in meters, the power of the surface in diopters is

$$P = \frac{(n-1)}{r}$$ (10.

This power is measured with an instrument called a dioptometer (Coleman et al., 1951). The power of the frontal (convex) surface is called the base power.

10.1.1 OPHTHALMIC MATERIALS

Ophthalmic lenses were originally made out of glass, but now most of them are made out of plastic. Their main optical characteristics are its refractive index and its Abbe number. The refractive index n can be defined by the Snell law as follows:

$$n = \frac{\sin \theta_1}{\sin \theta_2}$$ (10.

where
  θ_1 is the angle of incidence from vacuum (or air)
  θ_2 is the angle of refraction inside the glass or plastic

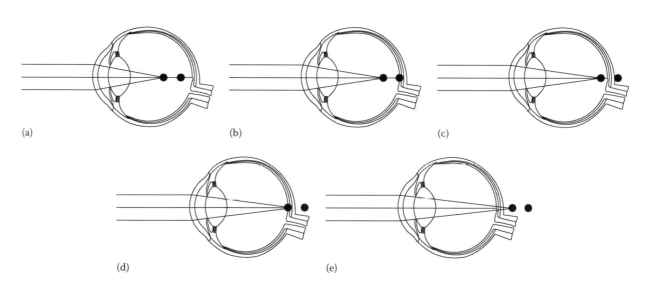

Figure 10.2 Combinations of myopia or hypermetropia and astigmatism.

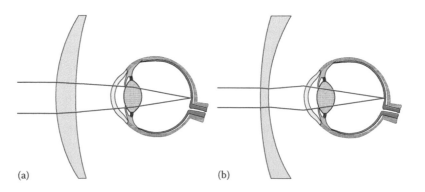

Figure 10.3 Correction of (a) hypermetropia and (b) myopia.

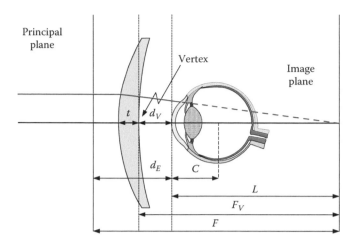

Figure 10.4 Some parameters used in the design of ophthalmic lenses.

The Abbe number is defined as

$$V_d = \frac{n_d - 1}{n_f - n_c} \qquad (10.3)$$

where n_d, n_β and n_c are the refractive indices for yellow, blue, and red light. This number represents the chromatic dispersion of the material. The larger the Abbe number is, the smaller the chromatic dispersion. A large chromatic dispersion introduces chromatic aberration in the image. In this sense, a large Abbe number is desirable.

Another important optical property is its chromatic transmission. The two important consequences of this transmission are its color and its ultraviolet (UV) cutoff wavelength. The color is determined by its chromatic transmission variations in the visible part of the spectrum. The chromatic transmission, even for white optical materials, where its chromatic transmission is nearly constant inside the visible spectrum, decreases fast in the UV region. This is beneficial, since UV radiation can damage the eye in several manners. The wavelength in the UV region, where the light transmission has fallen down to less than 1% for a 2 mm thick lens or glass plate, is called the UV cutoff.

Additional important properties of ophthalmic materials are its density and hardness. The density is important with high diopter powers, where the lens is thick, making it heavy. In this respect, plastic lenses have a great advantage over glass lenses. The hardness is an indication of the resistance to scratches. Glasses are more resistant than plastics.

10.1.1.1 Ophthalmic glasses

The refractive index for the most common ophthalmic glass is 1.523. Their main characteristic is that they are quite resistant to scratches. The main disadvantage is that they can break, but to eliminate this risk they can be hardened with a thermic process, making them even more resistant than most plastic lenses. Most glass manufacturers produce ophthalmic glasses with many different characteristics. Table 10.1 shows some of the most important glasses for ophthalmic purposes.

Besides the white color glasses, there are also tinted glasses in many colors, like gray, brown, green, or pink in several different shadows. Photochromic glasses are made with materials that react to the light, mainly UV, increasing its light absorption and making them either gray or brown. They appeared in 1962 but they are constantly improving. When the UV light disappears, they return to their original clear state.

10.1.1.2 Ophthalmic plastics

Most current ophthalmic lenses are made with plastics due to several factors, mainly that they are safer, since they do not break like glass and they are also lighter. The only disadvantage is that they scratch more easily. The main plastics for ophthalmic uses, whose characteristics are in Table 10.2, are as follows:

1. *CR-39.* This plastic is a polymerizable thermosetting resin, formed by allyl diglycol carbonate. The name comes from Columbia Resin No. 39, because it was the 39th formula of a thermosetting plastic developed by the Columbia Resins Project at the Pittsburgh Plate Glass Company in 1940. Their most important characteristic is a high degree of scratch resistance, but lower than that of glass.
2. *Trivex*, which is a urethane-based prepolymer. PPG named the material Trivex because of its three main performance properties, that is, superior optics, ultra-lightweight, and extreme strength. This plastic has the same UV blocking properties and shatter resistance of polycarbonate. Its low refractive index of 1.532 may result in slightly thicker lenses. It can be easily tinted.
3. *Polycarbonate* is a thermoplastic, lighter than other plastics. This material blocks UV rays and it is shatter resistant and it is better than CR-39, and it is used in sports glasses and glasses for children and teenagers. Because polycarbonate is soft and will scratch easily, scratch-resistant coating is typically applied after shaping and polishing the lens. Standard polycarbonate has a high Abbe value of 30, thus producing lenses with a high chromatic aberration.

Table 10.1 **Some ophthalmic glasses**

GLASS TYPE		V_d	N_c	N_d	N_f	DENSITY (g/cm³)	UV CUTOFF (nm)
Crown (most common)	D 0391, UV-W76, Unicrown 8214	58.8	1.5203	1.5230	1.5292	2.55	347
Crown (UV absorbing)	UV clear	59.7	1.5204	1.5231	1.5292	2.48	405
Flint	D 0290, HC-Weiss 0290	44.1	1.5967	1.6008	1.6103	2.67	348
Flint	D 0785, LaSF 1.8/35	35.0	1.7880	1.7946	1.8107	3.60	352
Flint	D 0082, LaSF 1.9/30	30.6	1.8776	1.8860	1.9066	4.02	363
Flint	Corning 80235	34.0	1.7952	1.8020	1.8184	3.63	363
Low-density flint	(D 0088)	30.8	1.6915	1.7010	1.7154	2.99	

Table 10.2 **Some ophthalmic plastics**

PLASTIC TYPE	V_d	N_d	DENSITY (g/cm³)	UV CUTOFF (nm)
CR-39	59.3	1.498	1.31	355
Trivex	43.0–45.0	1.532	1.1	380
Polycarbonate	30.0	1.586	1.2	385
Thiourethanes (high index)	42.0–32.0	1.600–1.740	1.3–1.5	380

4. *High-index plastics* (*thiourethanes*) are high-index plastics, producing thin lenses, but not lighter, than most plastics due to the high density. As most with high-index plastics, lenses have a high level of chromatic aberration. Another advantage of high-index plastics is their strength and shatter resistance, although not as shatter resistant as polycarbonate.

10.1.1.3 Ophthalmic lenses optical characteristics

Before the advent of plastics, the most common optical material was glass with an index of refraction equal to 1.523. The dioptric power of lenses made with this index of refraction and a normal thickness can be calculated with a reasonable accuracy with the thin lens formula to be described in the following, if an index of refraction equal to 1.53 was assumed; thus, for historical reasons and for practical convenience, the nominal dioptric power of a grinding or polishing tool for ophthalmic lenses, with radius of curvature r, is defined as $P_n = 0.530/r$. Thus, the real power P of the surface polished with a nominal power P_n is

$$P = \frac{(n-1)}{0.530} P_n \qquad (10.4)$$

Then, from Equation 10.3 we may show that the vertex power of the ophthalmic lens, which is defined as the inverse of the back focal length, is

$$P_V = \frac{P_1}{1 - P_1(t/1000n)} + P_2 \qquad (10.5)$$

where the thickness t is in millimeters. To make it simpler, this expression can be approximated by

$$P_V = P_1 + P_2 + \frac{P_1^2 t}{1000n} \qquad (10.6)$$

in order to make easier all hand calculations. With a slightly greater error we may also write

$$P_V = P_1 + P_2 \qquad (10.7)$$

To get a feeling for the error in these formulas, let us consider as an example a glass lens with a dioptric power $P_1 = 9$ D, a front base power $P_2 = -4$ D, and a thickness $t = 4$ mm. Then, we obtain

$P_V = 5.2179$ with the exact formula (10.5)
$P_V = 5.2127$ with the approximate formula (10.6)
$P_V = 5.0000$ with the approximate formula (10.7)

The effective power P_e is the inverse of the effective focal length in meters. Equation 10.3 may be written as

$$P_e = P_1 + \left(1 - \frac{P_1 t}{1000n}\right) P_2 \qquad (10.8)$$

The effective power of the lens in the last example is $P_e = 5.094$. Thus the relation between the effective power and the vertex power is

$$P_e = \left(1 - \frac{P_1 t}{1000n}\right) P_V \qquad (10.9)$$

where we can see that the vertex power and the effective power are almost equal when the power is small.

10.1.2 OPHTHALMIC LENS MAGNIFYING POWER

When an eye is larger than normal but the refractive components have the normal typical values, the image of an object at infinity is defocused on the retina. This refractive defect is called myopia. In hypermetropia, the opposite refractive defect, the eye is shorter than normal. An ophthalmic lens corrects this defect by shifting the real image formed by the optics of the eye to the correct position, a point in space conjugate to the retina. The magnification of the image in the retina is directly proportional to the effective focal length of the whole optical system, including the eyeglasses. If this combination of the optics of the eye with the ophthalmic lens preserves the original effective focal length, the size of the image is also preserved. Thus, the sizes of the images in a normal and in a corrected eye are equal only if the effective focal length is the same. In the case of myopia the elongation of the eye increases the focal length of the eye, but the separation between the eyeglasses and the cornea of the eye decreases the magnification. It may be easily shown that the effective focal length remains almost the same. This is only approximately true, because the average eye has a cornea-to-front principal plane distance of 16.0 mm, whereas a normal spectacle has a 14.5 mm distance.

In a myopic uncorrected eye, the image can change its size in a noticeable manner when the eyeglasses are worn, mainly if the dioptric power is medium or high. This change in magnifying power is given by

$$\Delta M = \left(\frac{1}{1 - (dP_e/1000)} - 1\right) \times 100\% \qquad (10.10)$$

where d is the distance from the principal plane of the lens to the cornea, in millimeters. This equation may be written in terms of the power P_1 of the base and the vertex power P_V as follows:

$$M = \left(\frac{1}{(1 - (dP_V)/1000)(1 - (P_1 t)/1000n)} - 1\right) \times 100\% \qquad (10.11)$$

Often, it is said that the first term in the denominator is due to the power of the lens and the second term to its shape. The effect of the first term is greater than that of the second. As an approximate rule, there is a magnification of about 1.4% for each diopter in the lens.

Another important property of ophthalmic lenses is that the functional or correcting power of the lens depends on its distance to the eye. If the lens-to-cornea distance is increased, the effective power decreases. Let the distance from the lens to the cornea be d_1, with its effective power P_1, and also the distance from the lens to the cornea be d_2, with its effective power P_2. Then, we have

$$\frac{P_2 - P_1}{P_1 P_2} = \frac{d_1 - d_2}{1000} \qquad (10.12)$$

For example, if a 5 D lens is moved 10 mm, the effective power changes by an amount 0.25 D. This effect is quite useful for old myopic people, because due to the presbyopia they can either remove their spectacles or just move them away from the eye, by sliding them along the nose. Thus, if the presbyopia is not extremely large and the myopia is medium or high, this simple shift of the eyeglasses permits the focusing of near objects. A myopic eye corrected with negative spectacles can be thought of as an emmetropic eye with an inverted Galilean telescope in front of it. It has been shown that this inverted telescope increases the depth of field (Malacara and Malacara, 1991), thus making easier near vision for presbyopic and myopic people.

10.2 SPHERICAL AND ASPHERICAL LENSES DESIGN

An optical light ray layout used for ophthalmic lens design is shown in Figure 10.5. The eye has a nearly spherical shape and moves in its cavity to observe objects in different directions. Thus, the stop is at the plane of the eye's pupil, which rotates about the center of rotation of the eye. The actual stop can be represented by an apparent stop located near the center of the eye. It is assumed that all observed objects in different directions are at the same distance from the eye. So, the object surface is spherical, with the center of curvature approximately at the stop. The image formed by the ophthalmic lens should also be at a constant distance for any object in the object surface. Thus, the image surface is also spherical, with the center of curvature at the stop.

The distance C from the cornea of the eye to the center of the eye globe is called the sighting center distance, and it has been found (Fry and Hill, 1962) to be a linear function of the refractive error of the eye, which may be expressed by

$$C = -\frac{P_V}{6} + 14.5 \text{ mm} \qquad (10.13)$$

The distance from the vertex of the lens to the cornea of the eye is not the same for all observers, but has small variations, with an average of about 14 mm. The thickness t is assumed to be constant for negative lenses and increasing linearly with the power for positive lenses, as shown in Figure 10.6. For positive lenses the edge thickness is taken as approximately constant, equal to 1 mm, as given by

$$t = -\frac{D^2 P_V}{8000(n-1)} + 2.0 \text{ mm} \qquad (10.14)$$

Since the vertex power of the lens and the stop position are fixed, the only degree of freedom we have for the correction of aberrations is the lens bending. The spherical aberration and the axial chromatic aberration are not a problem, because the eye pupil's diameter is very small compared with the focal length of the lens. The coma is not very important compared with the astigmatism, because of the large field and the small diameter of the eye's pupil. The remaining aberrations to be corrected are astigmatism, field curvature, distortion, and magnification chromatic aberration. Distortion and magnification chromatic aberrations cannot be corrected by just lens bending. Thus, we are left with the astigmatism and the field curvature, also sometimes called peripheral power error.

Ophthalmic lenses and their design techniques have been described by several authors, for example, Blaker (1983), Emsley (1956), Lueck (1965), Atchison (1985), Malacara and Malacara (1985a,b), Atchison (1992), Walker (2000), and Fowler and Petrre (2001). To design an ophthalmic lens is relatively easy if we plot the curvatures of the Petzval, sagittal, and tangential surfaces as a function of the power of the front lens surface, as shown in Figure 10.7. These curves were obtained for a thin lens ($t = 2$ mm) with a vertex power of three diopters and small field (5°), with a

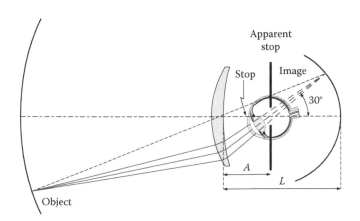

Figure 10.5 Optical schematics of an eye with its ophthalmic lens.

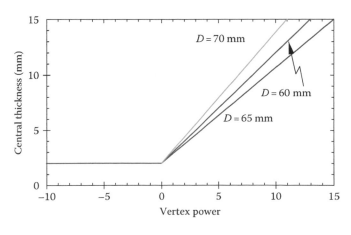

Figure 10.6 Central thickness in an ophthalmic lens.

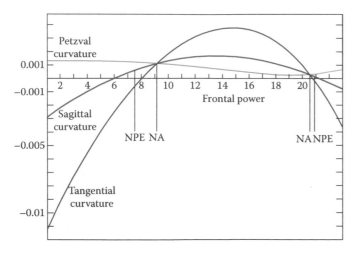

Figure 10.7 Change in the tangential and the sagittal curvatures versus the frontal power in an ophthalmic lens.

lens evaluation program that directly computes these curvatures. If we assume a relatively small field, so that only the primary aberrations are present, the sagittal astigmatism is one-third of the tangential astigmatism. In Figure 10.7 the zero for the vertical scale is the curvature of the focal surface with center of curvature at the stop. We may see small variations in the Petzval curvature because the vertex power is constant, but the effective focal length and hence the Petzval sum are not exactly constant. To analyze astigmatism in ophthalmic lenses with having to do a detailed exact ray tracing can be done by using the Coddington equations, specially modified for ophthalmic lenses, as shown by Landgrave and Moya-Cessa (1996).

The points where the three curves in Figure 10.7 meet are the solutions for no astigmatism and the points where the sagittal and tangential curves are symmetrical with respect to the horizontal axis are the solutions for no power error.

10.2.1 TSCHERNING ELLIPSES

We have seen that by bending we may correct either the astigmatism or the field curvature, but not both simultaneously. The second defect produces a defocusing of the objects observed through the edge of the lens. The observer may refocus the image by accommodation of the eye, but this introduces some eye strain, which may frequently be tolerated, especially by young persons. If the frontal lens power or base is used as a parameter for the bending, we may plot the total vertex power of the lens as a function of the base power that gives a lens without astigmatism and similarly for the peripheral power error. Thus, we obtain two ellipses as shown in Figure 10.8, called Tscherning ellipses. We see that for each ellipse there are two solutions, one with a low frontal power (Ostwald lenses) and the other with a higher frontal power (Wollaston lenses). The Tscherning ellipses were obtained using third-order theory, with a constant very small lens thickness, a very small field, and a constant distance from the vertex of the lens to the stop, equal to 29 mm.

When the restrictions of constant thickness, constant distance from the vertex of the lens to the stop, and small field are removed, the Tscherning ellipses deform as shown in Figure 10.9.

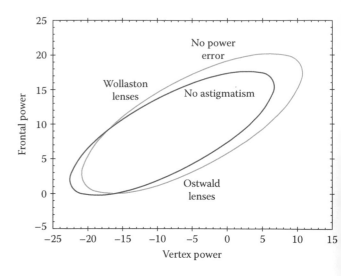

Figure 10.8 Tscherning ellipses for ophthalmic lenses free of astigmatism and power error.

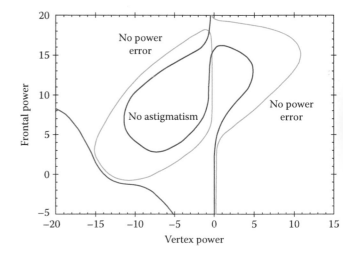

Figure 10.9 Tscherning ellipses deformed by the introduction of a finite lens thickness.

10.2.2 ASPHERICAL LENSES

We may see from the Tscherning ellipses that there are no solution without astigmatism or without peripheral power error for high lens vertex powers. In this case an aspheric surface may be used in the front surface. With aspheric surfaces, the Tscherning ellipses change their shape, extending the solution range to higher powers as shown in Figure 10.10 for the case of astigmatism and in Figure 10.11 for the case of zero power error. Aspherical surfaces for ophthalmic lenses have been studied by Smith and Atchison (1983), by Sun et al. (2000, 2002) using third-order theory, and b Malacara and Malacara (1985b) using exact ray tracing.

10.3 PRISMATIC LENSES

The two centers of curvature of the surfaces of the lens define th optical axis. Only when the two centers of curvature coincide because the surfaces are concentric, the optical axis is not define If the optical axis passes through the center of a round lens, the edge has a constant thickness all around.

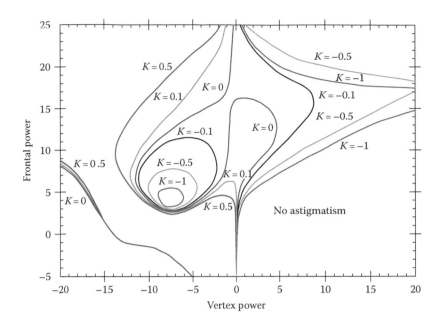

Figure 10.10 Tscherning ellipses for no astigmatism, deformed by the introduction of an aspheric surface.

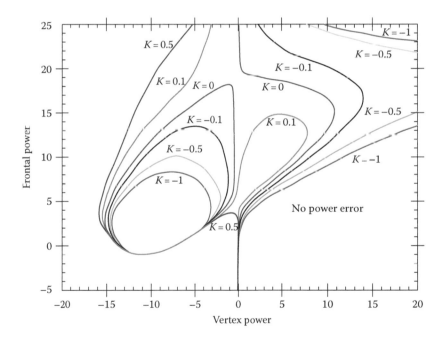

Figure 10.11 Tscherning ellipses for no power error, deformed by the introduction of an aspheric surface.

When the optical axis does not pass through the center of the lens, the lens is said to be prismatic because the two lens faces form an angle between them at the center of the lens. Then, a ray of light passing through the center of the lens is deviated by an angle ϕ. If a prism deviates a ray of light by an angle ϕ, as shown in Figure 10.12, the prismatic power P_P in diopters is given by

$$P_P = 100 \tan \phi \qquad (10.15)$$

thus, a prism has P_P diopters if a ray of light passing through the center of the lens is deviated P_P centimeters at a distance of 1 m.

If the angle between the two faces of the lens is θ, the angular deviation of the light ray is

$$\tan \phi = \frac{\sin \theta}{n - \cos \theta} \qquad (10.16)$$

or approximately, for thin prisms,

$$\phi = \frac{\theta}{n - 1} \qquad (10.17)$$

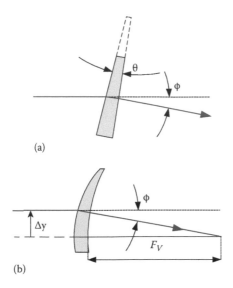

(a)

(b)

Figure 10.12 (a) Prism and (b) prismatic lens.

If two thin prisms with prismatic powers P_1 and P_2 are superimposed, forming at their bases an angle α between them, the resulting combination has a prismatic power P_R given by

$$P_R^2 = P_1^2 + P_2^2 + 2P_1P_2 \cos\alpha \tag{10.18}$$

and its orientation is

$$\sin\beta = \frac{P_2}{P_R}\sin\alpha \tag{10.19}$$

this result may also be obtained graphically, as shown in Figure 10.13.

A lens with vertex power P_V and prismatic power P_P is a lens whose optical axis is deviated from the center of the lens by an amount Δy given by

$$P_P = \frac{P_V \Delta y}{10} \tag{10.20}$$

where the decentration Δy is in millimeters.

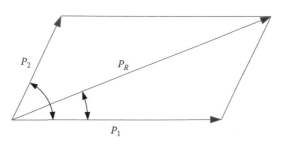

Figure 10.13 Vector addition of prisms.

10.4 SPHERO-CYLINDRICAL LENSES

A sphero-cylindrical lens has a toroidal or sphero-cylindrical surface. The lens does not then have rotational symmetry and an axial astigmatism is introduced, to compensate that of the eye. Optically, these lenses may be considered as the superposition of a spherical lens (with rotational symmetry) and a cylindrical lens (power in only one plane). As shown in Figure 10.14, a sphero-cylindrical lens is defined by (1) its spherical power, (2) its cylindrical power, and (3) its cylinder orientation. If a lens has a power P_1 in one diameter, at an angle ϕ with the horizontal, and a power P_2 in the perpendicular diameter, at an angle $\phi + 90°$, we may specify the lens as

　　Spherical power = P_1
　　Cylindrical power = $P_2 - P_1$
　　Axis orientation = ϕ

or as

　　Spherical power = P_2
　　Cylindrical power = $P_1 - P_2$
　　Axis orientation = $\phi + 90°$

The two specifications are identical. The only difference is that the spherical power is considered to be P_1 in one case and P_2 in the other. To pass from one form to the other is said to be to transpose the cylinder. A cylinder transposition is done in three steps as follows:

1. A new spherical power value is obtained by adding the spherical and cylindrical power values.
2. A new cylindrical power value is obtained by changing the sign of the old value.
3. The new axis orientation is obtained by rotating the old axis at an angle equal to 90°.

The power P_θ of a sphero-cylindrical lens along a diameter at an angle θ may be found with the expression

$$P_\theta = P_C \sin^2(\theta - \phi) + P_S \tag{10.21}$$

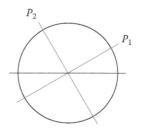

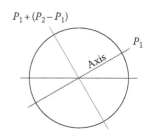

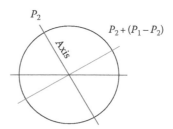

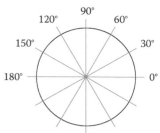

Figure 10.14 Powers and axis orientation in sphero-cylindrical lenses

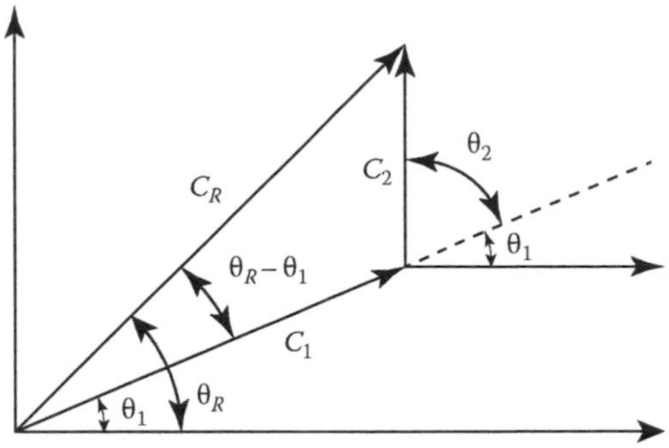

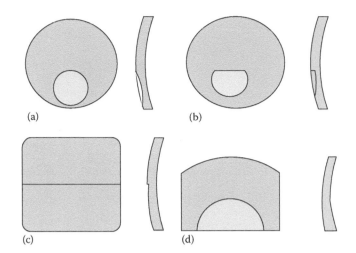

Figure 10.15 Vector addition of sphero-cylindrical lenses.

where

P_C is the cylindrical power
P_S is the spherical power
ϕ is the cylinder orientation

If two sphero-cylindrical lenses are superimposed, the combination has a cylindrical power P_{CR} given by

$$P_{CR}^2 = P_{C1}^2 + P_{C2}^2 + 2 P_{C1} P_{C2} \cos 2(\theta_2 - \theta_1) \qquad (10.22)$$

an axis orientation θ_R given by

$$\tan 2(\theta_R - \theta_1) = \frac{P_{C2} \sin 2(\theta_2 - \theta_1)}{P_{C1} + P_{C2} \cos 2(\theta_2 - \theta_1)} \qquad (10.23)$$

and a spherical power P_{SR} given by

$$P_{SR} = P_{S1} + P_{S2} + \frac{P_{C1} + P_{C2} - P_{CR}}{2} \qquad (10.24)$$

Graphically, these expressions may be represented as in Figure 10.15.

10.5 BIFOCAL AND MULTIFOCAL LENSES

When the eye lens cannot focus near objects due to age, eyeglasses with a positive dioptric power have to be used. However, the glasses to read a book or newspaper have to be removed in order to observe far objects. This may be troublesome if the change is very often. The problem can be solved if spectacles with lenses having a different dioptric power in the lower part of the lens are used. These lenses are called bifocal or multifocal and they are manufactured in a large variety of shapes and styles, some of them illustrated in Figure 10.16.

10.5.1 BIFOCAL LENSES

The first bifocal lenses were invented by Benjamin Franklin in 1784, by joining in a single ring two half lenses with different focal lengths. Bifocal lenses are made with two different concepts: (1) with two different materials, glasses or plastics having different refractive indices, as in Figure 10.16a and b, or (2) with a single material but with two different curvatures, as in Figure 10.16c and d.

Figure 10.16 Some common bifocal lenses.

Fused bifocal lenses are made with a convex segment fused on a concave cavity in the lens. The near vision segment should have a more positive dioptric power than the far vision zone in the lens. The difference of these dioptric powers is frequently called the addition. In fused lenses the near vision segment is made with a dense flint with $n = 1.625$, for addition powers smaller than 2.00 diopters, while for higher addition powers (2.00–4.00 diopters) a dense flint with refractive index $n = 1.654$ is used.

If the dioptric power of the convex (front face of the ophthalmic lens) with nominal refractive index ($n = 1.53$) is P_1 and that of the intermediate face of the segment is P_s, the addition power is

$$\text{Addition} = \left(\frac{n_p - n_s}{0.53} \right) P_1 + P_3 \qquad (10.25)$$

where n_p and n_s are the refractive indices of the ophthalmic lens and the addition segment, respectively.

Bifocal lenses have the problem that at the boundary between the near and the far zones a sudden change in the dioptric power is noticeable both for the person using the eyeglasses and for other persons. This problem has been solved, at least partially with several approaches, as it will now be described.

One solution to attenuate this sudden transition between the two dioptric powers is to design the addition segment in such a way that at the border between the two zones the power changes in a smooth manner in a band around it with a width of about 5 mm. Two examples of this kind of bifocal have the commercial names of *Beach Blend* and *Younger*. This solution is purely cosmetic, since the user of the eyeglasses notices the transition, but other persons don't.

10.5.2 PROGRESSIVE ADDITION LENSES

A second more popular solution is to manufacture the lenses with a shape such that the dioptric power varies in a smooth fashion over the whole surface of the lens, increasing from the upper to the lower zones. The first progressive addition lens was manufactured by Varilux and was designed by Maitenaz (1967). A review of progressive addition lenses is given by Bourdoncle et al. (1992). In the *Omnifocal* lenses the dioptric power varies

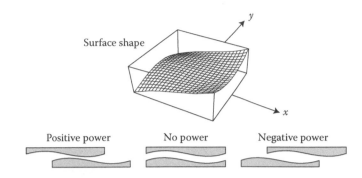

Figure 10.17 Schematics of the Alvarez lens.

quite slow but in others, like in the *Varilux*, the change is faster. In others, the variation is even faster in some regions of the lens. These lenses are known as progressive power lenses. Figure 10.17 shows one of these lenses.

Finally, another solution is by using lenses whose power changes by means of mechanism, for example, by moving the lenses either in a transverse or in an axial direction. One possible means is by means of liquids with a variable hydraulic pressure. Several of such systems had been described in the literature. One of the most interesting systems with moving lenses to change the dioptric power is invented in 1967 in California by Alvarez (1967), winner of the Nobel Prize in Physics in 1968. His optical system consists of a system of two identical aspheric lenses without rotational symmetry, one on top of the other, but one of them rotated 180° with respect to the other. The two lenses can be displaced with respect to each other. With the displacement of one of the two lenses along the *x*-axis, the spherical power of the combination changes.

To understand how these lenses work, let us think of each of the two lenses as a lens with a variable dioptric power, linearly increasing in one direction. The local curvatures on one of the surfaces should increase along the *x* direction. We will represent this surface by a polynomial $f(x, y)$. Since the lenses are thin compared with their diameter, the first partial derivatives, or slopes, are small. So, the local curvatures in the *x* and *y* directions are equal to the second partial derivatives in those directions.

Since the curvatures should change linearly along the *x*-axis, the polynomial should be of degree 3. To simplify this polynomial, we may impose some symmetry conditions. First, the surface is symmetrical with respect to the *x*-axis. Hence, all terms with an odd power in the variable *y* must be zero. Second, at the center of the lens ($x = y = 0$) the function has to be zero (no constant term), and there should be no slopes (zero tilt) and no curvatures (zero power). Hence, there should not be any constant term, and the coefficients of the terms x, y, x^2, and y^2 must be zero. Then the function $f(x, y)$ becomes

$$f(x, y) = a_1 y^2 x + a_2 x^3 \qquad (10.26)$$

Since it is a thin lens, the curvatures c_x and c_y along the *x*- and *y*-axes, respectively, are given by

$$c_x = \frac{\partial^2 f(x, y)}{\partial x^2} = 6a_2 x \qquad (10.27)$$

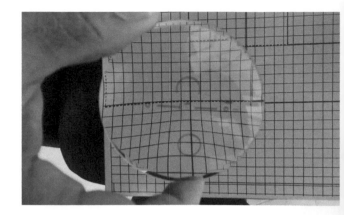

Figure 10.18 A progressive power lens.

and

$$c_y = \frac{\partial^2 f(x, y)}{\partial y^2} = 2a_1 x \qquad (10.2$$

Now, we require that these two curvatures are equal, so that the dioptric power is the same along the two axes, for all values of *y*, and also that the curvatures on the *y*-axis are zero, obtaining

$$a_2 = \frac{1}{3} a_1 \qquad (10.2$$

Hence, the function $f(x, y)$ is

$$f(x, y) = a_1 \left(y^2 + \frac{1}{3} x^2 \right) x \qquad (10.3$$

with the curvatures given by

$$c_x = c_y = 2a_1 x \qquad (10.3$$

If desired, in order to have a lens with a more uniform thickness, prismatic component can be added, without changing the dioptr power, by adding a linear term in *x* to the function $f(x, y)$.

In conclusion, each of the two Alvarez lenses is a progressive addition lens, whose power increases linearly with *x*. With two lenses, one on top of the other and one of them rotated 180° with respect to the other, the power of the combination can be changed by displacing one with respect to the other along the *x*-axis, as illustrated in Figure 10.17.

A typical progressive power lens is illustrated in Figure 10.18.

10.6 CONTACT LENSES

The most common contact lenses are the corneal lenses, which cover only the area of the cornea. These lenses have some important differences from an optical point of view from the normal spectacles:

1. There is no separation between the cornea and the lens.
2. The lens is not fixed to the face, since it moves with the eye.
3. The number of refracting surfaces is not increased, since the space between the contact lens and the cornea is filled with tears.

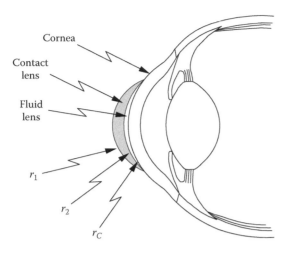

Figure 10.19 Contact lens on the eye.

The main advantages arising from these differences are the following:

1. No magnification of the image is introduced by the lens.
2. There are no off-axis aberrations in the lens, like coma, magnification chromatic aberration, and distortion.
3. Any possible corneal irregularities are eliminated, like astigmatism and keratoconus. This is true mainly for hard contact lenses.

Unfortunately, they also have some practical problems, for example, the following:

1. They require careful manipulation.
2. Not all persons can use them.
3. They can originate some medical problems in the eye if used without some special precautions.

In general, corneal lenses do not have the same curvature as the corneal surface in the concave side. To allow some ventilation and to avoid any sliding of the lens, the curvatures may be slightly different. There are some empirical criteria that must be followed for the proper fitting of the contact lenses.

There are two types of contact lenses, hard and soft and flexible. Hard lenses are frequently made with acrylic, and soft lenses are made with an extremely soft plastic material that can absorb water. An important difference between them is that hard lenses can eliminate corneal irregularities like astigmatism and keratoconus, but soft contact lenses cannot.

We can see that with hard contact lenses, optically we have two lenses, one is the contact lens and the other is the tear liquid lens between them, as illustrated in Figure 10.19.

10.7 INTRAOCULAR LENSES

In cataracts the eye lens loses its transparency. So, in cataract surgery the eye lens is removed, thus also removing approximately 19 D of power to the eye. The eye without the eye lens is called aphakic, from the Greek *phakos*, meaning lens. This power has to be restored by an intraocular lens to avoid a residual hypermetropy of about 19 D.

The first ophthalmologist to implant an intraocular lens was Sir Harold Ridley in November 1949 in London. However, the wide use of intraocular lenses in almost all cataract surgeries began in the 1980s. The first material that was used for these lenses was

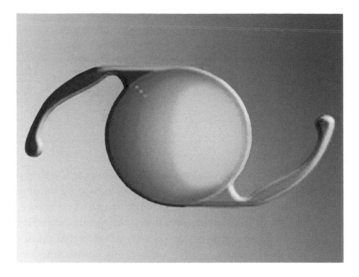

Figure 10.20 An intraocular lens.

polymethyl methacrylate, which is a rigid, nonfoldable material with refractive index equal to 1.49. Now silicone polymers have been used since 1984. This material is quite foldable and has a refractive index between 1.41 and 1.46. Some other materials, like hydrophobic foldable acrylic, are frequently used. Some intraocular lenses have in its periphery small hooks called haptics, in order to fix the lens in its proper place in the capsular bag inside the eye. Figure 10.20 shows an intraocular lens.

Intraocular lenses had been implanted since 1999 in order to correct high ammetropies, without removing the eye lens. This lens is called a *phakic intraocular lens*, since the eye lens remains in the eye. Naturally, an aphakic eye with an intraocular lens, cannot accommodate, like in very old people. Presently, bifocal or multifocal intraocular lenses are being implanted, to allow clear near vision, but many practical problems still have to be solved to find a perfect multifocal intraocular lens.

10.7.1 CALCULATION OF INTRAOCULAR LENSES

The first implanted intraocular lenses in the 1980s had the same average power of the eye lens. However, this approach did not give all persons the same quality of vision. For an intraocular lens to correct as well as possible all ammetropies, its dioptric power has to be properly calculated. Many methods had been developed over the years, constantly improving. A review of this subject has been published by Olsen (2007).

To calculate the intraocular lens, we need to know as precise as possible the following parameters:

1. The axial length L of the eye
2. The corneal dioptric power K
3. The posterior chamber constant A

The first two parameters have to be measured before the implantation of the lens. The position of the lens after the surgery is estimated with some available elements. The approximate value for the constant A is supplied by the lens manufacturer.

The axial length of the eye is the distance from the anterior or front surface of the cornea and the fovea. This is the most important parameter, since an error of about 1 mm will give an error of 2.35 D in the average human eye.

Table 10.3 Values of constant A for different locations of the intraocular lens

POSITION IN THE EYE		RANGE OF VALUES
Anterior chamber		115.0–115.3
	⌐ In the sulcus	115.9–117.2
Posterior chamber	│	
	⌐ In the bag	117.5–118.8

Table 10.4 Variation in the value of the constant A with the eye length

EYE LENGTH (mm)	SUM TO A CONSTANT
Smaller than 20.0	+3.0
20.0 to 20.9	+2.0
21.0 to 21.9	+1.0
22.0 to 24.5	0.0
Larger than 24.5	−0.5

The method currently used to measure the length of the eye is by means of A-scan ultrasonography, measuring the time that an ultrasound signal takes to go to the fovea and come back to the cornea. A practical problem is that the speed of sound is different for the different eye media. It has been accepted that a reasonable average speed is 1555 m/s, but the best instruments use different speed for each medium. Care must be taken to avoid pressuring too much on the cornea, with the measuring instrument, affecting the final results.

A method using partial coherence interferometry is more accurate for the measurement of the eye length, but the problem is that it cannot be used in an eye with the cataract.

The corneal dioptric power K is the power at the vertex of the cornea. Its radius of curvature is measured with keratometry or corneal topography, assuming a refractive index equal to 1.3375.

The value of the constant A depends on the position of the lens inside the eye. This constant depends on the material, on the shape of the lens, and mainly on its position inside of the eye. Its value is supplied by the lens manufacturer. Typical values for this constant are in Table 10.3.

This constant is closely related to two other frequently used parameters: (1) the distance from the iris of the eye pupil to the intraocular lens, also known as the surgeon factor, represented by SF (surgeon factor), and (2) the distance from the cornea to the intraocular lens, represented by ACD (anterior chamber depth). The formulas that relate these two quantities are

$$SF = 0.5663A - 65.6 \text{ mm}$$

$$ACD = \frac{SF + 3.595}{0.9704} \text{ mm} \tag{10.32}$$

10.7.1.1 Formulas for the calculation of the intraocular lens power

A few years ago, the dioptric power of the intraocular lens was estimated with relatively simple formulas that took into account some simple parameters and empirical results from a large number of previous implants. These methods were implemented with the regression formulas SRK and SRK II, whose names come from the researchers that proposed these formulas: Sanders and Kraf (1980), Retzlaff (1980), and Saunders et al. (1988). A review of these formulas was later published by Dang and Raj (1989).

The SRK formula works quite well for eyes with lengths between 22.0 and 24.5 mm, which are about 75% of all cases. Linear least squares fit of the parameters K and L with several thousand cases were made, using the constant A as a corrective additive value as follows:

$$P_0 = A - 0.9K - 2.5L \tag{10.33}$$

In order to improve this fitting as more results are available, the dioptric power that the intraocular lens should have had is calculated with the formula

$$P_0 = P_i + 1.5R_x \tag{10.34}$$

where

P_i is the implanted intraocular power
R_x is the residual ammetropic power

The original SRK formula was found in 1980. Several thousands of patients were studied and with the proper corrections, twenty years later, the following improved formula was proposed:

$$P_0 = 151.3 - 1.2K - 3.3L \tag{10.35}$$

The second formula, SRK II, assumes that the constant A is variable, adjusting its value, if the eye is too short or too large, as in Table 10.4.

Later, this formula was modified taking into account some theoretical calculations, in order to improve the results. It was called SRK II-T, where T stands for theoretical.

Other more recent formulas are based on the exact calculation using geometrical optics. The main ones are Holladay 1, Holladay 2, Hoffer-Q, Haigis-L, and Olsen (Holladay et al., 1988; Olsen 2007; Haigis 2008), all with different advantages and disadvantages. The more recent formulas are not empirical, but based on theoretical ray tracing. An example of this approach is given by the following formula:

$$P_0 = \frac{n\left(\dfrac{n}{K} - L\right)}{(L - ACD)\left(\dfrac{n}{K} - ACD\right)} \tag{10.36}$$

where the required parameters had already been defined.

10.8 LENSOMETERS OR VERTOMETERS

In optometric practice, it is essential to have an instrument to measure the dioptric powers in the simplest, accurate, and fastest manner. Some of these instruments are described in the book by Henson (1996). Herman Snellen around 1876 was one of the first to develop one of these instruments, using an optical bench. Unfortunately, its use required of persons with

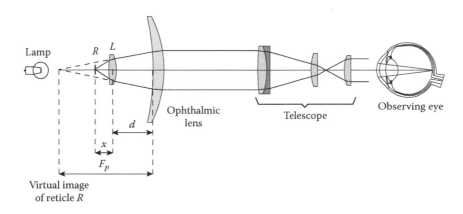

Figure 10.21 Optical schematics of a classic lens meter.

Figure 10.22 Two classic lensometers.

a good training. Troppman in 1912 simplified this instrument. The first commercial instruments developed with this purpose were patented in the 1920s in the twentieth century, first, by Bausch and Lomb, with the name of *vertometer* and later by American Optical with the name of *lensometer*. Later, the names of *focimeters* or *lensmeters* were used.

The lensometer or vertometer, whose optical schematics is in Figure 10.21, measures the back focal length F_p, or its inverse value, called vertex power P_v, of the ophthalmic lenses as it will be described. In this instrument, a light source illuminates a reticle, where a convergent lens with focal length f is at a distance x from the reticle. The lens, whose focal length is to be measured, is located at a distance d from the convergent lens.

Given a fixed distance d, the distance x is chosen such that the light coming out from the ophthalmic lens is collimated. This collimation is evaluated by means of a small telescope focused to an object at infinity. The distance F_p or its inverse P_v can be easily calculated if the value of x and the constants f and d are known. To prove this let us use the thin lens formula:

$$\frac{1}{f} = -\frac{1}{F_p - d} + \frac{1}{x} \tag{10.37}$$

Then, we can find the vertex power P_v:

$$P_v = \frac{1}{F_p} = \frac{f - x}{fx - fd - xd} \tag{10.38}$$

In order to have a linear variation of P_v with the distance x, we select

$$f = d \tag{10.39}$$

hence

$$P_v = \frac{1}{d} - \frac{x}{d^2} \tag{10.40}$$

Figure 10.22 shows two commercial lensometers of this type, manufactured between 1970 and 1990. They had a great success for more than 50 years, but they are not manufactured any more. They had been replaced by digital automatic instruments that work with similar but not identical methods.

REFERENCES

Alvarez, L. W., Two-element variable-power spherical lens, U.S. Patent No. 3,305,294 (1967).

Atchison, D. A., Modern optical design assessment and spectacle lenses, *Opt. Acta*, **32**, 607–634 (1985).

Atchison, D. A., Spectacle lens design: A review, *Appl. Opt.*, **31**, 3579–3585 (1992).

Blaker, J. W., Ophthalmic optics, in *Applied Optics and Optical Engineering*, R. R. Shannon and J. C. Wyant, eds., Vol. IX, Chap. 7, Academic Press, San Diego, CA, (1983).

Bourdoncle, B., J. O. Chauveau, and J. L. Mercier, Traps in displaying optical performances of a progressive addition lenses, *Appl. Opt.*, **31**, 3586–3593 (1992).

Coleman, H. S., M. F. Coleman, and D. S. Fridge, Theory and use of the dioptometer, *J. Opt. Soc. Am.*, **41**, 94–97 (1951).

Dang, M. S. and P. P. S. Raj, SRK formula in the calculation of intraocular lens power, *Br. J. Ophthalmol.*, **73**, 823–826 (1989).

Emsley, H. H., *Aberrations of Thin Lenses*, Constable and Co., London, U.K., (1956).

Fowler, C. and K. L. Petrte, *Spectacle Lenses: Theory and Practice*, Butterworth-Heinemann, Oxford, U.K., (2001).

Fry, G. A. and Hill, W. W., The center of rotation of the eye, *Am. J. Optom. Arch. Am. Acad. Optom.*, **39**, 581–595 (1962).

Haigis, W., Intraocular lens calculation after refractive surgery for myopia: Haigis-L formula, *J. Cataract Refract. Surg.*, **34**, 1658–1663 (2008).

Henson, D. B., *Optometric Instrumentation*, 2nd edn., Butterworth Heineman LTD, Oxford U.K., (1996).

Holladay, J. T., K. H. Musgrove, T. C. Prager, J. W. Lewis, T. Y. Chandler, and R. S. Ruiz, A three-part system for refining intraocular power calculations, *J. Cataract Refract. Surg.*, **14**, 17–24 (1988).

Landgrave, J. E. A. and J. R. Moya-Cessa, Generalized Coddington equations in ophthalmic lens design, *J. Opt. Soc. Am. A*, **13**, 1637–1644 (1996).

Lueck, I., Spectacle lenses, in *Applied Optics and Optical Engineering*, R. Kingslake, ed., Vol. III, Chap. 6, Academic Press, San Diego, CA, (1965).

Maitenaz, B., Image Retinienne donnée par un Verre Correcteur de Puissance Progressive, *Revu, Opt. Theor. Instrum.*, **46**, 233–241 (1967).

Malacara, D. and Z. Malacara, Tscherning ellipses and ray tracing in aspheric ophthalmic lenses, *Am. J. Opt. Phys. Opt.*, **62**, 456–462 (1985b).

Malacara, D. and Z. Malacara, An interesting property of inverted telescopes and their relation to myopic eyes, *Opt. Eng.*, **30**, 285–287 (1991).

Malacara, Z. and D. Malacara, Tscherning ellipses and ray tracing in aspheric ophthalmic lenses, *Am. J. Opt. Phys. Opt.*, **62**, 447–455 (1985a).

Olsen, T., Calculation of intraocular lens power: A review, *Acta Ophthalmol. Scand.*, **85**, 472–485 (2007).

Retzlaff, J., A new intraocular lens calculation formula, *J. Am. Intraocu Implant. Soc.*, **6**, 148–152 (1980).

Sanders, D. R. and M. C. Kraff, Improvement of intraocular lens powe calculation using empirical data, *J. Am. Intraocul. Implant. Soc.*, 263–267 (1980).

Sanders, D. R., J. A. Retzlaff, and M. C. Kraff, Intraocular lens power calculation, *J. Cataract Refract. Surg.*, **14**, 454–456 (1988).

Smith, G. and D. A. Atchison, Effect of conicoid asphericity on the tscherning ellipses of ophthalmic spectacle lenses, *J. Opt. Soc. Am* **73**, 441–145 (1983).

Sun, W.-S., H. Chang, C.-C. Sun, M.-W. Chang, C.-H. Lin, and C.-L. Tien, Design of high-power aspherical ophthalmic lenses with a reduced error budget, *Opt. Eng.*, **41**, 460–470 (2002).

Sun, W.-S., C.-L. Tien, C.-C. Sun, M.-W. Chang, and H. Chang, Ophthalmic lens design with the optimization of the aspherical coefficients, *Opt. Eng.*, **39**, 978–988 (2000).

Walker, B. H., *Optical Design for Visual Systems*, SPIE Press, Bellinghar WA, 2000.

Contact lenses

Ian Cox

Contents

11.1 A SHORT HISTORY OF CONTACT LENSES

Although the first practical contact lens was described in 1888 (Pearson et al. 1989), glass blown scleral shells formed individually to rest on the sclera and vault across the cornea were the norm until the 1930s. The advent of poly-methyl-methacrylate (PMMA or Perspex/Plexiglass) made it possible to make an all-plastic lens that could be fitted either by custom molding or trial fitting from a range of prelathed lenses (Mandell 1988). This reduced the weight and cost of lenses while improving comfort and wearing times beyond a couple of hours. It was not until 1948 that Kevin Tuohy made the first corneal contact lens (Tuohy 1948) by accidentally cutting through a scleral shell at the edge of the optic zone. Tuohy tried the small diameter lens that was left on his own eye and quickly realized that a lens fit within the cornea could be more comfortable and provide longer wearing times than a full scleral shell. This invention, combined with the realization that oxygen for corneal metabolism came directly from the atmosphere and not from the aqueous or limbal vasculature (Smelser and Ozanics 1953), led to a major shift in the industry to corneal contact lenses. These could be fit in a way that replenished the oxygenated tear film with every blink, thus extending comfortable wearing time to all waking hours. Through the 1950s and 1960s, the contact lens market expanded with commercially available corneal contact lens designs enabling the correction of myopia, hyperopia, astigmatism, and even novel bifocal designs for presbyopia correction.

In the mid-1960s, a Czech polymer chemist changed the industry by making the first "soft" contact lenses from his newly invented HEMA hydrogel material (Wichterle et al. 1960). This 38% water content material was highly flexible, oxygen permeable, and significantly more comfortable than the rigid PMMA corneal contact lenses that were available. This technology was licensed to Bausch + Lomb, a U.S.-based optics company, and after several years of development, the first soft contact lens approved by the FDA was launched in 1971 (Schaeffer and Beiting 2009). The dramatically improved comfort over rigid corneal contact lenses changed the contact lens industry in the United States, and ultimately the world, with rigid gas-permeable (RGP) contact lenses today accounting for only about 10% of the lenses that fit worldwide (Efron et al. 2013).

Initially only available in spherical powers to correct myopia and later hyperopia, soft lenses to correct astigmatism were first introduced in the United States in the early 1980s (Remba 1979, 1981). Unlike rigid lenses that "mask" the astigmatic component of the cornea, soft lenses conform to the underlying corneal shape, requiring a method of stabilization and orientation to be built into the physical shape of the lens. The most successful designs used an increasing thickness profile in the vertical meridian of the lens, allowing the natural squeezing force of the upper eyelid to orient and stabilize the lens on the eye between blinks (Richdale et al. 2007). Bifocal soft lenses designed to correct presbyopia were first introduced by major companies in 1982 (Bennett and Weissman 2005).

The first major challenge to soft contact lenses over the 1970s and 1980s was combatting adverse ocular responses related to deposition of protein and lipid on the lens surfaces from the tear film. This required daily cleaning and disinfection routines with specialized solutions and impacted the longevity of the lenses, prescribed as a single pair to be worn daily for as long as they lasted, which typically was a year or more. The second challenge was getting sufficient oxygen from the atmosphere through the hydrogel lenses to ensure an adequate physiological environment for the cornea. Many patients had their contact lens wear curtailed due to adverse corneal events precipitated by insufficient oxygen being available to the eyes during their wearing hours (Zantos and Holden 1978, Sweeney 1992). This was also the peak of "continuous wear," a modality where patients wore their contact lenses constantly, with removal as needed for cleaning (typically every 30 days in the early 1980s; Schaeffer et al. 2009). Although extremely convenient for patients, continuous wear (also known as extended wear) only exacerbated the issues of protein deposition, reduced lens life, and a significant increase in ocular adverse responses due to reduced oxygen availability to the cornea. In 1982, a small company in Denmark started cast molding hydrogel contact lenses and packaging them in small plastic blisters with foil covers. All other companies delivered their lenses individually and stored in a small glass serum vial, packaging that dated back to the original Bausch + Lomb lens in the 1970s. Danalens had produced the first "disposable" contact lens and created a major upheaval in the contact lens industry. Johnson & Johnson, sensing an opportunity to enter the lucrative contact lens market in the United States, acquired the Danalens production process and a small contact lens company called Frontier in 1981 who had a hydrogel lens material already approved by the FDA. Within 6 years, Vistakon launched the first readily available disposable lens in the United States (Rigel 2007). Launched initially as an extended wear lens to be replaced weekly, the marketplace soon dictated its use as a daily-wear-only (no overnight wear) lens with a replacement schedule of 2 weeks. Although the oxygen permeability of these new lenses was no better than previous lenses, the fact that patients could buy them for only a few dollars each (previously patients would typically pay over $200 for a pair of lenses) and replace them frequently rather than using heroic cleaning regimens to extend the life of their lenses made them a rapid success. Vistakon soon became the market leader in spherical soft contact lenses in the United States as the industry shifted to the new "disposable" paradigm. Within a few years, toric and multifocal options also became available as companies invested in the manufacturing capacity necessary to process these complex designs for a low cost, so no patient was excluded from a biweekly, or monthly, disposable lens. As manufacturing technology improved and cost of goods decreased into cents per lens, the option of a truly disposable lens—one that was worn once and then discarded with no overnight storage or care system—soon became a viable option. Vistakon again led the industry by launching the first daily-disposable contact lens in 1994, using a traditional hydrogel material. Although each lens was less than $1 to the patient, the high annual cost prohibited a rapid adoption of daily disposables, and it was another decade before this modality made any significant inroads into the marketplace.

In the intervening years, other companies were still chasing the elusive "pot of gold," a lens that was so physiologically compatible with the eye that it could be worn continuously for 30 days before removal without the risk of significant adverse ocular responses (Nilsson and Montan 1994). Lenses made from silicone elastomer were tried in the late 1970s and early 1980s, with the Dow Corning company being the most well-known manufacturer to try this alternative material (Bennett and Weissman 2005). Although inherently hydrophobic, the massive oxygen permeability of silicone elastomer materials made them very appealing to researchers; successful use of silicone in a contact lens would eliminate the myriad of adverse responses attributed to hypoxia of the cornea.

Unfortunately, silicone elastomer lenses had one undesirable and potentially dangerous flaw; their rubberlike nature generated negative pressure under the lens when worn and resulted in the lens sticking to the eye (Fanti and Holly 1980, Josephson and Caffery 1980). This phenomenon could not be overcome with design, and so the industry set about generating a hybrid material, a silicone–hydrogel: part silicone and part hydrogel. Although seemingly straightforward, the material scientists were essentially trying to mix "oil and water" and still maintain a transparent material. Bausch + Lomb, the first company to bring soft hydrogel contact lenses to the market in 1971, were also the first to develop a commercially viable silicone–hydrogel lens. This lens provided more than four times the oxygen transmission of hydrogel lenses, approved for up to 30 days of continuous wear and was first available in 1999. Clinicians immediately noted that the highly oxygen transmissive lenses eradicated significant adverse responses related to lack of oxygen at the cornea but were wary about using silicone–hydrogel lenses due to the up to 30 days of continuous wear indication awarded by the FDA. Experience over the years with extended wear of hydrogel lenses had shown that corneal ulcers or microbial keratitis was the single most significant adverse response associated with extended or continuous wear, with the FDA limiting wear time approval of all hydrogel lenses to seven nights maximum in 1989 due to their concern at the incidence levels (Schein et al. 1989). The industry hoped that the introduction of silicone hydrogels would eliminate the risk of microbial keratitis, but worldwide postmarket surveillance studies have shown that significantly increased oxygen transmission does not reduce the risk of microbial keratitis associated with the overnight wear of contact lenses (Schein et al. 2005). Clinicians and companies now recommend silicone–hydrogel lenses for daily wear or extended wear with monthly replacement, and the largest area of growth within the contact lens industry is in the daily-wear modality.

RGP lenses also underwent a material revolution over the years, with silicone-acrylates, and ultimately fluorosilicone polymers being developed to provide highly oxygen permeable and surface-wettable rigid lens materials. Although hampered by lens adhesion to the cornea when worn on a continuous or extended wear basis, RGP and ultimately scleral lenses have become the lens of choice when worn on a daily-wear basis for patients with high levels of ocular wavefront aberration caused by corneal pathology or postsurgical corneal shape changes. These patients can experience significantly improved vision under all lighting conditions for all waking hours, significantly improving their visual quality of life. However, for the more typical patient, rigid materials still provide significantly poorer comfort than lenses made from a rigid material, and so while currently available soft lens materials provide excellent physiological compatibility with the eye, and the cornea specifically, when worn in a daily-wear modality, the focus of the industry has been to improve end-of-day comfort through molecules embedded in the lens material during manufacture that release during lens wear and improve optical performance. This last development is ironic, since contact lenses are worn to correct the patient's vision, yet physiological compatibility was so important to successful contact lens wear that improved vision was a relatively low priority for the first three decades of soft lens development. The recent adoption of wavefront sensors has allowed measurement of the wavefront error of the eye, and of manufactured contact lenses, alone and in combination (Hong et al. 2001, Lu et al. 2003, Jiang et al. 2006). At least one manufacturer (Bausch + Lomb) is altering the inherent spherical aberration of their contact lens products to correct the average spherical aberration of the eye found in the general contact lens–wearing population (Porter et al. 2001) for both spherical and toric lenses and is employing sophisticated metrology and optical modeling in the development of all their contact lens products. Ideally, from an optics perspective, individual prescription contact lenses should be available for each eye based on wavefront measurements performed in the clinical setting, to enable correction of all higher-order aberrations for improved low-light vision; the challenge now for the industry is to deliver these in the same low-cost, disposable paradigm that patients are currently using.

11.2 VERTEX DISTANCE, MAGNIFICATION, ACCOMMODATION, AND CONVERGENCE

In a clinical setting, the optical power of the eye is typically determined by performing a subjective refraction. This procedure involves the iterative application of spherical and cylindrical lenses in front of the eye to determine the optimal visual acuity at distance. The lenses are mounted in front of the eye, in a trial frame, at a distance that replicates the position of a pair of spectacles to be worn by the patient. This is known as the spectacle plane, and the distance to the anterior corneal surface, typically 12–13 mm, is known as the vertex distance. When fitting a contact lens, clinicians must compensate for the vergence change that occurs from the spectacle plane to the anterior corneal plane to ensure the retinal image remains in optimal focus. While this can be calculated for each individual case, using the formula (Bennett 1974)

$$K = \frac{F}{1 - dF}$$

where

K is the refraction at the corneal plane
F is the refraction at the spectacle plane
d is the distance of the spectacle plane from the corneal plane

clinicians will typically use a lookup table or an electronic application to derive the appropriate compensated power(s) for a contact lens prescription, given that lenses can only be ordered in a minimum step size of 0.25 D.

This shift from spectacle plane to corneal plane will also impact the retinal image magnification, because the magnification (hyperopic corrections) or minification (myopic corrections) of the image caused by the anterior placement of the spectacle lenses will be reduced in magnitude when the correction is moved back to the corneal plane. Using the formula (Bennett 1974)

$$\text{Spectacle magnification} = \frac{1}{1 - aF}$$

where

F is the BVP of the lens
a is the distance from the lens to the entrance pupil of the eye

Spectacle magnification for spectacle lenses and contact lenses for the same ocular correction is shown in Table 11.1.

Clearly the change in magnification becomes greater as the magnitude of the spectacle correction increases.

Accommodation and convergence of the eye are also altered when moving from the spectacle plane to the corneal plane. Myopic patients wearing spectacles with the optical center aligned for distance viewing will experience base-in prism when converging to view a near object and will therefore need to converge less than if they were viewing the same object wearing contact lenses, which move with the eye as it turns and therefore have no prismatic effect for near viewing. Conversely, the hyperopic patient will experience base-out prism when viewing the same near object and will have to converge more wearing spectacles than with a contact lens correction (Bennett 1974, Mandell 1988).

Similarly, calculation of the vergence of light coming from distant and near objects reveals that myopic patients accommodate less when wearing spectacles than when corrected with contact lenses and viewing a near target, while hyperopic patients will accommodate more to bring a near target into focus wearing spectacles than with contacts.

These changes are important clinical considerations when fitting a patient with contact lenses; myopic patients with a

Table 11.1 **Spectacle magnification values for different back vertex power corrections when worn at the spectacle plane and the contact lens plane**

LENS BVP AT SPECTACLE PLANE	SPECTACLE MAGNIFICATION	CONTACT LENS MAGNIFICATION
+10.00 D	1.176	1.035
+5.00 D	1.081	1.016
−5.00 D	0.930	0.986
−10.00 D	0.870	0.974

Assumptions: 12 mm vertex distance, 3 mm distance to entrance pupil.

higher magnitude correction may appear to gain visual acuity over their spectacle prescription due to the reduced minification of the retinal image, but may also manifest asthenopic symptoms due to the increased accommodation and convergence necessary for near task viewing. For those approaching presbyopia, the increased accommodation and convergence demands may be sufficient to require a reading addition with their contact lenses, which is not necessary when wearing spectacles. Conversely, hyperopic patients may notice the reduction in magnification when moving from a spectacle correction to a contact lens correction but will benefit from the reduced accommodation and convergence demands at near.

11.3 CORRECTION OF MYOPIA AND HYPEROPIA WITH CONTACT LENSES

While spectacle lenses are the adopted method of vision correction for the majority of the world's ametropes, for many patients, the limitations of spectacles lead them to seek an alternate method of correction. Spectacles can be heavy and limit the field of view, particularly in higher prescriptions, limit the use of safety equipment such as helmets or visors, impact vision under environments of extreme temperature or humidity (fogging), are impractical for extreme or water sports, and may be less desirable from an appearance perspective (Walline et al. 2007, 2009). In general, contact lenses provide a vision solution for all of these shortcomings, and the fact that they are adopted by 36 million (approximately 24% of the U.S. vision-corrected population [Barr, 2006]) supports this claim. Contact lens designs for the correction of myopia and hyperopia must consider not only the optical needs of the patient but also the mechanical aspects of fitting the lens to the eye comfortably and securely, in addition to the physiological requirements of the cornea and conjunctiva. From an optical perspective, the ideal contact lens would center on the visual axis of the eye and remains stably fixed in that position under all positions of gaze and has an optical zone as large as the diameter of the contact lens itself. Unfortunately, this is physiologically unacceptable because the corneal epithelium receives its oxygen from the tear film as well as discharging metabolic waste products through the same medium. Contact lenses must demonstrate movement on the cornea with every blink (Young 1996, Wolffsohn et al. 2013), while maintaining a position over the visual axis for all positional movements of the eye. For comfort, the lenses must have a thin edge profile, regardless of the optical power of the lens correction, and the overall lens thickness must be controlled to ensure that the oxygen permeable lens material can supply sufficient oxygen to the post-lens tear film to meet the corneal epithelial metabolic needs. Therefore, contact lenses are typically a lenticular design, with anterior and posterior surface optic zones smaller than the overall lens diameter. For myopic corrections, particularly those of high power, this allows the relatively thick profile of the lens at the edge of the optic zone to taper down to an acceptable thickness at the edge of the lens. For hyperopic corrections, the same lenticular construction allows the center thickness of the lens to be minimized by reducing the lens thickness at the edge of the optic zone. Posterior lens curvature and lens diameter will control

how well the lens centers on the cornea and the magnitude of the lens movement with each eyelid blink (McNamara et al. 1999).

Soft lenses, typically made from hydrogel or silicone–hydrogel materials, demand that the lens diameter be larger than the diameter of the cornea and that the sagittal depth of the posterior lens surface be greater than the equivalent diameter of the cornea/scleral surface it is fitted to. This generates potential energy in the lens as it is squeezed into alignment with the corneal surface during the first blinks following placement on the eye, and this energy produces the forces that keep the soft lens aligned with and centered on the cornea in all positions of gaze. Insufficient sagittal depth may cause a soft lens to decenter excessively on a cornea during extreme gaze as well as move excessively with each blink, disturbing vision. Excessive sagittal depth may cause sufficient radial pressure on the central optic zone in myopic corrections (the lens will be thinnest in the center of the lens for these powers), where the lens surface will distort between blinks, again causing visual disturbance (Cox 2000). In extreme cases, the lens may buckle, causing a significant loss of vision (Figure 11.1).

Soft lenses are extremely comfortable and easy to fit and represent at least 90% of the contact lens fittings worldwide (Morgan et al. 2011b).

RGP lenses are typically smaller in diameter than the cornea and are purposely fitted to slightly misalign with the corneal shape to avoid lens adherence to the corneal surface. This misaligned fitting of the lens results in greater movement with the blink than a soft lens, and a less reliable centration of the lens in the same position on the cornea following each blink. RGP lenses are inherently less comfortable than soft lenses and are slightly more time consuming to fit, so they typically only represent about 10% of contact lens fittings worldwide (Efron et al. 2013).

Scleral lenses, made from the same materials as RGP lenses, are designed to vault the central corneal surface and rest on the scleral surface. These lenses show little movement with an eyelid blink and are very stable in terms of lens centration relative to the visual axis. Scleral lenses are the most time consuming to fit, particularly for an inexperienced clinician, and their size is daunting for many patients, so their use is very limited around the world (Nichols 2013).

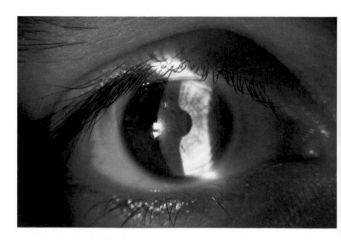

Figure 11.1 Thin soft contact lens fit with too great a sagittal depth (steep/short base curve) that is showing anterior surface buckling. Note that high molecular weight fluorescein has been instilled in the tears to show the surface deformation more clearly as the tears pool in the valleys of the distorted surface.

11.4 TEAR FILM AND ITS IMPACT ON VISUAL STABILITY WITH CONTACT LENSES

Soft lenses conform to the central corneal shape after the first couple of blinks, and the inherent energy within the lens keeps the surfaces aligned with the cornea if the lens is fitted correctly. Hence, the posterior tear film under a soft lens is thought to be very thin, perhaps less than 5 μm in thickness (Wang et al. 2003). The anterior surface tear film is also relatively thin, perhaps only a few microns thick, and typically does not demonstrate the same stability between blinks as the normal eye (King-Smith et al. 2000, Wang et al. 2003, Nichols et al. 2005). In fact, blink rate has been shown to increase when patients are wearing contact lenses, and one reason for this may be the instability of the anterior lens surface tear film (Pointer 1988, Jansen et al. 2010).

Since the anterior lens surface tear film represents the first refracting surface of the eye's optical system, and it also incurs the largest refractive index change (air to water), then, if the tear film thins and/or "breaks up" between blinks, vision will suffer due to the higher-order aberrations induced by this tear film surface variation (Figure 11.2).

Moreover, if the tear film disruption is occurring regularly before the typical interblink period for the patient, the anterior surface of the lens will partially dehydrate and require several forced blinks to rehydrate and stabilize the tear film and the retinal image quality for the patient. This instability can be reported by the patient as poor vision, but cannot be detected by the clinician using standard high-contrast letter acuity charts because the patient increases their blink rate during the testing procedure to achieve the best result (Pointer et al. 1985).

Conversely, RGP and scleral lenses can have considerably thicker posterior surface tear films due to the misalignment between the central posterior radius of the lens and that of the anterior cornea. This difference can contribute to the optical correction of the lens, adding positive or negative spherical or toric power to the overall prescription (Mandell 1988). While this can be calculated, most clinicians use a "rule of thumb," adding 0.50 D of correcting power to that of the lens itself for every 0.10 mm of radius disparity between the lens and the cornea, with flatter contact lens radii generating a negative power "tear lens" and steeper contact lens radii generating a positive power lens addition. Obviously, this optical correction technique also extends to more complex optical errors, such as astigmatism and higher-order aberrations, by substituting a spherically surfaced rigid contact lens for the nonspherical, and perhaps highly aberrated, corneal surface and allowing the tear film to "fill in the gap" between lens posterior surface and corneal anterior surface. With this technique, clinicians are not necessarily trying to fully correct the astigmatism or higher-order aberrations in the patient's ocular optical system, but rather reduce them significantly and improve retinal image quality over that provided by a soft lens or spectacle correction. Interestingly, RGP and scleral lens materials are inherently less wettable than soft lens materials, and hence, they are less capable of maintaining a stable tear film over the anterior surface of the lens compared to either soft lenses or the natural eye (Guillon 1998). In theory, this more rapid tear film breakup time should provide less stable vision than soft lenses. However, RGP and scleral lenses have a highly polished anterior lens surface whose quality is not impacted by dehydration. Therefore, while RGP and scleral lenses do demonstrate a shorter tear film breakup time (typically 4–6 seconds; Doane 1989, Creech 1998), patients wearing them do not observe the visual instability that they experience with soft lenses because the tear film clears rapidly at breakup, leaving a high-quality, stable optical surface until the next blink. This is one contributing factor to the clinical impression that RGP and scleral lenses give better visual quality than soft lenses (Timberlake et al. 1992, Fonn et al. 1995).

11.5 CORRECTION OF ASTIGMATISM WITH CONTACT LENSES

While we have demonstrated one mechanism that RGP and scleral lenses can reduce the impact of astigmatism on vision, namely, vaulting over the toric anterior corneal surface and using the tear film to fill in the void between the lens posterior surface and cornea (clinically known as "masking"), this technique only provides a full correction of astigmatism when the corneal surface toricity is the same magnitude and axis as the refractive astigmatism. Of course, soft lenses, being flexible and conforming to the corneal shape immediately after insertion, have no ability to "mask" astigmatism (Griffiths et al. 1998). To fully correct astigmatism, soft lenses all need to be fabricated with at least one toric surface and designed to maintain position in the correct orientation relative to the astigmatic axis of the eye during blinking motions and eye movements. Depending on the masking performance of the tear lens, RGP and scleral lenses may also have to have a similar toric design.

If the corneal astigmatism is sufficiently large in magnitude, one technique to stabilize the lens on the eye is to use a toric posterior lens surface designed to generally align with the anterior corneal surface (and in the case of a scleral lens, the anterior corneal surface and the scleral surface near the limbus), in much the same way that a saddle fits a horse (Mandell 1988). The anterior surface is also designed as a toric across the optic

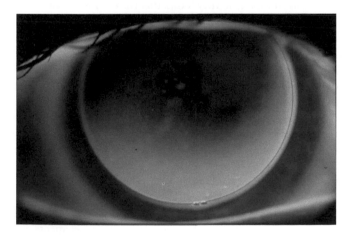

Figure 11.2 Anterior surface tear film break-up on a rigid gas permeable (RGP) lens showing the uneven surface (droplets) which will cause a reduction in vision until the patient renews the tear film with the next blink. Note that fluorescein has been instilled in the tear film and appropriate blue and yellow filters have been used to capture this photograph to make it easier to view the tear film.

portion of the lens to correct any residual sphere and cylinder correction of the eye. RGP lenses of this design are known as bitoric lenses. Clinicians find that this technique is successful using RGP or scleral lenses for eyes with 3.00 D or more of corneal astigmatism, with lens orientational stability becoming less than optimal as the magnitude of the corneal astigmatism decreases below this amount. Not surprisingly, due to their flexibility, this technique is not successful even with very large magnitudes of corneal astigmatism with soft lenses unless combined with additional design features to maintain lens orientation (Griffiths et al. 1998).

For soft lenses and patients fitted with RGP and scleral lenses that have corneal astigmatism less than 2.00 D, the lens must be designed with features that interact with the eyelids during a blink to orient the lens correctly relative to the corneal astigmatism axis. Ideally, the lens should maintain this orientation during the entire interblink period, regardless of horizontal, vertical, or cyclotorsional eye movements. In the early years of contact lenses, toric lenses were stabilized using prism ballasting, double "slab-off," or truncated designs (Holden 1975). Prism ballasting refers to a lens designed to gradually increase in thickness in the vertical meridian from the top to the bottom of the lens (i.e., the 90° [top] and 270° [bottom] meridians of the lens). This allows the superior eyelid to squeeze the thickest part of the lens to the bottom during the blink. The thick inferior portion of the lens comes into contact with the lower lid (which moves very little vertically during a blink), and these two interactions bring the lens into correct alignment no matter what orientation the lens was inserted onto the eye. Double "slab-off" designs start with a relatively thick lens and then remove the material across the top and bottom of the lens to provide sloped "ramps" for the eyelids to interact with and twist the lens into the correct orientation. Truncated lens designs rely on a straight edge formed across the bottom of the circular lens by removal of lens material (hence the term truncated) to interact with the lower eyelid during a blink, but have lost favor over the last decade due to the fact that they are the least reliable of all the design methods to orient a lens, and they create noticeable discomfort for the lens wearer. In more recent years the designs for orienting toric lenses have progressed from these fundamental designs to include quite complex peripheral anterior surface designs (known as peri-ballast designs) to provide better interaction with the upper and lower lids, while reducing the central thickness of the lens (improved oxygen permeability and physiology) and eliminating the prism across the optic zone and the associated higher-order aberrations that are inherent in a prism-ballasted design (Figure 11.3).

These designs provide very good orientational stability of toric soft lenses on eye with several of the most popular designs showing mean misorientation from the desired position of less than 10° (Young et al. 2009, Momeni-Moghaddam et al. 2014). Variation around this orientation position of equilibrium is typically less than 5° with each blink, and the clinician will provide the optimal astigmatism correction by compensating for the static misorientation value when prescribing the lenses (Teague and Gunderson 2006). For example, if the lens is orienting 10° nasal from the ideal position on the eye in primary gaze and the axis of astigmatism for this eye is 30°, the clinician will prescribe the contact lens with and axis of 20°, instead of 30°, knowing the 10°

of nasal misrotation will bring the optics of the lens into alignment when worn by the patient.

11.6 FLEXURE OF GP LENSES

The original corneal, or hard, contact lenses were made from PMMA, a relatively stiff material that would not warp or distort from the eyelid and tear film pressures inherent in contact lens wear. However, as manufacturers developed new oxygen permeable rigid lens materials, particularly those containing silicone, the stiffness of the materials decreased to the point that clinicians were noticing changes in the astigmatic correction of patients over time (Stevenson 1991, Carney et al. 1997). Patients whose corneal and refractive astigmatism was corrected using a spherical design RGP lens found that their vision were degrading over time, and clinicians were finding that the silicone-containing rigid lenses were deforming to partially match the astigmatic corneal shape, losing their ability to "mask" the corneal and refractive astigmatism of the patient's eye. Clinicians found that increasing the thickness of the lens provided a solution for some patients, but this often reduced the true benefit of silicon containing rigid lens materials—increased oxygen permeability and better physiology. Fortunately, manufacturers were able to improve their formulations and include fluorine in their formulations, which improved the stiffness and the wettability of the material while also improving the oxygen permeability. The vast majority of material used in RGP (corneal) and scleral lenses today is fluorosilicone acrylate. This has essentially eliminated the problem of lens flexure for the majority of patients.

11.7 CONFORMATION OF SOFT CONTACT LENSES

Based on their experience with hard lenses, early clinical adopters of soft spherical lenses expected that these lenses would correct (or partially correct) the refractive astigmatism of the eye by vaulting over the astigmatic surface of the anterior cornea (known as "masking"). They found that the early low water content lens designs with a center thicknesses of 300 µm or more did provide some correction, but the relatively rapid advancement of soft contact lens design toward thinner center thickness lenses (particularly for myopic corrections) and lower modulus higher water materials, in the desire to improve oxygen transmission to the corneal epithelium, soon made it clear that soft lenses conformed to the corneal surface and any corneal astigmatism (or other higher-order aberrations due to central corneal topography) are 'telegraphed' through to the front surface of a soft contact lens. It is now well understood that soft contact lenses, whether they are manufactured using hydrogel or silicone–hydrogel materials, conform to the corneal surface across the central optical portion of the lens (Plainis and Charman 1998). Soft contact lens designers use this feature to advantage, often placing elements of the optical correction on both the anterior and posterior surfaces of the lens (e.g., several toric lenses have the spherical component of the lens on the anterior surface and the cylindrical component on the posterior surface) to simplify and reduce the cost of manufacturing the lenses, knowing that in the clinical setting, the optical

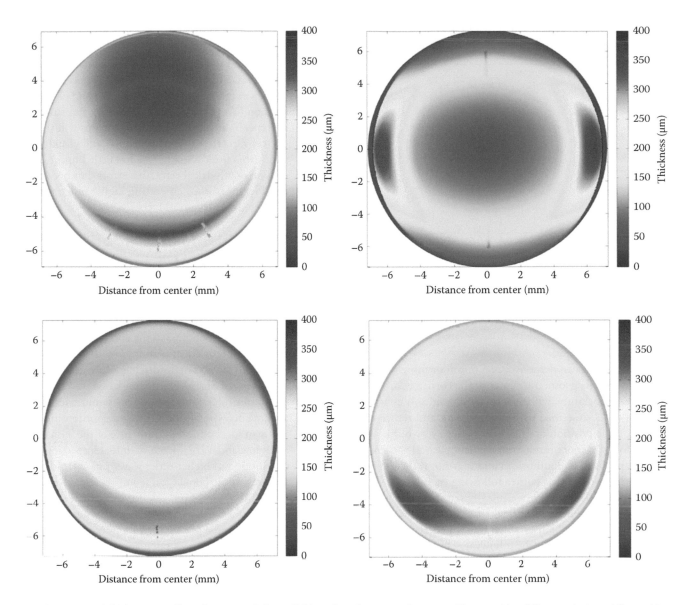

Figure 11.3 Measured thickness profiles of commercially available soft toric contact lenses to illustrate the different design philosophies used by manufacturers to stabilize the orientation of their lenses. Similar designs can also be used on RGP and scleral contact lenses to maintain the correct orientation of lenses with optical designs to correct astigmatism.

performance will be the same as a lens with both spherical and cylindrical components on the anterior surface.

11.8 CONTACT LENS MOVEMENT AND CENTRATION AND ITS IMPACT ON VISUAL PERFORMANCE

Ideally, from an optical perspective, a contact lens should be fixed in place and centered on the visual axis of the eye. From a corneal physiology and fitting perspective, a contact lens needs to move on the eye in response to the blinking mechanism of the eyelid (Mertz and Holden 1981, Brennan et al. 1994). The reasons for this are twofold: first, to provide a mechanism for the removal of epithelial metabolic waste products and other tear film debris, and second, to reduce the loss of comfort caused by the increase in surface friction between the eye lids and the lens

anterior surface as it dehydrates in low-humidity environments. Therefore, scleral lenses are fit with a small amount of movement with each blink (<0.2 mm; van der Worp 2010), soft lenses are fit with slightly more movement with the blink (0.2–0.5 mm; Brennan et al. 1994), and RGP lenses are fit with the most movement (1.0–2.0 mm; Hong et al. 2001) with each eye lid blink. The forces that control the amount of movement of a lens on the eye are also those that directly impact how well the contact lens centers on the cornea. Contact lenses fit with greater sagittal depth on the posterior surface generally move less and center better on the cornea, while lenses fit with lower sagittal depth will generally move more and have greater decentration on the cornea (Lowther and Tomlinson 1981). Scleral lenses will usually center well on the cornea, while soft contact lenses also center relatively well, with decentration values relative to the cornea typically in the 0.0–0.5 mm range for a lens fitted with minimal movement. RGP lenses, due to the fact that they are small in diameter and are fit intra-corneally, will typically stay within the corneal

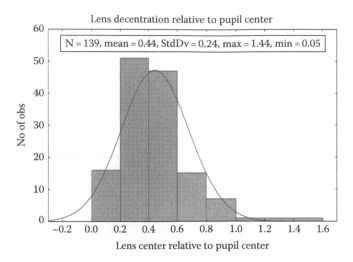

Figure 11.4 Distribution of lens decentration relative to the center of the pupil for 139 normal eyes fitted with a typical silicone hydrogel single vision lens design. Note that the distribution is skewed with most eyes showing relatively low amounts of decentration and only a few eyes showing considerably larger amounts of lens decentration.

diameter, but can demonstrate 1.0 mm or more of decentration relative to the geometric center of the cornea (Carney et al. 1997). More importantly, while scleral lenses and soft contact lenses tend to follow a prescribed path of movement and return to almost the exact same position on the cornea after every blink, RGP lenses will follow a more random movement path as the lens returns to its position of equilibrium and may not return to exactly the same location on the cornea following each blink. While this does not appear to have a clinically significant impact on vision for standard spherocylindrical corrections, it does reduce the potential for RGP lenses to be used for correcting higher-order aberrations, as seen in the following section.

This discussion of contact lens centration has been based on the mechanical fitting of the lens to the cornea and hence is described relative to the corneal center. However, the physiological pupil center is typically decentered nasally relative to the geometric center of the cornea and changes with pupil dilation (Yang et al. 2002). Therefore, not surprisingly, the best mechanically centered contact lens will still be decentered optically, relative to both the pupil and visual axis of the eye to which it is fitted. Again, this has little impact on the correction of spherocylindrical refractive errors, but becomes a significant consideration when correcting higher-order aberrations with contact lenses (Figure 11.4).

11.9 SPHERICAL ABERRATION AND CONTACT LENSES

Researchers have reported the presence of higher-order aberration in the eye for several decades (Koomen et al. 1949, Ivanoff 1953, Jenkins 1963), and until more recently, spherical aberration was primarily identified as the aberration most involved in the reduction of visual performance under large-pupil viewing conditions such as night driving. Only relatively recently with the advent of Shack–Hartmann-based wavefront sensors has the significant role of nonrotationally symmetrical aberrations such as coma been identified as a major contributing factor to the less than ideal optical performance of the human eye (Howland and Howland 1977). Even so, initial inspection of the average distribution of higher-order wavefront aberrations across a typical pre-presbyopic population requiring refractive correction shows a distinct deviation from zero for the spherical aberration, Zernike term while all other Zernike terms have an average close to zero, a value expected of a biological optical system attempting to optimize itself across a population (Figure 11.5).

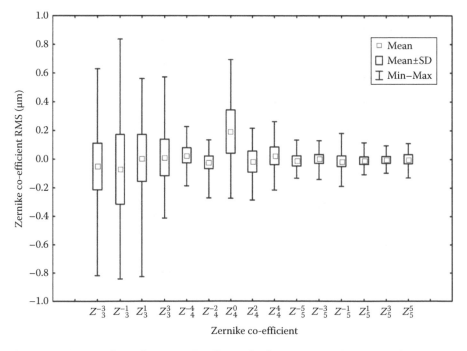

Figure 11.5 Mean values of higher order wavefront aberration Zernike modes for a 6mm pupil in a normal patient population aged 19 to 65 (n = 1325). The box represents ± one standard deviation from the mean value, while the error bars represent the range.

Contact lenses, by their design, have highly curved anterior and posterior surfaces to enable them to fit the eye in a stable and comfortable fashion and have a relatively constant alignment with the optical axis of the eye in all positions of gaze. This makes spherical aberration a potentially significant monochromatic aberration for contact lenses, unlike spectacles, which have shallow curvatures and off-axis positioning for all but central positions of gaze and are most affected by astigmatism and distortion aberrations (Westheimer 1961). These highly curved, and typically spherically shaped, surfaces inherently produce spherical aberration in contact lenses, regardless of whether they are soft lenses, RGP lenses, or scleral lenses. As lens power increases, spherical aberration also increases, with minus powered lenses producing negative spherical aberration and plus powers inducing positive spherical aberration (Dietze and Cox 2003).

If we consider the spherical aberration of the lens alone, patients with contact lens powers closer to zero should have the least visual impact from spherical aberration under large-pupil conditions, while patients with higher positive or negative corrections should notice a greater impact from spherical aberration when fit with contact lenses. However, clinically, the majority of contact lens–wearing patients are myopic and require a lens power between –1.00 D and –5.00 D, with the peak being close to –3.00 D. The average ocular spherical aberration for the population with single vision wearing contact lens is approximately 0.18 μm of positive spherical aberration over a 6 mm pupil (Kingston et al. 2013). Dietze and Cox (2003) showed that negative spherical aberration for spherically surfaced hydrogel soft contact lenses approached this magnitude for powers near –7.00 D, with substantial amounts of correction for powers as low as –4.00 D. Therefore, it appears that a portion of the contact lens–wearing population get some level of spherical aberration correction when wearing their spherically surfaced soft contact lenses under large-pupil conditions, by virtue of the inherent negative spherical aberration in their lens correction. Patients with very high negative or low negative defocus corrections may not experience the full benefit due to spherical aberration over or under correction. Patients with positive defocus corrections will actually experience greater levels of positive spherical aberration due to the combined effect of their inherent ocular spherical aberration and the positive spherical aberration induced by their contact lens correction.

This suggests that an appropriate aspheric correcting surface on a contact lens would reduce the spherical aberration of the eye significantly for all defocus corrections if it were incorporated into a contact lens. Indeed, this option for correcting higher-order aberrations with contact lenses has always been attractive to those involved in the fabrication of contact lenses, as adding an aspheric correction to a lens using a rotationally symmetrical manufacturing method such as lathing or cast molding is significantly less challenging than adding nonrotationally symmetrical aberration correcting optical surfaces. Interestingly, there is a wide range of spherical aberration values across the patient population, and it is not correlated to the magnitude of ametropia (Kingston et al. 2013).

This provides a dilemma to manufacturers of low-cost, high volume contact lens products, such as daily-disposable soft contact lenses, because it is not economically feasible to provide the optimum spherical aberration correction for each eye. The options for these companies are to provide a single spherical aberration correction based on the population average or to use spherical aberration correction as a fitting parameter and provide a slightly larger number of fitting alternatives (2 or 3 "bins" that cover different portions of the population range of spherical aberration). Smaller manufacturers using lathes to directly manufacture the lenses on a custom basis (i.e., each lens is manufactured based on the parameters of a patients eye supplied by the fitting clinician) are not bound by these economic restrictions and can provide a full correction of the spherical aberration present in that individual eye. Ray tracing modeling based on 446 individual eye models that incorporate the wavefront aberration profiles of actual patient eyes with 6 mm pupil sizes and refractive astigmatism of no greater than –0.50 D and programmed to estimate the retinal image resolution in terms of logMAR acuity shows that a single population average-based spherical aberration applied across a range of ametropia correction from +6.00 D to –9.00 D will provide some level of improved estimated logMAR acuity (by at least one letter) in 75% of eyes compared to the same population of eyes corrected with traditional spherically surfaced optics (Kingston et al. 2013). Reviewing the model results further showed that 67% of these 446 eyes demonstrated logMAR acuity of 0.00 or better with this single value spherical aberration correction incorporated into their spherical contact lens correction. Adding three potential spherical aberration corrections as a fitting parameter increases this group (predicted logMAR acuity of 0.00 or better) to 80% of the eyes, while custom optimization of the spherical aberration present in the population of model eyes with aspheric optics showed an improvement in logMAR acuity in 84% of eyes. These calculated results assumed that the lens was centered on the visual axis and that the residual defocus and astigmatism were corrected. Optimizing defocus only without correcting astigmatism reduced these group sizes to 50% for a single spherical aberration correction, 64% for three spherical aberration corrections, and 69% for a custom optimized spherical aberration correction.

Ultimately, correction of the measured population average value for spherical aberration with contact lenses provides the potential to improve visual performance to varying degrees under low-light, large-pupil conditions in a relatively large portion of the population. Additional degrees of freedom in the ability to refine this correction will increase both the magnitude of the visual improvement and the number of patients experiencing the benefit. However, other significant low- and high-order aberrations coexist with spherical aberration in the eye that cannot easily be corrected with an aspheric or aspheric toric lens as well as those generated by the decentration of the contact lens relative to the center of the pupil, and the correction of these aberrations needs to be addressed if the optical performance of the eye is to reach its maximum potential.

11.10 HIGHER-ORDER WAVEFRONT ABERRATIONS AND THEIR CORRECTION WITH CONTACT LENSES

Spherical aberration, while significant in the hierarchy of higher-order wavefront aberration of the eye, is typically not the dominant aberration in most eyes. Third-order Zernike terms, such as coma and trefoil, are the largest magnitude higher-order wavefront aberrations found in the general population (Porter et al. 2001). These aberrations must be corrected by a rotationally stable contact lens and manufactured using a process capable of creating nonrotationally symmetrical surfaces. In fact, Figure 11.6 shows the theoretically predicted distance visual acuity distribution calculated for a large sample of the population representing contact lens wearers (pupil size 6 mm) when corrected alternatively with sphere and cylinder only; with sphere, cylinder, and spherical aberration; and with sphere, cylinder, third-order Zernike terms, and spherical aberration. Eyes with 0.00 logMAR or better were 38%, 79%, and 99%, respectively, for these three levels of correction.

Clearly, while some patients benefit from correction of spherical aberration in addition to their myopia and astigmatism, there is a substantially larger increase in retinal image quality when the third-order Zernike terms are also corrected, and so a contact lens must be designed to correct both symmetrical and nonrotationally symmetrical higher-order aberrations of the eye if a true visual benefit is to be realized across a substantial proportion of the population.

This leads to the question, which type of contact lens would be ideal for neutralizing higher-order wavefront aberrations? RGP and scleral lenses have been utilized for years as a method for correcting eyes with pathological or postsurgical deformities of the anterior corneal surface, where standard spherocylindrical spectacle lenses do not provide adequate visual acuity. However, it is the rigid nature of the lens itself, rather than innovative optics that provides the wavefront aberration correction of these lenses. Vision of eyes with significantly distorted corneas can be improved with RGP lenses because the anterior surface of the contact lens forms the new refracting surface, with the tear film filling in the difference between the irregular cornea and the regular back surface of the lens. In this way, a large proportion of higher-order wavefront aberrations generated by the abnormal corneal surface are corrected, with the potential for any residual higher-order wavefront aberration from the ocular system to be corrected on the front surface of the RGP or scleral contact lens (Jinabhai et al. 2012). From a manufacturing perspective, generating a complex wavefront correcting surface on the front surface of an RGP or scleral lens is preferred, since the material is rigid during processing, and does not undergo any postprocessing expansion or contraction that is inherent in soft hydrophilic or silicone–hydrogel lenses.

Unfortunately, the physical nature of RGP lenses, and the way they must be fit to ensure tear exchange and mobility on the eye, renders them less desirable as a method for correcting wavefront aberrations of the eye through complex optical surfaces. GP lenses

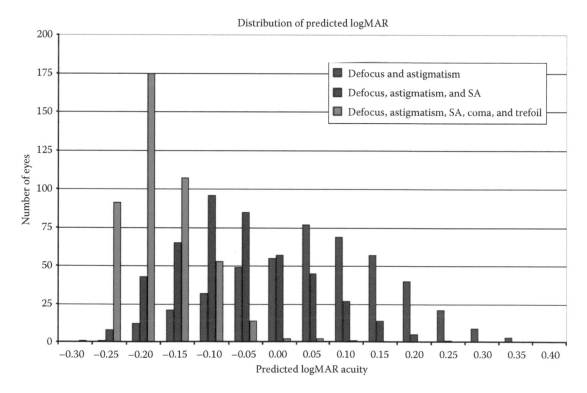

Figure 11.6 Theoretically predicted distance visual acuity distribution calculated for a large sample population (*n* = 446) representing contact lens wearers (pupil size 6 mm) when corrected alternatively with sphere and cylinder only; with sphere, cylinder and spherical aberration; and with sphere, cylinder, third order Zernike terms and spherical aberration. Eyes with 0.00 logMAR (6/6) visual acuity or better were 38%, 79% and 99% respectively for the three corrections.

are designed to be very mobile on the eye, moving at least 1.0 mm with each blink and finding a position of rest with up to 1 mm difference relative to the optical axis of the eye following each blink (Knoll and Conway 1987). Hence, stabilizing a conventional GP lens, such that it repeatedly returns to the same horizontal and vertical location relative to the visual axis, without rotating around this axis and maintaining physiologically desirable tear exchange behind the lens, is very difficult. Scleral lenses, which rest primarily on the scleral surface beyond the limbal junction with the cornea, hold potential promise, as they are very stable in their location on the eye. However, lenses of this type require a greater degree of fitting skill and clinician/patient interaction to generate a clinically acceptable fitting, and at this time the marketplace worldwide has shown they are not a preferred option for a mainstream ophthalmic correction.

Soft hydrophilic lenses are held in position relative to the cornea by an entirely different set of forces and are squeezed onto the cornea with the first blink following insertion and deformed to take the shape of the tissue beneath. The deformation of the lens that undergoes during this process generates radial stress in the lens, and it is this force combined with gravity that center the lens on the cornea at the position of equilibrium. Since there is only one optimal position on the corneal/scleral surface that provides the least radial stress, clinicians find that well-fitted soft lenses relocate to the same position on the cornea within 0.1–0.4 mm after every blink (Young 1992). Although the lens is not centered on the visual axis of the eye, it does relocate consistently relative to that axis after every blink. Control of rotation around the visual axis with soft lenses has been realized for over a decade, and the current generation of sophisticated prism-ballasted designs provides rotational stability within 5° between any series of blinks. It is this ability to relocate with great precision that makes soft lenses more clinically desirable than RGP lenses for correcting higher-order wavefront aberrations.

In theory, even slight changes in centration and rotation of a lens designed to correct up to fifth-order Zernike terms will significantly reduce the visual benefits experienced by that correction. Calculations to understand the tolerance/benefit ratio to decentration and rotational alignment of correcting surfaces typical of those found in the general ophthalmic population have been performed. Guirao et al. (2001) found that Zernike terms with higher (rotational) angular orders were more sensitive to the rotation of the correcting surface from the ideal position. Hence, coma is most tolerant to misrotation of the correcting surface, offering a visual benefit with misrotation up to 60°; astigmatism is the next most tolerant aberration, with visual benefit with misrotation up to 30°; trefoil is the next most tolerant aberration, with visual benefit with misrotation up to 20°; and so on. Rotation from the ideal correction position that is greater than these values would provide a retinal image quality that is poorer than leaving the Zernike term uncorrected. Decentration of the ideal correcting lens from the ideal correcting axis results in the higher-order aberrations generating a larger magnitude of lower-order aberrations. Hence, a correcting lens with coma will generate astigmatism, and defocus, spherical aberration will produce coma and tilt, while defocus or astigmatism will produce only tilt (prism). In general, higher-order Zernike terms are less tolerant to decentration than lower terms in relation

to their ability to improve retinal image quality. Guirao et al. calculated the reduction in visual benefit from lenses designed to correct wavefront aberrations of a 10-eye-sample population with increasing lens misrotation and decentration, respectively. They estimated that the lens would offer no better image quality than an uncorrected eye for a rotation of 45° or a translation of 1 mm and no improvement over a lens that corrects only defocus and astigmatism for a 17° rotation or a 0.6 mm translation. Clearly the visual benefit of a lens designed to correct both lower- and higher-order wavefront aberrations is dependent on repeatable lens centration and rotation following each blink or eye movement, the values generated by Guirao et al.'s analysis suggest that a soft lens is capable of remaining within these limits.

It is apparent that a contact lens designed to correct the higher-order wavefront aberrations of the eye needs to be a soft lens, prism-ballasted to control rotation, with the higher-order correction on the anterior surface of the lens (since lenses of this design may not perfectly align to the corneal surface and transfer subtle optical changes to the anterior surface of the lens). Since it would be extremely difficult to predict the manner in which the soft lens would distort as it is squeezed onto the cornea by the lid, the most pragmatic method would be to measure the wavefront aberrations with the contact lens in situ. Hence, a trial lens, with all the physical properties of the final custom-correcting lens but without the custom-correcting optical anterior surface, would be placed on the eye and allowed to settle, at which time wavefront sensor measurements would be taken through the lens–eye combination. In this way, the final custom-correcting lens will compensate for any variations in the eye's higher-order wavefront aberrations introduced by the lens itself or by the tear film between the lens and the cornea.

Alignment of the wavefront measurement is critical as the line of sight as defined by the center of the pupil will most likely be decentered relative to the geometric center of the lens. Therefore, the wavefront measurement must be performed with reference to both the center of the pupil and to the center of the lens. To achieve this, a marking scheme needs to be implemented on the trial lenses used for these measurements. These markings provide an indication of the center of the lens, as well as the rotational orientation of the lens during wavefront aberration measurements (Figure 11.7). The lathing parameters necessary to cut the nonrotationally symmetrical front surface using a 3-axis CNC lathe are calculated from the wavefront sensor measurements and the rotational position and decentration of the trial lens and downloaded to the lathe, generating the prescribed number of lenses, from a single trial lens to a 1 year supply (Figure 11.7).

Several research groups have attempted making customized higher-order wavefront aberration correcting soft contact lenses, and others are working on similar concept scleral lenses (Marsack et al. 2008, Katsoulos et al. 2009). Results are promising, with all investigators showing a significant reduction in ocular higher-order aberrations with custom-correcting contact lenses. Interestingly, however, the improvement in visual performance recorded by subjects corrected with these lenses, while measurable, does not correlate well to the magnitude of higher-order RMS reduction observed (Sabesan et al. 2007, Jinabhai et al. 2014). Sabesan and Yoon (2009) investigated

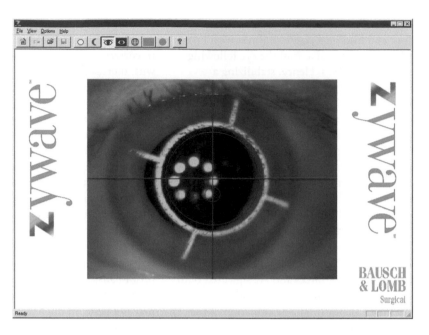

Figure 11.7 One possible marking scheme which could be used on a soft trial lens to locate the center of the lens and the rotational orientation of the lens during measurements of the wavefront aberration of the lens-eye combination.

this further by using an adaptive optics optical bench to correct higher-order wavefront aberration in a group of keratoconic subjects and a normal control group. Both groups had equivalent residual wavefront RMS values after correction with the adaptive optic system, but the keratoconic group demonstrated significantly lower corrected visual acuity scores compared to the control group. The researchers hypothesized that patients who are adapted to the retinal image distortion inherent to their ocular higher-order aberrations will not immediately benefit from a reduction in ocular higher-order RMS, and it is only after extended periods of wear that their visual system will adapt to the new retinal image quality. Sabesan and Yoon (2010) confirmed this hypothesis by conducting an experiment using an adaptive optic system to again alter the ocular aberrations of a group of keratoconic patients as well as a control group of normals. Both groups had their ocular aberrations corrected with the adaptive optic system, and then the residual aberration profile for the keratoconic patients with their habitual contact lens correction was added back into the system. Both groups of subjects were tested for high- and low-contrast visual acuity through this new higher-order aberration correction. Interestingly, the keratoconic subjects, who were habitually adapted to this level of higher-order aberration, showed significantly better visual performance over that achieved by the normals who were not adapted to this level of aberration, even though the measured aberration RMS was the same for both groups of subjects. This suggests that patients corrected with contact lenses customized to correct their individual ocular aberrations may not experience the full benefits of the correction until they have had enough time for their visual system to fully adapt to their new retinal image quality. It is unknown at this time how long this period of adaptation may be, but clinical experience with spectacle lens correction and patient adaptation to new aberration profiles induced by the spectacles suggests it will be a matter of days to weeks.

11.11 CORRECTION OF PRESBYOPIA WITH CONTACT LENSES

Although bifocal and multifocal contact lens designs have been available since the mass adoption of hard contact lenses as a form of optical correction in the late 1960s, there were typically custom designs available from small local laboratories, which required significant expertise on the part of the clinician to generate an acceptable visual performance at both distance and near for the patient. It was not until the early 1980s when the first mass-produced soft contact lens designs for presbyopia gained clearance for commercial release by the FDA that contact lenses were considered a viable option for the correction of presbyopia by a significant portion of clinicians. Indeed, even today, penetration is small, with only 37% of contact lens–wearing presbyopic patients receiving some form of presbyopic contact lens correction (Morgan et al. 2011a), the others wearing reading glasses over contact lenses.

First described in the late 1960s, one of the most common forms of contact lens correction for presbyopia used today is to correct one eye (typically the dominant eye) for distance vision, and the other (nondominant) eye for near viewing, with single vision contact lenses (Evans 2007). This technique is known as monovision, and although some binocular functions such as stereopsis are compromised over a full-distance binocular correction with reading glasses for near, research has shown that a monovision correction still provides superior binocular function over a true monocular correction, while providing visual acuity at distance and near equivalent to the patient's corrected monocular acuity. Indeed, research interest in monovision increased significantly during the late 1980s as market demands from the aging "baby boomer" segment of the population who had grown up with contact lens correction sought an acceptable presbyopic correction. Studies demonstrated that lower levels of optical defocus in the near corrected eye impacted binocular contrast sensitivity only at the higher spatial frequencies compared to a full-distance binocular correction (Loshin et al. 1982) and that patients with added powers greater than +1.50 D started to show poorer contrast sensitivity results than a monocular distance correction (Pardhan and Gilchrist 1990). A continuous field of through-focus vision is believed to be essential for a monovision correction to be successful as it allows the visual cortex to seamlessly move from visual input of one eye to the other. Gaps in the through-focus visual field (at intermediate distances for instance) may cause difficulties for the patient to unconsciously move from the dominant eye (distance) focus to the nondominant (near) focus or vice versa. Patients with strong ocular dominance for distant and near objects in the same eye will be not be successful in monovision (Schor and Erickson 1988).

Although very successful for early presbyopic patients, monovision quickly becomes unacceptable for most patients as their reading addition requirement increases above 1.50 D, and so a true bifocal or multifocal correction becomes the only viable contact lens alternative. Successful designs can be categorized into two main design categories: those based on the optical concept of having distinct optical zones on the lens, which cover the majority of the pupil for the desired object distance, that is, the distance zone covers the pupil when viewing an object at distance, and the near zone covers the pupil when viewing at near (alternating or translating vision lenses), and those that deliver light from distant and near objects to the pupil and onto the retina at the same time in all positions of gaze (known as simultaneous vision lenses).

Alternating vision lenses require a controlled movement of the lens on the eye (relative to the pupil) to translate the two different optical zones of the contact lens into their appropriate position. Investigators have found that the lower eyelid has little vertical movement during up and down gaze relative to the pupil (Borish and Perrigin 1987), and so most translating bifocal or multifocal contact lens design strategies have included features to provide increased interaction between the lens and the lower eyelid to hold the lens in position as the pupil of the eye moves from one optical zone on the lens to the other. These types of designs are more typically successful in RGP lenses because of their greater magnitude and ease of movement on the eye during down gaze,

although commercially successful soft contact lens designs have been introduced to the marketplace (Gussler et al. 1992; Figure 11.8).

Simultaneous vision lenses have light from all object distances reaching the retina at the same time, although only a portion of the light passing through the pupil will be in focus for the object distance being viewed by the patient at any particular time. Light that is not focused on the retina will, at best, reduce the contrast of the in focus image, but may also cause halo effects, ghosting, and secondary imaging that can be visually disturbing to the patient, particularly under high-contrast, low illumination viewing conditions, such as driving at night (Llorente-Guillemot et al. 2012). Simultaneous vision lens designs for contact lenses are typically rotationally symmetrical to avoid the need for lens stabilization and are either multifocal in nature, using a second- or third-order polynomial aspheric surface, or a bifocal or trifocal design with specific radial zones designed to correct light coming from distant, intermediate, and/or near objects. Lenses can be designed with either a higher positive power in the center of the optic zone relative to the power across the rest of the lens (center near) or a lower positive power in the center of the optic zone (center distance). As the surfacing capabilities of CNC lathes have improved in conjunction with more sophisticated metrology for nonspherical surfaces, designs have become more sophisticated and can include somewhat complex, rotationally symmetric, free-form surfaces to generate an extended depth of focus that provides adequate distance vision without significant secondary imaging while providing improved visual performance at intermediate and near working distances for presbyopic patients (Figure 11.9).

The performance of simultaneous vision lens designs is limited by a number of factors. The lens optical zone is frequently decentered relative to the pupil (unfortunately soft contact lenses tend to center well over the cornea, which is not usually centered well relative to the pupil, the magnitude of decentration varying greatly across the patient population). Patients have inherent higher-order aberrations of the eye that is being corrected, and these are usually only considered as a population average when the optical design of the presbyopic correcting lens is performed. Lens movement can cause visual instability as the purposely aberrated/aspheric optic of the lens moves upward with a blink,

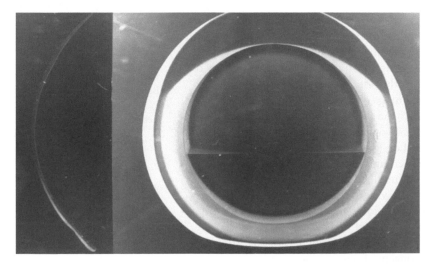

Figure 11.8 Face view using a contact lens comparator to show the design features of a soft translating, bifocal lens which was successfully commercialized in the mid 1980's. Lack of oxygen transmission and reduced comfort have led to the demise of these lens designs.

Power profiles

Figure 11.9 Face view of two commercially available simultaneous vision soft multifocal lenses imaged using Schlieren optics. Cartoons representing the power profiles for each lens design are shown below the lens images, with greater positive power represented by the vertical axis arrow direction. The horizontal axis represents the power profile moving from the optical center of the lens. The first lens is a bifocal with alternating distance and near zones surrounding the lens center. The second lens is a center near design, with an optical step at the junction between the central near and intermediate zone, with a gradual reduction in positive power over the intermediate zone until meeting the peripheral distance power zone.

and variations in pupil size occur with changes in ambient illumination. Attempts to mitigate this last variable have been made using refractive designs with alternating concentric distance and near zones (Kirschen et al. 1999) or diffractive designs (Young et al. 1990) that have been released to the market, but the significant ghosting and secondary imaging inherent to these abrupt transition designs have led to varying degrees of acceptance of these designs by the clinical community.

Based on commercial success, lenses with continuously changing surface radii tend to be more readily accepted due to reduced secondary imaging, although they are typically only suitable for presbyopic patients requiring lower power reading additions, as the reduction of image quality across all object distances becomes unacceptably poor as the asphericity, and hence depth of focus of the optical surface, is increased. Other nonoptical factors such as the desire for relatively low-cost options; monthly, or even daily, replacement; excellent comfort; and the requirement of a simple to follow and time efficient fitting procedure have all but eliminated the use of custom-designed simultaneous or alternating vision designs from the majority of the lenses prescribed worldwide.

Despite the shortcomings of current presbyopic contact lens optical designs, the demand for a successful contact lens solution to the correction of presbyopia has driven many clinicians to utilize combinations of designs from different manufacturers, with differing add powers and performance (some may be better for distance and intermediate vision; others may be better for intermediate and near vision) using empirical strategies to meet the individual visual requirements of each patient. This approach is known as "modified monovision" as it utilizes multifocal lenses rather than single vision lenses but is based in the fundamental concept of monovision, that is, dominant eye corrected for distance and the nondominant eye corrected

for near (Pence 1987). Utilizing multifocal lenses increases the depth of focus for each eye and allows the successful application of monovision to patients with higher add requirements than the +1.50 D limit after which binocularity with traditional single vision lens monovision begins to be significantly impacted.

Ultimately, the success rate of contact lens optical designs to correct presbyopia will only experience a significant improvement if lenses are prescribed on a more custom basis, taking additional individual measurements of ocular higher-order aberrations, pupil size, and position relative to the cornea and to the center of the lens when in situ, and designed to meet the visual demands of the patient while taking these variables into account. As with single vision lenses designed to correct ocular higher-order aberrations, this must be combined with a manufacturing process that provides the frequent replacement of lenses necessary to meet the physiological needs of the eye at a cost low enough to be acceptable to the marketplace.

REFERENCES

Barr, J., Contact lenses 2005. *Contact Lens Spectrum*, ePub (2006) Jan.

Bennett, A.G., *Optics of Contact Lenses*. Association of Dispensing Opticians, London, UK, 1974.

Bennett, E.S., Weissman, B.A., *Clinical Contact Lens Practice*. Williams & Wilkins; Lippincott, Philadelphia, PA, 2005.

Borish, I.M., Perrigin, D., Relative movement of lower lid and line of sight from distant to near fixation, *Am J Optom Physiol Opt* 64 (1987): 881–887.

Brennan, N.A., Lindsay, R.G., McCraw, K., Young, L., Bruce, A.S., Golding, T.R., Soft lens movement: Temporal characteristics, *Optom Vis Sci* 71 (1994): 359–363.

Carney, L.G., Mainstone, J.C., Carkeet, A., Quinn, T.G., Hill, R.M., Rigid lens dynamics: Lid effects, *CLAO J* 23 (1997): 69–77.

Cox, I.G., The whys and wherefore of soft lens visual performance, *Con Lens Ant Eye* 23 (2000): 3–9.

Creech, J.L., Do, L.T., Fatt, I., Radke, C.J., In vivo tear-film thickness determination and implications for tear-film stability, *Curr Eye Res* 17 (1998): 1058–1066.

Dietze, H.H., Cox, M.J., On- and off-eye spherical aberration of soft contact lenses and consequent changes of effective lens power, *Optom Vis Sci* 80 (2003): 126–134.

Doane, M.G., Interferometric measurement of in-vivo contact lens wetting, in *Proceedings of SPIE 1161, New Methods in Microscopy and Low Light Imaging*, San Diego, CA, 1989, p. 320.

Efron, N., Morgan, P.B., Woods, C.A., International contact lens prescribing survey consortium—International survey of rigid contact lens fitting, *Optom Vis Sci* 90 (2013): 113–118.

Evans, B.J., Monovision: A review, *Ophthalmic Physiol Opt* 27 (2007): 417–439.

Fanti, P., Holly, F.J., Silicone contact lens wear III. Physiology of poor tolerance, *Contact Intraocul Lens Med J* 6 (1980): 111–119.

Fonn, D., Gauthier, C.A., Pritchard, N., Patient preferences and comparative ocular responses to rigid and soft contact lenses, *Optom Vis Sci* 72 (1995): 857–863.

Griffiths, M., Zahner, K., Collins, M., Carney, L., Masking of irregular corneal topography with contact lenses, *CLAO J* 24 (1998): 76–81.

Guillon, J.P., Abnormal lipid layers. Observation, differential diagnosis, and classification, *Adv Exp Med Biol* 438 (1998): 309–313.

Guirao, A., Williams, D.R., Cox, I.G., Effect of rotation and translation on the expected benefit of an ideal method to correct the eye's higher order aberrations, *J Opt Soc Am (A)*, 18 (2001): 1003–1015.

Gussler, C.H., Solomon, K.D., Gussler, J.R., Litteral, G., Van Meter, W.S., A clinical evaluation of two multifocal soft contact lenses, *CLAO J* 18 (1992): 237–239.

Holden, B.A., The principles and practice of correcting astigmatism with soft contact lenses, *Aust J Optom* 58 (1975): 279–299.

Hong, X., Himebaugh, N., Thibos, L.N., On-eye evaluation of optical performance of rigid and soft contact lenses, *Optom Vis Sci* 78 (2001): 872–880.

Howland, H.C., Howland, B., A subjective method for the measurement of the monochromatic aberrations of the eye, *J Opt Soc Am* 67 (1977): 1508–1518.

Ivanoff, A., Letter to the editor: About the spherical aberration of the eye, *J Opt Soc Am* 46 (1953): 901–903.

Jansen, M.E., Begley, C.G., Himebaugh, N.H., Port, N.L., Effect of contact lens wear and a near task on tear film break-up, *Optom Vis Sci* 87 (2010): 350–357.

Jenkins, T.C.A., Aberrations of the eye and their effect upon vision. Part II, *Br J Physiol Opt* 20 (1963): 161–201.

Jiang, H., Wang, D., Yang, L., Xie, P., He, J.C., A comparison of wavefront aberrations in eyes wearing different types of soft contact lenses, *Optom Vis Sci* 83 (2006): 769–774.

Jinabhai, A., O'Donnel, C., Tromans, C., Radhakrishnan, H., Optical quality and visual performance with customised soft contact lenses for keratoconus, *Ophthal Physiol Opt* 34 (2014): 528–539.

Jinabhai, A., Radhakrishnan, H., Tromans, C., O'Donnell, C., Visual performance and optical quality with soft lenses in keratoconus patients, *Ophthal Physiol Opt* 32 (2012): 100–116.

Josephson, J.E., Caffery, B.E., Clinical experiences with the Tesicon silicone lens, *Int Contact Lens Clin* 7 (1980): 235–245.

Katsoulos, C., Karageorgiadis, L., Vasileiou, N., Mousafeiropoulos, T., Asimellis, G., Customized hydrogel contact lenses for keratoconus incorporating correction for vertical coma aberration, *Ophthal Physiol Opt* 29 (2009): 321–329.

King-Smith, P.E., Fink, B.A., Fogt, N., Nichols, K.K., Hill, R.M., Wilson, G.S., The thickness of the human precorneal tear film: Evidence from reflection spectra, *Invest Ophthalmol Vis Sci* 41 (2000): 3348–3359.

Kingston A.C., Cox I.G., Population spherical aberration: Associations with ametropia, age, corneal curvature, and image quality, *Clin Ophthalmol* 7 (2013): 933–938.

Kirschen, D.G., Hung, C.C., Nakano, T.R., Comparison of suppression, stereoacuity, and interocular differences in visual acuity in monovision and acuvue bifocal contact lenses, *Optom Vis Sci* 76 (1999): 832–837.

Knoll, H.A., Conway, H.D., Analysis of blink-induced vertical motion of contact lenses, *Am J Optom Physiol Opt* 64 (1987): 153–155.

Koomen, M., Tousey, R., Scolnik, R., The spherical aberration of the eye, *J Opt Soc Am* 39 (1949): 370–376.

Llorente-Guillemot, A., García-Lazaro, S., Ferrer-Blasco, T., Perez-Cambrodi, R. J., Cerviño, A., Visual performance with simultaneous vision multifocal contact lenses, *Clin Exp Optom* 95 (2012): 54–59.

Loshin, D.S., Loshin, M.S., Corner, G., Binocular summation with monovision contact lens correction for presbyopia, *Int Contact Lens Clin* 9 (1982): 161–165.

Lowther, G.E., Tomlinson, A., Critical base curve and diameter interval in the fitting of spherical soft contact lenses, *Am J Optom Phys Opt* 58 (1981): 355–360.

Lu, F., Mao, X., Qu, J., Xu, D., He, J.C., Monochromatic wavefront aberrations in the human eye with contact lenses, *Optom Vis Sci* 80 (2003): 135–141.

Mandell, R.N., *Contact Lens Practice*, 4th edn. Springfield, IL: Charles C. Thomas, 1988.

Marsack, J.D., Parker, K.E., Applegate, R.A., Performance of wavefront-guided soft lenses in three keratoconus subjects, *Optom Vis Sci* 85 (2008): 1172–1178.

McNamara, N.A., Polse, K.A., Brand, R.J., Graham, A.D., Chan, J.S., McKenne, C.D., Tear mixing under a soft contact lens: Effects of lens diameter, *Am J Ophthalmol* 127 (1999): 659–665.

Mertz, G.W., Holden, B.A., Clinical implications of extended wear research, *Can J Optom* 43 (1981): 203–205.

Momeni-Moghaddam, H., Naroo, S.A., Askarizadeh, F., Tahmasebi, F., Comparison of fitting stability of the different soft toric contact lenses, *Contact Lens Ant Eye* 37 (2014): 346–350.

Morgan, P.B., Efron, N., Woods, C.A., The international contact lens prescribing survey consortium, An international survey of contact lens prescribing for presbyopia, *Clin Exp Optom* 94 (2011a): 87–92.

Morgan, P.B., Woods, C.A., Tranoudis, I.G., Helland, M., Efron, N., Grupcheva, C.N. et al., International contact lens prescribing in 2010, *Contact Lens Spectrum*, ePub (2011b) Jan.

Nichols, J.J., Contact lenses 2012, *Contact Lens Spectrum*, ePub (2013) Jan.

Nichols, J.J., Mitchell, G.L., King-Smith, P.E., Thinning rate of the precorneal and prelens tear films, *Inv Ophthalmol Vis Sci* 46 (2005): 2353–2361.

Nilsson, S.E.G., Montan, P.G., The hospitalized cases of contact lens induced keratitis in Sweden and their relation to lens type and wear schedule: Results of a 3-year retrospective study, *CLAO J* 20 (1994): 97–101.

Pardhan, S., Gilchrist, J., The effect of monocular defocus on binocular contrast sensitivity, *Ophthal Physiol Opt* 10 (1990): 33–36.

Pearson, R.M., Efron, N., Hundredth anniversary of August Müller's inaugural dissertation on contact lenses, *Surv Ophthalmol* 34(2) (1989): 133–141.

Pence, N.A. Modified trivision: A modified monovision technique specifically for trifocal candidates, *ICLC* 14 (1987): 484–487.

Plainis, S., Charman, W.N., On-eye power characteristics of soft contact lenses, *Optom Vis Sci* 78 (1998) 78: 44–54.

Pointer, J.S., Eyeblink activity with hydrophilic contact lenses. A concise longitudinal study, *Acta Ophthalmol* 66 (1988): 498–504.

Pointer, J.S., Gilmartin, B., Larke, J.R., Visual performance with soft hydrophilic contact lenses, *Am J Optom Physiol Opt* (1985) 62: 694–701.

Porter, J., Guirao, A., Cox, I.G., Williams, D.R., Monochromatic aberrations of the human eye in a large population, *J Opt Soc Am A* 18 (2001): 1793–1803.

Remba, M.J., Clinical evaluation of FDA approved toric hydrophilic soft contact lenses (Part I), *J Am Optom Assoc* 50 (1979): 289–293.

Remba, M.J., Part II. Clinical evaluation of toric hydrophilic contact lenses, *J Am Optom Assoc* 52 (1981): 211–221.

Richdale, K., Berntsen, D.A., Mack, C.J., Merchea, M.M., Barr, J.T., Visual acuity with spherical and toric soft contact lenses in low- to moderate-astigmatic eyes, *Optom Vis Sci* 84 (2007): 969–975.

Rigel, L., A history of contact lens innovation, *Contact Lens Spectrum*, ePub (2007) Sept.

Sabesan, R., Jeong, T.M., Carvalho, L., Cox, I.G., Williams, D.R., Yoon, G., Vision improvement by correcting higher-order aberrations with customized soft contact lenses in keratoconic eyes, *Opt Lett* 15 (2007): 1000–1002.

Sabesan, R., Yoon, G., Visual performance after correcting higher order aberrations in keratoconic eyes, *J Vis* 9 (2009): 6.

Sabesan, R., Yoon, G., Neural compensation for long-term asymmetric optical blur to improve visual performance in keratoconic eyes, *Inv Ophthalmol Vis Sci* 51 (2010): 3835–3839.

Schaeffer, J., Beiting, J., Contact lens pioneers. The early history of contact lenses, *Rev Optom (Suppl)* 146 (2009): 3–23.

Schein, O.D., Glynn, R.K., Poggio, E.C., Seddon, J., Kenyon, K.R.; The Microbial Keratitis Study Group, The relative risk of ulcerative keratitis among users of daily-wear and extended-wear soft contact lenses. A case-control study, *N Engl J Med* 321 (1989): 773–778.

Schein, O.D., McNally, J.J., Katz, J., Chalmers, R.L., Tielsch, J.M., Alfonso, E., Bullimore, M., O'Day, D., Shovlin, J., The incidence of microbial keratitis among wearers of a 30-day silicone hydrogel extended-wear contact lens, *Ophthalmology* 112 (2005): 2172–2179.

Schor, C., Erickson, P., Patterns of binocular suppression and accommodation in monovision, *Am J Optom Physiol Opt* 65 (1988): 853–861.

Smelser, G.K., Ozanics, V., Structural changes in corneas of guinea pigs after wearing contact lenses, Arch Ophthalmol 49 (1953): 335–340.

Stevenson, R., Young's Modulus measurements of gas permeable contact lens materials, *Optom Vis Sci* 68 (1991): 142–145.

Sweeney, D.F., Corneal exhaustion syndrome with long-term wear of contact lenses, *Optom Vis Sci* 69 (1992): 601–608.

Teague, S., Gunderson, G., Soft contact lenses for astigmatism, in *Manual of Contact Lens Prescribing and Fitting*, M.M. Hom and A.S. Bruce, eds. St. Louis, MI: Butterworth Heinemann, 2006, pp. 341–361.

Timberlake, G.T., Doane, M.G., Bertera, J.H., Short-term, low-contrast visual acuity reduction associated with *in vivo* contact lens drying, *Optom Vis Sci* 69 (1992): 755–760.

Tuohy, K.M., Contact lens, U.S. Patent 2,510,438, filed February 28, 1948.

van der Worp, E., A guide to scleral lens fitting [monograph online]. Scleral Lens Education Society, 2010. Available from: http://commons.pacificu.edu/mono/4/.

Walline, J.J., Gaume, A., Jones, L.A., Rah, M.J., Manny, R.E., Berntsen, D.A., Chitkara, M., Kim, A., Quinn, N.; the Contact Lenses in Pediatrics (CLIP) Study Group, Benefits of contact lens wear for children and teens, *Eye Contact Lens* 33 (2007): 317–321.

Walline, J.J., Jones, L.A., Sinnott, L., Chitkara, M., Coffey, B., Jackson, J.M., Manny, R.E., Rah, M.J., Prinstein, M.J.; the ACHIEVE Study, Randomized trial of the effect of contact lens wear on self perception in children, *Optom Vis Sci* 86 (2009): 222–232.

Wang, J., Fonn, D., Simpson, T.L., Jones, L., Precorneal and pre- and postlens tear film thickness measured indirectly with optical coherence tomography, *Inv Ophthalmol Vis Sci* 44 (2003): 2524–2528.

Westheimer, G., Aberrations of contact lenses, *Am J Opt Arch Am Acad Optom* 38 (1961): 445–448.

Wichterle, O., Lim, D., Hydrophilic gels for biological use, *Nature* 185 (1960): 117–118.

Wolffsohn, J.S., Drew, T., Dhallu, S., Sheppard, A., Hofmann, G.J., Prince, M., Impact of soft contact lens edge design and midperipheral lens shape on the epithelium and its indentation with lens mobility, *Inv Ophthalmol Vis Sci* 54 (2013): 6190–6197.

Yang, Y., Thompson, K., Burns, S.A., Pupil location under mesopic, photopic, and pharmacologically dilated conditions, *Inv Ophthalmol Vis Sci* 43 (2002): 2508–2512.

Young, G., Soft lens fitting reassessed, *Contact Lens Spectrum* Dec 7(12) (1992): 56–61.

Young, G., Evaluation of soft contact lens fitting characteristics, *Optom Vis Sci* 73 (1996): 247–254.

Young, G., Grey, C.P., Papas, E.B., Simultaneous vision bifocal contact lenses: A comparative assessment of the in vitro optical performance, *Optom Vis Sci* 67 (1990): 339–345.

Young, G., McIlraith, R., Hunt, C., Clinical evaluation of factors affecting soft toric lens orientation, *Optom Vis Sci* 86 (2009): 1259–1266.

Zantos, S.G., Holden, B.A., Ocular changes associated with continuous wear of contact lenses, *Aust J Optom* 61 (1978): 418–426.

12 Corrections in highly aberrated eyes

Jason D. Marsack and Raymond A. Applegate

Contents

12.1 INTRODUCTION

The idea that an ophthalmic contact lens correction can be customized to correct the wavefront aberration of an individual eye has existed, at least in theory, for over half a century (Smirnov, 1961). While the benefit for normal eyes is limited to those with above-average higher-order aberrations and/or large pupils, the benefit to those with highly aberrated eyes can be significant (Williams, 2001). That said, the clinical availability of these "custom, wavefront-guided corrections" remains extremely limited at best and practically nonexistent. The underlying causes that lead to elevated levels of ocular aberration are varied (ocular disease, prior ocular surgery, trauma, etc.), but for patients with elevated levels of ocular aberration, current methods utilized to prescribe sphero-cylindrical refractive corrections do not provide normal levels of visual performance (e.g., Zadnik et al., 1998). This chapter defines the highly aberrated eye, discusses the state of the art associated with custom contact lens correction for the highly aberrated eye, and discusses the challenges that limit the widespread availability of these custom lenses.

12.2 HIGHLY ABERRATED EYE

12.2.1 DEFINING THE TERM "HIGHLY ABERRATED EYE"

The term "highly aberrated eye" is used frequently both in the literature and informally when discussing the optical performance of the eyes with reduced optical quality. But what is the property of the optics of the eye that is being described by the phrase "highly aberrated"? Certainly an eye suffering from a 12 D myopic refractive error could be considered to be "highly aberrated." So could an eye with 3 D of cylinder. But the colloquial use of the term "highly aberrated" is not typically applied to eyes that suffer from spherical or cylindrical refractive error, even when the magnitude of those errors is large. The reason is simple: these eyes can be well corrected with traditional sphero-cylindrical corrections, such as spectacles, soft contact lenses and rigid contact lenses.

The term "highly aberrated" is more commonly used to describe the eyes that suffer from elevated levels of optical error, where those errors are not well characterized as sphero-cylindrical error. Clinically, this type of non-sphero-cylindrical refractive error is sometimes referred to using the catch-all phrase "irregular astigmatism." But this term is misleading, as it implies that these optical errors are related to or a derivative form of astigmatism, which is inaccurate. The misconception can be better understood in the context of the prescriptions used to correct sphero-cylindrical errors of the eye. Traditional spectacle lenses for distance vision are commonly reported in minus cylinder (optometry) or plus cylinder (ophthalmology) form. If the refractive error is simple myopia, the wavefront error (WFE) is increasingly advanced with pupil diameter and rotationally symmetric (Figure 12.1 panel a). If the WFE is simple myopic astigmatism, the WFE is increasingly advanced in the meridian orthogonal to the axis of the astigmatism and is mirror symmetric along both principal meridians

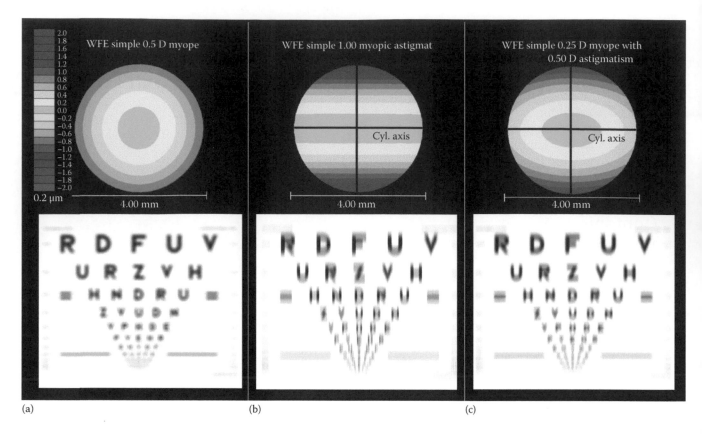

Figure 12.1 Panel a—Wavefront error (WFE) of a simple 0.50 D myope over a 4 mm pupil. Notice the rotational symmetry of the positive (advanced) WFE, which increases with increasing distance from the center of the 4 mm pupil. The lower portion of the figure is a corresponding retinal image simulation of a logMAR acuity chart. Panel b—WFE of a simple 1.00 D simple myopic astigmat (spherical equivalent = 0.50 D). Notice the perfect mirror symmetry of the positive (advanced) WFE along the power meridian and symmetry along the orthogonal axis meridian. The lower portion of panel b is a corresponding retinal image simulation of a logMAR acuity chart. Panel c—WFE of a compound myopic astigmat with 0.25 D of myopic spherical error and 0.50 D of myopic astigmatic error with the corresponding retinal simulation. Comparing the three simulations, notice how spherical error blurs in a rotationally symmetric manner (panel a), astigmatic error when the cylinder axis is at 180° draws the letters vertically, and the blur is in between simple spherical error and astigmatic error when there is both spherical and astigmatic errors. It is also interesting to note that all three errors demonstrated have a spherical equivalent error of 0.50 D, yet the resulting blur is markedly different.

(Figure 12.1 panel b). If the error is a combination of both simple spherical error and astigmatism, the WFE remains mirror symmetric along principal meridians (Figure 12.1 panel c). When astigmatism is present, the power difference in the two orthogonal principal meridians of the toric surface describes the level of astigmatism in the correcting lens. Considering this definition, what is meant when one refers to astigmatism as "irregular"? Certainly, the attribute being described is not the axis of the astigmatic correction or to the magnitude of the astigmatism.

Describing the optical performance of the highly aberrated eye requires a reporting method that is capable of representing optical errors other than sphere and cylinder. While several methods have been utilized historically, the normalized Zernike polynomial is the current ANSI standard, ANSI Z80.28-2004 utilized to report the optical aberrations in the eye. The Zernike polynomial decomposes the optical error of the eye into mathematically orthogonal terms, including defocus and astigmatism (which are the terms included in a sphero-cylindrical corrections), as well as higher-order terms such as trefoil, coma, and spherical aberration (Figure 12.2). It is the presence of elevated levels of these higher-order terms in differing magnitudes that leads to the application of the label "highly aberrated" to an eye and WFE maps that

are neither rotationally symmetric nor mirror symmetric (as is illustrated in Figure 12.3) and therefore cannot be well corrected with a simple sphero-cylindrical correction.

12.2.2 ORIGINS OF ELEVATED LEVELS OF OCULAR ABERRATION

In the context of contact lens correction, there are several conditions of the eye that tend to be frequently associated with high levels of optical aberration. Ocular disease processes (e.g., keratoconus and pellucid marginal degeneration) lead to rotationally asymmetric morphological changes in the cornea. Due to the fact that the refractive power of the eye is directly related to the shape of the eye and the fact that the anterior surface of the cornea is responsible for roughly 66% of the refracting power of the eye (Smith and Atchison, 1997), these rotationally asymmetric changes in shape induce rotationally asymmetric optical properties in the cornea, leading to an increased level of higher-order optical aberration in the eye. For example, keratoconus is a disease that is classically defined as resulting in an inferior steepening of the cornea, leading to an inferior displacement of the corneal apex. In the uncorrected eye, Pantanelli et al. showed that higher-order aberration in keratoconus was approximately 5.5 times higher

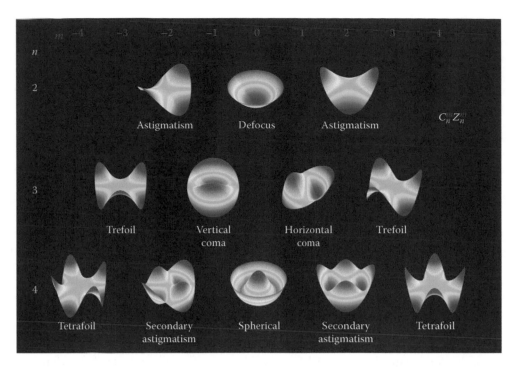

Figure 12.2 Zernike terms for the second to fourth radial orders (n = 2, 3, 4). The series is infinite. The zero and first radial orders are not shown because they do not influence the image quality. The zero radial order is a constant and the first radial order includes prism terms that impact image location. Numbers in green specify the radial order and numbers in red specify the angular frequency. The coefficient C specifies the magnitude of the particular Zernike term. Here the coefficients for each term are the same, so the relative shapes can be easily compared. When the terms are added together, they can form very asymmetric wavefront errors.

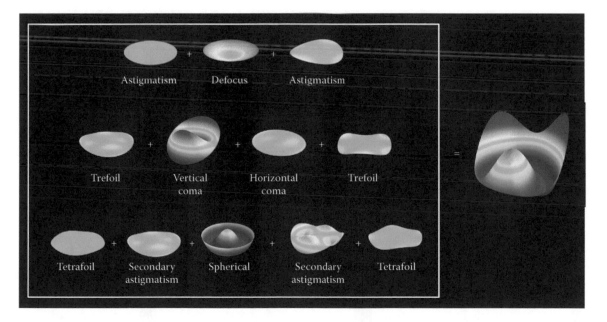

Figure 12.3 Illustrates the aberration structure of a highly eye with aberrations not correctable with sphere and cylinder. As can be seen in the figure, this eye is dominated by vertical coma and spherical aberration and has only small amounts of defocus (sphere) and astigmatism (cylinder). Since each Zernike term is mathematically independent (orthogonal) to all other Zernike terms, they can simply be summed up to visualize the total aberrations (total wavefront error [TWFE]). Notice that parceling the TWFE into orthogonal components specifies the major sources of wavefront error into components, giving a better insight as to why sphere and cylinder alone cannot adequately correct the optical errors of this eye.

than levels found in a normal population and that 53% of the HOA variance in the KC population can be accounted for by vertical coma (term C_3^{-1} in the Zernike polynomial) (Pantanelli et al., 2007). This result is optically consistent with the downward displacement of the corneal apex mentioned earlier. Along with

disease processes, prior ocular surgery, in particular, corneal transplantation (Pantanelli et al., 2012), and early corneal refractive procedures (Applegate and Howland, 1997) can lead to morphological asymmetry in the cornea and the induction of higher-order aberration. While the root cause of the change in corneal shape may differ, the presence of elevated levels of non-sphero-cylindrical errors poses challenges for the patient as well as the clinician attempting to provide the patient with a traditional refractive correction.

A first step in understanding the consequence of these aberrations present in the highly aberrated eye is to visualize them. Figure 12.4 panel a shows a high-order (HO) WFE map for an eye with mild to moderate keratoconus. It is called an HO WFE map because the defocus (sphere) and cylinder components have been removed so as not to mask the appearance of these HO aberrations. Unlike the WFE presented in Figure 12.1, this HO WFE map (panel a) displays a significant amount of rotational asymmetry, meaning that the error in this eye, in addition to the sphere and cylinder (displayed as the correction needed to reduce the sphere and cylinder components of the WFE to zero directly under the WFE map), is much more complex than simple sphero-cylindrical error displayed in Figure 12.1 and cannot be corrected with standard glasses. The inferior displacement of the corneal apex manifests as negative vertical coma, which can be seen in this picture as the intense region of blue in the inferior portion of the figure with a corresponding intense orange color in the top portion of the WFE map. That these higher-order WFEs cannot be eliminated by standard glasses is demonstrated in panel b where this same individual is wearing their glasses prescription during the WFE measurement. Notice that the HO WFE is essentially the same in both maps and the total WFE (TWFE)

is reduced. Further, notice that there is still a significant residual sphere and cylinder correction (displayed below the WFE map in panel b) despite the fact that the sphero-cylindrical correction that the individual is wearing provided the individual with the best possible subjective acuity. This is typical. All patients during a subjective refraction select a sphero-cylindrical lens to optimize their acuity. They do not select lenses to entirely eliminate their sphero-cylindrical errors. In panel c of Figure 12.4 the residual HO WFE of a normal individual wearing their best spectacle correction is provided for comparison. Notice again that measurement of the WFE of the eye reveals a residual correction, despite the fact that the individual is wearing their best subjective correction emphasizing the fact that patients select lenses to provide best acuity and not to eliminate all sphero-cylindrical WFE. The lower portion of each panel is a retinal image simulation for each case. By comparing the simulation of panel b to that of c, one can visualize why glasses do not provide the highly aberrated eye with a normal retinal image, even though in both cases the individual is wearing their best spectacle correction.

12.2.2.1 Whole eye WFE of one keratoconic eye

The full spectrum of the refractive error, notably the elevated levels of higher order aberration, would not be well compensated with a sphero-cylindrical correction (Figure 12.4 panel b) as compared to a normal eye (Figure 12.4 panel c). Significant levels of higher-order aberration in this keratoconic eye (notably coma)

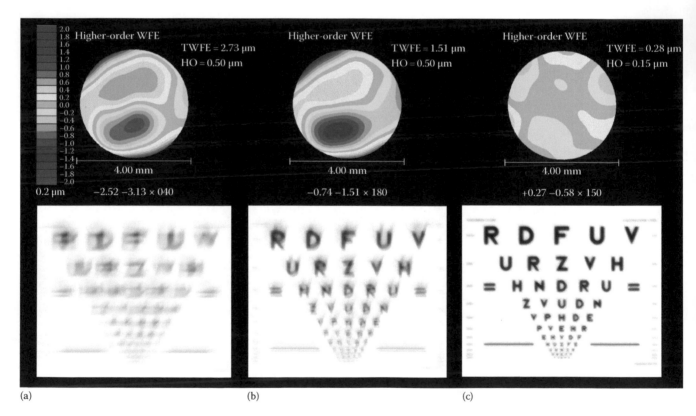

(a)   (b)   (c)

Figure 12.4 Panel a displays the higher-order wavefront error (WFE) of an uncorrected eye with moderate keratoconus, the total WFE, the high-order (HO) WFE, the sphero-cylindrical correction based on the measured second-order WFE, and a resulting retinal simulation of a logMAR visual acuity chart. Panel b displays the same components as panel a except the eye of panel a is now corrected with the individual's preferred sphero-cylindrical correction. Notice the HO WFE is not reduced. Panel c also displays the same components as Panel b except here for a normal eye wearing the individual's preferred sphero-cylindrical correction. Notice in both panel b and c there is a measureable and significant uncorrected sphero-cylindrical error illustrating that in both cases a subjective refraction does not typically eliminate sphero-cylindrical WFE. Instead the individual does as they are instructed and selects a sphero-cylindrical correction that optimizes acuity.

remain uncorrected and degrade retinal image quality. For this reason, spectacle correction does not typically provide the keratoconic subject with normal visual acuity (VA).

12.2.3 QUALITY OF LIFE IN INDIVIDUALS WITH ELEVATED LEVELS OF ABERRATION

The presence of elevated levels of uncorrected higher-order aberrations can have a significant impact on the quality of life. We rely heavily on the input obtained through our visual system to perform tasks of daily life, and the presence of elevated levels of aberration can have a significant impact on our ability to perform these tasks. This was made clear by the earliest forms of refractive surgery, where after surgery, ghosting of letters, halos particularly in low illumination (big pupil), and general distortion of the visual percept were reported (Applegate and Howland, 1989). These qualitative complaints were often observed even in cases where the reduction in sphero-cylindrical error is good if not excellent (as in Figure 12.3) and let patients seek retreatment (Salz, 2003). As an example, the patient may be able to read reasonably well on the acuity chart but continues to have subjective (qualitative) complaints about their vision. This type of qualitative complaint is common in patients with elevated levels of ocular aberration including individuals that suffer from disease or had previously undergone ocular surgery.

So what can this impact on the quality of life be measured? This burden can be put into context if one considers the point in life when the aberrations associated with keratoconus tend to impact the individual. The burden of keratoconus typically becomes apparent to the patient in the late teens or early twenties. This is the time in life associated with peak educational intensity, vocational preparation, and family formation. Simply put, current correction strategies leave an unmet need in a portion of the population with highly aberrated eyes, and this unmet need is driving the demand for increasingly sophisticated correction strategies, including improved contact lens corrections.

12.2.4 CURRENT CORRECTION STRATEGIES FOR HIGHLY ABERRATED EYES

For individuals suffering from elevated levels of higher-order aberration, rigid gas permeable (GP) contact lenses are the most common form of optical correction (Zadnik et al., 1998). The more symmetric anterior surface of the GP contact lens replaces the distorted cornea as the eye's first refracting surface, and the tear lens filling between the lens and cornea masks a portion of the higher-order aberration present in the eye through a process known as index matching. If the optical index of the tears was identical to that of the cornea, the anterior corneal aberrations would be eliminated and replaced with the aberrations of the new corneal surface defined by the contact lens. Unfortunately, the index match is not perfect, and while these lenses do reduce the level of aberration in the highly aberrated eye, they do not typically reduce them to normal levels. More importantly, these lenses that target only the sphero-cylindrical error do not provide the lens designer with a method to target individual higher-order aberration terms that may be elevated within an individual eye. The result is incomplete correction of higher-order aberration and reduced visual performance as compared to the normal eye (Kosaki et al., 2007; Marsack et al., 2007; Negishi et al., 2007).

12.3 WAVEFRONT-GUIDED CONTACT LENSES

12.3.1 BASIC CONCEPT BEHIND A WAVEFRONT-GUIDED CONTACT LENS CORRECTION

If the total aberration map of the highly aberrated eye can be measured and visualized, can this information also be utilized to construct a custom, wavefront-guided correction tailored to the unique needs of an individual eye? The answer is yes. To understand the philosophy employed in designing a wavefront-guided contact lens, it is helpful to first consider the process used in prescribing sphero-cylinder corrections and then contrast that process with the process associated with a wavefront-guided contact lens correction. Subjective refraction is the most common method utilized to identify a sphero-cylindrical correction in the clinic. That is to say, the patient is asked to make a series of two alternative forced choice judgments about the quality of their vision. The series of choices being compared guide the clinician to a sphero-cylindrical correction that provides the best subjective image quality, as judged by the patient. But what is the patient truly optimizing with this series of comparisons? The patient is not, as is commonly believed, identifying the sphero-cylindrical correction that best compensates their sphero-cylindrical error. In fact, the patient has no knowledge of their own level of sphero-cylindrical error or their level of higher-order aberration. The only information the patient has when making these decisions is the impact of those aberrations on their visual percept. Applegate et al. have demonstrated that aberrations interact to increase or decrease the visual percept (Applegate et al., 2003), and the fact that the patient is asked to optimize their percept (which is a result of all aberrations in their eye) with sphere and cylinder, it is most accurate to state that the patient is doing their best to choose the sphero-cylinder correction that optimizes their visual percept in the presence of their uncorrected higher-order aberrations.

In addition to the subjective process described earlier, clinicians currently have at their disposal objective methods to determine the sphero-cylindrical error of the eye. These instruments as a class are called autorefractors and utilize some objective optical metric to determine the eye's sphero-cylindrical correction. However, unlike the subjective method, the quality of vision is not assessed by the patient in this method. For this reason, objective refractions are typically used as a starting point for the subjective refraction process in the clinic. This is logical due to the fact that the patient's perceived quality is, in the end, the ultimate assessment of the quality of the correction.

The concept behind a wavefront-guided contact lens is different than the subjective and objective refraction processes described earlier, meaning a different process is being optimized. In the case of the wavefront-guided contact lens, the data needed to design the lens are quantified as the full monochromatic aberration profile of the eye when wearing a trial lens. This process does not rely on patient feedback to determine the optimal correction for the patient. The optimal correction in this case would be the correction that would correct the entire WFE (both lower- and higher-order aberrations) of the eye, which would have the effect of imaging a point of light in object space to a diffraction-limited spot on the retina. Under ideal conditions, the data describing the full WFE of the eye would be passed into the design process and used to implement a correction that fully compensated for the eye's error. While theoretically straightforward, achieving full correction of the WFE with a lens on the eye is not feasible for several reasons.

First, contact lenses move on the eye and are therefore constantly subject to some level of registration uncertainty. A full wavefront correction that would completely correct the WFE of the eye when aligned would induce WFEs when misaligned, and if the misalignment were great enough, it could induce aberrations that are larger than the level of uncorrected error in the eye (Guirao et al., 2001).

12.3.2 DESIGNING THE OPTICS OF A WAVEFRONT-GUIDED CONTACT LENS FOR THE HIGHLY ABERRATED EYE

When designing a wavefront-guided contact lens, it is important to first consider which aberrations will be integrated into the correcting optics. One factor that drives this decision is the expected (or better yet, measured) movement of the lens on the eye. Several approaches have been explored to examine the potential impact lens movement may have on the efficacy of the correction. Guirao et al. (2002) studied the impact of lens movement on the

optical performance of a theoretical wavefront-guided contact lens. This work laid out a method that allowed a percentage of each Zernike term to be corrected, such that there is always some benefit in the reduction of RMS WFE in the presence of lens movement. Shi et al. built on the work of Guirao et al., examining the impact of movement on the resultant optical and visual performance. The goal of the Shi et al.'s work was to identify a correcting aberration pattern that provided an increase in visual performance while keeping that visual performance within a defined level of variation (Shi et al., 2013). To accomplish this goal, Shi et al. examined the impact of the observed lens movement using visual image quality, due to the fact that a measured change in visual quality metrics (in particular visual Strehl ratio and neural sharpness) had been shown to be highly correlated to a measured change in visual performance as measured by acuity.

12.3.3 DEMONSTRATIONS OF CUSTOM WAVEFRONT GUIDED CONTACT LENS CORRECTIONS

The benefits associated with targeting higher-order aberration for correction in a custom contact lens have been demonstrated in a variety of forms on highly aberrated eyes. An example is presented in Figure 12.5. Panel a of Figure 12.5 displays an HO WFE map of an eye of an individual with severe keratoconus wearing

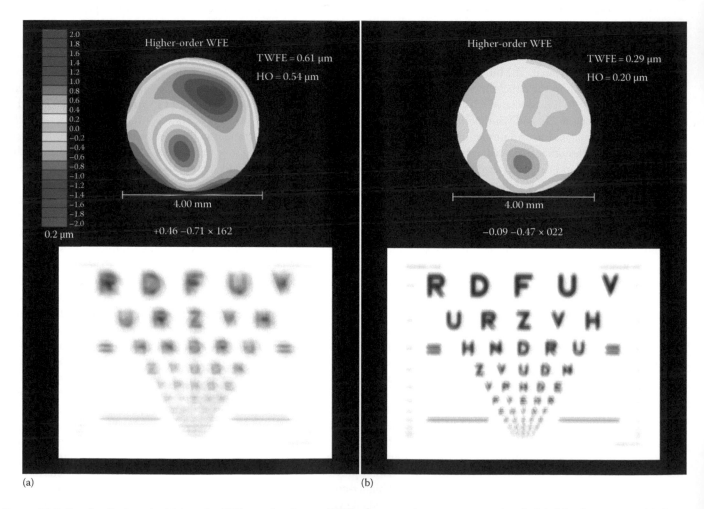

(a)   (b)

Figure 12.5 Panel a displays the high-order (HO) wavefront error (WFE) of a severe keratoconus wearing their habitual gas permeable lens, the total wavefront error (TWFE), the HO WFE, the residual sphero-cylindrical correction based on the measured second-order WFE, and a resulting retinal simulation of a logMAR visual acuity chart. Panel b displays the same components as panel a, except the eye of panel a is now wearing a wavefront-guided correction. Notice the HO WFE is markedly reduced.

their habitual (GP) lens. Notice the high level of residual coma is reversed from that seen in Figure 12.4 panels a and b. This is a result of the rigid GP contact lens masking the corneal first surface aberrations allowing the corneal posterior surface aberrations to manifest themselves more fully (Chen et al., 2007). Panel b displays an HO WFE map of the same eye wearing a wavefront-guided scleral lens. The residual Zernike refraction is displayed below the HO WFE map in both cases along with retinal image simulation of a logMAR chart resulting from the remaining residual TWFE.

The goal of a wavefront-guided correction is to achieve a result similar to that displayed in Figure 12.5—reduce HO WFE to normal levels. Sabesan et al. (2007) demonstrated a reduction in higher-order aberration in KC by a factor of 3× using a wavefront-guided soft contact lens and an improvement of 2.1 lines in VA, as compared to sphero-cylindrical correction. Marsack et al. (2008) compared correction of KC with a wavefront-guided soft contact lens to habitual GP corrections. In that experiment, all three subjects achieved photopic high-contrast (HC) logMAR VA equal to or better than values recorded with their habitual GP lenses. Both of these works were performed in soft contact lenses. Implementation of custom correction in the form of a soft contact lens is complicated by movement and flexure of the soft lens on the eye. Wavefront-guided soft contact lenses are desirable from the patient's point of view for the same reason typical eyes choose to wear soft lenses over rigid forms of correction (comfort being chief among these reasons). That said, soft lenses pose increased complexity in their construction and use because they must be hydrated after manufacture and tend to move significantly on the eye.

To reduce the impact of lens movement and flexure and to provide a more stable platform for customization, investigators began placing wavefront-guided optics in scleral contact lenses. Work by Sabesan et al. on keratoconus subjects wearing custom wavefront-guided scleral contact lenses demonstrated that higher-order RMS was reduced and both VA and contrast sensitivity improved with wavefront-guided scleral contact lenses. Sabesan et al. (2013) also reported that while improved, visual performance did not reach levels seen in subjects with normal levels of higher-order aberration. Work by Marsack et al. (2014) demonstrated that of 14 KC eyes studied while wearing wavefront-guided scleral lens corrections, 10 successfully reached residual HORMS levels within the range experienced by normal, age-matched subjects. Data in Figure 12.5 are derived from the data collected in the Marsack et al.'s work. However, visual performance as measured by HC acuity did not reach normal, age-matched levels, similar to the results of Sabesan. In summary, these small-scale demonstrations of wavefront-guided lenses show the promise of the technology, and this promise derives from the fact that its design can (1) personalize the correction to the aberration structure of an individual eye and in doing so (2) improve the image quality and resulting visual performance.

12.4 MOVING FORWARD WITH CUSTOM CORRECTIONS

12.4.1 CHALLENGES TO CLINICAL DEPLOYMENT OF WAVEFRONT-GUIDED CONTACT LENSES

While abstractly, delivery of a wavefront-guided lens may seem straightforward, it is in fact very challenging from technical, clinical, and data management points of view. One of the fundamental and unavoidable factors associated with achieving a good wavefront-guided correction is the collection of high-quality WFE data on the eye. Widespread implementation of wavefront-guided corrections will be predicated on an ability to quantify the aberrations present in an eye. Traditionally, wavefront sensors have been constructed with the typical clinical eye in mind, limiting their utility in the case of highly aberrated eyes. For this reason, the level of aberration present in these eyes can exceed the dynamic range of clinical instruments, making the collection of the needed WFE data difficult, if not impossible in some cases. Further, the availability of wavefront sensors in clinical use is relatively low when compared to other ophthalmic instruments. Simply put, there is limited use for these data in the standard eye care practice. This means that high-dynamic-range instruments capable of quantifying WFE data in these highly aberrated eyes will need to become more common in the clinic if these custom corrections are to become more available.

Yet another factor hampering the utility of the wavefront sensor in the clinic is the time constraints associated with the measurements needed to construct a wavefront-guided lens. The time constraint is particularly limiting when viewed in the context of today's clinic. Beyond the measurement of WFE data is the need to quantify other aspects of lens performance, such as the average position of a trial lens on the eye. This is necessary in order for the optics of the wavefront-guided lens to be placed correctly in front of the pupil. The process needed to manufacture an efficacious wavefront-guided lens often takes several visits, each lasting several hours. This clinical dilemma suggests that wavefront-guided contact lens will most likely first take hold in specialty clinics, perhaps those focused exclusively on fitting of these lenses or where the technology to capture the WFE of the eye is needed for other purposes (such as in a refractive surgery center).

Wavefront-guided lenses need to be cost effective and relatively easy for the eye care clinician to deploy. The next set of hurtles to overcome include development of reliable methods to insure lens stability on the eye, development of small footprint instrumentation that included accurate and precise high dynamic range wavefront sensing, corneal/scleral topography and automated lens position monitoring as well as development of easy to use design software capable of unique nontraditional modifications. The clinical success of this technology is a matter of making it practical and profitable for clinicians and manufacturers, while keeping it affordable for patients.

12.4.2 SUMMARY: WAVEFRONT-GUIDED CONTACT LENSES IN CONTEXT

What is the clinical potential of a wavefront-guided contact lens correction? That depends on the vantage point from which the question is asked. From the patient's point of view, the answer is a more complete optical correction of the aberrations that reduce retinal image quality and improve visual performance over their current mode of correction. But that speaks to only one aspect of what makes a patient "happy" with a contact lens correction. Based on the variability in need associated with individual patient preferences (comfort, wear time, cost, complexity of care regimen), it is clear that all modes of contact lens correction will continue to have a meaningful place in contact lens practice. From the practitioner's point of view, wavefront-guided corrections provide a tool to treat their optically most challenging patients, as well as a way to provide a unique service for their patients. Contact lenses that target patient-specific higher-order aberrations have been demonstrated, and investigations are currently looking to integrate them into the clinical environment. Highly aberrated eyes are particularly well suited for treatment with these wavefront-guided corrections.

REFERENCES

Applegate RA, Howland HC. Refractive surgery, optical aberrations, and visual performance. *J Refract Surg*. 1997;13:295–299.

Applegate RA, Marsack JD, Ramos R, Sarver EJ. Interaction between aberrations to improve or reduce visual performance. *J Cataract Refract Surg*. 2003;29:1487–1495.

Chen M, Sabesan R, Ahmad K, Yoon G. Correcting anterior corneal aberration and variability of lens movements in keratoconic eyes with back-surface customized soft contact lenses. *Opt Lett*. 2007;32:3203–3205.

Guirao A, Cox IG, Williams DR. Method for optimizing the correction of the eye's higher-order aberrations in the presence of decentrations. *J Opt Soc Am A*. 2002;19:126–128.

Guirao A, Williams DR, Cox IG. Effect of rotation and translation on the expected benefit of an ideal method to correct the eye's higher-order aberrations. *J Opt Soc Am A*. 2001;18:1003–1015.

Kosaki R, Maeda N, Bessho K, Hori Y, Nishida K, Suzaki A et al. Magnitude and orientation of Zernike terms in patients with keratoconus. *Invest Ophthalmol Vis Sci*. 2007;48:3062–3068.

Marsack JD, Parker KE, Applegate RA. Performance of wavefront-guided soft lenses in three keratoconus subjects. *Optom Vis Sci*. 2008;85:1172–1178.

Marsack JD, Parker KE, Pesudovs K, Donnelly WJ 3rd, Applegate RA. Uncorrected wavefront error and visual performance during RGP wear in keratoconus. *Optom Vis Sci*. 2007;84:463–470.

Marsack JD, Ravikumar A, Nguyen C, Ticak A, Koenig DE, Elswick JD, Applegate RA. Wavefront-guided scleral lens correction in keratoconus. *Optom Vis Sci*. 2014;91:1221–1230.

Negishi K, Kumanomido T, Utsumi Y, Tsubota K. Effect of higher-order aberrations on visual function in keratoconic eyes with a rigid gas permeable contact lens. *Am J Ophthalmol*. 2007;144:924–929.

Pantanelli S, Macrae S, Jeong TM, Yoon G. Characterizing the wave aberration in eyes with keratoconus or penetrating keratoplasty using a high-dynamic range wavefront sensor. *Ophthalmology*. 2007;114:2013–2021.

Pantanelli SM, Sabesan R, Ching SS, Yoon G, Hindman HB. Visual performance with wave aberration correction after penetrating, deep anterior lamellar, or endothelial keratoplasty. *Invest Ophthalmol Vis Sci*. 2012;53:4797–4804.

Sabesan R, Jeong TM, Carvalho L, Cox IG, Williams DR, Yoon G. Vision improvement by correcting higher-order aberrations with customized soft contact lenses in keratoconic eyes. *Opt Lett*. 2007;32:1000–1002.

Sabesan R, Johns L, Tomashevskaya O, Jacobs DS, Rosenthal P, Yoon G. Wavefront-guided scleral lens prosthetic device for keratoconus. *Optom Vis Sci*. 2013;90:314–323.

Salz JJ. Wavefront-guided treatment for previous laser in situ keratomileusis and photorefractive keratectomy: Case reports. *J Refract Surg*. 2003;19:S697–S702.

Shi Y, Queener HM, Marsack JD, Ravikumar A, Bedell HE, Applegate RA. Optimizing wavefront-guided corrections for highly aberrated eyes in the presence of registration uncertainty. *J Vis*. 2013;13(7):8.

Smirnov MS. Measurement of the wave aberration of the human eye. *Biofizika* 1961;6:776–795.

Smith G, Atchison DA. *The Eye and Visual Optical Instruments*, 1st ed. Cambridge, U.K.: Cambridge University Press; 1997, p. 291.

Williams DR. How far can we extend the limits of human vision? In: MacRae S, Krueger RR, Applegate RA, eds. *Customized Corneal Ablation: The Quest for Supervision*. Thorofare, NJ: Slack; 2001, pp. 11–32.

Zadnik K, Barr JT, Edrington TB, Everett DF, Jameson M, McMahon TT, Shin JA, Sterling JL, Wagner H, Gordon MO. Baseline findings in the collaborative longitudinal evaluation of keratoconus (CLEK) study. *Invest Ophthalmol Vis Sci*. 1998;39:2537–2546.

13 Accommodating intraocular lenses

Oliver Findl and Nino Hirnschall

Contents

13.1 INTRODUCTION

Within the last decades, patients' demand for spectacle independence after cataract surgery has increased rapidly. Modern intraocular lenses (IOLs) and modern IOL power calculation allow a significantly improved postoperative uncorrected distance visual acuity (and quality) by correcting for axial eye length, corneal astigmatism, and to some extent higher-order aberrations. But it is still not possible to fully restore accommodation after cataract surgery. Therefore, patients need to decide whether they would like to wear glasses for distance or instead for near work. Alternatively, spectacle dependence can be reduced by either monovision or multifocal IOLs. Both attempts include an optical compromise (as mentioned in the multifocal IOL chapter), in the case of the former reduced stereopsis and possible irritation due to the anisometropia and in the case of the latter mainly the loss in contrast vision, halos, and glare resulting in visual quality being imperfect at all distances.

There is no accommodating IOL on the market that allows full restoration of accommodation and recent studies showed that the accommodative effect is weak to not existing.[1] Another aspect that has to be taken into account is that studies investigating the effect of accommodating IOLs use different outcome parameters.

One question arises: "When is an accommodating IOL really an accommodating IOL?" On first sight the answer to this question appears to be simple. But how can this accommodation be measured? In many studies evaluating accommodating IOLs, the main outcome was spectacle independence, or patient satisfaction. As shown in the last section of this chapter, these parameters should be used with caution, as they can easily be biased, especially in nonrandomized trials. Furthermore, as mentioned in a recent Cochrane review, none of the investigated trials on so-called accommodating IOLs showed relevant accommodation and all studies were considered as having a severe performance bias.[2] However, although real accommodation was not proven for any of the listed so-called accommodating IOLs, we will keep the term "accommodating IOLs" within this chapter.

13.2 SINGLE-OPTIC ACCOMMODATING IOLs

Most accommodating IOLs have one single optic. All of these IOLs use the optic-shift principle. The proposed working principle is that the relaxation of the zonules induced by ciliary muscle contraction causes relaxation of the elastic capsular bag. As a consequence, circumferential compression is applied onto the haptics and a forward movement of the IOL is induced.

13.2.1 1CU

The 1CU single-piece accommodative IOL (HumanOptics AG, Erlangen, Germany) is the most excessively researched accommodating IOL. It consists of a 5.5 mm biconvex optic, measures 9.8 mm in diameter, and is made of foldable acrylic hydrophilic material. Four broad-based haptics, which are thinner toward the optic ("transmission element"), are attached to the optic and act as a hinge, as shown in Figure 13.1.[3]

The accommodative effect of the 1CU was controversially discussed in literature. One study group found distance-corrected near visual acuity (DCNVA) and refractive change results superior to monofocal IOLs[4,5] in a nonrandomized study and no fading effects were observed 1 year postoperatively.[6] DCNVA was significantly better with the 1CU compared to the respective control group in three further trials.[7–9] Findl et al.[10] conducted a randomized bilateral study with intraindividual comparison and found a mean forward shift of 314 µm. Using ray-tracing calculations for the individual eyes, this would have resulted in a refractive change of less than 0.5 D in most eyes and only few eyes reaching 1.0 D. Hancox et al.[11] observed little forward shifting of the 1CU with accommodation using partial coherence interferometry (PCI) in contrast to a backward movement of a standard control IOL, but the group did not detect any difference in DCNVA. Evidence was obtained that the accommodative potency of the IOL decreases over time in a trial where accommodation amplitude declined due to ACO and posterior capsule opacification (PCO) after a 1-year follow-up.[12] Two other investigations with observation periods of 2 years[13] and 3 years[14] had similar findings.

13.2.2 TETRAFLEX

This one-piece accommodating IOL (KH-3500; Lenstec, St. Petersburg, Florida, USA) is made from flexible hydrophilic hydroxylethylmethacrylate and has a 5.75 mm equiconvex square-edged optic for enhanced PCO prevention (Figure 13.2).[15] Two closed-loop haptics with a 5° angulation are designed to facilitate a forward shifting of the IOL as a whole with ciliary body contraction.[15] Additionally to this optic-shift principle, the Tetraflex was designed to utilize flexure changes to the optics to add dioptric power with accommodative effort.[16]

Sanders et al.[17] implanted a series of Tetraflex IOLs in 95 eyes of 59 patients. Six months postoperatively, 89.3% of cases with bilateral implants achieved a DCNVA of 20/40 or better, 67.9% of 20/32 or better, and 28.6% of 20/25 or better. Accommodative amplitude assessed using a push-up method was >1 D for all participants, >2 D for 96.2% of cases, and >3 D for 46.2% of cases 6 months postoperatively. Brown et al.[18] conducted an observational study comparing the near visual acuity of patients implanted bilaterally with either the Tetraflex or the Crystalens AT-50, and a superior functional reading performance was found in the Tetraflex cohort: the proportion of patients achieving a reading speed of at least 80 words/min throughout a range of different print sizes was significantly higher 1 year postoperatively. In a large trial,[19] results of a nonrandomized 1-year follow-up of 239 eyes with the Tetraflex implanted and 96 controls with monofocal IOL were reported and enhanced near reading ability and spectacle independence compared to a monofocal IOL was demonstrated. Two more recent trials investigated on the axial IOL movement with accommodation using AS-OCT. The first group[15] measured the IOL position in a sample of 13 eyes of 8 patients at least 2 years postoperatively. Surprisingly, no anterior movement of the lens optic was observed, but a slight (0.02 ± 0.05 mm) backward shift was seen with accommodation. However, a change in ocular aberrations (predominantly defocus, astigmatism, coma, and trefoil) was detected with increasing accommodative strain. These findings suggest that the Tetraflex' enhanced near function

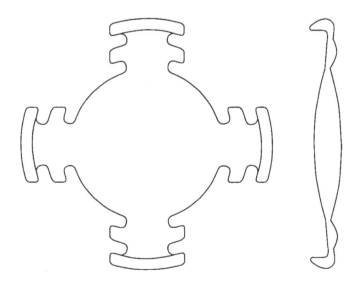

Figure 13.1 The 1CU intraocular lens (IOL) is a single-piece IOL with four flexible haptics. (From Menapace, R. et al., *Graefe Arch. Clin. Exp. Ophthalmol.*, 245(4), 473, 2007.)

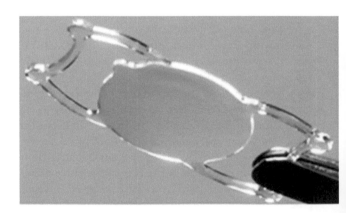

Figure 13.2 The Tetraflex intraocular lens consists of two flexible closed-loop haptics and a flexible optic. (From http://www.beyondlasi. co.uk/iol-vision-correction-over-45/accommodating-lenses, last visit: 15/08/2014.)

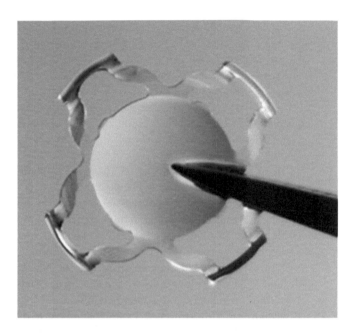

Figure 13.3 The OPAL-A intraocular lens (IOL) is an example of a single-piece IOL with four closed-loop haptics. (From Cleary, G. et al., *J. Cataract Refract. Surg.*, 36(5), 762, 2010.)

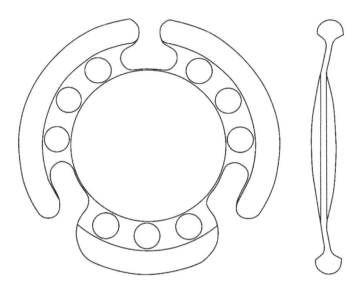

Figure 13.4 In contrast to other single-piece intraocular lenses (IOLs), the BioComFold IOL has a ring-shaped haptic. (From Menapace, R. et al., *Graefe Arch. Clin. Exp. Ophthalmol.*, 245(4), 473, 2007.)

is probably more attributable to a flexure of the optic during accommodation rather than to the presumed anterior shift of the IOL. In a second study, a mean forward shift of the Tetraflex of 337 ± 124 μm was detected and showed a strong correlation with the mean subjective accommodation of 0.94 ± 0.61 D.[20]

13.2.3 OPAL-A

The OPAL-A (HumanOptics AG, Germany) is a prototype one-piece acrylic focus shift accommodating IOL with a 5.5 mm single optic and a total diameter of 9.7 mm (Figure 13.3). The four flexible closed-loop haptics are designed to convey ciliary body contraction–induced circumferential compression into a forward shifting of the optic. Therefore, the optic–haptic junctions are designed to have a particularly large degree of flexibility. Cleary et al.[21] conducted a prospective nonrandomized pilot study including 22 patients that received the prototype IOL in one eye and were followed up for 6 months. The IOL displayed no clinically relevant forward shift with a mean pilocarpine-induced forward shift after 6 months of only 306 ± 161 μm. The highest mean optical change in response to a near stimulus of only 0.36 ± 0.38 D was observed 1 month postoperatively with a progressive decline in further follow-up visits.

13.2.4 BioComFold

The BioComFold (Morcher GmbH, Stuttgart, Germany) was the first commercially available accommodating IOL. The single-piece hydrophilic acrylic IOL has a 5.8 mm optic that is located anteriorly and is connected to an outer, discontinuous ring by broad, perforated, and angulated haptics (Figure 13.4). Two designs were marketed: model 43A was introduced in 1996, followed by model 43E in 1998, which had slight modifications in optic diameter, haptic design, and angulation. The proposed working mechanism of this IOL was centripetal compression of the haptics as a result of ciliary muscle contraction eventually

forcing the IOL to shift anteriorly. Only sparse clinical data on the performance of the BioComFold are available. In a randomized controlled trial by Legeais et al.[22] A-scan ultrasound was used to determine the anterior chamber depth of 15 eyes implanted with the BioComFold in cycloplegia and after instillation of pilocarpine 2%. A significantly higher forward shift of the study IOL (0.71 ± 0.55 mm) was observed one month postoperatively compared to a control IOL. Findl et al.[23] found mean anterior shifting of only 0.116 ± 0.11 mm for model 43A and 0.222 ± 0.24 mm for model 43E evaluated with PCI technique. No published reports on DCNVA results following implantation of the BioComFold IOL are as yet available.

13.2.5 CRYSTALENS

The Crystalens model AT-45 was introduced to the European market in 2002 by C&C Vision (later Eyeonics, now Bausch & Lomb Inc., Rochester, New York, USA). The newer model AT-50 is a three-piece construction and consists of an enlarged 5.0 mm biconvex silicone (Biosil) optic and two plate haptics with 50% thickness grooved hinges in vicinity to the optic–haptic junction, each with two laterally extending polyamide open loops at the end.[24] Overall diameter is 12.0 mm for IOLs <17.0 D and 11.5 mm for higher-powered IOLs (Figure 13.5). The IOL is designed to be implanted in a posteriorly vaulted manner with the posterior IOL surface resting on the lens capsule. A modified optic to increase depth of field was implemented in a further variation of the design, marketed as Crystalens HD.[25] This version is more of a multifocal IOL than an accommodating IOL.

In a randomized controlled trial, it was shown that the Crystalens IOL was not performing better compared to monofocal IOLs. In the same study it was also shown that the optic of the Crystalens IOL was moving in the wrong direction (backward instead of forward).[26] However, the Crystalens IOL is the only FDA-approved "accommodating" IOL.[27] Marchini et al.[28] performed a study including 20 eyes of 14 patients implanted with the AT-45. Fifteen percent and 45% of eyes achieved DCNVA of J1 and J3 or better, respectively.

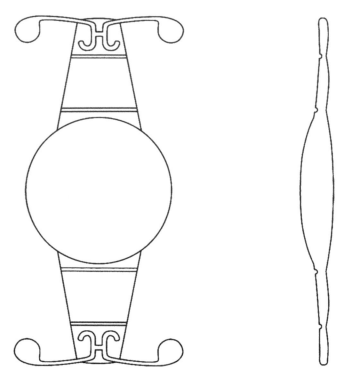

Figure 13.5 Although the design of the Crystalens intraocular lens (IOL) changed slightly over time, the principle of two flexible haptics adjusted to a single optic remained. In a more modern design ("HD"), the Crystalens IOL has a near add in the center. (From Menapace, R. et al., *Graefe Arch. Clin. Exp. Ophthalmol.*, 245(4), 473, 2007.)

13.3 ACCOMMODATIVE IOLs CHANGING OPTIC SHAPE

13.3.1 NuLens

The NuLens accommodating IOL's (NuLens Ltd, Herzliya Pituach, Israel) design features polymethyl methacrylate (PMMA) haptics with fine end plates that are secured to the ciliary sulcus without sutures, an anterior PMMA plane, a small chamber filled with flexible silicone gel, and a posterior piston operated by the empty capsular bag that applies pressure on the gel.[29] Consequently, the silicone gel bulges anteriorly through a round aperture in the anterior PMMA plane and—acting as a lens—adds refractive power.[30] Near focusing is established when the relaxed ciliary muscle stretches the capsular bag to push on the piston and to create this bulge. With ciliary body contraction, the capsular diaphragm is released resulting in disaccommodation[30] (Figure 13.6).

In a first feasibility trial,[30] the NuLens was implanted in primate eyes and, later, through a 9.0 mm incision in 10 eyes of 10 patients with cataracts and atrophic macular degeneration.[29] During a 12-month follow-up in humans, the IOL proved to be stable and well centered in all cases. A (subjectively determined) accommodative amplitude of 10 diopters was found.

However, there are some downsides to the NuLens design. It has been suggested by the developers that the "reversed" principle of ciliary body contraction–induced disaccommodation would necessitate an adaptive period of the visual system.[30] A further issue is the convergent eye movement associated with accommodation as near vision would be accompanied by divergence and distance vision by convergence. Potential difficulties with maintaining binocular single vision must not be neglected.[31] Furthermore, the developers annotated that a version of the NuLens suitable for small incision cataract surgery should be introduced with regard to the observed steep decline in endothelial cell count observed in the first human study as well as to reduce surgically induced astigmatism.[29]

13.3.1.1 FluidVision

Similar to the NuLens, the FluidVision IOL uses a fluid to change its optical shape, but in contrast to the NuLens, the FluidVision IOL stores the fluid in its haptic (Figure 13.7). With accommodative strain, the contracting ciliary body forces a refractive index-matched silicone oil from the outer haptics of the IOL through small channels into the optic, inducing an increase in optic thickness, IOL curvature, and, thus, refractive power.[32]

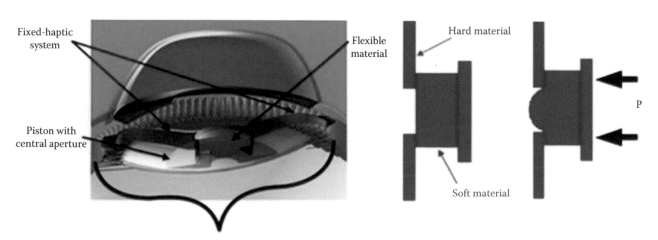

Figure 13.6 In contrast to flexible optic designs, the NuLens uses a small chamber filled with flexible silicone gel and a posterior piston operated by the empty capsular bag that applies pressure on the gel to allow accommodation. (From Alio, J.L. et al., *J. Cataract Refract. Surg.* 35(10), 1671, 2009; Ben-Nun, J. and Alio, J.L., *J. Cataract Refract. Surg.*, 31(9), 1802, 2005.)

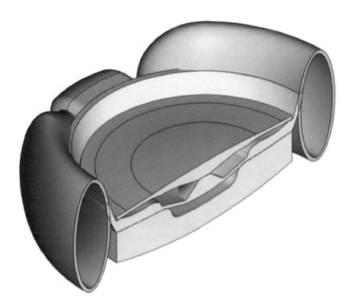

Figure 13.7 The FluidVision intraocular lens also uses a fluid to change the optical properties during accommodation. In contrast to the NuLens, the fluid is stored in the haptics. (From Roux, P., Early implantation results of the fluidvision AIOL in sighted eyes, *ASCRS*, Boston, MA, 2010.)

The FluidVision IOL (PowerVision Inc., Belmont, CA, USA) is still in development.

Results of a phase 1 clinical trial and early results of a phase 2 clinical trial have been presented.[32] The IOL, which was designed to be deliverable through a 4.0 mm incision, was implanted in three humans. All patients were within 1 D of emmetropia and an average accommodation of 5.6 D was found using a standard push-down test. Promising results of a recent study investigating on PCO formation 6 weeks after implantation of the FluidVision IOL in the animal model have been published.[33] Currently a multicenter study with an estimated enrollment of 115 participants is being conducted in 7 centers in Germany and South Africa.[34]

13.3.2 DUAL-OPTIC DESIGNS

Single-optic accommodating IOLs that utilize the shift of focus principle have the key disadvantage that the potential accommodative power is influenced by the refractive error and, thus, by IOL power. A dual-optic design with a high-powered, mobile anterior optic paired with a compensatory immobile negatively powered posterior optic does not exhibit this limitation. For example, a 1 mm anterior shifting of a 19.0 D single-optic accommodating IOL would result in an accommodation amplitude as high as 1.2 D. However, for a dual-optic IOL with a 32 D anterior lens and a posterior –12 D lens, the same forward shifting of the anterior optic produces a change of approximately 2.2 D.[35] Dual-optic devices occupy the entire capsular bag and the interoptic space is filled with aqueous humor. At rest, capsular tension causes a compression of the two optics, energy is stored in the interoptic articulations, and disaccommodation is established. Ciliary muscle contraction and consequent reduction in capsular tension induce a forward shifting of the anterior optic and an increased separation of the interoptic space.[36]

13.3.2.1 Synchrony IOL

The Synchrony IOL (Visiogen, Abbott Medical Optics, Santa Ana, CA, USA) received CE approval in 2006 and was the most commonly implanted dual-optic IOL to date. The one-piece IOL measures 9.5 mm in length and 9.8 mm in width, is made from foldable silicone, and can be injected through a 3.8 mm incision. A 32 D fixed power anterior optic measuring 5.5 mm is connected to a larger (6.0 mm) posterior optic of variable negative power via spring-loaded articulations (Figure 13.8).

Initial clinical results of 24 eyes implanted with the Synchrony IOL achieved a mean accommodative range of 3.22 ± 0.88 D as

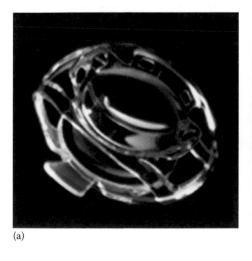

(a)

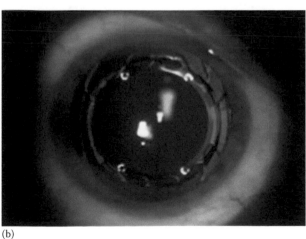

(b)

Figure 13.8 The first introduced dual-optic design accommodating intraocular lens (IOL) was the Synchrony IOL. (a) Shows the IOL ex vivo, (b) shows the implanted IOL. (From Ossma, I.L. et al., *J. Cataract Refract. Surg.*, 33(1), 47, 2007; Bohorquez, V. and Alarcon, R., *J. Cataract Refract. Surg.*, 36(11), 1880, 2010.)

calculated in defocus curve analysis.[37] Accommodative amplitude in a retrospective monofocal, single-optic control group was significantly lower (1.65 ± 0.58 D). Twenty-three eyes implanted with the accommodating IOL (96%) achieved a DCNVA of 20/40 or better. Moreover, Bohorquez et al.[38] reported stable or improved reading ability over time for patients with bilateral Synchrony implants. Distance-corrected mean reading acuity was 0.15 and 0.07 logMAR after 1 and 2 years, respectively. More recently, in 2012, Alio et al.[39] compared distance and near visual outcomes of 26 eyes implanted with a single-optic accommodating IOL (Crystalens HD) and 27 eyes with the Synchrony IOL implanted 6 months postoperatively. Both groups achieved good uncorrected and corrected distance visual acuity 6 months postoperatively (p = 0.79 and 0.26, respectively). In defocus curve testing, the Synchrony group showed significantly better median visual acuity at –3.00 and –3.50 D defocus. However, both IOLs demonstrated a steep drop of visual acuity with increasing negative defocus.

13.3.2.2 Lumina (AkkoLens)

This hydrophilic acrylic dual-optic IOL consists of an anterior element with a spherical lens to correct the overall refraction of the eye and a cubic optical surface for varifocal effect. The posterior element carries the second cubic surface.[40] As shown in Figure 13.9, the position of the two optical surfaces in regard to each other changes to change the optical profile of the IOL.

13.3.2.3 Sarfarazi (EA IOL, Bausch & Lomb, Rochester, NY, USA)

Sarfarazi is an elliptical dual-optic accommodating IOL (Figure 13.10). The anterior biconvex optic is 5.0 mm in diameter and the concave–convex posterior lens has a negative power. The overall diameter is 9.0 mm and is reduced to 8.5 mm during accommodation. The haptic system consists of three bands to allow fixation of the IOL in the capsular bag.[41]

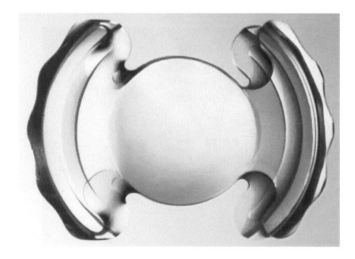

Figure 13.9 This hydrophilic acrylic dual-optic intraocular lens (AkkoLens) consists of an anterior spherical optic and a cubic optical surface. (From http://www.touchophthalmology.com/system/files/private/articles/951/pdf/rombach-2.pdf, last visit: 15/08/2014.)

Figure 13.10 The Sarfarazi intraocular lens (IOL) is an elliptical dual-optic accommodating IOL. The haptic system consists of three bands to allow fixation of the IOL in the capsular bag. (From http://www.chromatographyonline.com/lcgc/article/articleDetail.jsp?id=435050&sk=&date=&pageID=2; last visit: 15/08/2014.)

13.3.3 ELECTRICAL ACCOMMODATING IOLs

13.3.3.1 Sapphire (Elenza)

This accommodating IOL is still in development, but the concept is the following: the Sapphire IOL uses the change in pupil size as an accommodation stimulus (Figure 13.11). When the pupil decreases in size, as it is the case during accommodation, the sensors of the IOL detect these changes and change the optic accordingly. The pupillary changes due to accommodation are different concerning their amplitude and duration; therefore, it is stated that the sensors could distinguish between a pupillary response due to accommodation or due to light adjustments. Energy is supplied by a lithium-ion battery that is sealed in 24 carat gold to avoid toxicity. Recharging the battery appears to be one of the remaining problems and tests with recharges in pillows and special eye masks were tested. If the battery is not recharged, the IOL functions as a monofocal IOL. However, there are remaining safety issues and the IOL still is in a developmental state.[42]

13.3.4 LENS REFILLING

The principle of lens refilling is different compared to accommodating IOLs. The idea is to refill the lens capsule after removing the crystalline lens.[43] However, as shown in Figure 13.12, lens refilling changed significantly over the years. Primarily, the capsular bag was filled directly (Figure 13.12a), but leakage occurred in many cases. In a next step, different techniques were developed to overcome this problem, such as an endocapsular balloon (Figure 13.12b), which prevented silicone leakage with reasonable preservation of accommodation. Other methods, such as a specially designed sealing plug (Figure 13.12c), and later on an optic

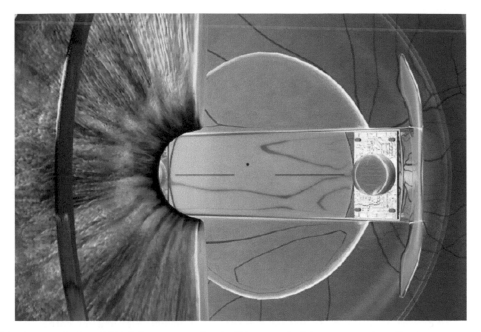

Figure 13.11 The Sapphire intraocular lens (IOL) follows a completely different principle compared to all other accommodating IOLs. It is an electric IOL that uses changes of the pupil as a stimulus for accommodation. Energy is supplied by the world's smallest existing lithium-ion battery that is covered in 24 carat gold. (From http://www.elenza.com/docs/Electronic_IOLS_EW_February_2012.pdf, last visit: 15/08/2014.)

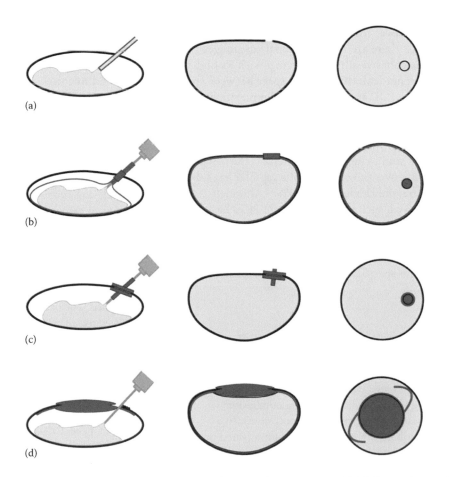

(a)

(b)

(c)

(d)

Figure 13.12 Schematic representation of lens refilling techniques used in animal experiments. (a) Mini capsulotomy by Haefliger and Parel,[26] Hettlich[27]; (b) endocapsular balloon by Nishi[31,32]; (c) capsule sealing plug by Nishi[35]; Koopmanns[5,37]; (d) optic plug by Nishi[37].   (Continued)

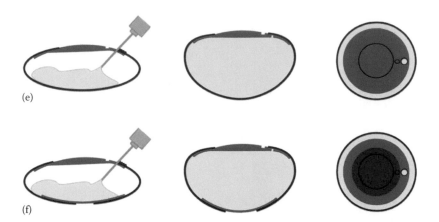

Figure 13.12 (*Continued*) Schematic representation of lens refilling techniques used in animal experiments. (e) accommodative intraocular len (IOL) by Nishi[39]; (f) accommodative IOL with posterior continuous curvilinear capsulorhexis with CCC optic by Nishi.[39] Diagrams depict filling technique (left) and sagittal (middle) and frontal (right) views of the filled capsular bag with injectable polymers (blue) and implanted devices (red). (From Nishi, Y. et al., *J. Cataract Refract. Surg.*, 35(2), 374, 2009.)

plug (Figure 13.12d,e) were also tested. To improve on the endocapsular balloon technique, a plug to seal a continuous curvilinear capsulorhexis was developed by Nishi and Nishi (Figure 13.12f).[44] This technique showed to be feasible due to a good outcome concerning centration and only little PCO. A more recent method uses an innovative foldable silicone IOL and a thin plate-disk haptic.[45] This technique allows the anterior capsule to cover the injection hole to prevent leakage of the silicone. There are two additional advantages of this technique. The cohesive silicone polymer with its high molecular weight does not leak through the space between the anterior capsule and the IOL and the injected liquid silicone polymerizes 2 h after injecting it; therefore, postoperative leakage is not be expected.[46]

The accommodative potential of lens refilling was shown to be promising in animal studies (up to six diopters).[47] The best accommodative potential was shown, when the capsular bag was not completely filled, but only about 60%–70%.[48] Although some problems were solved, such as the problem of PCO by using a posterior rhexis and a second optic that is in touch with the posterior lens capsule, some problems still remain, such as the possibility of leakage of the intracapsular injection fluid and fibrotic capsule opacification that leads to a reduced accommodative response.[46]

13.4 PATIENT MOTIVATION AND SPECTACLE INDEPENDENCE

One problem that still remains when presbyopic treatments are evaluated even under standardized reading conditions is the patient's and possibly also examiner's motivation. In a randomized trial, Leydolt et al.[49] observed that a motivated group of patients showed less spectacle dependence after cataract surgery and the implantation of a standard monofocal IOL compared to a not motivated control group. Patients could attain this through their large depth of focus, or pseudoaccommodation, which is influenced by various factors such as small pupil size, residual

myopic astigmatism, corneal multifocality, and higher-order aberrations of the eye.

When presbyopic treatments are evaluated, it should be taken into account that pseudoaccommodation strongly influences the subjective accommodative response measurements,[50] such as the push-up test and defocus curve testing. Additionally, macular and cortical function related to visual perception are known to play a role in extracting information from a defocused image, such as deciphering blurred optotypes from a reading chart.[51] These effects can result in good uncorrected distance and near vision in patient whether presbyopia was treated or not.

13.5 SUMMARY

Unfortunately, there is no evidence that any of the currently available accommodative IOLs offers any reproducible and clinically relevant accommodation. Additionally, accommodative IOLs appear to be not better compared to monofocal IOLs concerning near vision.[52]

Other concepts such as lens refilling are still in a developmental state and not ready for use in humans. It was shown that some patients that have undergone cataract surgery have good distance and near vision and do not need glasses for either distance. They attain this through a large depth of focus of the eye, or pseudoaccommodation, which is influenced by various factors. Therefore, visual perception also contributes to psychophysically assessed pseudoaccommodation. These effects can result in good uncorrected distance and near vision.[49]

REFERENCES

1. Findl O, Leydolt C. Meta-analysis of accommodating intraocula lenses. *Journal of Cataract and Refractive Surgery* 2007; **33**(3): 522–527.
2. Ong HS, Evans JR, Allan BD. Accommodative intraocular lens versus standard monofocal intraocular lens implantation in cataract surgery. *Cochrane Database of Systematic Reviews* 2014; 5 CD009667.

3. Menapace R, Findl O, Kriechbaum K, Leydolt-Koeppl C. Accommodating intraocular lenses: A critical review of present and future concepts. *Graefe's Archive for Clinical and Experimental Ophthalmology* 2007; **245**(4): 473–489.

4. Kuchle M, Seitz B, Langenbucher A, Gusek-Schneider GC, Martus P, Nguyen NX. Comparison of 6-month results of implantation of the 1CU accommodative intraocular lens with conventional intraocular lenses. *Ophthalmology* 2004; **111**(2): 318–324.

5. Langenbucher A, Huber S, Nguyen NX, Seitz B, Gusek-Schneider GC, Kuchle M. Measurement of accommodation after implantation of an accommodating posterior chamber intraocular lens. *Journal of Cataract and Refractive Surgery* 2003; **29**(4): 677–685.

6. Kuchle M, Seitz B, Langenbucher A, Martus P, Nguyen NX. Stability of refraction, accommodation, and lens position after implantation of the 1CU accommodating posterior chamber intraocular lens. *Journal of Cataract and Refractive Surgery* 2003; **29**(12): 2324–2329.

7. Heatley CJ, Spalton DJ, Hancox J, Kumar A, Marshall J. Fellow eye comparison between the 1CU accommodative intraocular lens and the Acrysof MA30 monofocal intraocular lens. *American Journal of Ophthalmology* 2005; **140**(2): 207–213.

8. Mastropasqua L, Toto L, Nubile M, Falconio G, Ballone E. Clinical study of the 1CU accommodating intraocular lens. *Journal of Cataract and Refractive Surgery* 2003; **29**(7): 1307–1312.

9. Sauder G, Degenring RF, Kamppeter B, Hugger P. Potential of the 1 CU accommodative intraocular lens. *The British Journal of Ophthalmology* 2005; **89**(10): 1289–1292.

10. Findl O, Kriechbaum K, Menapace R et al. Laserinterferometric assessment of pilocarpine-induced movement of an accommodating intraocular lens: A randomized trial. *Ophthalmology* 2004; **111**(8): 1515–1521.

11. Hancox J, Spalton D, Heatley C, Jayaram H, Marshall J. Objective measurement of intraocular lens movement and dioptric change with a focus shift accommodating intraocular lens. *Journal of Cataract and Refractive Surgery* 2006; **32**(7): 1098–1103.

12. Dogru M, Honda R, Omoto M et al. Early visual results with the 1CU accommodating intraocular lens. *Journal of Cataract and Refractive Surgery* 2005; **31**(5): 895–902.

13. Wolffsohn JS, Hunt OA, Naroo S et al. Objective accommodative amplitude and dynamics with the 1CU accommodative intraocular lens. *Investigative Ophthalmology & Visual Science* 2006; **47**(3): 1230–1235.

14. Liu XQ, Li MF, Guo MH, Chen L. Long-term clinical observation of an accommodative 1CU after implantation. (*Zhonghua yan ke za zhi*) *Chinese Journal of Ophthalmology* 2010; **46**(5): 415–418.

15. Wolffsohn JS, Naroo SA, Motwani NK et al. Subjective and objective performance of the Lenstec KH-3500 "accommodative" intraocular lens. *The British Journal of Ophthalmology* 2006; **90**(6): 693–696.

16. Wolffsohn JS, Davies LN, Gupta N et al. Mechanism of action of the tetraflex accommodative intraocular lens. *Journal of Refractive Surgery (Thorofare, NJ: 1995)* 2010; **26**(11): 858–862.

17. Sanders DR, Sanders ML. Visual performance results after Tetraflex accommodating intraocular lens implantation. *Ophthalmology* 2007; **114**(9): 1679–1684.

18. Brown D, Dougherty P, Gills JP, Hunkeler J, Sanders DR, Sanders ML. Functional reading acuity and performance: Comparison of 2 accommodating intraocular lenses. *Journal of Cataract and Refractive Surgery* 2009; **35**(10): 1711–1714.

19. Sanders DR, Sanders ML. US FDA clinical trial of the Tetraflex potentially accommodating IOL: Comparison to concurrent age-matched monofocal controls. *Journal of Refractive Surgery (Thorofare, NJ: 1995)* 2010; **26**(10): 723–730.

20. Dong Z, Wang NL, Li JH. Vision, subjective accommodation and lens mobility after TetraFlex accommodative intraocular lens implantation. *Chinese Medical Journal* 2010; **123**(16): 2221–2224.

21. Cleary G, Spalton DJ, Marshall J. Pilot study of new focus-shift accommodating intraocular lens. *Journal of Cataract and Refractive Surgery* 2010; **36**(5): 762–770.

22. Legeais JM, Werner L, Werner L, Abenhaim A, Renard G. Pseudoaccommodation: BioComFold versus a foldable silicone intraocular lens. *Journal of Cataract and Refractive Surgery* 1999; **25**(2): 262–267.

23. Findl O, Kiss B, Petternel V et al. Intraocular lens movement caused by ciliary muscle contraction. *Journal of Cataract and Refractive Surgery* 2003; **29**(4): 669–676.

24. Cumming JS, Slade SG, Chayet A. Clinical evaluation of the model AT-45 silicone accommodating intraocular lens: Results of feasibility and the initial phase of a Food and Drug Administration clinical trial. *Ophthalmology* 2001; **108**(11): 2005–2009; discussion 10.

25. Alio JL, Pinero DP, Plaza-Puche AB. Visual outcomes and optical performance with a monofocal intraocular lens and a new-generation single-optic accommodating intraocular lens. *Journal of Cataract and Refractive Surgery* 2010; **36**(10): 1656–1664.

26. Koeppl C, Findl O, Menapace R et al. Pilocarpine-induced shift of an accommodating intraocular lens: AT-45 Crystalens. *Journal of Cataract and Refractive Surgery* 2005; **31**(7): 1290–1297.

27. Cumming JS, Colvard DM, Dell SJ et al. Clinical evaluation of the Crystalens AT-45 accommodating intraocular lens: Results of the U.S. Food and Drug Administration clinical trial. *Journal of Cataract and Refractive Surgery* 2006; **32**(5): 812–825.

28. Marchini G, Pedrotti E, Sartori P, Tosi R. Ultrasound biomicroscopic changes during accommodation in eyes with accommodating intraocular lenses: Pilot study and hypothesis for the mechanism of accommodation. *Journal of Cataract and Refractive Surgery* 2004; **30**(12): 2476–2482.

29. Alio JL, Ben-nun J, Rodriguez-Prats JL, Plaza AB. Visual and accommodative outcomes 1 year after implantation of an accommodating intraocular lens based on a new concept. *Journal of Cataract and Refractive Surgery* 2009; **35**(10): 1671–1678.

30. Ben-Nun J, Alio JL. Feasibility and development of a high-power real accommodating intraocular lens. *Journal of Cataract and Refractive Surgery* 2005; **31**(9): 1802–1808.

31. Schor CM, Charles F. Prentice award lecture 2008: Surgical correction of presbyopia with intraocular lenses designed to accommodate. *Optometry and Vision Science: Official Publication of the American Academy of Optometry* 2009; **86**(9): E1028–E1041.

32. Roux P. Early implantation results of the fluidvision aiol in sighted eyes. ASCRS, Boston, MA, 2010.

33. Floyd AM, Werner L, Liu E et al. Capsular bag opacification with a new accommodating intraocular lens. *Journal of Cataract and Refractive Surgery* 2013; **39**(9): 1415–1420.

34. Auffarth G. NCT02049567. www.clinicaltrials.gov, last visit 02/10/2016.

35. McLeod SD, Portney V, Ting A. A dual optic accommodating foldable intraocular lens. *The British Journal of Ophthalmology* 2003; **87**(9): 1083–1085.

36. McLeod SD, Vargas LG, Portney V, Ting A. Synchrony dual-optic accommodating intraocular lens. Part 1: Optical and biomechanical principles and design considerations. *Journal of Cataract and Refractive Surgery* 2007; **33**(1): 37–46.

37. Ossma IL, Galvis A, Vargas LG, Trager MJ, Vagefi MR, McLeod SD. Synchrony dual-optic accommodating intraocular lens. Part 2: Pilot clinical evaluation. *Journal of Cataract and Refractive Surgery* 2007; **33**(1): 47–52.

38. Bohorquez V, Alarcon R. Long-term reading performance in patients with bilateral dual-optic accommodating intraocular lenses. *Journal of Cataract and Refractive Surgery* 2010; **36**(11): 1880–1886.

39. Alio JL, Plaza-Puche AB, Montalban R, Ortega P. Near visual outcomes with single-optic and dual-optic accommodating intraocular lenses. *Journal of Cataract and Refractive Surgery* 2012; **38**(9): 1568–1575.

40. Simonov AN, Vdovin G, Rombach MC. Cubic optical elements for an accommodative intraocular lens. *Optics Express* 2006; **14**(17): 7757–7775.

41. Natalini R. Twin optic elliptical IOL emulates natural accommodation. *Eye World* September 2003, p. 50.

42. http://www.elenza.com, last visit: 15/08/2014.

43. Kessler J. Experiments in refilling the lens. *Archives of Ophthalmology* 1964; **71**: 412–417.

44. Nishi O, Nishi K. Accommodation amplitude after lens refilling with injectable silicone by sealing the capsule with a plug in primates. *Archives of Ophthalmology* 1998; **116**(10): 1358–1361.

45. Nishi O, Nishi K, Nishi Y, Chang S. Capsular bag refilling using a new accommodating intraocular lens. *J Cataract Refract Surg* 2008; **34**(2): 302–309.

46. Nishi Y, Mireskandari K, Khaw P, Findl O. Lens refilling to restore accommodation. *Journal of Cataract and Refractive Surgery* 2009; **35**(2): 374–382.

47. Haefliger E, Parel JM, Fantes F et al. Accommodation of an endocapsular silicone lens (Phaco-Ersatz) in the nonhuman primate. *Ophthalmology* 1987; **94**(5): 471–477.

48. Nishi O, Nishi K, Mano C, Ichihara M, Honda T. Controlling the capsular shape in lens refilling. *Archives of Ophthalmology* 1997; **115**(4): 507–510.

49. Leydolt C, Neumayer T, Prinz A, Findl O. Effect of patient motivation on near vision in pseudophakic patients. *American Journal of Ophthalmology* 2009; **147**(3): 398–405, e3.

50. Win-Hall DM, Glasser A. Objective accommodation measurements in pseudophakic subjects using an autorefractor and an aberrometer. *Journal of Cataract and Refractive Surgery* 2009; **35**(2): 282–290.

51. Mon-Williams M, Tresilian JR, Strang NC, Kochhar P, Wann JP. Improving vision: Neural compensation for optical defocus. *Proceedings of the Royal Society B: Biological Sciences* 1998; **265**(1390): 71–77.

52. Findl O. Intraocular lenses for restoring accommodation: Hope and reality. *Journal of Refractive Surgery* 2005; **21**(4): 321–323.

53. http://www.beyondlasik.co.uk/iol-vision-correction-over-45/accommodating-lenses, last visit: 15/08/2014.

54. http://www.touchophthalmology.com/system/files/private/articles/951/pdf/rombach-2.pdf, last visit: 15/08/2014.

55. http://www.chromatographyonline.com/lcgc/article/articleDetail.jsp?id=435050&sk=&date=&pageID=2; last visit: 15/08/2014.

56. http://www.elenza.com/docs/Electronic_IOLS_EW_February_2012.pdf, last visit: 15/08/2014.

Adjustable intraocular lenses: The light adjustable lens

Christian A. Sandstedt

Contents

14.1 INTRODUCTION AND RATIONALE

An intraocular lens (IOL) is a surgically implanted, artificial lens designed to replace the natural crystalline lens in the human eye, typically in patients who have developed visually significant cataracts. Since their inception in the late 1940s, IOLs have provided improved uncorrected visual acuity (UCVA) compared to that of the cataractous or aphakic state; however, problems in predictably achieving emmetropia persist as most post–cataract surgery patients rely on spectacles or contact lenses for optimal distance vision.

The determination of IOL power required for a particular postoperative refraction is dependent on the axial length of the eye, the optical power of the cornea, and the predicted location of the IOL within the eye. Accurate calculation of IOL power is difficult because the determination of axial length, corneal curvature, and the predicted position of the IOL in the eye is inherently inaccurate.[1–4] Surgically induced cylinder and variable lens position following implantation will create refractive errors, even if preoperative measurements were completely accurate.[2] Currently, the options for IOL patients with less than optimal uncorrected vision consist of postoperative correction with spectacles, contact lenses, or refractive surgical procedures. Because IOL exchange procedures carry significant risk, secondary surgery to remove the IOL and replace it with a different power IOL is generally limited to severe postoperative refractive errors.

With current methods of IOL power determination, the vast majority of patients achieve a UCVA of 20/40 or better. A much smaller percentage achieves optimal vision without spectacle correction. Nearly all patients are within 2 diopters (D) of emmetropia.

In a study of 1676 patients, 1569 (93.6%) patients were within 2 D of the intended refractive outcome.[4] In 1320 cataract extractions on patients without ocular comorbidity, Murphy and coworkers found that 858 (65%) had UCVA ≥20/40.[4] A 2007 survey of cataract surgeons reported that incorrect IOL power remains a primary indication for foldable IOL explanation or exchange.[5,6]

In addition to imprecise IOL power determinations, postoperative UCVA is often limited by preexisting astigmatism. Several IOL manufacturers (e.g., Staar Surgical, Alcon, Bausch & Lomb [B&L], AMO, Carl-Zeiss) market toric IOLs that attempt to correct preexisting astigmatic errors. These IOLs are typically only available in discreet toric powers, and the axis must be precisely aligned at surgery. Other than surgical repositioning, there is no option to adjust the IOL's axis, which may shift postoperatively.[7] Furthermore, individualized correction of astigmatism is limited by the unavailability of multiple toric powers. An additional problem associated with using preimplantation corneal astigmatic errors to gauge the required axis and power of a toric IOL is the unpredictable effects of surgical wound healing on the final refractive error. After the refractive effect of the surgical wound stabilizes, there is often a shift in both magnitude and axis of astigmatism, offsetting the corrective effect of a toric IOL. Therefore, a means to postoperatively adjust (correct) astigmatic refractive errors after lens implantation and surgical wound healing is very desirable for an IOL. While limbal relaxing incision is a widely accepted technique for treating corneal astigmatism, the procedure is typically performed during cataract surgery and therefore does not address the effect of postimplantation wound healing.

Eyes undergoing corneal refractive procedures (approximately 1.0 million/year in the United States alone) that subsequently develop cataracts are challenging with respect to IOL power determination. Corneal topographic alterations induced by refractive surgery reduce the accuracy of keratometric measurements, often leading to significant postoperative ametropia.[8-15] Recent studies of patients who have had corneal refractive surgery (photorefractive keratectomy, laser in situ keratomileusis, radial keratotomy) and subsequently required cataract surgery frequently demonstrate refractive "surprises" postoperatively. As the refractive surgery population ages and develops cataracts, appropriate selection of IOL power for these patients has become an increasingly challenging clinical problem. The ability to address this problem with an adjustable IOL is valuable to patients seeking optimal vision after cataract surgery.

It is estimated that presbyopia currently affects more than 90 million people in the United States. Compounding this condition is the risk of cataract development as the patient ages. To treat both presbyopia and cataracts, several treatment approaches have been used such as monovision, multifocal IOLs, and accommodating IOLs. For standard monovision, the dominant eye is typically targeted for emmetropia postimplantation and the nondominant eye is targeted for –1.5 to –2.5 D of myopia. Research has shown, however, that if the refractive powers between each eye differ by >1.5 D, binocular summation does not occur and, in fact, binocular inhibition can cause the vision to degrade.[16] In addition, research has shown that higher binocular summation is strongly correlated with patient satisfaction and quality of vision.[17] Mini-monovision, while providing good quality of vision, does not provide for sufficient depth of focus (DOF) to enable spectacle independence. The two most widely adopted multifocal IOLs currently sold in the United States are the ReZoom (Abbott Medical Optics, Chicago, IL) and ReStor (Alcon, Fort Worth, TX) lenses. The ReZoom lens is comprised of five concentric, aspheric refractive zones.[18] Each zone is a multifocal element, and thus pupil size should play little or no role in determining the final image quality. However, the pupil size must be greater than 2.5 mm to be able to experience the multifocal effect. Image contrast is sacrificed at near and far distances to achieve the intermediate and has an associated loss equivalent to one line of visual acuity.[19] The ReStor lenses, both the 3.0 and 4.0 versions, provide simultaneous near and distance vision by a series of concentric, apodized diffractive rings in the central, 3 mm diameter of the lenses. The mechanism of diffractive optics should minimize the problems associated with variable pupil sizes and small amounts of decentration. The acceptance and implantation of both of these lenses have been limited by the difficulty experienced with glares, rings, halos, monocular diplopia, and the contraindication for patients with ≥2.0 D of astigmatism.[20,21] Recently, several IOL manufacturers have developed IOLs that attempt to take advantage of the existing accommodative apparatus of the eye in postimplantation patients to treat presbyopia. Two notable examples are the CrystaLens™ from B&L (Rochester, NY) and the AKKOMMODATIVE® 1CU from Human Optics AG (Erlangen, Germany). B&L's lens offers a plate haptic configured IOL with a flexible hinged optic (CrystaLens). Human Optics's lens (AKKOMMODATIVE 1CU) is similar in design but possesses four hinged haptics attached to the edge of the optic. The accommodative effect of these lenses is caused by the vaulting of the plate IOL by the contraction of the ciliary body. This vaulting may be a response of the ciliary body

contraction directly or by the associated anterior displacement of the vitreous body. Initial reports of the efficacy of these two lenses in clinical trials were quite high with dynamic wavefront measurement data showing as much as 2–3 D (measured at the exit pupil of the eye) of accommodation. However, the FDA Ophthalmic Devices panel review of B&L's clinical results concluded that only a 1 D accommodative response (at the spectacle plane) was significantly achieved by their lens. Regardless of the mechanism of action and design, all of these approaches suffer from the inherent inaccuracy of the preoperative biometry, surgical wound healing, potential shifts of the IOL during wound healing, and fibrosis of the capsular bag.

One of the newest concepts proposed to tackle the dual problems of cataracts and presbyopia is by increasing the DOF of the IOL using targeted amounts of the fourth-order spherical aberration. Two recent studies using adaptive optics simulators have been published and indicate that the induction of between –0.20 and –0.15 μm (referenced to a 4 mm pupil) of total ocular fourth-order spherical aberration along with some residual myopia, for example, –0.75 D, will provide patients with excellent vision from 40 cm to distance emmetropia.[22,23]

In light of the preceding discussion, a need exists for an IOL, which is adjusted postoperatively in vivo. Such a lens would remove the guesswork involved in presurgical power selection, overcome the wound healing response inherent to IOL implantation, and allow the vision to be customized to correspond to the patient's requirements at both distance and near.

14.2 LIGHT ADJUSTABLE LENS

Calhoun Vision has developed a silicone IOL that allows for noninvasive power adjustment postimplantation. Calhoun Vision's light adjustable lens (LAL) is based on the inclusion of a proprietary photoreactive silicone macromer and photoinitiator within a medical grade silicone polymer matrix. Postoperative in situ irradiation of the implanted Calhoun Vision LAL using the light delivery device to deliver targeted dosages of UV light (365 nm) produces modifications in the lens curvature resulting in predictable spherical, cylindrical, and aspheric changes.

14.2.1 MECHANISM OF ACTION

The material and optical design of the LAL is based upon the principles of photochemistry and diffusion whereby photoreactive components incorporated in the cross-linked silicone lens matrix are photopolymerized upon exposure to UV light (365 nm) of a select spatial irradiance profile. Figure 14.1 displays a cartoon picture of the LAL and its major constituents. The first component of note is shown as the long green strands and corresponds to the polymer matrix, which acts to give the LAL its basic optical and mechanical properties. The polymer matrix is composed of a high-molecular-weight (>200 K) polysiloxane that also possesses a covalently bonded UV blocker (not shown). Due to its cross-link density and inherent low glass transition temperature (~–125°C), the LAL's polymer matrix allows for relatively rapid diffusion throughout its polymer network. The second major constituent are the smaller, purple strands noted as macromer. The macromer is a low, relative to the matrix polymer, molecular weight polysiloxane. From a chemical standpoint, the majority of the macromer chain is

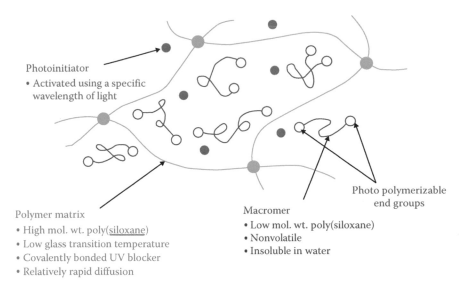

Photoinitiator
• Activated using a specific wavelength of light

Polymer matrix
• High mol. wt. poly(siloxane)
• Low glass transition temperature
• Covalently bonded UV blocker
• Relatively rapid diffusion

Macromer
• Low mol. wt. poly(siloxane)
• Nonvolatile
• Insoluble in water

Photo polymerizable end groups

Figure 14.1 Cartoon diagram of the major light adjustable lens chemical components.

the same as that of the polymer matrix, which allows for essentially infinite miscibility of the macromer within the polymer matrix. The fact that the macromer and polymer matrix are miscible with each other avoids the potential for phase separation and subsequent light scatter. The most unique aspect of the macromer molecule is the presence of symmetric, photopolymerizable methacrylate end groups at the end of each macromer chain. The final chemical moiety of note is listed as the photoinitiator, which acts to catalyze the photopolymerization reaction of the macromer end groups.

The mechanism upon which the LAL refractive power change is based is depicted graphically in Figure 14.2. Application of near UV light (365 nm) to the LAL induces the macromers to undergo photopolymerization forming an interpenetrating network within the irradiated target area of the lens. This action produces a change in the chemical potential between the irradiated and nonirradiated regions of the lens. To reestablish thermodynamic equilibrium, macromers from the unirradiated portion of the lens will diffuse along the concentration gradient into the photopolymerized portion of the lens, producing a swelling in the irradiated region that changes the lens curvature. The diffusion process will continue until equilibrium is reestablished, that is, 24–48 hours postirradiation.

As an example, if the central portion of the lens is irradiated and the peripheral portion is left nonirradiated, unreacted macromer diffuses into the central portion causing an increase

in the lens power (Figure 14.2b). Conversely, by irradiating the outer periphery of the lens, macromer migrates outward causing a decrease in the lens power. Cylindrical power adjustments are achieved in a similar manner by removing power in one meridian while adding power in the perpendicular meridian. By using a digitally generated beam profile, the axis of the cylindrical correction can be precisely aligned by digitally rotating the spatial irradiance profile. By controlling the irradiation dosage (i.e., beam intensity and duration), spatial irradiance profile, and target area, physical changes in the radius of curvature of the lens surface are achieved, thus modifying the refractive power of an implanted light adjustable IOL to either add or subtract spherical power, remove toricity, or adjust the amount of asphericity. Once the appropriate power adjustment and/or visual outcomes are achieved, the entire lens is irradiated to polymerize the remaining unreacted macromer to prevent any additional change in lens power. By irradiating the entire lens, macromer diffusion is prevented; thus, no additional change in lens power results. This second irradiation procedure is referred to as "lock-in" treatment.

The Calhoun Vision LAL (Figure 14.3) is a foldable posterior chamber, UV-absorbing, three-piece photoreactive silicone lens with blue polymethyl methacrylate (PMMA) modified-C haptics, a 6.0 mm biconvex optic with squared posterior edge, and an overall length of 13.0 mm. The LAL optic design incorporates

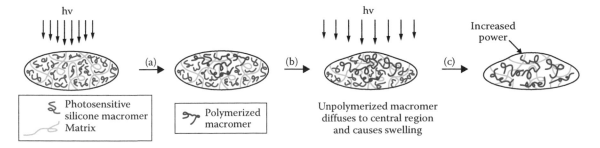

hv

Photosensitive silicone macromer
Matrix

Polymerized macromer

(a) → (b) →

hv

Unpolymerized macromer diffuses to central region and causes swelling

(c) →

Increased power

Figure 14.2 Schematic of the hyperopic power adjustment mechanism. (a) Selective irradiation of the central zone of light adjustable lens (LAL) polymerizes macromer creating a diffusion gradient between the irradiated and nonirradiated regions. (b) To reestablish equilibrium, excess macromer diffuses into the irradiated region (central zone) causing swelling. (c) Irradiation of the entire LAL completely photopolymerizes the macromer and "locks in" the shape change.

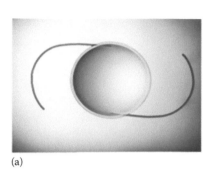

Posterior back layer

(a)   (b)

Figure 14.3 Calhoun Vision light adjustable intraocular lens (LAL). (a) Top view and (b) cross-sectional view of the optic showing LAL with and without a posterior layer. The posterior layer of up to 100 μm of a higher UV absorber concentration further enhances the UV-absorbing properties of the lens and safety of the retina during irradiation procedures.

a silicone posterior surface layer of 100 μm or less with a higher concentration of UV absorber than the photoreactive bulk lens material to further enhance the UV-absorbing properties of the LAL and provide additional retinal safety during the lens power adjustment and lock-in procedures.

A summary of the LAL design characteristics is presented as follows:

Lens optic
- Material: Photoreactive, UV-absorbing silicone
- Light transmission: UV cutoff at 10% T = 390 ± 2 nm for a +20 D lens
- Index of refraction: 1.43
- Diopter power: +10 to +16.0 D and +25.0 to +30 D in 1.0 D increments and +17.0 to +24.0 D in 0.5 D increments
- Optic type: Biconvex with posterior layer of 100 μm or less with a higher concentration of UV absorber than optic bulk
- Optic edge: Square on posterior surface and round on anterior surface
- Overall diameter: 13.0 mm
- Optic diameter: 6.0 mm

Haptics
- Configuration: Modified C
- Material: Blue PMMA
- Haptic angle: 10°

14.2.2 IN VITRO NOMOGRAM DEVELOPMENT

As alluded to in the sections regarding the mechanism of LAL power change, the type of refractive power correction (hyperopic, myopic, astigmatic, spherical aberration, etc.) is dependent upon the applied spatial irradiance profile, the average irradiance, and the time used for irradiation. To develop these nomograms, an experimental apparatus consisting of an irradiation system and a wavefront analysis instrument was employed. The irradiation system is composed of a mercury (Hg) arc lamp filtered to 365 nm (±5 nm FWHM), a critical illumination/projection system, and a digital mirror device (DMD). At the heart of this irradiation system is the DMD, which is a pixelated, micromechanical spatial light modulator formed monolithically on a silicon substrate. Typical DMD chips have dimensions of 15.1 mm × 12.7 mm. The individual micromirrors are ~14 μm on an edge and are covered with an aluminum coating. The micromirrors are arranged in an xy array, and the chips contain row

drivers, column drivers, and timing circuitry. The addressing circuitry under each mirrored pixel is a memory cell that drives two electrodes under the mirror with complimentary voltages. Depending on the state of the memory cell (a "1" or "0"), each mirror is electrostatically attracted by a combination of the bias and address voltages to one of the other address electrodes. Physically the mirror can rotate ±12°. A "1" in the memory causes the mirror to rotate +12°, while a "0" in the memory causes the mirror to rotate −12°. A mirror rotated to +12° reflects incoming light into the projection lens and onto the LAL. When the mirror is rotated −12°, the reflected light misses the projection lens. Thus, the great utility and advantage of the DMD device in its relation to the LAL is the ability of the researcher/ physician to easily define a specific spatial irradiance profile, program this into the DMD, and then irradiate the LAL. Because of the digital nature, the DMD technology offers greater resolution of the spatial light profile enabling the delivery of more precise, complex patterns to provide greater range and control to the LAL corrections.

The optical analysis portion of this instrument utilizes either a phase-shifting Fizeau interferometer (Wyko model 400) operating in double-pass configuration fitted with a 4 in. transmission sphere or an IOL Aberrometer (CrystalWave, Lumetrics, Rochester, NY) based upon a Shack–Hartmann wavefront sensor. In practice, a set of LALs are first mounted into the wet cell maintained at 35.0°C ± 0.5°C (simulated ocular temperature) and allowed to equilibrate for a minimum of 2 hours. For the experimental case when the interferometer is used, the wet cell is adjusted along the optical axis of the interferometer until the power in the wavefront across the full test aperture is minimized (≤0.010 wvs). A measurement of the wavefront in the exit pupil of the LAL and its position along the radius slide are recorded followed by irradiation of the lens. The total time for macromer diffusion is between 24 and 48 hours and depends upon the magnitude and the type, that is, myopic, hyperopic, astigmatic, and aspheric, of refractive power change. After the adjusted LAL has reached refractive stabilization (24–48 hours postirradiation), the LALs are returned to their original position on the radius slide followed by measurement of the LAL's adjusted wavefront. Analysis of the postirradiated wavefront along with subtraction of the pre- and postirradiated wavefronts gives direct information regarding the magnitude of the induced power change and any changes in the other aberrations induced by the irradiation procedure, for example, spherical aberration, coma, and astigmatism. Knowledge of the spatial irradiance profile applied to the LAL coupled with the analysis of the altered wavefront using both measurement methods allows guidance in the modification of the pattern to produce the desired changes. The procedures outlined earlier were followed in the development of the spherical spherocylindrical/aspheric nomograms.

Figure 14.4 displays in vitro examples of the type of spatial irradiance profiles used to generate specific refractive power changes and the resultant wavefronts produced by the irradiation. The wavefront picture on the left-hand side of the figure corresponds to that from a typical LAL at its preirradiation best focus position, that is, one focal distance away from the point source image formed by the interferometer's transmission sphere, along the optical axis of the interferometer in Figure 14.4. At this position, the wavefront exiting the LAL, which is subsequently retroreflected back into the interferometer, is collimated. The top, photopolymerization scheme in Figure 14.4 depicts

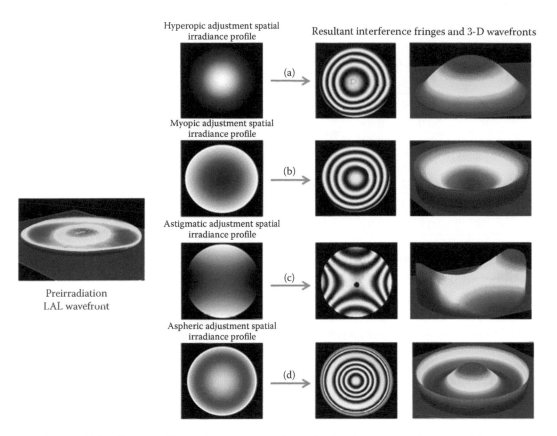

Figure 14.4 Representative spatial irradiance profiles and resultant wavefronts for (a) a hyperopic adjustment, (b) a myopic adjustment, (c) a toric adjustment, and (d) an aspheric adjustment.

irradiation of the LAL with a hyperopic adjustment profile. As evident in the grayscale image of the spatial irradiance profile and consistent with the description earlier regarding the LAL power change mechanism, this profile's irradiance is peaked in the center and falls off symmetrically toward the edges. The next two panels of the top scheme correspond to the raw interference fringes and resultant wavefront measured of an LAL at its preirradiation best focus position along the optical axis of the interferometer approximately 48 hours postirradiation. Clearly evident in this figure is the presence of four waves (~+1.25 D of added power) of optical path difference (OPD) induced by the photopolymerization process. The second photopolymerization scheme displays an example of spatial irradiance profile used to correct for myopic refractive errors. As displayed in this grayscale image, the myopic adjustment profile is peaked on the edges with corresponding reduced irradiance in the center. The impact of this applied profile is shown in the postirradiation interference fringes and wavefront that displays the induction of ~4 waves of OPD or ~1.25 D of subtracted power. The third reaction scheme displays the ability to create in situ toric lenses by combining hyperopic and myopic adjustment profiles perpendicular to each other. The postirradiation interference fringes and the resultant 3-D wavefront displays the induction of a classic, best focus astigmatic profile where power has been added along one meridian and removed along the other. The specific example shown in Figure 14.4c corresponds to the induction of 1.50 D of astigmatism. And finally the last reaction scheme depicts the ability of the LAL to be adjusted for aspheric adjustments. Qualitative inspection of the applied spatial irradiance profile

indicates that this profile possesses the mathematical form of the standard Zernike polynomial for fourth-order spherical aberration. The postirradiation interference fringes and corresponding 3-D wavefront clearly indicate the induction of asphericity in the LAL. The specific example shown corresponds to the induction of ~-0.30 μm of negative fourth-order asphericity.

The preceding description of the experimental methods and examples depicted in Figure 14.4 was provided to give the reader a conceptual idea of how the LAL responds to specific spatial irradiance profiles and how these specific profiles were generated. Tables 14.1 and 14.2 display actual power change and optical

Table 14.1 Summary of the in vitro hyperopic nomogram

NO. OF IRRADIATED LALs	ΔP (D) (SPECTACLE PLANE)	MTF AT 100 lp/mm
24	0.25 ± 0.05	0.51 ± 0.04
30	0.51 ± 0.05	0.56 ± 0.04
24	0.78 ± 0.08	0.55 ± 0.03
24	1.01 ± 0.04	0.49 ± 0.03
32	1.25 ± 0.05	0.49 ± 0.04
36	1.52 ± 0.09	0.48 ± 0.05
32	1.75 ± 0.13	0.48 ± 0.05
24	1.99 ± 0.12	0.51 ± 0.05

All power change data for the lenses are referenced to the spectacle plane and the MTF measurements were made in accordance with ISO 11979-2.

Table 14.2 **Summary of the in vitro myopic nomogram**

NO. OF IRRADIATED LALs	ΔP (D) (SPECTACLE PLANE)	MTF AT 100 lp/mm
24	−0.29 ± 0.05	0.48 ± 0.03
24	−0.48 ± 0.07	0.47 ± 0.04
24	−0.74 ± 0.05	0.49 ± 0.03
32	−0.95 ± 0.12	0.49 ± 0.03
24	−1.21 ± 0.08	0.49 ± 0.03
31	−1.52 ± 0.09	0.46 ± 0.02
28	−1.80 ± 0.09	0.46 ± 0.02
32	−1.94 ± 0.09	0.47 ± 0.03

All power change data for the lenses are referenced to the spectacle plane and the MTF measurements were made in accordance with ISO 11979-2.

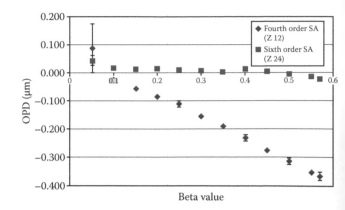

Figure 14.5 Induction of fourth (Z12)- and sixth (Z24)-order spherical aberration as a function of increasing β-value. The optical path difference data are referenced to a 4 mm aperture.

quality data for the in vitro developed hyperopic and myopic nomograms. The first column of each table refers to the number of individual LALs that were adjusted using a specific nomogram treatment condition. Column two lists the average power change (referenced to the spectacle plane) and first standard deviation for the groups of LALs adjusted with the specific nomogram condition. The third column displays the best focus modulation transfer function (MTF) values at a frequency of 100 lp/mm. All MTF measurements were made in the model eye described in ISO 11979-2: 1999. According to the ISO standard, MTF values ≥0.43 at 100 lp/mm meet the resolving capabilities of IOLs as established by ISO. Inspection of the power change results for both the hyperopic and myopic adjustments indicates that it is possible to adjust the refractive power of the LAL in discrete, 0.25 D increments from +2.0 D to −2.0 D. In addition, inspection of the final column indicates that from an optical quality standpoint, the adjusted LALs for a particular nomogram point pass the MTF imaging quality standards set forth in ISO document 11979-2: 1999. While not shown, combining the hyperopic and myopic nomogram spatial irradiance profiles permits the development of myopic, hyperopic, and mixed astigmatic power adjustments from ±2.0 D sphere in combination with −0.50 to −2.0 D of cylinder.

The presence of spherical aberration increases the DOF in the eye. In combination with a residual refractive error (defocus), induced spherical aberration can be used to provide patients with good contrast images both for distance and near objects. The key issue is to determine the required values of both fourth-order spherical aberration and defocus that provide good near and intermediate vision without deteriorating the image quality for distance objects. According to the work of Artal and Yoon, the presence −0.15 to −0.20 μm (referenced to a 4 mm pupil) of ocular fourth-order spherical aberration in combination with ~−0.75 D of spherical defocus will provide the eye with acceptable near (40 cm) and intermediate (60 cm) vision without significant reduction in distance vision.[22,23] In addition, a "normal" eye after implantation with an average power (+18 to +23 D) spherical IOL will have a resultant spherical aberration (referenced to a 4 mm pupil) of ~+0.10 to +0.12 μm. Therefore, in order for the LAL to be effective in producing sufficient DOF to the implanted eye, it is necessary to be able to induce a minimum of −0.30 μm of fourth-order spherical aberration.

The in vitro aspheric nomogram treatment conditions for the LAL were generated by linearly combining weighted amounts of the mathematical form of a fourth-order spherical aberration Zernike polynomial with a power neutral profile, that is, a profile that neither adds nor subtracts refractive power from the LAL. This is expressed mathematically as

$$\text{Profile}(\rho) = PN(\rho) + \beta Asph(\rho) \quad (14.$$

where

PN(ρ) represents the power neutral profile
β is a weighting factor that can range from 0 to 1
Asph(ρ) is given by

$$Asph(\rho) = A\rho^4 - B\rho^2 + 1 \quad (14.$$

Figure 14.5 displays a plot of the induced fourth- and sixth-order spherical aberration referenced to a 4 mm aperture as a function of increasing beta values, that is, the linear amount of the fourth-order polynomial added to the power neutral profile. Inspection of the data indicates a nearly linear, monotonic increase in the amount of induced fourth-order spherical aberration as a function of increasing beta value. In addition, these data show that it is possible to induce discrete amounts of fourth-order spherical aberration that are in excess of that necessary to create the required DOF. And finally, the results show that method employed does not induce sixth-order spherical aberration. This fact is important as considerations of the impact of sixth-order spherical aberration have shown that it can counteract the DOF induced by fourth-order spherical aberration.

14.3 LAL IN VIVO RESULTS

14.3.1 SPHERICAL/SPHEROCYLINDRICAL LAL ADJUSTMENTS

The preceding sections focused on the mechanism of action and in vitro response of the LAL to applied treatment protocols. However, several notable clinical studies utilizing the LAL to correct for postoperative spherical and spherocylindrical refractive errors have been performed and published. A prospective,

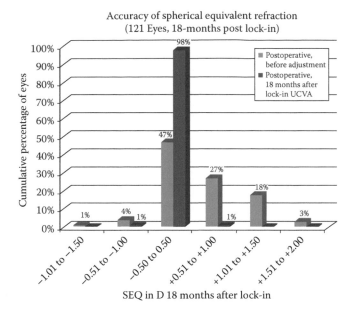

Figure 14.6 Comparison of the targeted manifest refraction spherical equivalent accuracy between the postoperative/preadjustment and 18 months post lock-in visits. (Reproduced with permission from Hengerer, F.H. et al., *Ophthalmology*, 118(12), 2382, 2011.)

nonrandomized clinical trial was conducted by Dr. Fritz Hengerer and Professor Burkhard Dick at the Center for Vision Science, Ruhr University Eye Clinic, in Bochum, Germany.[74] The study consisted of 122 eyes (91 subjects) followed up for 18 months postimplantation. In all cases, standard cataract procedures, including small incision phacoemulsification and implantation of the LAL, were applied. At 14–21 days postoperatively, the subjects returned to the clinic and were examined, refracted, and adjusted

with the required nomogram treatment protocol. At the time of primary adjustment, the average manifest refraction spherical equivalent (MRSE) value for all 122 eyes was +0.55 ± 0.77 D (range –1.38 to +1.75 D), the average sphere was +0.96 ± 0.85 D (range –0.75 to +2.25 D), and the average cylinder was measured at –0.98 ± 0.50 D (range 0.00 to –2.75 D). At the 18 months post lock-in visit, 99.2% of the eyes (121/122) returned for follow-up examination. At 18 months post lock-in, the mean MRSE was reduced to 0.03 ± 0.17 D (range –0.62 to 0.47 D). The average residual sphere and cylinder was measured at +0.10 ± 0.22 D (range –0.50 to +0.05 D) and –0.25 ± 0.22 D (range 0.00 to –0.75 D), respectively. Figure 14.6 summarizes the accuracy of the LAL refractive power change to intended MRSE target for the postoperative/preadjustment and 18-month exam visits. Inspection of the plot indicates that 98% of the treated eyes were within ±0.5 D of the targeted refraction as compared to only 47% at the postoperative/preadjustment visit.

Figure 14.7 compares the uncorrected distance visual acuities (UCDVAs) of the 121 eyes between the postoperative/preadjustment and 18 months post lock-in visits. At the preadjustment visit, 0% (0/121), 4% (5/121), and 22% (27/121) of the implanted eyes possessed an UCDVA of 20/16 or better, 20/20 or better, and 20/25 or better, respectively. Conversely, at 18 months postadjustment and post lock-in, these values have increased dramatically with 25% (30/121), 88% (107/121), and 100% (121/121) of the eyes presenting with an UCDVA of 20/16 or better, 20/20 or better, and 20/25 or better, respectively.

Figure 14.8 displays the achieved UCDVA results from the LAL study in Bochum with the FDA clinical results achieved using the commercially available ACRYSof IQ series toric IOL. As described in the figure, the ACRYSof IQ series toric lenses come in three discrete toric powers, that is, 1.0, 1.5, and 2.0 D.

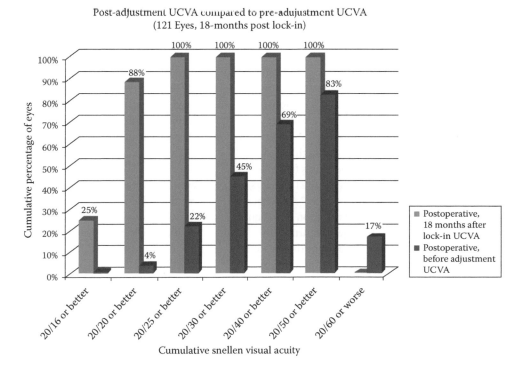

Figure 14.7 Comparison of the preadjustment and 18 months post lock-in uncorrected distance visual acuities. (Reproduced with the permission from Hengerer, F.H. et al., *Ophthalmology*, 118(12), 2382, 2011.)

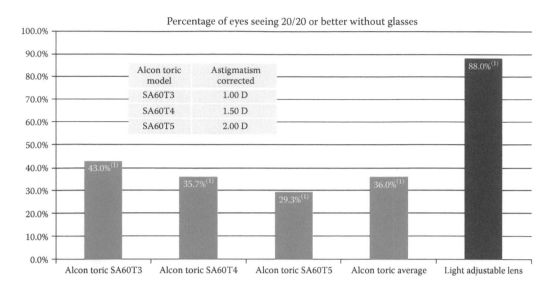

Figure 14.8 Comparison of the uncorrected distance visual acuity for eyes implanted with the Alcon toric lens and the light adjustable lens from work of Hengerer et al. (1) ACRYSof IQ Toric Product Information, Alcon Laboratories Inc., 2011; (2) Hengerer, F.H. et al., *Ophthalmology*, 118(12), 2382, 2011.

Inspection of the results indicates that the Alcon toric IOL achieved an average UCDVA of 20/20 or better in only 36% of the implanted eyes. In contrast, the LAL achieved an UCDVA of 20/20 or better in 88% of eyes. These results indicate that in clinical practice the LAL does an outstanding job in removing spherical/spherocylindrical errors and optimizing distance visual acuity postimplantation.

14.3.2 LAL AND POSTREFRACTIVE SURGERY EYES

As described in the introduction, eyes undergoing corneal refractive procedures (approximately 1.0 million/year in the United States alone) that subsequently develop cataracts are challenging with respect to IOL power determination. Corneal topographic alterations induced by refractive surgery reduce the accuracy of keratometric measurements, often leading to significant postoperative ametropia.[8–15] Recent studies of patients who have had corneal refractive surgery (photorefractive keratectomy, laser in situ keratomileusis, radial keratotomy) and subsequently required cataract surgery frequently demonstrate refractive "surprises" postoperatively. To evaluate whether postoperative refractive power adjustment of the LAL is effective in improving visual outcomes in cataract patients that had previously undergone corneal refractive procedures (either LASIK or PRK), a retrospective study of 34 eyes (21 patients) was recently performed by Dr. Lawrence Brierley.[25] All surgeries were performed using standard phacoemulsification techniques, and at 2 weeks postimplantation, each subject returned for examination and primary adjustments. Twenty of the eyes were targeted for emmetropia, while 14 were adjusted for monovision. For data analysis purposes and to ensure that the eyes targeted for monovision did not distort the MRSE data, all MRSE values immediately prior to adjustment and after lock-in were adjusted from clinical measurements to reflect MRSE to the predetermined refractive target. The range of MRSE to the eye's desired target immediately prior to the primary adjustment was from –1.0 to +2.875 D. At 1 week post lock-in, the range in MRSE dramatically reduced from –0.65 to +0.50 D. Further analysis

of the MRSE target between the preadjustment and 1 week post lock-in visits was performed by breaking up the cohort of 34 eyes into four preadjustment MRSE subgroups of ±0.25 D (n = 5; 15%), ±0.50 D (n = 12; 35%), ±1.0 D (n = 23; 68%), and >±1.0 D (n = 11; 32%). Comparison with the postadjustment and post lock-in MRSE values shows that the LAL was extremely effective in achieving the target refraction with 74% (25/34) and 97% (33/34) of eyes within ±0.25 and ±0.50 D, respectively. Based on the ability of the LAL adjustment procedure to hit the desired refractive target, there was also a dramatic improvement in the UCDVAs of the 20 eyes targeted for distance emmetropia. Figure 14.9 displays the UCDVAs for the 20 eyes targeted for distance emmetropia at the preadjustment and 1 week post lock-in visits. Inspection of this chart indicates that preadjustment, only 10% (2/20) and 30% (6/20) of the implanted eyes saw 20/20 and 20/25 or better, respectively. However, at 1 week post lock-in, these values have dramatically improved with 65% (13/20) and 95% (19/20) seeing 20/20 or better and 20/25 or better, respectively. The results of this study show that the LAL is an effective tool in ensuring that postrefractive surgery patients achieve their desired post–cataract surgery targets.

14.3.3 LAL AND THE CORRECTION OF PRESBYOPIA

Several different approaches have been attempted through the years to tackle the dual problems of cataracts and presbyopia. As discussed in the introduction of this chapter, the protocols and designs have ranged from classic monovision, multifocal IOLs, and accommodating IOLs. One of the newest approaches involves extending the DOF of one, or both, of the eyes using targeted amounts of fourth-order spherical aberration. Recent, independent adaptive optics researches by Professor Artal at the University of Murcia and Professor Yoon at the University of Rochester have indicated that the induction of fourth-order spherical aberration on the order of –0.20 µm (referenced to a 4 mm pupil) and residual myopia of ~–0.75 D will provide the human eye with enough DOF to increase the range of vision from 40 cm to distance emmetropia with tolerable losses in contrast sensitivity.[22,23]

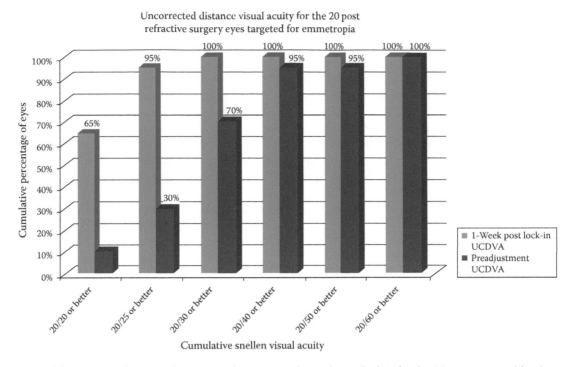

Figure 14.9 Uncorrected distance visual acuity values at preadjustment and 1 week post lock-in for the 20 eyes targeted for distance emmetropia.

As detailed earlier in the section on the in vitro nomogram development of the LAL, it is possible to induce targeted amounts of negative, fourth-order spherical aberration LAL by application of the appropriate spatial irradiance profile. In light of these facts, a prospective clinical trial of Calhoun Vision's LAL was conducted to evaluate the effectiveness of the LAL in subjects undergoing cataract extraction to provide distance, intermediate (60 cm), and near vision (40 cm). The study was conducted at CODET Vision Institute in Tijuana, Mexico, under the primary investigator Dr. Arturo Chayet. The required regulatory and IRB approvals were obtained prior to the initiation of the study.

In summary, the pool of subjects for this study included 20 bilateral patients that were implanted with the LAL with the ultimate intent of optimization of binocular intermediate (60 cm) and distance or near (40 cm) and distance vision. Preceding the delivery of any adjustment treatments, each subject's dominant and nondominant eye was determined. All dominant eyes were targeted for distance, and nondominant eyes were treated for near (40 cm) or intermediate (60 cm) vision. Of the 20 nondominant eyes, 8 received intermediate (60 cm) aspheric treatments while 12 received near (40 cm) aspheric treatments.

In summary, 95.0% (19/20) of the study subjects achieved a binocular UCDVA of 20/25 or better along with a corresponding binocular, uncorrected near (40 cm) visual acuity of J3 or better (20/40) in 90.0% (18/20) of the subjects. Figure 14.10 displays a plot of the preadjustment and 1 week post lock-in uncorrected,

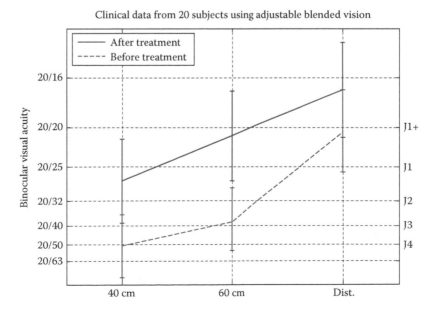

Figure 14.10 Comparison of the binocular, uncorrected visual acuities for all study subjects from 40 cm to distance emmetropia.

binocular visual acuity values from 40 cm to distance emmetropia. Inspection of the average preadjustment values indicates that prior to adjustment, the 20 study subjects possessed an average, uncorrected, binocular near visual acuity of ~20/50 (J4), an uncorrected binocular intermediate visual acuity of ~20/40 (J3), and a binocular UCDVA of ~20/20. However, after the required spherical/spherocylindrical and aspheric treatments, the near and intermediate binocular uncorrected visual acuities increased by ~2.5 lines (20/50 to 20/25– and 20/40 to 20/20–) and that for the distance visual acuity increased by ~1 line (20/20 to 20/16). These results indicate that the ability to correct for

residual spherocylindrical errors after cataract surgery followed by optimization of the specific eyes' asphericity will dramatically improve the uncorrected visual acuities from 40 cm to distance emmetropia.

Additionally, Figure 14.11 displays the preadjustment and post lock-in uncorrected visual acuities from 40 cm to distance emmetropia for the dominant and nondominant eyes as well as the binocular values. Comparison of the pre- and postadjustment visual acuities for the nondominant eyes shows a flattening of the visual acuity curve from 40 cm to distance emmetropia, indicating that application of the high and medium aspheric

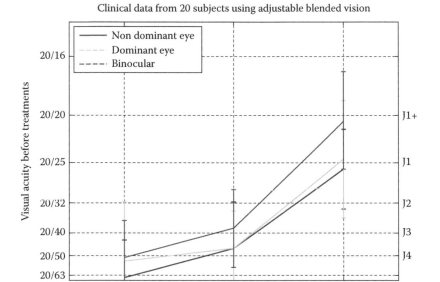

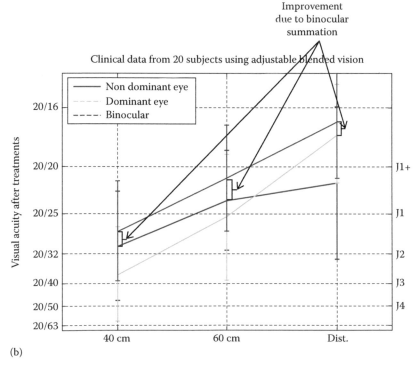

Figure 14.11 Plots of the (a) preadjustment and (b) post lock-in uncorrected visual acuities from 40 cm to distance emmetropia. The three plots appearing on each graph correspond to monocular measurements for the dominant and nondominant eyes as well as the combined binocular result.

treatment protocols induces a significant increase in the eyes' DOF. In addition, the pre- and postadjustment visual acuities for the dominant eyes indicate an average improvement of 1 line in VA. This result indicates that the ability to remove residual refractive error and minimize the fourth-order spherical aberration in the dominant eye postoperatively will significantly improve visual quality. As described in the introduction of this report, the main goal of the proposed binocular aspheric treatment protocol is to optimize near and distance visual acuity by the introduction of significant DOF in the nondominant eye, while at the same time maintaining binocular summation. The maintenance of binocular summation for the current process will result in better visual acuity and contrast sensitivity, particularly in low lighting conditions. As previous research has shown, binocular summation in subjects is highly correlated with patient satisfaction and quality of vision but will not occur if the difference in power between the two eyes is >1.5 D. The presence of binocular summation in this cohort of 20 subjects' eyes is highlighted in Figure 14.11b and clearly shows the improvement in binocular visual acuity versus either eye by itself.

14.4 SUMMARY AND CONCLUSIONS

Despite advances in cataract surgery, imprecise IOL power determination due to errors in biometry, preexisting corneal astigmatism, and unpredictable wound healing remain important clinical problems to address. In addition, the aging postrefractive surgery population and the desire for patients to see well at all distances require that a new cataract treatment modality be developed that addresses these patients' needs. Calhoun Vision has developed the LAL that provides cataract surgeons, for the first time, a method to effectively and noninvasively adjust the refractive error postimplantation and optimize visual outcomes. This novel lens represents a paradigm shift in the way intraocular technology is used to provide optimal vision that is truly customizable and highly predictable.

REFERENCES

1. Narvaez J, Zimmerman G, Stulting RD, Chang DH. Accuracy of intraocular lens power prediction using the Hoffer Q, Holladay 1, Holladay 2, and SRK/T formulas. *J Cataract Refract Surg.* 2006; 32:2050–2053.
2. Olsen T. Sources of error in intraocular-lens power calculation. *J Cataract Refract Surg.* 1992; 18:125–129.
3. Preussner PR, Wahl J, Weitzel D, Berthold S, Kriechbaum K, Findl O. Predicting postoperative intraocular lens position and refraction. *J Cataract Refract Surg.* 2004; 30:2077–2083.
4. Murphy C, Tuft SJ, Minassian DC. Refractive error and visual outcome after cataract extraction. *J Cataract Refract Surg.* 2002; 28(1):62–66.
5. Mamalis N, Brubaker J, David D, Espandar L, Werner L. Complications of foldable intraocular lenses requiring explanation or secondary intervention—2007 survey update. *J Cataract Refract Surg.* 2008; 34:1584–1591.
6. Jin GC, Crandall AS, Jones JJ. Intraocular lens exchange due to incorrect lens power. *Ophthalmology.* 2007; 114:417–424.
7. Sun XY, Vicary D, Montgomery P, Griffiths M. Toric intraocular lenses for correcting astigmatism in 130 eyes. *Ophthalmology.* 2000; 107(9):1776–1781.
8. Feiz V, Moshirfar M, Mannis MJ, Reilly CD, Garcia-Ferrer F, Caspar JJ, Lim MC. Nomogram-based intraocular lens power adjustment after myopic photorefractive keratectomy and LASIK. *Ophthalmology.* 2005; 112:1381–1387.
9. Wang L, Booth MA, Koch DD. Comparison of intraocular lens power calculation methods in eyes that have undergone LASIK. *Ophthalmology.* 2004; 111:1825–1831.
10. Latkany RA, Chokshi AR, Speaker MG, Abramson J, Soloway BD, Yu G. Intraocular lens calculations after refractive surgery. *J Cataract Refract Surg.* 2005; 31:562–570.
11. Mackool RJ, Ko W, Mackool R. Intraocular lens power calculation after laser in situ keratomileusis: Aphakic refraction technique. *J Cataract Refract Surg.* 2006; 32:435–437.
12. Packer M, Brown LK, Hoffman RS, Fine IH. Intraocular lens power calculation after incisional and thermal keratorefractive surgery. *J Cataract Refract Surg.* 2004; 30:1430–1434.
13. Fam HB, Lim KL. A comparative analysis of intraocular lens power calculation methods after myopic excimer laser surgery. *J Refract Surg.* 2008; 24:355–360.
14. Chokshi AR, Latkany RA, Speaker MG, Yu G. Intraocular lens calculations after hyperopic refractive surgery. *Ophthalmology.* 2007; 104(11):2044–2049.
15. Camellin M, Calossi A. A new formula for intraocular lens power calculation after refractive corneal surgery. *J Refract Surg.* 2006; 22(2):187–199.
16. Pardhan S et al. The effect of monocular defocus on binocular contrast sensitivity. *Ophthal Physiol Opt.* 1990; 10:33–36.
17. Wilkins MR et al. Randomized trial of multifocal intraocular lenses versus monovision after bilateral cataract surgery. *Ophthalmology.* 2013; 120(12):2449–2455.
18. Portney V. Multifocal ophthalmic lens, U.S. Patent No. 5,225,858, June 1991.
19. Steiner RF, Aler BL, Trentacost DJ, Smith PJ, Taratino NA. A prospective comparative study of the AMO array zonal-progressive multifocal silicone intraocular lens and a monofocal intraocular lens. *Ophthalmology.* 1999; 106(7):1243–1255.
20. Hansen TE, Corydon L, Krag S, Thim K. New multifocal intraocular lens design. *J Cataract Refract Surg.* 1990; 16:38–41.
21. Ellingson FT. Explanation of 3M diffractive intraocular lenses. *J Cataract Refract Surg.* 1990; 16:697–701.
22. Schwartz C, Canovas C, Manzanera S, Weeber H, Prieto PM, Piers P, Artal P. Binocular visual acuity for the correction of spherical aberration in polychromatic and monochromatic light. *J Vis.* 2013; 14(2):8, 1–11.
23. Zheleznyak L, Sabesan R, Oh JS, MacRae S, Yoon G. Modified monovision with spherical aberration to improve through-focus visual performance. *Investig Ophthalmol Vis Sci.* 2013; 54(5):3157–3165.
24. Hengerer FH, Dick B, Conrad-Hengerer I. Clinical evaluation of an ultraviolet light adjustable intraocular lens implanted after cataract removal: Eighteen months follow-up. *Ophthalmology.* 2011; 118(12):2382–2388.
25. Brierley L. Refractive results after implantation of a light adjustable intraocular lens in postrefractive surgery cataract patients. *Ophthalmology.* 2013; 120(10):1968–1972.

Laser refractive surgery

Jorge L. Alió* and Mohamed El Bahrawy

Contents

Our early personal experience with the excimer laser refractive surgery started in 1991, when we were the first to introduce the *VISX 20/20B* in Spain, and for nearly 25 years through which we experimented all generations of the technology, we again introduced the first sixth-generation excimer laser in Spain. As we look at the milestones of refractive surgery, we find that it had a rather interesting journey to reach our modern-day technologies, with a fair share of ups and downs; something a great number of eminent ophthalmologists and investigators had contributed to this journey, starting from the godfather of refractive surgery, Jose I. Barraquer, MD, developing his innovative studies of the foundations of keratomileusis in situ with the help of his apprentices such as Luis Ruiz, MD. The introduction of excimer laser to the field of refractive surgery

in the early 1980s was a major leap forward, to be first applied on a human eye by Marguerite McDonald, MD, under the supervision of Steve Kaufmann, MD, marking the beginning of photorefractive keratectomy (PRK) technique, as it is performed today. This was a benchmark for development in the field of epithelium removal concepts by different techniques such as laser subepithelial keratomileusis (LASEK), advanced surface ablation, and transepithelial ablation, with most of them showing limited results, to finally reaching the widely recognized and accepted laser in situ keratomileusis (LASIK), which was introduced by Ioannis Pallikaris, among others, nearly a decade after the description of Barraquer's concepts, and started its own fast development journey, passing through several generations of progressive high-performance systems, to reach our modern-day

* Dr. Jorge L. Alió is a consultant of SCHWIND® eye-tech-solutions and Carl Zeiss AG on issues related to technology.

sixth-generation excimer laser technology, with outstanding capabilities yet challenging limitations, setting the future plans of reaching perfection in refractive surgery.

15.1 MILESTONES OF REFRACTIVE SURGERY

15.1.1 EARLY UNDERSTANDING OF REFRACTIVE ERRORS

"The eye is like a mirror, and the visible objects is like the thing reflected in the mirror" as stated by Avicenna (Ibn Sina) in the early eleventh century[1]; since the beginning of civilization and vision has been the subject of conflicting theories by many ancient physicians as well as philosophers, as our understanding of it started to develop with the early definitions of Claudius Galenus (Galen) in the second century, describing the fundamentals of ocular anatomy and the peculiar physiological features of sight such as binocular vision,[2] this had a substantial influence on the prospectives of medieval Islamic medicine and philosophy, when Ibn Al-Haytham was able in his *Book of Optics* (Ketab Al-Manazir) to describe the extramission theory of sight, the effect of light on the eye, and how lenses were tools of improving sight.[2] Leonardo da Vinci in the early sixteenth century was the first to describe the refractive errors as being visual problems and contemplated them to be the source of visual disturbances.[3] Kepler in 1604 presented his idea of the retinal image as being reversed and then restored through the process of reflection and refraction, through placing greater emphasis on the fusion of both anatomy and geometry in studying the eye.[4]

Lenses have been known since the early ages of Egyptians, Greeks, and Romans as a tool of starting a fire, yet the concept of using it as a vision improvement device did not develop till the early eleventh century. The *reading stone*, a segment of a glass sphere that is laid against a reading material to magnify letters, started the nonsurgical innovation of a sharper eyesight sake.[5] In 1284, the first recorded pair of glasses were manufactured by the Italian Salvino D'Armate; were lenses made of glass- or crystal-like stone with a handle to hold them against the eyes. The following centuries witnessed the gradual growth in the craftsmanship of glasses, when Nicholas of Cusa was credited with discovering how to correct myopia with concave lenses, then Benjamin Franklin, with the idea of using separate lens segments in a rimmed frame to correct both myopia and presbyopia invented bifocals, and finally in the year 1825, the British astronomer George Airy created the first lens used to correct astigmatism.[6] The idea of contact lenses was suggested by the British Sir John Herschel, but the design and the origination of the name *contact lens or* (*contact shell*) were by the Swiss physician Dr. Adolf Eugen Fich who used to blow glass into bubbles and cut it in half, polish edges, and place it in the eye.[7]

15.1.2 CREATING THE CONCEPT OF REFRACTIVE SURGERY

Measurement of the anterior corneal surface by Scheiner in 1619 set the foundation of refractive surgery depending on the changes in the anterior corneal contour still recognized by ophthalmologists today. Later in 1823, the Purkinje principles and the four Purkinje images provided better understanding of keratometry and theories of visual accommodation, by helping create the *keratometer*, which was later used by surgeons to measure post cataract surgery astigmatism. The concepts of refractive surgery began as early as the year 1746 and led by a group of Dutch scientists, starting by the idea of removing the natural lens as a way of correcting high myopia when discussed Herman Boerhaave, followed by the proposition of using incisio across the steep meridian of the cornea to flatten it and neutraliz astigmatism by Hermann Snellen in 1869, an idea that was executed nearly two decades later by Faber, when he performed full thickness corneal incisions to decrease the naturally occurri astigmatism enabling a patient to pass his vision test in the royal military academy. It was stated by Leendert Jan Lans that by varying the number, direction, and shape of incisions, it is possible to manipulate the outcomes of visual correction and keratotomy principles that became the standards of refractive surgery. The following years were a phase of trial and error, till the Japanese surgeon Tsutomu Sato applied in 1939 anterior and posterior radial keratotomy based on the principles set by Lans to hundreds of pilots developing astigmatism after trauma and was able to treat up to 6 diopters (D) of astigmatism, but he was unaware of the effect of posterior keratotomy on injuring the endothelium, which led to subsequent development of corneal swelling in these patients; this was followed by several other tria that were focused only on astigmatism without regard to myopia or hyperopia.[8]

The year 1949 witnessed two historical events in ophthalmic surgery, the first when the British ophthalmologist Sir Harold Ridley envisioned the concept of intraocular lens and the other when another pioneer,[9] Professor José Barraquer (Figure 15.1), developed the idea of lamellar keratoplasty to flatten the curvatu of the cornea, significantly reducing myopia. His technique involved using a manually driven microkeratome, similar to a carpenter's plane to remove the anterior part of the cornea, freeze it, and then use a mechanical lathe called cryolathe to change the shape (Figure 15.2). His early trigonometric calculations were used to determine the volume of tissue removal required for a

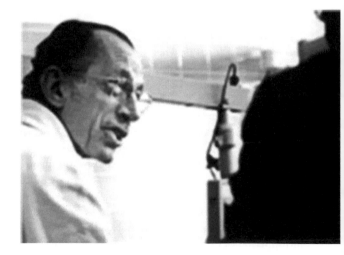

Figure 15.1 Professor Jose I. Barraquer, MD (1977–2007: Commemorating the ISRS/AAO and Global Refractive Surgery).

Figure 15.2 Cryolathe used for early lamellar surgery (1977–2007: Commemorating the ISRS/AAO and Global Refractive Surgery).

Figure 15.3 Barraquer–Krumeich–Swinger set (1977–2007: Commemorating the ISRS/AAO and Global Refractive Surgery).

particular refractive error correction; he coined the procedure as keratomileusis, which literally means sculpting of the cornea.[10]

Barraquer's law of thicknesses in 1964 described the corneal flatness with central tissue removal and steepness with peripheral tissue removal.[11] He then made trials of some other techniques to improve his methods as the *keratophakia*, which involves suturing a donor stromal disc under the initial cap, and then came the trials of Kaufman and Werblin in 1979, describing *epikeratophakia*, both aiming at overcoming the need to use the cryolathe, but the described techniques were reported to be neither predictable nor safe. Students of Barraquer raised the keratomileusis to its highest state of precision; the first was Swinger with the help of Krumeich, when they described the *Barraquer–Krumeich–Swinger* nonfreeze technique (Figure 15.3), a method of changing the shape of the cornea without freezing it, using a dye then a second pass of the microkeratome, with the aim of reducing surgical trauma and visual recovery time. The second was Luis Ruiz, who modified the principles of microkeratome by using an automated form called the *automated lamellar keratoplasty* (*ALK*), to correct high levels of myopia and hyperopia while avoiding irregular sections through a constant and reproducible speed of this automated microkeratome. It was Ruiz also who came up with the idea of passing the microkeratome a second time with the patient on table but with different suction setting, a procedure to be called *in situ keratomileusis*, and later demonstrated that by stopping the microkeratome before the end of the pass, it is possible to create a flap with a hinge that can be replaced again with no need for the previously required suture of the disc securing it with overnight patching.[12]

In the same era, the Russian ophthalmologist Dr. Svyatoslav Fyodorov (Figure 15.4) overcame the long-term problems that resulted from the disruption of corneal endothelium in Sato's

Figure 15.4 Dr. Svyatoslav Fyodorov, MD. (Courtesy of Stephen Trokel, MD.)

posterior keratotomies through a number of straight line incisions in a spoke-like fashion at the periphery confined to the anterior side of the cornea, called *radial keratotomy*. He was able to develop a system of controlling the degree of visual correction by varying the number of incisions and the amount of uncut clear central zones between them[13]; the idea was later introduced to the United States by Leo Bores, where the National Institutes of Health sponsored the *Prospective Evaluation of Radial Keratotomy* study.[14]

15.1.3 EARLY DEVELOPMENT OF EXCIMER LASER IN REFRACTIVE SURGERY

The Russian Nikolay Basov in 1970 using a xenon dimer gas introduced the name *excimer laser* as an abbreviation of *excited dimer*. Few years later the argon fluoride (ArF) excimer laser was developed, and it was first fired on an organic tissue by the IBM scientists Wynne, Blum, and Srinivasan in 1981 when they noticed that it did not cause damage to the surrounding tissue. They demonstrated that complex patterns could be made at a micronic level with each pulse removing a fraction of a micron, a process they called *ablative photodecomposition*, involving three main components: absorption, bond breaking, and ablation. In absorption, the high energy of a single photon of ultraviolet light, being 6.4 eV, was sufficient to exceed the energy required to break molecular bonds (~3.6 eV) and eject the particulate debris as when the concentration of photons or energy density exceeds a critical threshold value, sufficient molecular bonds are broken, enough to separate the tissue into microscopic fragments with sufficient kinetic energy to be ejected from the surface of the tissue, signifying the onset of ablation.[15]

The expectations of refractive surgery had an outstanding shift with the introduction of lasers as a tool for changing the corneal contour. First done in 1977 by Beckman and Peyman, they reported using a carbon dioxide laser to create a thermal shrinkage of the corneal tissue, where Dr. Peyman (Figure 15.5) continued his research at the department of physics at the University of Helsinki in 1984 with incentives of the LASIK using ArF and krypton fluoride excimer lasers and smooth corneal surface, maintaining Bowman membrane, and eliminating pain and potential infection, and the following years he continued to experiment other lasers such as Er:YAG laser as alternatives to excimer laser; in 1989 he reported corneal ablation under flap in rabbits with accepted patent of his method of corneal curvature modification in June 1989 (Figure 15.6) (Dr. Gholam Peyman, personal correspondence, August 2014); John Taboada, an investigator of ocular damage, reported the ability of ArF excimer laser to flatten the cornea, with no thermal damage to the remaining tissue after the application of a 248 nm excimer laser pulse to the epithelium, causing direct splitting of the molecular bonds with minimal adjacent heating—a process coined as "photoablation" (Figure 15.7).[12]

Influenced by the accuracy of Nd:YAG laser in cutting posterior capsule, Stephen Trokel (Figure 15.8), a professor of ophthalmology at Columbia University, started his trials applying different lasers such as Nd:YAG and carbon dioxide in the creation of keratotomies but found none of them suitable for the corneal application, but with the results of Taboada research, he was encouraged to use excimer for such purpose. Trokel formed a team with Srinivasan of IBM and was later joined by Ronald Krueger; this team was able determine the ablation threshold and the optimal ablation rate and to demonstrate the presence or lack of thermal tissue damage at certain wavelengths and fluences, revealing 193 nm to be the superior wavelength, with negligible thermal damage and lowest ablation threshold to be 50 mJ/cm². The greatest ablation efficiency ranged from 150 to 400 mJ/cm², and within this range, each pulse was determined to remove a depth of 1/4 to 1/3 of a micron of corneal stroma according to the beam shape.[12,16] John Marshall, of the institute of ophthalmology

Figure 15.5 Dr. Peyman awarded National Medal of Technology and Innovation for his ophthalmology innovations. (Courtesy of Gholam Peyman, MD.)

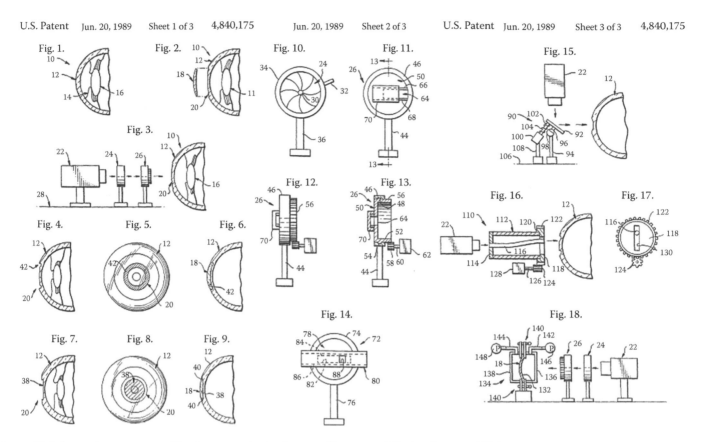

Figure 15.6 Corneal curvature modification patent by Dr. Peyman. (Courtesy of Gholam Peyman, MD.)

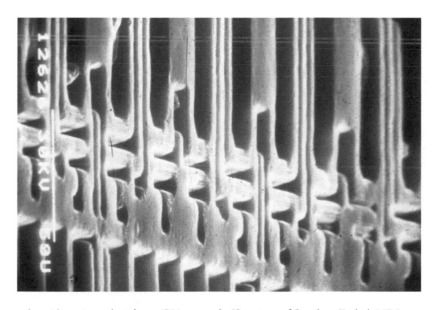

Figure 15.7 Early microlithography with excimer that drove IBM research. (Courtesy of Stephen Trokel, MD.)

in London, joined the team, working on the electron microscopy histopathology of the disc ablation samples. He verified the clarity of tissue interaction without formation of a corneal scar, describing a subepithelial pseudomembrane template on which the epithelium could regrow (Figure 15.9).[17] Marshall suggested a larger area of ablation to the center of the cornea as a feasible keratorefractive principle and described it as *photorefractive keratectomy* (*PRK*), with the need for further investigations on aspects of wound healing process, quality, clarity of the ablated surface, and finally the depth required for refractive error correction; Charles Munnerlyn, an optical engineer, used the early Barraquer's formulae to generate an algorithm of myopic ablation profiles, using progressive discs expanding or constricting in diameter.[18] McDonald and Marshall demonstrated long-term clarity, good healing, and dioptric stability in PRK animal trials up to 1 year after surgery.

First use on a human eye was in Germany, by Theo Seiler in 1985, who performed PRK on blind eyes, involving the

Figure 15.8 Dr. Stephen Trokel, MD. (Courtesy of Stephen Trokel, MD.)

Figure 15.10 Dr. Magritte McDonald, MD (1977–2007: Commemorating the ISRS/AAO and Global Refractive Surgery).

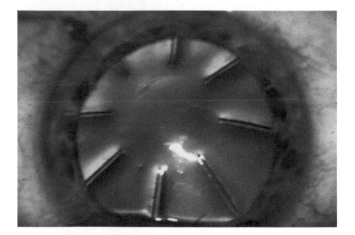

Figure 15.9 Early excimer radial keratotomies. (Courtesy of Stephen Trokel, MD.)

laser ablation of the anterior part of the cornea to flatten its central portion in order to correct myopia. Later that year the first international workshop on laser corneal surgery was held in Berlin. The first PRK on a sighted eye was performed by McDonald (Figure 15.10) in 1988, followed by L'Esperance who suggested small scanning spot excimer lasers, which could be controlled to ablate in specific patterns.[19]

The term LASIK was coined in 1990 by the Greek Ioannis Pallikaris (Figure 15.11) who with Lucio Buratto (Figure 15.12) in

Figure 15.11 Dr. Ioannis Pallikaris, MD (1977–2007: Commemorating the ISRS/AAO and Global Refractive Surgery).

Figure 15.12 Dr. Lucio Buratto and Dr. Jorge Alió.

the United States combined the moderately successful principles of ALK, originally set by Barraquer, through using the microkeratome, with the incredible accuracy of the excimer laser. The Russian ophthalmologist Razhev and his coworkers in Novosibirsk were performing similar technique and reporting their results of the very first LASIK[20]—a technique that avoided the anterior stromal haze and pain associated with PRK, because in LASIK, epithelium remains almost entirely intact, not exposing nerve ending and having a lower risk of infection and scarring. It was Michelson who first introduced the new technique to the United States in 1996, leading to the Food and Drug Administration (FDA) approval of the LASIK procedure in 1999.[21]

15.2 SUCCESSIVE PROGRESS ALONG CONSECUTIVE GENERATIONS OF EXCIMER LASER

The year 1983 witnessed a global race of developing excimer lasers for refractive use, with the cooperation of Trokel, Munnerlyn, and McDonald on one side in the United States and Seiler and his colleagues conducting studies on the other side in Germany.

Meditec introduced the first excimer laser prototype intended for clinical use in 1986 as a laser system with a delivery system into a box that was called *Meditec Excimer Laser 30 Hz* (MEL30); it was designed by Reinhardt Thyzel (Figure 15.13), a system with a big spot to hit mask radial keratotomy; instead of a cut, it used a mask to simulate cuts in the eye. The first cut was done on a blind human eye of a patient at the clinic of Prof. Dardenne in Bonn, to be later presented in Rome at the world congress as the first operational excimer laser based on an articulated arm attached to a laser console. Meditec at that time hired Kristian Hohla, MD, to start with him a company under the name of Technolas in Munich, Germany, with the sole purpose of building the laser heads for the technology; Meditec laser became Zeiss-Meditec and some of its former employees started WaveLight® company few years later (Reinhardt Thyzel, personal communication,

Figure 15.13 Mr. Reinhardt Thyzel. (Courtesy of Mr. Reinhardt Thyzel.)

April 6, 2014). At the same time, another prototype was designed by NIDEK® Co. of Japan, and both prototypes were later shifted to scanning slits to be used in PRK.

Munnerlyn in 1986 (Figure 15.14) introduced the first excimer laser using broad area ablation; a year later the first standard commercial prototype VISX laser, founded by Munnerlyn and Trokel, featured homogenized preprogrammed

Figure 15.14 Dr. Charles Munnerlyn, PhD. (From http://one.aao.org/munnerlyn-laser-surgery-center/munnerlyn-laser-center-landing-page.)

broad beam laser with intrinsic hot spots on colder areas. This prototype was first used on a human eye with malignant melanoma by McDonald on the same year. At the same time, Summit technologies started experimenting its first broad beam prototype with Seiler in Germany, who used both Summit and Meditec prototypes in his studies conducting PRK and PTK. McDonald under the supervision of Kaufmann led the phases I and II of the FDA approval of excimer PRK by 10, then 40 cases for each phase, respectively, to be followed by a 2100 case presentation of 3 excimer systems, Touton, VISX, and Summit in phase III, granting FDA approval to PRK in 1995 (Table 15.1).[22] International spread of technology started in 1989, with the installment of a VISX system in Jeddah, Saudi Arabia, by Dr. Akef El Maghraby and his colleagues, followed by a rather wider spread era in the years between 1989 and 1995.[15]

The following 25 years witnessed a rather remarkable and speed upgrades in the speed, precision, and safety profiles, evolving highly sophisticated instruments, from the first generation using a small optical zone, giving it the ability to correct only myopic refractive errors. The second generation still came with a fixed but larger optical zones with the ability to correct hyperopia in addition to myopia, with some of the platforms of this generation featuring incorporated passive eye tracker. The following generation still used broad beam laser but with fractional mask capabilities and aspheric ablation profiles, also featuring variable multizone optical zones. The fourth generation was the introduction of the flying spot, fractional rotation mask, and scanning slit capabilities; it incorporated active eye tracking system and was capable of some customized corrections that markedly enhanced hyperopic refractive correction. Wavefront technology was the next step, with high-speed flying spot laser and high-speed active tracking system of 340 Hz; platforms became able to perform optimized customized wavefront-guided corrections (Table 15.2 and Figure 15.15).

15.3 STANDARD OUTCOMES OF EXCIMER LASER REFRACTIVE SURGERY

Several ophthalmic authorities had set the benchmark for laser keratorefractive surgery, as the FDA, based on data presented by several evidence-based reviews, defined the correction limitation of excimer laser (Table 15.3).[30] The American Academy of Ophthalmology (AAO) reports[31] outlined the standard outcomes of excimer laser, using the substantial level II and level III evidence. It demonstrated that for low to moderate myopia (–1 to –6 D), results were highly effective and predicable in terms of obtaining very good to excellent uncorrected visual acuity, with mean postoperative refractive error –0.14 ± 0.32 with up to 93.5% UCVA > 20/20, 100% > 20/40, and up to 90% within 1 D of targeted refraction; the results were similar with low to moderate astigmatism less than 2 D. In moderate to high myopia (–6 to –15), results were much variable and similar to high astigmatism more than –5 D, the reports stated the results of randomized controlled clinical trials of moderate to high myopia (–6 to –15) to be less predictable with longer stability time; 66.6% within 1 D of target refraction and up to 57% achieved postoperative vision of 20/20, but with relatively higher overall percentage losing more than 2 lines of BCVA.[61] As with myopic LASIK, the results are superior for low amounts of hyperopia (<3 D) compared with higher amounts of hyperopia (>4–5 D), with limited outcomes in the correction of hyperopic astigmatism.[62] In the published studies in the report about the wavefront-guided LASIK, the mean preoperative manifest spherical equivalent (MSE) ranged from –3.16 to –7.30 D, and the mean postoperative MSE ranged from 0.14 to –0.40 D, with 72%–100% of these eyes being within 0.5 D of the intended postoperative target MSE, with evidence of improved contrast sensitivity and improvements in both lower- and higher-order aberrations in comparison to conventional LASIK.[63] Another report comparing the results of PRK to that of LASIK showed the superiority of LASIK outcomes to that of the PRK, in terms of early outcomes and patients' satisfaction, but with similarity in the long-term results, the mean predictability is –0.77 ± 1.0 and –1.43 ± 1.22 D, average regression from 1 to 6 months is 0.89 and 0.55 D, and 6 months BSCVA loss of 2 lines or more is 11.8% and 3.2% for PRK and LASIK, respectively[64] (Tables 15.4 and 15.5).

Table 15.1 **FDA-approved lasers for refractive surgeries**

DEVICE NAME	COMPANY	APPROVAL DATE	APPROVED USE
MEDITEC MEL 80 Excimer Laser System	Carl Zeiss, Inc.	March 28, 2011	LASIK
Star S4 IR Excimer Laser System with Wavescan System	VISX, Inc.	July 11, 2007	LASIK
NIDEK EC-5000 Excimer Laser System	NIDEK, Inc.	October 11, 2006	LASIK
MEL 80 Excimer Laser System	Carl Zeiss, Inc.	August 11, 2006	LASIK
WaveLight Allegretto Wave Excimer Laser System	WaveLight AG	July 26, 2006	LASIK
LADARVision 4000 Excimer Laser System And the Ladar 6000 Excimer Laser System	Alcon Laboratories, Inc.	May 2, 2006	LASIK
LADARVision 4000 Excimer Laser System	Alcon Laboratories, Inc.	May 1, 2006	LASIK
WaveLight Allegretto Wave Excimer Laser System	WaveLight AG	April 19, 2006	LASIK
Star S4 IR Excimer Laser System with Variable Spot Scanning (Vss) and Wavescan Wavefront System	VISX, Inc.	August 30, 2005	LASIK
Star S4 IR Excimer Laser System with Variable Spot Scanning (Vss)	VISX, Inc.	March 17, 2005	LASIK
Star S4 Excimer Laser System with Variable Spot Scanning (Vss) and Wavescan Wavefront System	VISX, Inc.	December 14, 2004	LASIK
LADARVision 4000 Excimer Laser System	Alcon Laboratories, Inc.	June 29, 2004	LASIK
WaveLight Allegretto Wave Excimer Laser System	Alcon Laboratories, Inc.	October 10, 2003	LASIK
TECHNOLAS 217Z Zyoptix System for Personalized Vision Correction	Bausch & Lomb Surgical, Inc.	October 10, 2003	LASIK
WaveLight Allegretto Wave Excimer Laser System	Alcon Laboratories, Inc	October 7, 2003	LASIK
Star S4 Activetrak Excimer Laser System and Wavescan Wavefront System	VISX, Inc.	May 23, 2003	LASIK
TECHNOLAS 217A Excimer Laser System	Bausch & Lomb Surgical, Inc.	February 25, 2003	LASIK
LADARVision 4000 Excimer Laser System	Alcon Laboratories, Inc.	October 18, 2002	LASIK
TECHNOLAS 217A Excimer Laser System	TECHNOLAS GMBH Perfect Vision	May 17, 2002	LASIK
VISX Star Excimer Laser System	VISX, Inc.	November 6, 2001	LASIK
Laserscan LSX Excimer Laser System for Laser-Assisted In-situ Keratomileusis (LASIK)	Lasersight Technologies, Inc.	September 28, 2001	LASIK
VISX Star Excimer Laser System	VISX, Inc.	April 27, 2001	LASIK
VISX Star S2 and S3 Excimer Laser System	VISX, Inc	October 18, 2000	PRK and Other
LADARVision Excimer Laser System (Hyperopia)	Summit Autonomous, Inc	September 22, 2000	LASIK
LADARVision Excimer Laser System	Summit Autonomous, Inc.	May 9, 2000	LASIK
NIDEK EC-5000 Excimer Laser System	NIDEK Technologies, Inc.	April 14, 2000	LASIK
TECHNOLAS 217A Excimer Laser System	TECHNOLAS GMBH Perfect Vision	February 23, 2000	LASIK
Laserscan LSX Excimer Laser System	Lasersight Technologies, Inc.	November 12, 1999	PRK and Other
SVS Apex Plus Excimer Laser Workstation	Summit Technology, Inc.	October 21, 1999	LASIK, PRK and Other
NIDEK EC-5000 Excimer Laser System (Park)	NIDEK Technologies, Inc	September 29, 1999	PRK and Other
KERACor 116 Ophthalmic Excimer Laser System	Bausch and Lomb Surgical	September 28, 1999	PRK and Other

(Continued)

Table 15.1 (*Continued*) **FDA-approved lasers for refractive surgeries**

DEVICE NAME	COMPANY	APPROVAL DATE	APPROVED USE
EC-5000 Excimer Laser System	NIDEK, Inc.	December 17, 1998	PRK and Other
VISX Excimer Laser System Model C "Star"	AMO Manufacturing USA, LLC	November 19, 1999	LASIK
LADARVision Excimer Laser System	Alcon Laboratories	November 2, 1998	PRK and Other
VISX Star S2 Excimer Laser System	VISX, Inc.	November 2, 1998	PRK and Other
Kremer Excimer Laser System	Lasersight Technologies, Inc.	July 30, 1998	LASIK
SVS Apex Plus Excimer Laser Workstation and Emphasis Disc	Summit Technology, Inc.	March 11, 1998	PRK and Other
VISX Excimer Laser System Models "B" and "C"	VISX, Inc.	January 29, 1998	PRK and Other
VISX Excimer Laser System Models "B" and "C"	VISX, Inc.	April 24, 1997	PRK and Other
VISX Excimer Laser System Models "B" and "C"	AMO Manufacturing USA, LLC	March 27, 1996	PRK and Other
SVS Apex Excimer Laser System	Summit Technology, Inc.	October 20, 1995	PRK and Other
VISX Model B /Model C Excimer Laser System	AMO Manufacturing USA, LLC	September 29, 1995	PRK and Other
Eximed UV200LA Excimer Laser System SVS Apex Excimer Laser System	Summit Technology, Inc.	March 10, 1995	PRK and Other

Source: U.S. Food and Drug Administration, FDA-Approved lasers for LASIK, PRK and other Refractive Surgeries, Sep 2011, Retrieved from: http://www.fda.gov/MedicalDevices/ProductsandMedicalProcedures/SurgeryandLifeSupport/LASIK/ucm168641.htm, May 1, 2014.

Table 15.2 **Features of the successive generations of excimer lasers**

First generation	Preclinical (Touton, VISX, Summit)
Second generation	Broad beam laser, fixed optical zone
Third generation	Broad beam laser, variable optical zone, multizone treatment
Fourth generation	Flying spot laser, built in tracker, hyperopic treatment
Fifth generation	Customized wavefront (guided, optimized) treatments
Sixth generation	• Faster ablation rates and tracking systems • Lower biological interaction • More variables under control • Pupil size • Advanced ablation profiles • Cyclotorsion control • Online pachymetry

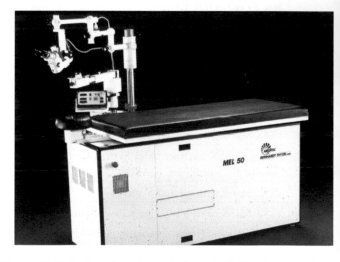

Figure 15.15 Early excimer laser platform; MEL 50. (Courtesy of Stephen Trokel, MD.)

15.4 LATEST SIXTH-GENERATION EXCIMER LASERS

15.4.1 INTRODUCTION

The modern-day trends of technology in excimer laser delivery system target the goal of minimally invasive surgery in LASIK, aiming at decreasing the amount of tissue ablation per diopter treated, increasing predictability, reducing the incidence of complications, decreasing the time required for the procedure, increasing the optical quality postoperatively, and using an advanced ablation wavefront-guided profiles under the control of key outcome variables as tracking, cyclotorsion, and online pachymetry.

15.4.2 TYPES

A number of manufacturers have been involved in the production of these state-of-the-art platforms, with differences in the specifications, but still with high-performance qualities. Here we discuss three market leaders: AMARIS® by SCHWIND eye-tech-solutions, Allegretto Wave® Eye-Q Laser by ALCON surgical: WaveLight® and NAVEX® Quest by NIDEK Co., Ltd.: NIDEK

Table 15.3 **FDA indications for LASIK and PRK**

REFRACTIVE ERROR	LASIK	PRK
Myopia	• Less than –14.0 D • With or without astigmatism between –0.50 and –5.00 D	• Up to –12.0 D • With or without astigmatism up to –4.00 D
Hyperopia	• Up to +5.00 • With or without astigmatism up to +3.00 D	• Up to +5.00 • With or without astigmatism up to +4.00 D
Mixed astigmatism	Astigmatism up to 6.00 D, cylinder is greater than sphere and of opposite sign.	

Source: AAO Refractive Management/Intervention PPP Panel, Hoskins Center for Quality Eye Care (July 2013), Refractive errors & refractive surgery PPP—2013, Retrieved from: http://one.aao.org/preferred-practice-pattern/refractive-errors—surgery-ppp-2013 (May 8, 2014).

advanced vision excimer laser system are three market technology leaders in the domain of the sixth-generation excimer lasers (Figure 15.16 and Table 15.6).[23–25]

15.4.3 CHARACTERISTICS OF THE NEW GENERATION

The following demonstrates the state-of-the-art features of the sixth-generation excimer lasers.

15.4.3.1 Faster excimer laser performance

Excimer laser systems, which can deliver more laser spots per second, are able to ablate more corneal tissue in a given time and thus result in a faster treatment time. The speed of existing sixth-generation laser platforms varies from 400 to 1050 Hz, 400 Hz (Eye-Q, WaveLight) and 1050 Hz (AMARIS, SCHWIND eye-tech-solutions). In this way, a 500 Hz platform will require less than 4 s per diopter to ablate a 6.5 mm optical zone compared to 7–10 s per diopter using previous laser platforms.[26]

15.4.3.2 Advanced fluence level adjustments

Depending on the planned refractive correction, 80% of corneal ablation is performed with high fluence level, while fine correction is performed with low fluence level improving resolution that significantly shortens laser treatment time, specially for higher refractions with remarkable exceptional precision.

15.4.3.3 Reduction of collateral thermal tissue damage

A high repetition rate may result in shorter intervals between laser pulses on the same area of the cornea. This may increase the thermal load on the cornea and result in thermal damage; some sixth-generation lasers use an Intelligent Thermal Effect Control to reduce the heating of the cornea significantly. This system ensures that the area around an applied laser spot is blocked for a certain time to let the cornea cool down, which will ensure thermally optimized, dynamically adapted distribution of laser pulses, allowing enough time for each area of the cornea to cool down between pulses and protect the stroma.[27]

15.4.3.4 Advanced eye trackers

The reduction of induced aberrations is a critical trend in modern laser refractive surgery; sixth-generation platform features advanced ablation profiles with the reduction of spot size as a key factor of control of the induced aberration, ranging between 0.54 and 0.65 mm, in comparison to a spot size of 0.8 mm or more for conventional excimer lasers. Also these profiles are able to correct preexisting optical aberration through integrated customized and wavefront ablation technology. The efficiency of the previous features required extremely accurate laser spot placement, in which the eye tracker should be at least twice as fast as the speed of the laser. A conventional laser platform eye tracker will have a capturing rate of 60–330 Hz, able to detect the pupil position at 4000 Hz generating a response time of 36 ms, which is clearly not fast enough for a high-speed laser platform of a speed reaching 700 Hz. The new five-dimensional turbo speed tracker has an acquisition speed of 1050 Hz generating a response rate less than 3 ms with unique rotational balance, tracking both the pupil and the limbus.[28]

Usually systems with eye tracker adjust linear eye movements in the x- and y-axes only, so lasers are able to follow "eye rolling" as it translates linear movements into rotations with the help of an eye model, so that horizontal and vertical rotations are followed and compensated; on the other hand, modern eye trackers not only track horizontal or vertical displacements of the eye but also track cyclotorsional rotations of the eye and rolling. The ability to track all movements of the eye is crucial to enable accurate place of the laser spots, as cyclotorsion movements of the eye can be classified as either static or dynamic cyclotorsion movements. Static cyclotorsion occurs when the patient moves from upright to supine position, while dynamic cyclotorsion occurs during the treatment procedure. Advanced eye trackers must be able to detect both types of cyclotorsion movements.[29]

15.4.3.5 Pupil size control

Automatic monitoring of pupil size ensures additional safety as illumination is automatically adjusted in such a way that the pupil is exactly the same size at the beginning of the treatment as it was at the preliminary examination.[23–25]

15.4.3.6 Newer ablation profiles

Most excimer laser delivery systems today come with ablation profiles designed to reduce the induction of aberrations. Customized or wavefront-guided ablation profiles are also available in order to correct preexisting individual optical aberrations. The size of the laser spot diameter is an important factor for the success of advanced surface profile; as most conventional excimer lasers have a spot size of 0.8 mm or more, a sixth generation like the SCHWIND AMARIS has a spot size of as small as 0.54 mm.[23–25]

15.4.3.7 Integrated online pachymetry

Some of the most recent platforms have introduced this feature as the AMARIS of SCHWIND, through which changes in corneal thickness are displayed in real time; the targeted measurements take place before flap preparation, after flap lifting, also during and after laser ablation to be documented in the treatment log.

Table 15.4 **LASIK selected standard results of the American Academy of Ophthalmology reports**

AUTHOR	NUMBER OF EYES	RANGE OF PREOP MYOPIA (D)	PREOP REFRACTION; MSE (D)	POSTOP REFRACTION; MSE (D)	PERCENT WITHIN ±0.50 D/1.0 D	POSTOP UCVA ≥20/20 (%)	POSTOP UCVA ≥ 20/40 (%)	LOSS OF ≥ 2 LINES BCVA (%)	FOLLOW-UP (MONTHS)
Low to moderate myopia (−1 to −6 D)									
Casebeer et al.[66]	911	−1.0 to −4.0 −4.0 to −7.0	NR	NR	75/90 52/73	NR	92 86	0 0	3 (100)
Reviglio et al.[67]	74 62	−1.0 to −4.0 −4.0 to −6.0	−2.21 ± 0.88 −4.59 ± 0.60	−0.09 ± 0.41 −0.26 ± 0.74	99.44 96.32	60.8 45.2	100 95.2	0 0	6 (100)
Morchen et al.[68]	35	−1.0 to −9.5	−4.8 ± 2.3	−0.22 ± 0.59	68/93.5	93.5	100	0	3 (88)
Moderate to high myopia (−6 to −25 D)									
Casebeer et al.[66]	911	−7.0 to −10	NR	NR	40/54	NR	68	0	3 (100)
McDonald et al.[69]	347	−1 to −11	NR	−0.29 ± 0.45	75.2/95.2	57	94	0.9	6 (94.4)
Reviglio et al.[67]	126	−6.0 to −10 −10 to −25	−7.63 ± 1.09 −12.70 ± 2.81	−0.37 ± 0.92 −0.64 ± 1.23	68.42 85.08	25.4 9.8	87.2 78.4	0 0	6 (100)
Hyperopia									
Rashad et al.[70]	85	1.25 to 5.00	3.31	0.43 ± 0.57	61/89	25	93	1.2	12
Tabbara et al.[71]	80	0.50 to 11.50	3.4	0.26 ± 0.80	58/84	44	98	1.3	6
Salz et al.[72]	143 117	0.88 to 6.00 0.50 to 5.75	2.56 2.84	0.05 ± 0.71 0.06 ± 0.78	74/91 73/89	54 58	94 94	3.4 1.4	12
Wavefront guided for primary myopia and astigmatism									
Durrie et al.[73]	30	NR	−4.66 ± 1.73	0.01 ± 0.34	83/NR	93	100	0	1
Pop and Payette[74]	71	NR	−4.4	−0.07	83/NR	92	100	0	3
Venter et al.[75]	93	NR	−3.72 ± 1.96	−0.07 ± 0.27	92/NR	88	100	0	6

Table 15.5 **PRK selected standard results of the American Academy of Ophthalmology reports**

AUTHOR	NUMBER OF EYES	RANGE OF PREOP MYOPIA (D)	PERCENT WITHIN ±1.0 D	POSTOP UCVA ≥20/20 (%)	POSTOP UCVA ≥20/40 (%)	LOSS OF ≥2 LINES BCVA (%)	FOLLOW-UP (MONTHS)
Alió et al.[76]	3000	–1 to –14	94	NR	NR	1	12
Higa et al.[77]	1218	–1 to –19	72	34	75	7	12
McCarty et al.[78]	347	<–5	87	47	87	4	12
	240	–5 to –10	65	25	71	9	

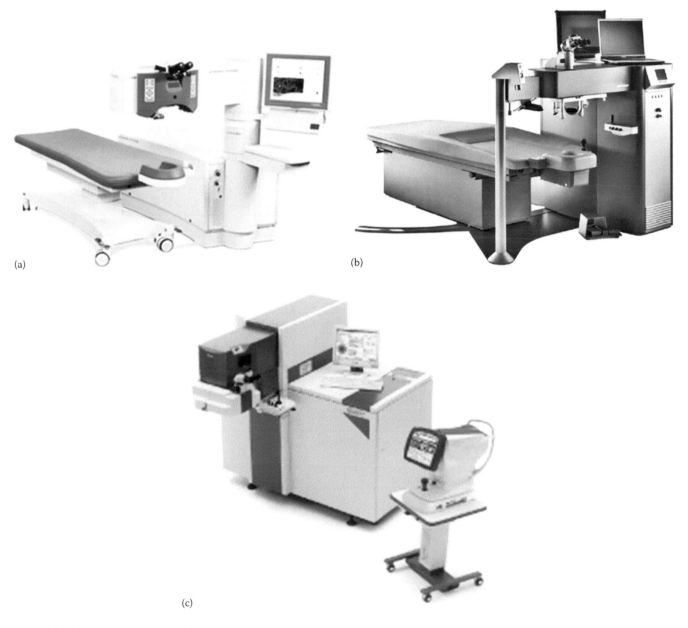

(a)

(b)

(c)

Figures 15.16 (a) AMARIS by SCHWIND. (From SCHWIND eye-tech-solutions: The SCHWIND AMARIS family, Retrieved from: http://www.schwind-amaris.com/en/home/, May 1, 2014.) (b) Allegretto Eye-Q by ALCON Labs. (From ALCON surgical: Wave Light® Allegretto Wave® Eye-Q Laser, Retrieved (May 1, 2014) from: http://www.alconsurgical.com/wavelight-allegretto-wave-eye-q-laser.aspx, May 1, 2014.) (c) NAVEX Quest by NIDEK. (NIDEK Co., Ltd., NIDEK advanced vision excimer laser system NAVEX Quest, Retrieved from: http://www.nidek-intl.com/products/ref_surgical/navex-quest.html, May 1, 2014.)

Table 15.6 **Comparison between sixth-generation excimer platforms**

Company	SCHWIND	NIDEK	WaveLight
Model	AMARIS	NAVEX Quest	Allegretto Eye-Q
Laser type	ArF	ArF	ArF
Laser beam	Flying spot	Slit scanning+ variable spot-size scanner	Flying spot
Beam profile	Super-Gaussian	Flat Top	Gaussian
Pulse rate	500–700 Hz	6 scans/s 60 Hz max	400–700 Hz
Pulse duration	10 ns	25 ns	10 ns
Peak fluence	160–450 mJ/cm^2	130 mJ/cm^2	400 mJ/cm^2
Beam size	0.54	10 × 2 mm scanning slit (1 mm for customized and hyperopia)	0.68 mm
Spot size (cornea)	0.54 mm	1.0 mm	0.95 mm (1.2 mm)
Optical zone (OZ)	4–10 mm	6.5 mm	4.5–8 mm (OZ)
Ablation zone	Optimized	8 mm	9 mm
Ablation profile	Aspheric (aberration-free)	Munnerlyn with aspheric transition zone	Aspheric (including Q-value)
Transition zones adjustable	No	Yes	Yes
Static cyclotorsion	Yes	Yes	Pseudo, yes
Dynamic cyclotorsion	Yes	Yes (TEC = torsion error correction)	Pseudo, yes
Cyclotorsion, sampling rate, Hz	36 Hz	30 Hz	NR
Ablation depth per shot (cornea)	0.42–0.68 µm	0.32 µm	0.65 µm
Ablation volume per shot (cornea)	110–220 pl	250 pl	0.38 µm^2
Ablation depth per diopter (6.5 mm OZ)	16.4 µm	15 µm	15.3 µm
Time per diopter (6.5 mm OZ)	<2.5 ms	5 s	3 s
Ablation depth –5 D/OZ = 6 mm	65 µm (12 s)	63 µm	65 µm
Eye tracking system	Active video tracking (SMI)	Active video tracking (SMI)	Active video tracking (SMI)
Sampling rate, Hz	1050 Hz	1050 Hz	400 Hz
Eye tracker response time	<3 ms	4 ms	4.0 m
Cyclotorsion, resolution	Static ±15°; dynamic ±7°	NR	NR
X–Y and Z-tracking	Active	Active	
Presbyopic treatment	No	Yes	No
Online pachymetry	Yes—integrated	No	No
Eye fixation	Green LED	Yes	LED
Centration of pupil	Automatic, user defined	Manual	User defined
Laser head/laser source	Coherent	Lambda	TUI
Fluence test needed every:	2 h	NR	Before every treatment day
Fluence and calibration	Automatic and objective	Manual and subjective	Manual and subjective
Capable of customized ablation	Yes	Yes	Yes
Ocular wavefront	Yes	Yes	Yes
Method used for wavefront	Hartmann–Shack	Yes	Tscherning principle

(Continued

Table 15.6 (*Continued*) Comparison between sixth-generation excimer platforms

Topographic system	Corneal Wavefront Analyzer/CSO	Topographer retinoscopy	Yes Oculus
Topographic link	Yes (Corneal wavefront)	Yes OPD-scan	Yes (topographic based on Zernike)
Dimensions (L × W × H)	264 × 144 × 136 (including patient bed)	137 × 151.6 × 147 cm	120 × 145 × 130 cm (without patient bed)
Weight (kg) (without patient bed)	550 kg	650 kg	265 kg (without bed and gas) patient bed 188 kg
Website: last visited May 2014	http://www.schwind-amaris.com/en/home/	http://www.nidek-intl.com/products/ref_surgical/navex-quest.html	http://www.alconsurgical.com/wavelight-allegretto-wave-eye-q-laser.aspx

Sources: SCHWIND eye-tech-solutions: The SCHWIND AMARIS family, Retrieved from: http://www.schwind-amaris.com/en/home/, May 1, 2014; ALCON surgical: Wave Light® Allegretto Wave® Eye-Q Laser, Retrieved from: http://www.alconsurgical.com/wavelight-allegretto-wave-eye-q-laser.aspx, May 1, 2014; NIDEK Co., Ltd., NIDEK advanced vision excimer laser system NAVEX Quest, Retrieved from: http://www.nidek-intl.com/products/ref_surgical/navex-quest.html, May 1, 2014.

15.5 OUTCOMES OF SIXTH-GENERATION EXCIMER LASER TECHNOLOGY

15.5.1 MYOPIC REFRACTIVE ERRORS

Our studies focused on the promised safety and accuracy of the new platforms in myopic correction through two separate reports of patients of high myopia of −8.50 D or more. In the first study, we studied 29 eyes of 17 patients, a mean age of 36.65 ± 10.80 years, with a mean spherical equivalent (MSE) of −8.39 ± 0.93 D, and followed up the results for 6 months. The efficacy of the treatment was 89.6% within 1.00 D of target refraction, and postoperative high-order aberrations (HOAs) was 0.95 ± 0.8 μm.[28] We confirmed the results with a larger sample in a second study to show an efficacy of 84.3% within 0.5 D of target refraction.[32] At the same time, Arba-Mosquera et al. published the results of a 3-month follow-up of 30 eyes with an MSE of −4.27 ± 1.62 D showing a mean residual spherical equivalent of −0.07 ± 0.25 D and a postoperative HOA of 0.425 ± 0.129 μm.[33] Later on Tomita et al. used a larger sample of 10,235 eyes of 5,191 patients; in his study the MSE was −5.02 ± 2.17 D, follow up was up to 3 months postoperatively, efficacy index was 1; with 88.4% within ±0.5 D and safety index was 1.03; 96.9% of patients had a vision of 0.0 on LogMAR, the HOA postoperatively were 0.70 ± 0.23 μm.[34] Most of the previous studies were conducted on AMARIS sixth-generation excimer laser platforms by SCHWIND eye-tech-solutions (Kleinostheim, Germany). Kanellopoulos et al. reported his results with the Alcon-WaveLight EX500 excimer laser by Alcon Laboratories (Fort Worth, TX), where he reported the results of the 12 months follow up of 58 patients; the preoperative MSE was −7.67 ± 1.55 D; his results showed 100% of patients were within 1.0 D defocus, and a keratometric stability of 0.22 D.[35]

15.5.2 HYPEROPIC REFRACTIVE ERRORS

The small spot hyperopic laser in situ keratomileusis, the so-called H-LASIK ablation at the periphery of the cornea designed to produce certain degrees of steepness. This treatment modality had several limitations such as decentration, decrease in best corrected visual acuity, high frequency for the need of retreatments, residual refractive errors, and induction of astigmatism due to the high levels of corneal aberrations as negative spherical aberration; this all causes loss of the efficiency of treatments and changes in biomechanics of the cornea.[36] Our studies reported the 6 month follow up of 51 eyes in 28 patients, of an MSE of +5.64 ± 0.93 D followed up for 6 months; we reported an efficacy of 70.37% within 0.5 D of target refraction, the HOA postoperatively; −0.44 ± 0.22 μm, with an efficacy and safety indices of 0.85 and 0.94, respectively,[37] another 6 months retrospective follow-up of 51 eyes with a spherical equivalent of more than 5.5 D; significant increase in corneal RMS high-order (RMS HO), spherical aberration (RMS SA), and coma (RMS COMA) aberration was observed 6 months after surgery (P < 0.01). Corneal asphericity for the 4.5 mm (Q45) and 8 mm (Q8) of corneal diameter also changed significantly during the postoperative period (P < 0.01). Strehl ratio change was not statistically significant (P = 0.77).[38] Arbelaez et al. published a study with an MSE of +3.02 ± 2.06 D (astigmatism was +1.36 ± 1.61 D); they reported a 6-month follow-up with a mean postoperative increase in HOA of 0.18 μm (P < 0.05); 89% were within 0.5 D of the target refraction and 94% ±0.5 of astigmatism, with comparable efficacy and safety indices of 0.89 and 1.1, respectively.[36] Kanellopoulos used a larger sample of 202 eyes with a longer follow-up of 2 years to demonstrate the safety and efficacy of topography-guided ablation using a 400 Hz WaveLight excimer laser by Alcon Laboratories (Fort Worth, TX); in his study the MSE was +3.04 ± 1.41 D; the results showed that the mean refraction spherical equivalent was ±0.50 D of target refraction in over 80% of cases, with an increase in the root mean square (RMS) of 15%.[39]

15.5.3 CORRECTION OF ASTIGMATISM

Patient satisfaction after refractive surgery, whether treated with wavefront guided treatment or not, is primarily dependent on the successful treatment of lower-order aberrations; sphere and cylinder of the eye, LASIK has been successful in the correction of mild to moderate myopic astigmatism, but with limited reports on the efficacy, predictability and safety of it, in higher myopic astigmatism in the terms of astigmatic correction of HOA, with the limitation of re-treatments if needed.[26] We reported a 3-month follow-up of 52 eyes of mixed astigmatism of more than 3 D, in which the efficacy was 65.3% within 1.0 D of target refraction.[40] Earlier Arbelaez et al. published a number of reports, with an average of 1.26 ± 3.29 D of astigmatism and with a 6-month follow-up. He found a HOA increase of 0.09 μm with a residual astigmatism of 0.50 ± 0.26 D (P < 0.0001).[26,41,42] The mean decrease of astigmatism magnitude in reported the most recent platforms was 93%, indicating a slight under correction of the preoperative astigmatism, which is marked improvements from the 36% to 91% reported with the use of older excimer laser platforms.[43,44] Also, recent reports showed a 100% efficacy within 0.25 D after 12 months.[35] Topography-guided hyperopic astigmatism correction showed a correction of a mean preoperative cylinder value of –1.24 ± 1.41 to the respective postoperative value of –0.35 ± 0.25[39] (Table 15.7).

15.5.4 RESULTS IN PHOTOREFRACTIVE KERATECTOMY

New platforms had shown similar satisfactory results with PRK. Aslanides et al., in their 2-year follow-up of 80 eyes with mild to high myopia and myopic astigmatism, found 91% to be within ±0.5 D of target refraction but reported a statistical increase in postoperative coma (+0.12 μm) and spherical aberration (+0.14 μm) compared to preoperative measurements (P < 0.001, both cases) as performed by the SCWIND AMARIS.[45] Also both VISX CustomVue® and WaveLight platforms performed equally in terms of visual acuity, safety, and predictability in PRK. The wavefront-guided group showed slightly improved contrast sensitivity. Both lasers induced a comparable degree of statistically significant spherical aberration and tended to increase other higher-order aberration measures as well.[46]

15.6 EXCIMER LASER CORRECTION OF PRESBYOPIA: PresbyLASIK TECHNIQUES

The idea of creating corneal multifocality for the treatment of presbyopia started with the application of radial keratotomy; also evidence of better subjective accommodation and near acuity was noticed in patients who underwent PRK myopic correction compared to their age group–matched controls. The use of LASIK, as a more controllable technique for corneal multifocality, avoiding the plastic compensatory effect of growing epithelium reactive to surface ablation profiles, is more adequate for presbyopia correction.[47]

Three different techniques have been proposed.

1. *Multifocal transitional profiles*: Started in the early 1980s, this technique is based on the creation of a transitional vertical multifocal ablation based on the creation of an intentional decentration of a hyperopic ablation profile; it created significant levels of vertical coma, making it less attractive to many surgeons, with very limited reports of the technique outcomes.
2. *Peripheral presbyLASIK*: The center of the cornea is left for distance while the rest is ablated in a way that a negative peripheral asphericity is created to increase the depth of focus.
3. *Central presbyLASIK*: A hyperpositive area is created for the near vision at the center, whereas the periphery is left for the far vision; this has a major advantage of performing the central hyperpositive area with minimal corneal excision associated with myopic, hyperopic profiles and also in emmetropes, but still with the main limitation of lack of adequate alignment making it prone to the induction of coma aberrations.

According to the reported results, both central and peripheral presbyLASIK obtain adequate spectacle independence simultaneously for far and for near. A neuroadaptation process is necessary for peripheral presbyLASIK. Epstein et al. reported a 4-year follow-up of 103 patients who underwent peripheral presbyLASIK; 89% of hyperopes and 92% of myopes were completely spectacle independent, with distance unaided visual acuity of 20/20 in 67.9% of hyperopes and 70.7% in myopes, also reporting that the overall increase in the HOAs was to a greater extent in hyperopic cases.[48] Results from central and peripheral presbyLASIK are shown in Table 15.8.

The level of scientific evidence from the literature is enough to consider that presbyLASIK is a useful tool in the correction of presbyopia. However, most of the techniques are still under development in clinical investigations, and further clinical data will validate the outcomes reported for the different techniques.

15.7 THE LIMITATIONS IN MODERN EXCIMER LASERS

In spite of the astonishing progress in the field of laser refractive surgery, a number of challenges are still facing the technology; one of the most critical is the limitations in the visual and optical outcomes in patients with high hyperopic refractive errors, as most authors reported a significant induction of the corneal HOAs, most significantly in the RMS coma.[36,37]

Ablation centration is a major issue in the excimer laser development; the decentration of ablation can lead to under correction and irregular astigmatism, which is most important in hyperopic patients,[49,50] who tend to have a larger angle kappa values.[49] There are four main methods of centration of laser refractive surgery that has been suggested in the literature: center of the pupil, coaxially sighted corneal light reflex (CSCLR), corneal vertex normal, and between the pupillary and visual axis.[51] Many reports had demonstrated the superiority of the CSCLR in terms of improving UCVA, safety profiles, and reduction of the HOA, specially in hyperopic patients.[39,52,53]

One of the main challenges is the limitations facing wavefront customized treatments, and in spite of the decrease in spot size to up to 5 mm and thermal damage control, limitations still

Table 15.7 Outcomes of sixth-generation excimer laser (AMARIS-SCHWIND)

AUTHOR	NUMBER OF PATIENTS	NUMBER OF EYES	MEAN AGE (YEARS)	GENDER (FEMALE/MALE)	MSE (D)	HOA (µm)	EFFICACY	EFFICACY INDEX	SAFETY INDEX	FOLLOW-UP (MONTHS)
Myopic patients										
Tomita et al.[79]	685	1,280	34 ± 8 (18:65)	371/314	−4.89 ± 2.12 (−0.5:−11.63)	0.66 ± 0.20	96.6%20/20 94.1% ±0.5 D	1.02	1.06	3
Tomita et al.[34]	5191	10,235	33.9 ± 7.89	2428/2763	−5.02 ± 2.17 (−2.75:−11.50)	0.70 ± 0.23	96.9% 0.0 LogMAR 88.4% ±0.5 D	1	1.03	3
Vega-Estrada et al.[28]	17	29	36.65 ± 10.80	N/A	−8.39 ± 0.93	0.95 ± 0.8	89.6% ±1.00 D 0.11 LogMAR ±0.26	NR	NR	6
Alió et al.[32]	32	51	23–61	N/A	≥−8.5	NR	84.3% ±0.5 D	NR	NR	6
Arba-Mosquera et al.[29]	NR	30	33(19–49)	53/47	−4.27 ± 1.62 (−7.38 to −1.38)	0.425 ± 0.129 (P < 0.01)	0.47 ± 0.72 lines (P < 0.05) −0.07 ± 0.25 (−0.63 to +0.50)	NR	NR	3
Hyperopic patients										
Alió et al.[37]	28	51	NR	NR	+5.64 ± 0.93 (3.50:7.88)	−0.44 ± 0.22 (P = 0.000)	70.37% ±0.5 D	0.85	0.94	6
Arbelaez et al.[36]	50	100	37 (21–59)	54% Female	+3.02 ± 2.06 (+0.13:+5.00) 1.36 ± 1.61 (Ast) (0.00:5.00)	↑ 0.18 (P < 0.05)	90% 20/20 89% ±0.5 94% ±0.5 A	0.89	1.1	6
Astigmatic patients										
Alió et al.[40]	36	52	21–53	NR	Mixed >3.0	NR	65.3% ±1.0 D	NR	NR	3
Arbelaez et al.[26]	25	50	NR	NR	−3.08 ± 2.32 1.26 ± 3.29	+0.57 (P < 0.005)	44% 20/20 0.12 ± 0.25 D P < 0.0001 0.50 ± 0.26 (As) P < 0.0001	NR	60% no changes	6
Arbelaez et al.[41]	NR	358	NR	NR	−3.13 ± 1.58 −0.69 ± 0.67	↑ 0.09	96% ±0.5 D 98% 20/20	NR	NR	6
Arbelaez et al.[42]	200	360	NR	NR	−0.14 ± 0.31 0.25 ± 0.37 (As)	↑ 0.09	97% ±0.5 D	NR	65% no changes	6

Table 15.8 **Results of peripheral versus central presbyLASIK**

AUTHOR	EYES	FOLLOW-UP (MONTHS)	PRE UCVA; 20/20 OR BETTER	PRE UCVA FOR NEAR	POST UCVA 20/20 OR BETTER	POST UCVA FOR NEAR	BSCVA LOSS OF TWO LINES OR MORE
Peripheral presbyLASIK							
Ghenassia[47]	6 hyperope 4 emmetrope 4 myopes	10	37.50%	12.5% J2 or better	50.00%	68.7% J2 or better	NR
Telandro et al.[80]	83 hyperope 77 myopes	3	NR	Hyperopes 29% J2 or better Myopes 52% J2 or better	NR	Hyperopes 58% J2 or better Myopes 73% J2 or better	1.00%
Pineli[47]	299	12	23.70%	11.3% J2 or better	97.30%	90.6% J2 or better	NR
Central presbyLASIK							
Alió et al.[81]	50 hyperopes	6	Mean, 0.39	Mean, 0.14 J1 or better	Mean, 0.97	Mean, 0.57 J1 or better	28.00%
Patel et al.[82]	26	6	Mean, 0.35	Mean, 0.15 J1 approx.	Mean, 0.68	Mean, 0.68 J2 approx.	NR
Jung et al.[83]	54 hyperopes	02/06/14	35.00%	100% 0.65 or worse	64.00%	100% 0.65 or better	4.00%

exist, mainly through the biomechanical changes induced by the wound healing patterns.[54] It is reported that 1 month after treatment, corneal hysteresis and the corneal resistance factor decreased significantly from 10.44 to 9.3 mm Hg and from 10.07 to 8.13 mm Hg, respectively.[55] So, when looking to customization as the planning of the most optimum ablation pattern specifically for each individual eye based on its diagnosis and visual demands, the best approach is a sophisticated pattern.[56] As wavefront-guided treatments may increase the HOA up to 100%. This induction is related to the baseline levels of HOAs, more significant in patients with less than 0.3 μm or greater than 0.3 μm of HOA, confirming that the customized ablation algorithms in all forms, is still not appropriate for the entire refractive surgery population. But results of these customized treatments still showed promising results as Arbeleaez et al. reported an average change in coma from 0.38 to 0.31 μm (–19%) (P = 0.04), in trefoil from 0.35 to 0.12 μm (–66%) (P = 0.0005), and in spherical aberration from +0.14 to +0.08 μm (–48%) (P = 0.02).[56] Furthermore, some authors compared a small spot scanning laser and a variable spot scanning laser, as Yu et al. reported in his study of 50 patients, in which one eye of each patient was treated by the small spot laser of Allegretto Wave Eye-Q system and the other with the variable spot scanning laser of VISX Star CustomVue S4 IR system; the small spot scanning laser group had significantly less spherical aberration (0.12 vs. 0.15) and significantly less mean total higher-order RMS (0.33 vs. 0.40 μm).[57]

The femtosecond assisted laser refractive surgery using sixth-generation excimer laser had reduced the magnitude of some of the previously mentioned limitations by creating thinner flaps, decreasing the amount of energy delivered to the cornea, decreasing the overall time required for the procedures, with increased precision while decreasing the biomechanical trauma

impact of the flap creation. This will produce more safety and almost near elimination of the risks of flap creations specially in high-risk eyes as in steep corneas. It has been reported that the postoperative aberrations induced when using the femtolaser in the flap creation is markedly decreased compared to the microkeratome, still with the main disadvantage being the prolonged time of the raster which is noticed in the 150 kHz fourth generation machines[58,59] (Table 15.9).

Patient satisfaction with the results of treatments with the recent excimer laser was remarkable. Kyprianou et al. reported satisfaction in 32 patients, with average age of 31.9 years, and a preoperative MSE of –3.05 D; it was evaluated by a questionnair consisting of 21 questions; he reported 100% satisfaction in his patients, being most satisfied in questions concerning quality of vision, distance vision, when watching TV and driving during daytime and during the night.[41]

15.8 THE FUTURE OF LASER REFRACTIVE SURGERY

Since the introduction of femtosecond refractive lasers in the year 2001, it had became a dominant part of the refractive techniques first as a flap fashioning modality, then as a dependent tool, in femtosecond refractive surgery, as Femtosecond Lenticular Extraction (FLEx) or Small Incision Lenticular Extraction (SMILE). The femtosecond laser is a focused infrared laser with a wavelength of 1053 nm that uses ultrafast pulses with a duration 100 fs (100×10^{-15} s). It is a solid-state Nd:Glass laser similar to a Nd:YAG laser, which operates on the principle of photoionization (laser-induced optical breakdown), producing photodisruption at its focal point, resulting in a rapidly expanding cloud of free

Table 15.9 Comparison between microkeratome and femtosecond laser created flaps

PARAMETER	MICROKERATOME	FEMTOSECOND LASER
Flap shape	Meniscus	Planar
Flap/hinge diameter	Keratometry dependent	Computer control
Flap thickness	Dependent on pachymetry, keratometry, intraocular pressure, blade quality and translational speed	Computer control
Thickness predictability	Moderate	High
Side cut	Shallow angled	Computer control
Epithelial ingrowth	More than femtosecond laser flaps	Less
Unique complications	Flap buttonhole	Opaque bubble layer, vertical gas breakthrough, transient light sensitivity syndrome, rainbow glare

Source: Kymionis, G.D. et al., *J. Refract. Surg.*, 28(12), 912, December 2012.

electrons and ionized molecules (plasma). Corneal flap creation in LASIK is the most common application of the femtosecond laser in corneal refractive surgery. More than 55% of all LASIK procedures in the United States were performed with femtosecond lasers in 2009 (Table 15.5). First described in 1996, laser intrastromal keratomileusis, was using picosecond laser, abd limiting the dependence on excimer laser as the sole tool of laser refractive surgery; these early trials required extensive manual dissection, leaving irregular interface. The shift to femtosecond laser, with the introduction of VisuMax in 2007, allowed much improved precision as an advantage over LASIK, as all of the potential variables associated with excimer laser ablation are avoided, such as stromal hydration, laser fluence, and environmental factors. Considering that the evolution of refractive surgery is heading closer to the preservation of corneal biomechanics to increase safety and achieve aberration-free results, also the new techniques had been proven to be more safe and effective in high myopic patients with limited postoperative dry eye syndrome and limited decrease in corneal sensation. So considering modern-day concepts of minimally invasive refractive surgery, FLEX and SMILE intrastromal keratomileusis techniques appear to be a reasonable technique worthy of future study and consideration.

ACKNOWLEDGMENTS

This study has been supported in part by a grant from the Spanish Ministry of Health, Instituto Carlos III, Red Temática de Investigación Cooperativa en Salud "Patología ocular del envejecimiento, calidad visual y calidad de vida," Subproyecto de Calidad Visual (RD07/0062) and a grant from the Spanish Ministry of Economy and Competitiveness, Instituto Carlos III, Red Temática de Investigación Cooperativa en Salud (RETICS) "Prevención, detección precoz y tratamiento de la patología ocular prevalente, degenerativa y crónica." Subprograma "dioptrio ocular y patologías frecuentes" (RD12/0034/0007).

REFERENCES

1. Nejabat M, Maleki B, Nimrouzi M, Mahbodi A, Salehi A. Avicenna and cataracts: A new analysis of contributions to diagnosis and treatment from the *Canon. Iran Red Cresc Med J* 2012; 14(5):265–270.
2. Splinter R, Hooper BA. Biomedical optics timeline. In: *Introduction to Biomedical Optics*, 1st edn. Boca Raton, FL: Taylor & Francis; 2007, pp. 5–6.
3. Ilardi V. From terrestrial to celestial vision. In: *Renaissance Vision from Spectacle to Telescopes*, 3rd edn. Philadelphia, PA: American Philosophical Society; 2007, pp. 239–241.
4. Lindberg D. Alhazen and the new intramission theory of vision. In: *Theories of Vision from Al-Kind to Kepler*, 1st edn. London, U.K.: University of Chicago Press; 1976, p. 86.
5. Knutson E (Sept 13, 2001). Vision and reading aids. Retrieved from: http://people.lis.illinois.edu/~chip/projects/timeline/1000knutson.html (Apr 15, 2014).
6. Glasses history (Date N/A). Salvino D'Armate inventor of glasses. Retrieved from: http://people.lis.illinois.edu/~chip/projects/timeline/1000knutson.html (Apr 15, 2014).
7. Goes FJ. Contact lens. In: *The Eye in History*, 1st edn. New Delhi, India: Jaypee Brothers; 2013, p. 147.
8. Wang M. History of refractive surgery. In: *LASIK Vision Correction*, 1st edn. Provo, UT: Medworld Publishing; 2000, pp. 15–20.
9. Apple DJ. The secret code: "Extra-Capsular Ext". In: *Sir Harold Ridley and His Fight for Sight*, 1st edn. Thorofare, NJ: SLACK Incorporated; 2006, pp. 2–10.
10. Barraquer JI. Quertoplasia refrativa. *Estudios e Informaciones Oftalmologicas* 1949;10:2–21.
11. Barraquer JI. Conducta de la cornea fronte a los cambios de espesor. *Arch Soc Oftal Optom* 1964;5:81–87.
12. Reinstein DZ, Archer TJ, Gobbe M. The history of LASIK. *J Refract Surg* 2012 Apr;28(4):291–298.
13. Tannebaum S, Svyatoslav Fyodorov MD. Innovative eye surgeon. *J Am Optom Assoc* 1995 Oct;66(10):652–654.
14. Bores LD. Radial keratotomy. *JAMA* 1986;256(2):212–213.
15. Krueger RR, Rabinowitz YS, Binder PS. The 25th anniversary of excimer lasers in refractive surgery: Historical review. *J Refract Surg* 2010 Oct;26(10):794–860.
16. Trokel SL, Srinivasan R, Braren B. Excimer laser surgery of the cornea. *Am J Ophthalmol* 1983;96(6):710–715.
17. Marshell J, Trokel S, Rothery S, Schubert H. An ultrastructural study of corneal incisions induced by an excimer laser at 193 nm. *Ophthalmology* 1985;92(6):749–758.
18. American Academy of Ophthalmology. (Nov 12, 2013). History of excimer laser in ophthalmology: Interview with Charles R. Munnerlyn, PhD. Retrieved from: one.aao.org/munnerlyn-laser-surgery-center/history-of-excimer-laser-in-ophthalmology (May 15, 2014).
19. Journal of Refractive Surgery. (Apr 16, 2012). History of LASIK. Retrieved from: video.healio.com/video/History-of-LASIK-2012 (May 15, 2014).

20. Gresham College. (Apr 17, 2013). Technology and vision. Retrieved from: www.gresham.ac.uk/lectures-and-events/technology-and-vision (September 15, 2014).

21. Zuberbuhler B, Tuft S. Introduction. In: *Corneal Surgery: Essential Techniques*, 1st edn. Heidelberg, Germany: Springer; 2013, pp. 1–2.

22. Azar DT, Koch D. History of LASIK. In: *LASIK (Laser in Situ Keratomileusis): Fundamentals, Surgical Techniques, and Complications*, 1st edn. New York: Marcel Dekker, Inc.; 2003, p. 23.

23. SCHWIND eye-tech-solutions: The SCHWIND AMARIS family. Retrieved from: http://www.schwind-amaris.com/en/home/ (May 1, 2014).

24. ALCON surgical: WaveLight® Allegretto Wave® Eye-Q Laser. Retrieved from: http://www.alconsurgical.com/wavelight-allegretto-wave-eye-q-laser.aspx (May 1, 2014).

25. NIDEK Co., Ltd. NIDEK advanced vision excimer laser system NAVEX Quest. Retrieved from: http://www.nidek-intl.com/products/ref_surgical/navex-quest.html (May 1, 2014).

26. Arbelaez MC, Vidal C, Arba-Mosquera S. Excimer laser correction of moderate to high astigmatism with a non-wavefront-guided aberration-free ablation profile: Six-month results. *J Cataract Refract Surg* 2009 Oct;35(10):1789–1798.

27. De Ortueta D, Magnago T, Triefenbach N, Arba Mosquera S, Sauer U, Brunsmann U. In vivo measurements of thermal load during ablation in high-speed laser corneal refractive surgery. *J Refract Surg* 2012 Jan;28(1):53–58.

28. Vega-Estrada A, Alió JL, Arba Mosquera S, Moreno LJ. Corneal higher order aberrations after LASIK for high myopia with a fast repetition rate excimer laser, optimized ablation profile, and femtosecond laser-assisted flap. *J Refract Surg* 2012 Oct;28(10):689–696.

29. Tomita M, Watabe M, Yukawa S, Nakamura N, Nakamura T, Magnago T. Supplementary effect of static cyclotorsion compensation with dynamic cyclotorsion compensation on the refractive and visual outcomes of laser in situ keratomileusis for myopic astigmatism. *J Cataract Refract Surg* 2013 May;39(5):752–758.

30. AAO Refractive Management/Intervention PPP Panel, Hoskins Center for Quality Eye Care (July 2013). Refractive errors & refractive surgery PPP—2013. Retrieved from: http://one.aao.org/preferred-practice-pattern/refractive-errors-surgery-ppp-2013 (May 8, 2014).

31. AAO Quality of Care Secretariat, Hoskins Center for Quality Eye Care (June 2013). Summary recommendations for keratorefractive laser surgery—2013. Retrieved from: http://one.aao.org/clinical-statement/summary-recommendations-lasik—january-2008 (May 8, 2014).

32. Alio JL, Vega-Estrada A, Piñero DP. Laser-assisted in situ keratomileusis in high levels of myopia with the amaris excimer laser using optimized aspherical profiles. *Am J Ophthalmol* 2011 Dec;152(6):954–963.e1.

33. Arba-Mosquera S, Arbelaez MC. Three-month clinical outcomes with static and dynamic cyclotorsion correction using the SCHWIND AMARIS. *Cornea* 2011 Sept;30(9):951–957.

34. Tomita M, Waring GO, 4th, Magnago T, Watabe M. Clinical results of using a high-repetition-rate excimer laser with an optimized ablation profile for myopic correction in 10235 eyes. *J Cataract Refract Surg* 2013 Oct;39(10):1543–1549.

35. Kanellopoulos AJ, Asimellis G. Refractive and keratometric stability in high myopic LASIK with high-frequency femtosecond and excimer lasers. *J Refract Surg* 2013 Dec;29(12):832–837.

36. Arbelaez MC, Vidal C, Arba Mosquera S. Six-month clinical outcomes after hyperopic correction with the SCHWIND AMARIS Total-Tech laser. *J Optom* 2010;3:198–205.

37. Alió JL, El Aswad A, Vega-Estrada A, Javaloy J. Laser in situ keratomileusis for high hyperopia (>5.0 diopters) using optimized aspheric profiles: Efficacy and safety. *J Cataract Refract Surg* 2013 Apr;39(4):519–527.

38. El Aswad A, Vega A, Arba Samuel, Alio J, Plaza-Puche A, Wróbel-Dudzińska D. Anterior corneal surface aberrations following excimer laser correction of high hyperopia; article submitted for publication.

39. Kanellopoulos AJ. Topography-guided hyperopic and hyperopic astigmatism femtosecond laser-assisted LASIK: Long-term experience with the 400 Hz eye-Q excimer platform. *Clin Ophthalmol* 2012;6:895–901.

40. Alio JL, El Aswad A, Plaza-Puche AB. Laser-assisted in situ keratomileusis in high mixed astigmatism with optimized, fast-repetition and cyclotorsion control excimer laser. *Am J Ophthalmol* 2013 May;155(5):829–836.

41. Arbelaez MC, Aslanides IM, Barraquer C, Carones F, Feuermannova A, Neuhann T, Rozsival P. LASIK for myopia and astigmatism using the SCWIND AMARIS excimer laser: An international multicenter trial. *J Refract Surg* 2010 Feb;26(2):88–98.

42. Arbelaez MC, Arba Mosquera S. The SCWIND AMARIS total-tech laser as an all-rounder in refractive surgery. *Middle East Afr J Ophthalmol* 2009 Jan;16(1):46–53.

43. Barraquer CC, Gutiérrez MAM. Results of laser in situ keratomileusis in hyperopic compound astigmatism. *J Cataract Refract Surg* 1999;25:1198–1204.

44. Payvar S, Hashemi H. Laser in situ keratomileusis for myopic astigmatism with the Nidek EC-5000 laser. *J Refract Surg* 2002;18:225–233.

45. Aslanides I, Padroni S, Arba-Mosquerab S. Aspheric photorefractive keratectomy for myopia and myopic astigmatism with the SCHWIND AMARIS laser: 2 years postoperative outcomes. *J Optom* 2013 Jan;6(1):9–17.

46. Moshirfar M, Churgin D, Betts B, Hsu M, Sikder S, Neuffer M, Church D, Mifflin M. Prospective, randomized, fellow eye comparison of WaveLight® Allegretto Wave® Eye-Q versus VISX CustomVue™ STAR S4 IRTM in photorefractive keratectomy: Analysis of visual outcomes and higher-order aberrations. *Clin Ophthalmol* 2011;5:1185–1193.

47. Alió JL, Amparo F, Ortiz D, Moreno L. Corneal multifocality with excimer laser for presbyopia correction. *Curr Opin Ophthalmol* 2009 July;20(4):264–271.

48. Epstein RL, Gurgos MA. Presbyopia treatment by monocular peripheral presbyLASIK. *J Refract Surg* 2009 June;25(6):516–52.

49. Basmak H, Sahin A, Yildirim N, Papakostas TD, Kanellopoulos AJ. Measurement of angle kappa with synoptophore and Orbscan II in a normal population. *J Refract Surg* 2007 May;23(5):456–460.

50. Pande M, Hillman JS. Optical zone centration in keratorefractive surgery. Entrance pupil center, visual axis, coaxially sighted corneal reflex, or geometric corneal center? *Ophthalmology* 1993 Aug;100(8):1230–1237.

51. Moshirfar M, Hoggan RN, Muthappan V. Angle Kappa and its importance in refractive surgery. *Oman J Ophthalmol* 2013 Sept;6(3):151–158.

52. Kermani O, Oberheide U, Schmiedt K, Gerten G, Bains HS. Outcomes of hyperopic LASIK with the NIDEK NAVEX platform centered on the visual axis or line of sight. *J Refract Surg* 2009 Jan;25(1 Suppl.):S98–S103.

53. Soler V, Benito A, Soler P, Triozon C, Arné JL, Madariaga V, Artal P, Malecaze F. A randomized comparison of pupil-centered versus vertex-centered ablation in LASIK correction of hyperopia. *Am J Ophthalmol* 2011 Oct;152(4):591–599.e2.

54. Dupps WJ Jr, Wilson SE. Biomechanics and wound healing in the cornea. *Exp Eye Res* 2006 Oct;83(4):709–720.

55. Ortiz D, Piñero D, Shabayek MH, Arnalich-Montiel F, Alió JL. Corneal biomechanical properties in normal, post-laser in situ keratomileusis, and keratoconiceyes. *J Cataract Refract Surg* 2007 Aug;33(8):1371–1375.

56. Arbelaez MC, Mosquera SA. The SCHWIND AMARIS total-tech laser as an all-rounder in refractive surgery. *Middle East Afr J Ophthalmol* [serial online] 2009 [cited 2014 May 5];16:46–53.

57. Yu CQ, Manche EE. Comparison of 2 wavefront-guided excimer lasers for myopic laser in situ keratomileusis: One-year results. *J Cataract Refract Surg* 2014 Mar;40(3):412–422.

58. Alio JL, Rosman M, Arba Mosquera S. Minimally invasive refractive surgery. In: *Minimally Invasive Ophthalmic Surgery.* Howard Fibe I and Mojon DS (Eds.). New York: Springer-Verlag; 2010, pp. 100–101.

59. Kyprianou G, Macháčková M, Feuermannová A, Rozsíval P, Langrová H. Subjective visual perception after laser treatment of myopia on two types of lasers. *Cesk Slov Oftalmol* 2010 Nov;66(5):213–219.

60. Kymionis GD, Kankariya VP, Plaka AD, Reinstein DZ. Femtosecond laser technology in corneal refractive surgery: A review. *J Refract Surg* 2012 Dec;28(12):912–920.

61. Sugar A, Rapuano CJ, Culbertson WW, Huang D, Varley GA, Agapitos PJ, de Luise VP, Koch DD. Laser in situ keratomileusis for myopia and astigmatism: Safety and efficacy: A report by the American Academy of Ophthalmology. *Ophthalmology* 2002 Jan;109(1):175–187.

62. Varley GA, Huang D, Rapuano CJ, Schallhorn S, Boxer Wachler BS, Sugar A. LASIK for hyperopia, hyperopic astigmatism, and mixed astigmatism: A report by the American Academy of Ophthalmology. *Ophthalmology* 2004 Aug;111(8):1604–1617.

63. Schallhorn SC, Farjo AA, Huang D, Boxer Wachler BS, Trattler WB, Tanzer DJ, Majmudar PA, Sugar A. Wavefront-guided LASIK for the correction of primary myopia and astigmatism a report by the American Academy of Ophthalmology. *Ophthalmology* 2008 July;115(7):1249–1261.

64. American Academy of Ophthalmology. Excimer laser photorefractive keratectomy (PRK) for myopia and astigmatism. *Ophthalmology* 1999;106:422–437.

65. U.S. Food and Drug Administration (Sept 2011). FDA—Approved lasers for LASIK, PRK and other refractive surgeries. Retrieved from: http://www.fda.gov/MedicalDevices/ProductsandMedicalProcedures/SurgeryandLifeSupport/LASIK/ucm168641.htm (May 1, 2014).

66. Casebeer JC, Kezirian GM. Outcomes of spherocylinder treatments in the comprehensive refractive surgery LASIK study. *Semin Ophthalmol* 1998 June;13(2):71–78.

67. Reviglio VE, Bossana EL, Luna JD, Muiño JC, Juarez CP. Laser in situ keratomileusis for myopia and hyperopia using the Lasersight 200 laser in 300 consecutive eyes. *J Refract Surg* 2000 Nov–Dec;16(6):716–723.

68. Morchen M, Kaemmerer M, Seiler T. Clinical results of wavefront-guided laser in situ keratomileusis 3 months after surgery. *J Cataract Refract Surg* 2001;27:201–207.

69. McDonald MB, Carr JD, Frantz JM, Kozarsky AM, Maguen E, Nesburn AB, Rabinowitz YS et al. Laser in situ keratomileusis for myopia up to –11 diopters with up to –5 diopters of astigmatism with the summit autonomous LADARVision excimer laser system. *Ophthalmology* 2001 Feb;108(2):309–316.

70. Rashad KM. Laser in situ keratomileusis for the correction of hyperopia from +1.25 to +5.00 diopters with the Technolas Keracor 117C laser. *J Refract Surg* 2001 Mar–Apr;17(2):113–122.

71. Tabbara KF, El-Sheikh HF, Islam SM. Laser in situ keratomileusis for the correction of hyperopia from +0.50 to +11.50 diopters with the Keracor 117C laser. *J Refract Surg* 2001 Mar–Apr;17(2):123–128.

72. Salz JJ, Stevens CA; LADARVision LASIK Hyperopia Study Group. LASIK correction of spherical hyperopia, hyperopic astigmatism, and mixed astigmatism with the LADARVision excimer laser system. *Ophthalmology* 2002 Sept;109(9):1647–1656; discussion 1657–1658.

73. Durrie DS, Stahl J. Randomized comparison of custom laser in situ keratomileusis with the Alcon CustomCornea and the Bausch & Lomb Zyoptix systems: One-month results. *J Refract Surg* 2004 Sept–Oct;20(5):S614–S618.

74. Pop M, Payette Y. Correlation of wavefront data and corneal asphericity with contrast sensitivity after laser in situ keratomileusis for myopia. *J Refract Surg* 2004 Sept–Oct;20(5 Suppl.):S678–S684.

75. Venter J. Wavefront-guided LASIK with the NIDEK NAVEX platform for the correction of myopia and myopic astigmatism with 6 month follow-up. *J Refract Surg* 2005 Sept–Oct;21(5 Suppl.):S640–S645.

76. Alió JL, Artola A, Claramonte PJ, Ayala MJ, Sánchez SP. Complications of photorefractive keratectomy for myopia: Two year follow-up of 3000 cases. *J Cataract Refract Surg* 1998 May;24(5):619–626.

77. Higa H, Liew M, McCarty C, Taylor H. Predictability of excimer laser treatment of myopia and astigmatism by the VISX Twenty-Twenty. Melbourne Excimer Laser Group. *J Cataract Refract Surg* 1997 Dec;23(10):1457–1464.

78. McCarty CA, Aldred GF, Taylor HR. Comparison of results of excimer laser correction of all degrees of myopia at 12 months postoperatively. The Melbourne Excimer Laser Group. *Am J Ophthalmol* 1996 Apr;121(4):372–383.

79. Tomita M, Watabe M, Yukawa S, Nakamura N, Nakamura T, Magnago T. Safety, efficacy, and predictability of laser in situ keratomileusis to correct myopia or myopic astigmatism with a 750 Hz scanning-spot laser system. *J Cataract Refract Surg* 2014 Feb;40(2):251–258.

80. Telandro A. Pseudo-accommodative cornea: A new concept for correction of presbyopia. *J Refract Surg* 2004 Sept–Oct;20(5 Suppl.):S714–S717.

81. Alió JL, Chaubard JJ, Caliz A, Sala E, Patel S. Correction of presbyopia by technovision central multifocal LASIK (presbyLASIK). *J Refract Surg* 2006 May;22(5):453–460.

82. Patel S, Alió JL, Feinbaum C. Comparison of Acri. Smart multifocal IOL, crystalens AT-45 accommodative IOL, and Technovision presbyLASIK for correcting presbyopia. *J Refract Surg* 2008 Mar;24(3):294–299.

83. Jung SW, Kim MJ, Park SH, Joo CK. Multifocal corneal ablation for hyperopic presbyopes. *J Refract Surg* 2008 Nov;24(9):903–910.

16

Nonlinear tissue processing in ophthalmic surgery

Holger Lubatschowski

Contents

16.1 LINEAR AND NONLINEAR LIGHT PROPAGATION

Light is an electromagnetic wave that is generated by the oscillation of an electric charge. This wave has both an electric and a transverse magnetic component and transports a quantum of energy through vacuum at a speed of $c = 2.998 \times 10^8$ m/s.

Any electromagnetic energy transport through material involves the absorption and reemission of wave energy by the atoms or molecules of the material. When the electromagnetic wave strikes an atom of the material, the energy of that wave is absorbed. The absorption of energy causes the electrons within the atom to oscillate. After a short period of oscillation, the electron generates a new electromagnetic wave with the same frequency and direction as the incoming electromagnetic wave. The reemitted quantum of energy travels through a short distance between the atoms and is absorbed again by the next atom. Since the absorption and reemission process causes a distinct delay, the net speed of the electromagnetic wave inside a medium is less than c. The delay is expressed in the index of refraction n and depends upon the optical density of that medium. As a result, the speed c_n of an electromagnetic wave inside a medium is given by $c_n = c/n$.

Table 16.1 lists the indices of refraction for a variety of media. Materials with an index of refraction close to 1 are those through which light travels fastest. These are the least optically dense materials in terms of phase delay. As the index of refraction increases, the optical density increases, and the speed of light in that material decreases.

Such a propagation of light in a medium is the normal case and is referred to as linear propagation. If the amplitudes of the electromagnetic waves are extremely high, which corresponds to an intense light field such as those provided by lasers, the electrons may respond nonlinearly to the electric field of the light. These nonlinearities occur in certain crystal materials. They are responsible for effects such as the oscillation of the electrons do not oscillate harmonically with the frequency of the incoming filed but with, for example, twice the frequency. As a consequence, the emitted field has the double frequency or half the wavelength of the incoming field. Such a process, called frequency doubling or second harmonic generation. It is used today routinely when, for example, a Nd:YAG laser is focused into a KDP (monopotassium phosphate) crystal and its 1064 nm infrared wavelength is turned into 532 nm visible green light.

Theoretically this process can also occur during surgical application at the patient's eye. Here the patient may recognize intense visible light on his retina although the incoming beam is infrared radiation. The well-structured stroma of the cornea, for example, has a huge potential to generate this doubling frequency, if irradiated with high-power infrared light.

Another nonlinear effect is the so-called optical Kerr effect. At light intensities, where the value of the electric field is comparable to interatomic electric fields ($\geq 10^8$ V/m), the refractive index of the material increases linearly with the intensity. For a typical laser beam with a Gaussian-shaped radial intensity profile, the refractive index becomes higher at the center of the beam. Such a distribution of the refractive index acts as a focusing lens. As a consequence, the laser itself is creating a focusing density profile that potentially leads to the collapse of a beam on itself (Figure 16.1) and, as a consequence, leads to damages of the material. This is commonly known as self-focusing.

For some clinical application this might be of relevance, since the focal spot of the laser pulse may be shifted toward the direction of the incoming laser beam leading to a misalignment in the intended position of the cut.

Table 16.1 **Index of refraction for a variety of media**

MATERIAL	INDEX OF REFRACTION
Vacuum	1.0000
Air	1.0003
Water	1.333
Cornea	1.376
Aqueous	1.337
Plexiglas	1.51
Crown glass	1.52
Diamond	2.417

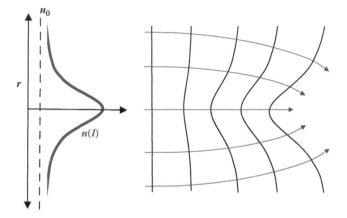

Figure 16.1 At high intensities the refractive index $n(I)$ increases with the intensity I. A Gaussian beam profile leads to a higher refractive index n and a smaller phase velocity c_n on the axis. Such a graded index profile acts as a focusing lens.

Multiphoton absorption is another nonlinear process where two or more photons are simultaneously absorbed, leading to an excitation for which single-photon energy would not be sufficient. For example, corneal tissue usually is transparent for visible or near-infrared radiation. At higher intensities, however, which can locally be achieved by focusing an ultrashort laser pulse, multiple infrared photons act as one UV photon and will be absorbed by the tissue (Figure 16.2). Multiphoton absorption is the basic absorption mechanism for tissue processing with femtosecond laser pulses.

16.2 THE PROCESS OF OPTICAL BREAKDOWN AND PHOTODISRUPTION

Photodisruption is a nonlinear procedure of tissue processing, which is—in contrast, for example, to photoablation— independent from the wavelength of the laser radiation. Only nonlinear absorption enables 3D tissue processing inside the medium since the laser radiation is not absorbed at the tissue surface directly by linear absorption [Lub 00, Noa 99, Vog 05].

Preferably the laser wavelength for photodisruptive tissue processing is chosen in a way that it penetrates the tissue

(low linear absorption, low scattering) and is safe for the eye in terms of phototoxicity. Near-infrared radiation ideally combines these two features and enables 3D processing of several ocular tissues.

The mechanism of photodisruption is best described as plasma-mediated ablation, or optical breakdown. It relies on the nonlinear absorption of laser energy in the tissue to generate a critical free electron density within the medium of at least 10^{21} cm^{-3} [Vog 05]. Fundamentally, optical breakdown is characterized by four successive events: (1) nonlinear absorption of the laser radiation, (2) generation of free electrons and plasma formation, (3) cavitation and shock wave generation, and (4) gas bubble formation (Figure 16.3).

As described earlier, nonlinear absorption is a multiphoton process and occurs when two or more photons are simultaneously absorbed, leading to an excitation for which single-photon energy would not be sufficient (Figure 16.2b and c). For example, the photon energy of a single photon of 1040 nm wavelength is 1.17 eV.* At this wavelength, which is the typical wavelength of ophthalmic femtosecond lasers, an extra six photons would be necessary to excite and release one electron from molecular bond such as water molecules.

Once the first free electrons are generated, they will increase their kinetic energy under the action of the strong electric field of the laser light. These accelerated electrons will again generate multiple free electrons in a cascade by impact ionization of other electrons in their neighborhood.

Optical breakdown occurs when the irradiance is sufficient to produce a critical density of free electrons beyond which the material substantially changes its chemical or physical properties (Figure 16.3a). At low laser pulse energy a so-called low-density plasma is produced. The cutting process is dominated by photochemically induced decomposition of the tissue and thermoelastic disruption. The free electrons transfer their energy to the tissue by locally increasing the temperature that, however, stays confined in the focal volume of the laser pulse. This is because the thermal diffusion is too slow to dissipate the energy during the pulse duration.

If the laser pulse energy is high enough, the material will be highly ionized and turns into a physical state that is called plasma. The created plasma first expands at supersonic velocity due to its high temperature and pressure and then slows down to the speed of the sound, emitting a shock wave (Figure 16.3b). The increase of the temperature creates a highly localized tensile stress that exceeds the critical tension for mechanical breakdown resulting in macroscopic tissue disruption and cavitation bubble formation (Figure 16.3c).

Depending on the incoming laser pulse energy and focusing condition, the maximum diameter of the cavitation bubble is in the range of several μm (nJ pulse energy) or up to 1 mm (mJ pulse energy). The cavitation bubble collapses within μs to ms and may oscillate for a few cycles if the viscosity of the surrounding medium is not too high (e.g., water) leaving a small gas bubble behind (Figure 16.3d). Mass spectroscopy analysis of the content

---

* 1040 nm is the typical wavelength of ophthalmic femtosecond lasers; eV = electron Volts is the common energy unit in atomic physics that corresponds to 1.6×10^{-19} J.

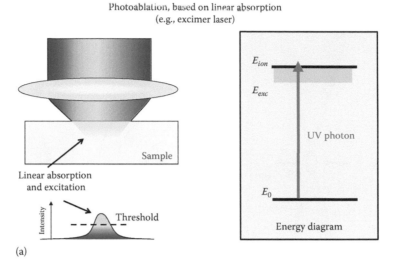

Photoablation, based on linear absorption
(e.g., excimer laser)

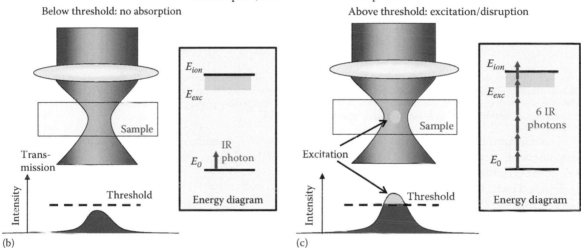

Photodisruption, based on nonlinear absorption

Below threshold: no absorption

Above threshold: excitation/disruption

Figure 16.2 (a) When corneal tissue is irradiated with UV-light (e.g., excimer laser), the photons will be absorbed at the surface. Due to the high energy of single UV-photons, they directly excite the tissue for fluorescence or ionization, which leads to ablation. (b) When tissue irradiated by IR-light (e.g., 1 μm wavelength) there is no absorption but transmission of the IR photons. (c) At higher intensities multiple infrared photons act as one UV photon and will be absorbed by the tissue and excite the tissue for fluorescence or ionization, which leads to disruption.

of the gas bubbles reveals a mixture of H_2, N_2, O_2, CO_2, methane, CH_4, water vapor, and carbon monoxide [Kai 94, Hei 02].

A single gas bubble usually cannot be seen with the naked eye. However, depending on the laser pulse energy and the spatial separation of the single laser spots, several individual gas bubbles aggregate to bigger bubbles, which, for example, can be observed as a thin intrastromal bubble layer after creating a flap with a femtosecond laser into the cornea for laser-assisted in situ keratomileusis (LASIK) treatment.

Even though femtosecond laser pulses are intended to cut corneal or crystalline lens tissue and light energy is transformed into heat and mechanical energy, one should keep in mind that a significant fraction of the laser pulse energy is transmitted through the focal region [Ham 97, Vog 99] and will be absorbed by the retinal pigment epithelium. On the retina the diameter of the beam has increased to several mm. As a consequence nonlinear interaction mechanisms are unlikely due to the decreased intensity. However, a potential thermal damage by linear absorption of the pigment epithelium should be taken

into consideration. As a matter of course, the laser power used in clinical applications such as corneal surgery or laser-assisted cataract surgery is well below any threshold of thermally induced damage [Sch 06, Wan 12].

16.3 FEMTOSECOND LASER TECHNOLOGY AND ITS APPLICATIONS IN OPHTHALMIC SURGERY

16.3.1 HISTORY OF FEMTOSECOND LASER APPLICATION

Photodisruption was already proposed in ophthalmology in the 1970s by Michail M. Krasnov using a Q-switched ruby laser. In a first paper [Kra 72], he targeted the trabecular meshwork to treat open-angle glaucoma and later on he suggested the mechanisms of photodisruption as a tool for cataract surgery [Kra 75]. In the

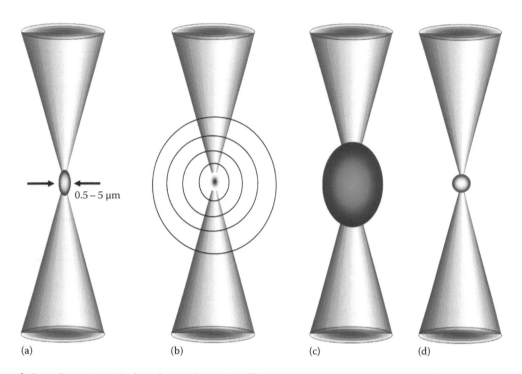

0.5 – 5 µm

(a)   (b)   (c)   (d)

Figure 16.3 Process of photodisruption: (a) when the irradiance is sufficient to produce a critical density of free electrons by nonlinear absorp-tion a plasma is produced. (b) The plasma expands at supersonic velocity emitting a shock wave. (c) The temperature rise creates a highly local-ized tensile stress that exceeds the critical tension for mechanical breakdown resulting in tissue disruption and cavitation. (d) The cavitation bubble collapses within µs to ms leaving a small gas bubble behind.

1980s, Danièle Aron-Rosa [Aro 80] and Franz Fankhauser [Fra 81] demonstrated Q-switched Nd:YAG lasers to cut the posterior capsule of the crystalline lens after secondary cataract.

Q-switched lasers have pulse durations in the nanosecond range (1 ns = 10^{-9} s). To induce multiphoton absorption, pulse energies in the millijoule (mJ) range are necessary. Because cavitation, shock wave emission, and residual gas bubbles scale with the pulse energy, the side effects at mJ pulse energy are tolerable for the disruption of the posterior lens capsule, but they are not acceptable for the remodeling of corneal tissue, where micron-level precision is required.

In the 1980s, Josef Bille and Stuart Brown first concluded that by shortening the pulse duration, high-intensity radiation would be available at considerable lower-pulse energy, leading to a higher precision in tissue processing. Moreover, in the late 1980s, laser technology could provide picosecond pulses (1 ps = 10^{-12} s) with a tolerably technical effort to realize a more or less tabletop machine for the doctor's office. In 1987, Drs. Bille and Brown founded Intelligent Surgical Lasers (ISL), a startup located in San Diego, California, with Tibor Juhasz, PhD, as a chief scientist. Their system, first demonstrated in 1989, operated at a 1053 µm wavelength and emitted pulses of several tens of picoseconds duration and around 100 µJ of pulse energy [Rem 92, Nie 93]. Although 100 µJ pulses dramatically decreased the unwanted mechanical side effects, this picosecond laser was not successful in performing reliable and reproducible intrastromal ablations.

Dr. Juhasz returned to the University of Michigan, where he met Ron Kurtz, MD, with whom he founded IntraLase Corp. (Irvine, CA) in 1997. Although their initial dream of performing refractive corneal surgery by intrastromal ablation and without excimer lasers [Kur 98] could not be realized completely, their femtosecond laser could replace the mechanical microkeratome and is still very successful.

16.3.2 CONCEPTS OF COMMERCIALLY AVAILABLE fs LASER SYSTEMS

Minimizing the pulse duration and the focal volume is the key parameter to reduce the energy threshold for optical breakdown that supply maximum precision of the laser cut and reduce the collateral damage.

Shortening the pulse duration is a basic physical problem, which is related to the maximum possible spectral bandwidth of the laser medium (so called time–bandwidth product). Titanium sapphire lasers, for example, have the broadest spectrum (>100 nm) and the shortest pulses (≪100 fs); however, they are very complex in their setup and relatively expensive. Ytterbium-doped fiber lasers or solid-state lasers, which emit around 1040 nm wavelength, are today the most reliable systems and also the cheapest way to produce femtosecond pulses. Their pulse duration is typically around 200–800 fs. In this range the energy threshold for optical breakdown increases almost linearly with pulse duration [Noa 99].

The focal volume of a Gaussian laser beam is determined by its axial extension, the so-called Rayleigh range ($z = \pi\omega_0^2/\lambda$) and the square of its beam waist $\omega_0 = f\lambda/\pi\omega_L$, where f is the focal length of the used focusing lens, ω_L is the radius of the beam at the focusing lens, and λ is the wavelength of the laser radiation. Accordingly the focal volume varies inversely with the power of four of the so-called numerical aperture NA $\propto \omega_L/f$ of the focusing optics. From this follows that, the larger the NA, the smaller the focal spot and finally the smaller is the energy threshold for disruption (Figure 16.4).

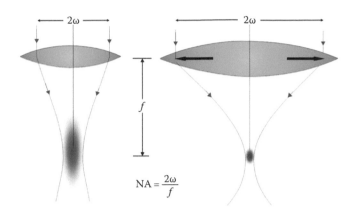

Figure 16.4 The focal volume of a focused Gaussian laser beam is determined by the numerical aperture NA which is the relationship of the diameter of the focusing lens and its focal length. The larger the NA, the smaller the focal volume.

$$NA = \frac{2\omega}{f}$$

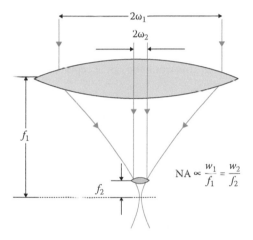

$$NA \propto \frac{w_1}{f_1} = \frac{w_2}{f_2}$$

Figure 16.5 Two different lenses producing the same focal spot size. While the large lens (f_1) is larger in its size, the distance to the focal spot is accordingly longer.

One option getting a large NA is of course by increasing the beam diameter at the focusing optics, which requires large and very expensive optical components. As an alternative, one can also decrease the focal length of the focusing objective, which on the other hand reduces the working distance of the laser system (Figure 16.5).

Since its beginning of corneal intrastromal laser ablation in 2001, the used pulse energy continuously has decreased from approximately 4 µJ to below 1 µJ by optimizing both parameters, the pulse duration and the numerical aperture. However, there is one concept in which the threshold was unprecedented trimmed down by maximizing the NA with an extraordinary short working distance (Figure 16.6). These are the Ziemer LDV systems, which require only a few nJ of pulse energy, resulting in an extremely precise cut, almost without any bubble formation. As a trade-off to this concept, there is a limited view to the operation field while the laser is operating because the focusing objective has to be moved very close to the patient's eye.

Based on its laser parameters, the nature of the cutting processes of the Ziemer laser systems differs to the other systems on the market (Figure 16.7). In all the other laser concepts, the cutting process is driven by mechanical forces, which are applied by the expanding bubbles and which disrupt the tissue. This cutting process is very efficient because the radius of disrupted tissue is larger than the laser spot itself. Hence, the spot separation of the scanned laser pulses can be larger than the spot diameter (Figure 16.7a). At the Ziemer laser, the cutting process is confined by the focal spot size of the laser pulse (Figure 16.7b). As a consequence, more pulses are needed to cut the same area. To keep the total operation time at the same level, higher pulse repetition rates (MHz) are required.

16.3.3 CORNEAL AND CRYSTALLINE LENS APPLICATION

Since the launch of the first IntraLase femtosecond laser systems for corneal surgery in 2001, there are now more than thousand systems sold, offered by five different manufacturers.

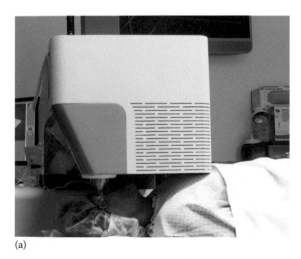

(a)

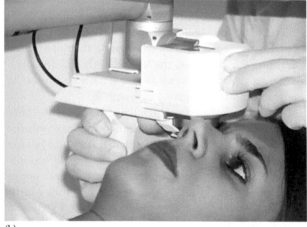

(b)

Figure 16.6 Different docking concepts to the patient's eye. (a) Conventional docking system where the patient is aligned to the beam delivery of the laser system by moving the patient bed. (b) Ziemers beam delivery concept where the complete optical focusing system is packed in a small handheld device.

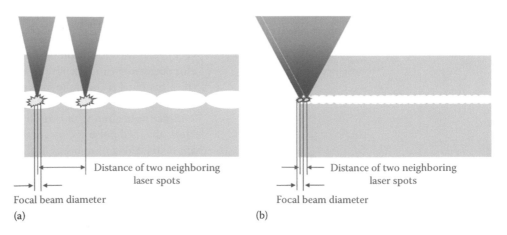

Figure 16.7 Different cutting mechanisms on photodisruption: (a) usually cutting process is driven by mechanical forces, which are generated by the expanding bubbles to disrupt the tissue. The radius of disruption is larger than the laser spot itself. Hence, the spot separation of the scanned laser pulses can be larger than the spot diameter. (b) At strong focusing (high NA), the cutting process is confined by the focal spot size of the laser pulse. As a consequence, more pulses are needed to cut the same area.

Besides IntraLase, now acquired by Abbott Medical Optics, USA, the other laser systems are the VICTUS (Bausch + Lomb), the VisuMax (Carl Zeiss, Germany), the WaveLight FS200 (Alcon, USA), and the Femto LDV (Ziemer Ophthalmic Systems, Switzerland).

Today these lasers have developed from a pure flap-maker for LASIK surgery [Rat 01] to a multiple tool for corneal surgery such as the generation of intrastromal pockets for corneal inlays and corneal rings, performing keratoplasties and arcuate incisions [Kym 12]. With the development of the SMILE® procedure (small incision lenticule extraction), the fs laser has even started to challenge the ArF-excimer laser technology since it uses only one femtosecond laser to complete the refractive surgery [Ang 12, Gan 14].

In 2009, a new era of fs laser application has begun with the introduction of the LenSx laser system for the assistance of cataract surgery [Nag 09]. Here the laser has extended its field of action to the crystalline lens where it is used to perform a capsulotomy, lens fragmentation, and corneal incisions. LenSx (today acquired by Alcon, USA) was immediately followed by Optimedica (today acquired by Abbott Medical Optics, USA) and Lensar (USA). Moreover, Bausch + Lomb as well as Ziemer followed with their system by redesigning their platforms for corneal surgery. Therefore, the VICTUS (B + L) and the LDV Z8 (Ziemer) are the only systems available for both corneal refractive surgery and cataract surgery.

Although both the corneal refractive lasers and the cataract devices are femtosecond lasers, it is important to emphasize that in fact a redesign of the systems were necessary and not just an upgrade to satisfy the requirements for both applications.

Laser systems for corneal refractive surgery are optimized to cut a smooth and precise area of about 10 mm width. For crystalline lens surgery, one has to target a volume of 7 mm in diameter and 4 mm in depth. Moreover, the lens is located significantly deeper inside the eye, passing through a number of refractive surfaces with different indices of refraction.

If a large volume has to be addressed by the laser focus, it is much easier to realize this with optics of lower numerical aperture (Figure 16.8). Accordingly the cutting precision will decrease, which is acceptable for cataract surgery, not only for lens fragmentation but also for capsulorhexis and corneal incisions. However, this would be not acceptable for corneal refractive surgery.

Besides the larger focal volume, higher pulse energy is required due to the strong scattering losses of laser power inside the sclerotic crystalline lens.

Finally, while processing the crystalline lens, a navigation system is essential, because the individual anatomy of the anterior chamber for each patient is different. Other than corneal procedures where usually the cutting depth is preset, a cataract laser system has to target the varying position of the anterior capsule and moreover has to keep an adequate safety margin to the posterior lens capsule.

Such a 3D imaging system can be realized by different optical techniques such as Scheimpflug imaging, confocal imaging, triangulation techniques, and optical coherence tomography (OCT) imaging. OCT imaging is implemented in most of the fs laser cataract surgery systems, because it is a scanning procedure that can be integrated pretty well into the delivery system of the fs laser beam.

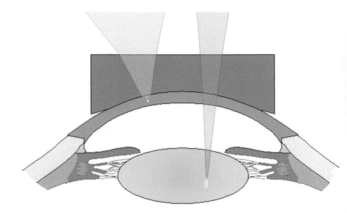

Figure 16.8 For corneal refractive surgery high precision over a wide lateral area is required. Here high numerical aperture focusing optics generating small focal spot sizes are ideal. For crystalline lens surgery a large volume has to be processed efficiently with optics of lower numerical aperture.

16.4 FUTURE APPLICATIONS

Ophthalmic femtosecond lasers enable safe surgery and short healing times because they can process tissue within a 3D space without altering its surface. The success of this platform in refractive and, more recently, cataract surgery is based on two unique characteristics: (1) the nonlinear absorption process and (2) extremely high precision and low side effects.

These excellent features in mind, it is obvious that refractive corneal surgery and cataract surgery will be just the beginning of a variety of impressively new applications.

Some of them, which are already discussed in the community, will be briefly described here.

Treatment of tractional vitreous attachments: Ultrashort laser pulses may be also used for the treatment of tractional vitreous attachments. However, this noninvasive strategy requires a new technique called adaptive optics to reduce the aberrations of the laser beam. Crystalline lens and vitreous cause these aberrations and deteriorate the focal spot which requires higher pulse energy, leading to unwanted side effects. However, if adaptive optics can be successfully incorporated into the beam delivery, these optical aberrations may be eliminated, thus achieving a well-focused, highly resolved laser spot (Figure 16.9) [Han 13].

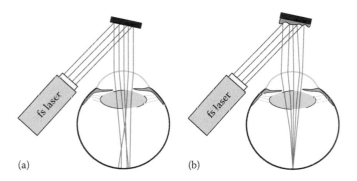

(a) (b)

Figure 16.9 (a) Crystalline lens and vitreous can cause aberrations which deteriorate the focal spot of the laser beam. (b) So called adaptive optics (here a deformable mirror) can preset the wavefront of the incident laser beam in that way, the aberrations will be compensated and a precise focal spot can be placed in the posterior segment of the eye.

Reversing presbyopia: Another promising application of the femtosecond laser is presbyopia reversal by restoring the flexibility of the crystalline lens. The hope here is that the femtosecond laser can be used to create microincisions inside the lens without surgically opening the eye (Figure 16.10). These microchannels could reduce the inner friction of the lens tissue, acting as sliding planes [Kru 05, Sch 09, Lub 10]. When delivered to rabbit eyes, these laser incisions did not cause cataract growth or wound-healing abnormalities. When applied to human autopsy eyes, an average increase of 100 μm in the anteroposterior lens thickness was seen, corresponding to a 1.00–2.00 D gain in accommodative amplitude.

Refractive index shaping: If the intensity of the femtosecond laser remains just below the threshold of optical breakdown, it is possible to create low-density plasma, which will allow free electrons to interact with the surrounding tissue. These chemical reactions could result in slight changes in the refractive index of optical media, and this phenomenon could be used to program diffractive lenses into the cornea and crystalline lens. In animal studies, refractive index shaping has been shown to be stable for several weeks or months [Din 08, Sav 14]. This principle could also be used to adjust the power of an IOL in situ [Bil 11].

Corneal collagen cross-linking (*CXL*): Ultrashort laser pulses applied to the posterior cornea or to scleral tissue may be possible using two-photon absorption. Therefore, surgeons could apply CXL to deeper areas of the eye for further beneficial effects in patients with keratoconus.

Reversing cataract: Photobleaching, or using multiphoton absorption to photochemically destroy absorbing, fluorescent, and scattering protein aggregates inside the nucleus, can remove the yellowing of the crystalline lens. In one experiment [Kes 11], human donor lenses were treated with an 800 nm infrared femtosecond pulsed laser. After treatment, the investigators found that the age-related yellow discoloration of the lens was reduced and the transmission of light increased. Finally, using coherent control, a quantum mechanical–based method for controlling dynamic light processes, it might be possible to selectively bleach the crystalline lens.

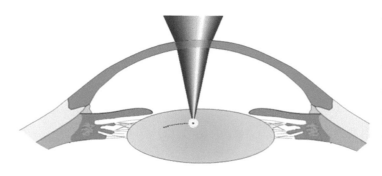

Figure 16.10 Femtosecond pulses can be used to create microincisions inside the crystalline lens. These microincisions act as sliding planes and might reduce the inner friction of the lens tissue. Right: Extracted porcine crystalline lens with sliding planes inside.

REFERENCES

[Ang 12] Ang M, Tan D, Mehta JS. (2012): Small incision lenticule extraction (SMILE) versus laser in-situ keratomileusis (LASIK): Study protocol for a randomized, non-inferiority trial. *Trials* 13(1), 75.

[Aro 80] Aron-Rosa D, Aron JJ, Griesemann JC, Thyzel R. (1980): Use of the neodymium-YAG laser to open the posterior capsule after lens implant surgery: A preliminary report. *J Am Intraocul Implant Soc* 6(4), 352–354.

[Bil 11] Bille JF. Generation and in situ modification of customized IOLs. In: Paper presented at *the ASCRS Symposium of Cataract, IOL and Refractive Surgery*, San Diego, CA, March 28, 2011.

[Din 08] Ding L, Knox WH, Bühren, Nagy LJ, Huxlin KR. (2008): Intratissue refractive index shaping (IRIS) of the cornea and lens using a low-pulse-energy femtosecond laser oscillator. *Invest Ophthalmol Vis Sci* 49(12), 5332–5339.

[Fra 81] Fankhauser F, Roussel P, Steffen J, Van der Zypen E, Chrenkova A. (1981): Clinical studies on the efficiency of high power laser radiation upon some structures of the anterior segment of the eye. First experiences of the treatment of some pathological conditions of the anterior segment of the human eye by means of a Q-switched laser. *Int Ophthalmol* 3(3), 129–139.

[Gan 14] Ganesh S, Gupta R. (2014): Comparison of visual and refractive outcomes following femtosecond laser-assisted lasik with smile in patients with myopia or myopic astigmatism. *J Refract Surg* 30(9), 590–596.

[Ham 97] Hammer DX, Jansen ED, Frenz M et al. (1997): Shielding properties of laser-induced breakdown in water for pulse durations from 5 ns to 125 fs. *Appl Opt* 36, 5630–5640.

[Han 13] Hansen A, Géneaux R, Günther A, Krüger A, Ripken T. (2013): Lowered threshold energy for femtosecond laser induced optical breakdown in a water based eye model by aberration correction with adaptive optics. *Biomed Opt Express* 4(6), 852.

[Hei 02] Heisterkamp A, Ripken T, Mamom T, Dommer W, Welling H, Lubatschowski H, Ertmer W. (2002): Nonlinear side effects of fs pulses inside corneal tissue during photodisruption. *Appl Phys B: Lasers Opt* 74, 419–425.

[Kai 94] Kaiser R, Habib M, Speaker M. (1994): Mass spectrometry analysis of cavitation bubbles generated by intrastromal ablation with the Nd:YLF picosecond laser. *Invest Ophthalmol Vis Sci* 35(Suppl.), 2026.

[Kes 11] Kessel L, Eskildsen L, van der Poel M, Larsen M. (2011): Non-invasive bleaching of the human lens by femtosecond laser photolysis. *PLoS ONE* 5(3), e9711.

[Kra 72] Krasnov MM. (1972): Laser puncture of the anterior chamber angle in glaucoma (a preliminary report). *Vestn Oftalmol* 3, 27–31.

[Kra 75] Krasnov MM. (1975): Laser-phakopuncture in the treatment of soft cataracts. *Br J Ophthalmol* 59(2), 96–98.

[Kru 05] Krueger RR, Kuszak J, Lubatschowski H, Myers RI, Ripken T, Heisterkamp A. (2005): First safety study of femtosecond laser photodisruption in animal lenses: Tissue morphology and cataratogenesis. *J Cataract Refract Surg* 31(12), 2386–2394.

[Kur 98] Kurtz RM, Horvath C, Liu HH, Krueger RR, Juhasz T. (1998): Lamellar refractive surgery with scanned intrastromal picosecond and femtosecond laser pulses in animal eyes. *J Refrac Surg* 14(5), 541–548.

[Kym 12] Kymionis GD, Kankariya VP, Plaka AD, Reinstein DZ. (2012): Femtosecond laser technology in corneal refractive surgery: A review. *J Refract Surg* 28(12), 912–920.

[Lub 00] Lubatschowski H, Maatz G, Heisterkamp, Hetzel U, Drommer W, Welling H, Ertmer W. (2000): Application of ultra short laser pulses for intrastromal refractive surgery. *Graefe's Arch Clin Exp Ophthal* 238, 33–39.

[Lub 10] Lubatschowski H, Schumacher S, Fromm M et al. (2010): Femtosecond lentotomy: Generating gliding planes inside the crystalline lens to regain accommodation ability. *J Biophotonics* 3(5–6), 265–268.

[Nag 09] Nagy Z, Takacs A, Filkorn T, Sarayba M. (2009): Initial clinical evaluation of an intraocular femtosecond laser in cataract surgery. *J Refract Surg* 25, 1053–1060.

[Nie 93] Niemz MH, Hoppeler TP, Juhasz T, Bille J. (1993): Intrastromal ablations for refractive corneal surgery using picosecond infrared laser pulses. *Laser Light Ophthalmol* 5(3):149–155.

[Noa 99] Noack J, Vogel A. (1999): Laser-induced plasma formation in water at nanosecond to femtosecond time scales: Calculation of thresholds, absorption coefficients, and energy density. *IEEE J Quantum Electron* 35(8), 1156.

[Rat 01] Ratkay I, Juhasz T, Kiss K, Ferencz I, Suarez C, Kurtz R. (2001): Ultrashort pulsed laser surgery: Initial application in LASIK. *Ophthalmol Clin North Am* 14, 347–355.

[Rem 92] Remmel R, Dardenne C, Bille J. (1992): Intrastromal tissue removal using an infrared picosecond Nd:YLF ophthalmic laser operating at 1053 nm. *Laser Light Ophthalmol* 4(3/4), 169–173.

[Sav 14] Savage DE, Brooks DR, DeMagistris M et al. (2014): First demonstration of ocular refractive change using Blue-IRIS in live cats. *Invest Ophthalmol Vis Sci* 55(7), 4603–4612.

[Sch 06] Schumacher S, Sander M, Stolte A. (2006): Investigation of possible fs-LASIK induced retinal damage. *Proc SPIE* 6138, 344–352.

[Sch 09] Schumacher S, Oberheide U, Fromm M et al. (2009): Femtosecond laser induced flexibility change of human donor lenses. *Vision Res* 49(14), 1853–1859.

[Vog 05] Vogel A, Noack J. (2005): Mechanism of femtosecond laser nanosurgery of cells and tissue. *Appl Phys B* 81(8), 1015.

[Vog 99] Vogel A, Noack J, Nahen K et al. (1999): Energy balance of optical breakdown in water at nanosecond to femtosecond time scales. *Appl Phys B: Lasers Opt* 68, 271–280.

[Wan 12] Wang J. (2012): Retinal safety of near-infrared lasers in cataract surgery. *J Biomed Opt* 17(9), 95001.

17 Corneal onlays and inlays

Corina van de Pol

Contents

17.1 EVOLUTION OF CORNEAL ONLAY AND INLAY TECHNOLOGIES

As the cornea is the primary refractive component of the eye and surgically the most accessible, providing surgical refractive correction at the corneal plane is a logical approach. The cornea can be reshaped using manual or laser incision techniques, such as radial keratotomy (RK) for the correction of myopia or astigmatic keratotomy (AK) for the correction of astigmatism. Additionally, femtosecond and excimer lasers have been used to reshape the cornea through the removal of tissue using techniques such as laser in situ keratomileusis (LASIK) and photorefractive keratectomy (PRK).

Corneal onlays and inlays present a means to correct refractive conditions of the eye through the addition of material rather than the removal of corneal tissue or permanent reshaping of the cornea. This provides a significant advantage in terms of removability of the inlay or replacement of correction as the refractive needs of the eye may change. The approaches to refractive correction in the corneal space include changing corneal curvature, implanting a "power" component, or implanting a "depth of focus" component.

Corneal onlays and inlays are not without their challenges, however. While the cornea is not a highly metabolic structure, it does have natural responses to materials placed within it and this presents the greatest challenge to a successful onlay or inlay procedure. This chapter will discuss the use of synthetic onlays and inlays implanted in the cornea to treat various conditions, specifically refractive errors.

17.1.1 EARLY EXPERIENCE

The implantation of synthetic materials into the eye is not a new concept. Intraocular lenses (IOLs) have been used to replace the eye's natural crystalline lens for over 60 years, with Ridley's first implantation of a polymethyl methacrylate (PMMA) IOL in 1949 [1]. Since this PMMA IOL was tolerated in the eye, researchers evaluated the possibility of implanting a synthetic material in the cornea but initially had limited success [2–9]. Using the same PMMA material as Ridley used in his IOLs, in 1963 Choyce implanted 32 patients' eyes with an 8 mm diameter, 0.2 mm thick acrylic corneal inlay just anterior to descemet's membrane for the management of corneal edema secondary to endothelial dystrophy [10]. These inlays, although nonporous, survived in the cornea with only a slight tendency toward opacification of the superficial cornea near the insertion point where the inlay was less deeply implanted. This technique was not carried forward to a refractive correction. Later developments in corneal inlays included various materials and designs intended to allow nutrient flow within the cornea, maintain corneal clarity, and deliver an optical correction.

The efforts toward development of a synthetic corneal onlay have been more recent [11–13]. A corneal onlay is intended to be placed at the very anterior aspect of the corneal stroma, often anterior to Bowman's membrane after removal of the epithelium, and the epithelium is either replaced or regrows over the surface of the onlay. Epikeratophakia is a technique where donor corneal stromal tissue is formed into a lenticule and added to the anterior corneal stroma to correct high refractive errors, such as those due to aphakia. Long-term viability of these lenticules, however, was often hampered by remodeling and epithelial tissue abnormalities [14–16]. Bioderived tissues were also developed and studied and found to have similar issues to donor corneal stroma [17,18]. In 1989, Colin described the development of synthetic keratophakia onlays that could be placed on the corneal surface without sutures [19]. Later development of the onlay was for the replacement of extended wear contact lenses, since this contact lens wear modality had been shown to cause corneal ulcers and other physiological issues. Synthetic onlays provide potential solutions for conditions such as corneal scarring, keratoconus, or high refractive error, in that they are essentially implantable contact lenses [11]. Evans et al. in 2002 described the development of a polymer that exhibited the required characteristics for an implantable contact lens [20]. Although significant work has been done in this area, to date corneal onlays have only been implanted in animal models and nonsighted human eyes with limited success.

17.1.2 OPTICAL CONSIDERATIONS

Generally, a corneal implant has to either increase or decrease the refractive power of the cornea to correct a hyperopic or myopic refractive error, respectively. Astigmatic errors would require a cylinder correction in addition to any other correction. In order to provide a correction with minimal aberrations, the implant has to provide a power change over a sufficiently large area of the cornea. This issue was dealt with in laser refractive procedures, such as LASIK or PRK, through the addition of a transition zone outside the central effective optical zone of the treatment [21–23]. Without this transition zone, the area around the treatment that is either too strong or too weak to bring the peripheral light rays into the same focus on the retina as the rays traveling through the central correction zone creates imperfections in the image. The effect is experienced as a halo around images or lights and is defined as "spherical aberration." If a refractive corneal implant has a diameter smaller than the patient's pupil, a similar effect might be experienced.

17.1.3 PHYSIOLOGICAL CONSIDERATIONS

Placing an inlay within the corneal stroma or an onlay within the more superficial layers of the cornea requires careful consideration of the cornea's physiological response to injury or the presence of a foreign body. The goal of these procedures is to minimize or control the cornea's natural response.

The cornea responds to injury with wound healing processes to maintain or stabilize normal corneal integrity and function [24–26]. When the cornea is cut, cells (keratocytes) within the tissue respond and create myofibroblasts, which act to remodel the injury, such as "stitching" together a lamellar resection when a flap or pocket is created in the cornea. The more aggressive the wound healing response, the more scar-type tissue remains within the cornea, which reduces corneal clarity. The cornea will continue to remodel over time eventually reducing the loss of transparency at the site of the injury.

When a foreign body is placed within the cornea, the cornea will respond in an effort to stabilize the foreign body. This often takes the form of encapsulation, a typical biological response to biologically "inert" materials [27]. The extent of encapsulation depends on the physical and chemical properties of the inlay and the location within the cornea. Specifically, the deeper stroma of the cornea is the least biologically active due to a lower keratocyte and nerve density [28–31].

Another physiological consideration is the body's need to maintain a smooth corneal surface. When refractive procedures change the profile of the cornea, the epithelium and stroma tend to either increase in thickness or decrease in thickness in regions of the corneal surface in order to create as smooth a surface as possible [32–37]. This remodeling of the corneal surface can work against the refractive effect desired, especially in cases of inlays and onlays that specifically alter corneal curvature for effect. Additionally, if the areas of irregularity are more extreme, as the eyelid sweeps across the cornea, the incomplete tear distribution over the corneal surface can result in desiccation of the epithelial layers (dellen formation).

17.1.4 MATERIAL CONSIDERATIONS

One of the primary considerations in material choice for a corneal implant is biocompatibility or at least biological "inert" properties that minimize corneal response. Nutrient transport and oxygen permeability are also considerations. For most implants, optical clarity is important. Additionally, the material should be able to be shaped based on the type of correction to be provided to the eye. In the case of corneal onlays, an important characteristic is its ability to support epithelial adherence, since often the epithelium is the only layer of tissue over the onlay.

17.2 REFRACTIVE ERROR CORRECTION

Hyperopia is a condition where the optics of the eye are not strong enough or the axial length of the eye is too short for images to come into focus on the retina without additional power. This additional power is either provided by positive (plus) powered lenses or contact lenses or may be achieved by strong accommodative effort. If a patient has had their natural crystalline lens removed and not had an IOL implanted, a condition known as "aphakia," the eye is left with a significant amount of hyperopia, or undercorrection. Positive spectacle lenses are also used to correct aphakia; however, they are extremely thick. Positive contact lenses are an option for aphakia; however, most patients who are aphakic are older and are less able to handle contact lenses or tolerate contact lens wear. For both hyperopia and aphakia, a change in the power of the eye can be achieved by either increasing the curvature of the cornea or introducing a positive powered implant.

Myopia is a condition of the eye where the optics of the eye are too strong or the axial length of the eye is too long for light to come into focus on the retina. Light from a distant object focuses in front of the retina in a myopic eye, necessitating a decrease in power in order to "push" the focus back to the retina. Negative (minus) powered spectacle or contact lenses are used to decrease the power of the eye. For myopia, a change in the power of the eye can be achieved by either decreasing the curvature of the cornea or introducing a negative powered implant.

17.2.1 CORNEAL SHAPE CHANGING

Shape-changing implants generally have a similar index of refraction as the cornea and primarily act to change the surface curvature of the cornea. In the case of hyperopic correction, these lenticules are thicker in the center and thinner toward the edge and need to be implanted either as onlays or as an inlay placed relatively shallow within the cornea in order to impart the shape change to the corneal surface (Figure 17.1).

The PermaVision® (Anamed, Inc., Lake Forest, California) intracorneal inlay was developed in the early 2000s using a material called Nutrapore®, a hydrogel substance with 78% water content and a similar index of refraction to the cornea (n = 1.376). The PermaVision inlay in the U.S. clinical trial had a diameter of 5 mm and a center thickness of 30–60 µm, depending on the amount of hyperopic correction being applied. While the PermaVision inlay showed effectiveness in the correction of hyperopia, especially in the first year after implantation, studies of long-term viability showed issues with corneal response to the presence of the inlay [38,39]. Specifically, researchers observed induced astigmatism, stromal opacification, perilenticular deposits and haze, and late onset ulceration of the cornea presumed to be neurotrophic [40–42]. The large diameter of the inlay and the proximity to the anterior cornea (most were implanted at 180 µm or less depth) may have contributed to the physiological responses seen.

Shape-changing inlays to correct hyperopia are not commonly used. However, the Raindrop™ inlay (ReVision Optics, Lake Forest, California), a successor to the PermaVision inlay, in development for the correction of presbyopia, has shown some effectiveness for the correction of low hyperopia. This inlay is described in Section 17.3.1.

A shape-changing implant that decreases corneal curvature, as would be required to correct myopia, could be achieved either through a lenticule with a thicker edge and thinner central region or through the use of rings implanted in the midperiphery of the cornea. The implants are implanted relatively deep within the peripheral cornea and are generally not wider than 1.0 mm. Nutrients flow around the inlay and disperse to the epithelium; therefore, these types of shape-changing implants do not need to necessarily transmit nutrients.

Channels or a pocket in the cornea for implantation of the rings or arcs can be created using special instruments or using a femtosecond laser, which can be programmed for precise depth, width, and arc. The implants are then slid into the channels or the pocket. In this mode of correction, the central cornea tends to retain a prolate profile, despite the overall flattening of the optical zone within the area of the rings [43–45]. This effect is due to the stiff nature of the implants and the fact that the collagen fibers in the cornea stretch from limbus to limbus across the cornea. The implantation of rings or arcs in the midperiphery of the cornea causes an elevation of the corneal fibers over the implants,

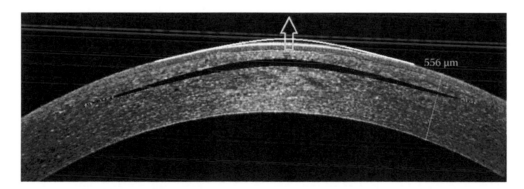

556 µm

Figure 17.1 Schematic representation of a positive-shaped corneal inlay (black) shown within an optical coherence tomography cross section of the cornea. The white line represents the forward bulging of the anterior corneal surface due to the space occupied by the inlay.

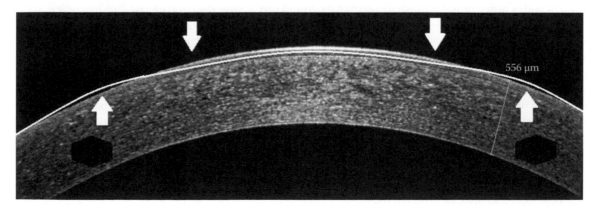

556 µm

Figure 17.2 Schematic representation of a corneal ring (black hexagon figures) shown within an optical coherence tomography cross section of the cornea. The white line represents the forward bulging of the periphery of the anterior corneal surface due to the space occupied by the ring and the relative central flattening.

a shortening of the arc length of the fibers, and a relative central flattening, as diagrammed in Figure 17.2.

There are a number of corneal ring implants that have been developed for the correction of myopia, including Keravision Intacs™ (Keravision, Inc., Fremont, California) and MyoRing® (DIOPTEX, GmbH, Linz, Austria). The Keravision Intacs were approved by the FDA in 1999 for the treatment of mild myopia (–1.00 to –3.00 diopters). Intacs, also called intrastromal corneal ring segments, consist of two segments or arcs made of PMMA material with a hexagonal cross section and a range of thicknesses from 0.21 to 0.45 mm [44,46,47]. Clinical studies of the Keravision Intacs showed effectiveness in the correction of myopia, relatively stable refractive endpoints, removability, and upgradability. However, at the same time, advances in laser refractive techniques, specifically LASIK technologies and wavefront-guided procedures, made keratorefractive laser techniques a more predictable procedure for the treatment of low myopia.

Rather than arcs or segments, the MyoRing corneal implant is a complete flexible ring made of PMMA with diameters ranging from 5 to 8 mm (www.dioptex.com). The MyoRing is implanted in a pocket with a 5.5 mm opening within the corneal stroma at a depth of 200–320 μm. The diameter and depth of the ring are determined based on the level of myopia being treated. The range of correction is indicated as –1.00 to –20.00 diopters. However, most published studies of the MyoRing corneal implant evaluate its use in the correction of keratoconus rather than myopia [48–52].

Shape-changing inlays are generally limited to the correction of low levels of refractive error and have less predictable outcomes than modern laser procedures, such as PRK or LASIK. After implantation, there can be epithelial and stromal remodeling which smooths out the surface of the cornea or there can be relaxing of the biomechanical properties of the cornea, resulting in some loss of effect. Therefore, these inlay technologies are less widely used for refractive error correction. In fact, currently very few corneal ring procedures are completed for the correction of myopia. Corneal ring technology has more recently been very successfully applied to the stabilization of the cornea in the treatment of keratoconus [51–57].

17.2.2 REFRACTIVE POWER ADDITION

An implant with a different index of refraction than the cornea can change the refractive power of the eye. Polysulfone is a synthetic material with a higher index of refraction (n = 1.633) than the cornea. Inlays made of polysulfone were studied for the correction of aphakia, hyperopia, and myopia starting in the mid-1980s [58–62]. These refractive implants could be very thin because of the high index of refraction and were shaped specifically to correct the eye's refractive error; in other words, a positive meniscus was shaped for aphakia or hyperopia correction and a negative meniscus for myopia correction. They were implanted deeper within the cornea in order to minimize anterior corneal shape changes. These higher index inlays provided refractive correction for very high levels of hyperopia and myopia; however, their reduced nutrient permeability eventually limited their use. Currently, this type of inlay is not being implanted due to issues with long-term viability of the inlay within the cornea [63,64].

17.3 PRESBYOPIC CORRECTION

Corneal inlays or onlays for refractive correction have not achieved widespread use in the refractive surgery market. However, an area where greater strides are being made is in the correction of presbyopia with corneal inlays.

Presbyopia is a condition of the eye associated with aging where the accommodative mechanism of the eye, consisting of the crystalline lens and ciliary muscle, is no longer able to increase the power of the eye to view near objects. Patients who do not require glasses to see distance (emmetropes) usually just use reading glasses to correct presbyopia. Patients who are myopic, hyperopic, or astigmatic (ametropes) generally use glasses to correct their distance vision with an addition (bifocals) or multifocal contact lenses in order to see clearly at near. Monovision is an approach where one eye is corrected for distance and the other eye is corrected for near. Monovision can be achieved with contact lenses, corneal refractive surgery (PRK or LASIK), or IOLs. A multifocal IOL is another option for the correction of presbyopia, either to replace a cataractous lens or a a refractive (clear lens) exchange procedure.

A corneal implant or laser procedure for presbyopia correction is less invasive than IOL implantation; therefore, it is a good option especially for patients without cataracts. There are three main types of presbyopia-correcting corneal inlays: shape changing, refractive, and small aperture. For all types of presbyopia-correcting inlays, the optics are designed to provide distance and near vision to the presbyopic eye and, in some case intermediate range vision as well. When the inlay is implanted monocularly in the nondominant eye, the optical system provide to the patient is a type of "modified monovision." In modified monovision, both the implanted and the unimplanted eye shoul see well at distance and the implanted eye provides most of the clarity of vision for intermediate and near viewing.

17.3.1 MULTIFOCAL CORNEAL SHAPE CHANGING

The Raindrop inlay (ReVision Optics, Lake Forest, California) is a hydrogel inlay with a similar index of refraction to the cornea. The inlay is 2 mm in diameter with an edge thickness of approximately 12 μm and a center thickness of approximately 32 μm. The inlay is implanted intrastromally in the cornea around 150 μm deep under a flap or within a pocket and centere over the pupil or first Purkinje reflex. The central thickness of the inlay acts to increase the central curvature of the anterior cornea, creating a "near" power zone. Near the edge of the inlay, in the transition zone where the cornea flap drapes down to mee the stromal bed, some intermediate power is provided. Finally, around the outside of the inlay, the natural cornea provides a distance vision zone. The net result is a hyperprolate anterior corneal curvature and an optical system that provides a central near zone that transitions smoothly to a peripheral distance zone

Figure 17.3 illustrates how light from a distant, intermediate, and near object would be focused on the retina by the Raindrop inlay. Light from a distant object is focused on the retina by the cornea surrounding the inlay, while the intermediate and near portions of the cornea/inlay combination bring the light into focus in front of the retina. These rays will fall unfocused onto the retina and be perceived as a "halo" or aberration. For light

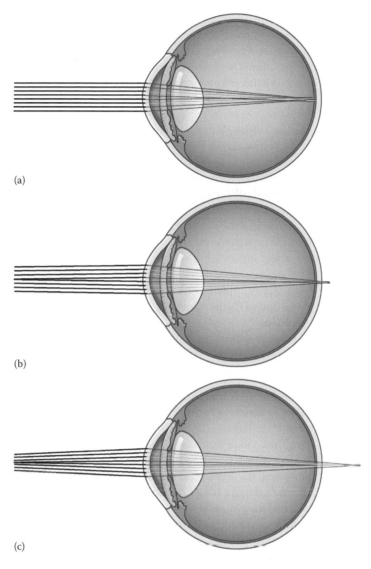

(a)

(b)

(c)

Figure 17.3 Demonstration of ray tracing for a distance (a), intermediate (b), and near (c) object through a cornea implanted with a Raindrop corneal inlay for presbyopia. Blue rays represent rays passing through the "distance" peripheral portion of the cornea/lens system; green rays represent rays passing through the "intermediate" transition zone of the cornea/lens system; and red rays represent rays passing through the "near" central portion of the cornea/lens system.

from an object at an intermediate distance (approximately 80 cm to 1 m), the transition portion of the corneal curvature (between the central near zone and the peripheral cornea) provides the focal power needed to focus these rays on the retina. The near portion of the cornea/inlay combination would focus the rays in front of the retina, while the peripheral portion of the cornea focuses the rays behind the retina. When viewing a near object, the light from the object is focused on the retina by the central portion of the cornea/inlay combination. Both the transition zone and the peripheral cornea would focus the light behind the retina. However, because of the pupillary constriction that occurs during accommodative effort, even in presbyopes, the peripheral unfocused rays may be blocked.

The Raindrop inlay has, in most cases, been implanted monocularly and has shown effectiveness in the treatment of presbyopia for emmetropic patients [65]. However, binocular

implantation has been shown to improve distance and near vision in low hyperopes with presbyopia [66].

17.3.2 MULTIFOCAL REFRACTIVE DESIGN

The refractive design multifocal inlays that are currently available have similar designs. Both the Flexivue™ (Presbia, Inc., Los Angeles, California) and the ICOLENS™ (Neoptics AG, Hunenberg, Switzerland) use a multifocal approach where the center of the lens provides distance vision and the peripheral portion of the lens provides the near addition power. The central distance zone is around 1.6 mm in diameter and the overall diameter of the inlay, which includes the peripheral near zone, is between 3.0 and 3.2 mm (Figure 17.4). Both lenses have a 150 µm central hole to facilitate nutrient flow and the edge thickness is approximately 15 µm. The inlays are made of a hydrogel with a higher index of refraction than the cornea. Since the inlays are refractive, corneal shape change is not necessary for effect. Both inlays are therefore implanted deeper in the cornea, around 280–300 µm, within a corneal pocket, usually created with a femtosecond laser system or special pocket-making keratome. The primary difference between the Flexivue and ICOLENS is that the Flexivue has zero power in the central zone, while the ICOLENS is available with some refractive power in the central zone to correct distance vision.

Figure 17.5 illustrates how light from a distant, intermediate, and near object would be focused on the retina by the Flexivue or ICOLENS inlay. Light from a distant object is focused on the retina through the central unpowered (Flexivue) or powered (ICOLENS) distance portion of the lens; the near powered peripheral portion of the lens bends these rays too much and they focus in front of the retina. These peripheral rays can be perceived as a "halo" or spherical aberration. When viewing an intermediate

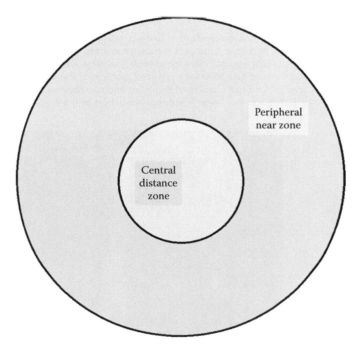

Figure 17.4 Schematic representation of the Flexivue and ICOLENS corneal inlay for presbyopia. The overall dimension is 3.0–3.2 mm, the inner distance portion of the inlay is 1.6 mm with a 150 µm central opening, and the peripheral ring provides the near addition power.

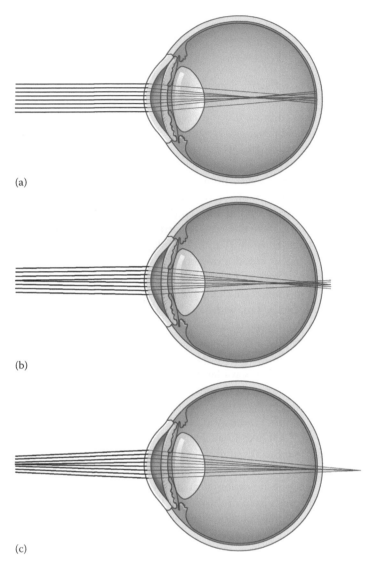

(a)

(b)

(c)

Figure 17.5 Demonstration of ray tracing for a distance (a), intermediate (b), and near (c) object through a cornea implanted with a Flexivue or ICOLEN corneal inlay for presbyopia. Blue rays represent rays passing through the "distance" central portion of the cornea/lens system and around the outside of the inlay and red rays represent rays passing through the "near" peripheral portion of the cornea/lens system.

object, the distance portion of the inlay focuses light behind the retina, while the near portion of the inlay focuses light in front of the retina. However, due to the small size of the distance portion of the inlay (1.6 mm), a certain amount of increased depth of focus is achieved and intermediate objects can be resolved at the retina. When a near object is viewed, rays that pass through the distance portion of the inlay focus behind the retina, while rays that pass through the peripheral near portion of the inlay focus at the retina. Again because of the "small aperture" effect of the central distance zone, the near object is imaged on the retina with minimal aberration. Additionally, pupil constriction with accommodative effort can help to decrease the impact of light that is focused by the cornea peripheral to the inlay.

Studies of the Flexivue and ICOLENS inlays show effectiveness in the correction of presbyopia with the implanted eye achieving an improvement in uncorrected near visual acuity and some decrease in uncorrected distance visual acuity [67–69]. The inlay is implanted monocularly, so the slight myopic shift in refraction of the implanted eye does not impact binocular vision. Additionally, morphological studies of the Flexivue implant indicate very little corneal response to the inlay [70].

17.3.3 INCREASING DEPTH OF FOCUS

The KAMRA™ inlay (AcuFocus, Inc., Irvine, California) uses a different approach to correct presbyopia. The inlay creates a small aperture to increase the depth of focus of the optical system of the eye. The concept is similar to the effect of changing the f-stop of a camera. When the f-stop is increased and the aperture of the camera decreased, the depth of field of objects that are clearly imaged increases (Figure 17.6). The depth of field in object space relates to depth of focus around the retina, which likewise increases with a smaller aperture. The increased depth of focus provides a range of near and intermediate vision with minimal effect on distance acuity.

The inlay is 3.8 mm in diameter with a 1.6 mm open aperture and is 4–6 µm thick. The dark outer ring of the inlay is made of carbon-impregnated polyvinylidene difluoride and has 8400 microperforations ranging from 5 to 11 µm in size arranged in a pseudorandom pattern to minimize unwanted optical effects and maximize nutrient flow (Figure 17.7). To minimize any corneal

(a)

(b)

Figure 17.6 Simulation of effect of aperture size on depth of field. (a) The camera is set for depth of focus (DOF) of f/5.6 that approximates a 4 mm pupil. (b) DOF of f/22 to approximate an aperture of 1.6 mm. (Images courtesy of Jack Holladay, MD.)

Figure 17.7 Schematic representation of the KAMRA corneal inlay for presbyopia. The overall dimension is 3.8 mm with a 1.6 mm central aperture with 8400 nutrient pores distributed in a pseudorandom pattern.

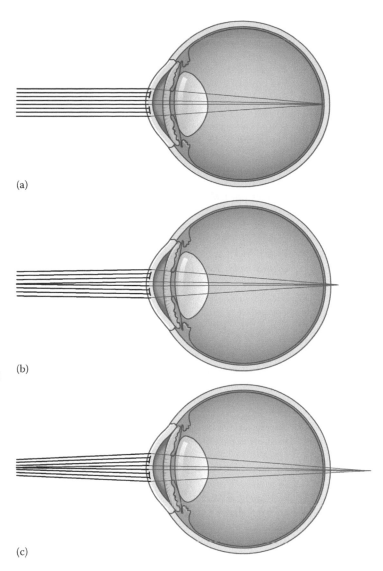

(a)

(b)

(c)

Figure 17.8 Demonstration of ray tracing for a distance (a), intermediate (b), and near (c) object through a cornea implanted with a KAMRA corneal inlay for presbyopia. Blue rays represent rays passing through the central aperture of the cornea/lens system and around the outside of the inlay.

surface curvature changes, the inlay is implanted around 200 or more microns depth within a corneal pocket created with a femtosecond laser system.

Figure 17.8 illustrates how light from a distant, intermediate, and near object would be focused on the retina by the KAMRA inlay. Light from a distant object is focused through the central aperture of the inlay, as well as around the inlay through the natural cornea; the inlay blocks the midperipheral rays from the distant object. When viewing an intermediate or near object, the rays are restricted by the aperture and still form a small blur circle resolvable on the retina. The decreasing pupil size with accommodative effort further restricts the peripheral rays, those passing around the outer dimension of the inlay, and improves image quality for intermediate and near objects.

Studies of the KAMRA corneal inlay have shown effectiveness in the improvement of reading performance, near and intermediate visual acuity, and having a minimal effect on distance visual acuity in presbyopes [71–75]. It has also been shown to be combinable with other refractive and ocular procedures, including LASIK, RK, and IOLs [76–80]. Biocompatibility and long-term viability of the inlay have also been shown [81,82].

17.4 SUMMARY

Corneal inlays and onlays provide an option for refractive and presbyopic correction that has the advantage of being reversible or even replaceable. The various approaches to correcting the optics of the eye by corneal shape changes, refractive implants, or increased depth of focus have all shown effectiveness. However, safety remains perhaps the biggest hurdle to overcome in this area. The cornea is a difficult space to work in, as demonstrated by the issues with biocompatibility and long-term viability

experienced with some of the inlay and onlay modalities presented. Yet, advances in material science, surgical procedure, and management of corneal physiological responses continue to advance this relatively noninvasive mode of visual correction.

REFERENCES

1. Apple, D.J. and J. Sims, Harold Ridley and the invention of the intraocular lens. *Surv Ophthalmol*, 1996. **40**(4): 279–292.
2. Krwawicz, T., M. Seidler-Dymitrowska, and A. Vorbrodt, Histochemical studies on the cornea in keratoplasty. *Klin Oczna*, 1953. **23**(3): 155–166.
3. Lieb, W.A. et al., Tissue tolerance of plastic resins: Experimental study of their physical, chemical and histopathological reactions. *Eye Ear Nose Throat Mon*, 1959. **38**(4): 303–307 passim concl.
4. Lieb, W.A. et al., Tissue tolerance of plastic resins; experimental study of their physical, chemical and histopathological reactions. *Eye Ear Nose Throat Mon*, 1959. **38**(3): 210–215 contd.

5. Krwawicz, T., Attempted modification of the corneal curvature by means of experimental plastic surgery. *Klin Oczna*, 1960. **30**: 229–236.

6. Krwawicz, T., New plastic operation for correcting the refractive error of aphakic eyes by changing the corneal curvature. Preliminary report. *Br J Ophthalmol*, 1961. **45**: 59–63.

7. Krwawicz, T., Experimental operations of partial lamellar resection of the corneal stroma for the equalization of myopia. *Klin Oczna*, 1963. **33**: 1–6.

8. Belau, P.G. et al., Correction of ametropia with intracorneal lenses. An experimental study. *Arch Ophthalmol*, 1964. **72**: 541–547.

9. Henderson, J.W. and P.G. Belau, Insertion of a plastic intracorneal lens: Report of case. *Mayo Clin Proc*, 1964. **39**: 772–774.

10. Choyce, P., Management of endothelial corneal dystrophy with acrylic corneal inlays. *Br J Ophthalmol*, 1965. **49**(8): 432–440.

11. Evans, M.D. et al., A review of the development of a synthetic corneal onlay for refractive correction. *Biomaterials*, 2001. **22**(24): 3319–3328.

12. Xie, R.Z., S. Stretton, and D.F. Sweeney, Artificial cornea: Towards a synthetic onlay for correction of refractive error. *Biosci Rep*, 2001. **21**(4): 513–536.

13. Sweeney, D.F., The Max Schapero Memorial Award Lecture 2004: Contact lenses on and in the cornea, what the eye needs. *Optom Vis Sci*, 2006. **83**(3): 133–142.

14. Gupta, S. and G.N. Rao, Morphologic features of clear and failed epikeratophakia lenticules. *Indian J Ophthalmol*, 1987. **35**(5–6): 183–185.

15. Rodrigues, M. et al., Clinical and histopathologic changes in the host cornea after epikeratoplasty for keratoconus. *Am J Ophthalmol*, 1992. **114**(2): 161–170.

16. Yamaguchi, T. et al., Histological study of epikeratophakia in primates. *Ophthalmic Surg*, 1984. **15**(3): 230–235.

17. Kornmehl, E.W. et al., In vivo evaluation of a collagen corneal allograft derived from rabbit dermis. *J Refract Surg*, 1995. **11**(6): 502–506.

18. Thompson, K.P. et al., Synthetic epikeratoplasty in rhesus monkeys with human type IV collagen. *Cornea*, 1993. **12**(1): 35–45.

19. Colin, J., Experimental sutureless synthetic epikeratophakia. *Arch Ophthalmol*, 1989. **107**(3): 318.

20. Evans, M.D. et al., Progress in the development of a synthetic corneal onlay. *Invest Ophthalmol Vis Sci*, 2002. **43**(10): 3196–3201.

21. Boxer Wachler, B.S. et al., Role of clearance and treatment zones in contrast sensitivity: Significance in refractive surgery. *J Cataract Refract Surg*, 1999. **25**(1): 16–23.

22. Cennamo, G. et al., Technical improvements in photorefractive keratectomy for correction of high myopia. *J Refract Surg*, 2003. **19**(4): 438–442.

23. Schipper, I., P. Senn, and A. Lechner, Tapered transition zone and surface smoothing ameliorate the results of excimer-laser photorefractive keratectomy for myopia. *Ger J Ophthalmol*, 1995. **4**(6): 368–373.

24. Eraslan, M. and Toker, E., Mechanisms of corneal wound healing and its modulation following refractive surgery. *Marmara Med J*, 2009. **22**(2): 169–178.

25. Netto, M.V. et al., Wound healing in the cornea: A review of refractive surgery complications and new prospects for therapy. *Cornea*, 2005. **24**(5): 509–522.

26. Binder, P.S., Barraquer lecture. What we have learned about corneal wound healing from refractive surgery. *Refract Corneal Surg*, 1989. **5**(2): 98–120.

27. Bryers, J.D., C.M. Giachelli, and B.D. Ratner, Engineering biomaterials to integrate and heal: The biocompatibility paradigm shifts. *Biotechnol Bioeng*, 2012. **109**(8): 1898–1911.

28. Patel, S. et al., Normal human keratocyte density and corneal thickness measurement by using confocal microscopy in vivo. *Invest Ophthalmol Vis Sci*, 2001. **42**(2): 333–339.

29. Patel, D.V. and C.N. McGhee, Mapping of the normal human corneal sub-Basal nerve plexus by in vivo laser scanning confocal microscopy. *Invest Ophthalmol Vis Sci*, 2005. **46**(12): 4485–4488.

30. Patel, D.V. and C.N. McGhee, In vivo confocal microscopy of human corneal nerves in health, in ocular and systemic disease, and following corneal surgery: A review. *Br J Ophthalmol*, 2009. **93**(7): 853–860.

31. Visser, N., C.N. McGhee, and D.V. Patel, Laser-scanning in vivo confocal microscopy reveals two morphologically distinct populations of stromal nerves in normal human corneas. *Br J Ophthalmol*, 2009. **93**(4): 506–509.

32. Abbas, U.L., P.S. Hersh, and P.R.K.S.G. Summit, Late natural history of corneal topography after excimer laser photorefractive keratectomy. *Ophthalmology*, 2001. **108**(5): 953–959.

33. Wilson, S.E. et al., The wound healing response after laser in situ keratomileusis and photorefractive keratectomy: Elusive control of biological variability and effect on custom laser vision correction. *Arch Ophthalmol*, 2001. **119**(6): 889–896.

34. Netto, M.V. and S.E. Wilson, Corneal wound healing relevance to wavefront guided laser treatments. *Ophthalmol Clin North Am*, 2004. **17**(2): 225–231, vii.

35. Kanellopoulos, A.J. and G. Asimellis, Epithelial remodeling following myopic LASIK. *J Refract Surg*, 2014. **30**(12): 802–805.

36. Kanellopoulos, A.J. and G. Asimellis, Epithelial remodeling after femtosecond laser-assisted high myopic LASIK: Comparison of stand-alone with LASIK combined with prophylactic high-fluence cross-linking. *Cornea*, 2014. **33**(5): 463–469.

37. Kanellopoulos, A.J. and G. Asimellis, Longitudinal postoperative lasik epithelial thickness profile changes in correlation with degree of myopia correction. *J Refract Surg*, 2014. **30**(3): 166–171.

38. Guell, J.L. et al., Confocal microscopy of corneas with an intracorneal lens for hyperopia. *J Refract Surg*, 2004. **20**(6): 778–782.

39. Michieletto, P. et al., PermaVision intracorneal lens for the correction of hyperopia. *J Cataract Refract Surg*, 2004. **30**(10): 2152–2157.

40. Ismail, M.M., Correction of hyperopia by intracorneal lenses: Two-year follow-up. *J Cataract Refract Surg*, 2006. **32**(10): 1657–1660.

41. Mulet, M.E., J.L. Alio, and M.C. Knorz, Hydrogel intracorneal inlays for the correction of hyperopia: Outcomes and complications after 5 years of follow-up. *Ophthalmology*, 2009. **116**(8): 1455–1460, 1460.e1.

42. Verity, S.M. et al., Outcomes of PermaVision intracorneal implants for the correction of hyperopia. *Am J Ophthalmol*, 2009. **147**(6): 973–977.

43. Holmes-Higgin, D.K. et al., Characterization of the aspheric corneal surface with intrastromal corneal ring segments. *J Refract Surg*, 1999. **15**(5): 520–528.

44. Twa, M.D. et al., One-year results from the phase III investigation of the KeraVision Intacs. *J Am Optom Assoc*, 1999. **70**(8): 515–524.

45. Silvestrini, T.A., M.L. Mathis, and B.E. Loomas, A geometric model to predict the change in corneal curvature from the intrastromal corneal ring (ICR). *Invest Ophthalmol Vis Sci*, 1994. **35**: 2023.

46. Burris, T.E., Intrastromal corneal ring technology: Results and indications. *Curr Opin Ophthalmol*, 1998. **9**(4): 9–14.

47. Linebarger, E.J. et al., Intacs: The intrastromal corneal ring. *Int Ophthalmol Clin*, 2000. **40**(3): 199–208.

48. Studeny, P., D. Krizova, and Z. Stranak, Clinical outcomes after complete intracorneal ring implantation and corneal collagen cross-linking in an intrastromal pocket in one session for keratoconus. *J Ophthalmol*, 2014. **2014**: 568128.

49. Jabbarvand, M. et al., Implantation of a complete intrastromal corneal ring at 2 different stromal depths in keratoconus. *Cornea*, 2014. **33**(2): 141–144.

50. Jabbarvand, M. et al., Continuous corneal intrastromal ring implantation for treatment of keratoconus in an Iranian population. *Am J Ophthalmol*, 2013. **155**(5): 837–842.

51. Alio, J.L., D.P. Pinero, and A. Daxer, Clinical outcomes after complete ring implantation in corneal ectasia using the femtosecond technology: A pilot study. *Ophthalmology*, 2011. **118**(7): 1282–1290.

52. Daxer, A., H. Mahmoud, and R.S. Venkateswaran, Intracorneal continuous ring implantation for keratoconus: One-year follow-up. *J Cataract Refract Surg*, 2010. **36**(8): 1296–1302.

53. Fahd, D.C., N.S. Jabbur, and S.T. Awwad, Intrastromal corneal ring segment SK for moderate to severe keratoconus: A case series. *J Refract Surg*, 2012. **28**(10): 701–705.

54. Guell, J.L., Are intracorneal rings still useful in refractive surgery? *Curr Opin Ophthalmol*, 2005. **16**(4): 260–265.

55. Kamburoglu, G. and A. Ertan, Intacs implantation with sequential collagen cross-linking treatment in postoperative LASIK ectasia. *J Refract Surg*, 2008. **24**(7): S726–S729.

56. Niknam, S. et al., Treatment of moderate to severe keratoconus with 6-mm Intacs SK. *Int J Ophthalmol*, 2012. **5**(4): 513–516.

57. Pinero, D.P. and J.L. Alio, Intracorneal ring segments in ectatic corneal disease—A review. *Clin Exp Ophthalmol*, 2010. **38**(2): 154–167.

58. Choyce, D.P., The correction of refractive errors with polysulfone corneal inlays. A new frontier to be explored? *Trans Ophthalmol Soc UK*, 1985. **104**(Pt 3): 332–342.

59. Choyce, D.P., The correction of high myopia. *Refract Corneal Surg*, 1992. **8**(3): 242–245.

60. Kirkness, C.M. and A.D. Steele, Polysulfone corneal inlays. *Lancet*, 1985. **1**(8432): 811.

61. Kirkness, C.M., A.D. Steele, and A. Garner, Polysulfone corneal inlays. Adverse reactions: A preliminary report. *Trans Ophthalmol Soc UK*, 1985. **104**(Pt 3): 343–350.

62. Werblin, T.P., R.L. Peiffer, and A.S. Patel, Synthetic keratophakia for the correction of aphakia. *Ophthalmology*, 1987. **94**(8): 926–934.

63. Lane, S.S. and R.L. Lindstrom, Polysulfone intracorneal lenses. *Int Ophthalmol Clin*, 1991. **31**(1): 37–46.

64. Horgan, S.E. et al., Twelve year follow-up of unfenestrated polysulfone intracorneal lenses in human sighted eyes. *J Cataract Refract Surg*, 1996. **22**(8): 1045–1051.

65. Garza, E.B. et al., One-year safety and efficacy results of a hydrogel inlay to improve near vision in patients with emmetropic presbyopia. *J Refract Surg*, 2013. **29**(3): 166–172.

66. Porter, T., Lang, A., Holliday, K., Sharma, G., Roy, A., Chayet, A., Favela, E., Barragan, E., and Gomez, S., Clinical performance of a hydrogel corneal inlay in hyperopic presbyopes, in *ARVO*, Orlando, FL, 2012.

67. Baily, C., T. Kohnen, and M. O'Keefe, Preloaded refractive-addition corneal inlay to compensate for presbyopia implanted using a femtosecond laser: One-year visual outcomes and safety. *J Cataract Refract Surg*, 2014. **40**(8): 1341–1348.

68. Limnopoulou, A.N. et al., Visual outcomes and safety of a refractive corneal inlay for presbyopia using femtosecond laser. *J Refract Surg*, 2013. **29**(1): 12–18.

69. Stojanovic, N.R., S.I. Panagopoulou, and I.G. Pallikaris, Refractive corneal inlay for near vision improvement after cataract surgery. *J Cataract Refract Surg*, 2014. **40**(7): 1232–1235.

70. Malandrini, A. et al., Morphologic study of the cornea by in vivo confocal microscopy and optical coherence tomography after bifocal refractive corneal inlay implantation. *J Cataract Refract Surg*, 2014. **40**(4): 545–557.

71. Dexl, A.K. et al., One-year visual outcomes and patient satisfaction after surgical correction of presbyopia with an intracorneal inlay of a new design. *J Cataract Refract Surg*, 2012. **38**(2): 262–269.

72. Seyeddain, O. et al., Femtosecond laser-assisted small-aperture corneal inlay implantation for corneal compensation of presbyopia: Two-year follow-up. *J Cataract Refract Surg*, 2013. **39**(2): 234–241.

73. Seyeddain, O., G. Grabner, and A.K. Dexl, Binocular distance visual acuity does not decrease with the Kamra intra-corneal inlay. *J Cataract Refract Surg*, 2012. **38**(11): 2062; author reply 2062–2064.

74. Waring, G.O.T., Correction of presbyopia with a small aperture corneal inlay. *J Refract Surg*, 2011. **27**(11): 842–845.

75. Dexl, A.K. et al., Reading performance and patient satisfaction after corneal inlay implantation for presbyopia correction: Two-year follow-up. *J Cataract Refract Surg*, 2012. **38**(10): 1808–1816.

76. Huseynova, T. et al., Small-aperture corneal inlay in patients with prior radial keratotomy surgeries. *Clin Ophthalmol*, 2013. **7**: 1937–1940.

77. Huseynova, T. et al., Small-aperture corneal inlay in presbyopic patients with prior phakic intraocular lens implantation surgery: 3-month results. *Clin Ophthalmol*, 2013. **7**: 1683–1686.

78. Tan, T.E. and J.S. Mehta, Cataract surgery following KAMRA presbyopic implant. *Clin Ophthalmol*, 2013. **7**: 1899–1903.

79. Tomita, M. et al., Small-aperture corneal inlay implantation to treat presbyopia after laser in situ keratomileusis. *J Cataract Refract Surg*, 2013. **39**(6): 898–905.

80. Tomita, M. et al., Simultaneous corneal inlay implantation and laser in situ keratomileusis for presbyopia in patients with hyperopia, myopia, or emmetropia: Six-month results. *J Cataract Refract Surg*, 2012. **38**(3): 495–506.

81. Santhiago, M.R. et al., Short-term cell death and inflammation after intracorneal inlay implantation in rabbits. *J Refract Surg*, 2012. **28**(2): 144–149.

82. Yilmaz, O.F. et al., Intracorneal inlay to correct presbyopia: Long-term results. *J Cataract Refract Surg*, 2011. **37**(7): 1275–1281.

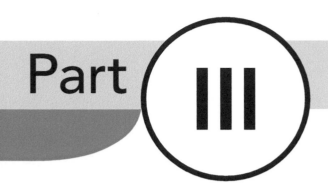

Part III

Impact of the eye's optics on vision

18

Optical and visual metrics

Antonio Guirao

Contents

18.1 INTRODUCTION

The eye is a complex imaging system whose optical components determine the first step of vision. Quantifying and measuring the *ocular quality*, and how this affects the quality of the retinal images, can allow us to better understand the visual process and to improve diagnostics and correction methods. The optical quality is lowered by aberrations (consequence of the shape of the refracting surfaces), diffraction at the pupil, and scattering (due to particles and nonuniformities localized in the media). The combined effect of these three factors is that light deviates from the ideal trajectory and spreads over a deteriorated retinal image that will limit the visual performance (Artal 2014). In general, a system with less aberrations, scattering, and diffraction will produce images with higher quality. Consequently, one expects to have a better vision when images at the retina are fine. However, although the optics, the retinal image stage, and the visual perception are dependent with each other, there is not a direct or obvious relationship between these three levels. In this context, it is necessary to distinguish between optical quality, optical performance (image quality), and visual performance and to be able to define *metrics* that quantitatively measure each of these matters.

The term *performance* is used to cover all aspects of an imaging system on its ability to perform a specific task. We talk about *optical performance* to refer to the efficacy or achievement with which the optical system produces good images. The term *image quality* is used in a more restrictive sense to describe the fineness

of detail that can be resolved in the image or the similarity between the image and their corresponding object. On the other hand, *visual performance* is defined by the speed and accuracy of processing visual information (CIE 2011), that is, by how well a visual task, for example, spatial vision, can be performed by a subject.

Traditional methods of assessing optical quality are defined in the *pupil plane* and account for the shape of the wave aberration (WA), for example, by means of global parameters such as the root-mean-square (RMS). On the other hand, a variety of metrics exist for specifying optical performance, based on either the *image plane* (spatial domain) or the *frequency domain*. Some of the most obvious image plane methods are intended to measure the quality of the point spread function (PSF) through scalar metrics as, for example, the Strehl ratio (SR) or the encircled energy (EE). Also, several image fidelity metrics have been developed whose goal is to measure the differences between two related images. In the Fourier or frequency domain, the modulation transfer function (MTF) represents a powerful tool for determining image quality for grating and extended objects. Meanwhile, visual performance or *subjective image quality* has been traditionally measured by means of the visual acuity and the contrast sensitivity function (CSF).

As a rule, optical quality ensures image quality and, therefore, a correct visual performance. If retinal images are good so will the vision under normal conditions (except in exceptional cases where images of extreme quality can cause visual artifacts such as aliasing). Likewise, if optics is poor, the visual perception

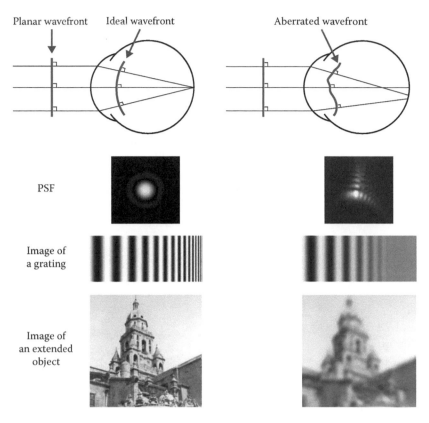

Figure 18.1 Optical wavefront, point spread function, image of a grating, and simulated image of a scene, for an ideal diffraction-limited eye and for a real aberrated eye. Different objective quality metrics may be defined based on the optical defects or based on the optical performance to form images.

will be largely limited. However, two facts should be taken into account: in some cases, a system with larger optical defects does not produce, paradoxically, worse images; and, after the optical stage of the visual processing, there are *retinal* and *neural factors* involved that explain how low-quality retinal images can still provide an acceptable visual performance. In other words, wavefront error is not always a good predictor of image quality or, for example, SR does not necessarily reflect subjective image quality. On the other hand, optical performance is relatively well matched to the capabilities of the neural network which it serves (Helmholtz 1885).

Measures of optical quality (aberrations) and optical performance (image quality, contrast transfer, etc.) are *objective methods* (Figure 18.1). While the final arbiter of image quality is the human viewer, these efforts to define objective metrics can be useful for many applications. Along with the eye, the optical performance of correcting lenses (contact, ophthalmic, and intraocular lenses), used increasingly more frequently, must be considered for understanding how the images are formed at the retina. Thus, for optical design the manufacturer needs some merit functions to achieve the quality requirements. The aberrations of the eye and their effects on retinal image quality have long been of interest. Further, the link between vision and ocular optical quality has enjoyed a renewed attention. How the complex interactions of aberrations impact visual performance is being systematically investigated by looking at correlations of objective and subjective estimates. Also, new objective metrics that

incorporate some neural characteristics of the visual system have been proposed recently that correlate better with clinical measures of visual performance.

In this chapter, we present an extensive list of metrics classified into five groups (Table 18.1). Some of them are functions, such as the MTF, but most are single-value or scalar metrics such as the entropy. The first group comprises parameters of optical quality that measure the aberrations. Then, three groups of image quality metrics are described: based on the PSF, the optical performance for grating objects, and metrics that compare extended objects with their images. All metrics in these four groups are objective measures. Finally, the visual metrics in the fifth group include subjective estimates of visual performance and objective metrics based on some neural feature. One metric alone is usually not enough to describe the system appropriately. Ultimately, optical and visual performance criteria should be selected to be appropriate for the application.

18.2 OPTICAL QUALITY (ABERRATION-BASED) METRICS

Optical properties are typically described by an aberration or wavefront error map. See Chapter 5 of *Handbook of Visual Optics: Fundamentals and Eye Optics, Volume One* for an introduction to optical aberrations. From the WA, several pupil plane metrics can be easily computed for quantifying the optical quality both of the eye and correcting lenses.

Table 18.1 **Summarized list of metrics**

OPTICAL QUALITY (ABERRATION BASED)	IMAGE QUALITY OF THE PSF	IMAGE QUALITY, FOURIER DOMAIN
WA	Strehl ratio	Contrast of grating image
Zernike coefficients	Intensity variance	Two-point optical resolution
Seidel coefficients	Peak autocorrelation	Grating optical resolution
RMS	Half width at half maximum	Cutoff frequency
Peak to valley	$1/e^2$ width	MTF
Pupil fractions	Width autocorrelation	Hopkins ratio
	Half energy diameter	Frequency for which MTF = 0.1 or 0.5
	D86 diameter	Area MFT
	Equivalent cylinder diameter	Area Hopkins ratio
	Encircled energy	Pseudo Strehl ratio on MTF
	Second moment	Volume OTF/volume MTF
	Entropy	

VISUAL METRICS	SIMILARITY OBJECT AND IMAGE
Minimum angle of resolution	Mean square error
Visual acuity	Peak signal-to-noise ratio
Snellen fraction	Correlation coefficient
Maximum resolved frequency	Mutual information
Threshold contrast	Receiver operator characteristic
CSF	Structural similarity index
Area CSF	
Area MFT/NTF	
Subjective quality factor	
Area OTF/NTF	
Area OTF/area MTF neurally weighted	
Visual Strehl ratio	
Neural sharpness	

18.2.1 POLYNOMIAL EXPANSION OF THE WAVE ABERRATION

The WA may be expressed as a polynomial expansion (Mahajan 1991), usually as a linear combination in the orthogonal variance-normalized Zernike base:

$$WA(\rho,\theta) = \sum_{n,\pm m} c_n^{\pm m} \cdot Z_n^{\pm m}(\rho,\theta) \tag{18.1}$$

where

$\rho = r/r_o$ is the normalized radial variable (with r_o the pupil radius)

θ is the angular variable over the pupil (Malacara 1992; Wang and Silva 1980; Wyant and Creath 1992)

$Z_n^{\pm m}(\rho,\theta)$ represent the *Zernike polynomials*

$c_n^{\pm m}$ the *aberration coefficients*

The subscript n indicates the order of aberration (second order, third order, fourth order, etc.)

The superscript m is the angular frequency denoting the times the wavefront pattern repeats itself

Equation 18.1 can be expanded by grouping the terms according to their order:

$$WA = c_0^0 \cdot Z_0^0 + \left\{c_1^1 \cdot Z_1^1 + c_1^{-1} \cdot Z_1^{-1}\right\} + \left\{c_2^0 \cdot Z_2^0 + c_2^2 \cdot Z_2^2 + c_2^{-2} \cdot Z_2^{-2}\right\}$$

$$+ \left\{c_3^1 \cdot Z_3^1 + c_3^{-1} \cdot Z_3^{-1} + c_3^3 \cdot Z_3^3 + c_3^{-3} \cdot Z_3^{-3}\right\} + \left\{c_4^0 \cdot Z_4^0 + \cdots\right\} + \cdots \tag{18.2}$$

Coefficients c_0^0 and $c_1^{\pm 1}$, piston and tilt, are not actually true optical aberrations, as they only shift the image but do not spread light. Defocus (c_2^0) and astigmatism ($c_2^{\pm 2}$) are the lowest-order true optical aberrations.

During the long history of optical design, one of the most familiar approaches to merit functions has been based on *Seidel aberrations*. These are fourth-order aberrations for a system with rotational symmetry that images light coming from both on-axis and off-axis points. The WA is

$$WA_{Seidel} = A_t \rho \cos\theta_t + A_d \rho^2 + A_a \rho^2 \cos^2\theta_a + A_c \rho^3 \cos\theta_c + A_s \rho^4 \tag{18.}$$

where the Seidel coefficients A_t, A_d, A_a, A_c, and A_s represent tilt (distortion), defocus (curvature of field), astigmatism, coma, and spherical aberration, respectively. Spherical aberration affects rays from on-axis points when the optical surfaces do not have the shape needed to bring all rays to a focus. The other aberrations affect rays from points off the axis. These rays pass through a system that, from their perspective, is tilted and asymmetric; so light is not brought to a sharp focus (astigmatism and coma), images are on a curved surface instead of a flat plane (curvature of field), and, finally, images lie at a nonproportional distance from the axis (distortion).

Even for object points located on the optical axis, the WA may have the form of Equation 18.3 if the optical system has no rotational symmetry. For extension, the Seidel coefficients (some of which depend on the off-axis distance) can be related t

Zernike coefficients for on-axis aberrations. If the WA in Zernike polynomials is taken up to fourth order, the relationships between Zernike and Seidel coefficients are

$$A_t = \sqrt{\left(2c_1^1 - 2\sqrt{8}c_3^1\right)^2 + \left(2c_1^{-1} - 2\sqrt{8}c_3^{-1}\right)^2}$$

$$A_d = 2\sqrt{3}c_2^0 - 6\sqrt{5}c_4^0 - \sqrt{6}\sqrt{\left(c_2^2\right)^2 + \left(c_2^{-2}\right)^2}$$

$$A_a = 2\sqrt{6}\sqrt{\left(c_2^2\right)^2 + \left(c_2^{-2}\right)^2}$$

(18.4)

$$A_c = 3\sqrt{8}\sqrt{\left(c_3^1\right)^2 + \left(c_3^{-1}\right)^2}, \quad A_s = 6\sqrt{5}c_4^0$$

18.2.2 ABERRATION COEFFICIENTS AS OPTICAL QUALITY METRICS

The simplest metric for optical quality is any of the aberration coefficients, which will inform of the amplitude of a particular aberration. A single coefficient does not capture the whole optical quality because other aberrations may also be present in varying degrees. However, this way is useful in several cases, for example, when a particular aberration is dominant and responsible for the greatest loss of optical quality (e.g., in keratoconic eyes), or when we want to compare an aberration in different eyes or subjects (e.g., the degree of coma across a population), the two eyes of the same subject (e.g., the spherical aberration of both eyes), or the evolution of the aberration before and after a process (age, surgery, etc.).

The advantage of Zernike polynomials for expressing the WA is that they are linearly independent; thus, individual aberration contributions to the wavefront may be isolated and quantified separately (Figure 18.2). Further, Zernike polynomials can be identified with *balanced aberrations*, which are aberrations of a certain order mixed with other of lower order such that the variance of the net aberration is minimized. For example, the term Z_4^0

includes spherical aberration balanced with defocus, so the coefficient yields a WA minimal in average. In this context, examples of optical quality metrics based on single aberration coefficients can be as follows:

- The Zernike spherical aberration coefficient: c_4^0.
- The coma amplitude from Zernike coefficients: $\sqrt{(c_3^1)^2 + (c_3^{-1})^2}$.
- In the case of nonrotationally symmetric aberrations (e.g., coma), the orientation angle is useful besides the aberration amplitude. In general, the amplitude of a particular aberration of order n and its orientation in the pupil is

$$A_n^m = \sqrt{\left(c_n^{+m}\right)^2 + \left(c_n^{-m}\right)^2}$$

$$\theta_n^m = \frac{1}{n}\arctan\left(\frac{c_n^{-m}}{c_n^{+m}}\right)$$

(18.5)

18.2.3 VARIANCE AND ROOT-MEAN-SQUARE OF THE WAVE ABERRATION

The *variance*, σ^2, is a statistical parameter that measures how far a set of numbers are spread out around the mean and from each other. The variance of the WA is the mean of the squares of the WA minus the square of the mean of the WA. Mathematically,

$$\sigma_{WA}^2 = \frac{1}{\pi}\int_0^{2\pi}\int_0^1 [WA(\rho,\theta)]^2 \rho\, d\rho\, d\theta - \left[\frac{1}{\pi}\int_0^{2\pi}\int_0^1 WA(\rho,\theta)\rho\, d\rho\, d\theta\right]^2$$

$$= \overline{WA^2} - (\overline{WA})^2$$

(18.6)

The square root of the variance is the *standard deviation*, σ, that measures the amount of dispersion of the WA values from the average.

One of the most practical and well-known optical quality metrics is the *root-mean-square* (RMS) of the WA. It is also a

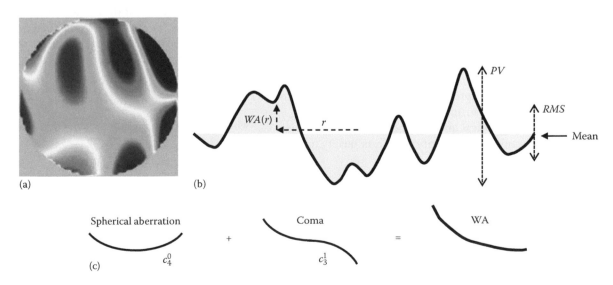

Figure 18.2 (a) Wave aberration (WA) map depicted on the pupil plane. (b) Cross section of the WA showing the deviation from the ideal aberration-free case. (c) Two different aberrations combine to give the WA. Optical quality metrics represent the magnitude of the aberrations, through polynomial expansion coefficients, variance, etc.

statistical parameter that is related to the variance and standard deviation. The name comes from its definition as the "square root of the mean of the squares" of the values. Since, in optics, the mean is always removed from the error (i.e., the WA is fixed to zero, piston term null), the variance is simply $\sigma_{WA}^2 = WA^2$, and the RMS will equal the standard deviation:

$$RMS = \sigma_{WA} = \sqrt{WA^2} \tag{18.7}$$

Because of the orthogonality of Zernike polynomials, the variance is given by the sum of the variance of each of the orthogonal components (Mahajan 1991). Thus, an advantage of Zernike base is that the RMS of the WA can then be obtained easily from the Zernike coefficients as

$$RMS^2 = \sum_{n, \pm m} \left(c_n^{\pm m} \right)^2 \tag{18.8}$$

In Equation 18.8, the piston term c_0^0 has been removed. Likewise, tilts (Zernike terms $c_1^{\pm 1}$) are removed when considering optics, since they only represent a displacement of the image, not a degradation.

The RMS metric of optical quality is based upon the principle of measuring the optical surfaces at many points and then arriving at a single number that is a statistical measure of the departure from the ideal form. This method puts into relative importance any isolated areas that may be in themselves highly deviated.

18.2.4 PEAK-TO-VALLEY WAVEFRONT ERROR

The *peak-to-valley* (PV) wavefront error is the maximum departure of the actual wavefront from the ideal wavefront in both positive and negative directions, that is, the distance between the maximum and the minimum values of the WA:

$$PV = \max(WA) - \min(WA) \tag{18.9}$$

This method is simple although it can be misleading because it looks at only two points and ignores the area over which the error is occurring. An optical system having a large PV error may actually perform better than other with smaller PV errors. The PV criterion became popular historically within the amateur community that used to express astronomical optics performance for mirrors, lenses, and instruments in PV terms. However, this method of rating optical quality is not representative. It is generally more meaningful to specify wavefront quality using the RMS error, as explained before.

18.2.5 PUPIL FRACTIONS

Some metrics of wavefront quality have been proposed based on the concept of *pupil fraction* (PF), which is defined as the fraction of the pupil area for which the optical quality is reasonably good but not necessarily diffraction limited (Thibos et al. 2004).

One of the methods for determining the area of the "good pupil" consists of taking a circular subaperture, concentric with the total pupil, within which some criterion of quality is reached (Corbin et al. 1999; Howland and Howland 1977).

This subaperture is called the *critical pupil* and has a critical radius r_c smaller than the pupil radius r_o. The PF is computed as

$$PF = \left(\frac{r_c}{r_o} \right)^2 \tag{18.10}$$

A different method is called the *tessellation method*. It consists of tessellating the entire pupil with small subapertures (e.g., about 1% of pupil diameter) and then rating each subaperture as "good" or "bad" according to some criterion. The sum of the good subapertures defines the area of the good pupil from which the PF is computed as

$$PF = \frac{\text{Area of good subapertures}}{\text{Area of the pupil}} \tag{18.11}$$

Both the critical pupil method and the tessellation method require criteria for deciding if the wavefront over a subaperture is good. The criterion for defining what is meant by good wavefront quality is somehow arbitrary and depends on the degree of optical performance expected or the specific application. For example, a criterion is that the RMS of the WA does not exceed the value of $\lambda/4$ or that a good subaperture meets $PV < \lambda/4$.

18.3 IMAGE QUALITY METRICS I: OPTICAL PERFORMANCE FOR POINT OBJECTS

In this section, we introduce image plane metrics that are intended to quantify the compactness, shape, and concentration of energy in the image of a point object.

18.3.1 THE POINT SPREAD FUNCTION

18.3.1.1 Image of a point object

The *point spread function* (PSF) is the distribution of light in the image of a point object formed by an optical system for a given aberration (Mahajan 1998). The PSF is a function of two variables (coordinates at the image plane) whose values are the light intensities for each point in the image. Mathematically, the PSF is the squared modulus of the *Fourier transform* (FT) of the generalized pupil function (Goodman 1996):

$$PSF(x, y) = \left| FT(P(\rho, \theta)) \right|^2 \tag{18.12}$$

where $P(\rho, \theta)$ is the *generalized pupil function* defined as

$$P(\rho, \theta) = A(\rho, \theta) \cdot \exp\left[\frac{2\pi i}{\lambda} WA(\rho, \theta) \right] \tag{18.13}$$

and

$A(\rho, \theta)$ denotes a circular aperture with a unit amplitude or an optional apodization function
$WA(\rho, \theta)$ is the WA
λ is the wavelength of the light

In general, the higher the aberrations, the more extensive the PSF and, therefore, the contrast decreases as the peak intensity drops because light spreads over a larger area. The PSF of a diffraction-limited system is the Airy spot. Each kind of aberration produces a characteristic PSF. The eye presents a combination of all aberrations that give a PSF more or less complicated and different from one subject to another.

18.3.1.2 Measuring the PSF

Today, it is relatively easy to measure the aberrations of an optical system, thanks to the aberrometry techniques, so the PSF can be calculated mathematically from the WA. Formerly, the PSF of the eye was estimated through the double-pass technique used extensively in physiological optics, which is based on recording images of a point source projected on the retina after retinal reflection and double-pass through the ocular media (Artal 2000; Iglesias et al. 1998).

For optical systems other than the eye, such as correcting lenses, the PSF can be calculated from aberration measurement or obtained directly by registration of passing light in an optical bench. However, in these cases the application of metrics based on the WA or the modulation transfer is more convenient, because when considering the effect of each element in a system as a whole, the degradation is additive in aberrations and multiplicative in contrast.

18.3.1.3 Moments of the PSF: Centroid and total intensity

A *moment* is a specific quantitative measure of the shape of a function. The nth moment of a real-valued continuous function $f(x)$ is

$$\int x^n f(x)dx \tag{18.14}$$

If the points represent intensity values of the PSF, then the zeroth moment is the total energy, the first moment divided by the total energy is the center of mass, and the second moment is the inertia.

The *centroid*, or center of mass, of the PSF is the point where the light of the image could be concentrated. The coordinates $(\bar{x}, \bar{y})$ of the centroid are computed as the arithmetic mean position of all the points in the PSF shape:

$$\bar{x} = \frac{\iint\limits_{image} x\, PSF(x,y)dxdy}{Vol(PSF)}, \quad \bar{y} = \frac{\iint\limits_{image} y\, PSF(x,y)dxdy}{Vol(PSF)} \tag{18.15}$$

where $Vol(PSF) = \iint\limits_{image} PSF(x,y)dxdy$ is the volume or total intensity energy in the image. In aberrated systems, the centroid of the PSF may not coincide with the coordinate origin.

The *normalized point spread function*, PSF_N, is defined with total intensity equal to 1:

$$PSF_N = \frac{PSF}{Vol(PSF)} \tag{18.16}$$

The normalized PSF may be seen as a probability density function. Then, the zeroth moment of PSF_N is the total probability (i.e., 1), the first moment is the mean, the second moment is the *variance*, and the third moment is the *skewness*.

18.3.1.4 The Airy disk of diffraction: Diffraction-limited PSF

The *Airy pattern* is the best focused spot of light that a perfect system with a circular pupil can make, limited by the diffraction of light, that is, the PSF for an unaberrated system. The *Airy disk* is the central core of the Airy pattern. It contains a maximum of 84% of the light entering the optical system. The angular radius of the Airy disk is

$$\theta_{Airy} = 1.22\frac{\lambda}{D} \tag{18.17}$$

where
  λ is the wavelength of the light
  D is the diameter of the pupil

18.3.2 STREHL RATIO AND INTENSITY SHARPNESS METRICS

18.3.2.1 Strehl ratio

The *Strehl ratio* (SR) is defined as the ratio of the intensity value at the center of the image with and without aberrations for the same pupil size:

$$SR = \frac{PSF(0,0)}{PSF_{d-l}(0,0)} \tag{18.18}$$

where PSF_{d-l} is the aberration-free diffraction-limited PSF. Strehl performance is usually expressed as a range of numbers from 1 to 0. A perfect system is 1, a completely imperfect system is 0, and acceptable standards occur somewhere in between.

The SR has been in regular use for many years and describes one of the most common and significant measures of optical performance. Although the mathematics involved in the calculation of the SR may be complex, they can be rather easily explained in concept. Once the PSF is available, the SR is really quite easy to relate to and is simple to understand. However, characterizing image quality by this single number will be meaningful only if the PSF is little distorted, which is true for a well-corrected system that operates close to the diffraction limit. For small aberrations, the SR and the RMS of the WA correlate well with each other. Several equations have been derived for expressing this relationship for low aberrations (SR down to values of about 0.5); one of the best known is (Mahajan 1982)

$$SR \approx e^{-\left(\frac{2\pi}{\lambda}RMS\right)^2} \approx 1 - \left(\frac{2\pi}{\lambda}RMS\right)^2 \tag{18.19}$$

where λ is the wavelength. However, only for large SR values the intensity is maximum at the point associated with minimum aberration variance. For large aberrations, there is no simple relationship between the SR and the aberrations.

An alternate definition of SR is often given in terms of the peak intensity (SR_{peaks}), so this metric is defined as the ratio of the maximum value of the PSF in the presence and absence of aberrations:

$$SR_{peaks} = \frac{\max(PSF)}{\max(PSF_{d-l})} \tag{18.20}$$

The peak of the *PSF* does not necessarily occur at the coordinate origin, specially for large and nonsymmetrical aberrations. Consequently, the SR_{peaks} is not equivalent in general to the actual SR.

18.3.2.2 Other intensity sharpness metrics

The SR refers to a single value of the PSF, its maximum value. It may happen that the PSF has a high peak value but the light is widely distributed (i.e., a bad image). Other metrics attempt to measure the sharpness of the PSF considering all the intensity values over the entire image (Figure 18.3). These metrics are, indirectly, an estimate of the spatial resolution since a sharper PSF will blur the image less. In particular, we will introduce here: intensity variance and autocorrelation. Other sharpness metrics can be found in the literature (Fienup and Miller 2003).

The *intensity variance of the PSF* summarizes in one number the histogram of intensity values of the PSF. Variance measures how far the values are spread out from the average. A variance of zero indicates that all the values are identical (i.e., a flat completely blurred PSF); on the contrary, a sharp image has a high variance. The intensity variance of the PSF is calculated as the average value of the squared PSF minus the average PSF squared:

$$\sigma^2_{PSF} = \overline{PSF^2} - \overline{PSF}^2 \tag{18.21}$$

The *standard deviation of the PSF*, σ_{PSF}, is the square root of the variance.

Peak of the autocorrelation of the PSF: The autocorrelation is t cross-correlation of the PSF with itself:

$$C(x,y) = PSF \otimes PSF = \iint\limits_{image} PSF(s,t)PSF(s-x,t-y)\,ds\,dt \tag{18.2}$$

A high spatial autocorrelation indicates that there are small differences between the values of the PSF in nearby points and large differences between distant points, that is, a smooth sharp PSF will have a narrow autocorrelation function. In this context a metric for sharpness is the peak of the autocorrelation function

18.3.3 METRICS FOR SPATIAL COMPACTNESS OF THE PSF

Some scalar metrics are designed to capture the attributes of spatial compactness of the PSF such that small values of the metric indicate a compact PSF of good quality. The most used a defined in the following (Figure 18.3).

18.3.3.1 Width metrics, referred to peak value and sharpness

The *full width at half maximum* (FWHM) is an expression of the extent of the PSF given by the distance between points on the curve at which the intensity reaches half its maximum value. *Half width at half maximum* (HWHM) is half of the FWHM (Charman and Jennings 1976; Westheimer and

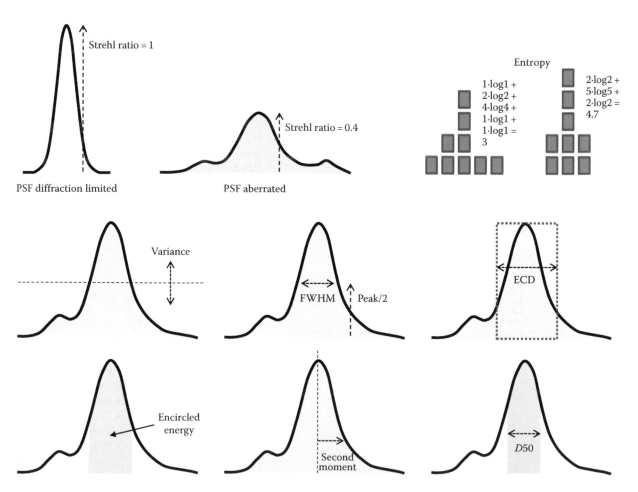

Figure 18.3 Cross sections of the point spread function (PSF) showing the idea behind different image quality metrics that account for the compactness, shape, and energy concentration of the PSF.

Campbell 1962). If the PSF is expressed in polar coordinates, the HWHM is computed as the average width of every cross section of the PSF:

$$HWHM = \sqrt{\frac{1}{\pi} \int_0^{2\pi} \int_0^\infty \Pi(r,\varphi) r\, dr\, d\varphi} \qquad (18.23)$$

where $\Pi(r,\varphi) = 1$ if $PSF(r,\varphi) > \max(PSF)/2$, otherwise $\Pi(r,\varphi) = 0$.

Similarly, the $1/e^2$ *width* is equal to the distance between the two points on the PSF distribution that are $1/e^2 = 0.135$ times the maximum value.

Width of the PSF autocorrelation: This metric is the HWHH of the autocorrelation $C(x,y)$ of the PSF. A similar metric is the *width of the correlation of the PSF with the diffraction-limited PSF* that is used as reference.

18.3.3.2 Diameter metrics, referred to encompassed energy

The *half energy diameter* (D50) is the diameter of a circular area centered on the PSF peak that captures 50% of the light energy. This diameter is computed from the implicit radius $r = D50$ that is the limit of the definite integral

$$\int_0^{D50} \int_0^{2\pi} PSF_N(r,\varphi) r\, dr\, d\varphi = 0.5 \qquad (18.24)$$

where PSF_N is the normalized PSF with intensity = 1 and peak value located at $r = 0$.

The *D86 diameter* is defined as the diameter of the circle that is centered at the centroid of the PSF profile and contains 86% of the total energy. In this case, we must do first a translation of the PSF so it is centered at the centroid. The D86 width is often used in applications that are concerned with knowing exactly how much power is in a given area, for example, applications of laser beams (Siegman 1998). The percentage of 86 is chosen because a circular Gaussian beam profile integrated down to $1/e^2$ of its peak value contains 86% of its total power.

The equivalent diameter (ECD) is the diameter of the circular base of that cylinder that has the same volume as the PSF and the same height. If the normalized PSF is taken, the value of ECD is given by

$$ECD = \sqrt{\frac{4}{\pi \max(PSF_N)}} \qquad (18.25)$$

18.3.3.3 Asymmetry of the PSF: Skewness

Skewness is a measure of the degree of asymmetry of the distribution of light in the PSF. If the left tail is longer than the right tail (the light of the distribution is concentrated on the right), the function has negative skewness and is said to be left-skewed or left-tailed. If the reverse is true, it has positive skewness (the right tail is longer; the light is concentrated on the left) and the distribution is said to be right-skewed or right-tailed.

The skewness is the third moment of the PSF. It can be calculated as the average of every cross section of the PSF centered at $r = 0$:

$$Skewness = \frac{1}{\pi} \int_0^\pi \left[\int_{-\infty}^\infty r^3\, PSF(r,\varphi) dr \right] d\varphi \qquad (18.26)$$

18.3.4 CONCENTRATION OF LIGHT

18.3.4.1 Encircled energy

The optics term *encircled energy* (EE) measures the fraction of the total energy in the PSF that lies within a circle of specified radius (Shannon 1997; Srisailam et al. 2011). It is calculated by first determining the total energy of the PSF over the full image plane and the centroid. Circles of increasing radius are then created at the centroid and the energy within each circle is calculated and divided by the total energy. As the circle increases in radius, more of the PSF energy is enclosed. The EE curve thus ranges from zero to one. The EE in a circle of radius R is

$$EE(R) = \int_0^R \int_0^{2\pi} PSF_N(r,\varphi) r\, dr\, d\varphi \qquad (18.27)$$

A typical criterion for EE is the radius of the PSF at which either 50% or 80% of the energy is encircled. Also, it is often considered the EE that falls within the core corresponding to the Airy disk; in this case the EE is called *light-in-the-bucket* (Thibos et al. 2004).

18.3.4.2 Second moment of intensity distribution

The *second moment of the PSF* (M2), also known as moment of inertia, measures the concentration of light in the near vicinity of the center (Bareket 1979). It represents the spatial variance of the PSF. If the normalized PSF is taken, the moment of inertia is

$$M2 = \int_0^\infty \int_0^{2\pi} r^2 PSF_N(r,\varphi) r\, dr\, d\varphi \qquad (18.28)$$

18.3.4.3 Entropy of the PSF

This metric is inspired by an information theory approach to optics (Barakat 1998; Bove 1993; Guirao and Williams 2003). The *entropy* (H) is mathematically calculated as follows (Shannon entropy):

$$H = -\sum_{x,y} PSF_N(x,y) \cdot \log PSF_N(x,y) \qquad (18.29)$$

where log is the decimal logarithm.

The entropy is a measure of the spatial variance of the PSF, that is, a measure of how the energy is distributed in the image. The aberration-free PSF shows the minimum entropy with the maximum concentration of light in the center. Aberrations increase the entropy because light tends to spread throughout the image. An image with an intensity constant level has the maximum entropy. Both the second moment and the entropy are metrics sensitive to the shape of the PSF tails (Figure 18.3).

18.4 IMAGE QUALITY METRICS II: OPTICAL PERFORMANCE FOR GRATING OBJECTS, METRICS BASED ON THE FOURIER DOMAIN

In the previous section, we have introduced image quality metrics based on the degraded image of point objects. However, in the real world, except when, for example, we look at stars, it is common to find extended objects with complicated light distributions. A method for studying the optical performance is based on the reduction of contrast that the optical system produces on the image of objects having a particular pattern detail. For that purpose, the Fourier analysis is a useful and powerful tool (Goodman 1996; Mahajan 1998; Williams 1998; Williams and Becklund 2002). By means of the so-called Fourier transform, complicated signals may be written as the sum of simple waves mathematically represented by sines and cosines. Significant simplification is often achieved by transforming spatial functions, such as intensity distributions, to the frequency domain of the Fourier space, which manifests the periodic information contained in the signal.

18.4.1 PERIODIC PATTERNS: CONTRAST AND OPTICAL RESOLUTION

18.4.1.1 Basics of gratings

In the study of visual perception, periodic patterns (sine and square-wave gratings) are frequently used to probe the capabilities of the visual system. A *grating* is a repeating sequence of light and dark. One adjacent pair of a light and a dark strip makes up one cycle. These cycles repeat over and over in a grating. A *sine-wave*, or sinusoidal, grating shows a smooth repetitive sine-shaped oscillation. A *square-wave* grating is a nonsinusoidal periodic waveform in which the amplitude alternates at a steady frequency between fixed minimum and maximum values, that is, it is composed of black and white bars.

Three parameters define periodic patterns (in our case, objects with periodic distributions of light and their images): spatial frequency, contrast, and phase.

The *spatial frequency* (f) is a measure of how often periodic components of the pattern repeat per unit of distance or angle.

Spatial frequency is often expressed in units of cycles per millimeter, or cycles per degree of visual angle.

The *contrast* (C), or modulation, is a measure of the difference between the extreme intensity values and the average intensity. *Michelson contrast* is defined as

$$C = \frac{I_{max} - I_{min}}{I_{max} + I_{min}} \tag{18.30}$$

where I_{max} and I_{min} are the maximum and minimum values of intensity in the pattern (object or image). A pattern with a flat or uniform intensity profile has contrast null. Conversely, a pattern with contrast unity presents minimum intensity values equal to zero.

The *phase* of the object, or image, is the shift (expressed in radians or degrees) of the location of the peak of the signal from the origin.

In general, extended objects are not as simple as a grid, but contain many spatial frequencies. In that case, the light distribution is decomposed by Fourier analysis as a combination of multiple gratings.

18.4.1.2 Contrast as a metric

If, in particular, one is considering the response of the optical system to a grating object of known contrast, a single-value metric is simply the Michelson contrast in the grating image as defined earlier, or the relative contrast between object and image (Figure 18.4). The generalization of this metric for grating objects with any spatial frequency will lead in the next section to introduce the MTF.

18.4.1.3 Two-point optical resolution

Often the quality of an imaging system is expressed in terms of *optical resolution*, which describes the ability of the system to resolve small details in the object that is being imaged (Born and Wolf 1999). The spatial, or angular, resolution is based on the minimum separation, or angle, at which two details (points, lines, stars, etc.) can be distinguished as individuals.

One requires empirical criteria to quantitatively calculate optical resolution. Several objective criteria have been used in classical optics, most taken from astronomy where they had a very practical significance. One of the standards is based on the *Rayleigh criterion of resolution* and states that two points are considered to be just resolved when the distance between them is such that, on the midpoint between the image of one point and

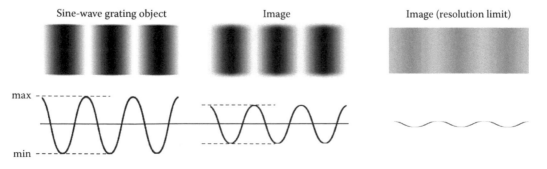

Sine-wave grating object   Image   Image (resolution limit)

Figure 18.4 Contrast of gratings. The contrast measures the difference between the extreme intensity values and the average intensity in an image. When the degradation is large, the contrast tends to zero until finally the stripes in the grating image become indistinguishable (concept of grating optical resolution).

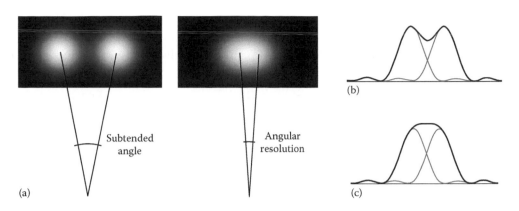

Figure 18.5 (a) Two-point optical resolution. (b) Rayleigh's resolution criterion. (c) Sparrow's resolution criterion.

the next, the intensity be 0.74% of the peak, that is, the contrast be 15% lower than the maximum.

The *Sparrow's resolution criterion* improves on this by saying that resolution is reached when the combined image of the two points no longer has a dip in brightness between them but instead has a roughly constant brightness from the peak of one image to the other (Figure 18.5). The Rayleigh criterion is more strict, while the alternate Sparrow criterion corresponds better with what the human eye can resolve.

The upper limit for two-point optical resolution is achieved for aberration-free optical systems. In this ideal case, the image of a point object is the diffraction Airy PSF and the angular resolution for the Rayleigh and the Sparrow criteria are given, respectively, by

$$\theta_{Rayleigh \atop d-l} = 1.22 \frac{\lambda}{D}, \quad \theta_{Sparrow \atop d-l} = 0.95 \frac{\lambda}{D} \quad (18.31)$$

where
  λ is the wavelength of the light
  D is the diameter of the lens's aperture
  The angle is expressed in radians
  The subscript $d - l$ means the diffraction-limited ideal case

18.4.1.4 Grating optical resolution and cutoff frequency

When considering gratings, resolution aims to the minimum spatial frequency in the grating object that can be resolved in its image. In this case, one often speaks of *grating optical resolution*. Thus, resolution is usually expressed in terms of lines per millimeter or cycles per degree, where a "line" or cycle is a sequence of one black strip and one white strip. The inverse of the frequency yields the spacing between two resolved details (two black or two white stripes). One can say, for example, that an optical system resolves 30 cycles per degree; the meaning of this is that two successive light stripes (or two successive dark stripes) separated by 1/30 degrees (= 2 minutes of arc) may be distinguished in the image, based on some objective criterion such as the Rayleigh or the Sparrow.

An upper limit to the objective estimate of grating optical resolution is given by the so-called cutoff frequency (f_{cutoff}), that is, the spatial frequency of a grating at which contrast in the image reaches zero and the details have disappeared completely. The maximum value for the cutoff frequency that can be achieved

corresponds to a perfectly corrected (diffraction-limited) optical system and is given by

$$f_{cutoff \atop d-l} = \frac{D}{\lambda} \text{ cycles/rad}, \quad \text{or} \quad f_{cutoff \atop d-l} = \frac{\pi}{180} \frac{D}{\lambda} \text{ cycles/degree} \quad (18.32)$$

One can see that the inverse of $f_{cutoff \atop d-l}$ is very close to the angular resolution $\theta_{Sparrow \atop d-l}$.

18.4.1.5 Drawbacks of the optical resolution methods

Resolution is a metric for measuring directly, by means of a single value, the performance of optical systems such as contact, intraocular, and ophthalmic lenses. In the case of the eye, the retinal image is not accessible and, then, resolution can only be estimated from the MTF measured indirectly or calculated from the ocular aberrations.

A weakness of the optical methods is that resolution is a threshold detection process. The criteria mentioned earlier are somehow arbitrary when applied to the human visual system. Ultimately, it is necessary to consider retinal and neural factors for knowing whether or not two objects may be resolved depending on their separation and contrast. This leads us to later define the visual resolution.

Another drawback is that objects in the real world are in general of relatively low contrast, so that the resolution limit obtained using high-contrast test patterns can be meaningless. The MTF and its psychophysical counterpart, the CSF, have a much complete information of the size of detail one might expect the eye to resolve in normal use.

18.4.2 OPTICAL TRANSFER FUNCTION AND MODULATION TRANSFER FUNCTION

18.4.2.1 Optical transfer function

The *optical transfer function* (OTF) specifies the translation and contrast reduction of a periodic sine pattern after passing through the optical system, as a function of its periodicity and orientation. Formally, the OTF is defined as the FT of the PSF (Goodman 1996; Mahajan 1998):

$$OTF(u,v) = FT(PSF(x,y)) \quad (18.33)$$

where (u, v) are the spatial frequencies of the pattern along two perpendicular directions.

Because of the relationship of Equation 18.12 between the PSF and the pupil function of the system, the OTF is also given by the autocorrelation of the pupil function:

$$OTF(u,v) = P \otimes P = \iint\limits_{pupil} P(s,t)P^*(s-u,t-v)ds\,dt \quad (18.34)$$

where the asterisk indicates a complex conjugate. Thus, the OTF of a system can be obtained from its WA without having to calculate the PSF.

The OTF is a complex valued function. The absolute value is commonly referred to as the *modulation transfer function* (MTF), which gives the relative contrast between the image and the object. On the other hand, when also the pattern translation is important, the complex argument of the OTF can be depicted as a second real-valued function, commonly referred to as the *phase transfer function* (PTF), that indicates a change in the location of the peak of the pattern. The OTF in terms of the MTF and PTF is

$$OTF(u,v) = MTF(u,v) \cdot e^{i\,PTF(u,v)} \quad (18.35)$$

A high-quality OTF is indicated by high MTF values and low PTF values.

The OTF provides a comprehensive and well-defined characterization of optical systems (Williams 1998). Any object can be conceived as the sum of gratings of various spatial frequencies, contrasts, phases, and orientations. The optical system of the eye is a filter that reduces the contrast and changes the relative position of each grating in the object spectrum.

18.4.2.2 Modulation transfer function

This function is a measurement of the ability of an optical system to transfer modulation or contrast at a particular spatial frequency from the object to the image (Figure 18.6). Mathematically, it is the modulus of the FT of the PSF:

$$MTF = |FT(PSF)| \quad (18.36)$$

The MTF is a function of a two-dimensional spatial frequency coordinate, (u,v). A one-dimensional MTF may be computed radially averaging the values across all angles:

$$MTF(f) = \frac{1}{2\pi} \int\limits_0^{2\pi} MTF(f,\varphi)d\varphi \quad (18.3$$

where the radial spatial frequency is $f = \sqrt{u^2 + v^2}$. In this one-dimensional MTF, the Y-axis is the contrast transferred by the eye's optics, and the X-axis represents sine waves with spatial frequency (f) varying from low (large spacing between adjacent white stripes) to high (fine gratings):

$$MTF(f) = \frac{C_{image}(f)}{C_{object}(f)} \quad (18.3$$

The MTF accounts for optical performance in terms of contrast. High contrasts at low spatial frequencies reproduce perfectly larg image details, while high contrasts at high frequencies drive how well smaller details are seen. For the human eye, the contrast of the image decreases as the spatial frequency increases.

The MTF is an important tool for the objective assessment of the image-forming capability of optical systems. Resolution metri represent the limit of the eye to resolve details, given that the imag is shown with 100% contrast. However, the MTF accounts for th contrast over the full range of spatial frequencies.

18.4.3 METRICS BASED ON THE MTF

18.4.3.1 Diffraction-limited MTF and Hopkins contrast ratio

The *diffraction-limited MTF* (MTF_{d-l}) is the MTF for a system in which the effects of optical aberrations are assumed to be negligible. An optical system cannot perform better than its MTF_{d-l} because any aberrations will pull the MTF curve down. The MTF_{d-l} represents the upper limit to the eye's performance, and it is based on the overall limiting pupil aperture and the wavelength of the light. The *cutoff frequency* for the MTF_{d-l} was defined before (Equation 18.32) as the spatial frequency for whi

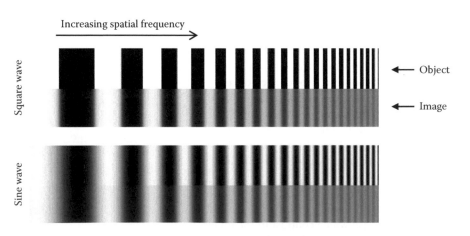

Increasing spatial frequency

Square wave

Sine wave

Object

Image

Figure 18.6 The modulation transfer function indicates the reduction of contrast from the object to the image for each spatial frequency. Thi reduction is shown in the figure for a square-wave grating and for a sine-wave grating.

the contrast becomes zero in the ideal case of an unaberrated system ($f_{cutoff MTFd-l} = D/\lambda$ cycles/rad).

The *Hopkins ratio* (HR), or Hopkins contrast ratio, is often used to characterize the image quality of an optical system. It is defined as the actual aberrated MTF divided by the diffraction-limited MTF as a function of spatial frequency (Hopkins 1966; Mahajan 1991):

$$HR(f) = \frac{MTF(f)}{MTF_{d-l}(f)} \quad (18.39)$$

Some authors use the term *optical quality factor* instead of HR when referring to this metric (Lisson and Mounts 1992).

18.4.3.2 Single values from the MTF

Instead of a complete curve MTF from zero to the cutoff frequency ($f_{cutoff MTF}$), in many cases, it is enough considering a limited range of spatial frequencies or, in some instances, picking out single MTF values at a few important frequencies.

A hint is to discard the MTF at low spatial frequencies because differentiation between high-performance and low-performance systems is difficult in this region. The high frequencies just under cutoff could be also avoided because even for well-corrected eyes the contrast is so small that erratic behavior in this region is usually of little consequence. Since the eye is relatively insensitive to detail at high spatial frequencies, a criterion consists of taking as an image quality metric the *spatial frequency at which MTF decreases to 10% level*. Experience has shown that the best indicators of image sharpness are the spatial frequencies at which the contrast attenuates by 50%, so an alternate metric is the *spatial frequency corresponding to the MTF equal to 0.5* (MTF50). Some works have considered the MTF value at the intermediate frequencies of 16 and 32 cycles/degree.

18.4.3.3 Area under the MFT or under the Hopkins ratio

While the MTF captures a lot of information about an imaging system, it is desirable to describe performance with a single figure of merit or scalar metric instead of a function. The area under the MTF has been one of the heavily researched metrics of image quality for the human eye. The image quality is directly linked to the integrated MTF curve between zero and the absolute limiting frequency, $f_{cutoff MTF}$, which means the richness of the information

contained in the image is a function of the area below the MTF curve. An alternative is to use the area of the HR curve.

Other approaches have been used. One is considering the area between spatial frequencies of 0 and 60 cycles/degree, or between 5 and 30 cycles/degree (Figure 18.7).

18.4.3.4 Strehl ratio from the Fourier domain

The SR defined in Equation 18.18 can be also calculated in the Fourier or frequency domain as the normalized volume under the OTF of the aberrated system:

$$SR = \frac{PSF(0,0)}{PSF_{d-l}(0,0)} = \frac{\displaystyle\int_{-\infty}^{\infty}\int_{-\infty}^{\infty} OTF(u,v)\,du\,dv}{\displaystyle\int_{-\infty}^{\infty}\int_{-\infty}^{\infty} OTF_{d-l}(u,v)\,du\,dv} \quad (18.40)$$

where OTF_{d-l} is the diffraction-limited OTF.

In many cases the PTF is unknown, which has led to the substitution of the MTF for the OTF in the previous calculation. Although this lacks rigorous justification, it is a popular method for defining a scalar metric (SR_{MTF}) once one knows the MTF. In particular, from the radially averaged MTF we can define (Driggers 2003)

$$SR_{MTF} = \frac{\displaystyle\int_{0}^{\infty} MTF(f)\,df}{\displaystyle\int_{0}^{\infty} MTF_{d-l}(f)\,df} \quad (18.41)$$

This metric is similar to the true SR but it is not the same, and it is also different to the SR_{peaks} defined in Equation 18.20.

18.4.3.5 Phase shifts and PTF

Most image quality metrics do not use phase in their calculation. The PTF tells how much the detail at each spatial frequency is shifted on the image relative to that detail on the object plane. In a nonsinusoidal extended object, the distribution of light can be broken down into sinusoidal components by Fourier methods. To preserve the exact appearance of the pattern, the sinusoidal

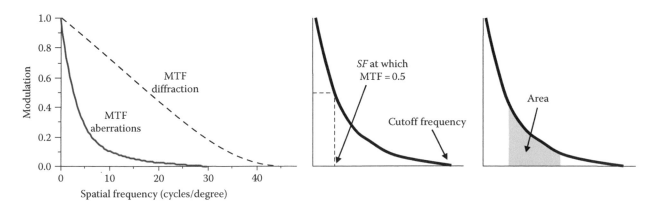

Figure 18.7 Modulation transfer function (MTF) for a real eye in comparison with the diffraction-limited MTF. Some examples of metrics based on the modulation transfer are spatial frequency at which the MTF decreases to 50%, cutoff frequency (*MTF* = 0), area under the MTF, and Hopkins ratio = quotient between MTF and diffraction MTF.

components must be kept in their original positions, which requires a linear PTF curve. Nonlinearities in the PTF, also called phase distortion, cause different spatial frequencies in the image to recombine with different relative phases.

No general statement can be made concerning the degradation of image quality caused the PTF. For certain cases, the PTF may supplement the information given by the MTF, but the PTF alone usually conveys little information. A metric that is intended to quantify phase shifts in the image is defined as the volume under the OTF normalized by the volume under the MTF (Thibos et al. 2004):

$$\frac{Vol(OTF)}{Vol(MTF)} = \frac{\int\limits_{-\infty}^{\infty}\int\limits_{-\infty}^{\infty} OTF(u,v)\,du\,dv}{\int\limits_{-\infty}^{\infty}\int\limits_{-\infty}^{\infty} MTF(u,v)\,du\,dv} \qquad (18.42)$$

18.4.4 MEASURING RESOLUTION AND MTF

A variety of test targets are available for measuring resolution (Smith 2000). These targets consist of equally spaced white and black bars. The widest bar the imager cannot discern, according to some resolution criteria, is the limitation of its resolving power. For example, the *Ronchi ruling*, or Ronchi grating, is a constant-interval alternating bar and space square-wave target that has a high edge definition and contrast ratio. Another of the existing test used for optical testing purposes is the *USAF 1951 resolution test target*. The pattern consists of groups of three bars with dimensions from big to small (covering a range of 0.25–228 cycles/mm) at different orientations.

The MTF for the eye can be obtained theoretically from WA measurements or double-pass PSF registration. In the case of correcting lenses, the MTF can be also measured directly by means of interferometry or by using sinusoidal grating objects.

As noted several times, one drawback of objective methods for assessing the visual performance is that they neglect the retinal and neural factors of the human visual system. However, a characteristic of the MTF is that it illustrates how well the eye could reproduce the contrast of the observed scene after the major limit imposed by the optical factors. In this sense, the MTF is objective and universal. Moreover, the MTF can be calculated from ocular aberration data giving scientists the ability to study the diffraction–aberration effects for different spatial frequencies and to design correcting lenses or surgery techniques for improving the optical quality. Another benefit is that the resulting MTF of a compound optical system is the product of all the MTF of its individual components, allowing researchers and manufacturers the comparison of the image quality with the expectations from the design stage and the prediction of system performance reliably.

18.5 IMAGE QUALITY METRICS III: SIMILARITY BETWEEN OBJECT AND IMAGE

As the ocular system forms a deteriorated image in the retina, the original object and its retinal image can be compared. The more similar these two images are, the better the optical performance.

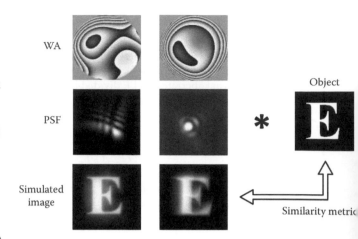

Figure 18.8 Convolution of a letter E with the point spread function calculated from the wave aberration data of two subjects. The simulated image and the reference image may be compared through several fidelity parameters such as the mean square error and the correlation coefficient.

The object can be a generated digital image of a real scene (O). Then, the corresponding retinal image (I) can be estimated computationally by convolving the eye's PSF with the object using the standard Fourier calculus. When the two images are available, the original and degraded, the following metrics may be used for measuring similarity and, therefore, estimating image quality (Figure 18.8).

18.5.1 SIMULATED IMAGES OF EXTENDED OBJECTS: CONVOLUTION

Mathematically, we may represent a complex or extended object as a sum over weighted impulse functions. The PSF may be independent of position in the object plane, in which case it is called shift invariant (Wandell 1995). In addition, if there is no distortion in the system, the image plane coordinates are linearly related to the object plane coordinates via the magnification M. Thus, the image of an extended object can be calculated as a superposition of weighted PSFs through a straightforward operation:

$$I(x,y) = \iint O(u,v) \cdot PSF(u-x/M, v-y/M)\,du\,dv \qquad (18.43)$$

where $O(u,v)$ and $I(x,y)$ represent the object and image, respectively. This integral is called *convolution*. So, the image of a complex object can be seen as a convolution of the true object and the PSF:

$$I = O * PSF \qquad (18.44)$$

18.5.2 FULL-REFERENCE METRICS

Full-reference metrics compute the quality difference by comparing every pixel of the distorted image to its corresponding pixel in the original object (Zhang et al. 2012).

18.5.2.1 Mean square error between object and image

The simplest and most widely used full-reference quality metric is the *mean squared error* (MSE), computed by averaging the squared intensity differences of distorted and reference image pixels (Gonzalez 2008):

$$MSE = \frac{1}{n} \sum_{i=1}^{N} (O_i - I_i)^2 \qquad (18.45)$$

where
- N is the number of pixels
- O_i and I_i are the intensity values in the pixel i

18.5.2.2 Peak signal-to-noise ratio

The related quantity of *peak signal-to-noise ratio* (PSNR) is an engineering term for the ratio between the maximum possible power of a signal and the power of corrupting noise that affects the fidelity of its representation. From the MSE, the PSNR is defined as

$$PSNR = 10 \cdot \log \frac{(MAX_O)^2}{MSE} \qquad (18.46)$$

with MAX_O being the maximum possible pixel value of the reference object O (e.g., when the pixels are represented using 8 bits per sample, this is 255).

Both MSE and PSNR can fail if the data diverge too much from the ideal case. It is well established that, in general, these simple approaches do not provide meaningful measures of image fidelity, and more sophisticated techniques are necessary.

18.5.2.3 Correlation coefficient

Pearson's correlation is widely used in image analysis (e.g., for comparing disparity between two images, for pattern recognition). Correlation coefficient (CR) measures linear covariation between two datasets, in this case between intensity values of the digital image of the reference object and its degraded image. Thus, higher correlation indicates that the two images have similar spatial patterns.

The *correlation coefficient* (CR) is defined as

$$CR = \frac{\sum_{i=1}^{N} (O_i - \bar{O})(I_i - \bar{I})}{\sqrt{\sum_{i=1}^{N} (O_i - \bar{O})^2} \sqrt{\sum_{i=1}^{N} (I_i - \bar{I})^2}} = \frac{\sigma_{OI}}{\sigma_O \sigma_I} \qquad (18.47)$$

where $\sigma_{OI} = (1/n) \sum_{i=1}^{N} (O_i - \bar{O})(I_i - \bar{I})$ is the covariance between object O and image I, and σ_O and σ_I are the standard deviation of O and I, respectively. Thus, CR is the covariance of the two variables divided by the product of their standard deviations. The coefficient has the value of 1 if the two images are absolutely identical and 0 if they are completely uncorrelated.

18.5.3 OTHER SIMILARITY MEASURES

18.5.3.1 Mutual information

The *mutual information* (MI) between two variables is a concept with roots in information theory and essentially measures the amount of information that one variable contains about another

(Papoulis 1991). It has emerged in recent years as an effective similarity measure for comparing images.

If the intensity distributions of the object and their simulated image are taken normalized to total energy equal to 1, the MI is defined as

$$MI = H_O + H_I - H_{OI} \qquad (18.48)$$

where H_O and H_I are the *marginal entropies* of object and image, respectively:

$$H_O = -\sum_{i=1}^{N} O_i \log O_i, \quad H_I = -\sum_{i=1}^{N} I_i \log I_i \qquad (18.49)$$

and H_{OI} is called the *joint entropy*:

$$H_{OI} = -\sum Hist(O, I) \cdot \log Hist(O, I) \qquad (18.50)$$

or entropy of the *joint histogram*, $Hist(O,I)$, which represents the probability distribution containing the number of simultaneous occurrences of intensities between the two maps.

18.5.3.2 Receiver operator methods

The *receiver operating characteristic* (ROC) is a method to evaluate the performance of a detection/imaging algorithm (Fawcett 2006). It has been a fundamental evaluation tool in clinical medicine. The ROC analysis consists of measuring the binary response of the system (target present or not) to one stimulus, for example, a reference image. First, both the reference and response images are transformed by applying a binary mask that labels pixels as background (value = 0) or foreground (value = 1). A certain threshold is fixed to decide which pixels should be considered as foreground or background. From these two binary images, the true-positive rate (TPR) and the false-positive rate (FPR) are defined as

$$TPR = \frac{TP}{F}, \quad FPR = \frac{FN}{B} \qquad (18.51)$$

where
- True positive (*TP*) is the number of foreground pixels, in the reference image, the system got right in the response image
- False negative (*FN*) are the foreground pixels wrongly identified as background
- *F* and *B* are the number of foreground and background pixels in the reference image, respectively

The ROC curve plots the TPR as a function of FPR for each decision threshold. Any point of this curve is a relationship between pixels correctly classified and pixels incorrectly classified. A curve located near of top left corner presents better performance compared to another one that is further away. The ROC curve is a two-dimensional depiction of imaging performance. A practical measure of the global performance by a single scalar value is given by the *area under the curve* of the ROC curve.

In particular, the ROC method can be applied to evaluate the performance of the optical system of the eye to produce faithful images. Let *MO* be the reference binary mask for an object (with *F* foreground pixels and *B* background pixels) and *MI* the response mask for its simulated retinal image at a certain threshold setting. Then the pixel-to-pixel comparison gives

$$TPR(t) = \frac{1}{F}\sum_{i=1}^{N} MO_i \cdot MI_i, \quad FPR(t) = \frac{1}{B}\sum_{i=1}^{N} MO_i \cdot (1 - MI_i) \tag{18.52}$$

where *t* is a decision/threshold parameter.

18.5.3.3 Structural similarity index

The *structural similarity* (SSIM) index is a method for measuring the similarity between two images (Wang et al. 2004). It is designed to improve on traditional methods like PSNR and MSE, which have proven to be inconsistent with human eye perception. SSIM is based on the hypothesis that the human visual system is highly adapted for extracting structural information.

SSIM considers image degradation as perceived change in *structural information*. Structural information is the idea that the pixels have strong interdependencies especially when they are spatially close. The SSIM metric is calculated as

$$SSIM = \frac{(2\bar{O}\bar{I} + c1)(2\sigma_{OI} + c2)}{(\bar{O}^2 + \bar{I}^2 + c1)\left(\sigma_O^2 + \sigma_I^2 + c2\right)} \tag{18.53}$$

where standard deviations and covariance are the same as defined before and *c*1 and *c*2 are two constants to avoid instability.

For image quality assessment, it is useful to apply the SSIM index locally rather than globally. Usually, it is computed within a local 8 × 8 square local window, which moves pixel by pixel over the entire image.

18.6 VISUAL METRICS

The human system contributes to the overall image-transfer process. The problem thus becomes psychophysical in addition to physical and requires a whole approach (Millodot 2009). In this section, we introduce metrics for measuring subjectively the visual performance and some objective metrics that take into account properties of the neural visual system.

18.6.1 VISUAL RESOLUTION

18.6.1.1 Minimum angle of resolution

The *minimum angle of resolution* (MAR) is the smallest separation between two closely high-contrast spaced details (points, lines, etc.) so that a subject is able to distinguish them as distinct. This measure does not correspond exactly with the spatial resolution defined previously because the MAR incorporates the visual factors.

When dealing with object patterns consisting of letters or gratings, the MAR is the smallest gap between letter strokes or grating bars that can be detected/resolved.

The MAR is often expressed in a logarithm scale. The *logMAR* is defined as the logarithm with base 10 of the MAR.

18.6.1.2 Visual acuity

Visual acuity is the spatial resolving capacity of the visual system, that is, the ability of the eye to see fine detail. There are various

ways to specify visual acuity, depending on the type of acuity ta. used. The most common is the *decimal visual acuity* (VA), define as the inverse of the MAR expressed in minutes of arc:

$$VA = \frac{1}{MAR} \tag{18.5}$$

For example, if the *MAR* = 1 minute of arc, the *VA* = 1. A subjec with *VA* = 2 can resolve 0.5 minutes of arc between two high-contrast points.

18.6.1.3 Snellen fraction

It is a measure of visual acuity based on a particular test pattern called the Snellen chart. The *Snellen fraction* (SF) is defined as

$$SF = \frac{D}{d} \tag{18.5}$$

where

    D is the distance at which the test is placed

    d is the distance at which the smallest resolved detail subtend an angle of 1 minute of arc (the smallest optotype of the Snellen chart identified subtends an angle of 5 minutes of arc)

For the small angles involved, we can replace the tangent by its argument, so the SF is practically equivalent to the decimal visu acuity:

$$SF \approx AV = \frac{1}{MAR} \tag{18.5}$$

In the most familiar acuity test, a Snellen chart is placed at a standard distance of 20 ft in the United States, or 6 m in the res of the world. At this distance, the letters representing normal acuity subtend an angle of 5 minutes of arc, and the thickness of the strokes and of the interspaces subtends 1 minute of arc. This level is designated 20/20 (in the United States) or 6/6 (in the res of world) and is the smallest line that a person with normal acui can read at a distance of 20 ft.

Again, the target image is displayed at high contrast (black letters on a white background). Difficulty may occur for decreased contrast (e.g., gray letters on a white background) in spite of normal visual acuity. This is because a subsequent contra sensitivity exam is required for describing complete visual function.

18.6.1.4 Grating acuity

The *maximum resolved frequency* is the maximum spatial frequency, f_{res}, detected when the eye observes sinusoidal or squa gratings of 100% contrast. It is related with the visual acuity and the MAR by

$$f_{res} = \frac{30}{MAR} = 30 \cdot VA \text{ cycles/degree} \tag{18.5}$$

The maximum resolved frequency by a subject is lower than the cutoff frequency, which only takes into account the optical

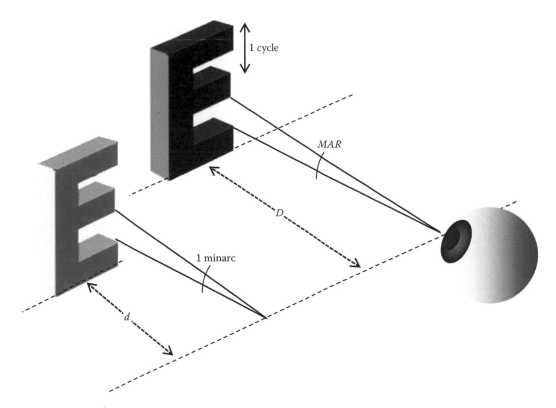

Figure 18.9 Concepts of minimum angle of resolution, visual acuity, and Snellen fraction.

factors. For instance, a subject with typical normal visual acuity of 1 can resolve a grating of 30 cycles/degree, although the MTF of its eye showed a cutoff frequency of 40 cycles/degree.

18.6.1.5 Measuring visual resolution and acuity: Charts and optotypes

A variety of test patterns are used for determining the visual resolution of the human eye (Duane et al. 1998). Most of them are based on single *optotypes*, which are standardized symbols (such as letters), or charts that contain several optotypes. A condition in all of these tests is that the image is shown with 100% contrast. The most common optotypes and charts are the two-point test, the Köning optotype, the Foucault grating, the Landolt C, the Snellen E, the logMAR chart, and the Snellen chart.

For example, a 20/20 Snellen letter has a bar/stroke width of 1 minute of arc, a letter height of 5 minutes of arc, 2.5 cycles, a grating period of 2 minutes of arc (1/30 degrees), and a grating spatial frequency of 30 cycles/degree (Figure 18.9).

18.6.2 CONTRAST SENSITIVITY FUNCTION

18.6.2.1 Threshold contrast

Suppose a visual target on a uniform background. The contrast of the target quantifies its relative difference in luminance from the background. It may be specified as *Michelson contrast*, as defined previously, $C = (I_{max} - I_{min})/(I_{max} + I_{min})$, or as *Weber contrast*. The Weber contrast is

$$C_W = \frac{I_{fore} - I_{back}}{I_{back}} \quad (18.58)$$

where I_{fore} and I_{back} are the intensity values for the foreground of the pattern and the background, respectively. This measure is also referred to as *Weber fraction*.

Weber contrast is claimed to be accurate for small dark symbols on a light background, where the viewer is assumed to be adapted to the background. Michelson contrast assumes the viewer is adapted to the sum of the background and foreground. Thus, Weber contrast is preferred for letter stimuli, whereas Michelson contrast is preferred for gratings.

The *threshold contrast* is the contrast required to see a target reliably. For gratings and periodic patterns, it is the minimum contrast that can be detected by the subject at a certain spatial frequency. The complete curve that plots the contrast threshold for all the spatial frequencies is the *contrast threshold function* (CTF).

18.6.2.2 Contrast sensitivity function

The reciprocal of the minimum perceptible contrast is called *sensitivity* (Pelli and Bex 2013). The *contrast sensitivity function* (CSF) tells us how sensitive we are to the various frequencies of visual stimuli or, stated equivalently, tells us what is the maximum spatial frequency that we can see for each possible value of contrast. The CSF is the reciprocal of the CTF.

If the frequency of visual stimuli is too high, we will not be able to recognize the stimuli pattern any more even if it has 100% contrast. The *cutoff frequency of the CSF* ($f_{cutoff\,CSF}$) is the maximum resolved frequency and, therefore, indicates the visual acuity for maximum contrast:

$$CSF(f_{cutoff\,CSF}) = 1 \quad (18.59)$$

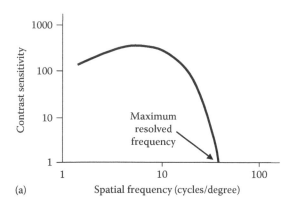
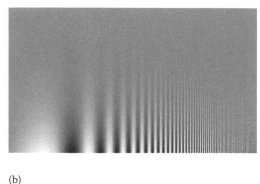

(a)   Spatial frequency (cycles/degree)   (b)

Figure 18.10 (a) From the contrast sensitivity function (CSF), different visual metrics can be defined: the maximum resolved frequency (which gives the visual acuity), the area under the CSF, the threshold contrast (reciprocal of the sensitivity) at some spatial frequency, etc. (b) The Campbell–Robson contrast sensitivity chart. Horizontal axis is spatial frequency; vertical axis is contrast.

The normal human contrast sensitivity shows a typical band-pass filter shape peaking at around 4 cycles/degree, and dropping off either side of the peak, and has a cutoff frequency between 25 and 50 cycles/degree.

For illustrating all the information carried by the CSF, we can discuss some points from the example shown in Figure 18.10. The maximum resolved frequency (sensitivity = 1) is 40 cycles/degree, which means a visual acuity of 1.33, Snellen acuity of 20/15, and minimum resolved angle of 0.75 minutes of arc. For a stimulus of 10 cycles/degree the sensitivity is 250, that is, the contrast threshold is 0.004. The example may be discussed in reverse: if the stimulus has contrast equal to 0.004, the maximum perceptible frequency is no longer 40 cycles/degree but 10 cycles/degree (i.e., we can speak of a *low-contrast VA* that would be 0.33 here).

As a rule, all contrasts and frequencies corresponding to points in the area enclosed by the CSF are visible.

Although the VA is a very important parameter, it is not sufficient to fully characterize the spatial vision. The usefulness of the CSF is that it reports the visual response for all conditions of contrast. From a clinical point of view, the CSF has great relevance because some diseases that are not detected by measuring the VA can be diagnosed and followed by means of the CSF.

18.6.2.3 Neural contrast functions

The *neural contrast sensitivity function* (NCSF), or neural CSF, reflects the contrast sensitivity of the neural visual system alone, without the optical effects (Campbell and Green 1965).

The *neural transfer function* (NTF), or *neural contrast threshold function*, is the reciprocal of the neural CSF and describes the contrast threshold or contrast transfer considering the neural factors alone.

The CSF can be calculated as the product of the optical MTF and the neural CSF:

$$CSF = MTF \cdot NCSF = \frac{MTF}{NTF} \quad (18.60)$$

18.6.2.4 Measuring contrast sensitivity and affecting factors

The CSF is measured using sinusoidal gratings of variable contrast and spatial frequency (Wandell 1995). This can be done using a computer/display system or with printed charts containing

stimuli of different frequencies and contrasts. In order to estimate a contrast threshold, the observer is tested over many trials, at various contrasts. Each trial is at some contrast and is scored right or wrong. The proportion of correct responses at each contrast is recorded. The observer's probability of correct response as a function of contrast is the *psychometric function*. The inverse of contrast threshold thus determined is the sensitivity. The NCSF can be measured subjectively by an interference fringe technique, which theoretically allows a sinusoidal pattern of very high contrast to be projected directly on the retina. In this way the results are no affected by aberration or diffraction from the eye's optics.

In addition to the optical factors that limit the ocular MTF and, therefore, the CSF (aberrations, pupil size, and wavelength), there are several others that affect the contrast sensitivity, such as luminance, adaptation, and visual field size. This means that determination of any visual metric must specify all these conditions.

18.6.3 MORE VISUAL METRICS

18.6.3.1 Area between the MFT and the contrast threshold

The intersection of the MTF and the NTF gives the highest spatial frequency for which the MTF is above the neural threshold. We call this the *crossover frequency* (f_{cross}) and is exactly the cutoff frequency of the CSF ($f_{cutoffCSF}$) defined before.

The area between the MFT and the neural contrast threshold function (or NTF) between zero and the crossover frequency is mathematically defined as

$$Area(MTF - NTF) = \int_{0}^{f_{cross}} [MTF(f) - NTF(f)]df \quad (18.61)$$

The rationale behind this metric is that it summarizes the extent of signal MTF in excess of the threshold requirement of the visual system over all usable frequencies.

When computing this metric, phase-reversed segments of the MTF curve count as positive area, which makes that spurious resolution be counted as beneficial when predicting visual performance for the task of contrast detection. This metric assumes that the area is homogeneous in image quality, that is,

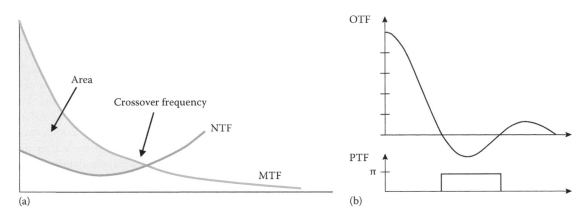

Figure 18.11 (a) Area between the modulation transfer function (MTF) and the neural contrast threshold. The rationale behind this metric is that it computes only the signal of the MTF over the usable frequencies. The crossover frequency is mathematically the cutoff frequency of the contrast sensitivity function. (b) Some visual metrics are based on the optical transfer function and are intended to discard phase-reversed contrasts.

that the excess of MTF over NTF is uniformly important for all spatial frequencies above the threshold requirement. Some authors suggested that the MTF be weighted by the spatial frequency power spectrum of the image to be seen to determine an effective area. On the other hand, it has been shown that it is important to have adequate signal above that minimally required for detection, but additional increases in this excess of MTF over NTF are of less value in many real-world tasks. This idea leads to define similar metrics by computing the MTF area only in a region near the threshold (Figure 18.11).

18.6.3.2 Subjective quality factor

Granger proposed a single metric that would correlate with perceived image quality (Granger and Cupery 1972). In his definition, *subjective quality factor* (SQF) defines a pass-band filter that is constructed by averaging the MTF between the peak frequency of the CSF and four times that frequency. More exactly, the SQF is the area of the MTF from $f1 = 0.5$ cycles/mm (1.7 cycle/degree) to $f2 = 2$ cycles/mm (6.7 cycles/degree):

$$SQF = \int_{f1}^{f2} MTF(f)d(\log f) \qquad (18.62)$$

This factor and the area between the MFT and the NTF are concepts originally developed for photographic systems, generalized to electro-optical system applications and image quality tasks.

18.6.3.3 Other metrics based on area

The *area between the OTF and the contrast threshold* is (Thibos et al. 2004)

$$Area(OTF - NTF) = \int_{0}^{fcross} [OTF(f) - NTF(f)]df \qquad (18.63)$$

where $OTF(f)$ is the radially averaged OTF, which is real valued because imaginary component vanishes after integration. The difference with the previous metric is that phase-reversed segments of the curve do not contribute to the area. Thus, this

metric would be appropriate for tasks in which phase-reversed contrasts (spurious resolution) actively interfere with performance.

The *area under the CSF* is calculated as

$$Area(CSF) = \int_{0}^{fcutoff} CSF(f)df = \int_{0}^{fcutoff} MTF(f) \cdot NCSF(f)df \qquad (18.64)$$

This is the same than the area under the MTF weighted by the neural CSF.

A metric that is intended to quantify the visually significant shifts in the image is the *area under the neurally weighted OTF divided by the area under the neurally weighted MTF:*

$$\frac{Area(OTF)_{NCSF}}{Area(CSF)} = \frac{\displaystyle\int_{0}^{fcutoff} OTF(f) \cdot NCSF(f)df}{\displaystyle\int_{0}^{fcutoff} MTF(f) \cdot NCSF(f)df}$$

$$= \frac{\displaystyle\int_{0}^{fcutoff} OTF(f) \cdot NCSF(f)df}{Area(CSF)} \qquad (18.65)$$

The *visual Strehl ratio* (VSR) is similar to the OTF method of computing the SR, except that the OTF is weighted by the NCSF:

$$VSR = \frac{\displaystyle\int_{-\infty}^{\infty}\int_{-\infty}^{\infty} OTF(u,v) \cdot NCSF(u,v)dudv}{\displaystyle\int_{-\infty}^{\infty}\int_{-\infty}^{\infty} OTF_{d-l}(u,v) \cdot NCSF(u,v)du\,dv} \qquad (18.66)$$

18.6.3.4 Neural sharpness

Neural sharpness (NS) was introduced as a way to capture the effectiveness of a PSF for stimulating the neural portion of the visual system (Williams et al. 2003). It is the maximum value of the convolution of the eye's PSF and a spatial sensitivity function with a Gaussian profile that represents the neural visual system:

$$NS = \max(PSF * g) \qquad (18.67)$$

where $g(x,y)$ is a bivariate Gaussian weighting function with a standard deviation of 1 minute of arc, which effectively ignores light outside of the central 4 minutes of arc of the PSF.

18.6.4 POLYCHROMATIC METRICS

The WA for each wavelength is treated separately because lights of different wavelengths are mutually incoherent. For this reason, metrics of wavefront quality do not generalize easily to the case of polychromatic light. However, there have been several attempts to characterize the image quality of the human eye including the longitudinal and the transverse chromatic aberrations (Font et al. 1994; Marcos et al. 1999; Ravikumar et al. 2008; Van Meeteren 1974).

The *polychromatic point spread function* may be computed as the superposition of the monochromatic PSFs for each wavelength defocused by axial chromatic aberration and shifted by lateral chromatic aberration:

$$PSF_{poly}(x, y) = \sum_{\lambda} PSF_{\lambda}(x, y) \qquad (18.68)$$

A different approach is to incorporate the effect of the human spectral sensitivity and calculate the polychromatic PSF as the weighted sum:

$$PSF_{poly} = \int_{\lambda 1}^{\lambda 2} PSF(\lambda) \cdot V(\lambda) d\lambda \qquad (18.69)$$

where $V(\lambda)$ is the luminous efficiency as function of wavelength (Thibos et al. 2004).

These definitions may be substituted in any of the equations given above to produce new polychromatic metrics of image quality. For example, the *polychromatic MTF* may be computed as the modulus of the FT of the polychromatic PSF. Also, simulated images in white light are easily derived from their monochromatic counterparts.

18.7 SUMMARY AND EXAMPLES OF APPLICATIONS

A problem of considering the aberration coefficients or the statistical parameters of the WA is that one does not know a priori what will be the appearance of the image formed by the system. As we have already noted, there is not a trivial relationship between optical quality and image quality. For example, balancing aberrations does not always lead to improved images. Also it occurs that some aberration terms may degrade the image more than others even if both have the same magnitude. However, optical metrics have proven to be useful in many applications in vision. For example, the analysis at the optical level of the astigmatism change or the coma induced by an incision of cataract surgery, or by refractive surgery, allows gaining insight

for improving these treatments that act on the optical surfaces of the eye. Some of the numerous works that have studied optical quality of the human eye in terms of the aberration coefficients or the RMS of the WA are, for instance, the relative contribution of the spherical aberration of cornea and lens (Artal et al. 2001), optical aberrations of the human cornea as a function of age (Guirao et al. 2000), spherical aberration versus accommodation (Cheng et al. 2004a; He et al. 2000), the limits to a perfect ideal customized wavefront correction due to the change of aberrations during accommodation (Artal et al. 2002), optical quality of intraocular lenses for pseudophakic eyes in terms of its spherical aberration (Piers et al. 2004), comparative study of changes in coma after cataract surgery (Guirao et al. 2004), the distribution of the eye's aberrations in the normal population (Porter et al. 2001), and a "virtual surgery" approach designed to predict optical performance in pseudophakic eyes (Tabernero et al. 2006).

Metrics defined on the image plane or in the frequency domain are intended to quantify optical performance regardless of optical quality. Some applications to vision of metrics that are based on the PSF are retinal image quality in patients implanted with intraocular lenses in terms of the PSF and the SR (Guirao et al. 2002a), off-axis aberrations estimated from double-pass measurements of the PSF (Guirao and Artal 1999), the HWHM of the double-pass PSF in subjects after correction of the ocular aberrations with adaptive optics (Logean et al. 2008), depth from focus based on entropy (Bove 1993), EE to investigate the optimum amount of spherical aberration for design of intraocular lenses (Wang and Koch 2007), etc. Also, entropy, SR, and intensity variance of the PSF have been used for predicting subjective image quality (Chen et al. 2005) and refractive errors from WA data (Guirao and Williams 2003).

Examples of metrics based on grating objects and the MTF can be found in the following works: the area under the MTF between 0 and 60 cycles/degree has been applied to predict subjective image quality (Chen et al. 2005) and refractive errors (Guirao and Williams 2003), a method has been proposed to determine intraocular lens power by maximizing the area under the MTF (Canovas and Artal 2011), the visual benefit of correcting higher-order aberrations of the eye was evaluated from the MTF at 16 and 32 cycles/degree (Guirao et al. 2002b) and from the HR at 16 cycles/degree (Williams et al. 2000), also the MTF at 16 cycles/degree was used to evaluate the "effective correction" of a laser beam in laser refractive surgery (Guirao et al. 2003), the pseudo SR computed from MTF areas (SR_{MTF}) was applied for quantifying optical performance of the aging eye (Guirao et al. 1999), and metrics based on the volume under the MTF have been used in studies of chromatic aberration (Marcos et al. 1999) and visual instrumentation (Mouroulis 1999).

The simulation of retinal images by computational convolution is a common method for estimating image quality. For example, the convolved images of the letter "E" based on the measured aberrations have been used to simulate vision in keratoconic eyes (Sabesan and Yoon 2009), in LASIK patients and vision through adaptive optics (Williams et al. 2000), or the peripheral image quality (Jaeken et al. 2013).

Objective image quality metrics for evaluating processed images are preferred to subjective evaluation, which is slow and inconvenient for practical usage. The most common computable objective measures of image quality are the mean square error and the PSNR. These distance metrics are often used for quality evaluation of medical images and retinal images (Nirmala et al. 2014). However, their predictions often do not agree well with the human visual perception. Image quality assessment can be improved by incorporating some models of human visual system. The SSIM index has been proposed, for example, to evaluate the subjective perception of quality of natural images (Wang et al. 2004). The MI was applied to registration of retinal images (Zhu 2007).

Faithful determination of visual performance depends on subjective or psychophysical measurements because of the visual system processing. Visual metrics such as the area under the CSF, or the polychromatic PSF and MTF, have been used, for example, to evaluate the effect of rotation and translation on the expected benefit of correcting the ocular aberrations (Guirao et al. 2001). The VSR accounted for visual acuity (Marsack et al. 2004) and predicted subjective refraction (Thibos et al. 2004). It has been measured the ability of the NS and the VSR to predict high- and low-contrast acuity (Applegate et al. 2006) and the impact of aberrations on subjective image quality by means of the NS (Chen et al. 2005).

The relationship between objective and subjective visual performance is being increasingly investigated. Just for mentioning a few works: the application of optical quality metrics to predict subjective quality of vision after laser in situ keratomileusis (Bühren et al. 2009), the SR of the PSF generated at rotated versions of aberrations support the hypothesis that the neural visual system is adapted to the eye's aberrations (Artal et al. 2004), the effect on visual performance of the interactions between aberrations (Applegate et al. 2003), the impact of higher-order aberrations on subjective best focus (Cheng et al. 2004b), estimation of visual quality from wavefront aberrations (Cheng et al. 2003), the impact of positive coupling of the eye's trefoil and coma in retinal image quality and visual acuity (Villegas et al. 2012), the effect of chromatic aberration on visual acuity (Campbell and Gubisch 1967) and visual performance (Thibos et al. 1991), etc.

REFERENCES

Applegate, R. A., Marsack, J. D., and Thibos, L. N., Metrics of retinal image quality predict visual performance in eyes with 20/17 or better visual acuity, *Opt. Vis. Sci.* 83 (2006): 635–640.

Applegate, R. A., Marsack, J. D., Ramos, R., and Sarver, E. J., Interactions between aberrations can improve or reduce visual performance, *J. Cataract Refract. Surg.* 29 (2003): 1487–1495.

Artal, P., Optics of the eye and its impact in vision: A tutorial, *Adv. Opt. Photon.* 6 (2014): 340–367.

Artal, P., Understanding aberrations by using double-pass techniques, *J. Refract. Surg.* 16 (2000): S560–S562.

Artal, P., Chen, L., Fernández, E. J. et al., Neural compensation for the eye's optical aberrations, *J. Vis.* 4 (2004): 281–287.

Artal, P., Fernández, E. J., and Manzanera, S., Are optical aberrations during accommodation a significant problem for refractive surgery? *J. Refract. Surg.* 18 (2002): S563–S566.

Artal, P., Guirao, A., Berrio, E., and Williams, D. R., Compensation of corneal aberrations by the internal optics in the human eye, *J. Vis.* 1 (2001): 1–8.

Barakat, R., Some entropic aspects of optical diffraction imagery, *Opt. Commun.* 156 (1998): 235–239.

Bareket, N., Second moment of the diffraction point spread function as an image quality criterion, *J. Opt. Soc. Am.* 69 (1979): 1311–1312.

Born, M. and Wolf, E., *Principles of Optics.* Cambridge, U.K.: Cambridge University Press, 1999.

Bove, V. M., Entropy-based depth from focus, *J. Opt. Soc. Am. A* 10 (1993): 561–566.

Bühren, J., Pesudovs, K., Martin, T., Strenger, A., Yoon, G., and Kohnen, T., Comparison of optical quality metrics to predict subjective quality of vision after laser in situ keratomileusis, *J. Cataract Refract. Surg.* 35 (2009): 846–855.

Campbell, F. W. and Green, D. G., Optical and retinal factors affecting visual resolution, *J. Physiol.* 181 (1965): 576–593.

Campbell, F. W. and Gubisch, R. W., The effect of chromatic aberration on visual acuity, *J. Physiol.* 192 (1967): 345–358.

Canovas, C. and Artal, P., Customized eye models for determining optimized intraocular lenses power, *Biomed. Opt. Express* 2 (2011): 1649–1662.

Charman, W. N. and Jennings, J. A. M., The optical quality of the monochromatic retinal image as a function of focus, *Br. J. Physiol. Opt.* 31 (1976): 119–134.

Chen, L., Singer, B., Guirao, A., Porter, J., and Williams, D. R., Image metrics for predicting subjective image quality, *Optom. Vis. Sci.* 82 (2005): 358–369.

Cheng, H., Barnett, J. K., Vilupuru, A. S. et al., A population study on changes in wave aberrations with accommodation, *J. Vis.* 4 (2004a): 272–280.

Cheng, X., Bradley, A., and Thibos, L. N., Impact of higher order aberration on subjective best focus, *J. Vis.* 4 (2004b): 310–321.

Cheng, X., Thibos, L. N., and Bradley, A., Estimating visual quality from wavefront aberration measurements, *J. Refract. Surg.* 19 (2003): 579–584.

CIE (Commission Internationale de l'Éclairage). *ILV: International Lighting Vocabulary.* Vienna, Austria: CIE S 017/E, 2011.

Corbin, J. A., Klein, S., and van de Pol, C., Measuring effects of refractive surgery on corneas using Taylor series polynomials, *Proc. SPIE,* 3591 (1999): 46–52.

Driggers, R. G., *Encyclopedia of Optical Engineering.* New York: Marcel Dekker, 2003.

Duane, T. D., Tasman, W., and Jaeger, E. A., *Duane's Foundations of Clinical Ophthalmology.* Philadelphia, PA: Lippincott, Williams & Wilkins, 1998.

Fawcett, T., An introduction to ROC analysis, *Pattern Recognit. Lett.* 27 (2006): 861–874.

Fienup, J. R. and Miller, J. J., Aberration correction by maximizing generalized sharpness metrics, *J. Opt. Soc. Am. A* 20 (2003): 609–620.

Font, C., Escalera, J. C., and Yzuel, M. J., Polychromatic point spread function: Calculation accuracy, *J. Modern Opt.* 41 (1994): 1401–1413.

Gonzalez, R. C., *Digital Image Processing.* Upper Saddle River, NJ: Pearson Prentice Hall, 2008.

Goodman, J. W., *Introduction to Fourier Optics.* New York: McGraw-Hill, 1996.

Granger, E. M. and Cupery, K. N., An optical merit function (SQF) which correlates with subjective image judgements, *Photogr. Sci. Eng.* 16 (1972): 221–230.

Guirao, A. and Artal, P., Off-axis monochromatic aberrations estimated from double pass measurements in the human eye, *Vis. Res.* 39 (1999): 207–217.

Guirao, A., González, C., Redondo, M., Geraghty, E., Norrby, S., and Artal, P., Average optical performance of the human eye as a function of age in a normal population, *Invest. Ophthalmol. Vis. Sci.* 40 (1999): 203–213.

Guirao, A., Porter, J., Williams, D. R., and Cox, I. G., Calculated impact of higher-order monochromatic aberrations on retinal image quality in a population of human eyes, *J. Opt. Soc. Am. A* 19 (2002b): 1–9.

Guirao, A., Redondo, M., and Artal, P., Optical aberrations of the human cornea as a function of age, *J. Opt. Soc. Am. A* 17 (2000): 1697–1702.

Guirao, A., Redondo, M., Geraghty, E., Piers, P., Norrby, S., and Artal, P., Corneal optical aberrations and retinal image quality in patients in whom monofocal intraocular lenses were implanted, *Arch. Ophthalmol.* 120 (2002a): 1143–1151.

Guirao, A., Tejedor, J., Artal, P., Corneal aberrations before and after small-incision cataract surgery, *Invest. Ophthalmol. Vis. Sci.* 45 (2004): 4312–4319.

Guirao, A. and Williams, D. R., A method to predict refractive errors from wave aberration data, *Optom. Vis. Sci.* 80 (2003): 36–42.

Guirao, A., Williams, D. R., and Cox, I. G., Effect of rotation and translation on the expected benefit of an ideal method to correct the eye's higher-order aberrations, *J. Opt. Soc. Am. A* 18 (2001): 1003–1015.

Guirao, A., Williams, D. R., and MacRae, S. M., Effect of beam size on the expected benefit of customized laser refractive surgery, *J. Refract. Surg.* 19 (2003): 15–23.

He, J. C., Burns, S. A., and Marcos, S., Monochromatic aberrations in the accommodated human eye, *Vis. Res.* 40 (2000): 41–48.

Helmholtz, H. V., *Popular Lectures on Scientific Subjects*. New York: Appleton, 1885.

Hopkins, H. H., The use of diffraction-based criteria of image quality in automatic optical design, *Opt. Acta* 13 (1966): 343–369.

Howland, H. C. and Howland, B., A subjective method for the measurement of monochromatic aberrations of the eye, *J. Opt. Soc. Am. A* 67 (1977): 1508–1518.

Iglesias, I., Lopez-Gil, N., and Artal, P., Reconstruction of the ocular PSF from a pair of double pass retinal images, *J. Opt. Soc. Am. A* 15 (1998): 326–339.

Jaeken, B., Mirabet, S., Marín, J. M., and Artal, P., Comparison of the optical image quality in the periphery of Phakic and Pseudophakic eyes, *Invest. Ophthalmol. Vis. Sci.* (2013) 54: 3594–3599.

Lisson, J. B., and Mounts, D. I., Estimation of imaging performance using local optical quality factor metrics, *Opt. Eng.* 31 (1992): 1038–1044.

Logean, E., Dalimier, E., and Dainty, C., Measured double-pass intensity point-spread function after adaptive optics correction of ocular aberrations, *Opt. Express* 16 (2008): 17348–17357.

Mahajan, V. N., *Aberration Theory Made Simple*. Bellingham, WA: SPIE Press, 1991.

Mahajan, V. N., *Optical Imaging and Aberrations*. Bellingham, WA: SPIE Press, 1998.

Mahajan, V. N., Strehl ratio for primary aberrations: Some analytical results for circular and annular pupils, *J. Opt. Soc. Am.* 72 (1982): 1258–1266.

Malacara, D., *Optical Shop Testing*. New York: John Wiley & Sons Inc., 1992.

Marcos, S., Burns, S. A., Moreno-Barriuso, E., and Navarro, R., A new approach to the study of ocular chromatic aberrations, *Vis. Res.* 39 (1999): 4309–4323.

Marsack, J. D., Thibos, L. N., and Applegate, R. A., Metrics of optical quality derived from wave aberrations predict visual performance, *J. Vis.* 4 (2004): 322–328.

Millodot, M., *Dictionary of Optometry and Visual Science*. Oxford, U.K.: Butterworth-Heinemann, 2009.

Mouroulis, P., *Visual Instrumentation: Optical Design & Engineering Principles*. New York: McGraw-Hill, 1999.

Nirmala, S. R., Dandapat, S., and Bora, P. K., Quality measures for retinal images, in *Ophthalmological Imaging and Applications*, E. Y. K. Ng et al., eds., Chapter 4. Boca Raton, FL: CRC Press, 2014.

Papoulis, A., *Probability, Random Variables, and Stochastic Processes*. New York: McGraw-Hill, 1991.

Pelli, D. G. and Bex, P., Measuring contrast sensitivity, *Vis. Res.* 90 (2013): 10–14.

Piers, P. A., Fernández, E. J., Manzanera, S., Norrby, S., and Artal, P., Adaptive optics simulation of intraocular lenses with modified spherical aberration, *Invest. Ophthalmol. Vis. Sci.* 45 (2004): 4601–4610.

Porter, J., Guirao, A., Cox, I. G., and Williams, D. R., Monochromatic aberrations of the human eye in a large population, *J. Opt. Soc. Am. A* 18 (2001): 1793–1803.

Ravikumar, S., Thibos, L. N., and Bradley, A., Calculation of retinal image quality for polychromatic light, *J. Opt. Soc. Am. A* 25 (2008): 2395–2407.

Sabesan, R. and Yoon, G., Visual performance after correcting higher order aberrations in keratoconic eyes, *J. Vis.* 9 (2009): 1–10.

Shannon, R. R., *The Art and Science of Optical Design*. Cambridge, U.K.: Cambridge University Press, 1997.

Siegman, A. E., How to (maybe) measure laser beam quality, in *DPSS (Diode Pumped Solid State) Lasers: Applications and Issues*, M. Dowley, ed., Vol. 17 of OSA Trends in Optics and Photonics, Washington D.C.: Optical Society of America 1998, paper MQ1.

Smith, W. J., *Modern Optical Engineering*. New York: McGraw-Hill Professional, 2000.

Srisailam, A., Dharmaiah, V., Ramanamurthy, M. V., and Mondal, P. K., Encircled energy factor as a point-image quality-assessment parameter, *Adv. Appl. Sci. Res.* 2 (2011): 145–154.

Tabernero, J., Piers, P., Benito, A., Redondo, M., and Artal, P., Predicting the optical performance of eyes implanted with IOLs to correct spherical aberration, *Invest. Ophthal. Visual Sci.* 47 (2006): 4651–4658.

Thibos, L. N., Bradley, A., and Zhang, X., Effect of ocular chromatic aberration on monocular visual performance, *Optom. Vis. Sci.* 68 (1991): 599–607.

Thibos, L. N., Hong, X., Bradley, A., and Applegate, R. A., Accuracy and precision of objective refraction from wavefront aberrations, *J. Vis.* 4 (2004): 329–351.

Van Meeteren, A., Calculations of the optical modulation transfer function of the human eye for white light, *Opt. Acta* 21 (1974): 395–412.

Villegas, E. A., Alcón, E., and Artal, P., Impact of positive coupling of the eye's trefoil and coma in retinal image quality and visual acuity, *J. Opt. Soc. Am. A* 29 (2012): 1667–1672.

Wandell, B. A., *Foundations of Vision*. Sunderland, MA: Sinauer Associates, 1995.

Wang, J. Y. and Silva, D. E., Wave-front interpretation with Zernike polynomials, *Appl. Opt.* 19 (1980): 1510–1519.

Wang, L. and Koch, D. D., Custom optimization of intraocular lens asphericity, *J. Cataract Refract. Surg.* 33 (2007): 1713–1720.

Wang, Z., Bovik, A. C., Sheikh, H. R., and Simoncelli, E. P., Image quality assessment: From error visibility to structural similarity, *IEEE Trans. Image Process.* 13 (2004): 600–612.

Westheimer, G. and Campbell, F. W., Light distribution in the image formed by the living human eye, *J. Opt. Soc. Am.* 52 (1962): 1040–1045.

Williams, C. S. and Becklund, O. A., *Introduction to the Optical Transfer Function*. Bellingham, WA: SPIE Press, 2002.

Williams, D. R., Applegate, R. A., and Thibos, L. N., Metrics to predict the subjective impact of the eye's wave aberration, in *Wavefront Customized Visual Correction: The Quest for Supervision II*, R.R. Krueger. R.A. Applegate, and S.M. MacRae, eds. Thorofare, NJ: Slack Inc., 2003, pp. 77–84.

Williams, D. R., Yoon, G. Y., Porter, J., Guirao, A., Hofer, H., and Cox, I. G., Visual benefit of correcting higher order aberrations of the eye, *J. Refract. Surg.* 16 (2000): 554–559.

Williams, T., *The Optical Transfer Function of Imaging Systems*. Bristol, U.K.: CRC Press, 1998.

Wyant, J. C. and Creath, K., Basic wavefront aberration theory for optical metrology, in *Applied Optics and Optical Engineering (Vol. XI)*, R.R. Shannon and J.C. Wyant, eds. Boston, MA: Academic Press, 1992, pp. 2–53.

Zhang, L., Zhang, L., Mou, X., and Zhang, D., A comprehensive evaluation of full reference image quality assessment algorithms, in *Proceedings of IEEE International Conference on Image Processing*, Orlando, FL, 2012, pp. 1477–1480.

Zhu, Y. M., Mutual information-based registration of temporal and stereo retinal images using constrained optimization, *Comput. Methods Prog. Biomed.* 86 (2007): 210–215.

19

Predicting visual acuity

Rafael Navarro

Contents

19.1 INTRODUCTION

Visual acuity (VA) is one of the most common clinical tests, since it is an easy, affordable, and reasonably fast method to find visual deficiencies due to either optical (refractive errors or aberrations) or neural (retinal or cortical diseases) causes. Both optical and neural factors limit visual performance (Banks et al. 1987). In particular, optical aberrations and refractive errors are present even in healthy young eyes, and hence, under photopic conditions, VA is used to be limited by optical factors in a majority of cases. This means that VA is the main criterion used in clinical subjective refraction, because VA is highly affected by optical defects. The new techniques for manipulating the optics of the eye have produced an increasing interest in the possibility of predicting VA under different levels of optical aberrations. On the one hand, the dynamic compensation of aberrations, using deformable mirrors (Liang and Williams 1997) and spatial light modulators (Vargas-Martín et al. 1998), or static compensation, using phase plates (Navarro et al. 2000), enabled enhanced or supernormal vision (Liang et al. 1997). On the other hand, cataract and refractive surgery modify the optical system of the eye in different ways, but often the result is that aberrations increase after surgery (Moreno-Barriuso et al. 2001). In both cases of decrease or increase of aberrations, it is important to be able to predict the impact of that change on visual performance, including VA.

From a historical perspective, Smith (1991) proposed a linear relationship between the minimum angle of resolution (predicted VA) with the refractive error and the pupil diameter. Even though it is not valid for low refractive errors, it provides a simple and powerful metric. Since then, a variety of metrics of either optical quality (wavefront), image quality, or visual quality were proposed. Initially, these metrics were applied to the problem of predicting the best refractive correction or clinical refraction (Guirao and Williams 2003; Cheng et al. 2004; Thibos et al. 2004; Navarro 2010; Martin et al. 2011). Closely related to the prediction of the best correction, these types of metrics were also used for predicting the accommodative response (Buchren and Collins 2006; López-Gil et al. 2009; Tarrant et al. 2010) or to the depth of focus (Yi et al. 2010) or the subjective visual quality (Chen et al. 2005; Legras and Rouger 2008; Bühren et al. 2009).

VA is a particular measure of visual quality, and hence the same metrics were also proposed for predicting VA (Marsack et al. 2004; Applegate et al. 2006; Cheng et al. 2010; Ravikumar et al. 2012, 2013; Young et al. 2013). The main advantages of these metrics are their simplicity and easy computation. The main limitations are, on the one hand, that in most cases they do not offer a prediction of the absolute VA and require a previous empirical calibration through linear regressions against experimental VA. On the other hand, even though they can provide good predictions of average trends, they often fail to predict VA of individual subjects.

The specific modeling of VA (Nestares et al. 2003) tries to simulate the VA exam in a realistic way. This includes the task (identification of optotypes), the main optical and neural stages limiting visual performance, and a decision-making stage through some types of template matching. The most common implementation of this crucial stage of guessing the input optotypes is based on the ideal observer paradigm (Geisler 1984). This type of simulation is opposite to the metrics mentioned earlier, in the sense that the model can be customized not only for the subject optical or neural stages but also for the type of task, different optotypes (Watson and Ahumada 2012), etc. Simplified versions of this type of VA model were also proposed (Dalimier and Dainty 2008; Watson and Ahumada 2008).

19.2 OPTICAL QUALITY AND VISUAL ACUITY

This section is devoted to briefly overview studies aimed to predicting the impact of optical quality on VA. First, the subject of optical, image, and visual quality metrics was extensively treated in Chapter 18. In natural viewing conditions, the eye's optical blur basically depends on the amount of aberrations. Note that VA is usually determined under fully photopic illumination, which means maximum neural response, and this will be assumed throughout most of this chapter. Figure 19.1 represents the simulated retinal image computed from the wavefront of a post-LASIK patient with high values of high-order aberrations, mainly coma and spherical aberration. Here we can observe the three main ways in which optical aberrations affect the image quality: (1) the contrast decreases with spatial frequency (increases with scale), (2) the resolution decreases due to a low-pass effect, and (3) phase distortions, which cause object deformations and even multiple images (monocular diplopia). In particular, contrast reversals may yield spurious resolution artifacts. A number of experimental studies assessed the impact of the different monochromatic aberrations that affect retinal image quality and VA (Liang and Williams 1997; Guirao et al. 2002; Applegate et al. 2003a,b; McLellan et al. 2006; Villegas et al. 2012) as well as chromatic aberrations (Marcos et al. 1999; Ravikumar et al. 2008). In addition, adaptive optics technology enabled studies aimed to predicting the visual benefit of correcting aberrations (Liang et al. 1997; Fernández et al. 2002; Yoon and Williams 2002; Piers et al. 2007; Rocha et al. 2007; Dalimier et al. 2008; Marcos et al. 2008).

Another type of studies evaluated the performance of different single-number metrics in predicting VA (Cheng et al. 2004; Marsack et al. 2004; Applegate et al. 2006; Cheng et al. 2010; Rouger et al. 2010; Ravikumar et al. 2012, 2013; Young et al. 2013). The overall conclusions of these studies are that image quality metrics, especially those including neural response, provide good predictions of the population average visual quality, but the prediction accuracy for individual subjects is modest. It was pointed out that neural effects such as blur adaptation may limit VA predictions relying only on the wavefront aberrations (Artal et al. 2004; Villegas et al. 2008). Examples of the most significant optical and image quality metrics are as follows (Thibos et al. 2004): the inverse of the root mean square (RMS) wavefront error (RMS_w^{-1}); the standard deviation of the intensity values of the point spread function (PSF) (STD); the standard deviation of the PSF spatial distribution (SM); the neural sharpness, defined as the Strehl ratio of the PSF weighted by a spatial neural sensitivity function (NS); the visual Strehl ratio obtained as the integral of the contrast sensitivity function computed as the product of the modulation transfer function (MTF) and the neural transfer function (NTF) (VSMTF); and the spatial frequency cutoff computed as the intersection of the radially averaged MTF and the inverse of the NTF (SFcMTF). Finally, the Smith's VA prediction, based on a simple formula by $VA_{Smith} = 1/(0.83 \times \Phi \times E)$ where Φ is pupil diameter (mm) and E is refractive error (D), provides good performance.

19.3 VISUAL ACUITY AND PATTERN RECOGNITION

Contrary to grating acuity, the optotypes used in standard clinical VA are complex stimulus, and the associated task is a decision making under several alternatives forced choice paradigm. Typically, the number of alternatives ranges from four (such as in Landolt C optotypes) to the number of letters in the alphabet (26 in English) for Snellen VA. In the last case, the task is letter identification quite similar to optical character recognition (OCR) artificial systems implemented in document scanners and other applications of artificial vision, such as automatic license plate recognition. The main difference is that in the VA test, the most important stage is when we are close to threshold. In that case, the size of the optotypes is similar to that of the optical blur patch, which means that the optotype image is strongly degraded (see bottom line of Figure 19.1). Interestingly, some of the algorithms proposed for robust pattern recognition in degraded images are inspired in human vision. In particular, image decomposition into frequency bands, similar to visual

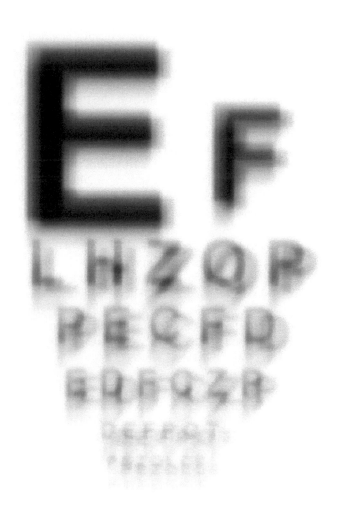

Figure 19.1 Simulated retinal image for a post-LASIK patient.

channels, was applied to invariant pattern recognition against defocus (Vargas et al. 2000). Subband decomposition was shown to be highly powerful in signal denoising (Cristóbal and Navarro 1994; Donoho 1995; Walker and Chen 2000). In summary, image decomposition into frequency bands seems a powerful strategy to avoid the two main causes of degradation in optical images: blur and noise.

This hypothesis was confirmed by implementing a biologically inspired Bayesian pattern recognition method for strongly degraded images (Navarro et al. 2004). In the Bayesian method, the choice among the set of possible alternatives is determined by that alternative providing the maximum *a posteriori* probability (MAP). The Bayesian method can incorporate previous knowledge in the form of an *a priori* probability, but here it is assumed that there is no *a priori* information. The probability is computed from an observation model. For the *i*th subband, the signal is

$$i_i(\mathbf{x}) = (o(\mathbf{x}) * h(\mathbf{x})) * g_i(\mathbf{x}) + n_i(\mathbf{x}). \quad (19.1)$$

This expression means that the image subband i_i (note that $\mathbf{x}$ is a vector of coordinates $\mathbf{x} = (x,y)$) is the spatial convolution with the *i*th filter g_i that selects the specific subband in the optical image plus some noise n_i. This optical image is the convolution of the object o with the PSF h. The noise n is assumed to be additive and independent for each subband. In this case, Gabor filters were used to obtain a multiscale (4 scales) and multiorientation (4 orientations) image representation (Nestares et al. 1998). The second stage in the procedure is to compute the probability of the input signal to correspond to every possible alternative (such as letters in the alphabet) object represented by its corresponding template. To this end, here we apply the main and strong simplifying assumption, namely, that the optical transfer function (OTF) is constant within each subband. In that case, the Fourier transform of h within the *i*th subband is a complex number. Its modulus h_i attenuates the contrast of the subband, and its phase ϕ_i produces a spatial shift $\mathbf{u}_i$ of the entire subband on the image plane. Now, the approximated observation mode is

$$i_i(\mathbf{x}) \approx h_i \delta(\mathbf{u}_i) * o_i(\mathbf{x}) + n_i(\mathbf{x}) \approx h_i o_i(\mathbf{x} - \mathbf{u}_i) + n_i(\mathbf{x}), \quad (19.2)$$

where

$\delta(\mathbf{u}_i)$ is a Dirac delta function placed at point $\mathbf{u}_i$
$o_i(\mathbf{x}) = o(\mathbf{x}) * g_i(\mathbf{x})$ is the *i*th subband of the object
h_i is the contrast modulation of the subband

This approximation is reasonable for moderate optical blur but fails for strong blur, and hence the failure of this approximation will be one of the main limitations of this method. In a similar way, the subband decomposition increases the robustness against noise, but the Bayesian method may fail when the noise is greater than the signal.

One key benefit of this approximation is that the subbands (4 × 4 Gabor filters in this particular implementation) can be realized as a discrete coarse sampling of the frequency domain of both the object and the OTF. This enables to estimate the

16 (4 × 4) complex samples of the OTF (one per subband), which allows us to have a rough estimation of the optical blur, as explained in Appendix A by Nestares et al. (2003). This provides robustness against blur of the Bayesian method, since having an estimation of the blur itself is somehow equivalent to perform an image restoration through a blind deconvolution. In fact, the pattern recognition procedure has two stages. First one estimates the optical degradation parameters (sampled OTF) h_i (contrast modulation) and $\mathbf{u}_i$ (phase shift) for each subband *i*. The values of these parameters can be estimated through minimization of the error function computed for all templates j and subbands i:

$$E_i^j = \sum_{\mathbf{x}} (i_i(\mathbf{x}) - h_i o_i(\mathbf{x} - \mathbf{u}_i))^2. \quad (19.3)$$

This yields the least squares estimation of the average degradation for each subband $\{\hat{h}_i, \hat{\mathbf{u}}_i\}$ (see Nestares et al. (2003) for a detailed formulation). Using the approximation mentioned earlier (Equation 19.2), and the estimated optical blur $\{\hat{h}_i, \hat{\mathbf{u}}_i\}$, the Bayesian formulation provides the expression for the maximum value of the probability that the pattern o contained in the input image is the *i*th template from the collection of possible patterns (i.e., the letters in the alphabet in the case of OCR):

$$P_M = \max\{p(o = o^j, \{\hat{h}_i, \hat{\mathbf{u}}_i\}\{i_i\})\}$$

$$\propto \exp\left(\frac{1}{2\sigma^2} \sum_{i=1}^{I} \frac{\left[\sum_{\mathbf{x}} i_i(\mathbf{x}) o_i^j\left(\mathbf{x} - \hat{\mathbf{u}}_i^j\right)\right]^2}{\sum_{\mathbf{x}} \left[o_i^j\left(\mathbf{x}\right)\right]^2}\right). \quad (19.4)$$

This means that for a collection of J possible templates, and after applying a frequency band decomposition, with I subbands, we compute the J probabilities $p(o = o^j, \{\hat{h}_i, \hat{\mathbf{u}}_i\}\{i_i\})$ that the unknown object o contained in the input image i (represented by the set of subbands $\{i_i\}$) and a given estimate of the optical blur $\{\hat{h}_i, \hat{\mathbf{u}}_i\}$ is object j. Then we take the maximum P_M of the set of probabilities $P_M = \max\{pj\}$ that provides the Bayesian response or decision for the task of object identification.

The algorithm was implemented using a multiscale (4 scales) and multiorientation (4 orientations) Gabor representation (Nestares et al. 1998), which is a plausible (although highly schematic) model of the early linear stages of image representation in the visual cortex (Landy and Movshon 1991). Figure 19.2 shows examples of the results obtained by a realistic simulation of strongly degraded images by blur caused by atmospheric turbulence, plus important amounts of additive noise (Navarro et al. 2004). Here the task was the species identification from a collection of eight different birds of prey (hawks and eagles), all of them having similar morphology. Identification results were highly satisfactory (5 correct out of 6) even for the strong degradation. The fact that there is 1 identification failure suggests that the amount of degradation is not far from the threshold value to allow identification. This would be similar to the scenario that one may find in a VA test.

Objects

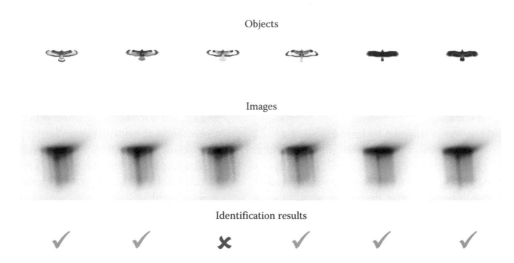

Images

Identification results

Figure 19.2 Results of object identification under high amounts of both blur and noise.

19.4 IDEAL OBSERVER MODEL OF VISUAL ACUITY

The pattern recognition method presented in the previous section can be used as the basis to develop a computational model of VA. This type of model is superior to simple metrics, as it allows us, on the one hand, implementing real tasks in an artificial vision application and, on the other hand, predicting VA of either a generic average model or custom models of individual subjects and specific optotypes (Watson and Ahumada 2012). The Bayesian formulation has been used before to develop ideal observer models (Geisler 1984, 1989; Banks et al. 1987). The ideal observer is able to use the available signal in an optimal way in decision-making tasks, such as forced choice between different alternatives, such as templates in object identification. The Bayesian approach means that the optimum response is the one maximizing the posterior

probability (MAP). In the optotype identification task, the Bayesian ideal observer always chooses the MAP alternative. The ideal observer model includes signal degradation both blur and noise (especially internal noise of the visual system) and the main stages of visual processing involved in the tasks. The ideal observer performs equal or better than real subjects, but it helps to explain important aspects of visual perception.

19.4.1 VISUAL ACUITY MODEL

The specific VA model explained here consists of a realistic simulation of a Snellen VA exam, where the optotypes are letters. The exam consists of applying the pattern recognition algorithm to a sequence of letters of different sizes. Usually, the letter scale is decreased progressively, and one counts the number of correct guesses for each scale. Figure 19.3 shows a block diagram of the main stages of the model. The degree of customization depends on the number

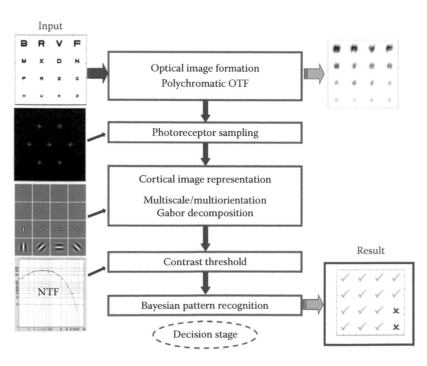

Figure 19.3 Block diagram showing the main stages of the VA model.

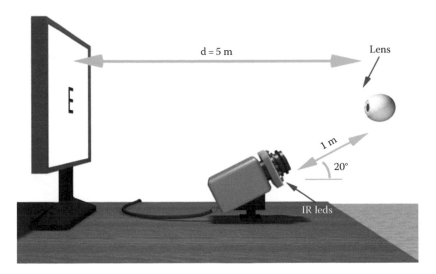

Figure 19.4 Setup for the simultaneous visual acuity tests and pupil diameter measurement.

of blocks that use generic or individual data. In this section only, the optical model is adapted to each eye (and pupil size).

The input image is processed by applying five blocks or moduli, which represent the most relevant stages of early visual processing for the particular case of VA:

1. The optical image formation is implemented as a polychromatic OTF linear filter, computed from the measured wavefront error for each individual eye.
2. The photoreceptor sampling grid is assumed to be hexagonal. A generic frequency of 120 cycles/deg was assumed. Its effect is to produce replicas of the OTF arranged along a hexagonal pattern with that frequency.
3. The cortical image representation linear stage was assumed to be a multiscale/multiorientation Gabor decomposition (Nestares et al. 1998). Generic Gabor frequency channels of one octave bandwidth; frequencies of 4, 8, 16, and 32 cycles/deg; and orientations of 0°, 45°, 90°, and 135° were used.
4. The contrast threshold given by the inverse of a generic NTF is related to the gain of the Gabor channels and the internal noise level. Fully photopic illumination level was assumed.
5. The Bayesian decision stage was applied as described in Section 19.3.

Therefore, this is a pattern recognition algorithm of degraded images where the set of alternative choices (templates) are the letters of the alphabet of different sizes according to the standard decimal VA scale. The output is a set of correct or wrong responses. One important feature of this type of model is that it provides absolute predictions without needing any parameter fit or calibration through linear regressions. Nevertheless, it is possible to adjust any of the generic neural stages (2, 3, and 4), that is, photoreceptor or Gabor channel frequencies or contrast threshold, to get further customization and hence better VA predictions in individual eyes.

19.4.2 EXPERIMENTAL VALIDATION

One of the most interesting properties of the model is that it can be customized not only to the specific subject but also to mimic experimental conditions of the VA test. A joint design of both experiment and simulation algorithm in such a way that the latter

can represent not only the overall experimental conditions but also the exact sequence of optotypes, or the pupil size measured during the VA exam, is possible. Figure 19.4 shows the experimental configuration for the validation study (Dalimier et al. 2009). The same optotypes, including fonts and scales, used for the simulation were displayed on a computer screen scaled for 5 m viewing distance. An infrared pupil meter took measurements at a few seconds rate during the VA exam, as pupil diameter is a critical parameter to estimate the OTF from the wavefront. The wavefront itself was previously measured (see Dalimier et al. 2009 for further details). The optotype presentation sequence was random, but it was stored to be later used for the simulation.

The average results for a group of 10 subjects are shown in Figure 19.5. The two curves represent experimental data (blue dots) and model predictions (black "X") of VA (decimal scale) as a function of positive defocus (so that it cannot be compensated by accommodation). The agreement is excellent for the whole range of 0 D to 2 D defocus studied, even though only the first block (optics) was customized. This result supports the biological plausibility of the model.

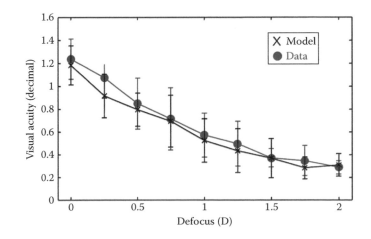

Figure 19.5 Model validation: Average results of experimental measurements (blue dots) and model predictions (black "X") of VA for different positive defocus.

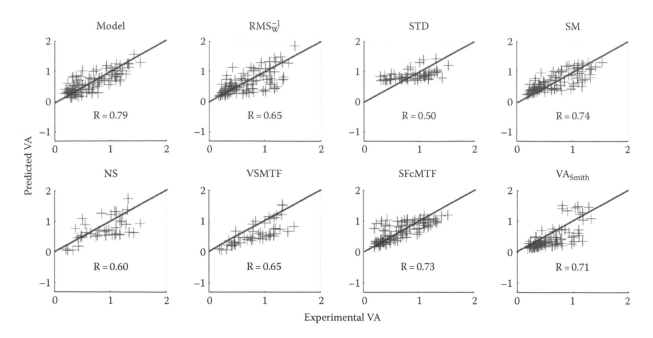

Figure 19.6 Correlations obtained between experimental data and predictions provided by different metrics and the present VA model (upper left plot).

The analysis of the whole data set is shown in Figure 19.6. Each panel shows a cloud of points comparing predicted versus experimental VA. Each cloud contains the complete data set for the different subjects and defocus conditions, for the VA model (upper left panel), and for the different metrics explained in Section 19.2. The red lines represent the ideal response (predicted = experimental) and R is the correlation coefficient. We can see that most metrics provide reasonable predictions after proper calibration through linear regressions (even the simplest Smith's formula). Nevertheless, the VA model gives the best performance (both highest correlation and closest to the ideal response) and what is more important without requiring any calibration or parameter optimization.

19.5 SIMPLIFIED VA MODELS

Other similar ideal observer template matching VA models were later proposed by different authors. They are simplified versions of the model mentioned earlier, which involve advantages and limitations. Here we overview two models.

19.5.1 WATSON AND AHUMADA MODEL

Watson and Ahumada (2008) proposed an extension of their model of contrast detection (Watson and Ahumada 2005) to include the Snellen VA task using "template matching" methods. The model includes two filtering stages: optical filtering by the OTF and neural filtering by the NTF to obtain a "neural image" (similar to blocks 1 and 4 in Figure 19.3). Then they add zero-mean Gaussian noise with standard deviation σ_n to that neural image. The model does not include photoreceptor sampling and visual frequency channel decomposition (blocks 2 and 3 in Figure 19.3). They compared different matching (or decision) rules, including the ideal observer paradigm, based on maximum probability, similar to the Bayesian model

described earlier. They found that other matching metrics such as minimum RMS (distance to the templates) provided similar results than those provided by the ideal observer approach. The main difference with the VA model of Section 19.4 is that this model does require calibration through 2 free parameters fit of the neural response for each subject. The initial generic NTF was customized by adjusting a frequency scale parameter ϕ, and the internal noise standard deviation σ_n was also adjusted to optimize model predictions for each subject. The frequency scale produces a shift of the NTF along the abscissa direction in a log–log plot, which is equivalent to adjust the high-frequency falloff. The noise level parameter σ_n would produce a sort of vertical shift (ordinate direction) of the neural sensitivity. The optimal values of both the noise level and the frequency scale parameter were found for each subject by minimizing the RMS difference between the model predictions and the experimental VA. This optimization process was implemented for each subject using multiple noise realizations. After this customized "calibration" by computing the optimal values of these two parameters for each subject, they were able to obtain a good correlation and low prediction errors with the measured VA using experimental data set of Cheng et al. (2004). After customization of both optics and neural response, they attained an RMS error of 0.056 logMAR and a correlation of 0.913 between model predictions and experimental logMAR VA for the complete data set of Cheng et al. (2004). More recently, they extended the model to a variety of optotypes (Watson and Ahumada 2012).

19.5.2 DALIMIER AND DAINTY MODEL

This is another ideal observer template matching model (Dalimier and Dainty 2008), similar to that of Watson and Ahumada (2008), but adapted for predicting the subject response for the task of orientation discrimination with Landolt C optotypes.

Thus, instead of letter identification, the model is specific for this four-alternative forced choice paradigm. The specific goal was to explain previous experimental results of VA when aberrations were corrected by adaptive optics (Dalimier et al. 2008). These results suggested a decrease of the benefit of correcting optical aberrations for low light levels. For this reason, here the emphasis is not on the effect of optical aberrations (the OTF is assumed to be diffraction limited after aberrations are corrected by an adaptive optics system) but on the impact of decreasing light level. For this reason, they included the differences in both pupil diameters and neural response, when the optotype illumination is decreased from photopic to low mesopic levels. This model is based on the simplified model of Watson and Ahumada, but it was further simplified, under the assumption that both optical and neural filters are known and generic (i.e., does not need customization). Since aberrations are corrected, the OTF is assumed to be diffraction limited and hence it is fully determined by the pupil diameter. The assumed NTF is standard and generic as well. However, this model does consider the changes of the NTF and OTF (through pupil diameter) with light level. This model also includes internal additive noise on the neural image (after optical and neural filtering) and selects the optotype template with the highest probability. The templates are filtered (by OTF and NTF) versions of the original optotypes. Even with this simplistic model, these authors were able to obtain a good prediction for the aberration correction benefit (or adaptive optics benefit) with a correlation of 0.83 between model predictions and experimental data, for the photopic light level. For the low mesopic condition, the correlation was lower but still consistent with the data.

19.6 CONCLUSIONS

Predicting VA is possible, even with simple metrics such as that proposed by Smith (1991), but the accuracy, reliability, and generality of these predictions depend on two main factors. On the one hand, to be realistic the model should incorporate the main stages of the visual process involved in the particular task of optotype identification, in the case of VA. In fact, only the most complete model was able to provide absolute predictions without any previous calibration. On the other hand, generic models provide good predictions for average responses, but predicting individual data would require customization not only of the optical image formation but also of the main stages of neural processing. Realistic customized models seem to do a good job for predicting VA.

REFERENCES

Applegate, R. A., C. Ballentine, H. Gross, E. J. Sarver, and C. A. Sarver, Visual acuity as a function of Zernike mode and level of root mean square error, *Optometry and Vision Science* 80 (2003a): 97–105.

Applegate, R. A., J. D. Marsack, R. Ramos, and E. J. Sarver, Interaction between aberrations to improve or reduce visual performance, *Journal of Cataract and Refractive Surgery* 29 (2003b): 1487–1495.

Applegate, R. A., J. D. Marsack, and L. N. Thibos, Metrics of retinal image quality predict visual performance in eyes with 20/17 or better visual acuity, *Optometry and Vision Science* 83 (2006): 635–640.

Artal, P., L. Chen, E. J. Fernandez, B. Singer, S. Manzanera, and D. R. Williams, Neural compensation for the eye's optical aberrations, *Journal of Vision* 4 (2004): 281–287.

Banks, M. S., W. S. Geisler, and P. J. Bennett, The physical limits of grating visibility, *Vision Research* 27 (1987): 1915–1924.

Buehren, J., K. Pesudovs, T. Martin, A. Strenger, G. Yoon, and T. Kohnen, Comparison of optical quality metrics to predict subjective quality of vision after laser in situ keratomileusis, *Journal of Cataract and Refractive Surgery* 35 (2009): 846–855.

Buehren, T. and M. J. Collins, Accommodation stimulus–response function and retinal image quality, *Vision Research* 46 (2006): 1633–1645.

Chen, L., B. Singer, A. Guirao, J. Porter, and D. R. Williams, Image metrics for predicting subjective image quality, *Optometry and Vision Science* 82 (2005): 358–369.

Cheng, X., A. Bradley, S. Ravikumar, and L. N. Thibos, The visual impact of Zernike and Seidel forms of monochromatic aberrations, *Optometry and Vision Science* 87 (2010): 300–312.

Cheng, X., A. Bradley, and L. N. Thibos, Predicting subjective judgment of best focus with objective image quality metrics, *Journal of Vision* 4 (2004): 310–321.

Cristóbal, G. and R. Navarro, Space and frequency variant image enhancement based on a Gabor representation, *Pattern Recognition Letters* 15 (1994): 273–277.

Dalimier, E. and C. Dainty, Use of a customized vision model to analyze the effects of higher-order ocular aberrations and neural filtering on contrast threshold performance, *Journal of the Optical Society of America A* 25 (2008): 2078–2087.

Dalimier, E., C. Dainty, and J. L. Barbur, Effects of higher-order aberrations on contrast acuity as a function of light level, *Journal of Modern Optics* 55 (2008): 791–803.

Dalimier, E., E. Pailos, R. Rivera, and R. Navarro, Experimental validation of a Bayesian model of visual acuity, *Journal of Vision* 9(12) (2009): 1–16.

Donoho, D. L., De-noising by soft-thresholding, *IEEE Transactions on Information Theory* 41 (1995): 613–627.

Fernández, E. J., S. Manzanera, P. Piers, and P. Artal, Adaptive optics visual simulator, *Journal of Refractive Surgery* 18 (2002): S634–S638.

Geisler, W. S., Physical limits of acuity and hyperacuity, *Journal of the Optical Society of America A* 1 (1984): 775–782.

Geisler, W. S., Sequential ideal-observer analysis of visual discriminations, *Psychological Review* 96 (1989): 267–314.

Guirao, A., J. Porter, D. R. Williams, and I. G. Cox, Calculated impact of higher-order monochromatic aberrations on retinal image quality in a population of human eyes, *Journal of the Optical Society of America A* 19 (2002): 1–9.

Guirao, A. and D. R. Williams, A method to predict refractive errors from wave aberration data, *Optometry and Vision Science* 80 (2003): 36–42.

Landy, M. S. and J. A. Movshon, Eds., *Computational Models of Visual Processing*. Cambridge, MA: MIT Press, 1991.

Legras, R. and H. Rouger, Just-noticeable levels of aberration correction, *Journal of Optometry* 1 (2008): 71–77.

Liang, J. and D. R. Williams, Aberrations and retinal image quality of the normal human eye, *Journal of the Optical Society of America A* 14 (1997): 2873–2883.

Liang, J., D. R. Williams, and D. T. Miller, Supernormal vision and high-resolution retinal imaging through adaptive optics, *Journal of the Optical Society of America A* 14 (1997): 2884–2892.

López-Gil, N., V. Fernández-Sánchez, L. N. Thibos, and R. Montés-Micó, Objective amplitude of accommodation computed from optical quality metrics applied to wavefront outcomes, *Journal of Optometry* 2 (2009): 223–234.

Marcos, S., S. A. Burns, E. Moreno-Barriuso, and R. Navarro, A new approach to the study of ocular chromatic aberrations, *Vision Research* 39 (1999): 4309–4323.

Marcos, S., L. Sawides, E. Gambra, and C. Dorronsoro, Influence of adaptive-optics ocular aberration correction on visual acuity at different luminances and contrast polarities, *Journal of Vision* 8(1) (2008): 1–12.

Marsack, J. D., L. N. Thibos, and R. A. Applegate, Metrics of optical quality derived from wave aberrations predict visual performance, *Journal of Vision* 4 (2004): 322–328.

Martin, J., B. Vasudevan, N. Himebaugh, A. Bradley, and L. Thibos, Unbiased estimation of refractive state of aberrated eyes, *Vision Research* 51 (2011): 1932–1940.

McLellan, J. S., P. M. Prieto, S. Marcos, and S. A. Burns, Effects of interactions among wave aberrations on optical image quality, *Vision Research* 46 (2006): 3009–3016.

Moreno-Barriuso, E., J. Merayo Lloves, S. Marcos, R. Navarro, L. Llorente, and S. Barbero, Ocular aberrations before and after myopic corneal refractive surgery: LASIK-induced changes measured with Laser Ray Tracing, *Investigative Ophthalmology and Visual Science* 42 (2001): 1396–1403.

Navarro, R., Refractive error sensing from wavefront slopes, *Journal of Vision* 10(3) (2010): 1–15.

Navarro, R., E. Moreno-Barriuso, S. Bará, and T. Mancebo, Phase plates for wave-aberration compensation in the human eye, *Optics Letters* 25 (2000): 236–238.

Navarro, R., O. Nestares, and J. J. Valles, Bayesian pattern recognition in optically degraded noisy images, *Journal of Optics A: Pure and Applied Optics* 6 (2004): 36–42.

Nestares, O., R. Navarro, and B. Antona, Bayesian model of Snellen visual acuity, *Journal of the Optical Society of America A* 20 (2003): 1371–1381.

Nestares, O., R. Navarro, J. Portilla, and A. Tabernero, Efficient spatial-domain implementation of a multiscale image representation based on Gabor functions, *Journal of Electronic Imaging* 7 (1998): 166–173.

Piers, P. A., S. Manzanera, P. M. Prieto, N. Gorceix, and P. Artal, Use of adaptive optics to determine the optimal ocular spherical aberration, *Journal of Cataract and Refractive Surgery* 33 (2007): 1721–1726.

Ravikumar, A., J. D. Marsack, H. E. Bedell, Y. Shi, and R. A. Applegate, Change in visual acuity is well correlated with change in image-quality metrics for both normal and keratoconic wavefront errors, *Journal of Vision* 13(28) (2013): 1–16.

Ravikumar, A., E. J. Sarver, and R. A. Applegate, Change in visual acuity is highly correlated with change in six image quality metrics independent of wavefront error and/or pupil diameter, *Journal of Vision* 12(11) (2012): 1–13.

Ravikumar, S., L. N. Thibos, and A. Bradley, Calculation of retinal image quality for polychromatic light, *Journal of the Optical Society of America A* 25 (2008): 2395–2407.

Rocha, K. M., L. Vabre, F. Harms, N. Chateau, and R. R. Krueger, Effects of Zernike wavefront aberrations on visual acuity measured using electromagnetic adaptive optics technology, *Journal of Refractive Surgery* 23 (2007): 953–959.

Rouger, H., Y. Benard, and R. Legras, Effect of monochromatic induced aberrations on visual performance measured by adaptive optics technology, *Journal of Refractive Surgery* 26 (2010): 578–587.

Smith, G., Relation between spherical refractive error and visual acuity, *Optometry and Vision Science* 68 (1991): 591–598.

Tarrant, J., A. Roorda, and C. F. Wildsoet, Determining the accommodative response from wavefront aberrations, *Journal of Vision* 10(4) (2010): 1–16.

Thibos, L. N., X. Hong, A. Bradley, and R. A. Applegate, Accuracy and precision of objective refraction from wavefront aberrations, *Journal of Vision* 4 (2004): 329–351.

Vargas, A., J. Campos, and R. Navarro, Invariant pattern recognition against defocus based on subband decomposition of the filter, *Optics Communications* 185 (2000): 33–40.

Vargas-Martín, F., P. M. Prieto, and P. Artal, Correction of the aberrations in the human eye with a liquid-crystal spatial light modulator: Limits to performance, *Journal of the Optical Society America A* 15 (1998): 2552–2562.

Villegas, E. A., E. Alcón, and P. Artal, Optical quality of the eye in subjects with normal and excellent visual acuity, *Investigative Ophthalmology and Visual Science* 49 (2008): 4688–4696.

Villegas, E. A., E. Alcón, and P. Artal, Impact of positive coupling of the eye's trefoil and coma in retinal image quality and visual acuity, *Journal of the Optical Society of America A* 29(8) (2012): 1667–1672.

Walker, J. S. and Y.-J. Chen, Image denoising using tree-based wavelet subband correlations and shrinkage, *Optical Engineering* 39 (2000): 2900–2908.

Watson, A. B. and A. J. Ahumada, A standard model for foveal detection of spatial contrast, *Journal of Vision* 5(9) (2005): 6.

Watson, A. B. and A. J. Ahumada, Predicting visual acuity from wavefront aberrations, *Journal of Vision* 8(17) (2008): 1–19.

Watson, A. B. and A. J. Ahumada, Modeling acuity for optotypes varying in complexity, *Journal of Vision* 12(19) (2012): 1–19.

Yi, F., D. R. Iskander, and M. J. Collins, Estimation of the depth of focus from wavefront measurements, *Journal of Vision* 10(3) (2010): 1–9.

Yoon, G.-Y. and D. R. Williams, Visual performance after correcting the monochromatic and chromatic aberrations of the eye, *Journal of the Optical Society of America A* 19 (2002): 266–275.

Young, L. K., G. D. Love, and H. E. Smithson, Accounting for the phase, spatial frequency and orientation demands of the task improves metrics based on the visual Strehl ratio, *Vision Research* 90 (2013): 57–67.

20 Neural adaptation to blur

Michael A. Webster and Susana Marcos

Contents

20.1 INTRODUCTION

A central premise of vision science is that the visual system is designed to efficiently encode and represent the properties of the visual environment (Simoncelli and Olshausen 2001). This optimization poses daunting challenges, in part because sensitivity is limited and the information available is often corrupted or incomplete. But it is also challenging because the environment and the observer are constantly changing, and thus the optimal coding schemes themselves must change. In this chapter, we examine how the visual system adapts to these changes, focusing on how vision adapts to image blur. Exposure to blur—introduced in the stimulus, by the optics of the eye, or by optical correction or ocular treatment—leads to relatively rapid and reversible changes in visual responses to blur. These neural adjustments affect both sensitivity (e.g., changing visual acuity) and perception (e.g., changing how blurry the world appears). We first review the general properties and functions of adaptation and then discuss how adaptation alters our perception of blur and how and where in the visual system these adjustments occur.

20.2 ADAPTATION

The optical design of the eye provides unique and dedicated processes for adjusting to blur, such as the focusing mechanism (accommodation) of the crystalline lens. In contrast, neural adaptation to blur is instead an example of adjustments that are very general and pervasive, affecting nearly all aspects of visual coding. In fact, all sensory systems exhibit profound adaptation and continuously recalibrate as the environment or the observer changes. To understand the nature and consequences of blur

adaptation, it is therefore instructive to first consider the general characteristics of perceptual adaptation. (For recent reviews detailing different aspects of visual adaptation, see Webster and MacLeod 2011; Clifford et al. 2007; Wark et al. 2007; Solomon and Kohn 2014; Webster 2011, 2015.)

Visual adaptation is one of many forms of neuroplasticity and refers to relatively rapid and temporary changes in perception when observers are exposed to a stimulus (Webster 2011). These are measured by monitoring how sensitivity or appearance is altered in the presence of an adapting stimulus, or by the lingering aftereffects when the adapting stimulus is removed. In a typical experiment, observers initially view an adapting stimulus for a short period and then make judgments about a probe stimulus (e.g., setting the contrast until it is just visible or judging its appearance such as its color, shape, or movement). The probe itself is shown briefly to avoid adapting to it and is often interleaved with reexposures or "top-ups" of the adapting stimulus to maintain a stable state of adaptation. In most cases, during or shortly after viewing the adapting pattern, the visibility of similar patterns is reduced, while patterns that are visible look different than they did before adapting, a change referred to as visual aftereffects (Thompson and Burr 2009). There are many classic examples of these aftereffects. Staring briefly at a red spot produces a greenish color afterimage; viewing a tilted line or grating causes a vertical grating to appear tilted in the opposite direction (Gibson and Radner 1937); and after watching downward motion (e.g., a waterfall), a static image appears to drift upward (Wohlgemuth 1911). As these examples illustrate, the aftereffects are typically "negative" in that a neutral stimulus (gray, vertical, or static) usually appears less like the adapting stimulus or in other words biased in the opposite direction.

Yet while the form of aftereffects is similar, the neural changes giving rise to them can occur at many different levels of the visual system, beginning at the retina (e.g., for color afterimages; Zaidi et al. 2012) and cascading throughout the visual stream (e.g., with different motion aftereffects tapping different types and levels of cortical processing; Mather et al. 2008).

Visual aftereffects result because adaptation produces selective changes in sensitivity to the adapting stimulus. For example, when viewing a grating of a particular size (spatial frequency) and orientation, sensitivity to that specific frequency and angle is reduced relative to other gratings, and the frequency or orientation of other gratings appears less like the adapting grating (Blakemore and Campbell 1969). The elevated thresholds presumably occur because the mechanisms responding to the adapting stimulus have become desensitized. On the other hand, the changes in appearance arise because these relative sensitivity changes bias the distribution of responses across the mechanisms (Webster 2011; Figure 20.1). Measurements of these effects have long been a central tool in psychophysical studies of the visual system. The selectivity of the sensitivity changes (e.g., how the strength of the adaptation depends on the similarity between the adapt and test patterns) provides a measure of the number and tuning of the underlying visual "channels" encoding the stimuli (Graham 1989). Such studies were fundamental to the development and characterization of "multiple channel" models of visual coding, for example, showing that the spatial contrast sensitivity function (CSF) reflects the envelope of multiple mechanisms each tuned to a narrow range of spatial frequencies.

As noted, these adaptation effects are ubiquitous in perception, and thus selective visual aftereffects can be demonstrated for most of the stimuli we experience. This includes not only simple features of color, form, or motion but also highly complex and abstract perceptual attributes. For example, the attributes by which we individuate or classify faces (e.g., their perceived identity, gender, or age) are also highly susceptible to adaptation (Webster and MacLeod 2011). Thus, after viewing a set of young faces, all faces appear older (O'Neil et al. 2014), while adaptation to a female face causes an androgynous face to appear more masculine (Webster et al. 2004). As noted, the basic pattern of adaptation and aftereffects is similar across different stimulus

domains (Clifford et al. 2000). These parallels suggest that adaptation is an intrinsic property of all sensory representations and thus is manifest at most, if not all, levels of visual coding. That is, adaptation may be a basic computational principle for how all neurons and neural systems operate (Webster 2015). Consequently, if the visual system represents a property of the world, then it probably adapts to that property. Finally, adaptation affects not only perception but also action. For example, sensory-motor learning and plasticity have been widely studied in the context of spatial distortions or displacements induced by wearing optical devices such as prisms (Shadmehr et al. 2010; Wolpert et al. 2011). In these cases, the adjustments involve recalibrations of both perceptual and motor responses.

The prevalence of adaptation implies that it plays an important role in sensory coding. A wide range of interrelated functions have been posited (Webster 2014). One of the most obvious involves sensitivity regulation. Neurons have very limited dynamic range yet often must operate over an enormous range of stimulus levels. Matching their responses to the prevailing stimulus is critical for protecting the system from response saturation and for maintaining discrimination. In vision, the need for continuous sensitivity adjustments is most obvious with regard to the 10^9 range of light levels we can be exposed to in the course of a single day. Other aspects of the world do not show such extreme variations, yet still require continuous sensitivity regulation. For instance, if the retinal image is strongly blurred, then mechanisms tuned to higher spatial frequencies will be chronically underexposed unless they increase their gain to be appropriate for the range of contrasts they can detect.

Sensitivity regulation is closely related to principles of coding efficiency. The limited available neural responses impose strong constraints on how much information the system can carry. This capacity can be optimized by tuning the responses within a neuron (Laughlin 1981), equating responses across neurons (Field 1987), and removing the correlations between them (Buchsbaum and Gottschalk 1983). Adaptation has been implicated in each of these roles. For example, in cortical cells, adaptation adjusts neural gain (Ohzawa et al. 1982), balances responses across cells (Benucci et al. 2013), and may also reduce redundancies between different neurons (Carandini et al. 1997). Applications of information theory have proven extremely powerful in predicting many of the characteristics of early visual coding and how it is controlled by adaptation (Atick 1990; Barlow 1990; Atick et al. 1993; Wainwright 1999; Simoncelli and Olshausen 2001; Wark et al. 2007). For example, we will see in the following that these ideas have been used to infer the encoding of blur from the spatial statistics of natural images.

Coding efficiency and adaptation are in turn closely tied to the concept of predictive coding, in which the visual system is calibrated for the expected value of the stimulus and thus resources are required only to represent the deviations or errors from the predictions. Thus, center-surround cells do not respond to uniform fields but only to the "unexpected" occurrence of edges (Srinivasan et al. 1982), while color-opponent mechanisms give no response to gray (von der Twer and MacLeod 2001). Adaptation defines these null responses because it normalizes neural activity according to prevailing stimulus. Thus, as we will see, what subjectively appears in focus, the "null" for blur, may correspond to a prediction

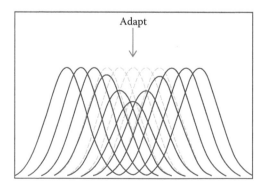

Figure 20.1 Channel models of selective adaptation. The stimulus dimension (e.g., spatial frequency or orientation) is represented by multiple narrowly tuned channels. Adaptation reduces responses in the channels tuned to the adapting stimulus and thus biases the distribution of responses. This increases thresholds for detecting similar stimuli and biases their appearance away from the adapting level. (From Webster, M.A., *J. Vis.*, 11(5), 3.1, 2011.)

about the structure of the world that is set by adaptation to the prevailing level of blur in the visual system.

A final proposed function of adaptation involves compensating visual coding for properties of the stimulus or the observer in order to maintain perceptual constancy. Adaptation is one of many diverse mechanisms contributing to constancy, by adapting out or removing irrelevant variations in the stimulus (Webster 2014). For example, in lightness or color constancy, the visual system maintains a stable representation of the color of objects (their intrinsic reflectance function) even when the lighting changes (Foster 2011). Adaptation can help to discount changes in the average spectrum of the lighting by adjusting sensitivity so that the average spectrum appears gray (Brainard and Wandell 1992). Similarly, if the scenes we were exposed to were reduced in contrast (e.g., because of haze or fog), then adaptation might function to recalibrate contrast coding to help maintain color constancy (Webster and Mollon 1995; Brown and MacLeod 1997). Similar adjustments could underlie perceptual constancy for blur if scenes became more blurry (Webster et al. 2002). In some cases, observers are exposed to environments with more blur, such as a diver underwater (Atchison et al. 2013) or a radiologist examining medical images (Burgess et al. 2001), and there is evidence that the visual system again adjusts to these stimulus changes so that the world appears to remain in focus (Ross 1974; Gislen et al. 2003; Kompaniez et al. 2013).

However, for the most part, "blur" is more a characteristic of the observer than the world, because it more often arises from limitations in the eye's optics or from neural limitations such as spatial summation. In such cases, the potential function of adaptation reflects a second vital aspect of perceptual constancy—in which perception is corrected for properties of the observer (Werner 1996; Webster 2011). There are again many examples of these compensatory adjustments. For example, sensitivity to wavelength varies dramatically over the lifespan and at different retinal locations, in part because of variations in lens and macular pigment. Both pigments reduce the short-wavelength light reaching the photoreceptors. Yet the world does not "look" yellower to observers with denser lens pigment (Werner and Schefrin 1993) nor yellower in the fovea compared to the periphery where macular pigment is absent (Webster et al. 2010). In such cases, color perception is instead almost completely compensated for the variations in the spectral sensitivity of the observer. We will see in the following that analogous adjustments in the spatial domain are fundamentally involved in compensating spatial vision for the inherent blur in the eye and neural coding.

20.3 SOURCES OF BLUR

Despite the fact that blur is a very prominent and widely studied aspect of image quality, the actual nature of blur as a visual stimulus is not fully understood (Watson and Ahumada 2011). To characterize it, it is helpful to distinguish three sources of blur, corresponding to the physical world, the optics of the eye, and the neural representation. Optical blur can arise from scattering (such as that produced by cataract) or by optical aberrations (including refractive errors—defocus and astigmatism—and high-order aberrations, prevailing in certain pathologies, increasing

with aging or induced in certain procedures). The sources of optical blur are the subject of many chapters in this book and are comparatively well understood. Here we note merely that its effects are "one-sided." Specifically, optical limits and imperfections of the eye degrade the image relative to the source but do not enhance it. "Best focused" corresponds to an upper limit of the sharpest image the visual system can achieve but not one that is too sharp. In contrast, the world itself, and its neural representation, can in principle vary from being overly sharp or overly blurred. Thus, in this case, "best focused" corresponds to the center of the scale, or a norm. For example, a scene itself can include either too little or too much fine structure so that high-frequency variations are under- or overrepresented. Similarly, the neural representation can either have too little or too much sensitivity to higher relative to lower frequencies, so that the neural responses are themselves too "blurred" or "sharp."

What does it mean for the world to be focused? An important insight into this question was the realization that natural images do not vary randomly but instead have characteristic structure. The intensities of nearby points tend to be correlated, and these correlations give rise to amplitude spectra that have more energy at lower spatial scales. In particular, the amplitude spectra of most scenes vary approximately inversely with spatial frequency or as $1/f$, which on a log-amplitude versus log-frequency scale plots as a slope of -1 (Field 1987; van der Schaaf and van Hateren 1996; Figure 20.2). The same relationship holds for square wave edges, in which contrast is again inversely proportional to the frequency of the higher harmonics. The characteristic $1/f$ structure of natural images might thus be thought to result because the world itself is composed of edges, yet is more likely to arise because the world is composed of multiple occluding objects with a power-law distribution of sizes (Ruderman 1997).

Brady and Field (1995) provided an intriguing demonstration that spatial vision appears matched to the $1/f$ spatial structure of natural scenes (Figure 20.3). An image of $1/f$ noise is physically dominated by the lower spatial frequencies, yet perceptually the image seems to have equal structure at all spatial scales—the finer details appear as salient as the coarse variations. In contrast, an image of white noise, in which the physical amplitude remains constant with frequency, instead appears completely dominated by the high-frequency structure and thus too sharp. Alternatively, if the image is filtered to a slope steeper than -1, it appears too blurred. The visual coding of the $1/f$ spectrum may serve to equate sensitivity across different spatial frequency ranges to optimize the information carried by each channel. In other words, the neural response to a $1/f$ spectrum is flat and thus unbiased. Field noted that this could in part be achieved by maintaining a constant peak sensitivity but allowing the channel bandwidth to increase with the preferred frequency (so that the bandwidths remained constant on a log or octave scale) (Field 1987). This scaling approximates the relationship between peak frequency and frequency bandwidth observed in cortical cells (De Valois et al. 1982). The falloff in the natural amplitude spectrum means that cells tuned to higher frequencies receive less energy at their single preferred frequency, but this sensitivity loss is offset by integrating the stimulus over a correspondingly wider bandwidth of frequencies. This predicts that while contrast sensitivity declines above ~4–6 cycles/deg when measured with

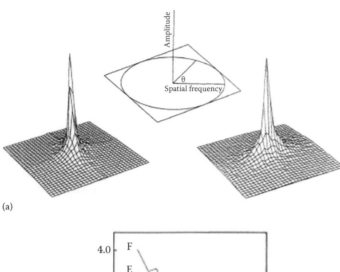

(a)

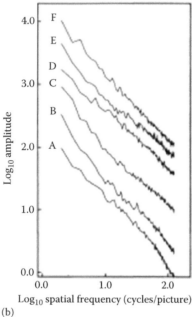

(b)

Figure 20.2 (a) Two-dimensional amplitude spectra of two natural images. (b) Plots of the amplitude averaged across orientation for 6 scenes. Amplitude falls in proportion to spatial frequency or as f^{-1}. (From Field, D.J., *J. Opt. Soc. Am. A*, 4(12), 2379, 1987.)

gratings, sensitivity remains constant for broadband images up t ~20 cycles/deg (Field and Brady 1997).

A caveat to this analysis is that many natural images are not 1/f, and there is instead significant variation in their amplitude spectra (Tolhurst et al. 1992). In fact, different types of scenes can be reliably classified by differences in their global amplitude spectra (Oliva and Torralba 2001). With regard to blur, Field and Brady (1997) noted that changes in the global spectrum could arise from two distinct sources. One is blur in the structure itself, so that higher frequencies have a lower local contrast. However, a second potential cause is variability in the density of the structure. In images with large uniform areas, there are fewe regions containing high-frequency edges or texture and thus again less total contrast at these frequencies. When the amplitud spectrum is computed for the local structure and thus corrected for structure density, the resulting spectrum predicts both physical blur and observers' subjective percepts of blur.

There are a number of additional aspects of blur that this analysis does not address. One is that there may be multiple potential cues to blur that the visual may exploit. For example, in a single edge the primary cue might be the spatial profiles (e.g., so that a blurrier edge has a broader "width") (Georgeson et al. 2007), while in noise (which has no edges) the dominant cue might be the contrast of texture at different scales. Moreover perceived blur can also be affected by changing the phase rather than the amplitude spectrum (Wang and Simoncelli 2003), and the phase changes introduced by optical aberrations can impact different measures of acuity in very different ways (Thorn and Schwartz 1990; Ravikumar et al. 2010). However, the local 1/f spectrum of images nevertheless provides a useful heuristic and potential actual perceptual code for image focus and one that th visual system might be normalized to represent by an unbiased response across spatial scale. Images with steeper or shallower spectra would appear blurred or sharper because this would bias the responses in favor of mechanisms coding coarser versus finer scales, respectively. Importantly, to the extent that the world has fairly characteristic structure, much of the variation in the spectrum of the retinal image of the world is the result of optical blur. Specifically, the presence of refractive errors (defocus,

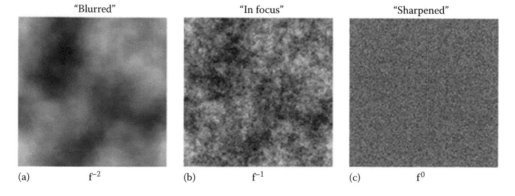

Figure 20.3 1/f noise (b) has more energy at lower frequencies yet appears to have equal contrast at all spatial scales. In contrast, in a white noise image, (c) the spectrum is flat yet appears to contain only high-frequency structure. Filtering with an exponent greater than −1 (a) causes the noise to instead appear too blurred. (After Brady, N. and Field, D., *Vision Res.*, 35(6), 739, 1995.)

astigmatism, or high-order aberrations) produces a filtering of the image with a modulation transfer function that is steeper than that of a diffraction-limited system, therefore producing a larger contrast reduction (particularly at higher spatial frequencies) and cutoff of higher spatial frequencies. Variations in the neural image additionally include marked changes in spatial summation and contrast sensitivity across the visual field. The function of adaptation to blur might therefore chiefly be to compensate for this optical and neural degradation in order to both optimize spatial resolution and also maintain percepts that are more closely tied to the expected properties of the world than to the sensitivity limits of the observer. In the following sections, we consider each of these consequences.

20.4 BLUR ADAPTATION AND VISUAL ACUITY

The initial studies of blur adaptation were motivated by the common anecdotal reports of spectacle wearers that their vision felt worse when they first removed their glasses but then improved over time. Pesudovs and Brennan formally evaluated changes in visual acuity in low-myopia subjects (<2.00 D) after a period of 90 min without their correction and found small (~0.04 logMAR) but significant improvements. These were unaccompanied by refractive changes and thus attributed to neural adaptation (Pesudovs and Brennan 1993). Subsequently, Mon-Williams et al. (1998) probed adaptation in emmetropes after inducing myopia with a +1.00 D lens worn for 30 min and found substantially larger improvements (~0.26 logMAR; Figure 20.4). These acuity changes have now been confirmed in a number of studies (e.g., George and Rosenfield 2004; Rosenfield et al. 2004; Cufflin et al. 2007; Rajeev and Metha 2010; Mankowska et al. 2012). The basic pattern of effects is similar in individuals with or without refractive errors or when the adaptation is assessed in the fovea or periphery, though some differences may occur because of the greater experience with

blurred images in individuals with poorer vision (Vera-Diaz et al. 2004; Jung and Kline 2010; Poulere et al. 2013). Neural adaptation to the observer's optical deficits is also suggested by findings that individuals with uncorrected errors show less impaired acuity than when the same blur is introduced in the eyes of normal observers (Sabesan and Yoon 2009, 2010; de Gracia et al. 2011).

The mechanisms mediating these improvements remain uncertain. For example, definitive tests have yet to be done to establish whether the acuity changes are more consistent with adaptation (~changes in neural sensitivity) or perceptual learning (~changes in how well observers can learn to utilize or interpret sensory information). Further, the specific contrast sensitivity changes that blur adaptation induces are not firmly established and instead appear to differ widely across studies. Understanding these contrast sensitivity changes is important for evaluating the implications of refractive errors and their corrections and for the development of novel corrections (Zalevsky et al. 2010). Mon-Williams et al. (1998) measured the effects of the adaptation on contrast sensitivity and found that the blur depressed sensitivity to medium spatial frequencies but not low or high. They suggested that the acuity changes occur because adaptation releases the higher frequencies from the masking effects normally exerted by the lower frequencies in the image. However, some studies have found that contrast sensitivity at higher frequencies is enhanced by blur adaptation (Rajeev and Metha 2010), while others have reported enhancements at low frequencies, which curiously were manifest only at suprathreshold contrasts (Ohlendorf and Schaeffel 2009). Importantly, relative to a uniform field, the effects of adaptation to both blurred and focused images on contrast sensitivity are strongest at lower spatial frequencies, where sensitivity is suppressed (Webster and Miyahara 1997; Bex et al. 2009). Different effects of adaptation on blur sensitivity have also been found, with some studies reporting improvements in sensitivity (Wang et al. 2006) and others finding deficits (Cufflin et al. 2007). Finally, it remains unclear how neural adjustments might impact sensitivity when a correction is applied. De Gracia et al. showed improvements in the CSF upon correction of high-order aberrations with adaptive optics (AO), but the improvement ratio in the CSF was much lower than that predicted by the MTF ratio (AO-corrected/ natural aberrations). Also, the improvement in CSF was meridional dependent (de Gracia et al. 2011). Rossi and Roorda (2010) found that when high-order aberrations were corrected with AO, visual resolution improved immediately and then showed little change despite extensive training. In contrast, Zhou et al. (2012) have reported that observers exhibited enhanced perceptual learning for contrast sensitivity after correcting their high-order aberrations.

While the improvements in visual acuity cannot be accounted for by optical adjustments and thus must involve neural factors, accommodative changes may also accompany the adaptation. For example, Vera-Diaz et al. (2004) found that myopes but not emmetropes showed an increase in their near accommodative response after brief exposures to blur. Cufflin and colleagues found no effect of adaptation on static accommodative responses

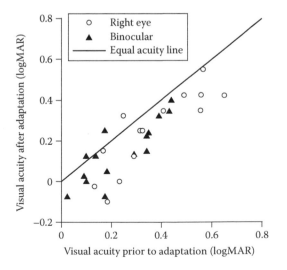

Figure 20.4 Comparison of visual acuity before or after adaptation to defocusing lenses. (From Mon-Williams, M. et al., *Proc. Biol. Sci.*, 265(1390), 71, 1998.)

(Cufflin et al. 2007) but changes in the dynamics for approaching targets (Cufflin and Mallen 2008). Interestingly, accommodative and pupillary changes are also involved in "adaptations" to vision underwater. Children of the Moken tribe of Southeast Asia harvest food from the seafloor and may achieve superior underwater acuity by constricting their pupil (but see Atchison et al. (2013)) and maximizing accommodation (Gislen et al. 2003), and these improvements can be trained relatively rapidly in other children (Gislen et al. 2006). However, it is not known whether their improved vision might also include the types of neural adjustments found with adaptation to defocusing lenses.

20.5 BLUR ADAPTATION AND PERCEIVED BLUR

The preceding studies explored how adaptation affects visual performance—the smallest stimulus changes we can detect or discriminate. Webster et al. (2002) instead measured the effects of blur adaptation on the subjective experience of blur. Their study was motivated by considerations, discussed previously, about the spatial structure of the world and how the visual system might be calibrated to represent it. Specifically, while vision might be approximately matched over evolutionary timescales to the amplitude spectra of images, for any given observer or time this match may be corrupted by optical blur or neural processing, or by how these factors change during development or aging. Thus, an "online" mechanism would be needed to fine-tune and maintain spatial coding to adjust for the idiosyncrasies of the observer's eye and brain.

To explore this, Webster et al. (2002) adapted observers to a blurred or sharpened image (created by varying the slope of the amplitude spectrum) and then adjusted the amplitude spectrum of the image until it appeared best focused (neither too blurred nor too sharp) (Figure 20.5). Prior exposure to a blurred image caused a physically focused image to appear oversharpened, and thus the image selected as best focused was moderately blurred. Adapting to sharpened images induced the opposite aftereffect. Unlike the changes in acuity, these perceptual aftereffects occur very rapidly, with only a few seconds of adaptation (Figure 20.6). The aftereffects show strong but incomplete transfer from one image to another, indicating that much of the adaptation is driven by the attribute of blur (or sharpness) independent of the specific image carrying it. Moreover, opposite aftereffects can be induced in images shown to either side of fixation and thus again cannot be accounted for by accommodative changes or general shifts in criterion. The results suggest that the perception of image focus can be rapidly and locally recalibrated in response to the blur experienced in the recent past.

The rapid and strong aftereffects of perceived blur stand in contrast to the slower and more subtle changes that adaptation induces in acuity, and there are likely a number of reasons for this.

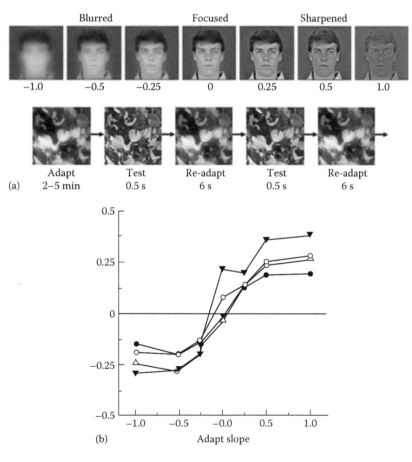

Figure 20.5 Blur aftereffects. (a) Images were filtered by varying the slope of the amplitude spectrum relative to the original image (0) to form a series ranging from too blurred to too sharp. Observers adapted to blurred or sharpened images and then adjusted the slope until it appeared correctly focused. (b) Aftereffects in one observer tested with 4 images. For each, adapting to blurred (or sharpened) images caused the original to appear too sharp (blurred), changing the physical slope that appeared best focused. (After Webster, M.A. et al., *Nat. Neurosci.*, 5(9), 839, 2002.)

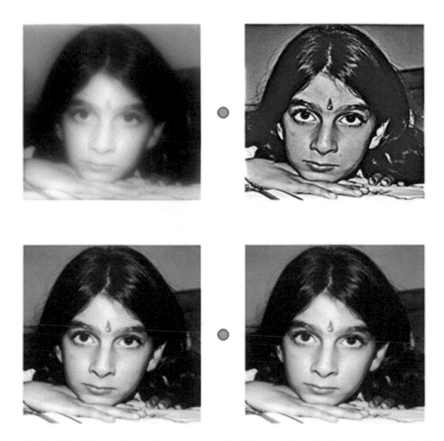

Figure 20.6 A demonstration of blur aftereffects. Fixate the upper spot for 30 s and then shift your gaze to the lower spot. The two lower images are both focused and should appear the same before adapting, but the lower right should appear more blurred after adapting.

First, as noted earlier, perceptual aftereffects (how things look) are generally much more salient than sensitivity changes (how well we can see them), even if they depend on the same response changes (Clifford et al. 2007). This is perhaps because the suprathreshold aftereffects reflect changes in the relative responses across mechanisms. Response changes may also be less apparent in measures of threshold sensitivity if the adaptation affects both signal and noise. A further issue is that acuity and perceived focus do depend at least in part on different mechanisms. Visual resolution is also limited by factors such as receptor sampling that cannot be compensated by adapting (Williams 1985). Moreover, acuity reflects the finest spatial frequencies we can resolve, while perceived focus instead depends on comparing the relative responses across a range of scales, and this range may not even include the highest visible frequencies. For example, adapting to images filtered to include only very high (or very low) frequencies does not bias the perceived focus of the original unfiltered images (Webster et al. 2001). Finally, the longer timescales associated with visual acuity changes could again also include processes such as perceptual learning.

Webster et al. (2002) also showed that the effects of context on perceived focus hold not only across time but across space. Surrounding a focused image by blurred images causes it to appear sharper or vice versa. For example, the middle row of Figure 20.7a includes only square wave edges, yet these edges appear sharpened when the abutting surround is blurrier and blurred when the surround is instead too sharp. Moreover, Figure 20.7b shows that these "simultaneous contrast" or "induction" effects occur even when the center and surround

are composed of images that do not contain aligned edges. As noted previously, adaptation effects are very general, and spatial contextual effects also occur for a wide range of stimulus dimensions (Schwartz et al. 2007). These tend to highlight the textural differences between different regions of an image and like adaptation reflect generic computations that again serve to normalize visual coding (Carandini and Heeger 2011).

Both adaptation and induction affect the perception of focus not only for spatial edges but also temporal edges (Bilson et al. 2005). Specifically, a step change in the brightness of a uniform field appears to have an overshoot after adapting to a gradual change in brightness, while appearing smeared in time following adaptation to sharpened transitions. Adaptation also biases the perceived blur in edges formed by pure color differences rather than luminance differences (Webster et al. 2006). This is despite the fact that equiluminant edges in general appear blurrier and chromatic blur is almost imperceptible when combined with luminance variations (Wandell 1995), though differences in chromatic blur are readily discriminable when the edges are shown in isolation (Wuerger et al. 2001). Notably, while spatial contrast effects for color are well known and striking, spatial induction of chromatic blur appears lacking (Webster et al. 2006).

As discussed earlier, one proposed function of adaptation is to promote perceptual constancy by compensating percepts for the sensitivity limits in the observer. Adaptation to blur could thus function to establish and maintain the correspondence between a focused world and its neural representation by discounting the blur introduced by the eye's optics or neural processing.

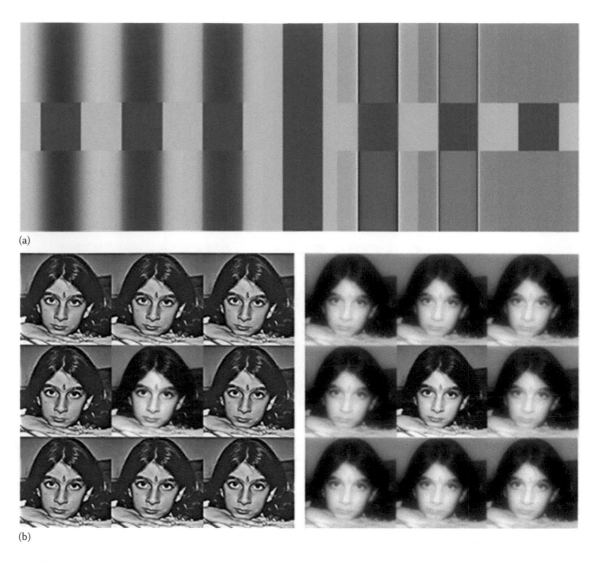

(a)

(b)

Figure 20.7 Examples of blur induction. (a) The central row of edges are identical square waves but appear sharpened when surrounded by blurred edges (left) while blurred when surrounded by sharpened edges (right). (b) The central face in each array is the same but differs in perceived blur because of the surround. These contrast effects may be more noticeable if the faces are viewed in the periphery. (From Webster, M.A. et al., *Nat. Neurosci.*, 5(9), 839, 2002.)

An important prediction of this functional account is that the blurry or sharpened images we adapt to should themselves appear more focused the longer we adapt to them. This change is known as normalization and is again found for many (but not all) adaptation effects. For example, a colored field appears less saturated (normalizing to gray) the longer we look at it (Brainard and Wandell 1992), while a distorted or distinctive face appears more average (Webster and MacLeod 2011). A similar normalization occurs with adaptation to blur and can in principle be accounted for by response changes in mechanisms tuned to different spatial scales that adapt to undo the low- or high-frequency biases in the adapting stimulus (Elliott et al. 2011; Haun and Peli 2013). Adaptation also normalizes the perception of images corrupted with high-frequency noise (Fairchild and Johnson 2007).

Similar processes may underlie the perceptual constancy for subjective focus we experience across the visual field despite the dramatic losses in spatial resolution with increasing eccentricity and again emphasize the role of adaptation in compensating not only for optical but also neural limitations. That is, the scene before us tends to look focused whether we are viewing it through our fovea or the periphery (even if we are aware that we cannot see it as well in the periphery). This could again occur if spatial perception at each retinal location is calibrated for the same physical spectrum and thus compensated for the differences in spatial sensitivity. Matches made between blurred edges in the fovea and periphery show that peripheral edges do appear sharper than predicted by the falloff in contrast sensitivity (Galvin et al. 1997). Moreover, blur induction from a surround does not occur when the surround itself is physically in focus, even though the visual resolution in the surround is much poorer (Webster et al. 2001).

20.6 BLUR ADAPTATION AND OPTICAL BLUR

The normalization of perceived blur suggests that the world may appear focused to us because we are habitually adapted or normalized for the intrinsic blur in the visual system, in effect removing this blur from our visual experience. Neural compensation for optical blur specific to our eyes is considered in detail in the following chapter by Artal. Here we note aspects of adaptation to actual optical blur that bear on the general properties of adaptation. The first, as the studies of visual acuity previously suggested, is that we do readily adapt to the patterns of blur introduced by optical aberrations. This was explored for low-order aberrations by Sawides et al. (2010), who examined adaptation to images filtered to simulate astigmatic blur. In their study, the image blur varied from horizontal to vertical with varying levels of simulated defocus added so that the blur strength remained constant (Figure 20.8). Adaptation to a given axis of the blur caused an isotropic image to appear more blurred along the orthogonal axis. These perceptual biases have also been found when observers are adapted by wearing astigmatic lenses (Anstis 2002; Yehezkel et al. 2010), and exposure to both simulated and optical astigmatism also leads to orientation-specific changes in visual acuity (Ohlendorf et al. 2011). Moreover, individuals with actual astigmatism show biases in their blur settings along their natural astigmatic axis (Vinas et al. 2012), again suggesting that they are adapted to their native blur.

Orientation selectivity for blur adaptation is not surprising, because contrast adaptation and the neural mechanisms that are thought to mediate it are themselves strongly orientation selective (Blakemore and Campbell 1969). However, the basis for the astigmatic aftereffects may be more complex. Blurring an image along one meridian not only introduces orientation-dependent changes in the blur or fuzziness of the image but can also change its perceived shape (e.g., turning a circle into an oriented ellipse). For example, the same face blurred with positive or negative astigmatism can look like images of different individuals (Sawides et al. 2010; Figure 20.9). Shape or figural aftereffects are also a well-known form of pattern aftereffects (e.g., after adapting to a vertical ellipse, a physical circle appears stretched horizontally (Kohler and Wallach 1944); and as we noted earlier, viewing a distorted face induces the opposite perceived distortion in the original face (Webster and MacLeod 2011). Thus, astigmatism could result in adaptation through multiple routes, driven directly by the blur or by the perceived spatial distortions or shape changes. To explore this, Sawides et al. (2010) compared how adaptation to a face shown at one size transferred to the image at a different size. In one case, the two images had the same level of blur in terms of visual angle (and thus less shape distortion in the larger image). In the other, the two images were magnified versions of each other and thus the same shape but more blur in the larger image. Aftereffects showed stronger transfer across the magnified images, suggesting that they were at least partly driven by the configural changes that the astigmatic blur introduced into the images. Thus, even though the stimulus for blur might arise early (in the retinal image), adaptation to it could include adjustments far along the visual hierarchy, in mechanisms encoding high-level form attributes of the image.

Adaptation to astigmatism has interesting implications for a perceptual puzzle known as the El Greco fallacy (Anstis 2002; Firestone 2013). The Spanish Renaissance artist painted highly elongated figures, and a recurring theory is that this was because he suffered from astigmatism and was merely recreating the world "as he saw it." The fallacy is that the same optical distortions should occur whether he was viewing the subject or

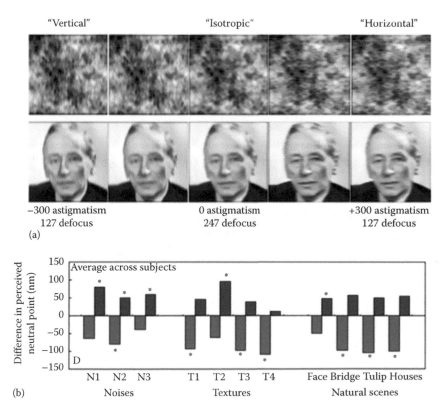

Figure 20.8 Adaptation to astigmatic blur. (a) Images were filtered to simulate different magnitudes of astigmatism with defocus added to maintain constant total blur. (b) Adapting to negative (blue) or positive (purple) astigmatism induced opposite biases in the astigmatic blur level that appeared isotropic. *p < 0.05.

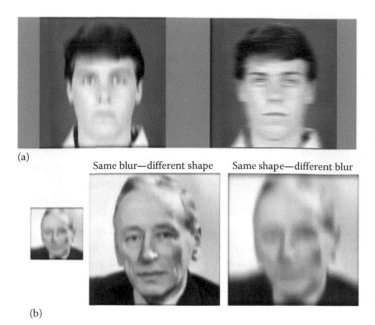

(a)

Same blur—different shape   Same shape—different blur

(b)

Figure 20.9 Figural distortions and blur. (a) Faces filtered with different axes of astigmatism (1.5 D) look like images of different individuals and thus adaptation could be driven by the shape changes in addition to the blur. (Images courtesy of Sowmya Ravikumar.) (b) A blurred image on the left scaled so that it has the same level of blur but different shape (middle) or magnified to have the same profile but greater blur (right). (After Sawides, L. et al., *J. Vis.*, 10(12), 22, 2010.)

their portrait, and thus to see them the same he would need to reproduce the same physical proportions. This illustrates the fact that observers in general have little awareness of their own visual limitations (Breitmeyer 2010). However, in the case of astigmatic blur, Sawides et al. (2012) noted a potential fallacy to the fallacy, for as described in the experiment previously, blur corresponds to a fixed visual angle and thus should have proportionally larger effects the farther the subject, or smaller their angle.

If the canvas was closer than the sitter, then the same blur could induce larger distortions in the smaller image. More importantly, adaptation to the astigmatic blur would in general be necessary to maintain stable shape percepts across changes in viewing distance.

Adaptation can also adjust to the blur introduced by high-order aberrations. Artal and colleagues (2004) used AO to show that observers experience images filtered through their own aberrations as better focused than rotated versions of the equivalent blur strength. Subsequent studies also using AO have demonstrated that images with a blur magnitude equivalent to an individual's native blur appear best focused, and moreover, that when observers are instead exposed to a smaller or larger magnitude of their own aberration pattern or to scaled versions of aberrations corresponding to others' eyes, this again induces blur aftereffects (Sawides et al. 2011a,b, 2012; Radhakrishnan et al. 2015; Figure 20.10). Such results strongly suggest that spatial vision is compensated for the individual's optical imperfections, though the extent to which focus percepts are calibrated for the specific aberration pattern remains uncertain. It also predicts that different observers will experience the same object as focused even though their optics may corrupt the retinal image in very different ways. This is because both observers are adapted to the same world and thus are applying different corrections to discount their different optics. Of course the extent to which adaptation actually converges the percepts of different observers depends in part on whether the processes of adaptation are themselves similar. Significant individual differences have been found in the pattern of blur aftereffects (Vera-Diaz et al. 2010), yet there are also strong similarities. For example, the pattern of blur adaptation appears very similar in younger and older adults (Elliott et al. 2007), despite the many optical and neural losses that occur with aging (Owsley 2011; see Chapter 69). It is possible that the mechanisms of adaptation survive aging in part because they are critically important for maintaining percepts for these losses.

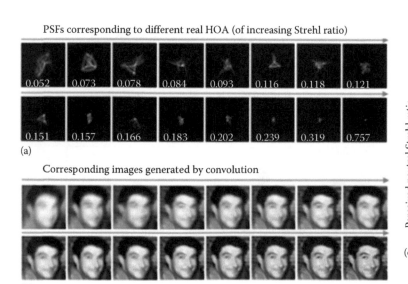

PSFs corresponding to different real HOA (of increasing Strehl ratio)

| 0.052 | 0.073 | 0.078 | 0.084 | 0.093 | 0.116 | 0.118 | 0.121 |

| 0.151 | 0.157 | 0.166 | 0.183 | 0.202 | 0.239 | 0.319 | 0.757 |

(a)

Corresponding images generated by convolution

(b)

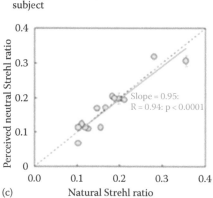

Correspondence between the image quality perceived as neutral and the retinal image quality produced by the aberrations of the subject

Slope = 0.95;
R = 0.94: p < 0.0001

Perceived neutral Strehl ratio

Natural Strehl ratio

(c)

Figure 20.10 (a) PSFs of different observers and (b) corresponding blur in an image. (c) The magnitude of blur that appears correctly focused to the observer corresponds closely to the magnitude of blur in the observer's own eyes.

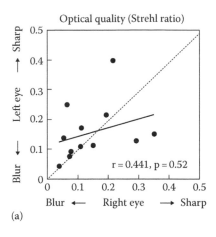 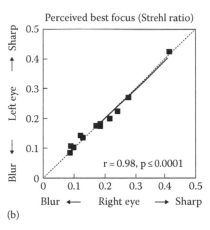 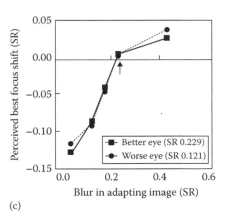

(a)   (b)   (c)

Figure 20.11 Interocular differences in blur perception and adaptation. (a) Differences in the magnitude of retinal image blur between the left and right eyes of a set of observers. (b) Despite these differences, the retinal blur that appears best focused is very similar in both eyes and determined by the eye with better optical quality. (c) Blur aftereffects are also very similar in both eyes and are calibrated for the blur of the better eye so that there is no shift in perceived focus when the adapt blur equals the better eye's blur. This holds even when the adaptation is assessed for the worse eye, suggesting that blur percepts through both eyes are set at a common binocular site calibrated for the eye with less blur. (From Radhakrishnan, A. et al., *Curr. Biol.*, in press.)

An important issue is whether this compensation reflects an actual calibration of visual sensitivity or is instead a "cognitive" adjustment in the observer's criterion or how they label stimuli. By the latter account, the blur level that is judged as focused may simply be because they represent the level of blur that we are accustomed to seeing. The fact that an image appears focused might therefore not correspond to a special state of the neural response but only to a learned response to special states of the image. This is again a general and fundamental question in perception. Many stimulus dimensions appear to be represented relative to a norm that itself appears neutral or unbiased. Thus, "focused" is a norm for blurred versus sharp in the same way that "gray" is for red versus green or "average" for tall versus short. The question is whether these psychologically neutral percepts correspond to physiologically neutral representations.

One way to test this is to ask whether the stimulus that an observer describes as neutral (e.g., focused) is also the stimulus that is neutral for adaptation (Webster and Leonard 2008). The blur level that does not produce an aftereffect is the level that visual sensitivity is calibrated for, because it is—uniquely—the stimulus that does not lead to a recalibration. In fact, this has been examined for blur and has revealed a close correspondence between subjective judgments of image focus and the neutral points for adaptation, with both closely tied to the level of native blur in the retinal image (Sawides et al. 2011a; Radhakrishnan et al. 2015; see Figure 20.11). The implication is that what we experience as focused does in fact seem to reflect an actual neutral or unbiased state in the neural representation, one that is adjusted to discount the individual's optical and neural imperfections.

20.7 THE NEURAL LOCUS OF BLUR ADAPTATION

At what stage of the pathway does the visual system adapt to blur? Contrast adaptation has long been assumed to reflect sensitivity changes at a cortical locus (Blakemore and Campbell 1969). The adaptation is strongly selective for orientation and spatial frequency and also shows strong interocular transfer (so that adapting to an image presented to one eye alters sensitivity to images seen by the other eye). This pattern selectivity and binocular convergence first arises in primary visual cortex (De Valois and De Valois 1980). Moreover, early physiological studies suggested that cortical cells were the earliest neurons in the visual system to adapt to contrast (Maffei et al. 1973). However, more recent work has revealed that the retina can adapt to a surprisingly wide range of image features including attributes like orientation or movement that the cells are not overtly tuned for (Gollisch and Meister 2010). Moreover, retinal cells of many species are now known to exhibit strong contrast adaptation (Kastner and Baccus 2014). In primates, these effects are mixed. Slower contrast adaptation of the type implicated by behavioral studies has been clearly demonstrated in some classes of retinal and geniculate cells (the magnocellular pathway) while it may remain absent in others (the parvocellular pathway) (Solomon et al. 2004). These pathways are thought to be precursors of different cortical streams subserving different visual functions such as "what" versus "where" or perception versus action (Goodale 2011; Ungerleider and Bell 2011). Thus, in principle, adaptation to blur could arise at a number of stages or as part of different visual subsystems.

Theoretically, there are also reasons for expecting that the visual system might position the adaptation so that each eye can be calibrated independently. Differences in optical aberrations often occur between the two eyes (Almeder et al. 1990; Marcos and Burns 2000), and early sites of adaptation could allow each to be compensated accordingly, similar to the way that emmetropization is achieved locally within the retina (Wallman and Winawer 2004). In fact, the compensation of color vision for the eyes' spectral sensitivity does occur very early in the retina and in large part may occur within the photoreceptors themselves (Webster and Leonard 2008). Moreover, even if the site of the adaptation is cortical, this does not preclude independent adjustment for each eye, for some aftereffects are strongly orientation selective yet also strongly monocular (such as the McCollough effect, in which color aftereffects are contingent on the orientation of the adapting colors; McCollough-Howard and Webster 2011).

Surprisingly, however, the adaptation of perceived blur is instead strongly binocular. Mon-Williams et al. (1998) showed that adapting to defocus in one eye affected visual acuity through the other, though the effect was smaller in the nonadapted eye. Kompaniez et al. (2013) instead examined the effects on subjective focus when the adapt and test images were shown to the same or different eyes. In this case, the aftereffects showed almost complete transfer between the eyes. Moreover, when observers were adapted to different blur levels in the two eyes, the aftereffects of perceived focus were similar through either eye and dominated by the sharper of the two images. This dominance is consistent with how we subjectively experience differences in blur between the two eyes or when focused and blurred images are superimposed. The sharper image determines the perceived focus and strongly suppresses the blur in the second image (Georgeson et al. 2007). This suppression is also found in binocular rivalry—where different images that cannot be fused are shown to each eye. The sharper image again strongly dominates (Fahle 1982; Arnold et al. 2007). Adaptation to these interocular differences may aid the utility of monovision corrections for presbyopia, where one eye is focused for far and the other for near (Radhakrishnan et al. 2014).

Radhakrishnan et al. (2015) recently examined the long-term consequences of normal interocular differences in optical aberrations. They used AO to correct the image in each eye and then introduce different magnitudes of blur (corresponding to the aberrations measured for a large set of real eyes). In one experiment, observers selected the blur level that appeared best focused through either eye. As shown in the previous studies described earlier, this level corresponded to the native blur that the individual was exposed to through their own eyes. Yet remarkably, their results also revealed that the native blur calibrating focus depended only on the eye with better optical quality, even when viewing the images monocularly through the eye with worse quality. In a second experiment, they measured blur aftereffects through either eye, by adapting to a range of blur levels that bracketed and included the native blur of each eye. In this case again, the aftereffects were completely governed by the better eye (Figure 20.11). Specifically, regardless of which eye was tested, whether the adaptation produced a sharper or blurrier aftereffect depended entirely on whether the adapting image was less or more blurred that the blur level of the better eye. These results point to a single "cyclopean" locus of the neural adaptation to blur, or in other words, suggest that the visual system adopts a single calibration for retinal image blur that is determined entirely by the stronger eye. Moreover, the effects of adaptation again reveal that this calibration reflects an actual norm in the neural coding of blur, one again set by the eye with superior optical quality.

20.8 BLUR ADAPTATION VERSUS SPATIAL FREQUENCY ADAPTATION

Thus far, we have illustrated blur adaptation by alluding to how blur adjusts to the spatial frequency spectrum and thus contrast sensitivity. However, the actual neural mechanisms for encoding blur, and thus its adaptation, are poorly understood. Is the adaptation acting on neurons or codes that represent blur directly, or only indirectly, for example, in spatial frequency channels that are instead simply encoding the contrast energy at different spatial scales? Functionally, this difference may not matter, for the fact that the visual system adjusts to calibrate for the image attribute of blur is important regardless of how that attribute is represented. However, mechanistically, it is of considerable interest to ask whether blur is encoded explicitly as an image feature or only implicitly by the variations in the spatial spectrum. While a number of studies have addressed this issue, the answer remains uncertain. For example, several studies have attempted to model blur discrimination and in particular the fact that discrimination thresholds follow a characteristic "dipper" function (in which the smallest discriminable increase in blur occurs not when the blur is added to focused images but to images that already have a modest level of "pedestal" blur). This effect and others (e.g., Georgeson et al. 2007; McIlhagga and May 2012) have been attributed to mechanisms sensitive to blur as a feature such as the spatial gradient of edges, yet some analyses suggest that discrimination (Watson and Ahumada 2011) and blur adaptation (Elliott et al. 2011; Haun and Peli 2013) can be accounted for by changes in contrast sensitivity without invoking explicit blur-encoding mechanisms.

However, other lines of evidence suggest that adaptation to blur does not follow in obvious ways from changes in spatial contrast sensitivity. One involves how contrast sensitivity itself is affected by adaptation. Webster and Miyahara (1997) measured changes in contrast thresholds and suprathreshold perceived contrast following adaptation to natural images or noise images with different power spectra (Figure 20.12). Surprisingly, focused images induced sensitivity losses that were strongly selective for lower spatial frequencies. That is, frequencies below ~4 cycles/deg became harder to detect, while thresholds for higher spatial frequencies were unaffected. These selective effects have been confirmed in recent psychophysical studies and even when the sensitivity changes are probed in single cortical cells (Sharpee et al. 2006; Bex et al. 2009). The low-frequency biases induced by the adaptation have important implications for spatial vision. The CSF is widely used for characterizing spatial sensitivity and predicting visibility. However, the standard function is based on measuring sensitivity, while observers are adapted to a uniform field. When instead adapted to natural images, the CSF is significantly more bandpass. For the present discussion, the point of note is that focused images produce large and selective sensitivity losses at lower frequencies, yet the images themselves do not appear sharper—adaptation to focused images does not produce a blur aftereffect.

Webster and Miyahara (1997) also found that the sensitivity losses depended only weakly on the actual power spectrum. That is, when adapting to blurred or sharpened images, the threshold changes remained similar unless the blur or sharpening became pronounced. The contrast sensitivity changes appear too gradual to predict the strong differential aftereffects in perceived blur. Finally, as we noted earlier, blur itself does not correspond simply to the global amplitude spectrum, and similarly, whether an image induces blur aftereffects depends on whether it appears focused, regardless of differences in the actual amount of energy at different scales.

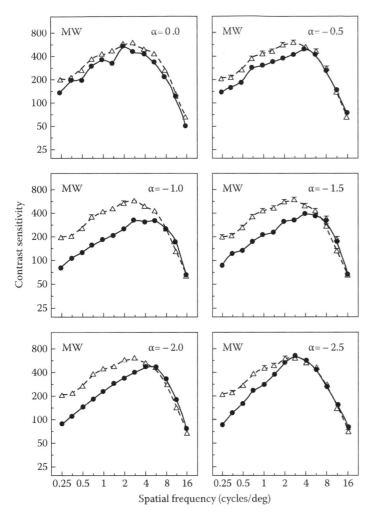

Figure 20.12 Effects of adaptation to noise images with different spectral slopes (α) on contrast sensitivity. Even images with "focused" slopes (α = −1) selectively reduce sensitivity at lower spatial frequencies. (From Webster, M.A. and Miyahara, E., *J. Opt. Soc. Am. A: Opt. Image Sci. Vis.*, 14(9), 2355, 1997.)

The low-frequency attenuation in contrast sensitivity with image adaptation is likely related to a phenomenon observed previously involving the interactions between spatial frequencies in adaptation effects. A square wave grating produces less adaptation at its third harmonic than when that harmonic is adapted to in isolation, as if the adaptation is really driven by the fundamental periodicity of the pattern rather than by the independent contrasts at each scale (Tolhurst 1972; Nachmias et al. 1973). A similar effect is found for complex image spectra. When 1/f noise is progressively high-pass filtered, the strongest adaptation occurs for the lowest frequency components (Webster et al. 2006). This suggests that with natural scenes much of the adaptation owes primarily to the DC or mean fluctuations in the overall images. Here again, however, the point to note is that in complex or naturalistic spectra the higher frequencies induce little change in threshold sensitivity, yet variations in the relative energy at different frequencies are a very potent stimulus for perceiving blur and for producing blur aftereffects (Webster et al. 2002).

We can also ask how perceived blur is affected by adapting out specific ranges of frequencies. Webster et al. (2001) examined this by filtering images into different one-octave bands and then measuring how adaptation to each altered the perceived blur of the original. Notably, the highest and lowest bands had negligible effect, even though these represented the most blurred or sharpened adaptors. Instead, the aftereffects depended primarily on frequency bands ranging over roughly 1–16 cycles/deg, suggesting that the adapted mechanisms are especially sensitive to—and might directly encode—the ratio of energy across this range.

20.9 TIMESCALES OF ADAPTATION

The timescales of adaptation are fundamentally important for understanding basic questions about what information and rates of change the visual system is tracking and also for answering clinically relevant questions such as how long it takes to adjust to a new pair of glasses. These timescales remain poorly characterized, yet it is increasingly evident that adaptation unfolds at multiple and widely varying rates. In the case of blur, we have noted that dramatic yet fleeting perceptual changes can occur within a few seconds of exposure, while induction effects are nearly instantaneous. At the other extreme, some adjustments may unfold only with very prolonged exposure but last indefinitely. Smallman et al. tracked perceived blur in an individual whose congenital cataracts were replaced at age 43 (Fine et al. 2002). Afterward, he perceived square wave edges as too sharp, and this persisted up to 2 years after surgery.

Most studies of visual adaptation have focused on relatively brief time intervals of seconds to minutes. Over this time, the buildup and decay of sensitivity changes tend to follow a power law of exposure duration (Greenlee et al. 1991; Bao and Engel 2012). However, a number of studies have extended adaptation to hours or days and have revealed a number of insights. First, longer durations seem more likely to reveal sensitivity improvements. For example, adaptation to an increase in contrast can quickly depress contrast sensitivity, yet when observers are exposed to reduced contrast, contrast sensitivity can increase but this may take hours to develop (Kwon et al. 2009; Zhang et al. 2009). In the case of blur adaptation and visual acuity, most investigators have reported improvements after adapting periods ranging from 30 min to few hours. Adjustments to corrections (e.g., refractive or cataract surgery) may take weeks or more to unfold (Pesudovs 2005; Parkosadze et al. 2013). Yet surprisingly, significant improvements in acuity have also been found with as little as 4–10 min of exposure to blurrier vision (Ohlendorf et al. 2011; Khan et al. 2013). Notably however, this is still orders of magnitude slower than the exposure time required to induce aftereffects of perceived focus.

A second important insight is that the adjustments over different timescales are distinct. One common test for demonstrating this is to show that aftereffects from a long exposure can be temporarily extinguished by a brief period of counteradaptation but then "spontaneously recover" (Shadmehr et al. 2010; Bao and Engel 2012; Mesik et al. 2013). This was also suggested for blur aftereffects in a study of senescent cataract patients by Parkosadze et al. (2013). Images continued to appear too sharp when tested up to 2 months after surgery, and brief exposures to blurred or sharpened images induced brief aftereffects that appeared to ride atop of the more persistent biases. As with other properties of adaptation, these characteristics are not specific to blur and instead reflect very general properties. For example, color vision and sensory-motor calibrations also include distinct short- and long-term changes (Delahunt et al. 2004; Belmore and Shevell 2010; Shadmehr et al. 2010). Moreover, adjustments over multiple timescales have a strong theoretical rationale—to optimally calibrate for changes that are brief and temporary requires adjustments that are themselves rapid but transient, while prolonged biases should be met with more permanent corrections (Kording et al. 2007).

An intriguing question is whether the timescales of adaptation are themselves adaptable. Individuals often report that a new lens prescription initially takes considerable time to adjust to and that a strong correction is poorly tolerated, prompting corrections in stages. Yet after a while, vision through the lens feels comfortable the moment it is worn. Yehezkel et al. (2010) examined the dynamics of such adjustments in observers who were adapted to astigmatic blur by wearing +1.00 D cylindrical lenses. Over sessions lasting 4 h, observers adjusted to the orientation biases introduced by the lenses, and these adjustments resulted in little aftereffect when the lens was removed, while immediately reviving when reworn on subsequent days. This suggests that the visual system might "learn to adapt" more quickly when a context is repeated or, alternatively, might be capable of storing different long-term blur calibrations that are contingent on the viewing context, similar to the contingent aftereffects that have been described for color and form (McCollough-Howard and Webster 2011). Vinas et al. (2012) explored the time course of blur percepts in habitually uncorrected astigmats after they received a correction. Significant shifts (in the image that appeared isotropic) occurred within 2 h and asymptoted within a week toward normal values. Thus, despite decades of exposure to their blur, they appeared to recalibrate rapidly to a correction. Corrected astigmats showed a different and surprising pattern—maintaining a bias toward their axis. A number of factors may account for this difference, but one possibility is that the subjects were tested with an AO system and thus without wearing their prescribed lenses. Like the subjects in the Yehezkel et al. (2010) study, this may have removed the contextual "frame" they were normally used to seeing clearly within and thus may have triggered their uncorrected adapted state.

ACKNOWLEDGMENTS

This work was supported by the National Eye Institute grant EY-10834 (MW), the European Research Council under the European Union's Seventh Framework Programme (FP7/2007–2013)/ERC Grant Agreement no. 294099, Spanish Government grant FIS2011-264605, and Seventh Framework Programme of the European Community through the Marie Curie Initial Training Network OpAL (OpAL is an Initial Training Network funded by the European Commission under the Seventh Framework Programme [PITN-GA-2010-264605]) (SM).

REFERENCES

Almeder, L. M., L. B. Peck, and H. C. Howland. 1990. Prevalence of anisometropia in volunteer laboratory and school screening populations. *Invest Ophthalmol Vis Sci* 31(11):2448–2455.

Anstis, S. M. 2002. Was El Greco astigmatic? *Leonardo* 35:208.

Arnold, D. H., P. M. Grove, and T. S. Wallis. 2007. Staying focused: A functional account of perceptual suppression during binocular rivalry. *J Vis* 7(7):1–8.

Artal, P., L. Chen, E. J. Fernandez, B. Singer, S. Manzanera, and D. R. Williams. 2004. Neural compensation for the eye's optical aberrations. *J Vis* 4(4):281–287.

Atchison, D. A., E. L. Valentine, G. Gibson, H. R. Thomas, S. Oh, Y. A. Pyo, P. Lacherez, and A. Mathur. 2013. Vision in water. *J V* 13(11):4-4. doi: 10.1167/13.11.4.

Atick, J. J. 1990. Could information-theory provide an ecological theory of sensory processing. *Network Comput Neural Syst* 3:213–251.

Atick, J. J., Z. Li, and A. N. Redlich. 1993. What does post-adaptation color appearance reveal about cortical color representation? *Vision Res* 33(1):123–129.

Bao, M. and S. A. Engel. 2012. Distinct mechanism for long-term contrast adaptation. *Proc Natl Acad Sci USA* 109(15):5898–5903.

Barlow, H. B. 1990. A theory about the functional role and synaptic mechanism of visual aftereffects. In *Visual Coding and Efficiency*, C. Blakemore (ed.), pp. 363–375. Cambridge, U.K.: Cambridge University Press.

Belmore, S. C. and S. K. Shevell. 2010. Very-long-term and short-term chromatic adaptation: Are their influences cumulative? *Vision Res* 51(3):362–366.

Benucci, A., A. B. Saleem, and M. Carandini. 2013. Adaptation maintains population homeostasis in primary visual cortex. *Nat Neurosci* 16(6):724–729.

Bex, P. J., S. G. Solomon, and S. C. Dakin. 2009. Contrast sensitivity in natural scenes depends on edge as well as spatial frequency structure. *J Vis* 9(10):1.1–1.19.

Bilson, A. C., Y. Mizokami, and M. A. Webster. 2005. Visual adjustments to temporal blur. *J Opt Soc Am A Opt Image Sci Vis* 22(10):2281–2288.

Blakemore, C. and F. W. Campbell. 1969. On the existence of neurones in the human visual system selectively sensitive to the orientation and size of retinal images. *J Physiol* 203(1):237–260.

Brady, N. and D. J. Field. 1995. What's constant in contrast constancy? The effects of scaling on the perceived contrast of bandpass patterns. *Vision Res* 35(6):739–756.

Brainard, D. H. and B. A. Wandell. 1992. Asymmetric color matching: How color appearance depends on the illuminant. *J Opt Soc Am A* 9(9):1433–1448.

Breitmeyer, B. 2010. *Blindspots: The Many Ways We Cannot See*. Oxford, U.K.: Oxford University Press.

Brown, R. O. and D. I. MacLeod. 1997. Color appearance depends on the variance of surround colors. *Curr Biol* 7(11):844–849.

Buchsbaum, G. and A. Gottschalk. 1983. Trichromacy, opponent colours coding and optimum colour information transmission in the retina. *Proc R Soc Lond B Biol Sci* 220(1218):89–113.

Burgess, A. E., F. L. Jacobson, and P. F. Judy. 2001. Human observer detection experiments with mammograms and power-law noise. *Med Phys* 28(4):419–437.

Carandini, M., H. B. Barlow, L. P. O'Keefe, A. B. Poirson, and J. A. Movshon. 1997. Adaptation to contingencies in macaque primary visual cortex. *Philos Trans R Soc Lond B Biol Sci* 352(1358):1149–1154.

Carandini, M. and D. J. Heeger. 2011. Normalization as a canonical neural computation. *Nat Rev Neurosci* 13(1):51–62.

Clifford, C. W., M. A. Webster, G. B. Stanley, A. A. Stocker, A. Kohn, T. O. Sharpee, and O. Schwartz. 2007. Visual adaptation: Neural, psychological and computational aspects. *Vis Res* 47(25):3125–3131.

Clifford, C. W. G., P. Wenderoth, and B. Spehar. 2000. A functional angle on some after-effects in cortical vision. *Proc R Soc B Biol Sci* 267(1454):1705–1710.

Cufflin, M. P., C. A. Hazel, and E. A. Mallen. 2007. Static accommodative responses following adaptation to differential levels of blur. *Ophthalmic Physiol Opt* 27(4):353–360.

Cufflin, M. P. and E. A. Mallen. 2008. Dynamic accommodation responses following adaptation to defocus. *Optom Vis Sci* 85(10):982–991.

Cufflin, M. P., A. Mankowska, and E. A. Mallen. 2007. Effect of blur adaptation on blur sensitivity and discrimination in emmetropes and myopes. *Invest Ophthalmol Vis Sci* 48(6):2932–2939.

de Gracia, P., C. Dorronsoro, G. Marin, M. Hernandez, and S. Marcos. 2011. Visual acuity under combined astigmatism and coma: Optical and neural adaptation effects. *J Vis* 11(2):5-5. doi: 11.2.5 [pii] 10.1167/11.2.5.

De Valois, R. L., D. G. Albrecht, and L. G. Thorell. 1982. Spatial frequency selectivity of cells in macaque visual cortex. *Vision Res* 22(5):545–59.

De Valois, R. L. and K. K. De Valois. 1980. Spatial vision. *Annu Rev Psychol* 31:309–341.

Delahunt, P. B., M. A. Webster, L. Ma, and J. S. Werner. 2004. Long-term renormalization of chromatic mechanisms following cataract surgery. *Visual Neurosci* 21(3):301–307.

Elliott, S. L., M. A. Georgeson, and M. A. Webster. 2011. Response normalization and blur adaptation: Data and multi-scale model. *J Vis* 11(2):7-7. doi: 11.2.7 [pii] 10.1167/11.2.7.

Elliott, S. L., J. L. Hardy, M. A. Webster, and J. S. Werner. 2007. Aging and blur adaptation. *J Vis* 7(6):8–8.

Fahle, M. 1982. Binocular rivalry: Suppression depends on orientation and spatial frequency. *Vision Res* 22(7):787–800.

Fairchild, M. D. and G. M. Johnson. 2007. Measurement and modeling of adaptation to noise in images. *J Soc Inf Disp* 15(9):639–647.

Field, D. J. 1987. Relations between the statistics of natural images and the response properties of cortical cells. *J Opt Soc Am A* 4(12):2379–2394.

Field, D. J. and N. Brady. 1997. Visual sensitivity, blur and the sources of variability in the amplitude spectra of natural scenes. *Vision Res* 37(23):3367–3383.

Fine, I., H. S. Smallman, P. Doyle, and D. I. MacLeod. 2002. Visual function before and after the removal of bilateral congenital cataracts in adulthood. *Vision Res* 42(2):191–210.

Firestone, C. 2013. On the origin and status of the El Greco fallacy. *Perception* 42(6):672–674.

Foster, D. H. 2011. Color constancy. *Vision Res* 51(7):674–700.

Galvin, S. J., R. P. O'Shea, A. M. Squire, and D. G. Govan. 1997. Sharpness overconstancy in peripheral vision. *Vision Res* 37(15):2035–2039.

George, S. and M. Rosenfield. 2004. Blur adaptation and myopia. *Optom Vis Sci* 81(7):543–547.

Georgeson, M. A., K. A. May, T. C. Freeman, and G. S. Hesse. 2007. From filters to features: Scale-space analysis of edge and blur coding in human vision. *J Vis* 7(13):7.1–7.21.

Gibson, J. J. and M Radner. 1937. Adaptation, after-effect and contrast in the perception of tilted lines. I. Quantitative studies. *J Exp Psychol* 20:453–467.

Gislen, A., M. Dacke, R. H. Kroger, M. Abrahamsson, D. E. Nilsson, and E. J. Warrant. 2003. Superior underwater vision in a human population of sea gypsies. *Curr Biol* 13(10):833–836.

Gislen, A., E. J. Warrant, M. Dacke, and R. H. Kroger. 2006. Visual training improves underwater vision in children. *Vision Res* 46(20):3443–3450.

Gollisch, T. and M. Meister. 2010. Eye smarter than scientists believed: Neural computations in circuits of the retina. *Neuron* 65(2):150–164.

Goodale, M. A. 2011. Transforming vision into action. *Vision Res* 51(13):1567–1587.

Graham, N. V. 1989. *Visual Pattern Analyzers*. Oxford, U.K.: Oxford University Press.

Greenlee, M. W., M. A. Georgeson, S. Magnussen, and J. P. Harris. 1991. The time course of adaptation to spatial contrast. *Vision Res* 31(2):223–236.

Haun, A. M. and E. Peli. 2013. Adaptation to blurred and sharpened video. *J Vis* 13(8):12-12. doi: 10.1167/13.8.12.

Jung, G. H. and D. W. Kline. 2010. Resolution of blur in the older eye: Neural compensation in addition to optics? *J Vis* 10(5):7.

Kastner, D. B. and S. A. Baccus. 2014. Insights from the retina into the diverse and general computations of adaptation, detection, and prediction. *Curr Opin Neurobiol* 25:63–69.

Khan, K. A., K. Dawson, A. Mankowska, M. P. Cufflin, and E. A. Mallen. 2013. The time course of blur adaptation in emmetropes and myopes. *Ophthalmic Physiol Opt* 33(3):305–310.

Kohler, W. and H. Wallach. 1944. Figural aftereffects: An investigation of visual processes. *Proc Am Philos Soc* 88:269–357.

Kompaniez, E., C. K. Abbey, J. M. Boone, and M. A. Webster. 2013. Adaptation aftereffects in the perception of radiological images. *PLoS ONE* 8(10):e76175.

Kompanicz, E., L. Sawides, S. Marcos, and M. A. Webster. 2013. Adaptation to interocular differences in blur. *J Vis* 13(6):19.

Kording, K. P., J. B. Tenenbaum, and R. Shadmehr. 2007. The dynamics of memory as a consequence of optimal adaptation to a changing body. *Nat Neurosci* 10(6):779–786.

Kwon, M., G. E. Legge, F. Fang, A. M. Cheong, and S. He. 2009. Adaptive changes in visual cortex following prolonged contrast reduction. *J Vis* 9(2):20.1–20.16.

Laughlin, S. 1981. A simple coding procedure enhances a neuron's information capacity. *Z Naturforsch C* 36(9–10):910–912.

Maffei, L., A. Fiorentini, and S. Bisti. 1973. Neural correlate of perceptual adaptation to gratings. *Science* 182(4116):1036–1038.

Mankowska, A., K. Aziz, M. P. Cufflin, D. Whitaker, and E. A. Mallen. 2012. Effect of blur adaptation on human parafoveal vision. *Invest Ophthalmol Vis Sci* 53(3):1145–1150.

Marcos, S. and S. A. Burns. 2000. On the symmetry between eyes of wavefront aberration and cone directionality. *Vision Res* 40(18):2437–2447.

Mather, G., A. Pavan, G. Campana, and C. Casco. 2008. The motion aftereffect reloaded. *Trends Cogn Sci* 12(12):481–487.

McCollough-Howard, C. and M. A. Webster. 2011. McCollough effect. *Scholarpedia* 6(2):8175.

McIlhagga, W. H. and K. A. May. 2012. Optimal edge filters explain human blur detection. *J Vis* 12(10):9.

Mesik, J., M. Bao, and S. A. Engel. 2013. Spontaneous recovery of motion and face aftereffects. *Vision Res* 89:72–78.

Mon-Williams, M., J. R. Tresilian, N. C. Strang, P. Kochhar, and J. P. Wann. 1998. Improving vision: Neural compensation for optical defocus. *Proc Biol Sci* 265(1390):71–77.

Nachmias, J., R. Sansbury, A. Vassilev, and A. Weber. 1973. Adaptation to square-wave gratings—In search of the illusive third-harmonic. *Vision Res* 13(7):1335–1342.

O'Neil, S. F., A. Mac, G. Rhodes, and M. A. Webster. 2014. Adding years to your life (or at least looking like it): A simple normalization underlies adaptation to facial age. *PLoS ONE* 9(12):e116105.

Ohlendorf, A. and F. Schaeffel. 2009. Contrast adaptation induced by defocus—A possible error signal for emmetropization? *Vision Res* 49(2):249–256.

Ohlendorf, A., J. Tabernero, and F. Schaeffel. 2011. Neuronal adaptation to simulated and optically-induced astigmatic defocus. *Vision Res* 51(6):529–534.

Ohzawa, I., G. Sclar, and R. D. Freeman. 1982. Contrast gain control in the cat visual cortex. *Nature* 298(5871):266–268.

Oliva, A. and A. Torralba. 2001. Modeling the shape of the scene: A holistic representation of the spatial envelope. *Int J Comput Vis* 42(3):145–175.

Owsley, C. 2011. Aging and vision. *Vision Res* 51(13):1610–1622.

Parkosadze, K., T. Kalmakhelidze, M. Tolmacheva, G. Chichua, A. Kezeli, M. A. Webster, and J. S. Werner. 2013. Persistent biases in subjective image focus following cataract surgery. *Vision Res* 89:10–17.

Pesudovs, K. 2005. Involvement of neural adaptation in the recovery of vision after laser refractive surgery. *J Refract Surg* 21(2):144–147.

Pesudovs, K. and N. A. Brennan. 1993. Decreased uncorrected vision after a period of distance fixation with spectacle wear. *Optom Vis Sci* 70(7):528–531.

Poulere, E., J. Moschandreas, G. A. Kontadakis, I. G. Pallikaris, and S. Plainis. 2013. Effect of blur and subsequent adaptation on visual acuity using letter and Landolt C charts: Differences between emmetropes and myopes. *Ophthalmic Physiol Opt* 33(2):130–137.

Radhakrishnan, A., C. Dorronsoro, L. Sawides, and S. Marcos. 2014. Short-term neural adaptation to simultaneous bifocal images. *PLoS One* 9(3):e93089.

Radhakrishnan, A., C. Dorronsoro, L. Sawides, M. A. Webster, and S. Marcos. 2015. A cyclopean neural mechanism compensating for optical differences between the eyes. *Curr Biol* 25(5):R188–R189.

Rajeev, N. and A. Metha. 2010. Enhanced contrast sensitivity confirms active compensation in blur adaptation. *Invest Ophthalmol Vis Sci* 51(2):1242–1246.

Ravikumar, S., A. Bradley, and L. Thibos. 2010. Phase changes induced by optical aberrations degrade letter and face acuity. *J Vis* 10(14):18.

Rosenfield, M., S. E. Hong, and S. George. 2004. Blur adaptation in myopes. *Optom Vis Sci* 81(9):657–662.

Ross, H. 1974. *Behavior and Perception in Strange Environments*. London, U.K.: George Allen & Unwin.

Rossi, E. A. and A. Roorda. 2010. Is visual resolution after adaptive optics correction susceptible to perceptual learning? *J Vis* 10(12):11.

Ruderman, D. L. 1997. Origins of scaling in natural images. *Vision Res* 37(23):3385–3398.

Sabesan, R. and G. Yoon. 2009. Visual performance after correcting higher order aberrations in keratoconic eyes. *J Vis* 9(5):6.1–6.10.

Sabesan, R. and G. Yoon. 2010. Neural compensation for long-term asymmetric optical blur to improve visual performance in keratoconic eyes. *Invest Ophthalmol Vis Sci* 51(7):3835–3839.

Sawides, L., P. de Gracia, C. Dorronsoro, M. Webster, and S. Marcos. 2011b. Adapting to blur produced by ocular high-order aberrations. *J Vis* 11(7):22-22. doi: 11.7.21 [pii] 10.1167/11.7.21.

Sawides, L., P. de Gracia, C. Dorronsoro, M. A. Webster, and S. Marcos. 2011a. Vision is adapted to the natural level of blur present in the retinal image. *PLoS One* 6(11):e27031.

Sawides, L., C. Dorronsoro, P. De Gracia, M. Vinas, M. Webster, and S. Marcos. 2012. Dependence of subjective image focus on the magnitude and pattern of high order aberrations. *J Vis* 12(8):4.1–4.12.

Sawides, L., S. Marcos, S. Ravikumar, L. Thibos, A. Bradley, and M. Webster. 2010. Adaptation to astigmatic blur. *J Vis* 10(12):22.

Schwartz, O., A. Hsu, and P. Dayan. 2007. Space and time in visual context. *Nat Rev Neurosci* 8(7):522–535.

Shadmehr, R., M. A. Smith, and J. W. Krakauer. 2010. Error correction, sensory prediction, and adaptation in motor control. *Annu Rev Neurosci* 33:89–108.

Sharpee, T. O., H. Sugihara, A. V. Kurgansky, S. P. Rebrik, M. P. Stryker, and K. D. Miller. 2006. Adaptive filtering enhances information transmission in visual cortex. *Nature* 439(7079):936–94:

Simoncelli, E. P. and B. A. Olshausen. 2001. Natural image statistics and neural representation. *Annu Rev Neurosci* 24:1193–1216.

Solomon, S. G. and A. Kohn. 2014. Moving sensory adaptation beyond suppressive effects in single neurons. *Curr Biol* 24(20):R1012–R1022.

Solomon, S. G., J. W. Peirce, N. T. Dhruv, and P. Lennie. 2004. Profound contrast adaptation early in the visual pathway. *Neuron* 42(1):155–162.

Srinivasan, M. V., S. B. Laughlin, and A. Dubs. 1982. Predictive coding: A fresh view of inhibition in the retina. *Proc R Soc Lond B Biol Sci* 216(1205):427–459.

Thompson, P. and D. Burr. 2009. Visual aftereffects. *Curr Biol* 19(1):R11–R14.

Thorn, F. and F. Schwartz. 1990. Effects of dioptric blur on Snellen and grating acuity. *Optom Vis Sci* 67(1):3–7.

Tolhurst, D. J. 1972. Adaptation to square-wave gratings: Inhibition between spatial frequency channels in the human visual system. *J Physiol* 226(1):231–248.

Tolhurst, D. J., Y. Tadmor, and T. Chao. 1992. Amplitude spectra of natural images. *Ophthalmic Physiol Opt* 12(2):229–232.

Ungerleider, L. G. and A. H. Bell. 2011. Uncovering the visual alphabet: Advances in our understanding of object perception. *Vision Res* 51(7):782–799.

van der Schaaf, A. and J. H. van Hateren. 1996. Modelling the power spectra of natural images: Statistics and information. *Vision Res* 36(17):2759–2770.

Vera-Diaz, F. A., J. Gwiazda, F. Thorn, and R. Held. 2004. Increased accommodation following adaptation to image blur in myopes. *J Vis* 4(12):1111–1119.

Vera-Diaz, F. A., R. L. Woods, and E. Peli. 2010. Shape and individual variability of the blur adaptation curve. *Vision Res* 50(15):1452–1461.

Vinas, M., L. Sawides, P. de Gracia, and S. Marcos. 2012. Perceptual adaptation to the correction of natural astigmatism. *PLoS ONE* 7(9):e46361.

von der Twer, T. and D. I. MacLeod. 2001. Optimal nonlinear codes for the perception of natural colours. *Network* 12(3):395–407.

Wainwright, M. J. 1999. Visual adaptation as optimal information transmission. *Vision Res* 39(23):3960–3974.

Wallman, J. and J. Winawer. 2004. Homeostasis of eye growth and the question of myopia. *Neuron* 43(4):447–468.

Wandell, B. A. 1995. *Foundations of Vision*. Sunderland, MA: Sinauer.

Wang, B., K. J. Ciuffreda, and B. Vasudevan. 2006. Effect of blur adaptation on blur sensitivity in myopes. *Vision Res* 46(21):3634–364:

Wang, Z. and E. P. Simoncelli. 2004. Local phase coherence and the perception of blur. In *Advances in Neural Information Processing Systems*, vol. 16, Vancouver BC: MIT Press.

Wark, B., B. N. Lundstrom, and A. Fairhall. 2007. Sensory adaptation. *Curr Opin Neurobiol* 17(4):423–429.

Watson, A. B. and A. J. Ahumada. 2011. Blur clarified: A review and synthesis of blur discrimination. *J Vis* 11(5):10-10. doi: 10.1167/11.5.1

Webster, M. A. 2011. Adaptation and visual coding. *J Vis* 11(5):3.1–3.2

Webster, M. A. 2014. Probing the functions of contextual modulation by adapting images rather than observers. *Vision Res.* 104:68–79.

Webster, M. A. 2015. Visual adaptation. *Annu Rev Vis Sci*, 1:547–567.

Webster, M. A., M. A. Georgeson, and S. M. Webster. 2002. Neural adjustments to image blur. *Nat Neurosci* 5(9):839–840.

Webster, M. A., K. Halen, A. J. Meyers, P. Winkler, and J. S. Werner. 2010. Colour appearance and compensation in the near periphery. *Proc R Soc B Biol Sci* 277(1689):1817–1825.

Webster, M. A., D. Kaping, Y. Mizokami, and P. Duhamel. 2004. Adaptation to natural facial categories. *Nature* 428(6982):557–56

Webster, M. A. and D. Leonard. 2008. Adaptation and perceptual norms in color vision. *J Opt Soc Am A* 25(11):2817–2825.

Webster, M. A. and D. I. A. MacLeod. 2011. Visual adaptation and face perception. *Philos Trans R Soc Lond B Biol Sci* 366(1571):1702–1725.

Webster, M. A. and E. Miyahara. 1997. Contrast adaptation and the spatial structure of natural images. *J Opt Soc Am A Opt Image Sci Vis* 14(9):2355–2366.

Webster, M. A., Y. Mizokami, L. A. Svec, and S. L. Elliott. 2006. Neural adjustments to chromatic blur. *Spat Vis* 19(2–4):111–132.

Webster, M. A. and J. D. Mollon. 1995. Colour constancy influenced by contrast adaptation. *Nature* 373(6516):694–698.

Webster, M. A., S. M. Webster, J. MacDonald, and S. R. Bharadwadj. 2001. Adaptation to blur. In *Human Vision and Electronic Imaging*, SPIE 4299, pp. 69–78, San Jose, CA.

Webster, S. M., M. A. Webster, J. Taylor, J. Jaikumar, and R. Verma. 2001. Simultaneous blur contrast. In *Human Vision and Electronic Imaging*, SPIE 4299, pp. 414–422, San Jose, CA.

Werner, J. S. 1996. Visual problems of the retina during ageing: Compensation mechanisms and colour constancy across the life span. *Prog Retin Eye Res* 15(2):621–645.

Werner, J. S. and B. E. Schefrin. 1993. Loci of achromatic points throughout the life span. *J Opt Soc Am A* 10(7):1509–1516.

Williams, D. R. 1985. Aliasing in human foveal vision. *Vision Res* 25(2):195–205.

Wohlgemuth, A. 1911. On the aftereffect of seen movement. *Br J Psychol Monogr Suppl* 1:1–117.

Wolpert, D. M., J. Diedrichsen, and J. R. Flanagan. 2011. Principles of sensorimotor learning. *Nat Rev Neurosci* 12(12):739–751.

Wuerger, S. M., H. Owens, and S. Westland. 2001. Blur tolerance for luminance and chromatic stimuli. *J Opt Soc Am A Opt Image Sci Vis* 18(6):1231–1239.

Yehezkel, O., D. Sagi, A. Sterkin, M. Belkin, and U. Polat. 2010. Learning to adapt: Dynamics of readaptation to geometrical distortions. *Vision Res* 50(16):1550–1558.

Zaidi, Q., Ennis, R., Cao, D., and Lee, B. 2012. Neural locus of color afterimages. *Curr Biol*, 22(3):220–224.

Zalevsky, Z., S. Ben Yaish, A. Zlotnik, O. Yehezkel, and M. Belkin. 2010. Cortical adaptation and visual enhancement. *Opt Lett* 35(18):3066–3068.

Zhang, P., M. Bao, M. Kwon, S. He, and S. A. Engel. 2009. Effects of orientation-specific visual deprivation induced with altered reality. *Curr Biol* 19(22):1956–1960.

Zhou, J., Y. Zhang, Y. Dai, H. Zhao, R. Liu, F. Hou, B. Liang, R. F. Hess, and Y. Zhou. 2012. The eye limits the brain's learning potential. *Sci Rep* 2:364.

Contrast adaptation

Frank Schaeffel

Contents

Contrast vision and contrast adaptation have been extensively studied over the past 50 years. More than 1000 peer-reviewed articles have been published on this topic (PubMed 2014). Certainly, seeing contrast is a fundamental requirement for spatial vision. But there are major challenges: contrast vision should be possible over a huge range of illuminance to which we are exposed, and very low contrasts should be detectable over a wide range of spatial frequencies. The neural mechanisms mediating contrast vision should take into account that retinal image contrast is generally very low at high spatial frequencies, which contain the information about fine details and are therefore of particular importance for high visual acuity. This chapter tries to summarize some aspects of contrast vision, its limits, and what can be gained by contrast adaptation. Important earlier reviews on contrast vision and contrast adaptation were provided by Shapley and Enroth-Cugell (1984) and on visual adaptation by Kohn (2007). Many details described in their reviews are not recapitulated here. Since the topic is multifaceted, it is hoped that the preselection of aspects does not offend those researchers whose studies have not been mentioned due to space limitations. The figures include portraits of the authors when available.

21.1 WHY CONTRAST VISION STARTS AT THE FIRST SYNAPSE IN THE RETINA

Contrast vision—the ability to distinguish potentially tiny differences in luminance between adjacent positions in the visual scene, or in time—starts already at the first synapses in the visual system that connects rods and cones to bipolar and horizontal cells. It is known since long that the surrounding photoreceptors inhibit the output of the photoreceptor in the center (lateral inhibition), which is considered a trick to enhance edges and to facilitate the detection of low contrasts. It becomes clear that, different from digital cameras that store the luminance values of each single pixel, the visual system does not transmit the output of single photoreceptors (pixels) to the visual cortex. This makes sense for two main reasons: (1) the visual system suffers from information overload anyway due to the graded outputs from 125 million photoreceptors and, in the case of cones, a huge amount of possible "gray levels," and (2) the correlation of "pixel values" in neighbored photoreceptors is very high and it would be clever to make use of this correlation to reduce the information

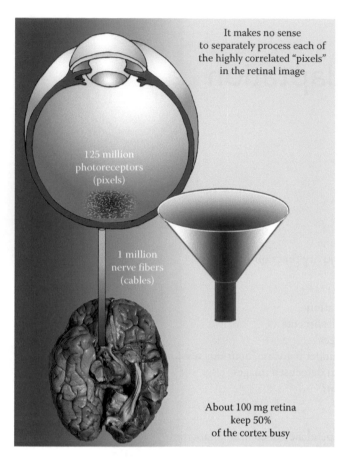

It makes no sense
to separately process each of
the highly correlated "pixels"
in the retinal image

125 million
photoreceptors
(pixels)

1 million
nerve fibers
(cables)

About 100 mg retina
keep 50%
of the cortex busy

Figure 21.1 Different from digital cameras, where the output of each single pixel is stored in a file, the visual system discards large amounts of information already at the first synapses in the retina. Humans have a 125 megapixel "sensor chip" in the eye but only 1 million "cables" to the brain, indicating that a compression of 100:1 is necessary at least. Furthermore, the bandwidth of the retinal ganglion cell axons is limited by the range of possible spike frequencies. This range covers only about 2 log units, while the ambient illuminance varies by 10 log units. A major task of the retina is therefore to cut down the visual information that is forwarded to the brain and one major step in this direction is to focus on spatiotemporal differences rather than absolute pixel values. Furthermore, high spatial resolution, requiring that each single pixel has a private line to the brain, is only available in the fovea. In the peripheral retina, the output of many thousands of photoreceptors may be pooled. The fovea covers only about 1° of the central visual field.

that needs to be transmitted through the optic nerve to the visual cortex. A central task of the retina is therefore to compress the output of 125 million photoreceptors into about 1 million nerve fibers that can be accommodated in the optic nerve (Figure 21.1).

A major trick is to discard the information about absolute luminances and to encode only differences in space (spatial contrast) or time (temporal contrast). As mentioned earlier, this starts already at the first synapse, which splits the photoreceptor outputs into ON and OFF bipolar cells. ON/OFF structures also show up in the receptive field structures of the output cells of the retina, the ganglion cells. Ganglion cells have circular receptive fields that can be divided into two basic classes, ON center and OFF surround, and OFF center and ON surround. In the first case, light in the center stimulates the ganglion cell and light in the periphery inhibits it. In the second case, light in the center has an inhibitory effect and light in the periphery is stimulatory to the

ganglion cell. As a result, homogenously illuminated areas scarcely stimulate ganglion cells and the information on absolute luminance is almost lost. Figure 21.2 shows a well-known demonstration that perceived brightness differs from absolute luminance cannot be reliably judged (Adelson's "checker shadow illusion").

21.2 HOW IS CONTRAST SENSITIVITY MEASURED: THE CONTRAST SENSITIVITY FUNCTION

There are two major variables to describe contrast: (1) Michelson contrast and (2) Weber contrast (Figure 21.3). Weber contrast is best suited to describe the contrast of a single object against a background. Michelson contrast is used to denote contrast of a periodic pattern, like a sine wave grating. It is important to realize that a contrast of 1 (or 100%) can scarcely be achieved in reality because it would require that no single photon comes from the darkest areas of a pattern, that is, its brightness should be zero—which is difficult to achieve. Examples of calculations Michelson contrast are shown in Figure 21.3.

The classical way to determine contrast sensitivity (CS) of a subject is to find the threshold contrast for detection of a pattern. To determine CS at different spatial frequencies (leading to the contrast sensitivity function [CSF]), sine wave gratings of different spatial frequencies are used. The reciprocal of the threshold contrast (as a fraction, not in percent) is defined as CS. Figure 21.4 illustrates that CS varies with spatial frequency. On the left, sine wave gratings are shown with increasing spatial frequency from the left to the right. Contrast drops equally for all spatial frequencies from the bottom to the top. It can be seen that, for lower spatial frequencies, the sine wave grating appears to vanish earlier than for the mid spatial frequencies. Obviously, CS is higher in the middle of the spatial frequency range. At higher spatial frequencies, CS declines again, generating a CSF with the shape of an inverted "U" (Figure 21.4, right). Above detection threshold (in the bottom part of the figure), contrast at different spatial frequencies is seen similar even though one might expect more contrast in the mid spatial frequency range. This phenomenon is termed "contrast constancy."

In humans, the CS peak is at 3–5 cyc/deg under photopic conditions. Two different factors account for the drop in CS at both ends of the tested spatial frequency range. At low spatial frequencies, the visual cortex may be responsible, or the fact that few receptive fields in the retina are large enough to respond to such a coarse pattern. On the high spatial frequency end, the optics of the eye is limiting. Like all optical imaging systems, it represents a low-pass filter for spatial frequencies. Fine details are imaged with lower contrast that elevates the detection threshold.

The CSF strongly depends on luminance (Figure 21.4, right). Its peak shifts toward lower spatial frequencies with decreasing luminance, and CS declines considerably down to 1/50 of its peak value from almost 500 under photopic conditions to less than 10 under scotopic conditions. Retinal illumination is measured in trolands. One troland is retinal illumination when a surface with a luminance of 1 cd/m² is observed through a one millimeter pupil. The unit ignores that retinal illumination also depends on the axial length of the eye

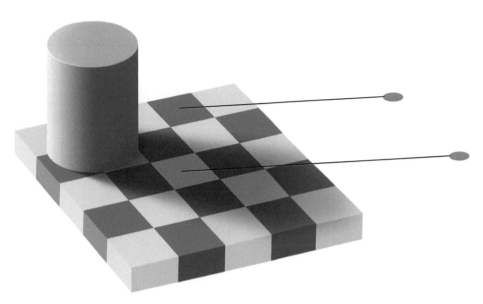

Figure 21.2 Absolute luminances are not carried on from the retina to the visual cortex, rather only relative brightness, as can be seen when the two checkerboard fields, denoted by the black lines, are compared. The lower field looks lighter than the upper one. However, if the pixel luminance in both fields is separately compared (as possible with the two isolated patches on the right), it becomes clear that they don't differ in luminance. (After Edward H. Adelson, web.mit.edu/persci/people/adelson/checkershadow_illusion.html, accessed October 3, 2016.)

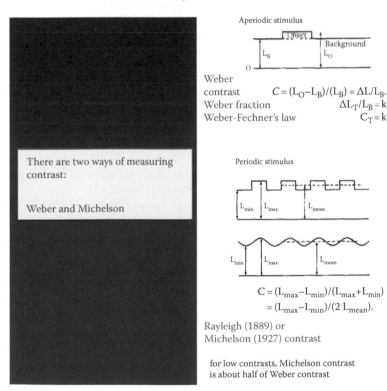

Shapley R, Enroth-Cugell C. Visual adaptation and retinal gain controls. In: *Progress in Retinal Research*, edited by Osborne N, Chader G. London: Pergamon Press, 1984, vol. 3, p. 263–346.

Visual Adaptation and Retinal Gain Controls

ROBERT SHAPLEY AND CHRISTINA ENROTH-CUGELL
Laboratory of Biophysics, Rockefeller University, New York, New York 10021, USA
and
Departments of Neurobiology & Physiology, and Engineering Sciences & Applied Mathematics, Northwestern University, Evanston, Illinois 60201, USA

Aperiodic stimulus

Weber contrast $C = (L_O - L_B)/(L_B) = \Delta L/L_B$.

Weber fraction $\Delta L_T/L_B = k$

Weber-Fechner's law $C_T = k$

There are two ways of measuring contrast:

Weber and Michelson

Periodic stimulus

$C = (L_{max} - L_{min})/(L_{max} + L_{min})$
$= (L_{max} - L_{min})/(2 L_{mean})$.

Rayleigh (1889) or Michelson (1927) contrast

for low contrasts, Michelson contrast is about half of Weber contrast

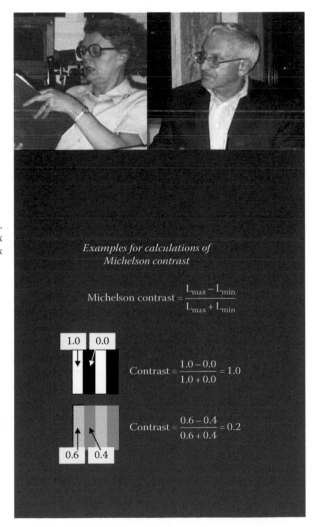

Examples for calculations of Michelson contrast

$$\text{Michelson contrast} = \frac{L_{max} - L_{min}}{L_{max} + L_{min}}$$

| 1.0 | 0.0 |

$$\text{Contrast} = \frac{1.0 - 0.0}{1.0 + 0.0} = 1.0$$

$$\text{Contrast} = \frac{0.6 - 0.4}{0.6 + 0.4} = 0.2$$

| 0.6 | 0.4 |

Figure 21.3 Description of Weber and Michelson contrast in Shapley's and Enroth-Cugell's classical review. On the right, examples for calculations of Michelson contrast are shown, although it must be noted that a contrast of 1.0 cannot be achieved in reality.

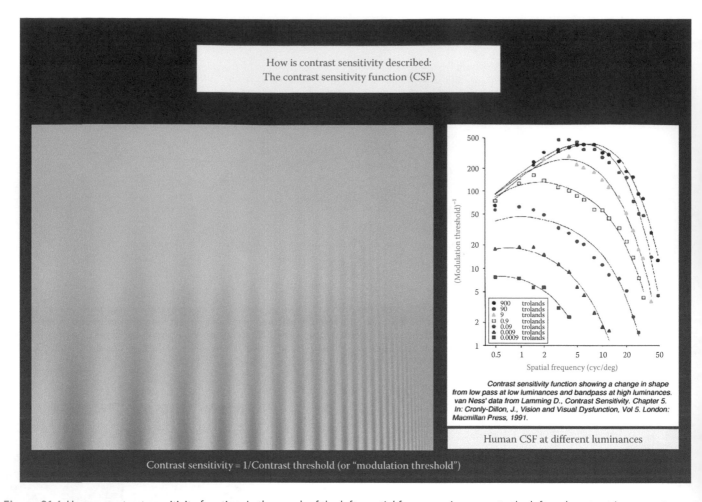

Figure 21.4 Human contrast sensitivity function. In the graph of the left, spatial frequency increases to the left and contrast increases to the top, exactly by the same amount for all spatial frequencies. Nevertheless, the sinusoidal modulation of luminance seems to disappear earlier both at low and high spatial frequencies (wide and narrow stripes, respectively) than in the mid spatial frequency range. Measuring the contrast (or modulation) at the detection threshold and taking the reciprocal of the threshold contrast lead to the contrast sensitivity functions which are shown on the right. These functions have the shape of an inverted "U." Contrast sensitivity declines with luminance and the spatial frequency range at which contrast sensitivity is maximal moves to the left to lower spatial frequencies.

since retinal image size is proportional to axial length and the same amount of photons is spread out over larger retinal areas.

21.3 LIMITS OF CONTRAST SENSITIVITY AT LOW LUMINANCES

Under low light conditions (scotopic vision, with rods only), our CS is limited by physics (e.g., Sterling 2003). Over the first three log units above the fully dark-adapted detection threshold of light, rods respond to single photons that can be seen when they are patch clamped and their "binary" responses are observed (single photons captured during their integration time of about 100 ms, the time period over which the rod outer segment membrane can summate the events). In this case, the statistics of arriving photons becomes important. At low luminances, photons arrive like single raindrops on pave stones. The probability of arrival is described by Poisson statistics. The standard deviation of the number of photon arrivals is then the square root of the number of photons captured during integration time. Therefore,

contrast vision at low luminances is limited by the variability of the "measured" luminances at each point. For instance, to detect a contrast of 0.1 (10%), at least n = 100 photons must be absorbed since luminance varies by the square root of n. To detect a contrast of 0.01 (1%), at least 10,000 photons must be absorbed. Therefore, over the first 2–3 log units of vision, CS rises with the square root of luminance (Figure 21.5).

There are also other factors that limit CS at low luminances. Even though each light-absorbing photopigment molecule, rhodopsin, has a half-life time of 317 years at body temperature, the packing of rhodopsin in the rod outer segments is so dense and the number of rhodopsin molecules so large that a significant number of photopigment molecules decay spontaneously and trigger the same events in the phototransduction cascade as if they had absorbed a photon. Furthermore, there is also "noise" in the biochemical steps of the phototransduction cascade itself. At highest light sensitivity (fully dark adapted), a single photon causes 2 mV hyperpolarization of the outer segment membrane. There is also a small chance of spontaneous opening of the cGMP-gated channels that are normally controlled in a light-dependent fashion. As a result of these factors, even in complete

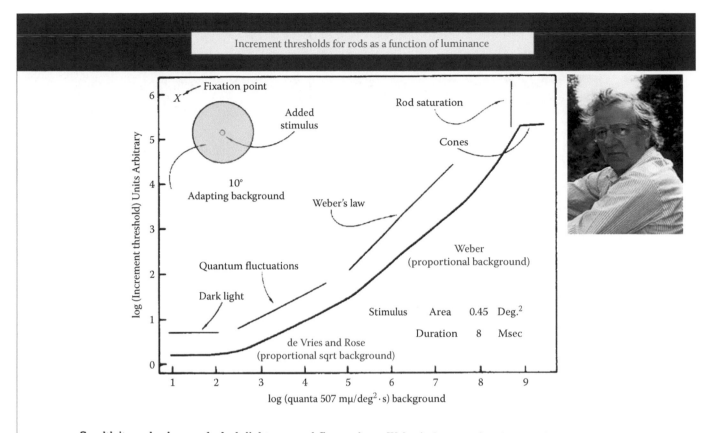

Increment thresholds for rods as a function of luminance

Sensitivity vs background: dark light, quantal fluctuations, Weber's Law, and rod saturation. The full repertoire of psychophysical laws is illustrated in this experimental curve which is the graph of sensitivity for a blue – green test spot, 0.75 deg in diameter, 8 ms in duration, presented 10 deg from the fovea on an orange colored background which was 10 deg in diameter. The different regimes are labeled in the figure. From Barlow(1965).

Figure 21.5 Factors limiting contrast sensitivity at different luminances. Below 2 log quanta/deg²·s at the corneal surface, the visual sensation is "dark light," determined by spontaneous decay of rhodopsin, and thermal noise in the steps of the phototransduction cascade. At higher luminances, contrast sensitivity is determined by the Poisson statistics, describing the probability of photon arrival in the photoreceptors and rises with the square root of the number of absorbed photons. Above about 5 log quanta/deg² s, contrast sensitivity rises proportional to the luminance of the observed pattern but contrast sensitivity finally reaches a maximum above 90 trolands (see Figure 21.4, right). (Modified from Shapley, R. and Enroth-Cugell, C., *Progr. Retinal Res.*, 3, 263, 1984.)

darkness, our visual sensation is not complete black but rather some noisy gray. At very low luminances (<1000 photons/deg² s on the corneal surface), sensation is limited by the "internal noise of the observer" (called "dark light"; Figure 21.5).

Body temperature also limits CS at very low luminances. Interestingly, poikilotherm animals (amphibia and reptiles), which have a body temperature similar to their environment, can gain CS from reduced thermal noise. It was shown that toads, hunting at night with a body temperature of 15°, have one log unit higher CS than humans, even after a correction was made for the difference in retinal image luminance in humans and toads (Aho et al. 1988).

Because the light sensitivity of humans rods is similar to the one of nocturnal mammals (also our rods can respond to a single photon), the only way for nocturnal animals to increase their CS at night is to increase retinal illuminance. This is achieved by lowering the aperture stop, that is, the ratio of focal length to pupil size (Figure 21.6). For instance, a barn owl has an aperture stop of less than 1, and a diurnal animal like a chameleon has

an aperture stop of about 5. The ratio of retinal illumination in both types of eyes is determined by the ratio of the squares of the two aperture stops, 1/25. Accordingly, the retinal image is 25 times brighter in the owl compared to the chameleon. Young children may have an aperture stop of around 2 (anterior focal length 16.7 mm, pupil size 8 mm), which means that their retinal image is only about four times darker than in an owl. Since CS rises with the square root of luminance in dim light, retinal illuminance can explain a difference in CS of a factor of 2. However, it was found that nocturnal mammals, like cat and owl, have about a six times higher CS at low light levels (Pasternak and Merigan 1981; Orlowski et al. 2012) than human subjects, and the difference cannot be fully explained only by optics. Cats and dogs have developed highly reflective layers behind the photoreceptors, the tapetum lucidum, to increase the chance that photons can be absorbed in a second pass. It has been calculated that the tapetum increases light sensitivity (and thereby CS) by further 29%. Because reflected photons are more scattered, this may be at the cost of visual acuity.

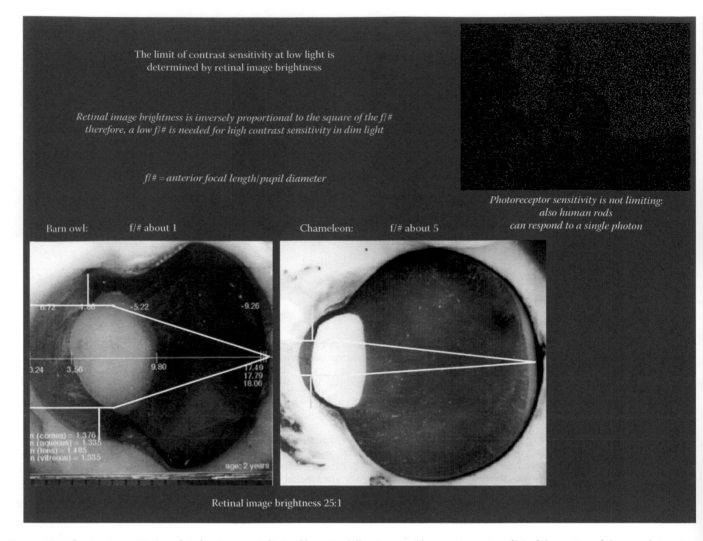

Figure 21.6 Contrast sensitivity at low luminances is limited by retinal illuminance. The aperture stop (f/#) of the optics of the eye determines contrast sensitivity at low luminances. It varies considerably among diurnal and nocturnal animals.

21.4 LIMITS OF CONTRAST SENSITIVITY AT HIGH LUMINANCES

Above the scotopic range (mediated by rod vision), CS follows Weber's law, which means that the ratio of object luminance to background luminance at detection threshold is proportional to the background luminance. The brighter the background, the larger must be the step in brightness between the object and the background to be detected. However, since (delta L)/L = constant (Weber's law), CS should also remain constant. This is not perfectly true (as can be seen in Figure 21.4, right) since CS does not further increase between 90 and 900 trolands. The reason might be light adaptation that reduces the gain of the phototransduction cascade and further neuronal mechanisms of "retinal gain control," like variations in the number of pooled photoreceptor signals. That photoreceptors can operate at a wide range of different luminances is a fundamental requirement for (spatial) vision. Without adaptation, the retina could be able to

encode only luminance differences of 1.5–2 log units (which matches the typical luminance differences in many visual scenes) because the bandwidth of spiking neurons (here the ganglion cell can vary their spike frequency only between 1 and 500, which is equivalent to 2.7 log units. Higher spike frequencies are not possible due to the refractory phase that follows after each action potential and has a duration of about 2 ms. It is clear that the bandwidth of spiking neurons would be too narrow to cover the full range of luminances over which we enjoy perfect spatial vision (a range of about 10 log units). Without adaptation, photoreceptors would saturate when luminance increases beyond 2 log units and further luminance differences could no longer be seen. The image would fade away. Saturation needs to be prevented already at the first step of vision, the phototransduction cascade, because there is no chance to recover a saturated signal any later. Changes in the gain of the phototransduction cascade are mediated by intracellular calcium levels. In bright light, most cGMP channels are closed and Ca influx is limited. Low Ca reduces the gain of the phototransduction cascade and shifts the photoreceptor response curve to higher luminances. Therefore, contrast vision remains functional over the wide range of luminances that occur in our visual environment. In the course of evolution, the human visual system has been optimized to function properly from very low light levels at night where only a few stars serve as light sources to the very bright conditions on the beach on a summer day. However, also light adaptation has its limits, for instance, at a glacier at noon on a sunny day, and glare may make it necessary to wear sun glasses to reduce retinal illuminance by a log unit or two.

21.5 CONTRAST SENSITIVITY IS NOT CONSTANT BUT SUBJECT TO ADAPTATION

The classical way of measuring CS is to determine threshold contrasts at different spatial frequencies, take their reciprocal, and plot the CSF (Figure 21.4, right). In this case, the threshold contrast was determined with a background of homogenous gray. Already 50 years ago, it was realized that CS is not stable but heavily dependent on the history of visual experience. If subjects look at a high-contrast sine wave grating with a defined spatial frequency, their CS for patterns of the respective spatial frequency is severely reduced (Figure 21.7a; see yellow arrow). Adaptation is moderately selective for the respective spatial frequency. For instance, if adaptation occurred to a sine wave grating with a spatial frequency of 7.1 cyc/deg, CS above 15 cyc/deg or below 2 cyc/deg remains unaffected (Blakemore and Campbell 1969, Figure 21.7a). Adaptation to phase-reversing sine wave gratings was also detected in cortical visually evoked potentials (VEPs; Figure 21.7b) since the amplitudes declined after adaptation but recovered after a period of normal visual experience. Contrast adaptation in the suprathreshold range was studied by Georgeson (Figure 21.7c) who found that "after adaptation to 32% contrast, a test contrast of 8% was matched with a (not adapted) matching contrast of 0.6%," indicating that CS had declined by more than an order of magnitude.

21.6 LEVEL OF CONTRAST ADAPTATION IN MORE NATURAL VISUAL ENVIRONMENTS

In reality, however, CS at a given spatial frequency is not determined by a homogenous gray background but rather by a continuously changing pattern of spatial details, which contains many different spatial frequency components. It is therefore more realistic to measure CS on top of patterned backgrounds. This was done by Bex, Solomon, and Dakin (Figure 21.8). In their experiments, the authors determined the threshold contrast for the subjects to detect a spatially scrambled, band-pass-filtered low-contrast circular pattern (visible in Figure 21.8a). The pattern was presented either on a "mean luminance background" of 50 cd/m² (Figure 21.8b, "unadapted"), after a movie was played for at least 5 s between the test intervals ("adapted"), or on top of the movie that was continuously played ("masked"). If CS was

measured with a movie in the background (either with correct phases among the spatial frequency components or with scrambled phases), CSFs varied considerably from the one measured with a "mean luminance background." It remained similar at high spatial frequencies under all three testing conditions but dropped in the spatial frequency range below 10 cyc/deg when the movie was played in the background (green curve, Figure 21.8). The drop reached an order of magnitude at spatial frequencies below 1 cyc/deg. Three conclusions can be drawn: (1) under natural viewing conditions, CS at lower spatial frequencies is reduced, probably because a lot of contrast is available at this spatial frequency range, (2) adaptation appears to equalize the perceived contrast across the spatial frequency spectrum ("whitening the spectrum"), and (3) at higher spatial frequencies, there is little adaptational change. Perhaps, because there is already little contrast available in the retinal image above 10 cyc/deg, CS is maintained high at all times (Figure 21.8, green arrow). It may be that the visual system operates already at its physical and neural limits of CS and that there is no room for further enhancement. It would not seem helpful to reduce CS by adaptation in the high spatial frequency range since fine details in the retinal image would no longer be seen. The effect of "whitening the spectrum" is illustrated in Figure 21.8c. When CS is reduced at low spatial frequencies by adaptation such that contrast appears similar across the entire spatial frequency spectrum (red lines in the bottom), the low-contrast edges in the image appear enhanced and the details become more clearly visible.

There is yet another mechanism that was proposed to help "whitening the spatial frequency spectrum": fixational eye movements. Rucci and his colleagues (Kuang et al., 2012) proposed that the spatiotemporal pattern of the miniature fixational eye movements that we do inadvertently during fixation of a target (and of which we are not aware) is such that it improves CS at high spatial frequencies at the cost of CS at lower spatial frequencies (Figure 21.9; compare "static" to "FEM" [= fixational eye movements considered]).

Apparently, the visual system has made every effort to optimize CS at higher spatial frequencies and the question is why. David Field has studied the contrast at different spatial frequencies in natural scenes (Figure 21.10). He found that the amplitude of Fourier components (spatial frequencies) in natural environment declines similarly for different scenes—that is, forest, bushes, rocks, and trees. As can be seen in the plot on the right, an almost linear decline in log amplitude can be observed, if plotted against the log of the spatial frequency. It is concluded that the amplitudes of the different spatial frequency components fall off with higher spatial frequencies (SF) like 1/SF, or their energy falls off with 1/SF². It becomes more clear why the visual system is optimized for high CS at high spatial frequencies: there is little energy in this range. In addition to the optics of the eye that already acts as a low-pass filter for spatial frequencies, also the input, the visual scene, has increasingly little energy at high spatial frequencies. These two factors combined, it becomes clear that the visual system has to develop as high CS in the high spatial frequency range as physically and neuronally possible.

Impact of the eye's optics on vision

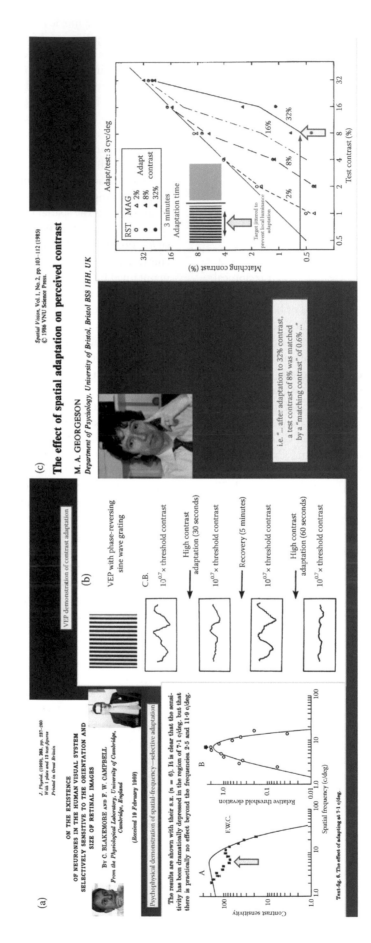

Figure 21.7 Contrast adaptation is selective for spatial frequency. (a) Contrast sensitivity was measured for different sine wave gratings in subject F.C.W. Contrast sensitivity declined by about half a log unit at 7 cyc/deg but remained unchanged below 2 cyc/deg and above 15 cyc/deg. The change is separately plotted on the adjacent figure on the right. (b) Adaptation to a phase-reversing sine wave grating could also be observed in recordings of visually evoked potentials. (c) Georgeson (1985) studied the amplitude of suprathreshold contrast adaptation and found that it can amount to more than one log unit.

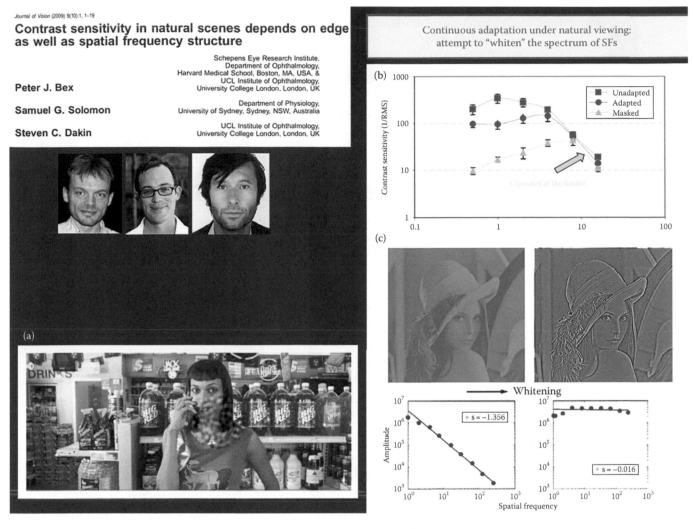

Figure 21.8 The contrast sensitivity function under natural viewing conditions varies from the one measured under conventional conditions (using sine wave gratings on homogenous backgrounds). In particular, contrast sensitivity is lower in the low spatial frequency range by an order of magnitude at 1 cyc/deg when a movie is played behind the test target. The authors measured contrast sensitivity of the subjects to detect a spatially scrambled, band-pass-filtered low-contrast circular pattern (visible in the picture in (a)) either on a homogeneous "mean luminance background" (generating the contrast sensitivity function shown as blue line in (b)), when a movie was presented just between the testing intervals (b red curve), or while the movie was continuously played in the background (b, green curve). In (c), the effects of equalizing contrast across the spatial frequency spectrum ("whitening the spectrum") are illustrated.

21.7 ADAPTATION TO THE SPATIAL FREQUENCY CONTENT OF REAL IMAGES

In 2002, Webster, Georgeson, and Webster studied how spatial filtering (reducing contrast of spatial frequency components in natural images) changes the perception of "sharpness" of their subjects. The spatial filter applied to a spatial frequency spectrum described by David Field (amplitude proportional to 1/SF) was a multiplication of each amplitude with a factor SF^S, where S was either –0.5 or +0.5. In the first case, the amplitude of spatial frequencies declined more steeply toward high spatial frequencies (a low-pass filter), and in the second case, the spatial frequency spectrum became more flat, enhancing the amplitudes at higher spatial frequencies (a high-pass filter).

Accordingly, a filtered image with the new spatial frequency spectra looked either blurry or sharp (Figure 21.11).

Adapting to a slope with s = zero (the original image) let them judge the original image as properly sharp. Adapting to a slope with s = –1 (low pass filtered) made them more tolerant against blur and even low-pass-filtered images looked sharp to them. Adapting them to a slope with s = +1 (high pass filtered) made them require more "sharpness." The "adaptation response curve" (Figure 21.12a) shows that the amplitude of adaptation is limited and that adaptation saturates at about 25% in both directions and that it was only about 50% of the actual difference in the exponent "S" (slope of function in Figure 21.12a about 0.5). Webster et al. (2002) also studied two further aspects, namely, whether the adapting and test images need to be identical (Figure 21.12b) and whether spatial long-ranging adaptational processes are involved (Figure 21.12c). They found that adaptation can

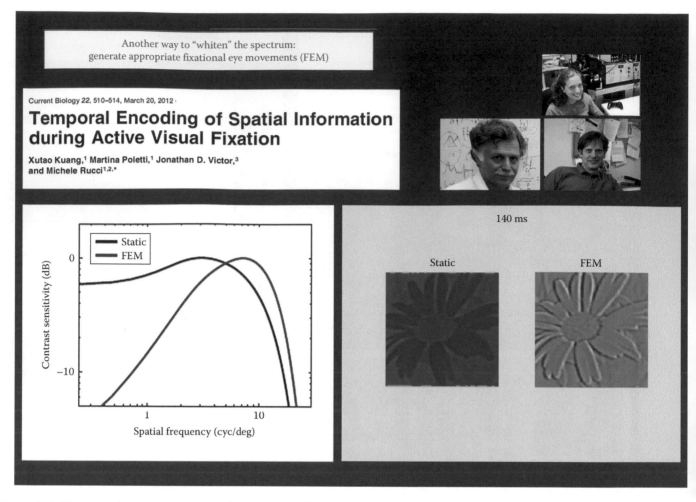

Figure 21.9 Change in the contrast sensitivity function due to fixational eye movements and their effects on the perception of a picture (static = no fixational eye movements, FEM = with fixational eye movements; a movie is available in the original publication).

easily be transferred from one image to the next, which may be somehow expected when the type of adaptation is specific for spatial frequencies but not to image content.

Webster et al. (2002) also found that the spatial frequency content in the surround affects the perceived sharpness of the fixated object in the center (Figure 21.12c). However, whether the level of sharpness in the periphery can affect the perceived sharpness of the fixated object remains uncertain as long as eye movements are possible to scan the image. In this case, spatial adaptation in the fovea is determined by the average sharpness of the image and not only the central target. This question could be resolved by using an eye tracker to present the images (Figure 21.12c) gaze contingently. The problem was recognized earlier. To reduce the effects of "mixing in" from the periphery, researchers used different presentation paradigms to elicit and test spatial adaptation: (1) images to be compared were presented sequentially (i.e., Webster et al. 2002; Figure 21.11), (2) images to be compared were presented next to each other but strict fixation was required on a fixation point between both images (i.e., Elliott et al. 2011; Haun and Peli, see Figure 21.19, below), (3) images were presented on top of each other but separated for each eye using a "custom-built stereoscope" (i.e., Kompaniez et al. 2013), and (4) interocular comparison of suprathreshold CS that assumes no interocular transfer (i.e., Ohlendorf et al. 2009).

21.8 CONTRAST SENSITIVITY FUNCTIONS AND CONTRAST ADAPTATION UNDER DEFOCUS: "SPURIOUS RESOLUTION"

An interesting feature of the modulation transfer of a defocused (diffraction-limited) optical system is that contrast does not completely fade away beyond the cutoff spatial frequency (Figure 21.13). Instead, there are a number of contrast reversals that become also visible when a sine wave grating is defocused, o a Siemens star (Figure 21.13c). This type of pseudoresolution is termed "spurious resolution." It also can confound measuremen of grating acuity with preferential looking techniques or VEPs (Bach et al. 1987).

As a result of spurious resolution, CSFs measured under defocus show conspicuous "notches" as can be seen in Figure 21.14. The position of the notches is dependent on the pupil size, the amount of defocus, and, in real eyes, the higher-order aberrations of the eye. They become particularly obvious if an eye is immersed in water and the loss of corneal refractive power generates large amounts of hyperopia (Figure 21.14, dotted blue line). Spurious resolution also needs to be taken into account when contrast adaptation is studied under

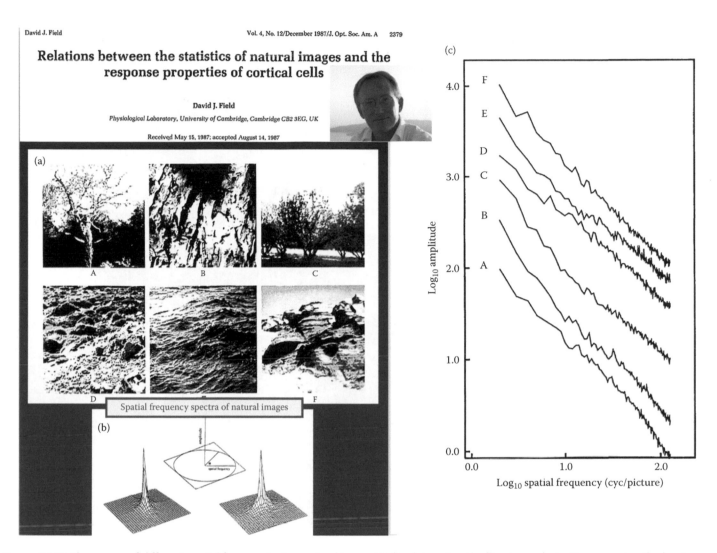

Figure 21.10 The energy of different spatial frequencies in a natural image can be determined by Fourier analysis. Decomposing the luminance profiles along radial lines from the center of the image to the edges into their Fourier components, their energy can be plotted as shown in (b) high energy in the center where the low spatial frequencies are depicted and declining energy for the higher spatial frequencies mapped out toward the edges. If the analysis is done for various natural scenes (a; see samples A–F) and the average amplitudes (the SQRT of the energy for a given radial distance from the center) are plotted for different spatial frequencies (c), it is obvious that the amplitudes decline linearly with increasing spatial frequency (SF) like 1/SF. (From Field, D.J., *J. Opt. Soc. Am. A.*, 4, 2379, 1987.)

defocus because there may be little adaptation at the spatial frequencies where the contrast sign reversal occurs.

21.9 THE NEXT STEP: ADAPTATION TO SPATIAL FREQUENCY SPECTRA IN DEFOCUSED IMAGES

Mon-Williams and colleagues (1998) were the first to study whether the claim made by myopic people that they can see sharper after they had taken off their glasses for a while has any justification. They tested visual acuity in emmetropic subjects right after they were made functionally myopic with a +1 diopter (D) trial lens. They compared visual acuity (as measured in logMAR, the logarithm of the minimum resolvable angular resolution in minutes of arc) right after the lens was put in place and 30 minutes later. The unit logMAR is counterintuitive since smaller values denote higher acuity. It can be seen in Figure 21.15,

left, that most of the data points are under the black line, which denotes equal acuity before and after adaptation. Visual acuity was better in most of the subjects after they had worn the +1 D lens for 30 minutes compared to their visual acuity right after the lens was put on.

The findings by Mon-Williams et al. (1998) show that visual acuity can improve by spatial adaptation, in the presence of persisting defocus. The findings were later confirmed by others. For instance, Rosenfield et al. (2004) measured visual acuity in fully corrected and moderately (mean –1.85 D) myopic subjects, immediately after they had taken off their glasses and 3 hours later. During the adaptation period, their visual acuity increased from 0.76 to 0.53 logMAR. Rosenfield et al. (2004) also confirmed that there were no changes in the optics of the eye using noncycloplegic autorefraction. They attributed the increase in visual acuity under defocus to "perceptual adaptation to the blurred images, which may occur at central sites within the visual cortex."

The topic was further studied by Rajeev and Metha (2010) who measured CS, rather than visual acuity, before and after

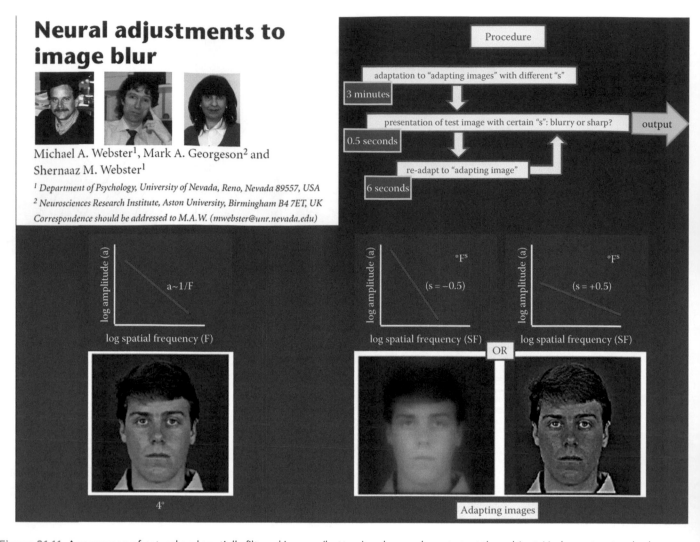

Neural adjustments to image blur

Michael A. Webster[1], Mark A. Georgeson[2] and Shernaaz M. Webster[1]

[1] Department of Psychology, University of Nevada, Reno, Nevada 89557, USA

[2] Neurosciences Research Institute, Aston University, Birmingham B4 7ET, UK

Correspondence should be addressed to M.A.W. (mwebster@unr.nevada.edu)

Figure 21.11 Appearance of natural and spatially filtered images (bottom) and procedures to test the subjects' judgment as to whether an image is "too sharp" or "too blurry." Bottom: images appear natural and in good focus when the amplitudes of spatial frequencies follow about the "1/SF rule" (see text). If amplitudes decline more steeply with increasing spatial frequency, the images appear blurry, and if they fall off more flat, the images appear too sharp. The judgments of the subjects were heavily affected by the previous exposure to other filtered images. Note that Webster et al. use "F" for spatial frequency, while "SF" is used in the current chapter.

adaptation to myopic defocus of 2 D for 30 minutes on a computer screen with a luminance of 46 cd/m² (903 trolands at 5 mm pupil size). They found that, under persisting defocus, CS between 8 and 12 cyc/deg increased by about 60% (=10^0.2; Figure 21.16b). No change was observed between 2 and 4 cyc/deg, which could have several reasons. It could be that defocus has too little effect on contrast in this spatial frequency range; however, this assumption can be refuted because the first Bessel function shows that there is almost no contrast at these spatial frequencies with 2 D of defocus; therefore, contrast adaptation should have been elicited (cutoff spatial frequency with a 5 mm pupil size and 2 D of defocus is at 2.1 cyc/deg). It also means that the range in which no adaptation was observed is already in the range of spurious resolution. It could be that contrast adaptation is generally limited in this spatial frequency range: humans have the highest CS (around 400–500) between 2 and 5 cyc/deg (Figure 21.16a). Perhaps the physical limit of CS is reached and no further improvement is possible since the measurements were done at about 900 trolands. Moreover, it remains puzzling why an increase in CS was found at 8–12 cyc/deg, far above the cutoff spatial

frequency. It is unclear how adaptation far above the cutoff spatial frequency can increase visual acuity. This leads to an interesting question: if adaptation would be limited above 10 cyc/deg as found in the study by Bex et al. (2009; Figure 21.8), there should be little room for adaptational improvement of visual acuity if it is already close to maximum. The amount of defocus for which contrast is zero at 10 cyc/deg is only 0.5 D (for a 5 mm pupil: defocus [D] = 21.3/(5 mm × 10 cyc/deg)). It is obvious that other published studies examined adaptational improvements of visual acuity with larger amounts of defocus, which affected contrast already at much lower spatial frequencies. Adaptation was studied in relatively low acuity ranges (logMAR > 0.6). Finally, it is not clear how consistent a defocus of 2 D was because the baseline refractions of the subjects have also some effect on retinal image defocus with a +2 D lenses.

Another surprising finding was that CS dropped at lower spatial frequencies after adaptation to defocus (Figure 21.16b). The advantage of this adaptational change is not clear. It could be that the total amount of contrast that can be transmitted over all spatial frequencies (basically the area under the CSF curve)

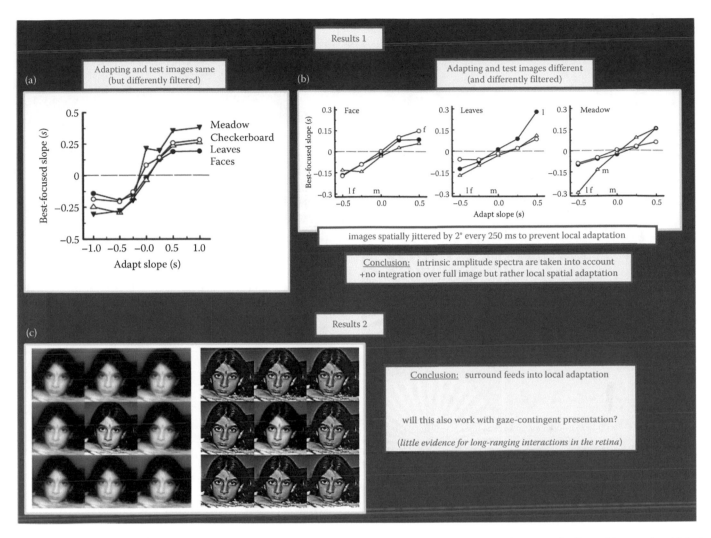

Figure 21.12 Adaptation to spatial frequency distributions in natural images. (a) Subjects were adapted to spatially filtered images in which the slopes of the spatial frequency distributions were manipulated (procedures and experimental protocol shown in Figure 21.11). Their judgment as to whether an image was blurry (s < 0) or too sharp (s > 0) was strongly influenced by the s value of the image that was previously used for adaptation. (b) Results were similar when adapting and test image contents were different. (c) The sharpness in the surround of a fixated object influences the judgment although it remains to be tested whether this holds true with gaze-contingent presentation of the image.

is limited by the bandwidth of the neurons, or the processing capacity in the visual cortex, and that the increase in CS at the high spatial frequency range can only be at the cost of the CS at the low spatial frequencies.

21.10 A RELATION OF CONTRAST ADAPTATION TO MYOPIA DEVELOPMENT?

It is interesting that the amplitudes of contrast adaptation vary among the different refractive groups. "Late onset myopes" (myopia developed during school years in early adolescence) show the largest amplitudes (i.e., Cufflin et al. 2007). It is unclear whether the larger amplitude of adaptation is the result of more defocused retinal images in myopia or whether it was there already before myopia developed but generated more tolerance to defocus and could therefore have stimulated its development. It is well established that poor retinal image contrast triggers the development of deprivation myopia in both animal models

and humans although it has not been resolved until today why positive defocus has a strong inhibitory effect on axial eye growth (i.e., Wallman and Winawer 2004).

For these reasons, Wallman suggested already in 1996 that the level of contrast adaptation (which changes with retinal image defocus) might represent a retinal error signal for the control of axial eye growth and myopia although it can only provide a unidirectional growth stimulus (Wallman and Schaeffel, 1996). The problem of emmetropization is that defocus of the retinal image varies rapidly over time depending on viewing distance and the level of accommodation. Therefore, momentary "measurements" of defocus cannot provide useful information for the control of axial eye growth. An integrator with long time constant is needed. A candidate would be the level of contrast adaptation that changes only slowly with the level of image sharpness. It can be observed in the chicken model of myopia that defocus imposed by spectacle lenses increases CS, by shifting the level of contrast adaptation (Diether et al. 2001; Figure 21.17).

Drugs that inhibit myopia development after intravitreal injection (i.e., atropine) also increase CS (Diether and Schaeffel 1999),

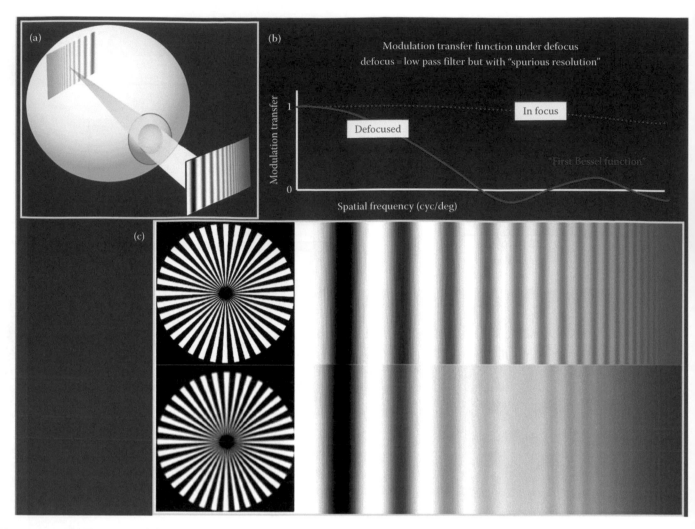

Figure 21.13 (a) The optics of the eye act as a low-pass filter for spatial frequencies even when the retinal image is in best focus, due to higher-order aberrations and diffraction at the pupil aperture. (b) When the image is defocused, contrast at higher spatial frequencies dro further as described by the first Bessel function. The spatial frequency at which contrast approaches zero (the first phase reversal) can be estimated by the simple equation SF = 21.3/(defocus * pupil size) (e.g., Schaeffel and de Queiroz 1990). Interestingly, beyond the cutoff spatial frequency, contrast returns again, although with reversed phase. Several phase reversals can be observed as illustrated in (c, sine wave pattern in the bottom).

suggesting a retinal mechanism. A possible biochemical mediator might be dopamine. Its release from dopaminergic amacrine cells is stimulated by atropine (Schwahn et al. 2000) and reduced when image contrast is poor (Feldkaemper et al. 1999). It is known for long that myopia development goes along with low retinal dopamine levels (Stone et al. 1989) and that intravitreal application of dopamine agonists can inhibit myopia development perhaps because they compensate for the drop in dopamine release.

21.11 ADAPTATION TO COMPLEX OPTICAL ABERRATIONS

In 2004, Artal and colleagues provided convincing evidence that there must be more than spatial frequency selective contrast adaptation. The wavefront errors in the pupil plane of subjects were measured and the test target presented to the subjects when wavefronts were rotated in steps of 30°. Since the root mean square (RMS) wavefront error remained always the same during rotation, the contrast at different spatial frequencies

was preserved. Subjects were asked to judge whether a stimulus, in this case a semirandom scrambled green spatial pattern (green circular pattern in the center of Figure 21.18), is perceived sharp or more blurred. It turned out that subjects judged the patterns as sharpest when the wavefront error map was oriented as in the natural eyes. Any other orientation caused the perception of mo blur. Since the contrast at different spatial frequencies did not change, contrast adaptation at different spatial frequencies coul not account for the effect. Obviously, this kind of adaptation must be more sophisticated; the question arises as to whether it could have taken place in the retina.

A similar study was performed by Sawides et al. (2011). They selected subjects with highly different levels of higher-order aberrations, with the RMS wavefront errors ranging from about 0.1 to 0.4 μm, and Strehl ratios (= the ratio of the area under the real modulation transfer function divided by the area under the diffraction-limited modulation transfer function) ranging from 0.097 to 0.193. The Strehl ratios show that the optical transfer functions were only about 10%–20% of the diffraction-limited transfer functions, which means not very good optics anyway.

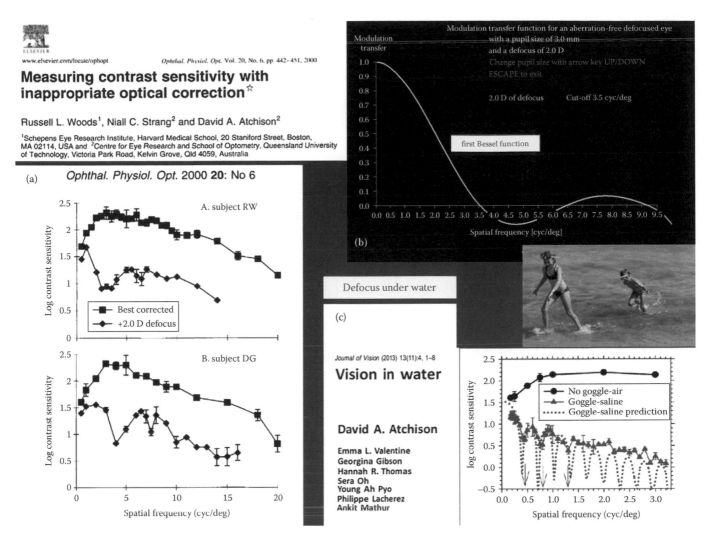

Figure 21.14 (a) Notches in the contrast sensitivity function in two subjects measured with 2 D of myopic defocus. In both subjects, contrast sensitivity is unexpectedly reduced between 3 and 4 cyc/deg, which is due to spurious resolution, (b) the first Bessel function for a pupil size of 3 mm and 2 D of defocus. The first phase reversal is at 3.5 cyc/deg, which matches the experimental results in (a). Under water, the eye becomes about 44 D hyperopic and contrast sensitivity is expected to show many notches (c, dotted blue line). In reality, some effects of spurious resolution are visible in the measured contrast sensitivity functions for the first phase reversals (continuous curve with blue triangles), but variations in pupil size and aberrations reduce the impact of the notches.

Two major results came out of these studies: (1) subjects with poorer optics judged images as sharp even if they were more blurry, compared to subjects with better optics who required images to contain more energy at higher spatial frequencies to judge them as sharp (this let the authors chose the title of the paper as *Vision Is Adapted to the Natural Level of Blur Present in the Retinal Image*), and (2) the adapting image may differ from the test image without that the amplitude of adaptation is reduced. This finding is in line with findings by Webster et al. (Figure 21.12) and excludes that afterimages played a role in the studied effects.

Haun and Peli (2013) no longer used stationary images for adaptation and testing but rather presented high- or low-pass-filtered video clips (Figure 21.19). Similar to Webster et al. (2002), they used a factor per spatial frequency component to either steepen (low-pass filter) or flatten (high-pass filter) the spatial frequency spectrum. The procedure made it possible to quantify the adaptation effects. Among other findings, they found that adaptation to certain spatial frequency spectra does not require stationary images but rather takes place in a highly dynamic way in a moving scene. Haun and Peli

(2013) also studied the gain of adaptation and found the largest gain around the position of best focus.

Ohlendorf et al. (2011) studied whether adaptation to aberrations, in this case astigmatism, can improve visual acuity. Subjects had to watch movies for 10 minutes, either with calculated (by convolution with an astigmatic point spread function) or optically induced astigmatism (induced by astigmatic lenses). Subsequently, their visual acuity was tested in astigmatically defocused letter charts, or letter charts including calculated astigmatic defocus. Subjects who had undergone adaptation to astigmatic defocus (either calculated or real) had gained about 0.1 logMAR or about 25% in visual acuity (Figure 21.20). Adaptation was highly selective for the axis of astigmatism; adaptation to another axis of astigmatism did not improve their visual acuity in the same letter chart.

Vinas et al. (2012) found that, in some subjects, the optimal adaptation to astigmatic defocus may take up to a week. Similar to Ohlendorf et al. (2011), they found that adaptation to astigmatic defocus was highly selective for the axis of astigmatism.

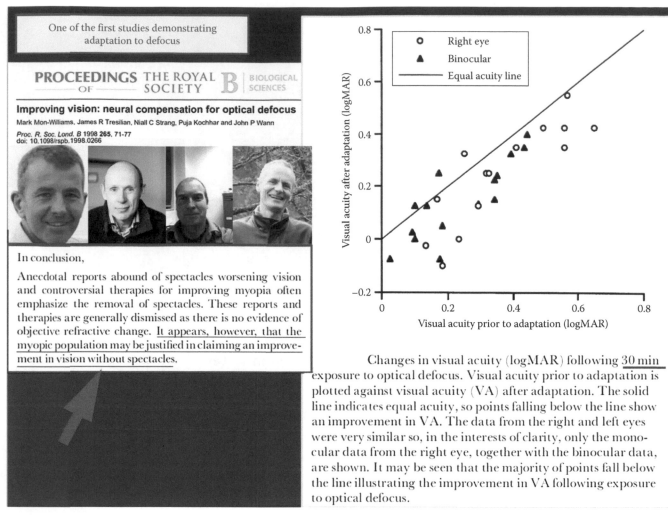

PROCEEDINGS OF THE ROYAL SOCIETY B BIOLOGICAL SCIENCES

Improving vision: neural compensation for optical defocus

Mark Mon-Williams, James R Tresilian, Niall C Strang, Puja Kochhar and John P Wann

Proc. R. Soc. Lond. B 1998 265, 71-77
doi: 10.1098/rspb.1998.0266

One of the first studies demonstrating adaptation to defocus

In conclusion,

Anecdotal reports abound of spectacles worsening vision and controversial therapies for improving myopia often emphasize the removal of spectacles. These reports and therapies are generally dismissed as there is no evidence of objective refractive change. It appears, however, that the myopic population may be justified in claiming an improvement in vision without spectacles.

Changes in visual acuity (logMAR) following 30 min exposure to optical defocus. Visual acuity prior to adaptation is plotted against visual acuity (VA) after adaptation. The solid line indicates equal acuity, so points falling below the line show an improvement in VA. The data from the right and left eyes were very similar so, in the interests of clarity, only the monocular data from the right eye, together with the binocular data, are shown. It may be seen that the majority of points fall below the line illustrating the improvement in VA following exposure to optical defocus.

Figure 21.15 Increase in visual acuity by adaptation to defocus of +1 D. When visual acuity is compared directly after a defocus of +1 D is imposed, and 30 minutes later, most of the subjects show an improvement of their visual acuity (their data are plotted under the black line, which indicates no change). Note that changes in acuity are plotted on the axes, not absolute acuity—it would be too striking if adaptation could occur in a range of maximal acuity (around logMAR 0.0).

21.12 WHERE DOES CONTRAST ADAPTATION OCCUR?

Researchers were always concerned about that they may not really study contrast adaptation but rather local luminance adaptation. The latter generates "afterimages" and a nice demonstration is a "Siemens star" that can be seen inverted in contrast when fixation moves (Figure 21.21). It interesting that the afterimage is visible almost instantaneously although the speed by which the biochemical gain in the phototransduction cascade is adjusted is limited, as one can observe when the light is switched on in the morning in a dark bedroom.

In most published studies, precautions were taken to exclude that the observed contrast adaptation was, in fact, just due to afterimages. For instance, Webster et al. (2002) jittered their images every 250 ms to prevent "local adaptation." However, there is evidence that contrast adaptation takes place already at the photoreceptor synapses. An example is that we don't see our retinal blood vessels, even though they cause shades on the photoreceptors behind them ("angioscotomas"). It can be nicely demonstrated

that local (contrast) adaptation prevents that they are seen. If one moves a small aperture (i.e., a pinhole in a little cardboard) in front of the pupil while looking at a white homogenous surface, the positions of the shades of the blood vessels move and adaptation starts failing. Suddenly, the retinal blood vessels become visible as black structures and one can also nicely see that the fovea is a circular field that is free of blood vessels. It would not make sense to inform the visual cortex about the spatial adaptation to the pattern of shades of the blood vessels and to involve interocular transfer, because the patterns are different in both eyes. It can therefore be assumed that this type of contrast adaptation is purely retinal and probably occurs in the photoreceptors themselves. Also the Siemens star (Figure 21.21) causes retinal adaptation and, if fixated with one eye for a while with the other covered, then switching to a homogenous gray surface like the background in Figure 21.21, exposing the other eye, no afterimages can be seen. The lack of interocular transfer suggests a retinal locus, or at least somewhere in the monocular pathway from retina to LGN and V1. If the retina can adapt to local contrasts, adaptation to low-pass-filtered or high-pass-filtered images can probably also

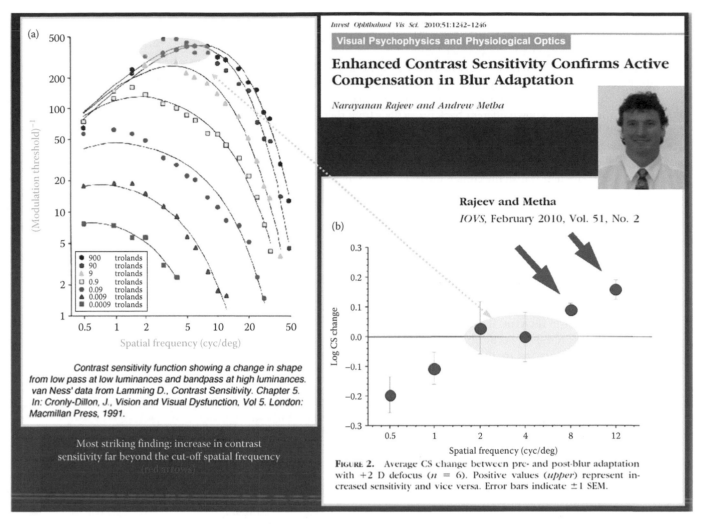

Figure 21.16 Changes in contrast sensitivity following exposure to myopia defocus of 2 D for 30 minutes and measured at 6 different spatial frequencies. Three findings were striking in (b): there was a reduction in contrast sensitivity at spatial frequencies below the estimated cutoff at around 2 cyc/deg, no change in the range where the retinal contrast might have been close to zero, and an increase in contrast sensitivity at spatial frequencies above the cutoff spatial frequency. It would be important to know whether the lack of a change in contrast sensitivity between 2 and 4 cyc/deg was due to the fact that contrast sensitivity was already maximal (it is the spatial frequency range with maximal contrast sensitivity in humans; see dotted yellow arrow pointing to (a)).

be explained by retinal mechanisms. If no transfer of the spatial adaptation can be found to the nonadapted eye, it can be assumed that the mechanism is retinal (or at least on the monocular part of the afference). There is no interocular transfer before the binocular neurons in layer 4 and beyond of the visual cortex.

There is ample evidence that the spatial filters in the retina can be shaped by visual factors or neurotransmitters in the retina. In chicks exposed to defocus, CS is enhanced after 30 minutes (Figure 21.17). The observation that CS can be similarly enhanced by a single intravitreal injection of a muscarinic antagonist (Diether and Schaeffel 1999) is in favor of the assumption of a retinal mechanism. Also retinal dopamine is well known to change horizontal cell (i.e., Kaneko and Stuart 1984) and amacrine cell coupling (i.e., Hampson et al. 1992), changing receptive field sizes and thereby the CSFs of the retina. Exposing the retina to low-pass-filtered images lowers dopamine release and increases receptive field sizes—again a retinal effect on CS at different spatial frequencies. In summary, adaptation to different spatial frequency spectra could well be mediated by just retinal mechanisms.

The more complex adaptations to optical aberrations and astigmatism, or adaptations to face distortions studied by Webster et al. (2002), are different. First, retinal ganglion cells and cells in the LGN show no orientation tuning. Adaptation to tilt of edges must therefore happen at higher levels. Second, one would expect interocular transfer when adaptations are taking place in visual cortex. In fact, Kompaniez et al. (2013) have found that adaptation to simulated optical defocus, simulated astigmatic defocus, or images in which the slope of the amplitude spectrum was varied to artificially sharpen or blur the image produced strong adaptation with pronounced interocular transfer of 60%–90% (Figure 21.22). It is striking that interocular transfer was so high given that the spatial filtering involved changing contrast at spatial frequencies—which could occur also just in the retina. It could be that the adaptation time of 120 s also triggered cortical changes in addition to retinal adaptation. Similar to Webster et al. (2002), Kompaniez et al. (2013) jittered their targets very 100 ms by ±16 pixels to avoid "local light adaptation" (retinal afterimages).

Another interesting observation was that when both eyes were adapted to different amounts of blur, the aftereffect was dominated by the eye with the sharper image.

The underlying pathways are illustrated in Figure 21.23 in simplified fashion. The (already perhaps partially adapted) output of the retina provides input to the contralateral and ipsilateral cortex. There are no known efferent back projections from there to the fellow eye. Therefore, binocular neurons in visual cortex must merge the adapted and the nonadapted inputs from both eyes, and a mixed signal must be forwarded to higher centers where the interocular transfer becomes apparent to the subject.

In summary, it is proposed that *more complex forms of adaptation can only occur at cortical levels where they are extracted and processed.* Since the retina itself does not extract forms, it also cannot adapt to them. This means that form adaptation, like to distorted faces, can only occur at a level where such features are extracted, that is, in the inferior temporal pathway (in the P pathway in the "ventral stream"). Adaptation to complex changes in shapes in the image that are generated by optical aberrations should also occur in the ventral stream. On the other hand, motion adaptation should occur in the M pathway in the dorsal stream. These conclusions are illustrated in Figure 21.24.

21.13 ELECTROPHYSIOLOGICAL CORRELATES AND SPATIAL FREQUENCY DEPENDENCY OF CONTRAST ADAPTATION

Heinrich and Bach (2001) adapted their subjects for 10 minutes to high-contrast phase-reversing checkerboards and studied the changes both in the retinal pattern ERG (PERG) and in the visually evoked cortical potentials (pattern VEP) (Figure 21.25). They found that effects of contrast adaptation to the checkerboard showed up at both levels. There were significant changes in amplitudes of the PERG (which decreased) and the VEP (which increased). The larger the changes, the higher the contrast to which the subjects were adapted. There were also changes in latency after each phase reversal during the course of adaptation, again in opposite directions in the PERG and the VEP (Figure 21.25c). The study shows that adaptation to high-contrast checkerboards involves changes both at the retinal and cortical levels.

Heinrich and Bach (2002) also studied the spatial frequency dependence of contrast adaptation, using phase-reversing sine wave gratings. They found no adaptation at a low spatial

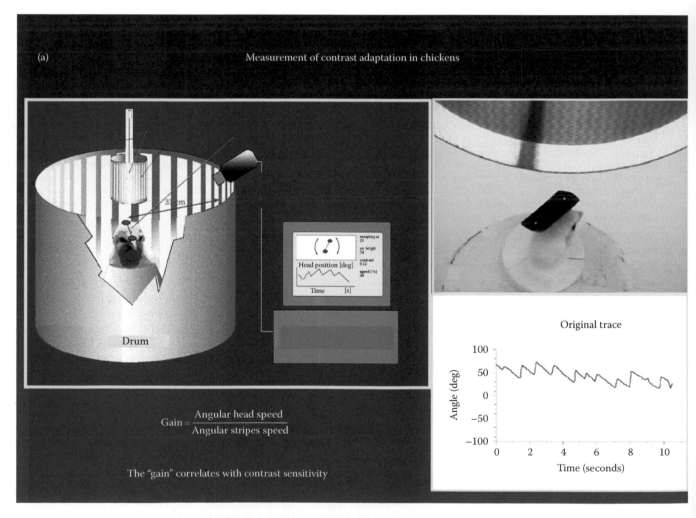

Figure 21.17 (a) Contrast sensitivity in a chicken can be measured by tracking its head movements when it is placed in the center of a drum with a drifting square wave grating inside. The "gain," defined as angular head speed divided by angular stripe speed, is proportional to the perceived contrast and represents therefore a measure of contrast sensitivity.   *(Continued)*

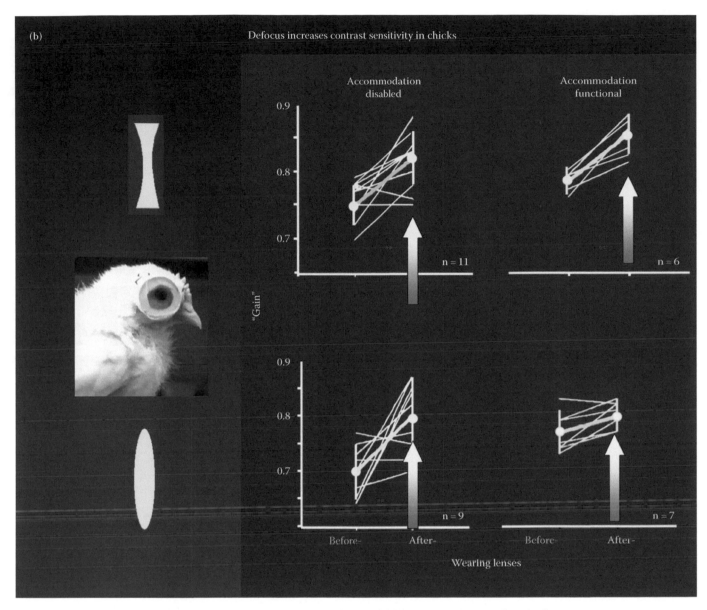

Figure 21.17 (*Continued*) (b) After chicks had been wearing spectacle lenses of −4 D or +4 D power for 1 h, their contrast sensitivity increased significantly, with little difference when accommodation was paralyzed (left) or intact (right).

frequency, 0.5 cyc/deg, neither in the PERG nor in the pattern VEP, but large effects at 5 cyc/deg (Figure 21.26). This is interesting, given that Rajeev and Metha (2010; Figure 21.16) found a decrease in CS during adaptation to defocus at 2 D at spatial frequencies below 5 cyc/deg, but no change at 5 cyc/deg. However, it needs to be kept in mind that defocus is not equivalent to a homogeneous reduction in image contrast but has a highly specific spatial frequency dependence as described by the first Bessel function.

It was found that interocular transfer also increases with increasing spatial frequency (Baker and Meese 2012; Figure 21.27). A trivial hypothesis could be that lower spatial frequencies elicit little contrast adaptation and contain little information about complex features in the visual scene; adaptation to them can therefore occur already at "lower centers" (retina?). They do not stimulate the P pathway and are therefore less represented at the cortical level where they would show up in interocular transfer.

It is interesting that binocular summation can improve visual performance only when the highest spatial frequencies are not represented in the retinal image. If retinal images are defocused in both eyes, binocular visual acuity is superior to monocular visual acuity. However, under fully corrected conditions, binocular summation does not provide additional benefit, neither during visual acuity testing nor in VEP recordings (Plainis et al. 2011; Figure 21.28). Similarly, if the retinal image is corrected for chromatic aberration and higher-order aberrations with adaptive optics, no further benefit can be gained when the input from both eyes is combined (Schwarz and Artal 2013, Schwarz et al. 2014; Figure 21.28).

21.14 NEURAL MECHANISMS

The initial mechanism of adaptation, perhaps partially responsible for afterimages, is light adaptation in the photoreceptors. It involves adjustment in the gain of the phototransduction cascade

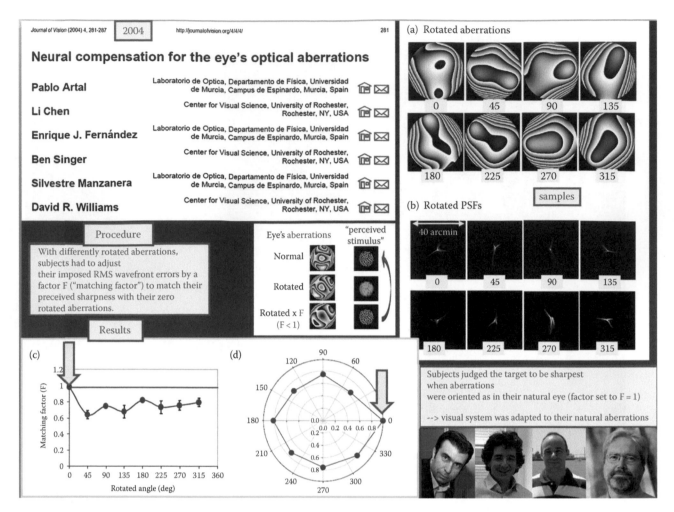

Figure 21.18 Effects of rotation of the wavefront error map on sharpness perceived by the subjects. The stimulus was a circular pattern shown in green (middle). The wavefront error maps in the subjects' pupils were determined (a) and, using adaptive optics, they were rotated in steps of 45° while the subjects had to adjust a matching factor F until the pattern seen through a rotated wavefront profile appeared equally sharp as with zero rotation. (b) Point spread functions associated with the rotated wavefront maps. The matching factor as a function of rotation angle, both as a linear plot (c) and a circular plot (d). The test pattern appeared sharpest when the wavefront map was at zero rotation and appeared less sharp at all other orientations, with subjective sharpness reduced to 60%–80%.

in the photoreceptor outer segments. It is controlled by calcium that comes into the cell through the light-controlled ion channels that close when photons are absorbed. High calcium levels increase the gain of phototransduction. The mechanism is not very fast (range of many seconds, as can be seen when we enter a dimly illuminated room, or come from a dark room into bright light). Therefore, it cannot explain "contrast gain control," the fast adjustment of CS when we adapt to a low- or high-pass-filtered images within fractions of a second (see experiments by Webster et al.; Figure 21.11). A fast mechanism that could account for these changes involves the synapses of the photoreceptors, which interact with neighbored photoreceptors through electrical and pH-mediated mechanisms that can respond very fast.

Demb (2008) has analyzed the different retinal mechanisms for contrast adaptation (Figure 21.29a). At the neural levels, there may be two mechanisms to preserve CS in the presence of high contrasts and to prevent saturation: (1) response gain changes and (2) baseline changes. There is electrophysiological evidence for both mechanisms. Demb has also described the synaptic contacts of bipolar, amacrine, and ganglion cells that mediate contrast adaptation (Figure 21.29b). The fast mechanisms (contrast gain control, partially by electrical

synapses, response much faster than a second, no involvement of GABA) occur at the photoreceptor synapses, while the slower mechanisms are mediated by neurotransmitters like GABA and dopamine and occur in the inner retina (see green boxes in Figure 21.29b). Contrast adaptation can be nicely demonstrated by paired recordings from bipolar and amacrine cells in mouse retinal slice preparations (Jarsky et al. 2011). Zaidi et al. (2012) have shown that afterimages involve adaptational changes in retinal ganglion cells in primates, with time constants in the range of seconds.

Both Solomon et al. (2004) and Hohberger et al. (2011) have provided evidence that contrast adaptation occurs more in the M pathway (responsible for luminance contrast detection, high temporal resolution but lower spatial resolution) and not in the P pathway (responsible for color and fine spatial details). This may be in contrast to the assumption made earlier that processing of complex spatial features like image distortions generated by higher-order aberrations should be a function of the P pathway. A complicating issue is that contrast adaptation and adaptation to more complex optical features like aberrations all occur at the same time and at different levels. It is therefore almost impossible to study one of the mechanisms in isolation.

Journal of Vision (2013) 13(8):12, 1–14   http://www.journalofvision.org/content/13/8/12

Adaptation to blurred and sharpened video

Andrew M. Haun   Schepens Eye Research Institute, Massachusetts Eye & Ear, Harvard Medical School, Boston, MA, USA

Eli Peli   Schepens Eye Research Institute, Massachusetts Eye & Ear, Harvard Medical School, Boston, MA, USA

Figure 21.19 Different from previous studies in which stationary images were used, Haun and Pelli (2013) studied spatial adaptation in movies. Subjects had to fixate a spot (red arrow), and high- or low-pass-filtered movie was presented to the left of the fovea and a blank adaptor or an unfiltered movie on the right. After 3 s of adaptation (30 s before the first trial), subjects had to judge in a 500 ms time window on which side a test video was sharper. Despite the variability of spatial and temporal features in the movies, the subjects adapted similar as previously found with stationary images.

21.15 TIME COURSES: HOW LONG DOES IT LAST?

Contrast adaptation includes mechanisms with very long time constants. For instance, Ohlendorf et al. (2009) have adapted their subjects to movies defocused by 3 D for 10 minutes (Figure 21.30).

Only one eye was adapted (viewing target at 1 m distance, trial lens +4 D). After the adaptation period, the subjects had to match the perceived contrast between both eyes in a sine wave grating of 3.2 cyc/deg, embedded in a Gabor patch. Interocular transfer was neglected in these measurements of "suprathreshold contrast sensitivity." It was found that CS in the eye that was previously exposed to defocus had increased by about 30% compared to the other eye that saw the movie in focus. After 10 minutes of adaptation, the effect lasted for about 3 minutes. After 5 minutes, CS in the previously defocused eye had returned to baseline (Figure 21.31a). An interesting and yet unresolved explained result was that contrast adaptation could only be elicited by myopic but not by hyperopic defocus (Figure 21.31b). Accommodation

was continuously monitored and could be excluded as a possible reason. Defocus of either sign should have similarly low pass filtered the retinal image since it was shown that accommodation did not change as it was controlled by the nondefocused eye. If adaptation was extended to 20 minutes, recovery was further slowed down—in two of the tested subjects there was no clear recovery over a period of 10 minutes (Figure 21.31c).

Khan et al. (2013) measured logMAR visual acuity in 12 emmetropic and 12 myopic young subjects while they were exposed to +1 D or +3 D of myopic defocus. While the adaptive improvement of visual acuity was highly significant in both groups (gain in acuity from about 1.1 logMAR to about 0.9 in the myopic group and from 1.17 to about 1.0 in the emmetropic group with the +3 D lenses, a small effect, between 1 and 2 lines), the improvement continued only over a time window of about 4 minutes and then reached a plateau.

That adaptational processes can be extremely long lasting has been shown in various other studies. Neitz et al. (2002) had subjects wear a red filter in front of one eye and a green filter in

Figure 21.20 Neuronal adaptation to astigmatism. When subjects watched an astigmatically defocused movie for 10 minutes (sample picture from unfiltered movie above, astigmatically defocused sample below), either with astigmatic trial lenses in front of their eyes or with astigmatic defocus calculated into the movie (bottom left), their visual acuity was about 20% better when they were tested in an astigmatically defocused letter chart, compared to when they were tested with previous adaptation to an unfiltered movie (top left).

front of the other. Since the filters attenuated light at different wavelength bands, the visual system increased sensitivity in the respective wavelength range to normalize color space. These adaptational processes can be measured by asking the subjects to match a single wavelength yellow ("unique yellow," 570 nm) with a mixture of red and green. For instance, if sensitivity was reduced by adaptation in the green range, more green was used by the subjects to match unique yellow. After about 3 days of adaptation, it also took about 3 days until subjects matched unique yellow with the same "dose" of red and green as untreated subjects.

Even longer-lasting effects were observed by Delahunt et al. (2004) in the changes in color perception. It is well known that, during aging, the crystalline lens becomes yellowish, with declining transmission in the short wavelength range. Figure 21.32a illustrates how an image should look like through an aging lens. However, even though much less energy is then in the retinal image in the short wavelength range, subjects are not aware of this loss due to their long-term color adaptation. If the lens is replaced by a plastic intraocular lens during cataract extraction, transmission in the blue is recovered and spectral sensitivity of

the subjects in the short wavelengths range is much enhanced (Figure 21.32b). Measuring the achromatic locus in those subjects in the CIE diagram (Figure 21.32d) shows that the recovery to "normal" (with a clear lens) takes many days (Figure 21.32c).

Another example of a very long-lasting adaptational change is the McCollough effect, where adaptation to black square wave grating, presented horizontally on a red background and vertically on a green background, leads to a greenish appearance when horizontal stripes are shown on a white background and to a reddish appearance when vertical stripes are shown on white background—color becomes linked to orientation. This contingent adaptational link can last for weeks.

Even though these studies deal with adaptation to changes in the spectral composition of light or spatial-color interactions, and not with contrast adaptation, they show that adaptational processes can be very long lasting. It is therefore not surprising that long-term spatial adaptation to low contrasts can increase CS for a long time. This feature has been used by companies to sell "vision training procedures" that are supposed to reduce the visual impact of myopia—but in fact to train the visual cortex to extract more spatial information from low-contrast retinal images.

Figure 21.21 A brief fixation of the green point generates prominent afterimages when one looks into the surrounding gray field. The question is whether the afterimage has only retinal origin or whether cortical adaptation is also involved.

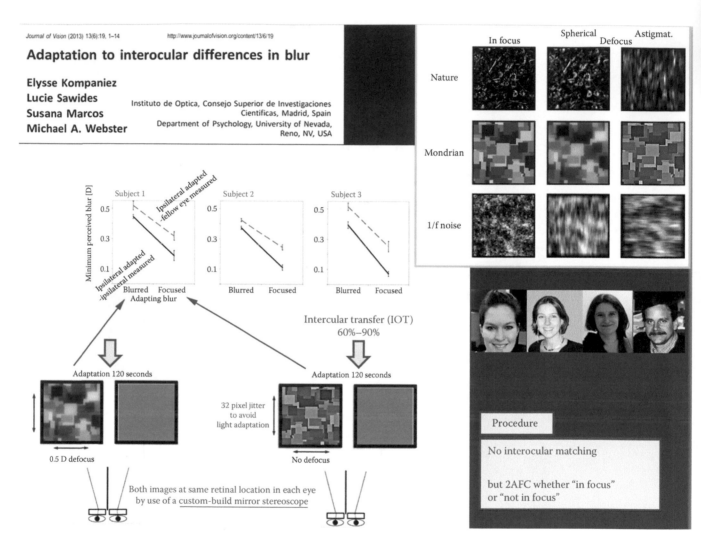

Figure 21.22 Adaptation to various types of simulated image blur shows strong interocular transfer. Subjects were monocularly adapted for 120 s to spatially filtered "nature," "Mondrian," and "1/f noise" targets and the adaptational changes were quantified in both eyes by a 2-alternative forced choice procedure. Interocular transfer of adaptation was surprisingly high, on average between 60% and 90%.

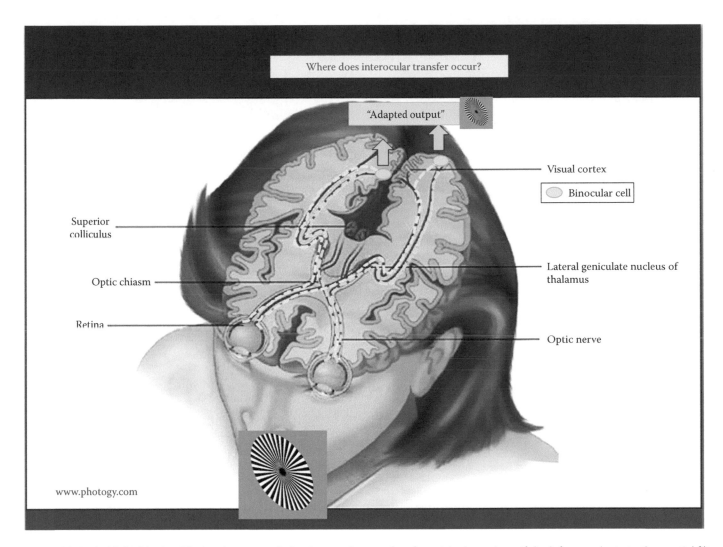

Figure 21.23 Probably highly simplified pathways mediating interocular transfer of spatial adaptations. If the left eye adapts to the spatial filter applied to a pattern, the adaptation may also be found in the fellow eye. This transfer can only happen at the level of binocular neurons in the visual cortex, so that the adapted output of these neurons represents a mixture of the inputs from the adapted and the nonadapted eye (green arrows). Note that there is in fact little evidence for interocular transfer of the afterimage of a Siemens star shown here as a stimulus.

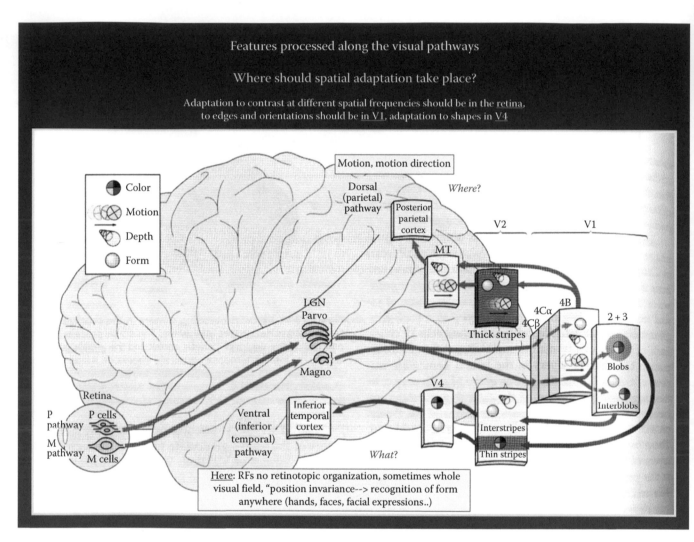

Figure 21.24 P and M pathways and features processed along the visual pathways. The assumption is that spatial adaptation can only occur at levels where the adapted features are processed in the brain. Adaptation to ocular aberrations requires most likely some kind of form vision that is processed in the P pathway in the ventral stream in V4 or upstream. At this level, receptive field size may cover the entire visual field and the visual features by which a cell can be triggered are highly specific, that is, certain faces. On the other hand, adaptation to motion and its direction should occur in the M pathway in the dorsal stream. (Adapted and expanded from Kandel et al., *Principles of Neural Science*, McGraw-Hill Professional, 4th Revised edition, 2000.)

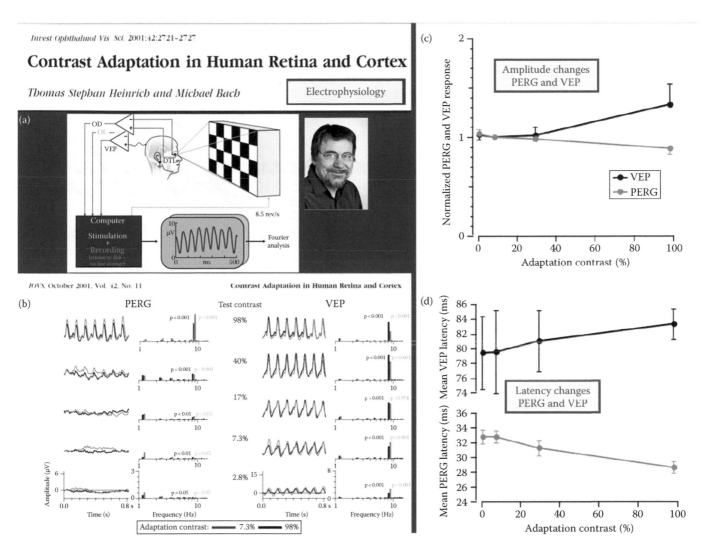

Figure 21.25 When subjects are adapted to phase-reversing checkerboards (a), changes can be observed both in the pattern electroretinogram (PERG) and the cortical visually evoked potentials (pattern VEP) (b). These changes are in opposite directions in both sites (c). VEP amplitudes increase and the PERG amplitudes decline, at least at high adaptation contrasts. Also latencies change, and the VEP latency increases, while the PERG amplitude declines with the adaptation contrast (d). All together, these data show that contrast adaptation occurs both at the retinal and cortical levels.

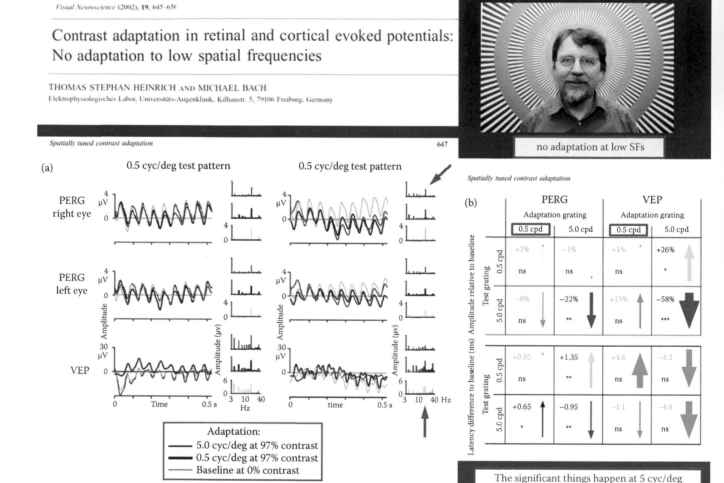

Visual Neuroscience (2002), **19**, 645–65C

Contrast adaptation in retinal and cortical evoked potentials: No adaptation to low spatial frequencies

THOMAS STEPHAN HEINRICH AND MICHAEL BACH

Elektrophysiologisches Labor, Universitäts-Augenklinik, Killianstr. 5, 79106 Freiburg, Germany

Spatially tuned contrast adaptation   647

no adaptation at low SFs

The significant things happen at 5 cyc/deg

Figure 21.26 Spatial frequency dependence of contrast adaptation, as measured by PERG and pattern VEP. Phase-reversing sine wave gratings were used. (a) Samples of original recordings for three different adaptation conditions and the respective Fourier spectra. The relevant Fourier component is at 17 reversals/s (red arrows). (b) The major changes that were observed in the PERG and the VEP are denoted b green (up) and red (down) arrows. Gray errors denote changes that did not achieve significance. There were no significant changes at the lowe spatial frequency of 0.5 cyc/deg.

Vision Research 63 (2012) 81–87

Contents lists available at SciVerse ScienceDirect

Vision Research

journal homepage: www.elsevier.com/locate/visres

Interocular transfer of spatial adaptation is weak at low spatial frequencies

Daniel H. Baker [*], Tim S. Meese

School of Life & Health Sciences, Aston University, Birmingham B4 7ET, UK

Fig. 1. Summary of interocular transfer effects reported by or obtained from previous studies. Symbols indicate the method used in each study: circles are for method of adjustment, stars for 2AFC, diamonds for yes/no and triangles for 4AFC or modified 2AFC. Each datum represents a single observer, except for the large grey diamond, which represents group data for six observers. Selby and Woodhouse (1981) also report IOT at lower spatial frequencies (0.5–2 c/deg), but the precise frequencies used for each observer are not clear from their manuscript so we include only their 8 c/deg data here. A further observer in the Meese and Baker (2011) study produced IOT of −45% (not shown). ●, Blakemore and Campbell (1969); ★, Gilinsky and Doherty (1969); △, Fiorentini, Sireteanu, and Spinelli (1976); ●, Lema and Blake (1977); ●, Hess (1978); ●; Levi, Harwerth, and Smith (1980); ○, Bjørklund and Magnussen (1981); ●, Blake, Overton, and Lema-Stern (1981); ○, Selby and Woodhouse (1981); ●, Sloane and Blake (1984); ●, Anderson and Movshon (1989); ◇, Snowden and Hammett (1996); ◇, Timney et al. (1996); ○, Falconbridge, Ware, and MacLeod (2010); ★, Meese and Baker (2011); △, Cass et al. (submitted for publication).

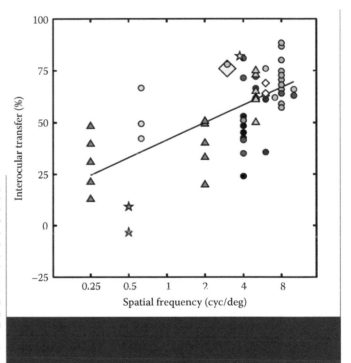

Figure 21.27 Interocular transfer of spatial adaptation as a function of spatial frequency. Data pooled from several studies (see legend of Baker and Meese's Figure 21.1). In general, interocular transfer increased with increasing spatial frequency.

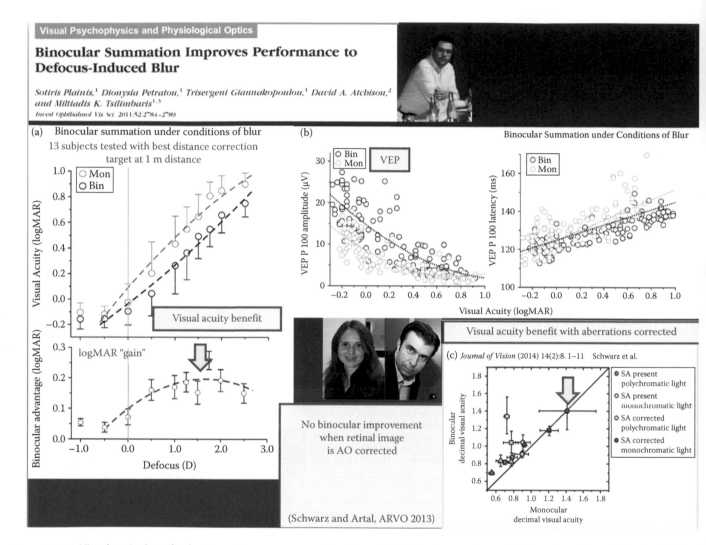

Figure 21.28 Visual acuity benefits for binocular versus monocular vision. (a) With best correction and high acuity of the subject (logMAR < 0) there is little benefit of binocular summation. However, with increasing defocus, also the benefit increases and reaches a peak at around 1.5 D of defocus (yellow arrow). (b) VEP P100 amplitudes reproduce the findings of subjective acuity testing. There were differences in VEP P100 latencies when monocular and binocular inputs were compared. (c) Binocular acuity (Snellen fraction) was superior to monocular acuity as long as aberrations (spherical and chromatic aberration) were present. If they were corrected, the binocular benefit disappeared and there was no longer any difference between monocular and binocular Snellen acuity (yellow arrow).

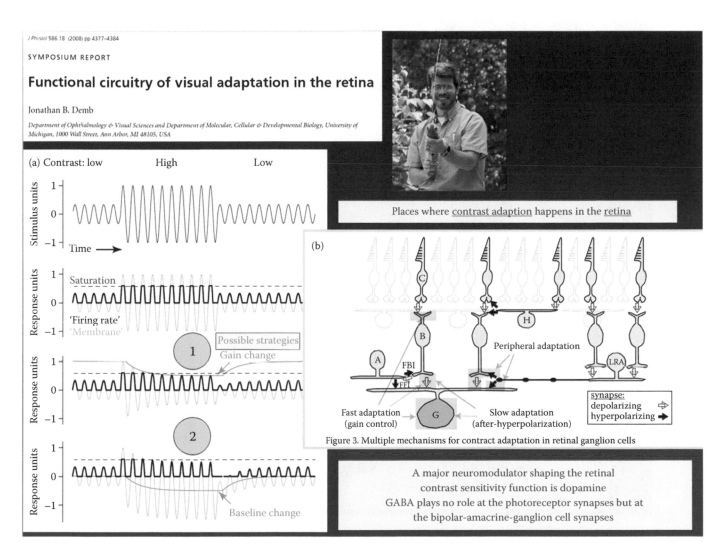

Figure 21.29 Mechanisms of contrast adaptation in the retina. (a) Both gain changes and baseline changes can serve to prevent saturation. (b) The major synapses mediating contrast adaptation in the retina are denoted by green boxes.

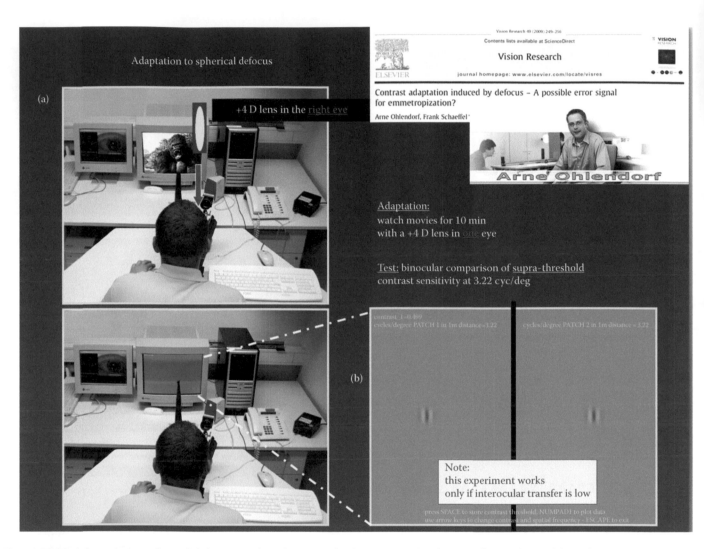

Figure 21.30 Adaptation to spherical defocus. (a) The subject watched a movie (top) with a +4 D lens in front of the right eye. Eye position was tracked to ensure that the subject looked at the screen. After 10 minutes, contrast sensitivity was compared in both eyes by an interocular contrast matching task (bottom). (b) Appearance of the screen during the subsequent interocular contrast matching task. The subject had to adjust the contrast of the pattern on the right to match the one on the left. A divider between both eyes prevented that the fixation could cross to the contralateral Gabor patch.

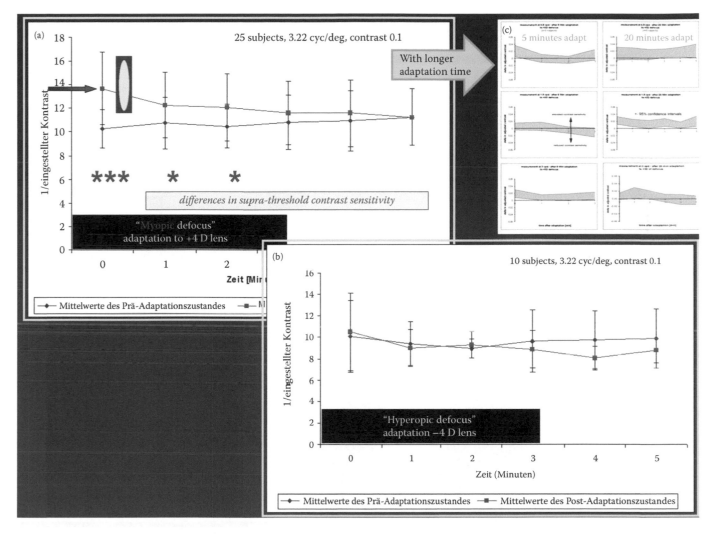

Figure 21.31 Time course of contrast adaptation in the experiment described in Figure 21.30. (a) After 10 minutes of adaptation to defocus, suprathreshold contrast sensitivity remained significantly elevated for further 3 minutes. No difference was detectable between both eyes after 5 minutes. (b) Surprisingly, no contrast adaptation could be elicited by hyperopic defocus imposed by a -4 D lenses, even though the defocus relative to the screen should have been even higher. (c) When the subjects watch the movie even longer than 10 minutes with the +4 D lens in front of one eye, the recovery from contrast adaptation was delayed. Some subjects did not recover over a 5 minute observation period.

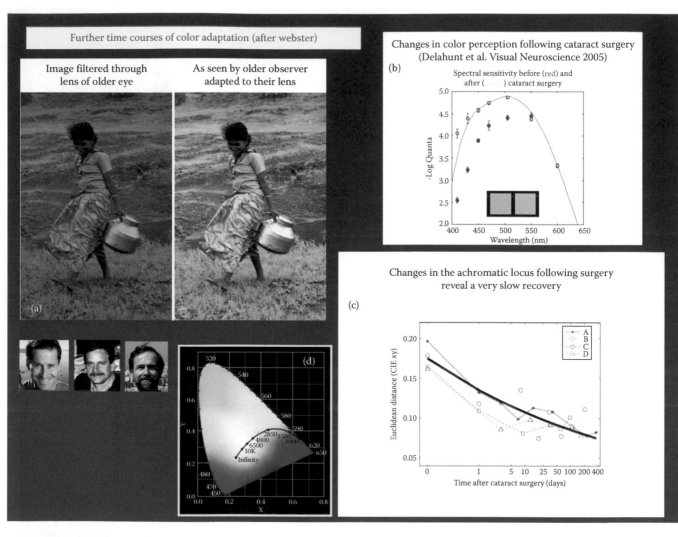

Figure 21.32 Effects of age-related yellowing of the crystalline lens on color vision. (a) The expected change in color vision is illustrated but not observed, (b) change in spectral transmission of the optics of the eye when the yellow lens is replaced by a clear plastic intraocular lens, (c) time course of the change of the position of the achromatic locus in the CIE diagram (d). (The figure was adapted from a presentation by M. Webster in Madrid 2012.)

ACKNOWLEDGMENTS

This chapter was prepared based on a key note lecture presented at the *Seventh Visual and Physiological Optics Meeting* in Wroclaw, August 25, 2014. I am grateful to Dr. Sven Heinrich, Freiburg, Professor Dr. Michael Bach, Freiburg, and Dr. Marita Feldkaemper, Tuebingen, for commenting on an earlier version of the manuscript.

REFERENCES

Aho AC, Donner K, Hydén C, Larsen LO, Reuter T, Low retinal noise in animals with low body temperature allows high visual sensitivity. *Nature*. 1988; 334: 348–350.

Artal P, Chen L, Fernández EJ, Singer B, Manzanera S, Williams DR, Neural compensation for the eye's optical aberrations. *J Vis*. 2004; 4: 281–287.

Bach M, Waltenspiel S, Schildwächter A, Detection of defocused gratings—Spurious resolution, a pitfall in the determination of visual acuity based on preferential looking or VEP. *Proceedings of the Third International Symposium of the Northern Eye Institute*, Manchester, U.K., 1987, pp. 562–565.

Baker DH, Meese TS, Interocular transfer of spatial adaptation is weak at low spatial frequencies. *Vision Res*. 2012; 63: 81–87.

Bex PJ, Solomon SG, Dakin SC, Contrast sensitivity in natural scenes depends on edge as well as spatial frequency structure. *J Vis*. 2009; 9(10): 1.1–1.19.

Blakemore C, Campbell FW, On the existence of neurones in the human visual system selectively sensitive to the orientation and size of retinal images. *J Physiol*. 1969; 203: 237–260.

Cufflin MP, Mankowska A, Mallen EA, Effect of blur adaptation on blur sensitivity and discrimination in emmetropes and myopes. *Invest Ophthalmol Vis Sci*. 2007; 48: 2932–2939.

Delahunt PB, Webster MA, Ma L, Werner JS, Long-term renormalization of chromatic mechanisms following cataract surgery. *Vis Neurosci*. 2004; 21: 301–307.

Demb JB, Functional circuitry of visual adaptation in the retina. *J Physiol*. 2008; 586: 4377–4384 (review).

Diether S, Gekeler F, Schaeffel F, Changes in contrast sensitivity induced by defocus and their possible relations to emmetropization in the chicken. *Invest Ophthalmol Vis Sci*. 2001; 42: 3072–3079.

Diether S, Schaeffel F, Long-term changes in retinal contrast sensitivity in chicks from frosted occluders and drugs: Relations to myopia? *Vision Res*. 1999; 39: 2499–2510.

Elliott SL, Georgeson MA, Webster MA, Response normalization and blur adaptation: Data and multi-scale model. *J Vis*. 2011; 11(2): pii: 7. doi: 10.1167/11.2.7.

Feldkaemper M, Diether S, Kleine G, Schaeffel F, Interactions of spatial and luminance information in the retina of chickens during myopia development. *Exp Eye Res*. 1999; 68: 105–115.

Field DJ, Relations between the statistics of natural images and the response properties of cortical cells. *J Opt Soc Am A*. 1987; 4: 2379–2394.

Georgeson MA, The effect of spatial adaptation on perceived contrast. *Spat Vis*. 1985; 1: 103–112.

Hampson EC, Vaney DI, Weiler R, Dopaminergic modulation of gap junction permeability between amacrine cells in mammalian retina. *J Neurosci*. 1992; 12(12): 4911–4922.

Haun AM, Peli E, Adaptation to blurred and sharpened video. *J Vis*. 2013; 13(8): pii: 12. doi: 10.1167/13.8.12.

Heinrich TS, Bach M, Contrast adaptation in human retina and cortex. *Invest Ophthalmol Vis Sci*. 2001; 42: 2721–2727.

Heinrich TS, Bach M, Contrast adaptation in retinal and cortical evoked potentials: No adaptation to low spatial frequencies. *Vis Neurosci*. 2002; 19: 645–650.

Hohberger B, Rössler CW, Jünemann AG, Horn FK, Kremers J, Frequency dependency of temporal contrast adaptation in normal subjects. *Vision Res*. 2011; 51: 1312–1317.

Jarsky T, Cembrowski M, Logan SM, Kath WL, Riecke H, Demb JB, Singer JH, A synaptic mechanism for retinal adaptation to luminance and contrast. *J Neurosci*. 2011; 31(30): 11003–11015.

Kandel ER, Schwartz JH, Jesell TM. *Principles of Neural Science*. 4th Edition 2000. McGraw-Hill Companies. Health Profession Division. New York, St. Louis, San Francisco, Auckland, Bogota (and 10 more).

Kaneko A, Stuart AE, Coupling between horizontal cells in the carp retina revealed by diffusion of Lucifer yellow. *Neurosci Lett*. 1984; 47(1): 1–7.

Khan KA, Dawson K, Mankowska A, Cufflin MP, Mallen EA, The time course of blur adaptation in emmetropes and myopes. *Ophthalmic Physiol Opt*. 2013; 33: 305–310.

Kohn A, Visual adaptation: Physiology, mechanisms, and functional benefits. *J Neurophysiol*. 2007; 97: 3155–3164 (review).

Kompaniez E, Sawides L, Marcos S, Webster MA, Adaptation to interocular differences in blur. *J Vis*. 2013; 13(6): 19.

Kuang X, Poletti M, Victor JD, Rucci M, Temporal encoding of spatial information during active visual fixation. *Curr Biol*. 2012; 22: 510–514.

Mon-Williams M, Tresilian JR, Strang NC, Kochhar P, Wann JP, Improving vision: Neural compensation for optical defocus. *Proc Biol Sci*. 1998; 265: 71–77.

Neitz J, Carroll J, Yamauchi Y, Neitz M, Williams DR, Color perception is mediated by a plastic neural mechanism that is adjustable in adults. *Neuron*. 2002; 35: 783–792.

Ohlendorf A, Schaeffel F, Contrast adaptation induced by defocus—A possible error signal for emmetropization? *Vision Res*. 2009; 49: 249–256.

Ohlendorf A, Tabernero J, Schaeffel F, Neuronal adaptation to simulated and optically-induced astigmatic defocus. *Vision Res*. 2011; 51: 529–534.

Orlowski J, Harmening W, Wagner H, Night vision in barn owls: Visual acuity and contrast sensitivity under dark adaptation. *J Vis*. 2012; 12: 4.

Pasternak T, Merigan WH, The luminance dependence of spatial vision in the cat. *Vision Res*. 1981; 21: 1333–1339.

Plainis S, Petratou D, Giannakopoulou T, Atchison DA, Tsilimbaris MK. Binocular summation improves performance to defocus-induced blur. *Invest Ophthalmol Vis Sci*. 2011; 52(5): 2784–2789.

Rajeev N, Metha A, Enhanced contrast sensitivity confirms active compensation in blur adaptation. *Invest Ophthalmol Vis Sci*. 2010; 51: 1242–1246.

Rosenfield M, Hong SE, George S, Blur adaptation in myopes. *Optom Vis Sci*. 2004; 81: 657–662.

Sawides L, de Gracia P, Dorronsoro C, Webster MA, Marcos S, Vision is adapted to the natural level of blur present in the retinal image. *PLoS ONE*. 2011; 6(11): e27031.

Schaeffel F, de Queiroz A, Alternative mechanisms of enhanced underwater vision in the garter snakes *Thamnophis melanogaster* and *T. couchii*. *Copeia* 1990; 1: 50–58.

Schwahn HN, Kaymak H, Schaeffel F, Effects of atropine on refractive development, dopamine release, and slow retinal potentials in the chick. *Vis Neurosci*. 2000; 17: 165–176.

Schwarz C, Cánovas C, Manzanera S, Weeber H, Prieto PM, Piers P, Artal P, Binocular visual acuity for the correction of spherical aberration in polychromatic and monochromatic light. *J Vis.* 2014; 14(2): pii: 8. doi: 10.1167/14.2.8.

Shapley R, Enroth-Cugell C, Visual adaptation and retinal gain control. *Progr Retinal Res.* 1984; 3: 263–343.

Solomon SG, Peirce JW, Dhruv NT, Lennie P, Profound contrast adaptation early in the visual pathway. *Neuron.* 2004; 42(1): 155–162.

Sterling P, How retinal circuits optimize the transfer of visual information. In: *The Visual Neurosciences*, Vol. 1, Chalupa LN and Werner JS (eds.), Cambridge, MA: MIT Press, 2003, pp. 234–259.

Stone RA, Lin T, Laties AM, Iuvone PM, Retinal dopamine and form-deprivation myopia. *Proc Natl Acad Sci USA.* 1989; 86: 704–706.

Vinas M, Sawides L, de Gracia P, Marcos S, Perceptual adaptation to the correction of natural astigmatism. *PLoS One.* 2012; 7(9): e46361.

Wallman J, Schaeffel F, Uses of blur to guide emmetropization. *Invest Ophthalmol Vis Sci.* 1996; 37: 2128 (ARVO abstract).

Wallman J, Winawer J, Homeostasis of eye growth and the question of myopia. *Neuron.* 2004; 43: 447–468 (review).

Webster MA, Georgeson MA, Webster SM, Neural adjustments to image blur. *Nat Neurosci.* 2002; 5: 839–840.

Zaidi Q, Ennis R, Cao D, Lee B, Neural locus of color afterimages. *Cu Biol.* 2012; 22: 220–224.

Visual changes with aging

Joanne Wood and Alex Black

Contents

22.1 INTRODUCTION

It is well established that many aspects of visual function decline with increasing age, both as a result of the normal structural and physiological changes that occur in the eye with age and the increased prevalence of ocular disease. This age-related decrease in visual function is reflected in population studies that consistently demonstrate an increase in the prevalence of visual impairment for those aged 65 years and above (Klein et al. 1991, Attebo et al. 1996, Rubin et al. 1997). For example, the prevalence of bilateral visual impairment, defined as visual acuity worse than the commonly adopted driver licensing standard of 6/12 (or 20/40), increases from 1% for those between 60 and 69 years to 26% for those aged 80 years and over (Wang et al. 2000). As the population ages, further appreciation and understanding of the aging changes in the visual system are critical, given that impaired visual function is associated with decreased quality of life and independence and can result in serious injuries from falls and motor vehicle collisions.

Normal aging is associated with a number of optical changes in the eye that reduce the amount of light reaching the retina (Werner et al. 2010); these factors are discussed extensively in a previous chapter of this handbook. In particular, the crystalline lens of the eye gradually loses its transparency (Said and Weale 1959, Weale 1992) and intraocular scatter and aberrations increase (Artal et al. 2003), while the pupil (the aperture of the eye) becomes more constricted (Winn et al. 1994) and less able to enlarge under low light levels. Together these changes reduce the amount of light reaching the retina of a visually normal 60-year-old to approximately one-third of that reaching the retina of a 20-year-old (Weale 1992). Thus, older adults often have problems under low-luminance conditions, particularly for nighttime driving (Kosnik et al. 1990), and need more external light to

achieve the same amount of retinal illumination as a younger person. Changes also occur at retinal and neural levels, including the age-related reduction in rod photoreceptors, where 20%–30% of rods are lost through the normal aging process (Curcio 2001), along with alterations in the integrity of the macular pigment, ganglion cells, retinal nerve fiber layer, and visual pathways (Weale 1992, Spear 1993, Bonnel et al. 2003, Lovasik et al. 2003).

Together, these age-related structural and physiological changes in the eye, as well as those along the visual pathways, lead to reductions in visual acuity, contrast sensitivity, visual field sensitivity, dark adaptation and motion sensitivity, a slowing of visual processing speeds, and increased glare sensitivity. In older populations, there is also greater interindividual variability in visual function, which may reflect the presence of subclinical ocular pathology, as well as the natural variability inherent in the normal aging process (Johnson and Choy 1987).

These aging changes in visual function are reflected by older adult's perceptions of their own visual abilities. Kline et al. (1992) demonstrated that the self-reported visual problems of older persons are associated with changes in their light sensitivity, dynamic vision, near vision, speed of visual processing, and visual search. The following sections discuss these changes in visual function with increased age, including a definition of each visual function, common measurement techniques, how that visual function changes with increasing age, potential underlying mechanisms, and the implications of these changes for performance of everyday activities.

22.2 VISUAL ACUITY

Visual acuity describes the ability to resolve fine spatial detail. It is the most common visual measure used for clinical assessment and in studies evaluating visual function in relation to everyday activities. Visual acuity is typically measured by determining the smallest high-contrast (black on white) target that is correctly recognized under photopic illumination levels while fixating centrally and can be measured relatively quickly using a variety of targets including letters, Landolt Cs, tumbling Es, pictures, or symbols. The progression of target size, from larger (at the top) to smaller targets at the bottom of the chart, also varies between different types of charts, including the nonstandard but more clinically popular Snellen progression, and the logarithmic size progression, where letter size, letter spacing and line spacing decrease in a uniform logarithmic fashion down the chart (Bailey and Lovie-Kitchin 2013). Visual acuity can be represented as a "Snellen fraction," where the numerator represents the testing distance, which is generally 6 m (or 20 ft in imperial units), and the denominator is an indication of the smallest letter, picture, or symbol that can be accurately recognized. Alternatively, the logarithm of the minimum angle of resolution (logMAR) is used, where a score of 0.00 on the logMAR scale is equivalent to a Snellen fraction of 6/6 (or 20/20).

The change in visual acuity with increasing age has been widely researched. These studies have primarily involved cross-sectional studies, though in more recent years there have been some longitudinal studies of larger population samples, with studies assessing visual acuity as either best corrected or habitual, with some study cohorts including those with various forms of age-related ocular pathology (Weale 1975, Pitts 1982, Owsley et al. 1983, Klein et al. 1991, Elliott et al. 1995, Wood and Bullimore 1995, Haegerstrom-Portnoy et al. 1999, Klein et al. 2001, Hong et al. 2013).

A compilation of visual acuity data from a range of studies involving normal healthy eyes indicated that mean logMAR visual acuity increased from −0.13 (Snellen equivalent 6/4.5) in 18- to 24-year-olds to −0.16 (Snellen equivalent 6/4⁻¹) in 25- to 29-year-olds, gradually becoming worse with increasing age to a mean value of −0.02 (Snellen equivalent 6/6⁺¹) for those aged 75 years and above (Elliott et al. 1995). The mean scores presented in that study were one line (0.10 log units) better than previously reported in less recent studies (Weale 1975, Pitts 1982), which is likely to have resulted from the use of truncated charts (which did not test beyond 6/6) and the inclusion of habitual rather than best-corrected visual acuity measures. The data reported by Owsley et al. (1983) were more similar to that of Elliott et al. (1995) for their younger subjects, but their older subjects reflected those found in the earlier studies (Weale 1975, Pitts 1982), possibly because of the use of different charts and less stringent exclusion criteria. Figure 22.1 represents data selected from some more recent studies that include best-corrected monocular visual acuity and from population samples that include binocular habitual visual acuity. While, collectively, these studies demonstrate a consistent decline in visual acuity with increasing age, the magnitude and time course of the decrease in visual acuity vary, with those assessing best-corrected monocular visual acuity showing a smaller age-related decline than those assessing habitual acuity.

Longitudinal studies of large population samples provide an opportunity to determine the change in visual acuity as a function of age in a single population, in contrast to the cross-sectional snapshots of different aged individuals within a population. Data from the Beaver Dam Eye Study indicate that the overall decrease in visual acuity over a 15-year period was just over a line (5.4 letters in the right and 7.1 letters in the left eye) (Klein et al. 2006), with similar data being reported from the Blue Mountains Eye Study, with an overall mean decrease of 6. and 6.8 letters for the right and left eye, respectively (Hong et al 2013). Both studies demonstrated that the mean decrease in the number of letters read correctly from baseline to the 15-year follow-up was greater with increasing age, but there were no significant gender effects. The higher incidence of visual acuity loss after 75 years of age in these population studies is consisten with the higher prevalence of ocular diseases such as age-related macular degeneration, diabetic retinopathy, and cataract in this age group (Attebo et al. 1996).

Which of the many reports on the age-related changes in visual acuity represent that of the "normal" aging process is unclear, given the inclusion of individuals with subclinical ocular disease

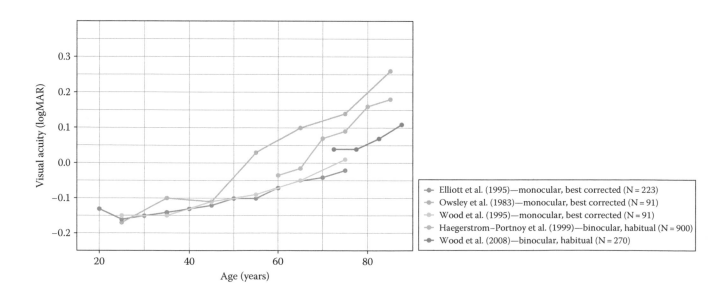

Figure 22.1 Visual acuity as a function of age. Data redrawn from various studies.

and failure to optimize refractions in many of these studies. Nevertheless, it is clear that many individuals can resolve much finer details, or higher spatial frequencies, than was suggested by many of the earlier studies and that some 80-year-olds in good ocular health can have acuity as high as 6/6 (Elliott et al. 1995).

The underlying cause of the decline in visual acuity seen with increasing age has been widely debated. Many authors have attributed the age-related acuity loss to optical factors including reduced retinal illumination and increased intraocular light scatter resulting from pupil miosis and changes in the crystalline lens (Owsley et al. 1983). However, others have suggested that optical factors do not fully account for this age-related loss in visual acuity, based upon calculations of the potential magnitude of acuity loss that should result from pupil miosis and increased light scatter (Weale 1975, 1992), and the fact that the age-related acuity loss is still evident in those with intraocular lenses following cataract extraction (Jay et al. 1987). The magnitude of the decline in acuity also becomes larger as luminance levels or the contrast of letters is reduced (Adams et al. 1988, Sturr et al. 1990) and the density of cone photoreceptors also remains relatively stable with increasing age (Curcio 2001). Collectively, these factors imply that changes in neural processing along the visual pathways are likely to underlie at least some of the acuity loss seen with increasing age (Spear 1993).

The age-related decrease in visual acuity has potential functional consequences for activities of everyday living, which has been explored in numerous driving- and falls-related studies. Visual acuity is the most commonly tested visual function in driving research, with numerous studies dating back to the 1960s having investigated the relationship between acuity and crash risk; however, the question of whether reduced acuity increases crash risk is far from clear. The earliest studies of visual acuity and crash risk examined the relationship between visual function and crash history in over 17,000 drivers (Burg 1967), finding only a weak correlation between acuity and crash rates, even when the sample was stratified for age (Hills and Burg 1977). Similarly, other studies using a range of sample sizes and methodologies have found only a weak relationship between acuity and crash risk (Hofstetter 1976, Ivers et al. 1999), while others failed to find any association (Decina and Staplin 1993, Cross et al. 2009). Driving performance studies have also shown that when visual acuity is reduced, driving performance, as assessed by sign recognition (Fonda 1989) and by the ability to undertake safety critical driving tasks (Lamble et al. 2002) or when rated as self-reported driving ability (Buyck et al. 1988), was not greatly impaired. Closed-road driving studies have also assessed driving performance under conditions of degraded acuity 6/12, 6/30, and 6/60 using optical blur (Higgins et al. 1998) and a simulated cataract condition, which reduced acuity to 6/12 (Higgins and Wood 2005). While acuity degradation significantly reduced some aspects of driving performance, other measures such as gap judgment and maneuvering ability were relatively unaffected. Interestingly, the simulated cataract condition produced selective deficits in sign and road hazard recognition that were not predictable from the modest loss in visual acuity associated with this condition (Higgins and Wood 2005).

Studies also report significant links between reduced visual acuity and increased risk of falls among older adults. Klein et al.

(1998) reported that older adults with binocular habitual visual acuity of 6/7.5 or worse were twice as likely to report more than two falls in the previous year. A longitudinal study showed that women aged over 65 years, whose visual acuity had decreased by two or more lines in the previous 4–6 years, were 43% more likely to have multiple falls in the subsequent year compared to those whose visual acuity had reduced by less than two lines over the same period (Coleman et al. 2004).

22.3 CONTRAST SENSITIVITY

As discussed in the previous section, visual acuity assesses the ability of an individual to resolve high-contrast and high-spatial-frequency targets, which are usually presented centrally and under optimal photopic lighting conditions. However, because the visual environment contains a wide range of spatial frequencies, contrast levels, and different lighting levels, standard measures of visual acuity may not necessarily represent functional visual performance for activities of daily living. Indeed, research has demonstrated that contrast sensitivity may provide a more complete assessment of spatial vision that is relevant to the capacity to complete everyday tasks, including walking and driving under day- and nighttime conditions, for both visually normal older adults and those with ocular disease (Cummings et al. 1995, Wood 2002, Wood and Carberry 2006, Wood et al. 2009).

Contrast sensitivity is typically determined by measuring the minimum contrast required to detect sinusoidally modulated gratings presented at a range of spatial frequencies to generate the contrast sensitivity function (CSF). The form of the CSF involves both optical and neural factors. Optically, retinal image quality is determined by the form of the modulation-transfer function, which shows an age-dependent reduction at high spatial frequencies (Artal et al. 1993), while at a neural level, the CSF is believed to reflect the sensitivity of multiple visual channels, which are selective to different bands of spatial frequencies (Campbell and Robson 1968). Alternative methods of assessing contrast sensitivity include chart-based techniques, which provide a measure of contrast thresholds at specific spatial frequencies across the CSF. For example, the Pelli–Robson chart includes large letters that decrease in contrast rather than size down the chart and measures contrast sensitivity at low spatial frequencies (Mantyjarvi and Laitinen 2001), while the Melbourne Edge Test assesses sensitivity closer to the peak of the CSF (Verbaken and Johnston 1986). An assessment with these chart-based contrast sensitivity charts in combination with a measure of visual acuity is believed to provide a good clinical compromise for describing the form of an individual's CSF (Woods and Wood 1995).

Spatial contrast sensitivity measured under photopic conditions declines with increasing age and is more sensitive to the effects of normal aging or ocular disease than high-contrast visual acuity (Eisner et al. 1987, Haegerstrom-Portnoy et al. 1999, Schneck et al. 2004). The decrease in contrast sensitivity generally becomes evident from the age of 65 years (Haegerstrom-Portnoy et al. 1999), although some studies have reported the decline to begin as early as 45 years of age (Sia et al. 2013). The decrease in contrast sensitivity is spatial frequency dependent, with significant age-related declines in sensitivity found at mid and high spatial frequencies, with the magnitude of the aging deficit increasing with increased spatial frequency (Derefeldt et al. 1979, Owsley et al. 1983, Higgins et al. 1988, Elliott et al. 1990). Lower spatial frequencies appear to be relatively robust to the effects of age; however, if low-spatial-frequency targets are temporally modulated (counter-phased), then an age-related decline is evident particularly as temporal frequency increases (Owsley et al. 1983, Elliott et al. 1990).

The magnitude and pattern of age-related loss in contrast sensitivity reported in various studies tend to vary and are likely to reflect variations in the lens density, retinal and neural health of the studied populations, and the presence of subclinical ocular disease and the inherent variability in older populations. In addition, the techniques used to measure contrast sensitivity also have an impact on the magnitude of the age-related trends, with those tests assessing contrast sensitivity at lower spatial frequencies tending to show smaller age-related declines than those at higher spatial frequencies (Figure 22.2).

The origin of the age-related decline in contrast sensitivity has been the subject of extensive research, with evidence suggesting

that a combination of optical and neural factors is likely to be involved. Age-related ocular media changes, including increased density of the crystalline lens, increased intraocular scatter, and the reduction in pupil size, all contribute to the age-related decline in the CSF. However, studies that have bypassed the optics of the eye either by measuring contrast sensitivity with laser interferometry (Burton et al. 1993) or when higher-order ocular aberrations are corrected with adaptive optics (Elliott et al. 2009) have demonstrated that the age-related losses in CSF are still evident for higher spatial frequencies, suggesting that changes in the neural pathways have a contributory, albeit relatively small role in the age-related decline in CSF. External noise paradigms recently used to further investigate the cause of the contrast sensitivity losses in normal healthy older adults have also suggested that the contributing factors are spatial frequency dependent (Allard et al. 2013).

The age-related decline in photopic contrast sensitivity is exaggerated when measured under reduced light levels (Sloane et al. 1988), with reductions in mesopic contrast sensitivity being

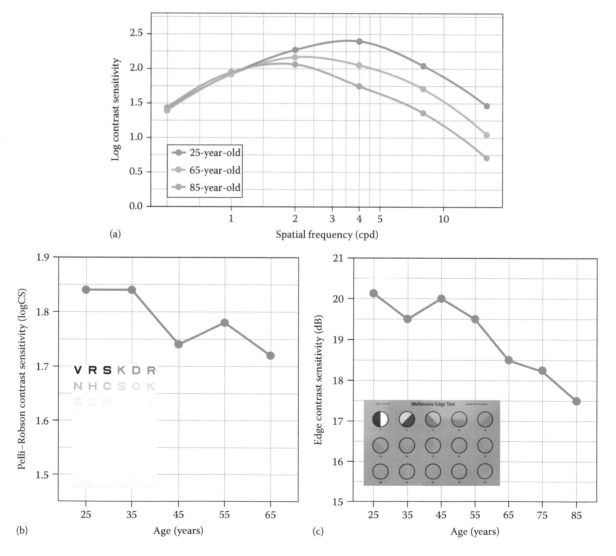

Figure 22.2 (a) Log contrast sensitivity as a function of spatial frequency for different age groups of observers (25, 65, and 85 years). (Adapted from Owsley, C. et al., *Vis. Res.*, 23(7), 689, 1983.) (b) Pelli–Robson contrast sensitivity as a function of age. (Adapted from Mantyjarvi, M. and Laitinen, T., *J. Cataract Refract. Surg.*, 27(2), 261, 2001.) (c) Melbourne Edge contrast sensitivity as a function of age. (Adapted from Verbaken, J.H. and Johnston, A.W., *Am. J. Optom. Physiol. Opt.*, 63(9), 724, 1986.)

evident 10 years before the onset of declines in photopic contrast sensitivity (Puell et al. 2004a). These age-related changes are likely to underlie the problems that older adults report under low light levels or at night (Kosnik et al. 1988, Owsley et al. 2006) and are likely to be a contributory factor to the age-related avoidance of nighttime driving (Ball et al. 1998, Brabyn et al. 2005).

While optical changes are potentially the main contributing factor for contrast sensitivity losses at photopic levels, optical factors are unlikely to be responsible for older adults' accentuated loss in contrast sensitivity at low luminance (Sloane et al. 1988), implying that neural factors must contribute to these changes. Anatomical evidence supports this suggestion of the involvement of neural changes, including the age-related reduction in the number of rods, but not cones, in the parafoveal region (Gao and Hollyfield 1992, Curcio et al. 1993).

22.4 VISUAL FIELDS

The visual field is the spatial extent or area over which a person can detect the presence of a given target at any one time. Visual fields can be measured using crude, manual techniques such as confrontation, or more sophisticated automated perimeters, which have become widely available in clinical practice following improvements in testing strategies in terms of speed and reliability. Given that information from across the visual field comprises a large component of visual sensory input, visual field loss has been linked to difficulties in mobility, driving, and other important tasks of daily living.

Visual fields are generally measured with either static targets of varying brightness presented at different locations across the visual field or kinetically, where the targets move and are of fixed brightness. For the purpose of detection and diagnosis of ocular disease, visual fields are generally tested monocularly (with one eye occluded), whereas for functional purposes, such as assessment for driving capacity, binocular visual fields are more relevant. Standard automated perimetry strategies use a white target against a white background at photopic light levels and show the highest sensitivity at the fovea with a gradual decline in sensitivity with increasing eccentricity.

Importantly, the extent of the visual field (Drance et al. 1967) and the level of sensitivity across the visual field (Johnson et al. 1989) decrease with age. There are many factors that may contribute to this decline, including age-related preretinal factors, such as media opacities and pupil miosis, which reduce retinal image quality, and the neural cell losses that occur in the retina and visual pathways with increasing age. Even if stimulus characteristics are altered to minimize the aging effects of the pupil and crystalline lens, visual field sensitivity is still reduced in individuals free of ocular disease (Johnson et al. 1989), suggesting that loss of visual field sensitivity is driven more by neural than optical changes in the aging eye.

In a large study of people aged between 10 and 89 years without ocular disease, Spry and Johnson (2001) showed significant age-related reductions in mean sensitivity within the central 30° of the visual field using a Goldmann size III target. Regression analysis demonstrated a nonlinear reduction in sensitivity with age, with a small reduction up to the age of around 50 years (–0.45 dB/decade), followed by a greater reduction in later life (–1.00 dB/decade). Figure 22.3 represents these data, along with that from a study of older adults including some with early or subclinical ocular disease and hence lower levels of sensitivity (Wood et al. 2008). This age-related trend is also reflected in earlier studies, which showed an overall sensitivity loss of about 0.8 dB/decade, with smaller sensitivity changes occurring in the central 10° regions, while greater age-related sensitivity losses were evident in more peripheral regions (Johnson et al. 1989).

More recently introduced visual field testing strategies, such as frequency doubling technology (FDT) and short-wavelength automated perimetry (SWAP), are becoming increasingly popular, given evidence that they can detect pathology earlier than standard white-on-white strategies. Importantly, these tests also demonstrate significant changes in visual field sensitivity with age. Adams et al. (1999) showed a decrease in sensitivity for detection of FDT stimuli of around 0.6 dB/decade from the age of 15 years and an acceleration of this decline over the age of 60 years. Similarly, Johnson et al. (1988) showed that SWAP sensitivity decreased with age, at a rate of approximately 0.15 log unit/decade.

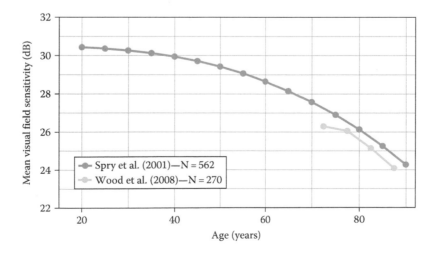

Figure 22.3 Decrease in visual field sensitivity with age. (Data from Spry, P.G. and Johnson, C.A., *Optom. Vis. Sci.*, 78(6), 436, 2001; Wood, J.M. et al., *J. Am. Geriatr. Soc.*, 56(6), 986, 2008.)

Visual fields are widely believed to be important for safe driving and are included in driver licensing standards in many countries, although the exact field requirements vary between states and countries. Visual field tests commonly used for assessing fitness to drive include the binocular Esterman visual field test (EVFT) and Medmont driving test, both of which extend beyond the central 30° and involve suprathreshold targets. Merging two monocular threshold fields, known as the integrated visual field (IVF), may be as useful as the EVFT in assessing fitness to drive in patients with a range of field loss (Crabb et al. 2004, Chisholm et al. 2008), particularly given that monocular fields are routinely assessed in patients with ocular disease; however, the link between IVF and crash risk has yet to be determined.

Johnson and Keltner (1983) in a large-scale study of 10,000 drivers showed that binocular field loss more than doubled crash rates compared to controls; however, monocular field loss was not associated with increased crash risk. Conversely, other studies using a range of experimental designs failed to find a significant relationship between visual field loss and crash risk (Burg 1967, Owsley et al. 1998). More recently, a population-based study (Rubin et al. 2007) and smaller-scale studies of drivers with glaucomatous visual field defects (McGwin et al. 2005, Haymes et al. 2007) provide support for Johnson and Keltner's (1983) original findings, demonstrating that only those with more extensive field loss demonstrate impaired driving performance and increased crash risk.

There is also a growing body of evidence that shows that visual field loss increases the risk of mobility difficulties and falls among older adults, particularly as visual fields contribute to efficient navigation within the environment (Turano et al. 2005). Studies show that visual field loss is linked with reduced mobility, particularly increased numbers of obstacle collisions and slower times to complete a mobility course (Turano et al. 2004). Furthermore, visual field loss among older adults has been linked to increased risk of falls in population studies (Coleman et al. 2007, Freeman et al. 2007), as well as in those with ocular conditions causing visual field loss, such as glaucoma (Black et al. 2011).

22.5 SPEED OF VISUAL PROCESSING AND DIVIDED ATTENTION: USEFUL FIELD OF VIEW

Assessment of functional or attentional fields provides another approach to the visual field testing typically conducted in clinical settings (as outlined in the previous section). Functional fields represent the extent of the visual field across which information can be processed during a single fixation, or the speed of visual processing within a given visual area, and have been most commonly measured using a technique known as the useful field of view (UFOV) (Ball et al. 1988, Ball and Owsley 1992), which was designed to better reflect the everyday visual difficulties experienced by older adults that are not captured by conventional field testing (Ball et al. 1990). The UFOV is a computer-generated task that involves participants identifying a central target, while at the same time having to identify peripherally presented targets, either in the presence or absence of distractors. The test is believed to assess higher-order cognitive

abilities and relies on selective and divided attention and rapid visual processing speeds, but performance also depends upon visual sensory function since targets must be visible in order to be attended to (Wood and Owsley 2014).

The slowing of visual processing speeds with increasing age is well established, where older adults require more time to detect and correctly identify targets, and comprises an important contributory factor to the problems with higher-order processing evident in many older adults, even in those without dementia (Salthouse 1992). This age-related slowing is exacerbated when individuals are required to divide attention between central and peripheral tasks (divided attention) and in the presence of distractor tasks (selective attention) as is the case in the UFOV data presented in Figure 22.4.

The extent of the UFOV has been shown to decrease through slowing of visual processing speeds with increasing age, particularly in the presence of distractor targets and at more peripheral eccentricities (Ball et al. 1988), although there are significant interindividual differences (Rubin et al. 2007). This finding was confirmed by a large-scale study of older adults (58–102 years), which used an alternative attentional field task that required participants to focus on a central target (counting the

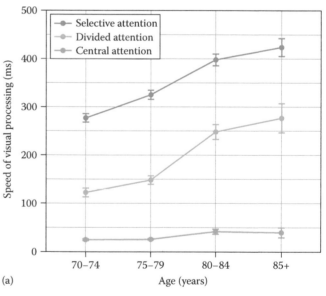

(a)

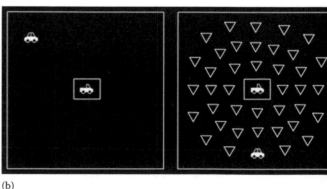

(b)

Figure 22.4 (a) Speed of visual processing as a function of age. (From Wood, J.M. et al., J. Am. Geriatr. Soc., 56(6), 986, 2008.) (b) Schematic of the useful field of view test illustrating the divided attention task (on the left) and selective attention task (on the right).

Figure 22.5 Schematic of the useful field of view in a real-world driving situation.

number of times it flashed on and off) while detecting peripherally presented suprathreshold targets (Haegerstrom-Portnoy et al. 1999). Interestingly, it was found that while conventional perimetric sensitivity measured for suprathreshold targets showed little decrease with increasing age, the attentional fields showed a dramatic reduction with increasing age. This restriction of the attentional/functional field is thus likely to be due to higher-order declines rather than simply optical or retinal factors.

A growing body of evidence suggests that a reduction in the UFOV, expressed either in terms of extent or speed of visual processing, is a strong predictor of driving ability and safety. A schematic representation of the potential impact of a reduction in the UFOV on perception of the driving environment is given in Figure 22.5. Reduced UFOV performance is a strong predictor of both retrospective (Owsley et al. 1991, Ball et al. 1993) and prospective crashes in general populations of older adults (Owsley et al. 1998, Rubin et al. 2007, Cross et al. 2009), as well as in those with ocular disease (Haymes et al. 2007). For example, a 40% reduction in the UFOV is associated with a 16.3 times higher risk of an injurious crash compared to those with no impairment (Owsley et al. 1998). Studies have also reported strong associations between a reduction in the UFOV and unsafe on-road performance (Myers et al. 2000, Wood 2002) and driving simulator performance (Roenker et al. 2003). The UFOV is also effective at predicting prospective crash risk when administered in a driver licensing setting (Ball et al. 2006), providing support for its inclusion in older driver screening.

The potential for the UFOV to predict problems with other functional activities of daily living has also been explored. These studies have demonstrated that impaired UFOV performance is associated with difficulties with mobility (Owsley and McGwin 2004), balance (Reed-Jones et al. 2012), and increased falls risk (Vance et al. 2006), as well as a range of other everyday activities (Edwards et al. 2005).

Importantly, there is growing evidence that visual processing speeds can be improved in older adults through computer-based training programs administered in either a laboratory or clinic-based setting, as well as in a home-based situation

(Wadley et al. 2006, Ball et al. 2007). These improvements in visual processing speed have been shown to reduce crash risk, where at-fault crash risk was halved following training (Ball et al. 2010), and have other positive benefits for health and functional well-being, including faster completion of everyday visual tasks (Edwards et al. 2005), reduced risk of depression (Wolinsky et al. 2009), and improvements in self-rated health (Wolinsky et al. 2010) and in health-related quality of life (Wolinsky et al. 2006).

22.6 DARK ADAPTATION

Dark adaptation represents the ability of the visual system to adjust sensitivity when ambient light levels change from bright light (photopic) to low light (scotopic) conditions. Measurement of dark adaptation is therefore an important clinical tool for the assessment of change in visual sensitivity over time in the dark and is commonly used to assess retinal photoreceptor function. A variety of techniques are available to measure dark adaptation, but the most common approach is to assess sensitivity over time following an initial photobleach exposure, using an achromatic stimulus, either at the fovea or around 5°–12° from the fovea (Jackson et al. 1999, Dimitrov et al. 2008, Owsley et al. 2014).

The dark adaptation curve, as shown in Figure 22.6, plots the improvement in sensitivity as the visual system adapts to the dark and is comprised of several components. The first component represents the rapid reduction in thresholds corresponding to cone-mediated recovery, which reaches a plateau at around 2 log units below the initial threshold. The rod–cone break occurs approximately 10–15 min into dark adaptation, where visual sensitivity switches from cone to rod mediated. Impaired dark adaptation can impact upon various activities of daily living, including mobility and driving, particularly when walking from very bright to dim room lighting or recovering from bright oncoming headlights when driving at night. The recent interest in dark adaptation has also been driven by its potential as a biomarker for early macular disease among older adults (Dimitrov et al. 2008, Gaffney et al. 2012).

Studies have demonstrated significant changes in dark adaptation with increasing age, which is reflected by older adults'

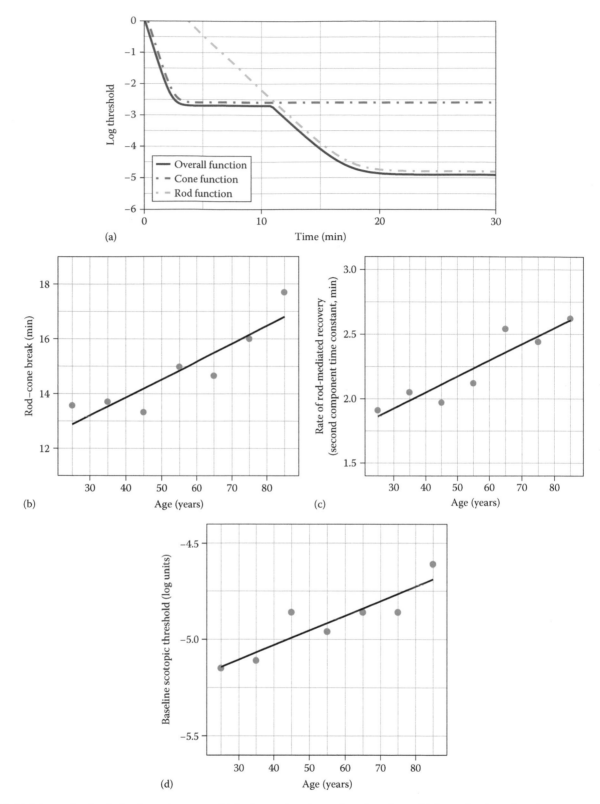

Figure 22.6 (a) Schematic of a dark adaptation curve. (b) Rod–cone break as a function of age. (c) Rate of rod-mediated recovery as a function of age. (d) Baseline scotopic threshold as a function of age. (Data redrawn from Jackson, G.R. et al., *Vis. Res.*, 39(23), 3975, 1999.)

reports of difficulty in seeing under low illumination and at night, even in the absence of ocular disease (Kline et al. 1992). These age-related changes occur in both the rod- and cone-mediated components of dark adaptation and may be due to the optical changes with age, such as pupil miosis and media opacities, as well as retinal and neural changes.

The cone-mediated component of dark adaptation slows with age, where the cone adaptation time for adults over 70 with no ocular disease was reported to be around twice that of those aged 10–30 years, despite the fact that final cone thresholds remain relatively stable across age (Gaffney et al. 2012). Investigations have also demonstrated age-related changes in the rod-mediated component of dark adaptation, which reflect impairments in rod photopigment regeneration and neural cell loss. Jackson et al. (1999) showed that the rod–cone break was delayed with age, with the rod–cone break for adults in their 70s being around 2.5 min slower relative to those in their 20s. In addition, age significantly slowed the rod recovery components and reduced final baseline scotopic sensitivity. Even older adults with normal macular health can exhibit abnormal rod-mediated dark adaptation dynamics, which may indicate preclinical changes consistent with conditions such as macular degeneration (Owsley et al. 2014).

Impaired dark adaptation is likely to significantly impact on performance across a range of everyday activities, particularly when moving into areas of reduced light levels. Impaired dark adaptation has been linked to an increased risk of falls (McMurdo and Gaskell 1991) and an increased risk of nighttime crashes among older drivers (Mortimer and Fell 1989).

22.7 DISABILITY GLARE

Disability glare describes the reduction in visual performance that occurs in the presence of a bright glare source, which induces forward intraocular light scatter. In particular, the straylight from intraocular scatter causes a reduction in the retinal image quality, affecting a range of visual functions including visual acuity and contrast sensitivity. Disability glare can affect everyday tasks such as driving; driving situations in which disability glare may cause

Figure 22.7 Example of glare at night from oncoming traffic.

transient loss of visual performance include nighttime driving as a result of approaching headlights (Figure 22.7) and at dawn and dusk from the sun positioned low on the horizon. Discomfort glare is a psychological consequence of glare that does not cause impairment in visual performance yet can cause significant distress and distraction due to the dazzling effect of the glare source.

While there are no universally accepted measures of disability glare, it is typically assessed by determining the extent to which visual performance is impaired, for example, high- or low-contrast acuity or contrast sensitivity, in the presence of an off-axis glare source (Elliott and Bullimore 1993, Bailey and Bullimore 1997). It has been challenging to compare between disability glare studies due to the lack of standardized measures, as the amount of intraocular scatter induced varies according to the intensity and direction of the glare source, as well as in differences in the visual function being assessed. Two commonly used measures include the Berkeley Glare Test (BGT) and the Brightness Acuity Tester (BAT) (Figure 22.8). The BGT assesses the decrease in binocular low-contrast (10%) acuity that results when presented under different levels of surrounding glare of 300, 800, and 3000 cd/m² (Bailey and Bullimore 1997). Conversely, the BAT assesses monocular disability glare and consists of a handheld internally illuminated hemispheric bowl of around 350 cm/m², which is held close to one eye, while visual function (either contrast sensitivity or visual acuity) is assessed through the central 12 mm aperture.

Older adults often experience higher levels of disability glare than younger adults due to increased intraocular scatter, particularly in the presence of media opacities such as cataracts (Elliott and Bullimore 1993). In addition, the reduction in pupil size with age may exacerbate disability glare by reducing retinal illumination. Straylight meters can provide an indirect measure of disability glare by quantifying the amount of forward intraocular scatter (van Rijn et al. 2005). In healthy eyes, retinal straylight remains relatively constant until the age of 45, followed by a gradual increase with advancing age (Rozema et al. 2010). Intraocular stray light is significantly related to cataract severity to a much greater extent than standard measures of visual acuity and contrast sensitivity (Michael et al. 2009).

Several studies have shown higher levels of disability glare with age, even in the absence of significant eye disease. Bailey

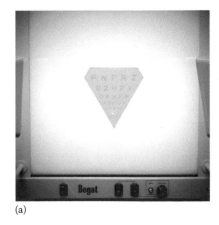

(a)

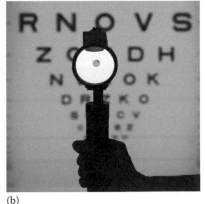

(b)

Figure 22.8 Images of two commonly used techniques used for the measurement of disability glare including the Berkeley Glare Test (a) and the Brightness Acuity Tester (b).

and Bullimore (1997) reported greater decrements in low-contrast acuity on the BGT for older adults without eye disease in the high-glare versus no-glare condition (10.2 ± 4.8 letters lost) compared to a younger group (2.3 ± 1.8 letters lost). Similarly, Elliott and Bullimore (1993) reported greater disability glare in older adults with no significant eye disease in the medium glare relative to the no-glare condition on the BGT (7 ± 4 letters lost) compared to the younger group (1 ± 2 letters lost). In a large population study, Rubin et al. (1997) reported a significant increase in glare sensitivity with age using the Pelli–Robson contrast sensitivity chart with the BAT; for every decade increase in age, an additional Pelli–Robson letter was lost in the presence of glare.

In older adults with cataracts, studies consistently show greater disability glare compared to their age-matched counterparts without cataracts (Elliott and Bullimore 1993, Wood and Carberry 2006) and significant improvements following cataract removal (Elliott et al. 1997, Wood and Carberry 2006). Similarly, older adults with various eye conditions, such as glaucoma, are also strongly affected by disability glare, which relates to their perceived visual disability (Nelson et al. 2003).

While most disability glare tests are conducted under photopic conditions, some have been developed for low-luminance conditions to reflect lighting conditions when driving at night. Puell et al. (2004b) assessed mesopic contrast sensitivity in the absence and presence of a glare source using the Mesotest II in 297 drivers aged from 21 to over 70 years. Contrast sensitivity measured without glare decreased gradually from the age of 51 to 60 years, while contrast sensitivity measured with glare started to decline at the earlier age of 41–50 years. Both conditions showed a decline of around 0.1 log contrast sensitivity per decade from 50 years onward.

Many of these age-related changes in disability glare have been shown to be related to the performance of everyday tasks. For example, older drivers with higher levels of disability glare self-regulate and reduce their driving exposure, particularly at night (Brabyn et al. 2005). While previous research has failed to find significant links between disability glare and at-fault crash risk (Owsley et al. 2001), there is some evidence that increased glare sensitivity is linked with crash involvement in older drivers (Rubin et al. 2007). This is consistent with findings from a simulator study of older adults, which demonstrated poorer driving safety when performing turns against oncoming traffic in the presence of glare, particularly for low-contrast vehicles (Gray and Regan 2007). The effects of headlamp glare halved the likelihood that drivers detected the presence of pedestrians at nighttime in on-road studies and also significantly decreased the distances at which pedestrians were first recognized (Wood et al. 2012). Similarly, Theeuwes et al. (2002) found that the ability of drivers to detect simulated pedestrians at night was significantly decreased in the presence of a simulated glare source mounted on the vehicle bonnet, even when the brightness of the simulated glare source was relatively low. These studies are consistent with the findings of Ranney et al. (2000) using a night driving simulator, which demonstrated that the presence of glare delayed the detection of pedestrians.

Studies also show significant associations between disability glare and mobility performance in older adults. In a large population study, West et al. (2002) reported that greater disability glare, as measured using the BGT, was significantly associated with a greater odds of self-reported mobility problems and failing physical performance tests (walking test and chair stand).

22.8 MOTION PERCEPTION

Motion perception describes the ability to detect or discriminate the motion of targets within the environment and has been studied using a wide range of approaches. Importantly, the ability to perceive motion and discriminate one's own motion is considered to be critical for the performance of a range of everyday tasks, including driving and navigating through the environment.

The impact of aging on motion perception has been a particular focus of research over many years; however, the extent to which aging has been reported to affect motion sensitivity varies widely between studies and depends on the contrast, speed, and the type of targets employed to assess motion thresholds (Gilmore et al. 1992, Norman et al. 2003, Snowden and Kavanagh 2006, Bennett et al. 2007). Random-dot kinematograms (RDK) have been commonly used to explore age-related changes in motion sensitivity (Wood and Bullimore 1995, Snowden and Kavanagh 2006, Bennett et al. 2007, Billino et al. 2008, Roudaia et al. 2010). Some of these studies have reported that motion sensitivity is reduced with increasing age (Trick and Silverman 1991, Wood and Bullimore 1995, Billino et al. 2008, Allen et al. 2010), while others have reported no differences (Gilmore et al. 1992, Atchley and Andersen 1998, Snowden and Kavanagh 2006). Figure 22.9 represents the decline in the lower displacement limit as measured with RDK stimuli from one of our previous studies (Wood and Bullimore 1995).

Studies also suggest that age-related differences in motion perception are more evident for low-contrast targets (Allen et al. 2010) and are dependent on the target speed (Atchley and Andersen 1998, Snowden and Kavanagh 2006, Billino et al. 2008) and duration of stimuli, where extended viewing times have been shown to improve performance for older adults (Bennett et al. 2007). Evidence also suggests that these age-related deficits in motion perception become evident in later

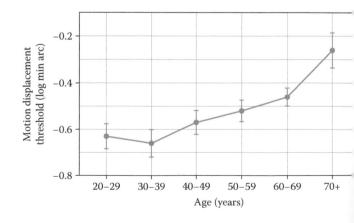

Figure 22.9 Motion displacement thresholds as a function of age. (Adapted from Wood, J.M. and Bullimore, M.A., *Ophthalmic Physiol Opt.*, 15(1), 31, 1995.)

adulthood rather than having a gradual onset as do many other age-related changes in visual function (Bennett et al. 2007) described elsewhere in this chapter.

Age-related preretinal changes including pupil miosis and light scatter have been shown to have only a minimal effect on the age-related decline in motion sensitivity evident for motion coherence thresholds (Norman et al. 2003, Betts et al. 2009). This is supported to some extent by the finding that motion sensitivity is not affected by increasing blur (Ball and Sekuler 1986), although reduced contrast sensitivity has been shown to impact on some aspects of motion sensitivity such as speed perception (Thompson et al. 2006). Interestingly, recent work has suggested that when the RDK display aperture is small, detection of the direction of translational global motion is actually better in older compared to younger observers (Hutchinson et al. 2014). The mechanisms underpinning this finding and other age-related changes seen in higher-level motion processing are believed to involve neural factors including decreases in spatial summation, visual attention, reduced sampling efficiency, cortical inhibition, and increased neural noise (Karas and McKendrick 2012, Bogfjellmo et al. 2013, Hutchinson et al. 2014).

A number of studies have also explored the impact of aging on measures of motion perception related to performance of everyday activities, particularly those related to driving. Increased age has been shown to be associated with deficits in collision detection (Andersen and Enriquez 2006, Andersen et al. 2000), less accurate time-to-contact judgments and difficulties in the perception of speed and heading (Conlon and Herkes 2008, DeLucia and Mather 2006, Poulter and Wann 2013), which are critical for enabling smooth and rapid responses to road hazards, particularly at intersections. Older adults also exhibit difficulties in perceiving shape from motion (Norman et al. 2004a, Blake et al. 2008) and the ability to discriminate the motion of point-light walkers, known as biological motion (Norman et al. 2004b, Billino et al. 2008, Pilz et al. 2010).

Many of these age-related deficits in motion sensitivity have been shown to relate to the performance of everyday tasks. For example, age-related deficits in various aspects of motion perception have been shown to be predictive of indices of driving performance and safety, particularly in older adults (Wood 2002, Raghuram and Lakshminarayanan 2006, Wood et al. 2008, Lacherez et al. 2014a,b). Motion perception has also been strongly linked to self-reported failures of attention using established questionnaire measures (Henderson and Donderi 2005, Raghuram and Lakshminarayanan 2006, Henderson et al. 2010). In one of our recent nighttime studies, drivers' capacity to recognize the presence of pedestrians wearing reflective clothing at night (in terms of recognition distances) was shown to significantly decrease with age and was most strongly predicted by the lower displacement threshold in RDK displays after controlling for age (Wood et al. 2014; Figure 22.10). This is in accord with the suggestion by previous authors that visual motion provides an important cue for discriminating the movement of hazards, such as pedestrians, against their background (Warren et al. 1989, Cutting et al. 1995, Loomis and Beall 1998).

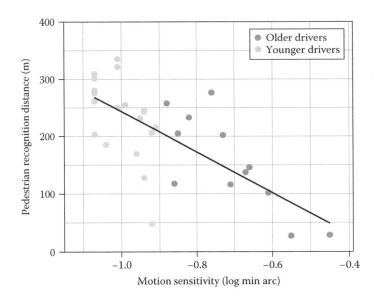

Figure 22.10 Relationship between motion sensitivity and overall pedestrian recognition distances for younger and older drivers. (Adapted from Wood, J.M. et al., *Ophthalmic Physiol. Opt.*, 34(4), 452, 2014.)

REFERENCES

Adams, A. J., L. S. Wong, L. Wong, and B. Gould. 1988. Visual acuity changes with age: Some new perspectives. *American Journal of Optometry and Physiological Optics* 65 (5):403–406.

Adams, C. W., M. A. Bullimore, M. Wall, M. Fingeret, and C. A. Johnson. 1999. Normal aging effects for frequency doubling technology perimetry. *Optometry and Vision Science* 76 (8):582–587.

Allard, R., J. Renaud, S. Molinatti, and J. Faubert. 2013. Contrast sensitivity, healthy aging and noise. *Vision Research* 92:47–52.

Allen, H. A., C. V. Hutchinson, T. Ledgeway, and P. Gayle. 2010. The role of contrast sensitivity in global motion processing deficits in the elderly. *Journal of Vision* 10 (10):15, 1–10.

Andersen, G. J., J. Cisneros, A. Saidpour, and P. Atchley. 2000. Age-related differences in collision detection during deceleration. *Psychology and Aging* 15 (2):241–252.

Andersen, G. J. and A. Enriquez. 2006. Aging and the detection of observer and moving object collisions. *Psychology and Aging* 21 (1):74–85.

Artal, P., M. Ferro, I. Miranda, and R. Navarro. 1993. Effects of aging in retinal image quality. *Journal of the Optical Society of America. A, Optics and Image Science* 10 (7):1656–1662.

Artal, P., A. Guirao, E. Berrio, P. Piers, and S. Norrby. 2003. Optical aberrations and the aging eye. *International Ophthalmology Clinics* 43 (2):63–77.

Atchley, P. and G. J. Andersen. 1998. The effect of age, retinal eccentricity, and speed on the detection of optic flow components. *Psychology and Aging* 13 (2):297–308.

Attebo, K., P. Mitchell, and W. Smith. 1996. Visual acuity and the causes of visual loss in Australia: The Blue Mountains Eye Study. *Ophthalmology* 103 (3):357–364.

Bailey, I. L. and M. A. Bullimore. 1997. A new test for the evaluation of disability glare. *Optometry and Vision Science* 68 (12):911–917.

Bailey, I. L. and J. E. Lovie-Kitchin. 2013. Visual acuity testing. From the laboratory to the clinic. *Vision Research* 90:2–9.

Ball, K., J. D. Edwards, and L. A. Ross. 2007. The impact of speed of processing training on cognitive and everyday functions. *Journals of Gerontology. Series B: Psychological Sciences and Social Sciences* 62 (Spec No. 1):19–31.

Ball, K., J.D. Edwards, L.A. Ross, G.Jr. McGwin. 2010. Cognitive training decreases motor vehicle collision involvement of older drivers. *Journal of the American Geriatrics Society* 58:2107–2113.

Ball, K. and C. Owsley. 1992. The useful field of view test: A new technique for evaluating age-related declines in visual function. *Journal of the American Optometric Association* 63:71–79.

Ball, K., C. Owsley, and B. Beard. 1990. Clinical visual perimetry underestimates peripheral field problems in older adults. *Clinical Vision Sciences* 5:113–125.

Ball, K., C. Owsley, M. E. Sloan, D. L. Roenker, and J. R. Bruni. 1993. Visual attention problems as a predictor of vehicle crashes in older drivers. *Investigative Ophthalmology and Visual Science* 34 (11):3110–3123.

Ball, K., C. Owsley, B. Stavey et al. 1998. Driving avoidance and functional impairment in older drivers. *Accident Analysis and Prevention* 30 (3):313–322.

Ball, K., D. L. Roenker, V. G. Wadley et al. 2006. Can high-risk older drivers be identified through performance-based measures in a department of motor vehicles setting? *Journal of the American Geriatrics Society* 54:77–84.

Ball, K. and R. Sekuler. 1986. Improving visual perception in older observers. *Journal of Gerontology* 41 (2):176–182.

Ball, K. K., B. L. Beard, D. L. Roenker, R. L. Miller, and D. S. Griggs. 1988. Age and visual search: Expanding the useful field of view. *Journal of the American Optometric Association* 5:2210–2219.

Bennett, P. J., R. Sekuler, and A. B. Sekuler. 2007. The effects of aging on motion detection and direction identification. *Vision Research* 47 (6):799–809.

Betts, L. R., A. B. Sekuler, and P. J. Bennett. 2009. Spatial characteristics of center-surround antagonism in younger and older adults. *Journal of Vision* 9 (1):25.1–15.

Billino, J., F. Bremmer, and K. R. Gegenfurtner. 2008. Differential aging of motion processing mechanisms: Evidence against general perceptual decline. *Vision Research* 48 (10):1254–1261.

Black, A. A., J. M. Wood, and J. E. Lovie-Kitchin. 2011. Inferior field loss increases rate of falls in older adults with glaucoma. *Optometry and Vision Science* 88 (11):1275–1282.

Blake, R., M. Rizzo, and S. McEvoy. 2008. Aging and perception of visual form from temporal structure. *Psychology and Aging* 23 (1):181–189.

Bogfjellmo, L. G., P. J. Bex, and H. K. Falkenberg. 2013. Reduction in direction discrimination with age and slow speed is due to both increased internal noise and reduced sampling efficiency. *Investigative Ophthalmology and Visual Science* 54 (8):5204–5210.

Bonnel, S., S. Mohand-Said, and J. A. Sahel. 2003. The aging of the retina. *Experimental Gerontology* 38 (8):825–831.

Brabyn, J A., M E. Schneck, L A. Lott, and G. Haegerstrom-Portnoy. 2005. Night driving self-restriction: Vision function and gender differences. *Optometry and Vision Science* 82:755–764.

Burg, A. 1967. *The Relationship between Vision Test Scores and Driving Record: General Findings.* Los Angeles, CA: Department of Engineering, University of California, Los Angeles.

Burton, K. B., C. Owsley, and M. E. Sloane. 1993. Aging and neural spatial contrast sensitivity: Photopic vision. *Vision Research* 33 (7):939–946.

Buyck, A., L. Missotten, M. J. Maes, and H. Van de Voorde. 1988. Assessment of the driving behaviour of visually handicapped persons. In *Vision in Vehicles II*, eds. Gale, A.G., Haslegrave, C.M., Smith, P., Taylor, S.P., pp. 131–142. Elsevier Science Publishers BV (North Holland).

Campbell, F. W. and J. G. Robson. 1968. Application of Fourier analysis to the visibility of gratings. *Journal of Physiology* 197 (3):551–566.

Chisholm, C. M., F. G. Rauscher, D. C. Crabb et al. 2008. Assessing visual fields for driving in patients with paracentral scotomata. *British Journal of Ophthalmology* 92 (2):225–230.

Coleman, A. L., S. R. Cummings, F. Yu et al. 2007. Binocular visual-field loss increases the risk of future falls in older white women. *Journal of the American Geriatrics Society* 55 (3):357–364.

Coleman, A. L., K. Stone, S. K. Ewing et al. 2004. Higher risk of multiple falls among elderly women who lose visual acuity. *Ophthalmology* 111 (5):857–862.

Conlon, E. and K. Herkes. 2008. Spatial and temporal processing in healthy aging: Implications for perceptions of driving skills. *Neuropsychology, Development, and Cognition. Section B: Aging, Neuropsychology and Cognition* 15 (4):446–470.

Crabb, D. P., F. W. Fitzke, R. A. Hitchings, and A. C. Viswanathan. 2004. A practical approach to measuring the visual field component of fitness to drive. *British Journal of Ophthalmology* 88 (9):1191–1196.

Cross, J. M., G. McGwin, Jr., G. S. Rubin et al. 2009. Visual and medical risk factors for motor vehicle collision involvement among older drivers. *British Journal of Ophthalmology* 93 (3):400–404.

Cummings, S. R., M. C. Nevitt, W. S. Browner et al. 1995. Risk factors for hip fracture in white women. Study of Osteoporotic Fractures Research Group. *New England Journal of Medicine* 332 (12):767–773.

Curcio, C. A. 2001. Photoreceptor topography in ageing and age-related maculopathy. *Eye* 15 (Pt 3):376–383.

Curcio, C. A., C. L. Millican, K. A. Allen, and R. E. Kalina. 1993. Aging of the human photoreceptor mosaic: Evidence for selective vulnerability of rods in central retina. *Investigative Ophthalmology and Visual Science* 34 (12):3278–3296.

Cutting, J. E., P. M. Vishton, and P. A. Braren. 1995. How we avoid collisions with stationary and moving objects. *Psychological Review* 102 (4):627–651.

Decina, L. E. and L. Staplin. 1993. Retrospective evaluation of alternative vision screening criteria for older and younger drivers. *Accident Analysis and Prevention* 25 (3):267–275.

DeLucia, P. R. and R. D. Mather. 2006. Motion extrapolation of car-following scenes in younger and older drivers. *Human Factors* 48 (4):666–674.

Derefeldt, G., G. Lennerstrand, and B. Lundh. 1979. Age variations in normal human contrast sensitivity. *Acta Ophthalmologica* 57 (4):679–690.

Dimitrov, P. N., R. H. Guymer, A. J. Zele, A. J. Anderson, and A. J. Vingrys. 2008. Measuring rod and cone dynamics in age-related maculopathy. *Investigative Ophthalmology and Visual Science* 49 (1):55–65.

Drance, S. M., V. Berry, and A. Hughes. 1967. Studies on the effects of age on the central and peripheral isopters of the visual field in normal subjects. *American Journal of Ophthalmology* 63 (6):1667–1672.

Edwards, J. D., V. G. Wadley, D. E. Vance et al. 2005. The impact of speed of processing training on cognitive and everyday performance. *Aging and Mental Health* 9 (3):262–271.

Eisner, A., S. A. Fleming, M. L. Klein, and W. M. Mauldin. 1987. Sensitivities in older eyes with good acuity: Cross-sectional norms. *Investigative Ophthalmology and Visual Science* 28 (11):1824–1831.

Elliott, D., D. Whitaker, and D. MacVeigh. 1990. Neural contribution to spatiotemporal contrast sensitivity decline in healthy ageing eyes. *Vision Research* 30 (4):541–547.

Elliott, D. B. and M. A. Bullimore. 1993. Assessing the reliability, discriminative ability, and validity of disability glare tests. *Investigative Ophthalmology and Visual Science* 34 (1):108–119.

Elliott, D. B., A. Patla, and M. A. Bullimore. 1997. Improvements in clinical and functional vision and perceived visual disability after first and second eye surgery. *British Journal of Ophthalmology* 81:889–895.

Elliott, D. B., K. C. Yang, and D. Whitaker. 1995. Visual acuity changes throughout adulthood in normal, healthy eyes: Seeing beyond 6/6. *Optometry and Vision Science* 72 (3):186–191.

Elliott, S. L., S. S. Choi, N. Doble et al. 2009. Role of high-order aberrations in senescent changes in spatial vision. *Journal of Vision* 9 (2):24.1–16.

Fonda, G. 1989. Legal blindness can be compatible with safe driving. *Opthalmology* 96:1457–1459.

Freeman, E. E., B. Munoz, G. Rubin, and S. K. West. 2007. Visual field loss increases the risk of falls in older adults: The Salisbury Eye Evaluation. *Investigative Ophthalmology and Visual Science* 48 (10):4445–4450.

Gaffney, A. J., A. M. Binns, and T. H. Margrain. 2012. Aging and cone dark adaptation. *Optometry and Vision Science* 89 (8):1219–1224.

Gao, H. and J. G. Hollyfield. 1992. Aging of the human retina. Differential loss of neurons and retinal pigment epithelial cells. *Investigative Ophthalmology and Visual Science* 33 (1):1–17.

Gilmore, G. C., H. E. Wenk, L. A. Naylor, and T. A. Stuve. 1992. Motion perception and aging. *Psychology and Aging* 7 (4):654–660.

Gray, R. and D. Regan. 2007. Glare susceptibility test results correlate with temporal safety margin when executing turns across approaching vehicles in simulated low-sun conditions. *Ophthalmic and Physiological Optics* 27 (5):440–450.

Haegerstrom-Portnoy, G., M. E. Schneck, and J. A. Brabyn. 1999. Seeing into old age: Vision function beyond acuity. *Optometry and Vision Science* 76 (3):141–158.

Haymes, S. A., R. P. Leblanc, M. T. Nicolela, L. A. Chiasson, and B. C. Chauhan. 2007. Risk of falls and motor vehicle collisions in glaucoma. *Investigative Ophthalmology and Visual Science* 48 (3):1149–1155.

Henderson, S. and D. C. Donderi. 2005. Peripheral motion contrast sensitivity and older drivers' detection failure accident risk. In *Proceedings of the Third International Driving Symposium on Human Factors in Driver Assessment, Training and Vehicle Design*, Rockport, Maine.

Henderson, S., S. Gagnon, A. Bélanger, R. Tabone, and C. Collin. 2010. Near peripheral motion detection threshold correlates with self-reported failures of attention in younger and older drivers. *Accident Analysis and Prevention* 42 (4):1189–1194.

Higgins, K. E., M. J. Jaffe, R. C. Caruso, and F. M. deMonasterio. 1988. Spatial contrast sensitivity: Effects of age, test-retest, and psychophysical method. *Journal of the Optical Society of America. A, Optics and Image Science* 5 (12):2173–2180.

Higgins, K. E. and J. M. Wood. 2005. Predicting components of closed road driving performance from vision tests. *Optometry and Vision Science* 82 (8):647–656.

Higgins, K. E., J. Wood, and A. Tait. 1998. Vision and driving: Selective effect of optical blur on different driving tasks. *Human Factors* 40 (2):224–232.

Hills, B. L. and A. Burg. 1977. *A Reanalysis of California Driver Vision Data: General Findings*. Crowthorne, Berkshire: Transport and Road Research Laboratories.

Hofstetter, H. W. 1976. Visual acuity and highway accidents. *Journal of the American Optometric Association* 47 (7):887–893.

Hong, T., P. Mitchell, E. Rochtchina et al. 2013. Long-term changes in visual acuity in an older population over a 15-year period: The Blue Mountains Eye Study. *Ophthalmology* 120 (10):2091–2099.

Hutchinson, C. V., T. Ledgeway, and H. A. Allen. 2014. The ups and downs of global motion perception: A paradoxical advantage for smaller stimuli in the aging visual system. *Frontiers in Aging Neuroscience* 6:199.

Ivers, R. Q., P. Mitchell, and R. G. Cumming. 1999. Sensory impairment and driving: The Blue Mountains Eye Study. *American Journal of Public Health* 89 (1):85–87.

Jackson, G. R., C. Owsley, and G. McGwin, Jr. 1999. Aging and dark adaptation. *Vision Research* 39 (23):3975–3982.

Jay, J. L., R. B. Mammo, and D. Allan. 1987. Effect of age on visual acuity after cataract extraction. *British Journal of Ophthalmology* 71 (2):112–115.

Johnson, C. A., A. J. Adams, and R. A. Lewis. 1989. Evidence for a neural basis of age-related visual field loss in normal observers. *Investigative Ophthalmology and Visual Science* 30 (9):2056–2064.

Johnson, C. A., A. J. Adams, J. D. Twelker, and J. M. Quigg. 1988. Age-related changes in the central visual field for short-wavelength-sensitive pathways. *Journal of the Optical Society of America. A, Optics and Image Science* 5 (12):2131–2139.

Johnson, C. A. and J. L. Keltner. 1983. Incidence of visual field loss in 20,000 eyes and its relationship to driving performance. *Archives of Ophthalmology* 101 (3):371–375.

Johnson, M. A. and D. Choy. 1987. On the definition of age-related norms for visual function testing. *Applied Optics* 26 (8):1449–1454.

Karas, R. and A. M. McKendrick. 2012. Age related changes to perceptual surround suppression of moving stimuli. *Seeing Perceiving* 25 (5):409–424.

Klein, B. E., R. Klein, K. E. Lee, and K. J. Cruickshanks. 1998. Performance-based and self-assessed measures of visual function as related to history of falls, hip fractures, and measured gait time. *The Beaver Dam Eye Study. Ophthalmology* 105 (1):160–164.

Klein, R., B. E. Klein, K. E. Lee, K. J. Cruickshanks, and R. J. Chappell. 2001. Changes in visual acuity in a population over a 10-year period: The Beaver Dam Eye Study. *Ophthalmology* 108 (10):1757–1766.

Klein, R., B. E. Klein, K. E. Lee, K. J. Cruickshanks, and R. E. Gangnon. 2006. Changes in visual acuity in a population over a 15-year period: The Beaver Dam Eye Study. *American Journal of Ophthalmology* 142 (4):539–549.

Klein, R., B. E. Klein, K. L. Linton, and D. L. De Mets. 1991. The Beaver Dam Eye Study: Visual acuity. *Ophthalmology* 98 (8):1310–1315.

Kline, D. W., T. J. B. Kline, J. L. Fozard et al. 1992. Vision, aging, and driving: The problems of older drivers. *Journal of Gerontology: Psychological Sciences* 47:27–34.

Kosnik, W., L. Winslow, D. Kline, K. Rasinski, and R. Sekuler. 1988. Visual changes in daily life throughout adulthood. *Journal of Gerontology* 43 (3):P63–P70.

Kosnik, W. D., R. Sekuler, and D. W. Kline. 1990. Self-reported visual problems of older drivers. *Human Factors* 32:597–608.

Lacherez, P., S. Au, and J. M. Wood. 2014. Visual motion perception predicts driving hazard perception ability. *Acta Ophthalmologica* 92 (1):88–93.

Lacherez, P., J. M. Wood, K. J. Anstey, and S. R. Lord. 2014. Sensorimotor and postural control factors associated with driving safety in a community-dwelling older driver population. *Journals of Gerontology. Series A: Biological Sciences and Medical Sciences* 69 (2):240–244.

Lamble, D., H. Summala, and L. Hyvarinen. 2002. Driving performance of drivers with impaired central visual field acuity. *Accident Analysis and Prevention* 34 (5):711–716.

Loomis, J. M. and A. C. Beall. 1998. Visually controlled locomotion: Its dependence on optic flow, three-dimensional space perception, and cognition. *Ecological Psychology* 10 (3–4):271–285.

Lovasik, J. V., M. J. Kergoat, L. Justino, and H. Kergoat. 2003. Neuroretinal basis of visual impairment in the very elderly. *Graefes Archive for Clinical and Experimental Ophthalmology* 241 (1):48–55.

Mantyjarvi, M. and T. Laitinen. 2001. Normal values for the Pelli-Robson contrast sensitivity test. *Journal of Cataract and Refractive Surgery* 27 (2):261–266.

McGwin, G., Jr., A. Xie, A. Mays et al. 2005. Visual field defects and the risk of motor vehicle collisions among patients with glaucoma. *Investigative Ophthalmology and Visual Science* 46 (12):4437–4441.

McMurdo, M. E. and A. Gaskell. 1991. Dark adaptation and falls in the elderly. *Gerontology* 37 (4):221–224.

Michael, R., L. J. van Rijn, T. J. van den Berg et al. 2009. Association of lens opacities, intraocular straylight, contrast sensitivity and visual acuity in European drivers. *Acta Ophthalmologica* 87 (6):666–671.

Mortimer, R. G. and J. C. Fell. 1989. Older drivers: Their night fatal crash involvement and risk. *Accident Analysis and Prevention* 21 (3):273–282.

Myers, R. S., K. K. Ball, T. D. Kalina, D. L. Roth, and K. T. Goode. 2000. Relation of useful field of view and other screening tests to on-road driving performance. *Perceptual and Motor Skills* 91 (1):279–290.

Nelson, P., P. Aspinall, O. Papasouliotis, B. Worton, and C. O'Brien. 2003. Quality of life in glaucoma and its relationship with visual function. *Journal of Glaucoma* 12 (2):139–150.

Norman, J. F., A. M. Clayton, C. F. Shular, and S. R. Thompson. 2004a. Aging and the perception of depth and 3-D shape from motion parallax. *Psychology and Aging* 19 (3):506–514.

Norman, J. F., S. M. Payton, J. R. Long, and L. M. Hawkes. 2004b. Aging and the perception of biological motion. *Psychology and Aging* 19 (1):219–225.

Norman, J. F., H. E. Ross, L. M. Hawkes, and J. R. Long. 2003. Aging and the perception of speed. *Perception* 32 (1):85–96.

Owsley, C., K. Ball, G. McGwin et al. 1998. Visual processing impairment and risk of motor vehicle crash among older adults. *Journal of the American Medical Association* 279 (14):1083–1088.

Owsley, C., K. Ball, M. E. Sloane, D. L. Roenker, and J. R. Bruni. 1991. Visual/cognitive correlates of vehicle accidents in older drivers. *Psychology and Aging* 6 (3):403–415.

Owsley, C., C. Huisingh, G. R. Jackson et al. 2014. Associations between abnormal rod-mediated dark adaptation and health and functioning in older adults with normal macular health. *Investigative Ophthalmology and Visual Science* 55 (8):4776–4789.

Owsley, C. and G. McGwin. 2004. Association between visual attention and mobility in older adults. *Journal of the American Geriatrics Society* 52 (11):1901–1906.

Owsley, C., G. McGwin, and K. Ball. 1998. Vision impairment, eye disease, and injurious motor vehicle crashes in the elderly. *Ophthalmic Epidemiology* 5 (2):101–113.

Owsley, C., G. McGwin, Jr., K. Scilley, and K. Kallies. 2006. Development of a questionnaire to assess vision problems under low luminance in age-related maculopathy. *Investigative Ophthalmology and Visual Science* 47 (2):528–535.

Owsley, C., R. Sekuler, and D. Siemsen. 1983. Contrast sensitivity throughout adulthood. *Vision Research* 23 (7):689–699.

Owsley, C., B. T. Stalvey, J. Wells, M. E. Sloane, and G. McGwin, Jr. 2001. Visual risk factors for crash involvement in older drivers with cataract. *Archives of Ophthalmology* 119 (6):881–887.

Pilz, K. S., P. J. Bennett, and A. B. Sekuler. 2010. Effects of aging on biological motion discrimination. *Vision Research* 50 (2):211–219.

Pitts, D. G. 1982. Visual acuity as a function of age. *Journal of the American Optometric Association* 53 (2):117–124.

Poulter, D. R. and J. P. Wann. 2013. Errors in motion processing amongst older drivers may increase accident risk. *Accident Analysis and Prevention* 57:150–156.

Puell, M. C., C. Palomo, C. Sanchez-Ramos, and C. Villena. 2004a. Normal values for photopic and mesopic letter contrast sensitivity. *Journal of Refractive Surgery* 20 (5):484–488.

Puell, M. C., C. Palomo, C. Sanchez-Ramos, and C. Villena. 2004b. Mesopic contrast sensitivity in the presence or absence of glare in a large driver population. *Graefes Archive for Clinical and Experimental Ophthalmology* 242 (9):755–761.

Raghuram, A. and V. Lakshminarayanan. 2006. Motion perception tasks as potential correlates to driving difficulty in the elderly. *Journal of Modern Optics* 53 (9):1343–1362.

Ranney, T. A., L. A. Simmons, and A. J. Masalonis. 2000. The immediate effects of glare and electrochromic glare-reducing mirrors in simulated truck driving. *Human Factors* 42 (2):337–34

Reed-Jones, R. J., S. Dorgo, M. K. Hitchings, and J. O. Bader. 2012. WiiFit Plus balance test scores for the assessment of balance and mobility in older adults. *Gait and Posture* 36 (3):430–433.

Roenker, D. L., G. M. Cissell, K. K. Ball, V. G. Wadley, and J. D. Edward 2003. Speed-of-processing and driving simulator training result in improved driving performance. *Human Factors* 45 (2):218–233.

Roudaia, E., P. J. Bennett, A. B. Sekuler, and K. S. Pilz. 2010. Spatiotemporal properties of apparent motion perception and aging. *Journal of Vision* 10 (14):5, 1–15.

Rozema, J. J., T. J. Van den Berg, and M. J. Tassignon. 2010. Retinal straylight as a function of age and ocular biometry in healthy ey *Investigative Ophthalmology and Visual Science* 51 (5):2795–2799

Rubin, G. S., E. S. Ng, K. Bandeen-Roche et al. 2007. A prospective, population-based study of the role of visual impairment in moto vehicle crashes among older drivers: The SEE study. *Investigative Ophthalmology and Visual Science* 48 (4):1483–1491.

Rubin, G. S., S. K. West, B. Munoz et al. 1997. A comprehensive assessment of visual impairment in a population of older Americans. The SEE Study. Salisbury Eye Evaluation Project. *Investigative Ophthalmology and Visual Science* 38 (3):557–568.

Said, F. S. and R. A. Weale. 1959. Variation with age of the spectral transmissivity of the living human crystalline lens. *Gerontologica* 3:1213–1231.

Salthouse, T. A. 1992. Influence of processing speed on adult age differences in working memory. *Acta Psychologica* 79 (2):155–170

Schneck, M. E., G. Haegerstrom-Portnoy, L. A. Lott, J. A. Brabyn, and G. Gildengorin. 2004. Low contrast vision function predict subsequent acuity loss in an aged population: The SKI study. *Vision Research* 44 (20):2317–2325.

Sia, D. I., S. Martin, G. Wittert, and R. J. Casson. 2013. Age-related change in contrast sensitivity among Australian male adults: Florey Adult Male Ageing Study. *Acta Ophthalmologica* 91 (4):312–317.

Sloane, M. E., C. Owsley, and S. L. Alvarez. 1988. Aging, senile miosi and spatial contrast sensitivity at low luminance. *Vision Research* 28 (11):1235–1246.

Snowden, R. J. and E. Kavanagh. 2006. Motion perception in the ageing visual system: Minimum motion, motion coherence, and speed discrimination thresholds. *Perception* 35 (1):9–24.

Spear, P. D. 1993. Neural bases of visual deficits during aging. *Vision Research* 33 (18):2589–2609.

Spry, P. G. and C. A. Johnson. 2001. Senescent changes of the normal visual field: An age-old problem. *Optometry and Vision Science* 78 (6):436–441.

Sturr, J. F., G. E. Kline, and H. A. Taub. 1990. Performance of young and older drivers on a static acuity test under photopic and mesopic luminance conditions. *Human Factors* 32 (1):1–8.

Theeuwes, J., J. W. Alferdinck, and M. Perel. 2002. Relation between glare and driving performance. *Human Factors* 44 (1):95–107.

Thompson, P., K. Brooks, and S. T. Hammett. 2006. Speed can go up as well as down at low contrast: Implications for models of motic perception. *Vision Research* 46 (6–7):782–786.

Trick, G. L. and S. E. Silverman. 1991. Visual sensitivity to motion: Age-related changes and deficits in senile dementia of the Alzheimer type. *Neurology* 41 (9):1437–1440.

Turano, K. A., A. T. Broman, K. Bandeen-Roche et al. 2004. Association of visual field loss and mobility performance in older adults: Salisbury Eye Evaluation Study. *Optometry and Vision Science* 81 (5):298–307.

Turano, K. A., D. Yu, L. Hao, and J. C. Hicks. 2005. Optic-flow and egocentric-direction strategies in walking: Central vs peripheral visual field. *Vision Research* 45 (25–26):3117–3132.

van Rijn, L. J., C. Nischler, D. Gamer et al. 2005. Measurement of stray light and glare: Comparison of Nyktotest, Mesotest, stray light meter, and computer implemented stray light meter. *British Journal of Ophthalmology* 89 (3):345–351.

Vance, D. E., K. K. Ball, D. L. Roenker et al. 2006. Predictors of falling in older Maryland drivers: A structural-equation model. *Journal of Aging and Physical Activity* 14 (3):254–269.

Verbaken, J. H. and A. W. Johnston. 1986. Population norms for edge contrast sensitivity. *American Journal of Optometry and Physiological Optics* 63 (9):724–732.

Wadley, V. G., R. L. Benz, K. K. Ball et al. 2006. Development and evaluation of home-based speed-of-processing training for older adults. *Archives of Physical Medicine and Rehabilitation* 87 (6):757–763.

Wang, J. J., S. Foran, and P. Mitchell. 2000. Age-specific prevalence and causes of bilateral and unilateral visual impairment in older Australians: The Blue Mountains Eye Study. *Clinical and Experimental Ophthalmology* 28 (4):268–273.

Warren, W. H., Jr., A. W. Blackwell, and M. W. Morris. 1989. Age differences in perceiving the direction of self-motion from optical flow. *Journal of Gerontology* 44 (5):P147–P153.

Weale, R. A. 1975. Senile changes in visual acuity. *Transactions of the Ophthalmological Societies of the United Kingdom* 95 (1):36–38.

Weale, R. A. 1992. *The senescence of human vision*. Oxford: Oxford University Press.

Werner, J. S., Schefrin, B. E., Bradley, A. 2010. Optics and vision of the aging eye. In *Handbook of Optics: Vision and Vision Optics*, eds. Bass, M., Enoch, J. M., Lakshminarayanan, V., pp. 14.11–14.38. New York: McGraw-Hill.

West, C. G., G. Gildengorin, G. Haegerstrom-Portnoy et al. 2002. Is vision function related to physical functional ability in older adults? *Journal of the American Geriatrics Society* 50 (1):136–145.

Winn, B., D. Whitaker, D. B. Elliott, and N. J. Phillips. 1994. Factors affecting light-adapted pupil size in normal human subjects. *Investigative Ophthalmology and Visual Science* 35 (3):1132–1137.

Wolinsky, F. D., F. W. Unverzagt, D. M. Smith et al. 2006. The effects of the ACTIVE cognitive training trial on clinically relevant declines in health-related quality of life. *Journals of Gerontology. Series B: Psychological Sciences and Social Sciences* 61 (5):S281–S287.

Wolinsky, F. D., M. W. Vander Weg, R. Martin et al. 2009. The effect of speed-of-processing training on depressive symptoms in ACTIVE. *The Journals of Gerontology. Series A, Biological Sciences and Medical Sciences* 64 (4):468–472.

Wolinsky, F. D., M. W. Vander Weg, R. Martin et al. 2010. Does cognitive training improve internal locus of control among older adults? *Journals of Gerontology. Series B: Psychological Sciences and Social Sciences* 65 (5):591–598.

Wood, J. M. 2002. Age and visual impairment decrease driving performance as measured on a closed-road circuit. *Human Factors* 44 (3):482–494.

Wood, J. M., K. J. Anstey, G. K. Kerr, P. F. Lacherez, and S. Lord. 2008. A multidomain approach for predicting older driver safety under in-traffic road conditions. *Journal of the American Geriatrics Society* 56 (6):986–993.

Wood, J. M. and M. A. Bullimore. 1995. Changes in the lower displacement limit for motion with age. *Ophthalmic and Physiological Optics* 15 (1):31–36.

Wood, J. M. and T. P. Carberry. 2006. Bilateral cataract surgery and driving performance. *British Journal of Ophthalmology* 90:1277–1280.

Wood, J. M., P. F. Lacherez, A. A. Black et al. 2009. Postural stability and gait among older adults with age-related maculopathy. *Investigative Ophthalmology and Visual Science* 50 (1):482–487.

Wood, J. M., P. Lacherez, and R. A. Tyrrell. 2014. Seeing pedestrians at night: Effect of driver age and visual abilities. *Ophthalmic and Physiological Optics* 34 (4):452–458.

Wood, J. M. and C. Owsley. 2014. Useful field of view test. *Gerontology* 60 (4):315–318.

Wood, J. M., R. A. Tyrrell, A. Chaparro et al. 2012. Even moderate visual impairments degrade drivers' ability to see pedestrians at night. *Investigative Ophthalmology and Visual Science* 53 (6):2586–2592.

Woods, R. L. and J. M. Wood. 1995. The role of contrast sensitivity charts and contrast letter charts in clinical practice. *Clinical and Experimental Optometry* 78:43–57.

23 Stereoacuity and optics

José Ramón Jiménez

Contents

23.1 STEREOPSIS

Stereopsis is one of the most advanced functions of our visual system, as it enables us to distinguish spatial (3D) locations of visual objects around us (Fielder and Moseley 1996; Howard and Rogers 1995, 2002; Reading 1983; Solomon 1978; Steinman et al. 2000; Stidwill and Flecher 2011), an essential property in our species, for which it is fundamental to have 3D vision in order to adapt to the demands of our surroundings, that is, accurately calculate positions and distances (especially short distances) and apprehend complex visual contexts (Fielder and Moseley 1996; Mazyn et al. 2004; O'Connor et al. 2010).

With only one eye, depth perception is possible but not so effective as with stereopsis. The information on depth without stereopsis is derived from monocular cues (Howard and Rogers 1995, 2002; Reading 1983; Solomon 1978; Steinman et al. 2000; Stidwill and Flecher 2011), which are founded on learning. Thus, the relative movement of the observer with respect to the object (movement parallax), the relative movement of objects with respect to the observer, shadows, relative size, aerial perspective, lights, and texture are monocular cues by which the subject judges depth, although not with the quality provided by stereopsis. Not all monocular cues are equally effective, and thus relative movement of the observer with respect to the object and shadows, for example, are powerful monocular cues, while aerial perspective and lights are less effective monocular cues. Also, the mechanism of convergence–accommodation offers minor information to deduce objects' position (Howard and Rogers 1995, 2002; Reading 1983; Solomon 1978; Steinman et al. 2000; Stidwill and Flecher 2011).

Stereoscopic vision is based on disparity. This is a geometric function that depends on the spatial positions of objects with respect to the fixation point. Thus, two points situated in different spatial positions imply unequal angles for eyes and different projections on the retina (Figure 23.1). The difference between the two angles is known as disparity.

The detection of disparity has its neurophysiological basis. Our visual system has disparity-sensitive neurons. Different models are based on the notion that we have neurons syntonized to positive, negative, and zero disparity (Cumming and De Angelis 2001; Hibbard 2008; Lehky and Sejnowski 1990; Orban et al. 2006; Poggio 1977). Our visual system obtains 3D information from the visual scene being viewed, gathering information provided by these families of disparity-sensitive neurons. Failures in detection of one or more of these families provoke problems or hamper the interpreting of our 3D information (Howard and Rogers 1995, 2002; Reading 1983; Richards 1971; Solomon 1978; Steinman et al. 2000; Stidwill and Flecher 2011).

The classification of disparity in zero, positive, or negative disparity has its correspondence with the geometry of the visual scene that we are seeing. Thus, the circle defined by the circumference passing through the centers of the two eye pupils and the point of fixation is called the Vieth–Müller circle. This circle divides the space into two regions: interior of the circle or zone with positive disparities and the exterior of the circle or zone with negative disparities. The Vieth–Müller circumference would be the

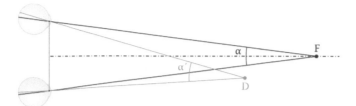

Figure 23.1 Different spatial positions of points D and F give rise to different projections and angles on the retina and a disparity value, η, of η = α′ − α, this disparity being used by the visual system to estimate the positions of D and F with respect to the observer.

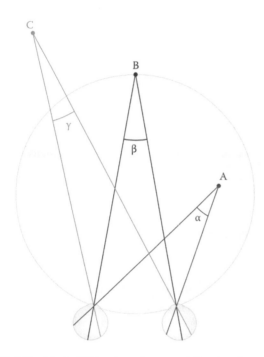

Figure 23.2 The Vieth–Müller circumference, defined as the circumference that passes through the fixation point (B) and the centers of the pupils of the eyes, dividing the space into two regions: the interior region of the circle, in which the points have positive disparities, and the exterior region of the circle, in which the points have negative disparities. The points of the circumference have zero disparity.

zero-disparity zone (Figure 23.2) (for more information about the Vieth–Müller circle and other related concepts such as horopter, see Howard and Rogers 1995, 2002; Ogle 1950; Reading 1983; Solomon 1978; Steinman et al. 2000; Stidwill and Flecher 2011).

To elucidate how the visual system can calculate the disparity map of a visual scene is an extraordinarily complex problem in which the final mechanisms are still unclear. The problem of calculating the disparity map of a visual scene is also known as the stereo correspondence problem (Howard and Rogers 1995, 2002; Marr 1980; Marr and Poggio 1976), since to know the disparity map is equivalent to determining, given a point on the retina, to what point on the other retina it corresponds in order to produce a simple, unified vision. In a normal scene, from the points of the image on a retina, there may be millions of pairs corresponding to those of the other retina but only one real valid correspondence that generates the disparity map associated with the visual image being viewed. The problem of stereo correspondence is clearly manifested with random-dot stereograms (RDS) (explained

immediately in the following). In the solution of the stereo correspondence problem, the visual information at different spatial frequencies constitutes an essential point in the solution of this problem (Howard and Rogers 1995, 2002; Marr 1980).

Usually, studies on stereoscopic vision from the standpoint of visual or optical quality have been very limited, among other reasons, because the problems of having reduced or no stereopsis are not considered by many clinicians as problematic as other symptoms, such as blurred vision, diplopia, and headaches, these being symptoms with which the patient is very familiar, and often give rise to more complaints. Many people are not even aware that they lack stereopsis, despite that they may be conscious of having difficulties in manipulating objects, for example, often dropping things being handled manually. In fact, some emmetropization techniques, such as monovision (Evans 2007; Jain 1996), cancel or partially limit stereopsis to give the patient acceptable near and far vision even at the cost of reduced 3D vision.

The range or extent of disparities that can be detected by the observer is a key question, as it enables us to know the spatial region or its size where the observer can perceive stereoscopically, reflecting the quality of stereoscopic vision. A greater disparity range permits a larger spatial region in which to perceive stereoscopically (Howard and Rogers 1995, 2002; Jiménez et al. 1997, 2008a,b). A narrower extent of perceived disparities indicates that the region where the observer perceives stereoscopically is smaller and therefore judgments on the spatial position of objects prove more limited. The disparity range of each observer enables the definition of two parameters of stereoscopic vision quality: the minimum disparity perceived, or stereoacuity, and the upper disparity limit, or maximum disparity.

Stereoacuity indicates the region from the fixation point where we can perceive stereoscopically. A lower stereoacuity value would indicate higher stereoscopic vision quality, as it permits sharper discrimination of depth for points near the fixation point (Howard and Rogers 1995, 2002).

On the other hand, a high maximum disparity value (upper disparity limit) would also indicate better stereoscopic vision, as greater maximum disparity indicates a larger spatial region around the fixation point where judgments can be made concerning the 3D position of objects based on stereopsis, providing a more effective depth discrimination (Jiménez et al. 1997, 2008a,b).

Concerning the two limits of stereoscopic vision, stereoacuity has traditionally been the parameter most studied when exploring the limits of stereoscopic vision, especially in clinical practice, where different tests have been developed to evaluate it, though not with high accuracy (Reading 1983; Solomon 1978; Steinman et al. 2000; Stidwill and Flecher 2011).

23.2 CONCEPT AND CHARACTERISTICS OF STEREOACUITY

As defined earlier, stereoacuity is the minimum disparity that an observer can detect and this minimum disparity corresponds with the smallest depth difference we can see stereoscopically, which is the depth-discrimination threshold. Different methods and different devices can be used to measure stereoacuity (see

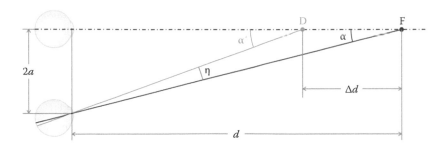

Figure 23.3 Schematic of binocular vision: two aligned points D and F (fixation point) separated by distance Δd and giving rise to disparity η.

Section 23.1.3). Small values of a few seconds of arc (2–6 arc sec) under laboratory conditions can be reached (Howard and Rogers 1995, 2002; Westheimer and McKee 1980; Wong et al. 2002). In clinical optometric practice, the minimum stereoacuity values provided by tests are usually around 40 arc sec (the TNO test can reach 15 arc sec), while 60 arc sec is a standard of normal stereoacuity for clinical purposes (Stidwill and Fletcher 2011).

As indicated previously, when stereopsis or its quality is being evaluated, especially in clinical practice, only stereoacuity is assessed. However, good stereoacuity is not necessarily indicative of good stereopsis, as in everyday life, larger disparities are used for depth-discrimination tasks. In fact, the sensitivity of the disparity range can differ markedly from one person to another, each being sensitive to a particular disparity range, with difficulties sometimes appearing even in detecting positive versus negative disparities, depending on whether a neuronal anomaly occurs in one of the families syntonized to positive, negative, or zero disparity (Howard and Rogers 1995, 2002; Reading 1983; Richards 1971; Solomon 1978; Steinman et al. 2000; Stidwill and Fletcher 2011). Making a comparison with the contrast sensitivity function and visual acuity, we note that good visual acuity does not provide information of the different ranges of sensitivity to spatial frequency, just as knowing the stereoacuity value does not tell us the range of the different disparities detected by an observer.

From the scheme shown in Figure 23.3, Equation 23.1 shows the relation between disparity and the depth-discrimination increment (the depth difference between fixation and test rod or point):

$$\eta = \frac{2a\Delta d}{d^2 + d\Delta d} \cong \frac{2a\Delta d}{d^2} \tag{23.1}$$

where

η is the disparity
d is the fixation distance
Δd is the distance between the fixation point (F) and point to calculate disparity (D)
$2a$ is the interpupillary distance

Stereoacuity (or minimum detected disparity) is proportional to interpupillary distance and depth-discrimination threshold and is inversely proportional to the square of fixation distance.

From Equation 23.1, we can deduce the limit distance, d_l, from which we cannot discriminate stereoscopically. If we make the distant target tend to infinity and we make the angle subtended between the fixation target and the distant target equal to stereoacuity, η_t, we can easily calculate the value from which perception is no longer stereoscopic, this being $d_l = 2a/\eta_t$.

If we introduce standard values of the parameters $\eta_t = 10$ arc sec and $2a = 60$ mm, we find that the limit distance is $d_l \cong 1240$ m. This value is extremely high, as the value used for stereoacuity ($\eta_t = 10$ arc sec) is determined under laboratory conditions. Under normal observation conditions, where numerous factors limit the stereoscopic discrimination capacity, beyond 5–10 m, our practical capacity for stereoscopic discrimination becomes sharply diminished.

23.3 DEVICES AND PSYCHOPHYSICAL METHODS FOR MEASURING STEREOACUITY

The instruments and psychophysical methods used to measure stereoacuity are very diverse. The devices can be classified as real-depth and projected-depth devices, although currently mainly the latter are used.

23.3.1 REAL-DEPTH DEVICES

In a real-depth device, the stereoscopic test is a real object that is usually moved outside the fixation plane, a plane where one or more objects may be encountered. These types of devices and tests are usually easy to use for observers and require no experience to detect depth. The only requirements usually refer to the alignment and the way to sit in order to avoid monocular cues and to ensure that stereoacuity measurements do not change over time if the alignment and the way to sit is not stable.

Two-rod or three-rod tests are often used. The most commonly used design (Figure 23.4) is that of the Howard-Dolman apparatus (Howard 1919; Howard and Rogers 1995, 2002; Ogle 1950; Reading 1983; Solomon 1978; Steinman et al. 2000; Stidwill and Flecher 2011). With this device, the subject's task is to move the adjustable rod until perceiving the depth difference with respect to the fixed rod. The minimum magnitude of displacement to see in depth allows the stereoacuity to be calculated from Equation 23.1. Normally, the method used to measure stereoacuity with these devices is the adjustment method, although at times, the constant-stimuli method is more precise despite being more time consuming (see Reading 1983 for more psychophysical methods).

The adjustment method (Reading 1983) requires the person directing the experiment to place the movable rod in a position far from the fixed rod and ask the observer to move the rod to the location judged equidistant from the fixed rod, d_i. This process is repeated a number of times with different offset positions in both directions (far and near). For each direction (far and near), the

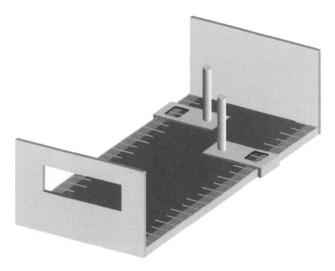

Figure 23.4 Scheme of Howard-Dolman stereoscope.

different d_i are taken to calculate the depth-increment threshold and, from Equation 23.1, the stereoacuity.

In the constant-stimuli method (Reading 1983), a position is sought so that the movable rod is clearly seen as being closer to the observer than the fixed rod. Conversely, a position is also sought so that the movable rod is clearly seen to be farther than the fixed rod. This distance range is divided into a number of intervals and the movable rod is randomly presented to the observer in these positions a number of times. From the percentage of positive responses in each case (near or far), stereoacuity is estimated (Reading 1983).

A simple modification of the Howard-Dolman apparatus is also used, the three-needle test (Howard and Rogers 1995, 2002; Ogle 1950; Reading 1983). In this real test, two vertical lines are presented in a front-parallel plane and a third vertical line between them is moved nearer or farther from front-parallel plane. With the adjustment or constant-stimuli method, stereoacuity can be measured.

As indicated earlier, real-depth tests are falling into steadily declining use against the projection systems. With real-depth devices, the possibilities of the geometry of the system (rod size, separation) are clearly limited. Furthermore, the real tests can be computer simulated.

23.3.2 PROJECTED-DEPTH DEVICES (STEREOPAIRS)

In this type of test, stereopsis is studied by presenting the observer with stereopairs (a pattern to each eye independently) with the help of some system of separation, and the observer comfortably fuses the stereopair (Howard and Rogers 1995, 2002; Ogle 1950; Reading 1983; Solomon 1978; Steinman et al. 2000; Stidwill and Flecher 2011). Different stereoscopic displays are based on different approaches that allow the left and right images to be separated so that each image is seen only by one eye (Javidi et al. 2009; Onural 2011):

- *Temporal stereoscopic devices*: Both images (left and right) are transmitted from the display alternately in time. Usually, the observer wears glasses with shutters that switch in time so that light is allowed to an eye only when the display is showing the appropriate image. If the switch frequency is high (above 60 Hz per eye), the flicker is not seen under most conditions (Javidi et al. 2009; Onural 2011).

- *Polarization stereoscopic devices*: Polarization (linear or circular) is used to encode the two images with different polarized light. Observers wear glasses with different filters in each eye that allow only the appropriate polarization image to be seen. Linear polarization tends to provide higher luminance but causes problems when patients tilt their heads when examining the image. Circular polarization avoids head-tilt problems but involves other optical problems such as lower image luminance (Javidi et al. 2009; Onural 2011).

- *Wavelength stereoscopic devices*: In this traditional system, two colors are chosen, the left image encoded in one and the right in the other. The filters used on the glasses allow the images to be seen by the appropriate eye. An extended example is the traditional anaglyph (red-green or red-blue encoding). Great research and advances in coding and filters have extended the use of these devices although they are not usually used in research (Javidi et al. 2009; Onural 2011).

- *Spatial stereoscopic devices*: These devices (Javidi et al. 2009; Onural 2011) spatially separate the left and right images, providing a different display in front of each eye. Some of these devices use a head-mounted display or a chin rest to help this separation. The Wheatstone stereoscope (Figure 23.5a), the first stereoscope (Wheatstone 1838, 1852), belongs to this category. This stereoscope allows the images for the right and left eyes to be presented by reflection through mirrors placed at 45° to the line of sight. The mirror for the observer's right eye reflects an image on the observer's right and analogously for the left eye. There are slight modifications of this stereoscope that use four mirrors (see Figure 23.5b). Brewster's stereoscope, based on the prismatic effect of lenses, has also been a traditional spatial stereoscopic device (Brewster 1856).

- *Autostereoscopic displays*: These devices ensure that each eye sees the correct image without the observer using any device. To achieve this, the display needs to incorporate an optical element, which serves to direct different images in different directions to the two eyes. Autostereoscopic displays are quickly evolving and they are used steadily more although not usually for research tasks in stereoscopic vision. With autostereoscopic devices, the convergence–accommodation mechanism is altered and some observers may have difficulties in fusing stereograms and thus long use in a single session is not recommended (Howard and Rogers 2002; Stidwill and Flecher 2011).

Stereoacuity measured both for real and for projected tests depends largely on the experimental conditions and stimulus parameters. Thus, for example, angle size of the stimuli, luminance stimuli, background luminance, time exposure of stimuli, time between stimuli, eccentricing, fixation distance, and other variables influence stereoacuity (Howard and Rogers 1995, 2002; Ogle 1950; Reading 1983; Solomon 1978; Steinman et al. 2000; Stidwill and Flecher 2011), and therefore they need to be well controlled and fixed during experiments.

23.3.3 RANDOM-DOT STEREOGRAMS

It is worth mentioning a test, RDS, which is widely used in projected tests. The introduction of this test in 1960 (Juslez 1960, 1971) triggered a revolution in studies on binocular vision (Howard and Rogers 1995, 2002). An RDS (see Figure 23.6) is a set of black and white points randomly arrayed over a

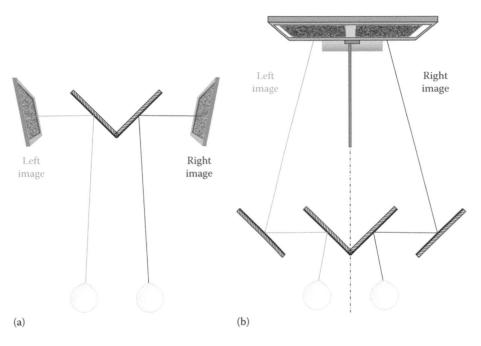

(a)   (b)

Figure 23.5 Scheme of Wheatstone stereoscope. (a) Two mirrors. (b) Four mirrors.

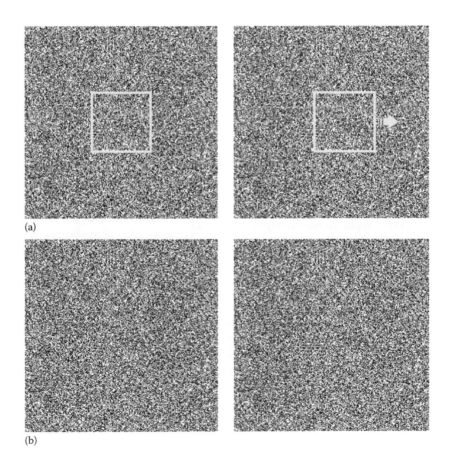

(a)

(b)

Figure 23.6 Random-dot stereograms: (b) The distribution of random dots in the two images of the pattern makes it so that monocularly there is no monocular cue nor recognizable monocular information that provides any 3D information. (a) The shift of a small square generates the disparity that causes the perception of a small square in depth.

pattern, generally a square. The pattern is duplicated to produce an identical image for each eye. Some of the dots, usually with the shape of a small square, are displaced laterally in one of the two images to produce disparity (Juslez 1960). The displacement of the small square to generate disparity can also occur in both squares but in opposite directions.

When the stereogram is fused, the small square is seen in depth (in front or behind, depending on the sense of the displacement of the small square) against the nonshifted background. However, there are no monocular cues nor recognizable monocular information, and stereoscopic perception can appear only when there is binocular fusion. If we see only one pattern of the stereoscopic pair, we cannot find the small square. The fused figure appears only in the binocular image; this has been also termed as cyclopean depth perception. For perception, RDS involves cortical binocular processes because binocular vision takes place in the visual cortex (Howard and Rogers 1995, 2002). As there are no recognizable monocular information nor monocular cues that can help to perceive stereoscopically, some observers have difficulty perceiving RDS and many need a long training or instruction time to perceive the test.

Stereopsis perceived with tests such as RDS with no monocular recognizable information is known as global stereopsis. Local stereopsis is stereopsis perceived with stereoscopic targets containing monocularly recognizable forms such as contours or lines (Stidwill and Flecher 2011).

23.4 STEREOACUITY AND OPTICAL FACTORS

Adaptive optics applied to the human eye has rapidly developed since the end of the 1990s (see related chapters in this handbook). The application of the measurement and control techniques of aberrations has been widely used in the development of laser-scanning ophthalmoscope, fundus imaging, visual simulation, optical coherence tomography, and surgical techniques of emmetropization (refractive surgery, intraocular lenses, multifocal corrections, etc.), among other applications, although these studies were initially limited to monocular aspects despite that normal human vision is binocular. Devices have appeared for measuring and monitoring visual aberrations under binocular conditions (Fernández et al. 2009, 2010; Hampson et al. 2008; Kobayashi et al. 2008; Sabesan et al. 2012; Schwarz et al. 2011). From the standpoint of visual function, the binocular devices that use adaptive optics constitute an enormous leap for fundamental and practical research, as they enable the study of functions linked to binocular vision, which is the normal state of vision for any person.

The study of the influence of optical factors in stereoacuity has traditionally been limited to questions concerning interocular differences in low-order aberrations (defocus and astigmatism), whether natural interocular differences in ametropia or interocular differences induced by different emmetropization techniques (lenses, contact lenses, monovision, surgery, etc.) and their effects on visual quality (Howard and Rogers 1995, 2002; Westheimer and McKee 1979, 1980; Wood 1983). To analyze the influence of higher-order aberrations on binocular vision is now possible with adaptive optics devices.

In this section, we will limit ourselves to the study of the optical aspects that affect one of the most thoroughly studied binocular functions, stereoacuity, though we examine some optical aspects with major implications for upper disparity, which can help us to understand the influence of optical aspects in stereopsis and also provide evidence regarding factors that may be important for a better characterization of stereoacuity.

23.4.1 STEREOACUITY AND INTEROCULAR DIFFERENCES IN LOW-ORDER ABERRATIONS: MONOCULAR AND BINOCULAR DEFOCUS

The change in stereoacuity with defocus (monocular or binocular) has been measured by a large variety of tests. Most experiments confirm that monocular or binocular defocus deteriorates stereoacuity and this deterioration is proportional to the magnitude of the defocus (Lovasik and Szymkiw 1985; Schmidt 1994; Schor and Heckmann 1989) although monocular defocus generates worse stereoacuity than binocular defocus (Cormack et al. 1997; Halpern and Blake 1988; Legge and Gu 1989). This deterioration has been confirmed with real stereoscopes such as the Howard-Dolman stereoscope, projected-test devices, and even binocular adaptive optics devices, as will be shown in the following.

The numerical relationship that quantifies the deterioration between defocus and stereoacuity depends on the method and experimental device used (Howard and Rogers 1995, 2002). Also, the degree of tolerance for monocular defocus differs between global and local stereopsis, the tolerance to defocus being greater for global stereopsis (Wood 1983). Deterioration due to unilateral defocus increases as image blur and the eye approaches the fovea suppression limit, a value reached at around 2 D (Geib and Baumann 1990). Some authors have shown that a stereoacuity of 40 s is maintained for 0.5–1.0 D of monocular defocusing, whereas other authors (Peters 1969) have found that 80% of the subjects undergo a complete loss of stereoacuity for 1.0 D of monocular defocusing.

This effect of monocular defocus on stereoacuity can be appreciated in the technique of monovision (Evans 2007; Jain et al. 1996), one of the refractive error correction techniques used for the correction of presbyopia. Monovision consists basically of setting the refraction of one eye for far vision (usually, the dominant eye) while the other eye is corrected for near vision (usually, the nondominant eye), generating an interocular difference in defocus that deteriorates stereoacuity for different distances, as many studies demonstrate (see review, Alarcón et al. 2011; Evans 2007).

23.4.2 STEREOACUITY AND HIGHER-ORDER EYE ABERRATIONS

23.4.2.1 Evidence from the upper disparity limit

Before the development of binocular adaptive optics visual analyzers, which enable an exhaustive control of binocular eye aberrations, different studies showed certain evidence that interocular differences in higher-order aberrations could influence stereopsis (Jiménez et al. 2008a,b).

For normal observers, one study (Jiménez et al. 2008a) showed that interocular differences in higher-order eye aberrations closely correlate with the upper disparity limit, a parameter that characterizes the limits of stereoscopic vision together with stereoacuity, as shown before. This study of 30 normal emmetropic subjects found a significant descending correlation between maximum disparity and interocular differences in higher-order eye aberrations (total RMS, spherical and coma aberrations) showing that the higher interocular differences in eye aberrations, the lower the upper disparity. The results show the sensitivity of the upper disparity limit to interocular differences in higher-order interocular aberrations, stereo correspondence being more effective with lower higher-order interocular differences in eye aberrations. If a dependence with interocular differences in higher-order aberrations is found for maximum disparity, a certain dependence for stereoacuity could be expected, as both parameters represent the limits of stereoscopic vision.

Experiments with patients operated on with LASIK (laser in situ keratomileusis) confirm this tendency found for normal observers (Jiménez et al. 2008b). It is well known that LASIK generates profound changes in the cornea, this usually increasing eye aberrations. The results for 23 people indicate that the upper disparity limit declines from 41.1 min of arc on average (with best presurgery correction) to 31.3 min of arc after successful LASIK, being significant in 83% of the patients. This deterioration is significantly correlated with an increase in the postsurgical interocular differences in higher-order aberrations (RMS, spherical and coma) (Figure 23.7)—the more postsurgical interocular differences in higher-order aberrations, the lower the upper disparity limit.

Jiménez et al. (2008b) also measured stereoacuity pre- (best corrected) and post-LASIK but with a non-high-accuracy test, the random stereotest (including Titmus stereofly, Wirt rings, and random-dot targets). Presurgery, all the patients could perform the tests, the majority of the patients reaching 20 s of arc for minimum disparity; after LASIK, only eight patients failed to reach the minimum disparity detected before LASIK with the randot test, and they needed one or two more disparity steps. Although the randot stereotest is not accurate enough for measuring the minimum disparity perceived by an observer,

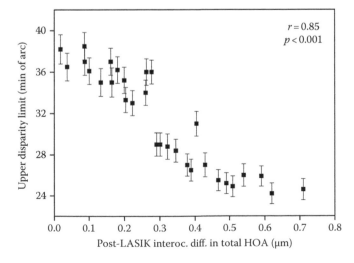

Figure 23.7 Upper disparity limit in function of post-LASIK interocular differences in higher-order aberrations. (From Jiménez, J.R., Castro, J.J., Hita, E., and Anera, R.G., Upper disparity limit after LASIK, *J. Opt. Soc. Am. A*, 25, 1227, 2008b. With permission of the Optical Society of America.)

the evidence for eight patients showed that stereoacuity could be influenced by higher-order aberration changes when a higher-accuracy test is used.

23.4.2.2 Stereoacuity and adaptive optics control

Fernández et al. (2010) developed a binocular adaptive optics simulator to simultaneously control eye aberrations and the effect on stereopsis by studying the variable stereoacuity. They used an instrument (Figure 23.8) capable of manipulating the aberrations of each eye separately while subjects performed visual tests. The correction device is a liquid-crystal-on-silicon spatial light modulator permitting the control of aberrations in both eyes of the observer simultaneously in an open loop. The device can work as an electro-optical binocular phoropter with two microdisplays projecting different scenes to each eye.

In their experiments, they used two normal subjects to test a number of binocular aberration combinations under natural viewing conditions (normal accommodation and natural pupil). Clearly, the results of the experiments for two observers cannot be generalized, as it is extremely small and nonsignificant sample, but it does show the usefulness of binocular adaptive optics simulators in studying the binocular function and the influence of eye-aberration changes on stereoacuity. The stereoscopic stimuli used were the three-needle test and RDS. In the three-needle test, the two outer wires are in the fixation plane and the central one is displaced backward and forward to generate disparity. All disparity control was undertaken computationally.

Different experiments were performed. First, different values of defocus were induced in each eye. As an example, for one observer, they determined stereoacuity with full refraction-correction results (when low-order aberrations, defocus and astigmatism, were corrected) giving a value of 4 s for stereoacuity. When equal defocus was introduced in both eyes, 1 D on a 4 mm pupil, the stereoacuity slightly deteriorated to 6 s. Finally, when 1 D of defocus was induced on the right eye (simulating anisometropia), the stereoacuity deteriorated to 8 s. Defocus in all situations was shifted toward the hyperopic direction, with the image formed behind the observer's retina, avoiding accommodation to compensate for the blur. These experiments on monocular and binocular defocus provide stereoacuity deterioration, as did results shown in previous experiments (Cormack et al. 1997; Halpern and Blake 1988; Legge and Gu 1989; Lovasik and Szymkiw 1985; Schmidt 1994; Schor and Heckmann 1989).

The adaptive optics simulator was used to test more complex optical conditions using RDS to estimate stereoacuity in the presence of high-order aberrations for one subject. The aberration selected by Fernández et al. (2010) for analyzing the influence of higher-order eye aberrations on stereoacuity was trefoil. For the observer tested, the stereoacuity when low-order aberrations, defocus and astigmatism, were corrected was 4 s. A bilateral addition of 1 μm pure trefoil deteriorated stereopsis, increasing stereoacuity to 18 s. When 1 μm of trefoil was unilaterally induced in the right eye, stereoacuity was also deteriorated with respect to low-order aberration correction, rendering a value in this case of 13 s.

These results constitute evidence of the effects on stereoacuity by inducing aberrations that reveal a deterioration when

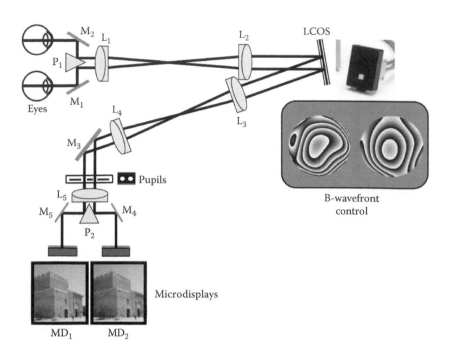

Figure 23.8 Scheme of a binocular adaptive optics simulator. (From Fernández, E.J., Prieto, P.M., and Artal, P., Adaptive optics binocular visual simulator to study stereopsis in the presence of aberrations, *J. Opt. Soc. Am. A*, 27, A48–A55, 2010. With permission of the Optical Society of America.)

higher-order aberrations are increased in one or two eyes, although the present study was limited to the trefoil aberration and a single observer. It would be necessary to study other higher-order aberrations in order to draw firmer conclusions. This has far-reaching implications from the clinical standpoint, as many emmetropization techniques induce and increase higher-order aberrations and therefore can negatively influence binocular vision and stereopsis, as results on maximum disparity have shown (Jiménez et al. 2008b).

23.4.2.3 Stereoacuity with improved optics

Other researchers (Vlaskamp et al. 2011) have investigated the way in which the optics of the eye affects stereopsis, studying the effect in a stereoacuity task. For this, they have compared stereoscopic performance with normal, well-focused optics (6 and 4 mm) and with optics improved by eliminating chromatic aberration and correcting higher-order aberrations.

For three emmetropic observers (spherical and cylindrical refractive errors both smaller than 0.5 D), they improved the optical quality of the retinal image by filtering the stimuli at 550 nm to cancel chromatic aberrations, reducing the effective pupil diameter to 2.5 mm with an artificial pupil, and higher-order aberrations were reduced by using phase plates that cancel the aberrations after measuring eye aberrations using a Shack–Hartmann sensor.

For stereoacuity measurements, they used a two-line stereogram for determining minimum disparity detected. In the experiment, the stimuli were white lines on a dark background, performing the tests with different contrasts between the lines and background. Results showed nonsignificant differences for stereoacuity between three conditions tested (normal, well-focused optics for 6 mm pupil; normal, well-focused for 4 mm pupil and with improved optics)

providing that optics improvements had no effect on stereo performance even for different contrasts. Thus, the resolution (stereoacuity) of human stereopsis is not limited by the optics of the well-focused eye. However, in these experiments, contrast sensitivity and visual acuity were also measured with normal well-focused optics and improved optics, resulting in better performance for contrast sensitivity and visual acuity with improved optics.

To explain this difference, the authors hypothesize that the difference is due to how eye movements affect performance in visual resolution and stereo resolution tasks (stereoacuity). They propose that small eye movements during fixation more negatively affect stereo resolution than visual resolution (i.e., visual acuity, contrast sensitivity). These minor eye movements bring about changes in disparity computation that are similar to spatial blur. According to these authors, the blur due to eye movements may be the limit to stereo performance rather than the blur inherent to the optics of normal, well-focused eyes. Stereoacuity therefore shows certain asymmetry, and optical improvement in the retinal image in both normal eyes does not improve stereo resolution but an increase in eye aberrations for both eyes seems to deteriorate stereoacuity.

23.4.2.4 Stereoacuity with binocular adaptive optics simulations for clinical applications

As indicated earlier, the binocular adaptive optics simulators have enormous potential for simulating different surgical techniques of emmetropization that may indicate whether or not they are appropriate for a patient, including binocular function. The usefulness of these simulators in some cases is essential because some surgical techniques of emmetropization can be irreversible or have a reversibility that can alter the function and structure

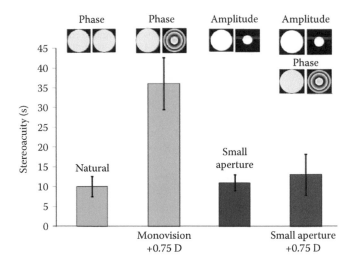

Figure 23.9 Average stereoacuity under different conditions: natural vision, 0.75 D monovision, small pupil in one eye, and small pupil with monovision in one eye. (From Fernández, E.J., Scharwz, C., Prieto, P.M., Manzanera, S., and Artal, P., Impact on stereoacuity of two presbyopia correction approaches: Monovision and small aperture inlay, *Biomed. Opt. Express*, 4, 822–830, 2013. With permission of the Optical Society of America.)

of the eye, and therefore a prior simulation can provide essential information for visual function, including binocular aspects, such as stereoacuity.

As an example of clinical application, the paper by Fernández et al. (2013) indicates the utility of simulators in studying the effect on stereoacuity of a correction of presbyopia with the monovision technique and that of a small aperture inlay in one eye. In this work, the experimental conditions were assayed in a binocular adaptive optics vision analyzer modified for such experiments. Stereoacuity was measured with two methods: monovision and small aperture inlay in one eye using the three-needle test. These researchers tested different cases for the stimulus placed at distance: natural vision, 0.75 D monovision, natural vision and small pupil, and 0.75 D monovision and small pupil. In all cases, the standard pupil diameter was 4 mm and the small pupil diameter (simulated inlay) was 1.6 mm. Four subjects took part in the experiments.

The results are presented in Figure 23.9, reflecting that monovision with 0.75 D added caused a deterioration in stereoacuity by a factor of close to 4. However, the use of a small aperture in one of the eyes significantly reduced the negative effect of monovision on stereoacuity. The results of the experiment showed that a small aperture (e.g., an intrastromal corneal inlay) can provide values of stereoacuity similar to those found under normal binocular vision. We should mention that these experiments were performed under photopic conditions; mesopic and scotopic illumination conditions were not studied in this work. Therefore, more works are needed to draw conclusions for a wider range of experimental conditions.

This experiment is a good example of the importance of binocular adaptive optics for testing binocular vision before treatment (surgical or not). It is hoped that the clinical use and research on these devices will enable better characterization of stereopsis and binocular vision from the optical standpoint,

contributing knowledge on the visual system that has been limited until now to monocular studies, as only in the last few years some research has begun to delve into the binocular aspects, including stereopsis.

REFERENCES

Alarcón A., Anera R.G., Villa C., Jiménez del Barco L., and Gutiérrez L., Visual quality after monovision correction by laser in situ keratomileusis in presbyopic patients, *Journal of Cataract and Refractive Surgery* 37 (2011):1629–1635.

Brewster D. *The Stereoscope: Its History, Theory and Construction, with Its Application to the Fine and Useful Arts and to Education.* London, U.K.: J. Murray, 1856.

Cormack L.K., Stevenson S.B., and Landers D.D., Interactions of spatial frequency and unequal monocular contrasts in stereopsis, *Perception* 26 (1997):1121–1135.

Cumming B.G. and De Angelis G.C., The physiology of stereopsis, *Annual Revision Neuroscience* 24 (2001):203–238.

Evans B.J.W., Monovision: A review, *Ophthalmic and Physiological Optics* 27 (2007):417–439.

Fernández E.J., Prieto P.M., and Artal P., Binocular adaptive optics visual simulator, *Optics Letters* 34 (2009):2628–2630.

Fernández E.J., Prieto P.M., and Artal P., Adaptive optics binocular visual simulator to study stereopsis in the presence of aberrations, *Journal of the Optical Society of American A* 27 (2010): A48–A55.

Fernández E.J., Scharwz C., Prieto P.M., Manzanera S., and Artal P., Impact on stereo-acuity of two presbyopia correction approaches: Monovision and small aperture inlay, *Biomedical Optics Express* 4 (2013):822–830.

Fielder R. and Moseley M.J., Does stereopsis matter in humans?, *Eye* 10 (1996):233–238.

Geib T. and Baumann C., Effect of luminance and contrast on stereoscopic acuity, *Graefe's Archive Clinical Experimental Ophthalmology* 228 (1990):310–315.

Halpern D.L. and Blake R.R., How contrast affects stereoacuity, *Perception* 17 (1988):483–495.

Hampson K.M., Chin S.S., and Mallen E.A.H., Binocular Shack–Hartmann sensor for the human eye, *Journal of Modern Optics* 55 (2008):703–716.

Hibbard P.B., Binocular energy responses to natural images, *Vision Research* 48 (2008):1427–1439.

Howard I.P., A test for the judgment of distance, *American Journal of Psychology* 2 (1919):656–675.

Howard I.P. and Rogers B.J., *Binocular Vision and Stereopsis.* Oxford, U.K.: Oxford University Press, 1995.

Howard I.P. and Rogers B.J., *Seeing in Depth.* Toronto, Ontario, Canada: Toronto University Press, 2002.

Jain S., Arora I., and Azar D.T., Success of monovision in presbyopes: Review of the literature and potential applications to refractive surgery, *Survey of Ophthalmology* 40 (1996):491–499.

Javidi B., Okano F., and Son, J., *Three-Dimensional Imaging, Visualization and Display.* New York: Springer, 2009.

Jiménez J.R., Castro J.J., Hita E., and Anera R.G., Upper disparity limit after LASIK, *Journal of the Optical Society of America A* 25 (2008b):1227–1231.

Jiménez J.R., Castro J.J., Jiménez R., and Hita E., Interocular differences in higher-order aberrations on binocular visual performance, *Optometry and Vision Science* 85 (2008a):174–179.

Jiménez J.R., Rubiño M., Hita E., and Jiménez del Barco L., Influence of the luminance and opponent chromatic channels on stereopsis with random-dot stereograms, *Vision Research* 37 (1997):591–596.

Juslez B., Binocular depth perception of computer-generated patterns, *Bell Systems Technical Journal* 39 (1960):1125–1162.

Juslez B., *Foundations of Cyclopean Perception*. Chicago, IL: University of Chicago Press, 1971.

Kobayashi M., Nakazawa N., Yamaguchi T., Otaki T., Hirohara Y., and Mihashi T., Binocular open-view Shack–Hartmann wavefront sensor with consecutive measurements of near triad and spherical aberration, *Applied Optics* 47 (2008):4619–4626.

Legge G. and Gu Y., Stereopsis and contrast, *Vision Research* 29 (1989):989–1004.

Lehky S.R. and Sejnowski T.J., Neural model of stereoacuity and depth interpolation based on a distributed representation of the stereo disparity, *Journal of Neuroscience* 10 (1990):2281–2299.

Lovasik J.V. and Szymkiw M., Effects of aniseikonia, anisometropia, accommodation, retinal illuminance, and pupil size on stereopsis, *Investigative Ophthalmology and Visual Science* 26 (1985):741–750.

Marr D., *Vision*. San Francisco, CA: W.H. Freeman and Company, 1980.

Marr D. and Poggio T., Cooperative computation of stereo disparity, *Science* 194 (1976):283–287.

Mazyn L.I.N., Lenoir M., Montagne G., and Savelsbergh G.J.P., The contribution of stereo vision to one handed catching, *Experimental Brain Research* 157 (2004):383–390.

O'Connor R., Birch E.E., Anderson S. et al., The functional significance of stereopsis, *Investigative Ophthalmology and Visual Science* 51 (2010):2019–2023.

Ogle K.N., *Researches in Binocular Vision*. Philadelphia, PA: W.B. Saunders Company, 1950.

Onural L., *3D Video Technologies: An Overview of Research Trends*. Bellingham, WA: Spie Press, 2011.

Orban G.A., Jansenn P., and Vogels R., Extracting 3D structure from disparity, *Trends in Neuroscience* 29 (2006):466–472.

Peters H.B., The influence of anisometropia on stereosensitivity, *American Journal of Optometry* 46 (1969):120–123.

Poggio G.F. and Fisher B., Binocular interaction and depth sensitivity in striate and prestriate cortex of behaving rhesus monkey, *Journal of Neurophysiology* 40 (1977):1392–1405.

Reading R.W., *Binocular Vision: Foundations and Applications*. Boston, MA: Butterworth Publishers, 1983.

Richards W., Anomalous stereoscopic depth perception, *Journal of the Optical Society of America A* 61 (1971):410–414.

Sabesan R., Zheleznyak L., and Yoon G., Binocular visual performance and summation after correcting higher order aberrations, *Biomedical Optics Express* 3 (2012):3176–3189.

Schmidt P.P., Sensitivity of random-dot stereoacuity and Snellen acuity to optical blur, *Optometry and Vision Science* 71 (1994):466–471.

Schwarz C., Prieto P.M., Fernández E.J., and Artal P., Binocular adaptive optics vision analyzer with full control over the complex pupil functions, *Optics Letters* 36 (2011):4779–4781.

Schor C. and Heckmann T., Interocular differences in contrast and spatial frequency: Effects on stereopsis and fusion, *Vision Research* 29 (1989):837–847.

Solomons H., *Binocular Vision: A Programmed Text*. London, U.K.: William Heinemann Medical Books Ltd., 1978.

Steinman S.B., Steinman B.A., and Garzia R.P., *Foundations of Binocular Vision: A Clinical Perspective*. New York: McGraw-Hill, 2000.

Stidwill D. and Fletcher R., *Normal Binocular Vision: Theory, Investigation and Practical Aspects*. Oxford, U.K.: Wiley-Blackwell, 2011.

Vlaskamp B.N.S., Yoon G., and Banks M.S., Human stereopsis is not limited by the optics of the well-focused eye, *Journal of Neuroscience* 31 (2011):9814–9818.

Westheimer G. and McKee S.P., What prior uniocular processing is necessary for stereopsis?, *Investigative Ophthalmology and Visual Science* 18 (1979):614–621.

Westheimer G. and McKee S.P., Stereoscopic acuity with defocused and spatially filtered retinal images, *Journal of the Optical Society of America A* 70 (1980):772–778.

Wheatstone C., Contributions to the physiology of vision-part the first. On some remarkable, and hitherto unobserved, phenomena of binocular vision, *Philosophical Transactions of the Royal Society* 12 (1838):371–394.

Wheatstone C., Contributions to the physiology of vision-part the second. On some remarkable, and hitherto unobserved phenomena of binocular vision, *Philosophical Transactions of the Royal Society* 142 (1852):1–17.

Wong B.P., Woods R.L., and Peli E., Stereoacuity at distance and near, *Optometry and Vision Science* 79 (2002):771–778.

Wood I.C., Stereopsis with spatially degraded images, *Investigative Ophthalmology and Visual Science* 3 (1983):337–340.

Index

T - #0199 - 111024 - C404 - 279/216/19 - PB - 9780367869939 - Gloss Lamination